Bio-Inspired Robotics

Bio-Inspired Robotics

Special Issue Editors

Toshio Fukuda
Fei Chen
Qing Shi

MDPI • Basel • Beijing • Wuhan • Barcelona • Belgrade

Special Issue Editors
Toshio Fukuda
Nagoya University
Japan

Fei Chen
Italian Institute of Technology
Italy

Qing Shi
Beijing Institute of Technology
China

Editorial Office
MDPI
St. Alban-Anlage 66
Basel, Switzerland

This is a reprint of articles from the Special Issue published online in the open access journal *Applied Sciences* (ISSN 2076-3417) from 2017 to 2018 (available at: http://www.mdpi.com/journal/applsci/special_issues/bio_inspired_robotics)

For citation purposes, cite each article independently as indicated on the article page online and as indicated below:

LastName, A.A.; LastName, B.B.; LastName, C.C. Article Title. *Journal Name* **Year**, *Article Number*, Page Range.

ISBN 978-3-03897-045-3 (Pbk)
ISBN 978-3-03897-046-0 (PDF)

Cover image courtesy of UBTECH.

Contents

About the Special Issue Editors

Toshio Fukuda, Professor, received his B.S. degree from Waseda University, Tokyo, Japan, in 1971 and his M.S. and Ph.D. degrees from the University of Tokyo, Tokyo, Japan, in 1973 and 1977, respectively. From 1977 to 1982, he was with the National Mechanical Engineering Laboratory, Tsukuba, Japan. From 1982 to 1989, he was with the Science University of Tokyo. Starting in 1989, he was with Nagoya University, Nagoya, Japan, where he was a Professor with the Department of Micro System Engineering, and a Professor at Meijo University, Nagoya Japan. He is currently a Professor (1000 Foreign Experts Plan) with the Beijing Institute of Technology, Beijing, China, where he is mainly engaged in the research fields of intelligent robotic systems, cellular robotic systems, mechatronics, and micro/nano robotics. Prof. Fukuda was the President of IEEE Robotics and Automation Society (1998–1999), Editor-in Chief of IEEE/ASME Transactions on Mechatronics (2000–002), Director Division X: Systems and Control (2001–2002), IEEE Founding President of Nanotechnology Council (2002–2003, 2005), Director Region 10 (2013–2014), Director Division X: Systems and Control (2017–2018). He is 2019 IEEE President-elect.

Fei Chen is a researcher in the department of advanced robotics with the Italian Institute of Technology (IIT), Italy. He received the B.S. degree in computer science from Xi'an Jiaotong University (XJTU), Xi'an, China, in 2006, the M.S. degree in computer science from Harbin Institute of Technology (HIT), Harbin, China, in 2008, and the Dr. Eng. degree from Nagoya University, Japan, in 2012. Between 2012 and 2013, he was working as a Japanese COE Program Researcher in Nagoya University. After moving to IIT, he is leading the mobile manipulation group focusing on robot learning and manipulation. He was the PI of AutoMAP project funded within FP7 EUROC, developing the novel mobile manipulation system for the maintenance of the Large Hadron Collider (LHC) with the European Organization for Nuclear Research (CERN). He is currently working on a legged mobile manipulation system for grape vine pruning automation. He received the best automation paper finalist award in ICMA2016, and the best cognition paper award in ICARM2016. His main research interest is robotic mobile manipulation, intelligent sensing and learning, human–robot collaboration.

Qing Shi, Associate Professor, received his B.S. degree in Mechatronics from the Beijing Institute of Technology, Beijing, China, in 2006, and his Ph.D. degree in Biomedical Engineering from Waseda University, Japan, in 2012. He was a Research Associate at GCOE Global Robot Academia of Waseda University from 2009 to 2013. Since 2013, he has been a full-time Lecturer, and is now an Associate Professor at Beijing Institute of Technology. His research interests are focused on biorobotics, mechatronic systems, computer vision, and micro/nano robotics. Dr. Shi has published more than 50 international journal/conference papers, and has received the Best Journal Paper Award of Advanced Robotics in 2015, and was the Best Cognitive Robotics Paper Finalist of ICRA 2014. He is currently the Associate Editor of ROBOMECH Journal. Additionally, he has served as Associate Editor of IROS and ICRA, Guest Editor of IEEE Transactions on Nanotechnology and Applied Sciences, and was the Program Co-Chair of IEEE CBS 2017.

Editorial

Special Feature on Bio-Inspired Robotics

Toshio Fukuda [1,2,3], Fei Chen [4,*] and Qing Shi [3]

[1] Institute for Advanced Research, Nagoya University, Chikusa-ku, Nagoya 464-8601, Japan; tofukuda@nifty.com
[2] Department of Mechatronics Engineering, Meijo University, Nagoya, Aichi Prefecture 468-0073, Japan
[3] Intelligent Robotics Institute, School of Mechatronic Engineering, Beijing Institute of Technology, Beijing 100081, China; shiqing@bit.edu.cn
[4] Department of Advanced Robotics, Istituto Italiano di Tecnologia, Via Morego 30, 16163 Genoa, Italy
* Correspondence: fei.chen@iit.it; Tel.: +39-10-71781-217

Received: 4 May 2018; Accepted: 14 May 2018; Published: 18 May 2018

1. Introduction

Modern robotic technologies have enabled robots to operate in a variety of unstructured and dynamically-changing environments, in addition to traditional structured environments. Robots have, thus, become an important element in our everyday lives. One key approach to develop such intelligent and autonomous robots is to draw inspiration from biological systems. Biological structure, mechanisms, and underlying principles have the potential to feed new ideas to support the improvement of conventional robotic designs and control. Such biological principles usually originate from animal or even plant models for robots, which can sense, think, walk, swim, crawl, jump or even fly. Thus, it is believed that these bio-inspired methods are becoming increasingly important in the face of complex applications. Bio-inspired robotics is leading to the study of innovative structures and computing with sensory-motor coordination and learning to achieve intelligence, flexibility, stability, and adaptation for emergent robotic applications, such as manipulation, learning, and control.

This Special Issue invites original papers of innovative ideas and concepts, new discoveries and improvements, and novel applications and business models relevant to the selected topics of "Bio-Inspired Robotics". Bio-Inspired Robotics is a broad topic and an ongoing expanding field. This Special Issue collects 30 papers that address some of the important challenges and opportunities in the broad and expanding field. We group these papers as follows.

2. Legged and Mobile Robots

Developing legged moving robots or wheeled mobile robots inspired by animals or human beings is one of the major research direction in bio-inspired robotics. This section collects papers in this domain by grouping the papers as follows: biped robots, quadruped robots, multi-legged robots and wheeled robots.

2.1. Biped Robots

Yang et al. perform a review summarizing the chronological historical development of bipedal robots and introducing some current popular bipedal robots. The basic theory-stability control and key technology-motion planning of bipedal robots are introduced and analyzed [1]. Hu and Mombaur present an optimal control-based approach for the humanoid robot iCub that allows to generate optimized walking motions using a precise whole-body dynamic model of the robot. The optimal control problem is formulated to minimize a set of desired objective functions with respect to physical constraints of the robot and contact constraints of the walking phases [2]. Jiang et al. propose a jumping control scheme for a bipedal robot to perform a high jump taking into account the motion during the launching phase. The half-body of the robot is modeled as three planar links. A geometrically simple motion is conducted through which the gear reduction ratio that matches the maximum motor output for high jumping is selected, followed by the strategies to further exploit the motor output performance via criteria including the location of ZMP (zero moment point) and the torque limit [3]. Zang et al. introduce a novel foot system with passively adjustable stiffness for biped robots which is adaptable to small-sized bumps on the ground. The robotic foot is attached by eight pneumatic variable stiffness units. Each variable stiffness unit mainly consists of a pneumatic bladder and a mechanical reversing valve. When walking on rough ground, the pneumatic bladders in contact with bumps are compressed, and the corresponding reversing valves are triggered to expel out the air, enabling the pneumatic bladders to adapt to the bumps with low stiffness, while the other pneumatic bladders remain rigid and maintain stable contact with the ground, providing support to the biped robot [4]. Gerardo Muscolo et al. present a novel method to determine the center of mass position of each link for human-like multibody biped robots. They optimize the formulation to determine the total center of mass position tested in other works on a biped platform with human-like dimensions. This formulation is able to give as output the center of mass positions of each link of the platform [5]. Liu et al. describe a hybrid control approach aiming to integrate the main characteristics of human walking into a simulated seven-link biped robot. Three bipedal gaits are considered, including a fully actuated single support phase with the stance heel supporting the body, an under-actuated single support phase, with the stance toe supporting the body, and an instantaneous double support phase when the two legs exchange their roles. The walking controller combines virtual force control and foot placement control, which are applied to the stance leg and the swing leg, respectively [6]. Otani et al. propose a stabilizing control method for humanoids during the stance phase while hopping and running by considering swinging the legs rapidly during the flight phase to prevent rotation in the yaw direction. They develop an angular momentum control method based on human motion for a humanoid upper body. The method involves calculation of the angular momentum generated by the movement of the humanoid legs and calculation of the torso and arm motions required to compensate for the angular momentum of the legs in the yaw direction. They also develop a humanoid upper-body mechanism having human link length and mass properties, using carbon-fiber-reinforced plastic and a symmetric structure for generating large angular momentum [7].

2.2. Quadruped Robots

In order to effectively plan the robot gaits and foot workspace trajectory (WT) synchronously, Zeng et al. present a novel biologically inspired control strategy for the locomotion of a quadruped robot based on central pattern generator—neural network—workspace trajectory (CPG-NN-WT). A neural network is designed and trained to convert the CPG output to the preplanned WT, which can make full use of the advantages of the CPG-based method in gait planning and the WT-based method in foot trajectory planning simultaneously [8]. Ba et al. propose a position-based impedance control method for the hydraulic drive unit on the joints of bionic legged robots. They propose a first-order sensitivity matrix to analyze the dynamic sensitivity of four main control parameters under four working conditions. Two sensitivity indexes are defined and verified by experiments to study the parameter sensitivity quantificationally [9].

2.3. Multi-Legged Robots

To find a common approach for the development of an efficient system that is able to achieve an omnidirectional jump, Zhu et al. present a jumping kinematic of a legged robot based on the behavior mechanism of a jumping spider. To satisfy the diversity of motion forms in robot jumping, a kind of 4 degrees of freedom (DOF) mechanical leg is designed. Taking the change of joint angle as inspiration by observing the behavior of the jumping spider during the acceleration phase, a redundant constraint to solve the kinematic is obtained. A series of experiments on three types of jumping—vertical jumping, sideways jumping and forward jumping—is carried out, while the initial attitude and path planning of the robot is studied [10]. To better understand how animals control locomotion, Rubeo et al. model animals and their nervous systems with dynamical simulations, namely synthetic nervous systems (SNS). In order to pick up the parameter values that produce the intended dynamics, they introduce a design process that solves this problem without the need for global optimization, by test this method on SimRoach2, a dynamical model of a cockroach. Each leg joint of SimRoach2 is actuated by an antagonistic pair of Hill muscles. A distributed SNS is designed based on pathways known to exist in insects, as well as hypothetical pathways that produced insect-like motion. Each joint's controller is designed to function as a proportional-integral (PI) feedback loop and tuned with numerical optimization [11].

2.4. Wheeled Robots

Villaseñor et al. use a Germinal Center Optimization algorithm (GCO) which implements temporal leadership through modeling a non-uniform competitive-based distribution for particle selection. GCO is used to find an optimal set of parameters for a neural inverse optimal control applied to all-terrain tracked robot. In the Neural Inverse Optimal Control scheme, a neural identifier, based on Recurrent High Orden Neural Network trained with an extended kalman filter algorithm, is used to obtain a model of the system, then, a control law is designed using such model with the inverse optimal control approach [12]. Wang et al. address trajectory tracking of an omni-directional mobile robot (OMR) with three mecanum wheels and a fully symmetrical configuration. The omni-directional wheeled robot outperforms the non-holonomic wheeled robot due to its ability to rotate and translate independently and simultaneously. A kinematics model of the OMR is established and a model predictive control algorithm with control and system constraints is designed to achieve point stabilization and trajectory tracking [13].

3. Animal and Plant Inspired Robots

This section introduce the collect papers on developing animal and plant robots, including monkey, snake, fish, flying and plant robot.

3.1. Monkey Inspired Robots

Zhu et al. study member-to-member transition and its utility in global path searching for biped climbing robots. To compute operational regions for transition, hierarchical inspection of safety, reachability, and accessibility of grips is taken into account. A novel global path rapid determination approach is subsequently proposed based on the transition analysis. This scheme is efficient for finding feasible routes with respect to the overall structural environment, which also benefits the subsequent grip and motion planning [14]. Lo et al. report a model-based development of a monkey robot that can perform continuous brachiation locomotion on swingable rod, as the intermediate step toward studying brachiation on the soft rope or on horizontal ropes with both ends fixed. The model, which is composed of two rigid links, is inspired by the dynamic motion of primates. The model further serves as the design guideline for a robot that has 5 DOF: two on each arm for rod changing and one on the waist to initiate a swing motion [15].

3.2. Snake Inspired Robots

In nature, snakes can gracefully traverse a wide range of different and complex environments. Snake robots that can mimic this behaviour could be fitted with sensors and transport tools to hazardous or confined areas that other robots and humans are unable to access. In order to carry out such tasks, snake robots must have a high degree of awareness of their surroundings (i.e., perception-driven locomotion) and be capable of efficient obstacle exploitation (i.e., obstacle-aided locomotion) to gain propulsion. Sanfilippo et al. survey and discuss the state-of-the-art, challenges, and possibilities of perception-driven obstacle-aided locomotion for snake robots. To this end, different levels of autonomy are identified for snake robots and categorised into environmental complexity, mission complexity, and external system independence. From this perspective, they present a step-wise approach on how to increment snake robot abilities within guidance, navigation, and control in order to target the different levels of autonomy. They put obstacle-aided locomotion into the context of perception and mapping. Finally, they present an overview of relevant key technologies and methods within environment perception, mapping, and representation that constitute important aspects of perception-driven obstacle-aided locomotion [16]. Kelasidi et al. take into account both the minimization of the power consumption and the maximization of the achieved forward velocity in order to investigate the optimal gait parameters for bio-inspired snake robots using lateral undulation and eel-like motion patterns. They consider possible negative work effects in the calculation of average power consumption of underwater snake robots. To solve the multi-objective optimization problem, they propose transforming the two objective functions into a single one using a weighted-sum method. In this way, they are able to obtain some observations about the optimal values of the gait parameters, which provide very important insights for future control design of bio-inspired snake robots [17]. Wang et al. develop a novel snake-like robot which can perform common gaits to adapt to different environments. A multi-gait is established and used as a reference for the articulation design. A non-snake-like mechanism with linear articulation is combined with the classical swing joint. A prototype is designed and constructed for verification and analysis. Two basic main gaits, namely, serpentine and rectilinear locomotion, are fused, and a novel obstacle-aided locomotion based on rectilinear motion is developed [18].

3.3. Fish Inspired Robots

In order to efficiently harness tidal flow energy in a cost-efficient manner, development of a mechanism that is inherently resistant to these harsh conditions is required. Yamamoto et al. develop a simple oscillatory-type mechanism based on robotic fish tail fin technology. This uses the physical phenomenon of vortex-induced oscillation, in which water currents flowing around an object induce transverse motion. They consider two specific types of oscillators, firstly a wing-type oscillator, in which the optimal elastic modulus is being sort, and secondly, the optimal selection of shape from 6 basic shapes for a reciprocating oscillating head-type oscillator. Analysis of the flow field clearly showed that the discontinuous flow caused by a square-headed oscillator results in higher lift coefficients due to intense vortex shedding, and that stable operation can be achieved by selecting the optimum length to width ratio [19]. Koca et al. develop a complete non-linear dynamic model comprising entirely kinematic and hydrodynamic effects of Carangiform locomotion based on the Lagrange approach by adapting the parameters and behaviors of a real carp. In order to imitate biological features, swimming patterns of a real carp for forward, turning and up-down motions are analyzed by using the Kineova 8.20 software. The proportional optimum link lengths according to actual size, swimming speed, flapping frequency, proportional physical parameters and different swimming motions of the real carp are investigated with the designed robotic fish model. Three-dimensional locomotion is evaluated by tracking two trajectories in a MATLAB environment. A Reaching Law Control approach for inner loop (Euler angles-speed control) and a guidance system for the outer loop (orientation control) are proposed to provide an effective closed-loop control performance [20]. Xing et al. present a novel, multiply gaited, vectored water-jet, hybrid locomotion-capable, amphibious spherical robot III (termed ASR-III) featuring a wheel-legged, water-jet composite driving system incorporating a lifting and supporting wheel mechanism (LSWM) and mechanical legs with a water-jet thruster. The LSWM allows ASR-III to support the body and slide flexibly on smooth (flat) terrain. The composite driving system facilitates two on-land locomotion modes (sliding and walking) and underwater locomotion mode with vectored thrusters, improving adaptability to the amphibious environment [21]. Gu and Guo present a novel propulsion system for the third-generation Spherical Underwater Robot (SURIII), the improved propulsion system is designed and analyzed to verify its increased stability compared to the second-generation Spherical Underwater Robot (SURII). With the new propulsion system, the robot is not only symmetric on the X axis but also on the Y axis, which increases the flexibility of its movement. The new arrangement also reduces the space constraints of servomotors and vectored water-jet thrusters. The experimental results demonstrate the propulsive force is better than a previous version [22].

3.4. Plant Inspired Robots

Del Dottore et al. present a plant root behavior-based approach to define the control architecture of a plant-root-inspired robot, which is composed of three root-agents for nutrient uptake and one shoot-agent for nutrient redistribution. By taking inspiration and extracting key principles from the uptake of nutrient, movements and communication strategies adopted by plant roots, they develop an uptake–kinetics feedback control for the robotic roots. Exploiting the proposed control, each root is able to regulate the growth direction, towards the nutrients that are most needed, and to adjust nutrient uptake, by decreasing the absorption rate of the most plentiful one [23].

3.5. Flying Robots

This Special Issue also collects one paper for UAV robot, although it is for data processing not for flying robots design. This work aims at the reconnaissance task by a UAV to survey an area and retrieve strategic information. Cisneros et al. present a data-foraging-oriented reconnaissance algorithm based on bio-inspired indirect communication. The approach establishes several paths that overlap to identify valuable data sources. Inspired by the stigmergy principle, the aerial vehicles indirectly communicate through artificial pheromones. The aerial vehicles traverse the environment using a heuristic algorithm that uses the artificial pheromones as feedback [24].

4. Bio-Inspired Robotic Components

Three papers in this section present some interesting work on the design of robotic parts or components which are inspired by living animals or human beings.

Fuller and Schultz present the modeling, characterization and validation for a discrete muscle-like actuator system composed of individual on–off motor units with complex dynamics inherent to the architecture. The dynamics include innate hardening behavior in the actuator with increased length. A series elastic actuator model is used as the plant model for an observer used in feedback control of the actuator [25]. Wang et al. present the design of a legged robot with gecko-mimicking mechanism and mushroom-shaped adhesive microstructure (MSAMS) that can climb surfaces under reduced gravity. The design principle, adhesion performance and roles of different toes of footpad are explored and discussed in this paper. The effect of the preload velocity, peeling velocity and thickness of backing layering on the reliability of the robot are investigated. Results show that pull-force is independent of preload velocity, while the peeling force is relying on peeling velocity, and the peel strength increased with the increasing thickness of the backing layer [26]. Aiming at the inspection of rough stone and concrete wall surfaces, Jiang and Xu design a grasping module of cross-arranged claw which can attach onto rough wall surfaces by hooking or grasping walls. Based on the interaction mechanism of hooks and rough wall surfaces, the hook structures in claw tips are developed. Then, the size of the hook tip is calculated and the failure mode is analyzed. The effectiveness and reliability of the mechanism are verified through simulation and finite element analysis. Afterwards, the prototype of the grasping module of claw is established to carry out grasping experiment on vibrating walls. The experimental results demonstrate that the proposed cross-arranged claw is able to stably grasp static wall surfaces and perform well in grasping vibrating walls, with certain anti-rollover capability [27].

5. Bio-inspired Medical and Rehabilitation Robotic Technology

The rest papers collected by the Special Issue are summarized here for medical and rehabilitation application.

Jiang et al. propose a robot inverse kinematics solver based on a particle swarm optimization (PSO) back propagation (BP) neural network algorithm to solve the inverse kinematics problem of a 6 DOF UR3 robot, overcoming some disadvantages of BP neural networks. The back propagation neural network improves the convergence precision, convergence speed, and generalization ability. The results show that the position error is solved by the research method with respect to the UR3 robot inverse kinematics with the joint angle less than 0.1 degrees and the output end tool less than 0.1 mm, achieving the required positioning for medical puncture surgery, which demands precise positioning of the robot to less than 1 mm [28]. Human locomotion is a synergetic process of the musculoskeletal system characterized by smoothness, high nonlinearity, and quasi-periodicity. Duan et al. use previous and current inertial measurement unit readings to predict human locomotion based on their kinematic properties. Takens' reconstruction method is used to characterize quasi-periodicity and nonlinear systems. With Takens' reconstruction framework, they develop following methods, including Gaussian coefficient weighting and offset correction which is based on the smoothness of human locomotion, Kalman fusion with complementary joint data prediction and united source of historical embedding generation which is synergy-inspired, and Kalman fusion with the Newton-based method with a velocity and acceleration high-gain observer also based on smoothness [29]. Jiang et al. present a hardware-based method that utilizes a shielded drive circuit to eliminate extraneous interferences on biopotential signal recordings, while also preserving all useful components of the target signal. The performance of the proposed method is evaluated by comparing the results with conventional hardware and software filtering methods in different biopotential signal recording experiments [30].

Acknowledgments: This special issue receives technical support by IEEE Robotics and Automation Society (RAS) Technical Committee on Neuro-Robotics Systems, and Technical Committee on Cyborg and Bionic Systems and IEEE Systems, Man, and Cybernetics Society (SMC) Technical Committee on Bio-Mechatronics and Bio-Robotics Systems.

Conflicts of Interest: The authors declare no conflict of interest.

References

1. Yang, X.; She, H.; Lu, H.; Fukuda, T.; Shen, Y. State of the Art: Bipedal Robots for Lower Limb Rehabilitation. *Appl. Sci.* **2017**, *7*, 1182. [CrossRef]
2. Hu, Y.; Mombaur, K. Bio-Inspired Optimal Control Framework to Generate Walking Motions for the Humanoid Robot iCub Using Whole Body Models. *Appl. Sci.* **2018**, *8*, 278. [CrossRef]
3. Jiang, X.; Chen, X.; Yu, Z.; Zhang, W.; Meng, L.; Huang, Q. Motion Planning for Bipedal Robot to Perform Jump Maneuver. *Appl. Sci.* **2018**, *8*, 139. [CrossRef]
4. Zang, X.; Liu, Y.; Li, W.; Lin, Z.; Zhao, J. Design and Experimental Development of a Pneumatic Stiffness Adjustable Foot System for Biped Robots Adaptable to Bumps on the Ground. *Appl. Sci.* **2017**, *7*, 1005. [CrossRef]
5. Muscolo, G.G.; Caldwell, D.; Cannella, F. Calculation of the Center of Mass Position of Each Link of Multibody Biped Robots. *Appl. Sci.* **2017**, *7*, 724. [CrossRef]
6. Liu, Y.; Zang, X.; Heng, S.; Lin, Z.; Zhao, J. Human-Like Walking with Heel Off and Toe Support for Biped Robot. *Appl. Sci.* **2017**, *7*, 499. [CrossRef]
7. Otani, T.; Hashimoto, K.; Miyamae, S.; Ueta, H.; Natsuhara, A.; Sakaguchi, M.; Kawakami, Y.; Lim, H.-O.; Takanishi, A. Upper-Body Control and Mechanism of Humanoids to Compensate for Angular Momentum in the Yaw Direction Based on Human Running. *Appl. Sci.* **2018**, *8*, 44. [CrossRef]
8. Zeng, Y.; Li, J.; Yang, S.X.; Ren, E. A Bio-Inspired Control Strategy for Locomotion of a Quadruped Robot. *Appl. Sci.* **2018**, *8*, 56. [CrossRef]
9. Ba, K.; Yu, B.; Gao, Z.; Li, W.; Ma, G.; Kong, X. Parameters Sensitivity Analysis of Position-Based Impedance Control for Bionic Legged Robots' HDU. *Appl. Sci.* **2017**, *7*, 1035. [CrossRef]
10. Zhu, Y.; Chen, L.; Liu, Q.; Qin, R.; Jin, B. Omnidirectional Jump of a Legged Robot Based on the Behavior Mechanism of a Jumping Spider. *Appl. Sci.* **2018**, *8*, 51. [CrossRef]
11. Rubeo, S.; Szczecinski, N.; Quinn, R. A Synthetic Nervous System Controls a Simulated Cockroach. *Appl. Sci.* **2018**, *8*, 6. [CrossRef]
12. Villaseñor, C.; Rios, J.D.; Arana-Daniel, N.; Alanis, A.Y.; Lopez-Franco, C.; Hernandez-Vargas, E.A. Germinal Center Optimization Applied to Neural Inverse Optimal Control for an All-Terrain Tracked Robot. *Appl. Sci.* **2018**, *8*, 31. [CrossRef]

13. Wang, C.; Liu, X.; Yang, X.; Hu, F.; Jiang, A.; Yang, C. Trajectory Tracking of an Omni-Directional Wheeled Mobile Robot Using a Model Predictive Control Strategy. *Appl. Sci.* **2018**, *8*, 231. [CrossRef]
14. Zhu, H.; Gu, S.; He, L.; Guan, Y.; Zhang, H. Transition Analysis and Its Application to Global Path Determination for a Biped Climbing Robot. *Appl. Sci.* **2018**, *8*, 122. [CrossRef]
15. Lo, A.K.-Y.; Yang, Y.-H.; Lin, T.-C.; Chu, C.-W.; Lin, P.-C. Model-Based Design and Evaluation of a Brachiating Monkey Robot with an Active Waist. *Appl. Sci.* **2017**, *7*, 947. [CrossRef]
16. Sanfilippo, F.; Azpiazu, J.; Marafioti, G.; Transeth, A.A.; Stavdahl, Ø.; Liljebäck, P. Perception-Driven Obstacle-Aided Locomotion for Snake Robots: The State of the Art, Challenges and Possibilities. *Appl. Sci.* **2017**, *7*, 336. [CrossRef]
17. Kelasidi, E.; Jesmani, M.; Pettersen, K.Y.; Gravdahl, J.T. Locomotion Efficiency Optimization of Biologically Inspired Snake Robots. *Appl. Sci.* **2018**, *8*, 80. [CrossRef]
18. Wang, K.; Gao, W.; Ma, S. Snake-Like Robot with Fusion Gait for High Environmental Adaptability: Design, Modeling, and Experiment. *Appl. Sci.* **2017**, *7*, 1133. [CrossRef]
19. Yamamoto, I.; Rong, G.; Shimomoto, Y.; Lawn, M. Numerical Simulation of an Oscillatory-Type Tidal Current Powered Generator Based on Robotic Fish Technology. *Appl. Sci.* **2017**, *7*, 1070. [CrossRef]
20. Ozmen Koca, G.; Bal, C.; Korkmaz, D.; Bingol, M.C.; Ay, M.; Akpolat, Z.H.; Yetkin, S. Three-Dimensional Modeling of a Robotic Fish Based on Real Carp Locomotion. *Appl. Sci.* **2018**, *8*, 180. [CrossRef]
21. Xing, H.; Guo, S.; Shi, L.; He, Y.; Su, S.; Chen, Z.; Hou, X. Hybrid Locomotion Evaluation for a Novel Amphibious Spherical Robot. *Appl. Sci.* **2018**, *8*, 156. [CrossRef]
22. Gu, S.; Guo, S. Performance Evaluation of a Novel Propulsion System for the Spherical Underwater Robot (SURIII). *Appl. Sci.* **2017**, *7*, 1196. [CrossRef]
23. Del Dottore, E.; Mondini, A.; Sadeghi, A.; Mazzolai, B. Swarming Behavior Emerging from the Uptake–Kinetics Feedback Control in a Plant-Root-Inspired Robot. *Appl. Sci.* **2018**, *8*, 47. [CrossRef]
24. Castañeda Cisneros, J.; Pomares Hernandez, S.E.; Perez Cruz, J.R.; Rodríguez-Henríquez, L.M.; Gonzalez Bernal, J.A. Data-Foraging-Oriented Reconnaissance Based on Bio-Inspired Indirect Communication for Aerial Vehicles. *Appl. Sci.* **2017**, *7*, 729. [CrossRef]
25. Fuller, C.; Schultz, J. Characterization of Control-Dependent Variable Stiffness Behavior in Discrete Muscle-Like Actuators. *Appl. Sci.* **2018**, *8*, 346. [CrossRef]
26. Wang, Z.; Wang, Z.; Dai, Z.; Gorb, S.N. Bio-Inspired Adhesive Footpad for Legged Robot Climbing under Reduced Gravity: Multiple Toes Facilitate Stable Attachment. *Appl. Sci.* **2018**, *8*, 114. [CrossRef]
27. Jiang, Q.; Xu, F. Grasping Claws of Bionic Climbing Robot for Rough Wall Surface: Modeling and Analysis. *Appl. Sci.* **2018**, *8*, 14. [CrossRef]
28. Jiang, G.; Luo, M.; Bai, K.; Chen, S. A Precise Positioning Method for a Puncture Robot Based on a PSO-Optimized BP Neural Network Algorithm. *Appl. Sci.* **2017**, *7*, 969. [CrossRef]
29. Duan, P.; Li, S.; Duan, Z.; Chen, Y. Bio-Inspired Real-Time Prediction of Human Locomotion for Exoskeletal Robot Control. *Appl. Sci.* **2017**, *7*, 1130. [CrossRef]
30. Jiang, Y.; Samuel, O.W.; Liu, X.; Wang, X.; Idowu, P.O.; Li, P.; Chen, F.; Zhu, M.; Geng, Y.; Wu, F.; et al. Effective Biopotential Signal Acquisition: Comparison of Different Shielded Drive Technologies. *Appl. Sci.* **2018**, *8*, 276. [CrossRef]

Review

State of the Art: Bipedal Robots for Lower Limb Rehabilitation

Xiong Yang [1,†], Haotian She [2,†], Haojian Lu [1], Toshio Fukuda [2] and Yajing Shen [1,3,*]

1 Department of Mechanical and Biomedical Engineering, University of Hong Kong, Tat Chee Avenue, Kowloon, Hong Kong 999077, China; xiongyang2-c@my.cityu.edu.hk (X.Y.); haojianlu2-c@my.cityu.edu.hk (H.L.)

2 Beijing Institute of Technology, No.5 Yard, Zhong Guan Cun South Street, Haidian District, Beijing 100000, China; 3120150094@bit.edu.cn (H.S.); tofukuda@nifty.com (T.F.)

3 Centre for Robotics and Automation, CityU Shenzhen Research Institute, Shenzhen 518000, China

* Correspondence: yajishen@cityu.edu.hk; Tel.: +852-3442-2045

† These authors contributed equally to this work.

Received: 5 October 2017; Accepted: 3 November 2017; Published: 16 November 2017

Abstract: The bipedal robot is one of the most attractive robots types given its similarity to the locomotion of human beings and its ability to assist people to walk during rehabilitation. This review summarizes the chronological historical development of bipedal robots and introduces some current popular bipedal robots age. Then, the basic theory-stability control and key technology-motion planning of bipedal robots are introduced and analyzed. Bipedal robots have a wide range of applications in the service, education, entertainment, and other industries. After that, we specifically discuss the applications of bipedal robots in lower limb rehabilitation, including wearable exoskeleton robots, rehabilitation equipment, soft exoskeleton robots, and unpowered exoskeleton robots, and their control methods. Lastly, the future development and the challenges in this field are discussed.

Keywords: bipedal robot; bipedal locomotion; rehabilitation; robot review

---

1. Introduction

Bipedal robots are robots with two legs; the main difference between them and other robots is their bipedal locomotion. The research on bipedal robots began in the 1960s. In 1968, Smo-Sher, who worked at General Motors in the U.S., developed a controlled bipedal walking robot named Rig [1]. This started the research on bipedal robots. From 1968 to 1969, Yugoslavia's famous scientist Vukobratovic proposed an important theory on bipedal robots, the Zero Moment Point (ZMP) stability criteria [2]. Since then, scholars and research institutions in various countries have started research work on bipedal robots. The main development course of the bipedal robot is shown in Table 1 and Figure 1.

Table 1. The main development of the bipedal robot [3–35].

Time	Scientist/Institution	Achievement
1968	R. Smo-Sher, United States	Rig
1969	M. Vukobratovic, Yugoslavia	ZMP stability criteria
1969	Kato Ichiro, Japan	WAP-1
1970	Witt, United Kingdom	"Witt" type robot
1984	Kato Ichiro, Japan	WL-10RD
1986	Honda, Japan	E0
1988	National University of Defense Technology, China	KDW-I
1989	Mogeer, United States	Passive Dynamic Waking
1990	Y. F. Zheng et al., United States	Neural networks, SD-1

Table 1. *Cont.*

Time	Scientist/Institution	Achievement
1993	Honda, Japan	P-1
1997	Honda, Japan	P-3
1997	J. E. Pratt and G..A. Pratt, United States	Virtual model control
1999	MIT, United States	COG
2000	Honda, Japan	ASIMO
2000	Sony, Japan	SDR-3X
2002	Beijing Institute of Technology, China	BHR
2004	South Korea	HUBO
2004	RobotCub Consortium, Italy	iCUB
2005	University of Florida, United States	Rabbit
2005	MIT, America	Domo
2007	Aldebaran Robotics, France	NAO
2008	University of Tehran, Iran	Surena I
2009	Technical University of Munich, German	LOLA
2009	Aldebaran Robotics, France	Romeo
2010	AIST, Japan	HRP-4C
2012	NRL, United States	SAFFiR
2013	Institute of Robotics and Mechatronics, German	TORO
2013	Boston Dynamics, United States	PETMAN
2015	University of Tehran, Iran	Surena III
2016	Boston Dynamics, United States	Atlas

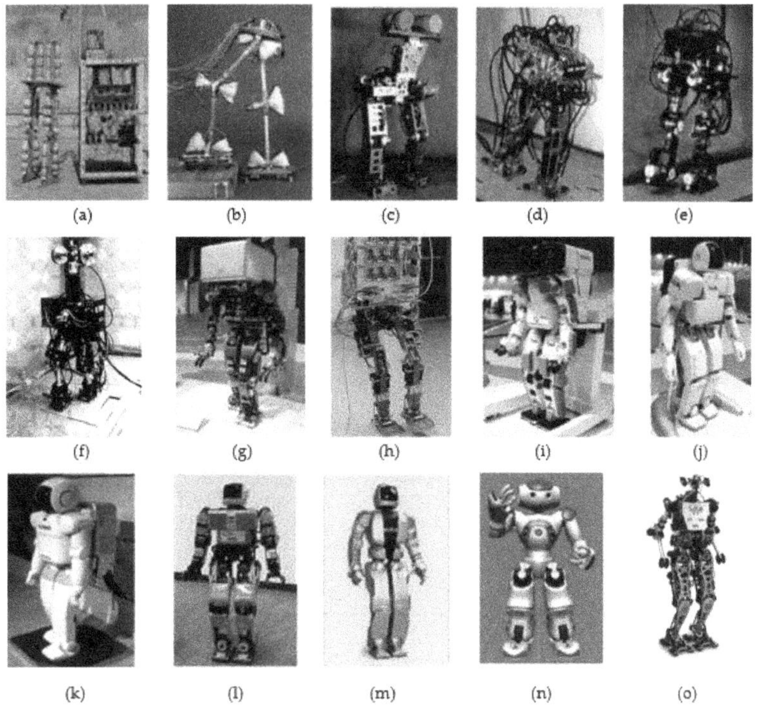

Figure 1. The historical development of bipedal robots. (**a**) WAP-1; (**b**) WAP-3; (**c**) WL-5; (**d**) WL-9DR; (**e**) WL-10RD; (**f**) WL-12; (**g**) Honda P-1; (**h**) BIP2000; (**i**) Honda P-2; (**j**) Honda P-3; (**k**) ASIMO; (**l**) KHR-2; (**m**) HUBO; (**n**) NAO; (**o**) LOLA. ((**a–e**,**g**,**i–m**) were reproduced with permission from [36]; (**f**) was reproduced with permission from [37], Copyright Springer, 1997; (**h**) was reproduced with permission from [38], Copyright Elsevier, 2004; (**n**) was reproduced with permission from [39], Copyright Springer, 2014; (**o**) was reproduced with permission from [40], Copyright Springer, 2013).

Japan has a prominent position in the field of bipedal robot research and has made many milestone achievements. In 1968, Professor Kato Ichiro of Waseda University first started the development of bipedal robots in Japan. From 1969 to 1984, Kato's laboratory launched more than 10 bipedal robots. The walking speed of the robot has developed from WL-5 with 45 s per step, to WL-10RD with 1.3 s per step [36]. As a pioneer in bipedal robot research, Kato has made a significant contribution. Since 1986, Honda has launched seven E series robots, three P series robots, and an intelligent robot called ASIMO. Honda's research work, especially the P-3 and ASIMO, pushed the development of bipedal robots to a new level, making the development, production, engineering, and marketing of biped robots practical. Japan has many other scientific research institutions engaged in bipedal robot development and theoretical research work, such as Matsushita Electric Works, Fujitsu, Farrah, and Hitachi. They have performed significant research on bipedal robots and have achieved some success.

In parallel with the Japanese developments, Mogeer put forward the "Passive Dynamic Waking" theory in 1989 [41,42]. This theory improved the system of bipedal robots. The next year, Zheng et al. proposed the use of neural networks to achieve stable dynamic walking and achieved their goal with the SD series bipedal robots [43,44]. This was the first time intelligent algorithms were integrated with gait planning. In 1997, Pratt and Pratt et al. of the Massachusetts Institute of Technology proposed a virtual model control (VMC) strategy in the research of the bipedal robot Spring Flamingo [45]. The use of VMC effectively avoided the cumbersome robot dynamics and inverse kinematics calculations. This was helpful for the research of bipedal robots. In 2005, the University of Florida developed a bipedal walking robot, Rabbit, to show the world its running ability [46]. The walking ability of bipedal robots was again improved.

At present, the research and development of bipedal robots are mainly concentrated in Japan, the U.S., China, South Korea, and France. However, the focus of each country's research is different. Japanese bipedal robots are biased toward simulating human movements and living characteristics. European countries developed bipedal robots biased toward medical services, and the American research bipedal robot has focused on military applications. Over the past decade, new bipedal robot products were being constantly developed. In 2010, Japan's Advanced Industrial Science and Technology (AIST) released an entertaining bipedal robot HRP-4C [20]. Although HRP-4C is an entertainment robot meant for singing and dancing, it is technologically advanced. PETMAN robots were developed by Boston Dynamics in 2013 [47], a lifelike humanoid robot mainly used to test protective clothing. Surena III was the latest generation of bipedal robots launched by Tehran University in 2015 [24]. Its body can move backward while standing on one foot. In 2016, Boston Dynamics showcased Atlas, developed for the U.S. military [25]. Atlas can not only walk like a human body with its legs, but can also adapt to a variety of outdoor terrain. These bipedal robots, as shown in Figure 2, represent the top level of bipedal robot development. The theory of bipedal locomotion has basically matured, although room for improvement exists in stability. The researchers should allow the bipedal robot to identify the external environment information, so that it knows how and where to go. The implementing agency, cooperating with the bipedal locomotion, needs to be improved and perfected. Once this is attained, bipedal robots will be able to complete more tasks.

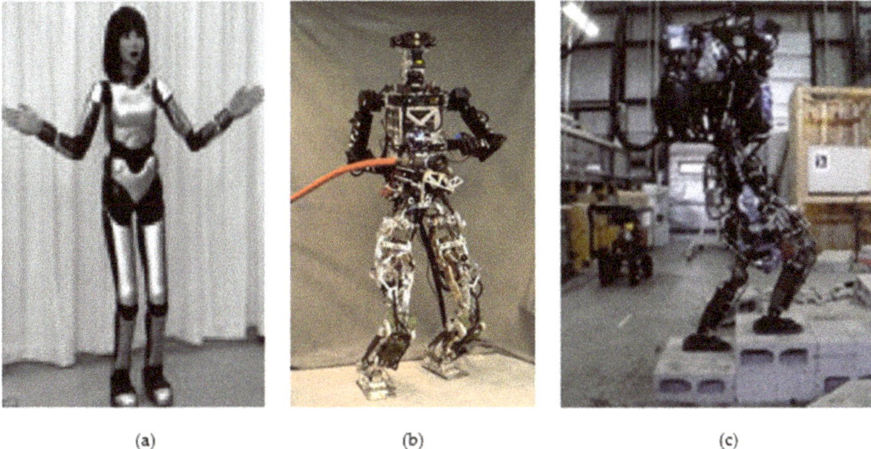

Figure 2. Popular bipedal robots at current stage. (**a**) HRP-4C used for entertainment; (**b**) SAFFiR used for firefighting; (**c**) Atlas walking on an uneven ground. ((**a**) was reproduced with permission from [36]; (**b**) was reproduced with permission from [48], Copyright Elsevier, 2015; (**c**) was reproduced with permission from [49], Copyright Springer, 2016.)

Conversely, the movement of bipedal robots is similar to that of humans, and many bipedal robot technologies can be applied to rehabilitation, such as structural design, stability control, and gait planning. So, the development of bipedal robots will promote the development of rehabilitation, and rehabilitation will become one of the important applications of bipedal robots.

2. Bipedal Locomotion

2.1. Hardware and Structure

The bipedal locomotion of the robot imitates human walking, so the robot also mimics the human leg in the design of the mechanical structure, including the hip, knee, and ankle [50]. Figure 3a shows the classic lower limb structure with 12 degrees of freedom, where each leg has 6 degrees of freedom. Usually, the rotation pair (Figure 3b) will be used to simulate the joints of the human leg. The actuator is used instead of the muscle to drive the various joints. Currently, widely-used drives include motors, hydraulic pumps, and pneumatic muscles. A large number of sensors installed on the robot are equivalent to the human senses used to obtain their own state of motion and external environment information, for example, torque sensors, angle sensors, and contact switches [51]. This provides the basis for the control of the bipedal locomotion. This hardware provides the possibility for the bipedal robot to walk on the flat, slopes, or even up the stairs like a human. The imitation of human lower limb structure laid the foundation for the application of bipedal robot in rehabilitation.

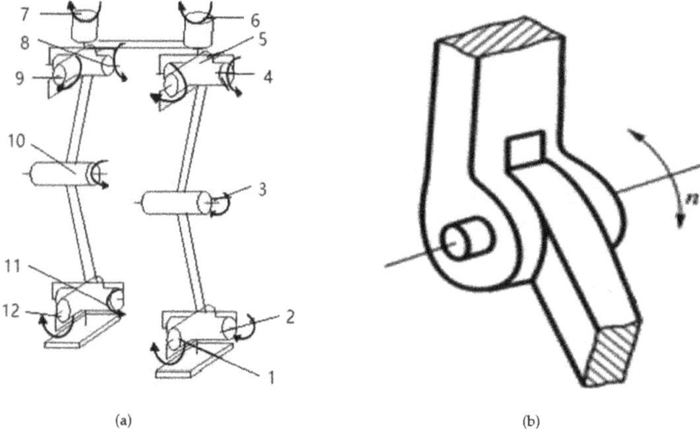

Figure 3. The mechanical structure of the bipedal locomotion. (**a**) The classic mechanical structure with 12 degrees of freedom; (**b**) The rotating hinge structure.

2.2. Stability Control

Stability control is an important part of bipedal locomotion. When people walk on uneven surfaces or are disturbed, the upper limbs and lower limbs ware adjusted accordingly to maintain body balance. Due to the role of inertia, the robot's state is very unstable when it is influenced by outside interference. To allow bipedal robots to be as stable and balanced as humans, researchers created stability control based on sensory reflexes, including the zero moment point (ZMP) reflex, the landing-phase reflex, and the body-posture reflex [52]. The three sensory re-flexes are independent and work when their conditions are satisfied.

- ZMP Reflex

ZMP is an abbreviation for zero moment point. ZMP reflex does not work when ZMP is in a stable area. The ZMP reflex will not be active until ZMP is at a stable area boundary and exceeds the stability range. By changing the corner of the ankle joint, the ZMP can be returned to the stable range. Therefore, the use of ZMP reflex can ensure that ZMP is always in a stable area during walking. The angle of the supporting ankle can be calculated by Equations (1) and (2) [53]:

$$\Delta\theta_a(nT_s) = \sum_{j=1}^{j=n} \delta\theta_a(jT_s) \tag{1}$$

$$\delta\theta_a(jT_s) = \begin{cases} K_{ac} * d_{ZMP}(jT_s), \ F_z > 0 \\ -K_{as}\Delta\theta_a((j-1)T_s), \ F_z = 0 \end{cases} \tag{2}$$

where T_s is the sampling period, and nT_s is current time. $d_{ZMP}(jT_s)$ is the distance between the actual ZMP position and the expected stable area boundary. K_{ac} and K_{as} are coefficients. F_z is the force between the supporting foot and the ground.

- Landing-Phase Reflex

When the ground level is uneven, the contact time between the foot and the ground will be corresponding in advance or delay. If the gait is not corrected in time, the robot is likely to fall. Landing-phase reflex can effectively solve this problem. Landing-phase reflex will increase the height of the leg lift when the contact time between the foot and the ground is advanced. The increase in the height of the leg lift can be calculated according to the Formulas(3) and (4) [53]:

$$\Delta Z_f(nT_s) = \sum_{j=1}^{j=n} \delta Z_f(jT_s) \tag{3}$$

$$\delta Z_f(jT_s) = \begin{cases} K_{fc} * F_z,\ 0 < F_z < Mr \\ -K_{fs}\Delta Z_f((j-1)T_s),\ F_z \geq Mr \end{cases} \tag{4}$$

Similarly, when the contact time between the foot and the ground is delayed, the height of leg lift will be reduced accordingly, as per Equations (5) and (6) [53]:

$$\Delta Z_f(nT_s) = \sum_{j=1}^{j=n} \delta Z_f(jT_s) \tag{5}$$

$$\delta Z_f(jT_s) = \begin{cases} -h,\ F_z = 0 \\ -K_{fs}\Delta Z_f((j-1)T_s),\ F_z \geq Mr \end{cases} \tag{6}$$

where K_{fc}, K_{fs} are coefficients, and M_r is estimated body weight of robot. h is a constant variable. F_z is the force between the supporting foot and the ground. The change of foot height $\Delta Z_f(nT_s)$ is achieved by controlling the hip and knee joints.

- Body Posture Reflex

Changes in body posture also affect bipedal locomotion. Reasonable body posture contributes to the stability of the system. When the robot is tilted due to outside interference, the body posture deviates from the expected target. The role of the body posture reflex is to change the hip to adjust the body posture. This ensures the body posture remains in a reasonable range. From the current actual posture to the target posture, the angle of the hip can be calculated from Equations (7) and (8) [53]:

$$\Delta\theta_h(nT_s) = \sum_{j=1}^{j=n} \delta\theta_h(jT_s) \tag{7}$$

$$\delta\theta_h(jT_s) = \begin{cases} \Delta\theta_{actb}(jT_s),\ F_z > 0 \\ -K_{hs}\Delta\theta_h((j-1)T_s),\ F_z = 0 \end{cases} \tag{8}$$

where $\Delta\theta_{actb}(jT_s)$ is the deviation between the actual body posture and the required body posture, and K_{hs} is the coefficient. F_z is the force between the supporting foot and the ground.

- Intelligent Algorithm

The above control ideas build a reasonable research model and theoretical background for the actual control of the robot. On this basis, the correct introduction of an intelligent control algorithm will improve the applicability and robustness of the algorithm. The traditional intelligent control algorithms, such as connectionism, fuzzy logic, and genetic algorithms, have strong self-learning, adaptive, and fault-tolerant capabilities, and are gradually introduced into the control of each model. The author of the literature [54] is one of the most successful scholars who used the neural network to control the robot. Initially, they used supervised learning to train three CMAC neural networks for adaptive learning, and then used them to control the front to back balance, the left to right side pendulum, and the continuous changing pose. In addition, according to the needs of actual robot control, the combination of intelligent control algorithms, such as fuzzy neural network, neuron fuzzy logic, and neuron genetic algorithm, is also appearing in the control of biped robots [55].

- Torque Control

The torque-controlled robot has been proposed as a new concept. This method is based on the force balance at the joints. Generally, a torque sensor is installed at the joints of the robot. The motion of the robot can be obtained by processing the sensor data, and the location and size of the external force can be calculated. On this basis, the robot can achieve touch stop, drag teaching, impedance control, and other functions using a reasonable force control algorithm. Unlike position-controlled robots, this flexibility allows a greater degree of safety when interacting with people, as well as greater robustness in contact with the environment. In 2013, DLR introduced a torque-controlled humanoid robot TORO. TORO is a research platform for scientific topics addressing bipedal movement and dynamics [56], which will promote the development and popularization of torque-control methods. As torque-control has a unique advantage, it will attract more attention in the future.

2.3. Motion Planning

Like humans, the bipedal robot walks by alternating legs. To achieve a stable walk, a reasonable gait must be planned. The quality of the gait will directly affect the stability of walking, the size of the driving torque, and aesthetics. Similarly, in the application of rehabilitation robots, motion planning is also needed for coordination with the movement of people. At present, five kinds of gait planning methods exist.

- Gait Planning Based on Bionic Kinematics

The purpose of bipedal locomotion is to imitate human walking characteristics, so gait planning can learn from the human bionic gait. Human motion capture data (HMCD) [57,58] can plan complex and diverse actions. As the reference action is generated by the human body, the planning method becomes simpler. Honda's research team generated ASIMO's gait based on the analysis of the mutual restraint and coordination between the joints of human lower limbs. However, the feasibility of the method entirely depends on the walking data, and accurately and completely measuring or recording the characteristics of human walking data is difficult with the existing instrument. So obtaining gait data for robots is challenging.

- Gait Planning Based on Stability

The study of human dynamic walking shows that the balance of human walking is not due to the relatively large soles of the feet, but the complex coordination of the body. To achieve a stable gait, a large number of research scholars have proposed stabilization criteria for different models, such as center of pressure criterion (CoP) [59], foot rotation indicator (FRI) [60], and zero moment point (ZMP) criterion. Among them, the ZMP stability criterion proposed by Vukobratovic is one of the most widely used. Honda's ASIMO used the ZMP theory to achieve walking and balance of bipedal robots.

- Model-Based Gait Planning

The model-based gait planning method mainly includes the multi-link model (Figure 4a) and the inverted pendulum model (Figure 4b) [61,62]. The gait planning of the multi-link model has been successfully applied to the Honda robot. This gait planning method is based on the ideal model while ignoring the radial and lateral motion of the coupling. Therefore, when the planned gait is applied to the actual prototype, an unavoidable error occurs. However, this is one of the most widely used methods. The adaptability of the data to the actual environment, and the robustness to the interference, need to be improved in the future.

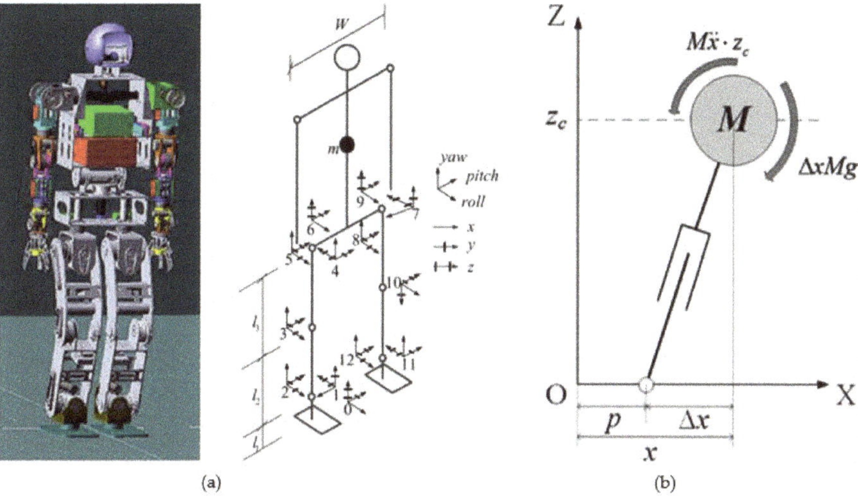

Figure 4. The model-based gait planning (**a**) The multi-link model; (**b**) The inverted pendulum model. ((**a**) was reproduced with permission from [63]; (**b**) was reproduced with permission from [64], Copyright Cambridge University Press, 2016.)

- Gait Planning Based on Energy Consumption Optimization

People walk with the lowest energy consumption and remain stable as result of the evolutionary process. Therefore, planning a gait with the energy-optimized method can make bipedal locomotion more human. Vukboratovic first analyzed the torque required to achieve the joint movements from the energy point of view [65]. Capi generates the walking gait of the biped robot Bonten-Maru for the optimization target with the lowest energy and the minimum variation in joint torque [66]. The energy optimization method can fully develop the performance of bipedal robots and reduce the system requirements. However, the calculation of the optimal planning is complex and cannot be implemented in real time.

- Gait Planning Based on Intelligent Algorithm

Intelligent techniques, such as neural networks, fuzzy logic, and genetic algorithms, have powerful self-learning, adaptive, and fault-tolerant capabilities that have attracted many robotics researchers to apply them to gait planning. In 1992, each joint of robot SD-2 was represented by a joint neuron [67]. In the study, the neural network obtained the relationship between the foot force and the angle of the corresponding joint angle adjustment. In addition to the neural network, due to the advantages of fuzzy logic in terms of knowledge expression, many scholars apply it to the generation and control of the biped gait. The literature [68] uses a fuzzy gait parameter adjustment algorithm to dynamically control the gait parameters, including the step size and rotation speed. In the literature [69], the concept of fuzzy logic has been combined with Linear Quadratic Regulator (LQR) controller theory to design the best method to allow the biped robot system to have a balanced and stable gait. However, intelligent algorithms generally require a large number of training samples and long-term calculations, and the spatial structure and data convergence of the samples have not yet been fully solved. Much work remains to be done in applying the intelligent algorithm to gait planning.

3. Rehabilitation Application

The lower limbs of the human body are mainly responsible for standing, maintaining balance, and walking. The requirements of these functions are the same in bipedal robots. Conversely, bipedal robots are designed and controlled by the principle of realistically imitating people. As a result, bipedal robots and related technologies, such as stability control and motion planning, can be applied in human lower limb rehabilitation training and assisted walking. The application of bipedal robot technology to human lower limb rehabilitation training is limited. In 2000, with the support of the U.S. Defense Advanced Research Projects Agency, Berkeley's Human Engineering Laboratory (HEL), SARCOS Robotics, Oak Ridge National Laboratory (ORNL), and Millennium Jet began research on bipedal exoskeleton robots. Most bipedal exoskeleton robots are still in the initial stages of development. However, the technical achievements on bipedal robots have contributed to the development of lower limb rehabilitation robots, particularly for gait planning, stability control, and sensor applications. Of course, a re-examination of the theoretical approach helps to improve the robots.

However, bipedal robots and rehabilitation robots cannot be equated. For rehabilitation robots, humans can act as a link in control, whereas the existing bipedal robot technology often does not consider this aspect. Therefore, developing and improving the control scheme from the unique perspective of the rehabilitation robot is necessary. Overall, the bipedal robot technology has a large volume of reference material for lower limb rehabilitation robots. Regardless of the type of lower limb rehabilitation robot, they all use bipedal robot control theory, especially for stability control and gait planning. Four common types of lower limb rehabilitation robots exist.

- Wearable Exoskeleton Robot

The wearable exoskeleton robot is one of the earliest proposed robots to assist with walking. Its biggest feature is that it allows the user's range of activities to be less restrictive. It can be used to assist patients with hemiplegia to walk, and to improve the body's normal function. This field has the highest number of researchers. The Berkeley lower extremity exoskeleton (BLEEX) (Figure 5a), developed by the University of California at Berkeley and the Ergonomics Laboratory, already has two versions [70]. The mechanical system of the bipedal exoskeleton robot was designed according to the biological characteristics of the normal adult man. A plurality of sensors and actuators mounted on the exoskeleton legs provide the necessary control information to the control center. The computer accepts this information while detecting the user's real-time status. By adjusting the drive torque to assist joints, people can walk long distances while carrying external load. Experiments show that when the load on BLEEX-1 is 34 kg, walking speed can reach 1.3 m/s. However, the human body feels only 2 kg of the external weight. When the load on BLEEX-2 is up to 45 kg, walking speed can be increased to 2 m/s. Cyberdyne HAL [71] (Figure 5b), Israel ReWalk (Figure 5c), and Rex Bionics' Rex (Figure 5d) are all wearable exoskeletons. Due to the complexity of the mechanical structure and the energy consumption of the equipment, the current wearable exoskeleton robot efficiency is low.

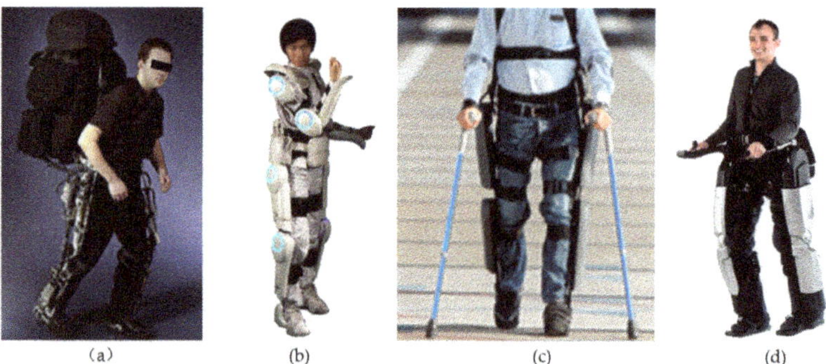

Figure 5. The wearable exoskeleton robot. (**a**) BLEEX; (**b**) Cyberdyne HAL; (**c**); Israel ReWalk; (**d**) Rex Bionics' Rex. ((**a**,**c**) were reproduced with permission from [72]; (**b**,**d**) were reproduced with permission from [73], Copyright Springer, 2013.)

- Rehabilitation Equipment

Rehabilitation equipment also uses bipedal robot technology. Unlike the wearable exoskeleton robot, rehabilitation equipment is generally larger and heavier in quality. Rehabilitation equipment usually requires more sensors to obtain human motion data, and have specialized data processing equipment. Although rehabilitation robots can provide better service as well as a rehabilitation effect, they are expensive and cumbersome. So, rehabilitation robots can only exist in hospitals or large rehabilitation centers. Toyota demonstrated the rehabilitation equipment Welwalk-1000 in 2017 and will lease the rehabilitation system to various medical institutions by the end of the year [74]. The patient's thighs, knees, ankles, and feet are fixed to the equipment and then the patient walks on a treadmill. In the course of walking, the internal sensor detects the action and continually adjusts the robot. This system contributes to the recovery of patients with slow progress, and training can accelerate rehabilitation. Simultaneously, the system helps the therapist to monitor the patient's progress and better grasp the patient's condition. LOKOMAT (Figure 6a) and LokoHelp (Figure 6b) have a function similar to Welwalk-1000. Being bulky and hard to carry are their common shortcomings.

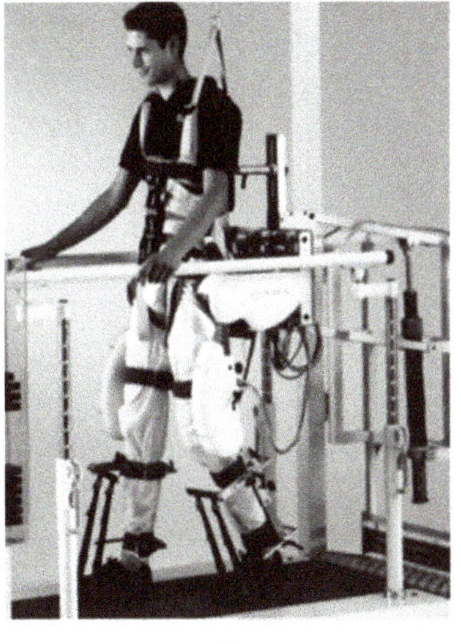

(a)

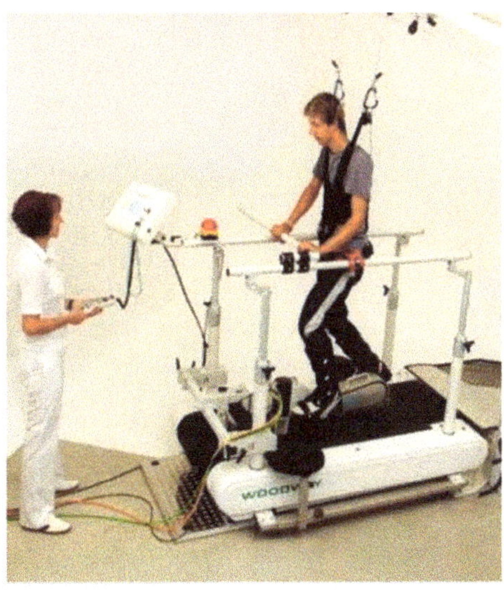

(b)

Figure 6. Rehabilitation equipment. (**a**) LOKOMAT; (**b**) LokoHelp. (reproduced with permission from [75].)

- Other Emerging Exoskeleton Robots

To minimize the burden on the human body due to the weight of the lower limb auxiliaries, exoskeleton robots with soft structures have been proposed by researchers. The advantages of the soft material include the weight and the ability to overcome the constraints of a rigid structure. The soft exoskeleton can better adapt to the deformation of the human body during walking. However, how to effectively drive a soft material and pass the force on to the human body becomes a challenge. In 2013, Harvard University developed an exoskeleton robot Exosuit [76] (Figure 7a). The robot employs a flexible design that allows the wearer to remove the constraints of the rigid material and move more naturally. In realizing the assistive walk function, the robot weighs only 7.5 kg. The video of the stroke on the treadmill proves that it can improve a person's ability to walk.

When the soft exoskeleton is worn, the patients no longer need crutches, and their footsteps become faster and more confident.

The unpowered exoskeleton robot is a new concept. Similar to passive walking in bipedal robots, robotic joints do not require a driver, using the body's own gravity or spring potential to assist people to walk. Although the principle of the unpowered exoskeleton robot is simple and does not require a control system, the design must consider the constraint of the degree of freedom, gait planning, and force analysis. Researchers from Carnegie Mellon University and the University of North Carolina designed an unpowered exoskeleton system, Exoboot [77] (Figure 7b). This exoskeleton system uses carbon fiber material, so it is very light at about 500 g. The use of springs to store energy reduces the energy consumed in walking by nearly 7%. Advanced exoskeleton robots have many advantages over traditional exoskeleton robots. However, the research on advanced exoskeleton robots has just started and remains a good research prospect. Table 2 summarizes and compares the characteristics of different exoskeleton robots.

(a)

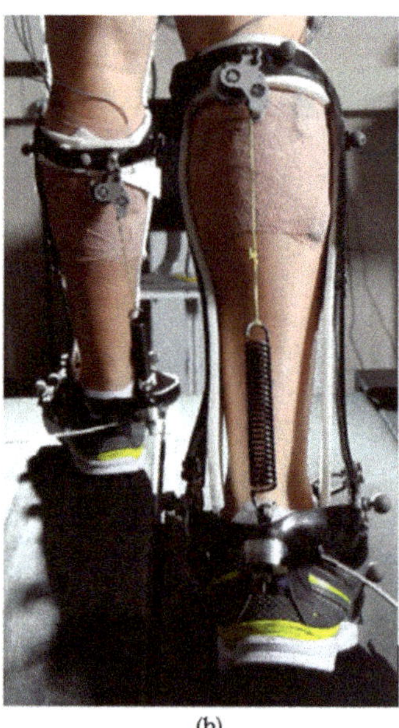
(b)

Figure 7. Advanced exoskeleton robot. (**a**) Soft exoskeleton robot; (**b**) Unpowered exoskeleton robot Exoboot. ((**a**) was reproduced with permission from [78]; (**b**) was reproduced with permission from [77], Copyright Nature, 2015.)

Table 2. Comparison of various types of rehabilitation robots.

Type	Advantages	Disadvantages	Example
Wearable exoskeleton robot	Wearable, the range of use is wide	Uncomfortable, Rigid structure	BLEEX, HAL, ReWalk
Rehabilitation equipment	Functional, Rehabilitation effect is good	Bulky	Welwalk-1000, LOKOMAT
Soft exoskeleton robot	Light quality, Unconstrained	Auxiliary force is relatively small	Exosuit
Unpowered exoskeleton robot	Unpowered	Structural design is difficult	Exoboot

For bipedal robots, the gait planning based on bionic kinematics requires the collection of human motion data. Since the assistance of exoskeleton robots is based on the accurate acquisition of human motion intent, these signal acquisition techniques can be applied to the control of exoskeleton robots. The current exoskeleton robot control methods are: manipulator control, force feedback control, EMG signal control, pre-programmed control, master-slave control, ZMP control, and sensitivity amplification control. The hybrid auxiliary limb (HAL) [71] (Figure 5b), developed by Japanese robot manufacturer Cyberdyne, is controlled based on the EMG signal. HAL is also the world's first bipedal skeleton robot that can be controlled by human minds. In 2013, the HAL bipedal exoskeleton robot received global safety certification in Japan. Obtaining a Global Certification means that HAL will be delivered to the market worldwide, which is good news for people with disabilities. ReWalk [79] (Figure 5c), designed and manufactured by Israel ReWalk, is adaptive, unlike other common exoskeletons or prostheses. It self-learns and walks at the habitual pace of the user with an intelligent control method. In 2017, Zhang proposed using the person as a part of the control of the exoskeleton robot by measuring the body's metabolism of the gas to obtain the real-time state of the human body [80]. Based on this, the control of exoskeleton robots was optimized. Figure 8 shows the operating principle of the entire system. This theory has raised the l man-machine cooperation of exoskeleton robots to a new level.

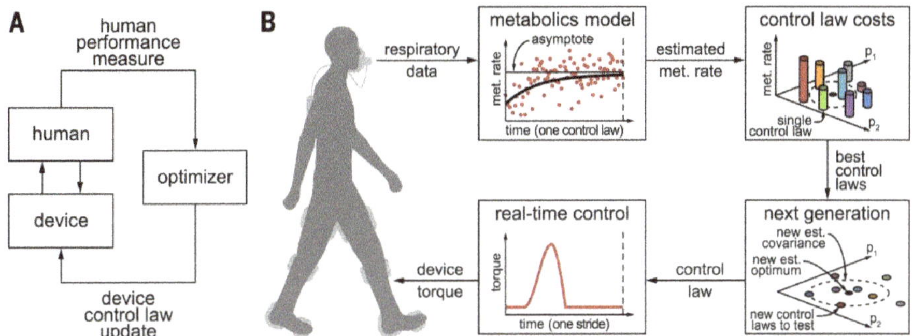

Figure 8. Human-in-the-loop optimization of exoskeleton assistance during walking (reproduced with permission from [80], Copyright Science, 2017).

The number of elderly people suffering from hemiplegia or paraplegia for whom walking is an inconvenience or impossibility is huge, and this number continues to rise. With an aging society, the number of elderly people will increase sharply in the future, and the number of elderly people incapacitated will inevitably increase. With the increase of these social needs, rehabilitation will become one of the important applications of the bipedal robot.

4. Challenges and Outlooks

The application of bipedal robots to rehabilitation has become a research hotspot, and its application prospects are very broad. However, there is still a long way to go for the commercialization of the rehabilitation equipment. First, the rehabilitation effect of the equipment is the aspect that people most concerned. Although rehabilitation equipment can significantly reduce the workload of physical therapist and contribute to the popularity of rehabilitation training, the rehabilitation effect of equipment has not yet reached the level of physical therapist. Practicality is also one of the factors that rehabilitation equipment needs to consider. The structure of traditional rigid rehabilitation equipment is too complex and cumbersome, which is not only detrimental to wear but also not conducive to the late maintenance and repair. The new soft rehabilitation robot greatly simplifies the structure and weight of the equipment, but the auxiliary force is also reduced. The commercialization of rehabilitation equipment must take the cost into account. At present, the price of rehabilitation robot, such as HAL and ReWalk, is very high. This is not all patients can afford. In order to achieve commercialization in the future, the cost of rehabilitation equipment must be well controlled. Comprehensive research status, there are still many challenges in theoretical and technical aspects:

- Mechanical Structure

The rehabilitation robot is a complex system with multiple joints and redundant degrees of freedom. The complex and heavy structure is presently one of the important factors limiting the development of the rehabilitation robot. A fully functional intelligent rehabilitation robot must have a compact and disposable mechanical body. How to achieve the desired function and optimize the structure is worth exploring. The use of advanced materials, such as carbon fiber and titanium alloy, is the usual method used to reduce the weight. However, the types of advanced materials are limited and the cost is usually high, resulting in it being difficult to meet the demand. However, with the progress in processing and manufacturing, integration and modular design will be the development trend in the future. In general, the mechanical structure should be designed to meet people's requirements for flexibility, stability, and comfort.

- Kinematics and Dynamics

The control of the rehabilitation robot is performed based on kinematics and dynamics. The accuracy of the model determines the final control effect. Due to the high order, strong coupling, and non-linearity of the rehabilitation robot system, completely solving the kinematics and dynamics is difficult. Some scholars have introduced methods such as neural networks, genetic algorithms, fuzzy logic, chaos theory, and other methods. However, these methods are not yet mature and have their own advantages and disadvantages. With the introduction of new theoretical methods and the continuous improvement of computer performance, better solutions should be presented. In the future, accurate kinematics and dynamics simulation on a computer will be possible.

- New Drives

The drive is an important part of a rehabilitation robot. As performance requirements for rehabilitation equipment become increasingly higher, the drive requirements are following suit. Rehabilitation equipment must provide power to help people walk. To provide sufficient power, the drive used by the robot is often cumbersome. This increases the weight of the equipment and wastes energy. The ideal drive must be lightweight, small, and provide enough drive torque, while having good heat dissipation. The development requirements of the robot are not being met by the traditional drive method, so people are attempting to find new drives. Shape memory alloy drive and piezoelectric ceramic drive are being gradually applied to the field of robots. These drives are faster and more capable of load, making them good prospects for the field of rehabilitation robots.

- Energy Resources

Most rehabilitation robots are mainly powered by batteries, and the range of movement is limited by the capacity and efficiency of the battery. The ideal energy source should have a very high energy density, high-temperature resistance, corrosion resistance, be renewable, and inexpensive. Therefore, improving the drive source to make it small, lightweight, and have a large capacity is a problem that must be solved in the development of bipedal robots. Researchers in the future can seek new energy technologies, including solar energy and bioenergy, to address the energy development bottleneck. With the development of nanotechnology, high-density batteries are expected to solve the energy problem, such as with graphene batteries. Wireless charging is also a means of solving the energy problem with rehabilitation robots. However, the current wireless charging technology is not mature enough and cannot achieve long-distance transmission.

- Sensing Technology

To accurately obtain the user's motion information, many sensors, such as force sensors, torque sensors, gyroscopes, vision sensors, and acoustic sensors, are installed in bipedal robots. The control of the robot is sensor-based. However, the sensor accuracy can be improved and the cost can be decreased. Additionally, the traditional rigid sensors tend to be relatively large and limit the movement of the human body. Many new types of sensors based on soft material have been presented, which are well-suited to human body detection. A soft structure not only fits the surface of the human body, but is also more comfortable for the user. A micro-sensor is a new type of sensor based on semiconductor technology. Micro-sensors are increasingly popular because of their miniaturization, intelligence, low power, and high integration. Although these new sensors are still facing problems in manufacturing, sensitivity, and stability, they have potential for robot sensing, especially for application in rehabilitation.

- Human-Robot Co-Fusion

In the traditional training process, rehabilitation is the interaction between a patient and a physical therapist. When using robots, the ideal robot should have the same or similar ability as physical therapists in terms of being able to detect the rehabilitation condition of the patient, design a personalized training plan, and dynamically optimize the training process. However, current robotic systems are still in a very early stage. The systems can perform some basic tasks through programming, but the results are far from intelligent. A new concept called "human-in-the-loop" has been proposed by the research team at CMU, where the human and robot are integrated in a closed loop and the robot is able to adjust the working conditions according to the human's walking condition. This type of human-robot co-fusion could be a trend for the development of rehabilitation robots.

5. Conclusions

Although the research on bipedal robots has had many achievements, many shortcomings in stability, man-machine coordination, and cooperation still exist, limiting the large scale entrance of the bipedal robot in our lives. The application of bipedal robots to rehabilitation places a higher requirement on the structure, control, stability, and human-computer interaction. With the development of biotechnology, electromechanical integration technology, and control technology, bipedal robot technology is also maturing. In the near future, bipedal robots might be better suited for application to rehabilitation and provide better service to people.

Acknowledgments: This work was partly supported by National Science Foundation of China (61773326), Shenzhen (China) Basic Research Project (JCYJ20160329150236426), ITF of Hong Kong (GHP/017/14SZ), and Hong Kong RGC General Research Fund (CityU 11278716).

Author Contributions: This review is drafted by X.Y., H.S. and H.L. Toshio Fukuda and Yajing Shen gave some advice and ideas about the organization of the article. X.Y. and H.S. contributed equally to this work.

Conflicts of Interest: The authors declare no conflict of interest.

References

1. Chen, B.; Luo, M.; Guo, F.; Chen, S. Walking mechanism and kinematic analysis of humanoid robot. In Proceedings of the 2013 International Conference on Advanced Mechatronic Systems (ICAMechS), Luoyang, China, 25–27 September 2013; pp. 491–494.
2. Vukobratović, M.; Branislav, B. Zero-moment point—Thirty five years of its life. *Int. J. Humanoid Robot.* **2004**, *1*, 157–173. [CrossRef]
3. Akhtaruzzaman, M.; Shafie, A.A. Evolution of humanoid robot and contribution of various countries in advancing the research and development of the platform. In Proceedings of the 2010 International Conference on Control Automation and Systems (ICCAS), Gyeonggi-do, Korea, 27–30 October 2010; pp. 1021–1028.
4. Takanishi, A. Quasi dynamic walking of the biped walking robot. *J. Robot. Soc. Jpn.* **1983**, *1*, 196–203. [CrossRef]

5. Lim, H.-O.; Atsuo, T. Biped walking robots created at Waseda University: WL and WABIAN family. *Philos. Trans. R. Soc. Lond. A* **2007**, *365*, 49–64. [CrossRef] [PubMed]
6. Lim, P.; Al Kushi, A.; Gilks, B.; Wong, F.; Aquino-Parsons, C. Early stage uterine papillary serous carcinoma of the endometrium. *Cancer* **2001**, *91*, 752–757. [CrossRef]
7. Kato, T.; Takanishi, A.; Jishikawa, H.; Kato, I. The realization of the quasi-dynamic walking by the biped walking machine. In Proceedings of the Fourth Symposium on Theory and Practice of Robots and Manipulators, Zaborów, Poland, 8–12 September 1981.
8. Takanishi, A.; Naito, G.; Ishida, M.; Kato, I. Realization of plane walking by the biped walking robot WL-10R. In *Theory and Practice of Robots and Manipulators*; Springer: Colfax County, MX, USA, 1985; pp. 383–393.
9. Kajita, S.; Kanehiro, F.; Kaneko, K.; Fujiwara, K.; Yokoi, K.; Hirukawa, H. A realtime pattern generator for biped walking. In Proceedings of the IEEE International Conference on Robotics and Automation, Washington, DC, USA, 11–15 May 2002; Volume 1, pp. 31–37.
10. Zheng, Y.F.; Shen, J. Gait synthesis for the SD-2 biped robot to climb sloping surface. *IEEE Trans. Robot. Autom.* **1990**, *6*, 86–96. [CrossRef]
11. Espiau, B.; Sardain, P. The anthropomorphic biped robot BIP2000. In Proceedings of the IEEE International Conference on Robotics and Automation, San Francisco, CA, USA, 24–28 April 2000; Volume 4, pp. 3996–4001.
12. Hirai, K.; Hirose, M.; Haikawa, Y.; Takenaka, T. The development of Honda humanoid robot. In Proceedings of the IEEE International Conference on Robotics and Automation, Leuven, Belgium, 16–20 May 1998; Volume 2, pp. 1321–1326.
13. Pratt, J.E.; Pratt, G.A. Exploiting natural dynamics in the control of a planar bipedal walking robot. In Proceedings of the Annual Allerton Conference on Communication Control and Computing, Monticello, IL, USA, 23–25 September 1998; Volume 36, pp. 739–748.
14. Sakagami, Y.; Watanabe, R.; Aoyama, C.; Matsunaga, S.; Higaki, N.; Fujimura, K. The intelligent ASIMO: System overview and integration. In Proceedings of the IEEE/RSJ International Conference on Intelligent Robots and Systems, Lausanne, Switzerland, 30 September–4 October 2002; Volume 3.
15. Kuroki, Y.; Fujita, M.; Ishida, T.; Nagasaka, K.I.; Yamaguchi, J.I. A small biped entertainment robot exploring attractive applications. In Proceedings of the IEEE International Conference on Robotics and Automation, Taipei, Taiwan, 14–19 September 2003; Volume 1, pp. 471–476.
16. Huang, Q.; Li, K.; Wang, T. Control and mechanical design of humanoid robot BHR-01. In Proceedings of the 3rd IARP International Workshop on Humanoid and Human Friendly Robotics, Tsukuba, Japan, 11–12 December 2002; pp. 10–13.
17. Akachi, K.; Kaneko, K.; Kanehira, N.; Ota, S.; Miyamori, G.; Hirata, M.; Kanehiro, F. Development of humanoid robot HRP-3P. In Proceedings of the 5th IEEE-RAS International Conference on Humanoid Robots, Tsukuba, Japan, 5–7 December 2005; pp. 50–55.
18. Park, I.W.; Kim, J.Y.; Lee, J.; Oh, J.H. Online free walking trajectory generation for biped humanoid robot KHR-3 (HUBO). In Proceedings of the IEEE International Conference on Robotics and Automation, Orlando, FL, USA, 15–19 May 2006; pp. 1231–1236.
19. Lohmeier, S.; Buschmann, T.; Ulbrich, H. Humanoid robot LOLA. In Proceedings of the IEEE International Conference on Robotics and Automation, Kobe, Japan, 12–17 May 2009; pp. 775–780.
20. Kaneko, K.; Kanehiro, F.; Morisawa, M.; Miura, K.; Nakaoka, S.I.; Kajita, S. Cybernetic human HRP-4C. In Proceedings of the 9th IEEE-RAS International Conference on Humanoid Robots, Paris, France, 7–10 December 2009; pp. 7–14.
21. Kaneko, K.; Kanehiro, F.; Morisawa, M.; Akachi, K.; Miyamori, G.; Hayashi, A.; Kanehira, N. Humanoid robot hrp-4-humanoid robotics platform with lightweight and slim body. In Proceedings of the IEEE/RSJ International Conference on Intelligent Robots and Systems (IROS), San Francisco, CA, USA, 25–30 September 2011.
22. Henze, B.; Werner, A.; Roa, M.A.; Garofalo, G.; Englsberger, J.; Ott, C. Control applications of TORO—A torque controlled humanoid robot. In Proceedings of the 14th IEEE-RAS International Conference on Humanoid Robots (Humanoids), Madrid, Spain, 18–20 November 2014; p. 841.
23. Edwards, L. *PETMAN Robot to Closely Simulate Soldiers*; Technical Report; PhysOrg: Douglas, UK, 2010.
24. Sadedel, M.; Yousefi-koma, A.; Khadiv, M. Offline path planning, dynamic modeling and gait optimization of a 2D humanoid robot. In Proceedings of the Second RSI/ISM International Conference on Robotics and Mechatronics (ICRoM), Tehran, Iran, 15–17 October 2014; pp. 131–136.

25. De Santis, A.; Siciliano, B.; De Luca, A.; Bicchi, A. An atlas of physical human–robot interaction. *Mech. Mach. Theory* **2008**, *43*, 253–270. [CrossRef]
26. Azevedo, C. Control architecture and algorithms of the anthropomorphic biped robot BIP2000. In Proceedings of the International Conference on Climbing and Walking Robots, Madrid, Spain, 2–4 October 2000.
27. Pratt, J.; Chew, C.M.; Torres, A.; Dilworth, P.; Pratt, G. Virtual model control: An intuitive approach for bipedal locomotion. *Int. J. Robot. Res.* **2001**, *20*, 129–143. [CrossRef]
28. Chestnutt, J.; Lau, M.; Cheung, G.; Kuffner, J.; Hodgins, J.; Kanade, T. Footstep planning for the honda asimo humanoid. In Proceedings of the 2005 IEEE International Conference on Robotics and Automation, Barcelona, Spain, 18–22 April 2005; pp. 629–634.
29. Kuroki, Y. A small biped entertainment robot. In Proceedings of the 2001 International Symposium on Micromechatronics and Human Science, Nagoya, Japan, 9–12 September 2001; pp. 3–4.
30. Zhang, L.; Huang, Q.; Lu, Y.; Xiao, T.; Yang, J.; Keerio, M. A visual tele-operation system for the humanoid robot bhr-02. In Proceedings of the International Conference on Intelligent Robots and Systems, Beijing, China, 9–15 October 2006; pp. 1110–1114.
31. Fujita, M.; Kuroki, Y.; Ishida, T.; Doi, T.T. A small humanoid robot sdr-4x for entertainment applications. In Proceedings of the International Conference on Advanced Intelligent Mechatronics, Kobe, Japan, 20–24 July 2003; Volume 2, pp. 938–943.
32. Kim, J.Y.; Park, I.W.; Lee, J.; Kim, M.S.; Cho, B.K.; Oh, J.H. System design and dynamic walking of humanoid robot KHR-2. In Proceedings of the 2005 IEEE International Conference on Robotics and Automation, Barcelona, Spain, 18–22 April 2005; pp. 1431–1436.
33. Park, I.W.; Kim, J.Y.; Lee, J.; Oh, J.H. Mechanical design of humanoid robot platform KHR-3 (KAIST humanoid robot 3: HUBO). In Proceedings of the 5th IEEE-RAS International Conference on Humanoid Robots, Tsukuba, Japan, 5–7 December 2005; pp. 321–326.
34. Shamsuddin, S.; Ismail, L.I.; Yussof, H.; Zahari, N.I.; Bahari, S.; Hashim, H.; Jaffar, A. Humanoid robot NAO: Review of control and motion exploration. In Proceedings of the 2011 IEEE International Conference on Control System, Computing and Engineering (ICCSCE), Penang, Malaysia, 25–27 November 2011; pp. 511–516.
35. SoftBank Robotics. Available online: https://www.ald.softbankrobotics.com/en/robots/romeo (accessed on 8 September 2017).
36. Akhter, A.; Shafie, A.A. Advancement of android and contribution of various countries in the research and development of the humanoid platform. *Int. J. Robot. Autom.* **2010**, *1*, 43–57.
37. Yamaguchi, J.I.; Takanishi, A. Development of a leg part of a humanoid robot—Development of a biped walking robot adapting to the humans' normal living floor. *Auton. Robot.* **1997**, *4*, 369–385. [CrossRef]
38. Azevedo, C.; Poignet, P.; Espiau, B. Artificial locomotion control: From human to robots. *Robot. Auton. Syst.* **2004**, *47*, 203–223. [CrossRef]
39. Kristiina, J.; Wilcock, G. Multimodal open-domain conversations with the Nao robot. In *Natural Interaction with Robots, Knowbots and Smartphones*; Springer: New York, NY, USA, 2014; pp. 213–224.
40. Buschmann, T.; Favot, V.; Schwienbacher, M.; Ewald, A.; Ulbrich, H. Dynamics and control of the biped robot lola. In *Multibody System Dynamics, Robotics and Control*; Springer: Vienna, Austria, 2013; pp. 161–173.
41. McGeer, T. Passive dynamic walking. *Int. J. Robot. Res.* **1990**, *9*, 62–82. [CrossRef]
42. McGeer, T. Passive walking with knees. In Proceedings of the IEEE International Conference on Robotics and Automation, Detroit, MI, USA, 10–15 May 1990; pp. 1640–1645.
43. Salatian, A.W.; Zheng, Y.F. Gait synthesis for a biped robot climbing sloping surfaces using neural networks. I. Static learning. In Proceedings of the IEEE International Conference on Robotics and Automation, Nice, France, 12–14 May 1992; pp. 2601–2606.
44. Zheng, Y.F. A neural gait synthesizer for autonomous biped robots. In Proceedings of the IEEE International Workshop on Intelligent Robots and Systems, Towards a New Frontier of Applications, Ibaraki, Japan, 3–6 July 1990; pp. 601–608.
45. Pratt, J.; Dilworth, P.; Pratt, G. Virtual model control of a bipedal walking robot. In Proceedings of the IEEE International Conference on Robotics and Automation, Albuquerque, NM, USA, 20–25 April 1997; Volume 1, pp. 193–198.

46. Kac, E. *Telepresence & Bio Art: Networking Humans, Rabbits, & Robots*; University of Michigan Press: Ann Arbor, MI, USA, 2005.
47. Nelson, G.; Saunders, A.; Neville, N.; Swilling, B.; Bondaryk, J.; Billings, D.; Lee, C.; Playter, R.; Raibert, M. Petman: A humanoid robot for testing chemical protective clothing. *J. Robot. Soc. Jpn.* **2012**, *30*, 372–377. [CrossRef]
48. Kim, J.H.; Lattimer, B.Y. Real-time probabilistic classification of fire and smoke using thermal imagery for intelligent firefighting robot. *Fire Saf. J.* **2015**, *72*, 40–49. [CrossRef]
49. Kuindersma, S.; Deits, R.; Fallon, M.; Valenzuela, A.; Dai, H.; Permenter, F.; Koolen, T.; Marion, P.; Tedrake, R. Optimization-based locomotion planning, estimation, and control design for the atlas humanoid robot. *Auton. Robots* **2016**, *40*, 429–455. [CrossRef]
50. Chen, C.Y.; Shih, B.Y.; Shih, C.H.; Wang, L.H. RETRACTED: Design, modeling and stability control for an actuated dynamic walking planar bipedal robot. *J. Vib. Control* **2013**, *19*, 376–384. [CrossRef]
51. Ramezani, A.; Hurst, J.W.; Hamed, K.A.; Grizzle, J.W. Performance analysis and feedback control of ATRIAS, a three-dimensional bipedal robot. *J. Dyn. Syst. Meas. Control* **2014**, *136*, 021012. [CrossRef]
52. Hong, Y.-D.; Kim, J.H. 3-D command state-based modifiable bipedal walking on uneven terrain. *IEEE/ASME Trans. Mech.* **2013**, *18*, 657–663. [CrossRef]
53. Huang, Q.; Yoshihiko, N. Sensory reflex control for humanoid walking. *IEEE Trans. Robot.* **2005**, *21*, 977–984. [CrossRef]
54. Miller, W.T. Real-time neural network control of a biped walking robot. *IEEE Control Syst.* **1994**, *14*, 41–48. [CrossRef]
55. Kuindersma, S.; Deits, R.; Fallon, M.; Valenzuela, A.; Dai, H.; Permenter, F.; Tedrake, R. Optimization-based locomotion planning, estimation, and control design for the atlas humanoid robot. *Auton. Robots* **2016**, *40*, 429–455. [CrossRef]
56. Englsberger, J.; Werner, A.; Ott, C.; Henze, B.; Roa, M.A.; Garofalo, G.; Albu-Schäffer, A. Overview of the torque-controlled humanoid robot TORO. In Proceedings of the 14th IEEE-RAS International Conference on Humanoid Robots (Humanoids), Madrid, Spain, 18–20 November 2014; pp. 916–923.
57. Shotton, J.; Sharp, T.; Kipman, A.; Fitzgibbon, A.; Finocchio, M.; Blake, A.; Kipman, A.; Moore, R. Real-time human pose recognition in parts from single depth images. *Commun. ACM* **2013**, *56*, 116–124. [CrossRef]
58. Semwal, V.B.; Gora, C.N. Toward developing a computational model for bipedal push recovery—A brief. *IEEE Sens. J.* **2015**, *15*, 2021–2022. [CrossRef]
59. Sardain, P.; Guy, B. Forces acting on a biped robot. Center of pressure-zero moment point. *IEEE Trans. Syst. Man Cybern. Part A* **2004**, *34*, 630–637. [CrossRef]
60. Choi, J.H.; Grizzle, J.W. Planar bipedal walking with foot rotation. In Proceedings of the American Control Conference, Portland, OR, USA, 8–10 June 2005; pp. 4909–4916.
61. Sakaino, S.; Tomoya, S.; Kouhei, O. Multi-DOF micro-macro bilateral controller using oblique coordinate control. *IEEE Trans. Ind. Inf.* **2011**, *7*, 446–454. [CrossRef]
62. Zhao, Y.; Sentis, L. A three dimensional foot placement planner for locomotion in very rough terrains. In Proceedings of the IEEE-RAS International Conference on Humanoid Robots (Humanoids), Osaka, Japan, 29 November–1 December 2012; pp. 726–733.
63. Zeng, J.; Chen, H.; Yin, Y.; Yin, Y. A Humanoid Robot Gait Planning and Its Stability Validation. *J. Comput. Commun.* **2014**, *2*, 68. [CrossRef]
64. Ho, Y.F.; Li, T.H.S.; Kuo, P.H.; Ye, Y.T. Parameterized gait pattern generator based on linear inverted pendulum model with natural ZMP references. *Knowl. Eng. Rev.* **2016**, *32*, e3. [CrossRef]
65. Vukobratovic, M.; Manja, K. *Kinematics and Trajectory Synthesis of Manipulation Robots*; Springer: Berlin, Germany, 2013; Volume 3.
66. Capi, G.; Nasu, Y.; Barolli, L.; Mitobe, K. Real time gait generation for autonomous humanoid robots: A case study for walking. *Robot. Auton. Syst.* **2003**, *42*, 107–116. [CrossRef]
67. Salatian, A.W.; Yi, K.Y.; Zheng, Y.F. Reinforcement learning for a biped robot to climb sloping surfaces. *J. Robot. Syst.* **1997**, *14*, 283–296. [CrossRef]
68. Shi, H.; Li, X.; Liang, W.; Chen, H.; Wang, S. A novel fuzzy omni-directional gait planning algorithm for biped robot. In Proceedings of the 17th IEEE/ACIS International Conference on Software Engineering, Artificial Intelligence, Networking and Parallel/Distributed Computing (SNPD), Shanghai, China, 30 May–1 June 2016; pp. 71–76.

69. Lee, H.W. A study of the use of fuzzy control theory to stabilize the gait of biped robots. *Robotica* **2016**, *34*, 777–790. [CrossRef]
70. Zoss, A.B.; Hami, K.; Andrew, C. Biomechanical design of the Berkeley lower extremity exoskeleton (BLEEX). *IEEE/ASME Trans. Mechatron.* **2006**, *11*, 128–138. [CrossRef]
71. Sankai, Y. Leading edge of cybernics: Robot suit hal. In Proceedings of the International Joint Conference, Orlando, FL, USA, 12–17 August 2007; pp. 1–2.
72. Chen, B.; Ma, H.; Qin, L.Y.; Gao, F.; Chan, K.M.; Law, S.W.; Qin, L.; Liao, W.H. Recent developments and challenges of lower extremity exoskeletons. *J. Orthop. Transl.* **2016**, *5*, 26–37.
73. Ferrati, F.; Bortoletto, R.; Pagello, E. Virtual modelling of a real exoskeleton constrained to a human musculoskeletal model. In *Conference on Biomimetic and Biohybrid Systems*; Springer: Berlin/Heidelberg, Germany, 2013; pp. 96–107.
74. TOYOTA Global Newsroom. Available online: http://newsroom.toyota.co.jp/en/detail/15989382 (accessed on 30 October 2017).
75. Dzahir, M.A.M.; Yamamoto, S.I. Recent trends in lower-limb robotic rehabilitation orthosis: Control scheme and strategy for pneumatic muscle actuated gait trainers. *Robotics* **2014**, *3*, 120–148. [CrossRef]
76. Wehner, M.; Quinlivan, B.; Aubin, P.M.; Martinez-Villalpando, E.; Baumann, M.; Stirling, L.; Walsh, C. A lightweight soft exosuit for gait assistance. In Proceedings of the 2013 IEEE International Conference on Robotics and Automation (ICRA), Karlsruhe, Germany, 6–10 May 2013; pp. 3362–3369.
77. Collins, S.H.; Wiggin, M.B.; Sawicki, G.S. Reducing the energy cost of human walking using an unpowered exoskeleton. *Nature* **2015**, *522*, 212–215. [CrossRef] [PubMed]
78. Panizzolo, F.A.; Galiana, I.; Asbeck, A.T.; Siviy, C.; Schmidt, K.; Holt, K.G.; Walsh, C.J. A biologically-inspired multi-joint soft exosuit that can reduce the energy cost of loaded walking. *J. Neuroeng. Rehabil.* **2016**, *13*, 43. [CrossRef] [PubMed]
79. Zeilig, G.; Weingarden, H.; Zwecker, M.; Dudkiewicz, I.; Bloch, A.; Esquenazi, A. Safety and tolerance of the ReWalk™ exoskeleton suit for ambulation by people with complete spinal cord injury: A pilot study. *J. Spinal Cord Med.* **2012**, *35*, 96–101. [CrossRef] [PubMed]
80. Zhang, J.; Fiers, P.; Witte, K.A.; Jackson, R.W.; Poggensee, K.L.; Atkeson, C.G.; Collins, S.H. Human-in-the-loop optimization of exoskeleton assistance during walking. *Science* **2017**, *356*, 1280–1284. [CrossRef] [PubMed]

Article

Bio-Inspired Optimal Control Framework to Generate Walking Motions for the Humanoid Robot iCub Using Whole Body Models

Yue Hu[1,2] *and Katja Mombaur[1,2]

[1] Optimization, Robotics & Biomechanics, Institute of Computer Engineering (ZITI), Heidelberg University, Berliner Str. 45, 69120 Heidelberg, Germany; katja.mombaur@ziti.uni-heidelberg.de

[2] Interdisciplinary Center for Scientific Computing (IWR), Heidelberg University, INF205, 69120 Heidelberg, Germany

* Correspondence: yue.hu@ziti.uni-heidelberg.de; Tel.: +49-6221-54-14868

Received: 16 November 2017; Accepted: 6 February 2018; Published: 12 February 2018

Abstract: Bipedal locomotion remains one of the major open challenges of humanoid robotics. The common approaches are based on simple reduced model dynamics to generate walking trajectories, often neglecting the whole-body dynamics of the robots. As motions in nature are often considered as optimal with respect to certain criteria, in this work, we present an optimal control-based approach that allows us to generate optimized walking motions using a precise whole-body dynamic model of the robot, in contrast with the common approaches. The optimal control problem is formulated to minimize a set of desired objective functions with respect to physical constraints of the robot and contact constraints of the walking phases; the problem is then solved with a direct multiple shooting method. We apply the formulation with combinations of different objective criteria to the model of a reduced version of the iCub humanoid robot of 15 internal DOF. The obtained trajectories are executed on the real robot, and we carry out a discussion on the differences between the outcomes of this approach with the classic approaches.

Keywords: humanoid robot; bipedal locomotion; optimal control; optimization

1. Introduction

Humanoid robots have been designed and built for decades, inspired by nature and in particular human structure and behavior. The goal is to use humanoid robots in environments that are made for and populated by humans, either supporting the human in his/her activities if they are too heavy or too boring, or replacing him/her in situations that are too dangerous. Furthermore, replacing humans with respect to work in a space is a common target for humanoid robotics. Therefore, there is a general expectation that humanoid robots should be able to perform the same tasks and have the same motion capabilities as humans and, in the future, even outperform humans in these tasks.

Ever since the first appearance of humanoid robots, many of them have been equipped with legs. However, the bipedal locomotion capabilities of these robots are still far behind what humans are able to perform. To date, walking remains one of the major open challenges of bipedal humanoid robotics. The difficulty of achieving stable, dynamic and versatile walking for bipedal robots is due to many reasons: the anthropomorphic shape of the humanoid robot exhibits a high redundancy with respect to most motion tasks, but at the same time, it is also difficult to determine feasible solutions, since many of the joint angle combinations result in infeasible motions. Humanoids are generally underactuated with no direct actuators for the overall position and orientation in space, which have to be moved instead by a coordinated action of many other actuators considering feasibility. It should be noted that for most current humanoid robots, a typical human walking behavior is much too challenging in terms of speed, joint motion ranges, required torques and impacts. Humanoid locomotion also involves changing contacts with the environment, which have to be properly chosen. Stability control of humanoid locomotion is one of the biggest problems, since bio-inspired bipedal walking cannot be performed in a statically stable way, but must be dynamically stable. The precise definition of dynamic stability of human walking is still unknown, and up to now, many humanoid robots still apply the well-known Zero Moment Point (ZMP) criterion [1] instead, which is very convenient to use, but results in more conservative gaits.

Motion generation for humanoid robots can be divided into two main categories: motion remapping and model-based generation. In the first case, walking motions from humans are recorded via motion capturing systems, then they can be transferred to the robot by taking into account robot specific constraints such as physical constraints and stability criteria. The recorded motions can be first processed as motion primitives and then concatenated to generate complex motions, such as achieved on the robot HRP-2 (Humanoid Robotics Projects, Kawada) [2], or directly transferred to the robot by means of reduced models and adjustments of end effector positions, as achieved on the JAXON and CHIDORI robots [3]. In the case of model-based motion generation, the most common method is to use reduced models together with the ZMP criterion. Reduced models are for example the Linear Inverted Pendulum (LIPM) [4] or the table cart [5], which allowed many robots to achieve stable walking, such as on HRP-2 [6], by means of Nonlinear Model Predictive Control (NMPC), on the compliant humanoid robot CoMan [7] and iCub [8]. In all these cases, the whole-body posture is computed by means of inverse kinematics matching the desired end effector positions computed with the dynamics of the reduced models. The main reason for which reduced models are still widely used is because they allow for fast online control. However, it is known that due to the many assumptions that are introduced due to the use of such models, the whole-body dynamics of the robot is not well exploited, and the motions often appear over-constrained and not very natural.

Another, more bio-inspired way of motion generation is to use optimization, or to be more precise, optimal control, based on whole-body models of the robots. Optimality is often considered to be a guiding principle of nature, and movements that are performed frequently such as everyday motions or trained motions tend to be optimal with respect to some optimality criterion. By the use of mathematical optimization in the generation of motions for humanoid robots, this biological optimization process is mimicked. The optimization criteria used can be either biologically inspired (by analyzing the optimality criteria of humans or other biological systems, e.g., by inverse optimal control [9]), or technically motivated. In addition, optimization helps to solve the feasibility and the redundancy issues of walking. It generates motions that are feasible with respect to all constraints that can be formulated in a detailed whole-body model of the robot and of the motion task. From all redundant solutions of a particular task, it selects the one that is optimal with respect to the chosen optimization criterion. In addition, optimization techniques can be used to kinematically adjust humanoid motions to recorded human motions, taking the different geometry into account (see, e.g., [10]). See [11] for a general overview of optimization in the context of humanoids.

Optimal control has previously been used to generate challenging motions using whole-body models for robots such as HRP-2 [12–18]. In an approach that is situated between whole-body optimization and motion generation by reduced models, optimization of template models has been used for generating walking motion of humanoid robots in challenging scenarios [19]. Collocation has been a popular method to compute walking motions for humanoid robots using whole-body models and contact constraints, such as in [20,21], where motions could be obtained in a time in the range of minutes to hours.

In previous works, we have achieved the first walking motions on one of the most disseminated complex humanoid robots iCub [22], by means of the table cart model and ZMP criterion [8], allowing the robot to achieve walking on level ground and, for the first time, on slopes and stairs, with the aim of measuring the walking capabilities of iCub. In particular, this work has been performed on the reduced version of the iCub available in our lab with neither head and arms (named the HeiCub

(iCub of Heidelberg University)), but the results are transferable to the full iCub models. We later improved the walking performances by means of a closed-loop control scheme based on NMPC, which allowed the robot to be able to change walking directions online [23].

In this work, instead, we want to use the whole-body dynamic model of the robot to generate walking motions on level ground. Similar to the setup of the problem for HRP-2 in [15], here, we formulate the optimal control problem for walking using the detailed model of the the humanoid robot iCub/HeiCub. In the optimal control problem, we set several objective functions and constraints to describe the hybrid dynamics of walking and to respect the physical constraints of the robot itself. In contrast to [20,21], the problem is solved using a direct multiple direct shooting method [24], which allows obtaining precise and accurate results and has been used previously to obtain open-loop stable robot motions [25]. The obtained motions are then transferred to the robot, and the outcomes are discussed with a comparison with the motions we had previously achieved on the same robot with reduced models. The main contributions of this paper consist of:

- Formulation of the whole-body dynamic model for walking problems for the humanoid robot iCub.
- Formulation of the optimal control problem for a complex walking sequence involving cyclic, as well as non-cyclic motion phases.
- The use of a direct multiple shooting method, which allows solving the equations of motion at a high precision at all times in the gait and to precisely take into account the kinematic and dynamic constraints of the robot.
- Implementation and experimental validation on the HeiCub humanoid, a reduced iCub, with computation of performance indicators as a comparison with motions generated using reduced models.

This paper is organized as follows: In Section 2, we give a description of the iCub humanoid robot and its reduced version HeiCub. Section 3 describes in detail the different walking phases involved in the optimal control problem, as well as the dynamics equations in all cases. In Section 4, we illustrate the optimal control problem formulation giving the details of states, controls, parameters, objective functions and constraints. Section 5 shows the numerical results and the software tools used, followed by the experimental results obtained on the robot in Section 6. The paper concludes in Section 7 with a final discussion of the work and conclusions, as well as possible future developments.

2. The (He)iCub Humanoid Robot

The iCub [22] is a humanoid robot designed and built by the Fondazione Istituto Italiano di Tecnologia (IIT) and distributed among more than 30 research institutions and universities all over the world. The robot was originally designed with the aim of conducting cognitive studies; therefore, it was meant to resemble a 3–4-year-old child. The latest version of the robot consists of a whole-body humanoid with head, arms and legs, for a total of 53 degrees of freedom (DOF), including dexterous hands.

The iCub is mostly known for its grasping [26,27] and balancing [28] capabilities, which have been widely exploited by several works in past and recent years, while in the first versions of iCub, the robot legs had weak actuators and small feet that did not allow the robot to perform walking, but only crawling. In the latest available version, the legs have been redesigned [29] inspired by the mechanical design of the CoMan humanoid robot, which has demonstrated several walking capabilities [30–32].

In the context of the European Project KoroiBot, IIT has delivered to Heidelberg University a reduced version of the standard iCub, consisting of 15 internal DOF: three in the torso and six in each leg, as shown in Figure 1. The robot is furthermore equipped with four Force Torque sensors (F/T), two in the upper legs and two in the feet. The robot has an on-board PC104 with a dual core, but has no battery; therefore, it needs to be connected to an external power supply by means of cables, which serve also as network communication cables, allowing one to use external computers to carry out bulky computations. The robot has also four custom Series Elastic Actuators (SEA) [33] with springs that can be unmounted to obtain rigid actuators. In the context of this paper, the springs are unmounted, and therefore, the joints are considered perfectly rigid.

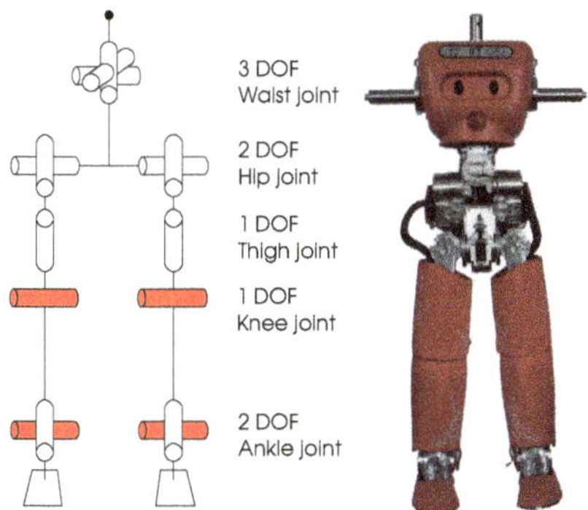

Figure 1. The HeiCub (iCub of Heidelberg University) humanoid robot. In red, the series elastic actuators, which are not considered in the context of this work.

The robot has a weight of 26.4 kg, and it is 0.97 m tall. The leg length from the hip axis is 0.51 m, and the feet are 0.2 m long and 0.1 m wide. The weight distribution of the robot is about 6 kg for the torso, 5 kg for the pelvis and 7.5 kg each leg. All the kinematic and dynamic parameters of the robot are described in a URDF (Unified Robot Description Format) file, which was extracted directly from the original CAD model. The hardware limitations, including joint limits, joint velocity limits and torque limits, are as in Table 1. Please note that the velocity and torque limits have been obtained via experiments; therefore, they are not perfectly precise, i.e., these limits might be conservative, as they are set to guarantee the safety of the robot.

Table 1. HeiCub joint limits.

Joint	Range Limits (deg)	Velocity Limits (deg/s)	Torque Limits (Nm/s)
l_hip_pitch, r_hip_pitch	[−33, 100]	[−100, 100]	[−50, 50]
l_hip_roll, r_hip_roll	[−19, 90]	[−150, 150]	[−50, 50]
l_hip_yaw, r_hip_yaw	[−75, 75]	[−150, 150]	[−50, 50]
l_knee, r_knee	[−100, 0]	[−150, 150]	[−50, 50]
l_ankle_pitch, r_ankle_pitch	[−36, 27]	[−150, 150]	[−50, 50]
l_ankle_roll, r_ankle_roll	[−24, 24]	[−150, 150]	[−50, 50]
torso_pitch	[−20, 60]	[−150, 150]	[−50, 50]
torso_roll	[−26, 26]	[−150, 150]	[−50, 50]
torso_yaw	[−50, 50]	[−150, 150]	[−50, 50]

To distinguish this robot from the standard iCub, we will use the name HeiCub (iCub of Heidelberg University) to refer to it from now on. The legs of HeiCub have the exact same mechanical design as any other standard iCub, and it also uses the same software infrastructure as the iCub. These features allow us to transfer control frameworks developed for the iCub to HeiCub and vice versa, by just adapting the number of DOF and the upper body structure.

3. Model and Dynamics

As during walking, different contacts are involved, we define walking as a hybrid dynamics system, where the dynamics switches according to contact conditions, i.e., hybrid and non-smooth dynamics. To have a precise description of the dynamics of walking, in the following, we first list the phases that involve the different contacts and then the dynamics equations that will be used in the optimal control problem.

3.1. Walking Phases

Different phases can be identified for a walking sequence. This is usually done also for human walking motion analysis [34], where the different feet contacts are described and the phases might change according to the walking environment [35]. Differently from humans, HeiCub has a rigid flat foot, common among humanoid robots. Therefore, the walking phases for such a humanoid are different from human walking, as we have to consider completely flat contacts between the feet and the ground.

As shown in Figure 2, the walking sequence can be seen as three sub-sequences:

- Starting step, the robot starts from a complete stop (i.e., all velocities to zero) and takes the first step, leading to the periodic motion.
- Periodic steps, which are the steps that the robot can repeat during walking. In this case, we assume single-step periodic, i.e., the left and right leg configurations can be mirrored, as the robot is symmetric. The periodicity is enforced on touchdown.
- Ending step, the final step where the robot comes to a complete stop from the periodic motion.

It is to be noted that in this case, we assume that one step is enough to lead the motion into a periodic motion and to lead the motion to an end. This assumption might not be valid for every system and situation; however, we have verified that in the cases we considered for HeiCub in this paper, this assumption can be used. A further discussion is carried out in Section 7.

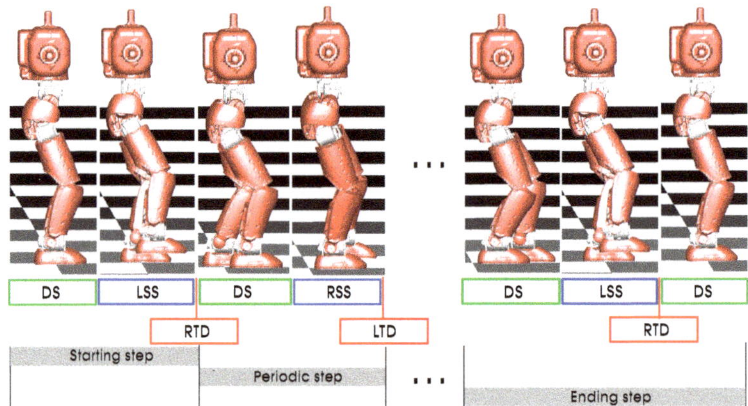

Figure 2. Walking phases of HeiCub. DS = Double Support, LSS = Left Single Support, RSS = Right Single Support, LTD = Left Touch Down, RTD = Right Touch Down. The whole sequence can be seen as three sub-sequences. The periodic step can be repeated for a desired number of times that do not need to be further modeled.

The walking phases involving different contacts are described as follows:

- DS: Double Support, where both feet are on the ground.
- LSS: Left Single Support, where the left foot is on the ground and the right leg swings to the next support location.
- RSS: Right Single Support, as for LSS, the right foot is on the ground and the left leg is swinging.

In addition, there are also two impacts that follow each of the single-support phases and precede the double-support phases:

- RTD: Right Touch Down, when the left foot is in single support and the right foot strikes the ground, we assume that when the foot of the robot touches the ground, it is completely flat.
- LTD: Left Touch Down, the left foot strikes the ground when the right foot is in single support.

To define a flat contact, it is enough to define three contact points on each of the feet, as shown in Figure 3.

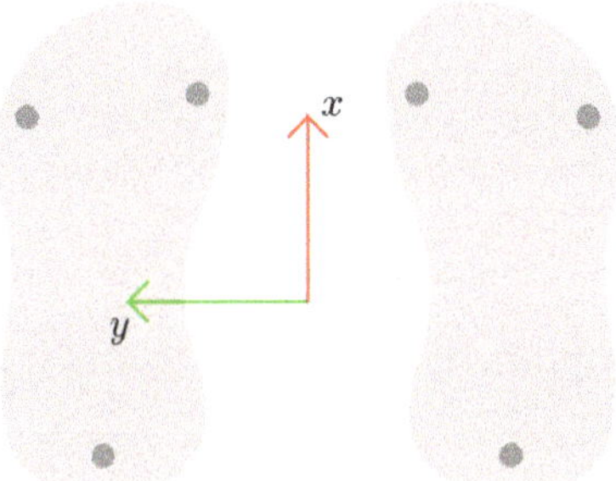

Figure 3. Feet shapes of the HeiCub robot. Three points are defined for contact modeling on each foot.

Given the description of the walking phases, we illustrate in the following the dynamics equations based on few assumptions.

3.2. Dynamics

The robot is described with the generalized coordinates $\mathbf{q} \in \mathbb{R}^{n_{dof}}$, where $n_{dof} = 21$, with 15 internal DOF and six external DOF describing the floating base, with three translations along x, y, z and three rotations about the same axis represented with Euler angles.

The dynamics of a rigid multi-body system such as a robot can be described by the equations of motion:

$$\mathbf{H}(\mathbf{q})\ddot{\mathbf{q}} + \mathbf{C}(\mathbf{q}, \dot{\mathbf{q}}) = \boldsymbol{\tau}, \tag{1}$$

where the matrix $\mathbf{H}$ is the joint space inertia matrix, $\ddot{\mathbf{q}}$ is the joint acceleration vector, $\mathbf{C}$ the Coriolis, centrifugal and gravitational term and $\boldsymbol{\tau}$ are the joint torques. Given that the robot is a floating base system, the vector of joint torques is assumed to be:

$$\boldsymbol{\tau} = \begin{bmatrix} 0 \\ \boldsymbol{\tau}_A \end{bmatrix}, \tag{2}$$

where $\boldsymbol{\tau}_A \in \mathbb{R}^{15}$ is the vector of active joint torques.

We introduce the following set of constraints due to contacts:

$$\mathbf{g}(\mathbf{q}) = \mathbf{0}, \tag{3}$$

which depend on the walking phases, as previously described.

In our formulation, we assume that contacts are perfectly rigid and non-sliding and impacts are instantaneous and inelastic. Therefore, taking into account (3) in (1), we obtain:

$$\mathbf{H}(\mathbf{q})\ddot{\mathbf{q}} + \mathbf{C}(\mathbf{q}, \dot{\mathbf{q}}) = \boldsymbol{\tau} + \mathbf{G}^T\boldsymbol{\lambda}, \tag{4}$$

where $\boldsymbol{\lambda}$ is the vector of external forces due to the contacts and $\mathbf{G} == (\partial \mathbf{g} / \partial \mathbf{q})$ is the contact Jacobian.

An additional set of equations for the contact constraints is obtained by differentiation twice of Equation (3):

$$\mathbf{G}(\mathbf{q})\ddot{\mathbf{q}} + \dot{\mathbf{G}}(\mathbf{q})\dot{\mathbf{q}} = \mathbf{0}. \tag{5}$$

Combining Equations (4) and (5), we obtain the following system of equations for unknown $\ddot{\mathbf{q}}$ and λ:

$$\begin{bmatrix} \mathbf{H} & \mathbf{G}^T \\ \mathbf{G} & \mathbf{0} \end{bmatrix} \begin{bmatrix} \ddot{\mathbf{q}} \\ -\lambda \end{bmatrix} = \begin{bmatrix} \tau - \mathbf{C} \\ -\gamma \end{bmatrix}, \tag{6}$$

where $\gamma = \dot{\mathbf{G}}(\mathbf{q})\dot{\mathbf{q}}$. This system describes the dynamics of the continuous phases of single and double supports, i.e., DS, LSS and RSS. In the case of DS, the number of contacts is six, i.e., both feet have flat contacts with the ground, while in the case of LSS and RSS, the number of contacts is three, i.e., only one foot has flat contact with the ground.

Due to the assumptions on the impacts, the dynamics describing the instantaneous change in the generalized velocities can be obtained by integrating Equations (4) and (5) over a time singleton. The following system of equations can be written for the unknown generalized velocities after the impact $\dot{\mathbf{q}}^+$ and the impulses at each constraint $\mathbf{\Lambda}$:

$$\begin{bmatrix} \mathbf{H} & \mathbf{G}^T \\ \mathbf{G} & \mathbf{0} \end{bmatrix} \begin{bmatrix} \dot{\mathbf{q}}^+ \\ -\mathbf{\Lambda} \end{bmatrix} = \begin{bmatrix} \mathbf{H}\dot{\mathbf{q}}^- \\ \mathbf{v}^+ \end{bmatrix}, \tag{7}$$

where $\dot{\mathbf{q}}^-$ are the generalized velocities before the impact, and $\mathbf{v}^+$ the desired velocity of contact points after the impact, which is $\mathbf{0}$ due to the previously-described assumptions. The system as in Equation (7) describes the dynamics of the impacts LTD and RTD as in Figure 2.

4. Optimal Control Problem

The optimal control problem is formulated to treat the hybrid dynamic system described in the previous section; therefore, it results in a multiple-phase optimal control problem, where each of the phases are as described in Figure 2.

The general formulation of a multiple phases optimal control problem can be described as follows:

$$\min_{\mathbf{x},\mathbf{u},\mathbf{p},s_0,s_1,\dots,s_{n_{ph}}} \sum_{j=0}^{n_{ph}-1} \int_{s_j}^{s_{j+1}} \Phi_L(t,\mathbf{x}(t),\mathbf{u}(t),\mathbf{p})dt + \Phi_M(s_{j+1},\mathbf{x}(s_{j+1}),\mathbf{p}) \tag{8}$$

subject to:

$$\begin{aligned} \dot{\mathbf{x}}(t) &= \mathbf{f}_j(t,\mathbf{x}(t),\mathbf{u}(t),\mathbf{p}), \\ & \quad t \in [s_j, s_{j+1}] \\ & \quad j = 0,\dots,n_{ph}-1, \\ & \quad s_0 = 0, s_{n_{ph}} = T \\ \mathbf{x}(s_j^+) &= \widetilde{\mathbf{J}}_j(\mathbf{x}(s_j^-),\mathbf{p}),\ j = 0,..,n_{ph}-1 \\ g_j(t,\mathbf{x}(t),\mathbf{u}(t),\mathbf{p}) &\geq \mathbf{0},\ t \in [s_j, s_{j+1}] \\ r^{eq}(\mathbf{x}(0),..,\mathbf{x}(T),\mathbf{p}) &= \mathbf{0} \\ r^{ineq}(\mathbf{x}(0),..,\mathbf{x}(T),\mathbf{p}) &\geq \mathbf{0}. \end{aligned} \tag{9}$$

In the problem formulation, objective functions of Lagrangian type $\Phi_L(\cdot)$ and/or Mayer type $\Phi_M(\cdot)$ are minimized with respect to the states $\mathbf{x}(t)$, the controls $\mathbf{u}(t)$ and the model parameters $\mathbf{p}$. These objective functions can be different for each phase, as described by the sum over the number of phases n_{ph}. The constraints of the optimal control problem are defined in Equation (9), where $\mathbf{f}_j(\cdot)$ is the differential equation describing the phase j, with the duration of phase j being s_j. The transition function between two consecutive phases is $\widetilde{J}(\mathbf{x}(s_j^-,\mathbf{p}))$. The path constraints are represented by $g_j(\cdot)$, $r^{eq}(\cdot)$ and $r^{ineq}(\cdot)$ being the boundary conditions and point equality and inequality constraints.

In the case of the walking problem, the number of phases is $n_{ph} = 10$, of which seven are continuous phases and three are the impacts, which are modeled as phases with zero time.

4.1. States, Controls and Parameters

The states of the optimal control problem are the generalized positions and velocities of the robot in the case of rigid actuation:

$$\mathbf{x}(t) = \begin{bmatrix} \mathbf{q}(t) \\ \dot{\mathbf{q}}(t) \end{bmatrix}, \tag{10}$$

where $\mathbf{q}, \dot{\mathbf{q}} \in \mathbb{R}^{n_{dof}}$; which means that the right-hand sides of the differential states are:

$$\dot{\mathbf{x}}(t) = \begin{bmatrix} \dot{\mathbf{q}}(t) \\ \ddot{\mathbf{q}}(t) \end{bmatrix} = \begin{bmatrix} \dot{\mathbf{q}}(t) \\ FD(\mathbf{q}, \dot{\mathbf{q}}, \boldsymbol{\tau}) \end{bmatrix}, \tag{11}$$

where $FD(\cdot)$ is a forward dynamics computation. The right-hand side of each continuous phase is described by the equations as in Equation (6) with specific sets of contacts as per Figure 2. The transition functions $\widetilde{J}(\mathbf{x}(s_j^-, \mathbf{p}))$ associated with state variable discontinuities consist of the impact dynamics equations as in Equation (7).

The controls are represented by the active joint torques:

$$\mathbf{u}(t) = \boldsymbol{\tau}_A(t). \tag{12}$$

Therefore, $\mathbf{u} \in \mathbb{R}^{15}$.

The set of parameters includes the step length and the step width, expressed in m. The former is left free to the optimization, while the latter is kept fixed to 0.14 m for the time being:

$$\mathbf{p} = \begin{bmatrix} p_{steplength} \\ p_{stepwidth} \end{bmatrix}. \tag{13}$$

The step width is the distance between the feet at touchdown, i.e., when both feet are on the ground, and does not apply for the single-support phases when one of the legs is swinging.

The duration of each phase s_j, and therefore also the total time T, is left free to the optimization to find for the best value for the specific parameters and objective functions.

4.2. Constraints

As we want to obtain feasible motions for an existing humanoid robot, boundary constraints need to be set according to the physical limits of the robot on:

- Joint angles range,
- Joint velocities,
- Torques.

Ground reaction forces and contact positions are checked according to the contact points of each phase by means of the contact points of each foot as described in Figure 3, modeled as equality and inequality constraints. Depending on the phase, the ground reaction forces have to be positive when the contact is established, and the positions have to be zero on z, e.g., during RSS, the right foot reaction forces have to be positive and the foot flat on the floor, while the left foot positions have to be above the floor level to avoid penetration issues.

Collision avoidance also represents a key constraint that needs to be enforced in motion generation for robots. In this work, it is modeled as a set of constraints based on geometric capsules, i.e., cylinders with rounded caps [36], as shown in Figure 4. The cylinders approximate the limits of each of the links of the legs, such as the upper leg, the lower leg and the foot, for both legs. The distance $d(i, j)$ between the capsules i and j is used as a constraint to avoid collision between links, i.e., $d(i, j) > 0$, where i and j are couples of different links of the two legs.

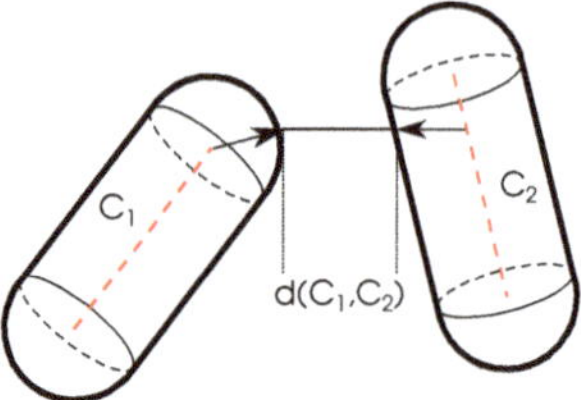

Figure 4. Collision avoidance with rounded caps. The distance between the two is calculated as d.

Orientation constraints are enforced on the feet and torso of the robot only at the start and end of the phases as equality constraints, such that after touchdown, the feet are straight aligned and the torso upright with respect to the world reference frame.

As the stability criterion, we use the ZMP, which we ensure lies inside the support polygon, which is approximated by a polygon that approximates the shape of the foot during single-support phases and an enlarged polygon including both feet and the area between the feet during double-support phases, as depicted in Figure 5. The ZMP is computed from the ground reaction forces using the whole-body model of the robot.

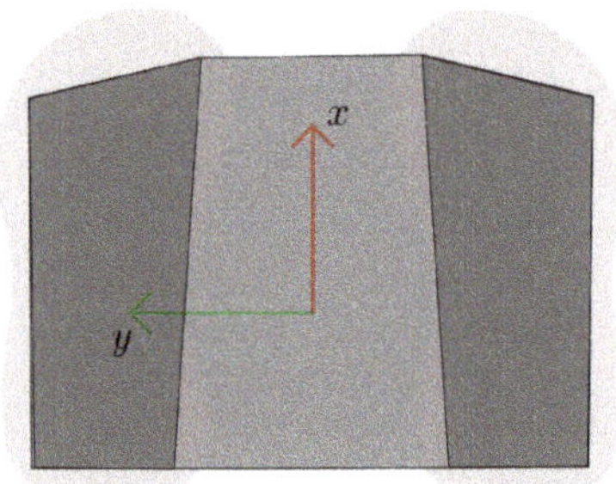

Figure 5. Approximation of the support area of the foot. The darker grey zones delimit the support areas for single support cases. The lighter grey zone changes according to the positions of the feet.

Periodicity constraints are enforced on the intermediate periodic step as depicted in Figure 2. We impose a one-step periodicity, i.e., at the end of the step, the state of the robot is mirrored with respect to the beginning, meaning that left and right legs have a mirrored configuration, as well as the torso, such that by mirroring the obtained periodic step, we can obtain a full sequence of multiple steps. The periodicity is imposed on states $\mathbf{x}$ and controls $\mathbf{u}$.

The joint velocities are constrained to be zero at the beginning of the first phase and at the end of the last phase as equality constraints, to ensure that the motion starts from a complete stop and ends with a complete stop, i.e., $\dot{\mathbf{q}}(0) = \mathbf{0}, \dot{\mathbf{q}}(T) = \mathbf{0}$.

4.3. Objective Functions

Different objective functions have been defined for the problem of walking, which can be combined by means of weighting factors:

- Minimization of joint torques squared, which is always included to ensure smooth torques (with small weighting factor):
$$\Phi_{\tau}(\cdot) = \mathbf{W} \cdot \mathbf{u}^T \mathbf{u}. \tag{14}$$

- Minimization of absolute mechanical work:
$$\Phi_{work}(\cdot) = \mathbf{W} \cdot \sum_{i=0}^{n_{dof}-6} \mid \dot{q}_{i-6} \cdot u_i \mid . \tag{15}$$

- Minimization of joint accelerations squared in order to obtain smooth velocity trajectories (with small weighting factor):
$$\Phi_{\ddot{q}}(\cdot) = \mathbf{W} \cdot \ddot{\mathbf{q}}^T \ddot{\mathbf{q}}. \tag{16}$$

- Torso orientation minimization (ort. min., which is sometimes also referred to as torso stabilization) in terms of torso movements with respect to the world reference:
$$\Phi_{torso}(\cdot) = \theta_{torso} - \theta_{world}. \tag{17}$$

The matrix **W** is a weighting matrix that allows one to weight each joint differently, as not all joints have an equal contribution and an equal order of magnitude. For the time being, the same **W** is used for all objective functions.

The reason for which the torso orientation minimization has been included as one of the objectives is that it has been observed that the torso performs large movements in order to compensate for the angular momentum, given that the robot does not have arms. Since the robot also has cameras on the torso, which faces toward the front, it is desired that the movements of the torso with respect to the world reference not be too large; this is similar to the necessity of head stabilization in many humanoid robots.

The above-listed objective functions can then be combined as one weighted objective:

$$\Phi_L(\cdot) = c_{\tau}\Phi_{\tau} + c_{work}\Phi_{work} + c_{\ddot{q}}\Phi_{\ddot{q}} + c_{torso}\Phi_{torso}. \tag{18}$$

The weighting factors serve to scale the contribution of each objective, but also to take into account the difference in order of magnitudes. Different combinations of the weighting factors are explored in the next section.

5. Results

In this section, we first briefly describe the software tools that have been used to achieve the results of the optimal control framework, then the numerical results are shown.

5.1. Software Tools

The multiple-phase optimal control problem is solved with the software package MUSCOD-II [37] of IWR (Interdisciplinary Center for Scientific Computing), Heidelberg University. The solution of the optimal control problem is carried out using the direct multiple-shooting method as described in [24], in which controls and states are discretized. The phase times $s_{n_{ph}}$ are divided into m_{ph} intervals, over which a simple function (constant or linear) is chosen for the controls. State variables are parametrized with the multiple shooting method, which in this cases uses the same grid as the control discretization. From these two discretizations, we obtain a large, but structured NLP (Nonlinear Programming problem), which is solved using an adapted SQP (Sequential Quadratic Programming) method. The dynamics of the system is handled on all multiple shooting intervals in parallel with the NLP solution by means of an efficient integrator at a desired precision.

In this paper, we choose to use piecewise constant discretization for the controls, with 50 intervals for the single-support phases and 25 for the double-support phases, as they are usually shorter than the single-support phases. The number of shooting intervals is chosen the same as the control intervals.

The dynamics quantities are computed using the Rigid Body Dynamics Library (RBDL) [38], which allows for the computation of forward dynamics with contacts, as well as impact dynamics.

5.2. Numerical Results

We generated different motions with combinations of different optimality criteria. In particular, we have used the following combinations, which are summarized in Table 2:

1. Minimization of joint torques, with a small weight on minimization of joint accelerations.
2. Minimization of joint torques and torso orientation minimization, with a small weight on minimization of joint accelerations.
3. Minimization of absolute work, with a small weight on minimization of joint torques and joint accelerations.
4. Minimization of absolute work and torso orientation, with a small weight on minimization of joint torques and joint accelerations.

Table 2. Combination of objective functions. Minimization of torques and joint accelerations is always included with small weighting factors to ensure smooth trajectories.

Objectives	$\Phi_{\tau}(\cdot)$	$\Phi_{work}(\cdot)$	$\Phi_{\ddot{q}}(\cdot)$	$\Phi_{torso}(\cdot)$
1	1	0	10^{-4}	0
2	1	0	10^{-4}	10
3	10^{-4}	1	10^{-4}	0
4	10^{-4}	1	10^{-4}	10

The computations have been carried out using MUSCOD-II and RBDL on a standard desktop PC with the i7 CPU running at 3.60 GHz with eight cores. For the whole sequence to be computed, the average time is about 12–15 h, due to the chosen high number of intervals.

The box constraints on states and controls are set as per the hardware limits of the robot as in Table 1. The initial guess is generated from a sequence of walking motion obtained with the reduced models from our previous works and set as the same for all combinations of objectives considered, as setting the initial guesses all to zeros might lead to longer computation times in such a large and complex nonlinear optimization problem. We cannot ensure that the obtained results are completely independent of the initial guess, but the outcomes have shown that the resulting motions show differences with respect to the ones generated with reduced models. This can be seen in a more intuitive way from the example of a motion sequence of a single step shown in Figure 6. The first sequence is a motion generated with the 3D Linear Inverted Pendulum Model (LIPM), in this case, the whole-body motion was obtained applying inverse kinematics to the given center of mass (CoM) and feet trajectories from a pattern generator. In order to generate feasible trajectories, the orientations of feet and torso have been constrained with respect to the world reference frame. In contrast to the first sequence, the second sequence, which is generated with optimal control with minimization of torques (Case 1 in Table 2), shows a clear change in the orientation of both torso and swing foot. As also discussed in Section 4, the stabilization of the torso might be a desired feature; therefore, in the third sequence, it is possible to see how the torso is stabilized using the set of objectives as per Case 2 in Table 2.

In order to show the results in a more consistent way, the intermediate periodic step has been concatenated into nine steps, resulting in a sequence of a total of 11 steps for each of the cases as listed in Table 2. The resulting CoM trajectories are shown in Figure 7, and the whole-body joint trajectories are shown in Figure 8. The effect of the torso orientation minimization can be seen in a clear way in

Figure 9, which shows the orientation of the torso with respect to the world reference frame during the whole walking sequence.

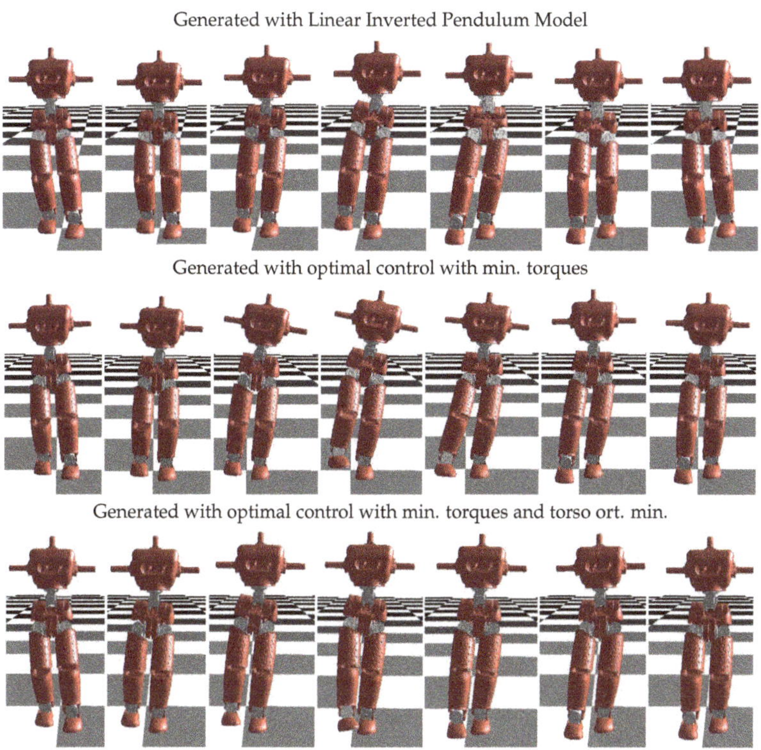

Figure 6. Example sequences of one step motions from using the reduced model, the whole-body model minimizing torques and the whole-body model with torso orientation minimization.

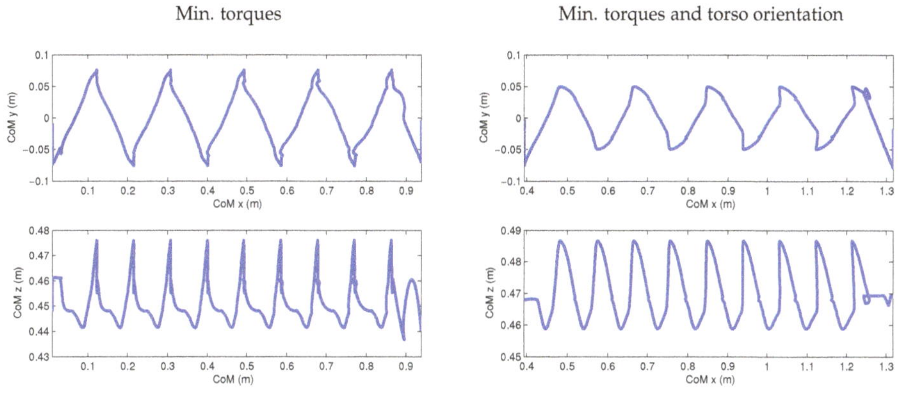

Figure 7. Cont.

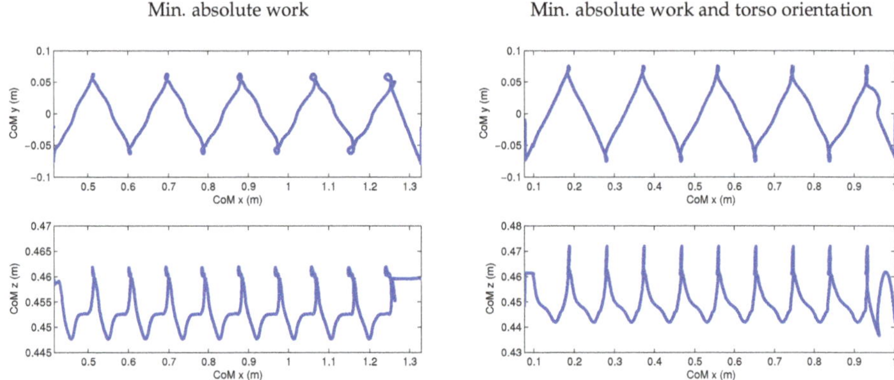

Figure 7. Center of mass trajectories of a full sequence with nine periodic steps combinations of objectives as in Table 2.

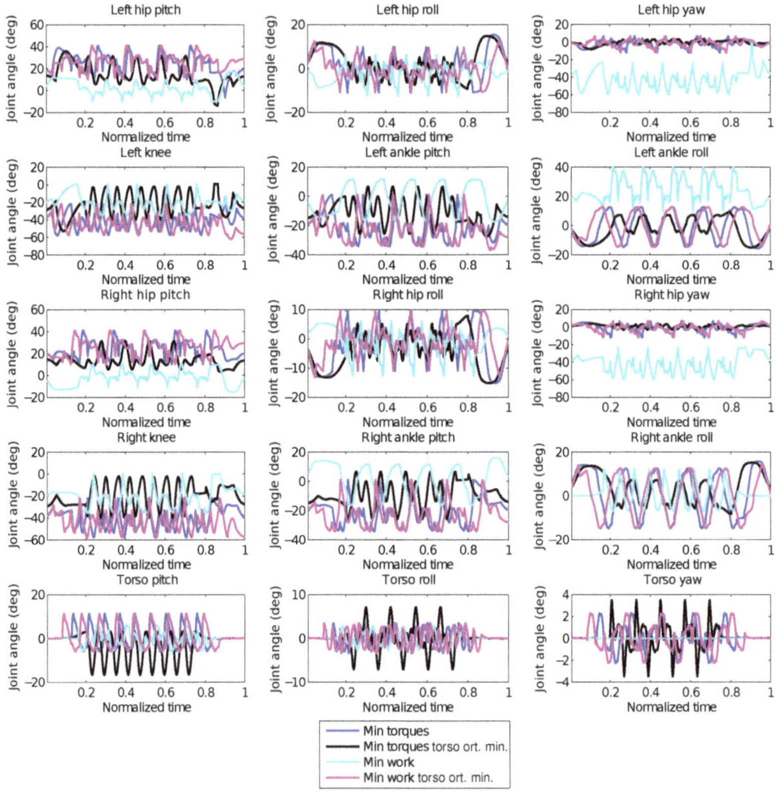

Figure 8. Joint angle trajectories of all the 15 DOF of the robot, obtained for the four combinations of objective functions. Time is normalized for all trajectories for comparison purposes. It is to be noted that torso orientation minimization does not mean reduction of the movements of the torso joints, but the orientation w.r.t. the world reference.

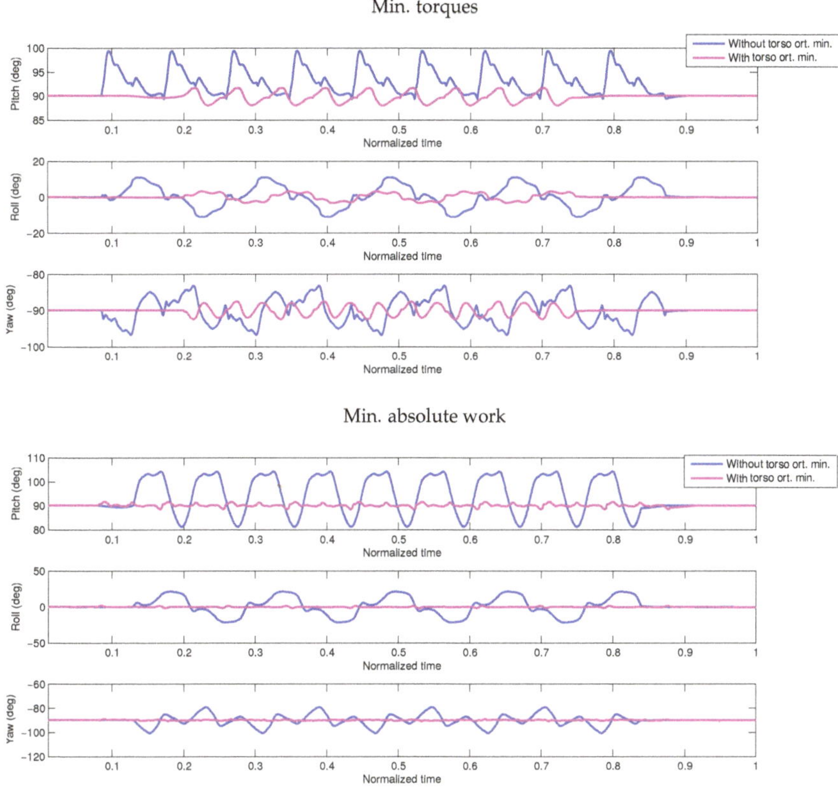

Figure 9. Torso orientation trajectories for the two cases of minimization of torques and absolute work. The orientation is expressed with respect to the world reference frame. The nominal reference (torso upright) is [90, 0, −90]. Time is normalized for all trajectories for comparison purposes.

From the results, with respect to methods based on reduced models, a major difference that can be observed is the variation in height of the CoM, which is usually kept fixed due to the linearization of the ZMP equations. Using optimal control instead, we have not introduced such a constraint, and as we can observe from Figure 8, the knee joint has big variations and reaches also completely stretched configurations. The CoM height variation presents however spikes when switching to the next step; these spikes are due to the impacts, which are set with zero velocity as per our assumptions in Section 3. In future work, they could be reduced by minimizing the impact forces. We can also observe that the CoM projection on the ground (xy plane) is still very similar to what is usually obtained with the reduced models as in [8,23]. This is mainly due to mechanical design and restrictions of the robot itself, such as the rigid flat feet and also the missing arms. The reason for which the robot swings the torso without orientation minimization is to compensate for the angular momentum. If the robot had arms, they could be used to achieve angular momentum compensation, but this is not the case for HeiCub. As a matter of fact, we can observe from Figure 6 that when the torso orientation minimization is introduced, the robot takes steps with a smaller step width during the swing phase with respect to the case when there is no torso orientation minimization. From Figures 7 and 8, we can also observe the differences between the motion obtained with minimization of torques and absolute work, where the latter shows a more constant variation in the CoM trajectory and the legs are turned inwards.

6. Experimental Validation

In order to test the obtained motion sequences on the robot, the periodic step has been concatenated to obtain a longer sequence of walking for nine periodic steps. The motions are executed on the robot by means of position control in open loop with closed loop joint angle tracking. This choice is due to the fact that torque control for walking on the iCub family of robots still cannot ensure good torque tracking, as the robot does not have joint torque sensors; the torques are estimated via force torque sensors that often incur in saturation issues at impacts and cannot therefore ensure precise torque tracking for motions such as walking.

The trajectories obtained from the optimal control framework are discretized as per the discretization grid chosen for MUSCOD-II, which does not correspond to the thread rate required by the robot controller. Due to the choice of letting the time be free in the optimization, it is not possible to choose a discretization grid that corresponds exactly to the thread rate. Furthermore, the number of shooting nodes would be too high for the problem to be solvable in a reasonable time. Therefore, all the trajectories have been interpolated with a spline interpolation in order to obtain the correct number of samples required by the robot controller. Furthermore, it is to be noted that currently on the robot, there is no proper implementation of the floating base estimation; therefore, the sliding effects that might occur during the execution of the motions are not compensated. The CoM trajectories of the robot are obtained with the computed floating based and the joint angle trajectories from the encoders.

The robot has executed all four obtained motions, proving the feasibility of the motions and of the periodicity constraints, as can be seen in the Supplementary Video S1. We can see the resulting CoM trajectories in red in Figure 10. We can observe from the results that the CoM trajectories could be followed closely by the robot; however, when the torso orientation minimization is not included, a bigger deviation can be observed in the height of the CoM trajectories in correspondence with the spikes; this is due to the impact forces as mentioned in the previous section. It seems however that the introduction of torso orientation minimization has a damping effect on the error when executed on the robot.

To compare in a more systematic way the outcomes of the optimal control framework on the robot with the results obtained in our previous works with reduced models, we computed the key performance indicators, which are illustrated in detail in the following section.

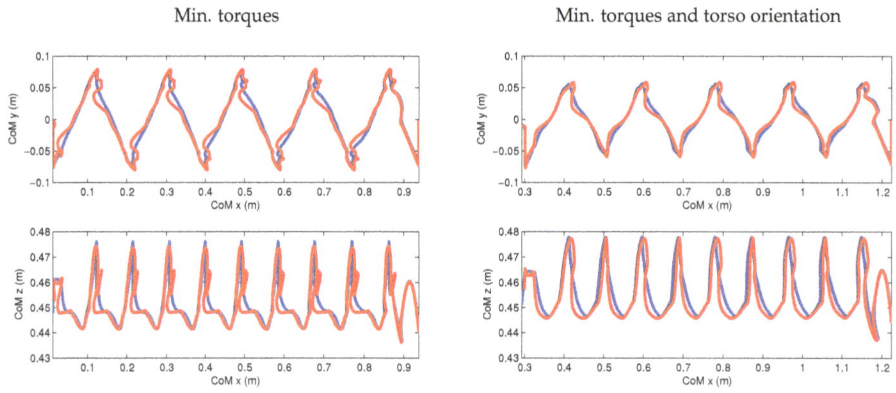

Figure 10. Cont.

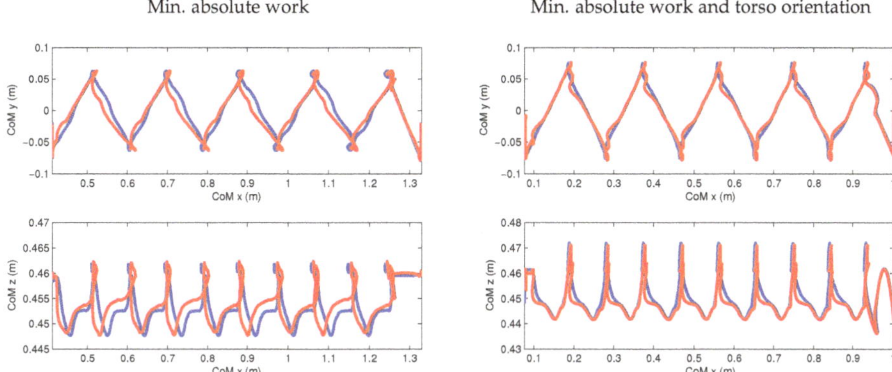

Figure 10. [Hardware experiment result.] Center of mass trajectories obtained on the robot (in red) with all four combinations of objective functions as in Table 2.

Key Performance Indicators

The Key Performance Indicators (KPIs) are a series of indicators developed in the European Project KoroiBot (http://www.koroibot.eu) designed to compare the performances of different methods applied on humanoid robots. These indicators are meant to compare the walking capabilities of the same humanoid robot with different control approaches and/or methodologies, and not to compare different robots, as it is highly difficult to compare robots that are very different in size and mechanical structure.

For the HeiCub, a subset of the KPIs has been defined in our previous works [8,23]. In the following, we give a brief explanation of the measured indicators, which were chosen based on quantities that are measurable from the robot sensors:

- Walking velocity v_{max}:
  The maximum achieved walking velocity; in the case of motions generated with optimal control, this corresponds to the velocity of the resulting sequence.
- Walking timings t^{ss}/t^{ds} and step period:
  Single- and double-support times of a single step, the whole duration of which is indicated as the step period. In the case of optimal control, we consider the timings of the periodic step.
- Cost of transport:

$$\mathbf{E}_{CT} = \frac{\sum_{m=1}^{M} \int_{t_0}^{t_f} I_m(t) V_m(t) dt}{m_{robot} g \cdot d} \tag{19}$$

  The cost of transport is defined as a unitless quantity, where M is the total number of motors, I_m and V_m are the current and voltage measurements of the motor m, m_{robot} is the mass of the robot and d is the traveled distance.
- Froude number:

$$Fr = v_{max}/\sqrt{g \cdot h}, \text{ for } h = l_{leg} \tag{20}$$

  where l_{leg} is the robot leg length. The Froude number is a dimensionless number used in fluid mechanics to characterize the resistance of an object moving through water. Alexander used it to characterize animal locomotion, given the that also legged locomotion is a dynamic motion in gravity [39]. A given Froude number can be assigned to a certain walking style. Running, for example, starts at a Froude number of approximately 1.0

- Precision of task execution:
  Expressed in terms of tracking errors. The root mean square error is computed by summing the squared difference between the measurement and the desired position over all points of the whole trajectory.

The KPIs computed for each of the combinations of objective functions as in Table 2 are reported in Table 3 for the whole sequence, while in Tables 4–6, a few KPIs referred to the three steps of the optimized sequence: starting step, periodic step and ending step. In these tables, we report also the ones obtained from previous works [8,23] for comparison purposes. As we can see from these tables, the cost of transport could be reduced with the minimization of torques, while the minimization of mechanical work does not show significant improvements in these indicators with respect to the reduced models. In particular, when minimizing work, the cost of transport is higher with respect to the LIPM model, which seems to contradict the wanted objective. It has to be noted that this might be due to the different formulations used for the objective function and the computation of the cost of transport. The cost of transport is computed with data from the robot motors, which could also have some measurement errors.

Table 3. [Hardware experiment result.] Key Performance Indicators (KPIs) measured for all 4 combinations of objective functions, in comparison with results with reduced models. The support times and step period refer to the ones of the periodic step. NMPC, Nonlinear Model Predictive Control.

KPIs	Cart-Table	NMPC (LIPM)	Min Torques	Min Torques and Torso ort.	Min Work	Min Work and Torso ort.
v_{max} (m/s)	0.037	0.065	0.053	0.079	0.043	0.053
t^{ss}/t^{ds} (s)	1.5/1.0	0.6/0.6	1.06/0.98	0.7/0.37	1.06/1	0.8/0.8
step period (s)	2.5	1.2	2.04	1.07	2.06	1.6
Cost of Transport	4.27	2.99	2.69	1.97	3.32	3.09
Froude Number	0.017	0.029	0.024	0.035	0.019	0.024
Joint error (deg)	1.45	1.21	1.27	1.29	1.11	0.9
CoM error (cm)	0.61	0.44	0.29	0.31	0.39	0.29

Table 4. [Hardware experiment result.] Key Performance Indicators (KPIs) for the starting step.

KPIs	Cart-Table	NMPC (LIPM)	Min Torques	Min Torques and Torso ort.	Min Work	Min Work and Torso ort.
Cost of Transport	3.99	2.54	2.49	2.30	2.56	2.67
Joint error (deg)	1.35	1.12	1.19	1.22	1.08	0.9
CoM error (cm)	0.59	0.42	0.25	0.28	0.36	0.27

Table 5. [Hardware experiment result.] Key Performance Indicators (KPIs) for the periodic step. Note that in the case of motions obtained with the cart-table and NMPC, no explicit periodicity constraints were introduced, so the data are approximated on an intermediate step that would correspond to a periodic step.

KPIs	Cart-Table	NMPC (LIPM)	Min Torques	Min Torques and Torso ort.	Min Work	Min Work and Torso ort.
Cost of Transport	4.33	3.10	2.89	1.90	3.31	3.19
Joint error (deg)	1.65	1.20	1.37	1.39	1.20	1.0
CoM error (cm)	0.63	0.46	0.31	0.35	0.40	0.32

Table 6. [Hardware experiment result.] Key Performance Indicators (KPIs) for the ending step.

KPIs	Cart-Table	NMPC (LIPM)	Min Torques	Min Torques & Torso ort.	Min Work	Min Work & Torso ort.
Cost of Transport	4.05	2.64	2.54	2.20	2.61	2.62
Joint error (deg)	1.35	1.15	1.20	1.20	1.08	0.85
CoM error (cm)	0.61	0.43	0.26	0.30	0.38	0.27

When stabilizing the torso orientation, the motion generated using optimal control outperforms the reduced model in walking speed as the double-support time is much shorter; as a matter of fact, also the cost of transport and Froude number are better in this case with respect to the other cases. A higher Froude number indicates basically faster walking both in terms of speed and covered distance; in fact, we can observe that in the case of minimizing torque and introducing torso orientation minimization, the Froude number is the highest, while in the case of minimum work only, it is quite low due to the longer step timings. Tracking errors, such as joint tracking and CoM tracking errors, appear to be overall smaller with respect the case of reduced models, with all objective function combinations. These errors are slightly bigger with the faster walking motion, i.e., the case with minimization of torques and torso orientation minimization, which are still smaller than the case of reduced models. In our objective functions, we did not aim at maximizing walking velocity, but rather left the optimization to find the best solution for the given objective; therefore, it was not expected that the walking speed would necessarily outperform the reduced models. The results however show that the robot can achieve faster walking within its physical limitations, which was not possible with the reduced models.

7. Discussion and Conclusions

In this paper, we have presented the first results of optimal control methods applied to whole-body models of an iCub robot for generating walking motions using a direct multiple shooting method. We have also described the underlying framework and model in much detail. In contrast to other approaches to motion generation, optimization is an all-at-once approach that determines all characteristics of a motion simultaneously to optimize a chosen performance criterion called the objective or cost function and to satisfy all important constraints related to the robot and the task description. For this research, we have focused on the HeiCub robot, which is a reduced version of the iCub with no head and no arms, but the same approach could be applied to the full iCub by just using the corresponding model, which is also available.

In our previous research, the HeiCub had already achieved walking, using methods based on reduced models such as the LIPM and the table cart, and these approaches also have been transferred to the full iCub. These experiences also allowed us to identify parameters and physical constraints that were now included in the optimal control problem formulations. Even though the simplified approaches allowed one to generate a variety of walking motions on level ground in different directions (forward, backward, sidewards) and up and down stairs and slopes, the motions showed some of the typical characteristics of walking based on simplified models, e.g., keeping the pelvis at (nearly) constant height, resulting in a less dynamic appearance of the gait. In addition, the weight distribution of the HeiCub robot violates some of the center of mass assumptions taken for the simple models (see Section 2), because its legs are comparatively heavy. Walking generation methods taking into account the precise mass distribution of the robot therefore seemed to be very promising to generate more suitable motions on the robot, especially since we already had positive experiences with such methods with generating squatting motions and simple push recovery steps for the same robot [40,41], as well as complex walking motions for other humanoid robots, such as HRP-2 [15].

The results of this first optimization study are very encouraging. We have shown that

- It is possible to formulate optimal control problems for the iCub/HeiCub robots that allow one to simultaneously generate periodic walking motions, as well as the necessary starting and stopping steps that take the robot from standing position to the periodic cycle, as well as from back to standing position. So far, only single starting and stopping steps have been included in the formulation, but it is straightforward to extend this to multiple steps, which may be required for faster walking.

- Different objective functions result in visibly different walking styles for the robot. In particular we have compared a minimization of torques squared and of absolute mechanical work, in both cases with and without a combined term on torso stabilization. A very small term on joint accelerations was present in all objective functions. Further objective functions are the subject of current research. The fact that we are able to generate walking in different styles in an automated way already presents a significant difference from the classical walking generation methods using the simple models for which all outcomes look very similar.
- The resulting motions are still not close to biological motion, but they have made significant progress in the right direction. In particular, the motions show variations in the height of the CoM. See the discussion below for making optimized walking motions more biological or "human-like".

As discussed in the Introduction, optimization can be a very helpful tool to bring humanoid robots to their technical limits, i.e., to make the most out of given robot hardware. This requires that suitable optimization criteria targeting an improvement of performance be used, which has not been done so far in this study. In current research, we are also looking at step length, step frequency and walking speed maximization. The objective functions used in this study targeted the smoothness of motions, as well as measures of efficiency. We are also interested in using human-inspired criteria for generating humanoid motions. For this, we can rely on the results of inverse optimal control studies, which identify criteria underlying human motions from motion capture data. These criteria can then be applied to humanoid models in optimal control problem formulations as the one presented in this paper. We have demonstrated a similar approach based on simpler template models of human and humanoid walking in [19], also for the HeiCub, but it would be new to apply this concept of transferring bio-inspired optimization criteria to whole-body models of the HeiCub robot.

There are several characteristics of the robot and its inherent control approaches that limit the performance. The fact that the robot has no arms and all counteraction to the lower limb angular momentum has to be done by the torso already prevents it from acting in a fully-human-like manner. While humans exhibit a strongly stabilized head and torso while walking forward on flat ground, such a criterion induces strong limitations to the motion if no actions of the arms are possible. In this context, also the extension of the work to the full iCub model would be interesting to see how the arms are used for angular momentum compensation.

As discussed before, the HeiCub and iCub robots also have several limitations in the joint angles, the angular velocities and the joint torques that prevent them from coming close to human performance. In addition, the current foot shape and controllers impose that the robot, as most contemporary robots, walks with flat feet and cannot, e.g., have heel only or toe only contact as humans can, resulting in limited step lengths and step heights (the latter on stairs). The optimal control approach presented in this paper could also serve to generate walking motions with partial or more flexible foot contact, as soon as the hardware allows it. While optimization can help to exploit the capabilities of the robot hardware, it is best to simultaneously also improve the hardware to have an overall performance improvement.

As mentioned in Section 2, the robot is equipped with series elastic actuators. The model of the robot with a reduced number of DOF, but including elasticity has been used in optimal control frameworks for the generation of squat and push recovery motions [40,41]; the current framework for walking will be extended to include also the elasticity of the elastic actuators, which will be mounted for the experimental validations. The optimal control problem including elasticity is a larger nonlinear problem with respect to the one described in this work, which will require longer computation times. On the hardware side, a proper control for the SEA needs to be implemented.

In addition to the extensions already discussed, future work on whole-body optimal control of the HeiCub robot will also include more complex walking scenarios like stairs and slopes for which the approach based on simple models is even further from the real behavior than for motion on flat ground and for which many details of the motion such as the foot trajectories in the air had to be inserted manually. We expect that for these motions, the gain in performance and walking quality by the whole-body optimal control approach will be more significant than for walking.

The computation times for the method presented here are very long for multiple reasons. First, it should be noted that the whole-body dynamic equations for a complex multi-phase system have to be solved (and derivatives have to be computed) precisely in every optimization step. Second, the discretization chosen for the controls and the multiple shooting intervals in the present study is very accurate, and from the optimal control perspective, a much coarser grid would be possible without any problem. This was done in the attempt to get closer to the robots' control intervals, but could also be solved by means of interpolation in the interest of higher efficiency. However what is most important is that solving such optimal control problems online each time the robot has to perform a motion can be avoided. In previous research, we developed an approach based on a combination of optimal control with movement primitives [18]. In this method, optimal control serves to generate multiple training data in an offline process on the basis of which movement primitives are learned. This takes some time, but is not time-critical and only has to be done once for a class of motions. New versatile motions can then be efficiently generated online, just using the movement primitives without the need to solve any optimal control problems. As has been demonstrated in the case of another robot, the resulting motions are close to optimal and dynamic feasibility and work well on the real system. We therefore expect this to be a very interesting approach for the iCub/HeiCub robot, which makes this a very important argument to further explore whole-body optimal control in the context of bio-inspired walking generation for these robots.

Supplementary Materials: The following are available online at www.mdpi.com/2076-3417/8/2/278/s1, Video S1: Experimental Results.

Acknowledgments: The work carried out in this paper was funded by the Heidelberg Graduate School of Mathematical and Computational Methods for the Sciences (HGS MathComp), founded by DFG Grant GSC 220 in the German Universities Excellence Initiative. We acknowledge financial support by Deutsche Forschungsgemeinschaft within the funding programme Open Access Publishing, by the Baden-Württemberg Ministry of Science, Research and the Arts and by Ruprecht-Karls-Universität Heidelberg. This work is based on work that was carried out for the European Project KoroiBot concluded in September 2016, which received funding from the European Union Seventh Framework Programme (FP7/2013) under Grant Agreement No. 611909. We wish to thank the Simulation and Optimization group of H. G. Bock at the Heidelberg University for providing the optimal control code MUSCOD.

Author Contributions: Yue Hu and Katja Mombaur jointly planned and designed the study and contributed to analyzing the results and writing the paper. Yue Hu has carried out the computations and the experiments.

Conflicts of Interest: The authors declare no conflict of interest. The funding sponsors had no role in the design of the study; in the collection, analyses or interpretation of data; in the writing of the manuscript; nor in the decision to publish the results.

References

1. Vukobratović, M.; Borovac, B. Zero-moment point—Thirty five years of its life. *Int. J. Humanoid Robot.* **2004**, *1*, 157–173.
2. Mukovskiy, A.; Vassallo, C.; Naveau, M.; Stasse, O.; Soueres, P.; Giese, M.A. Adaptive synthesis of dynamically feasible full-body movements for the humanoid robot HRP-2 by flexible combination of learned dynamic movement primitives. *Robot. Auton. Syst.* **2017**, *91*, 270–283.
3. Ishiguro, Y.; Kojima, K.; Sugai, F.; Nozawa, S.; Kakiuchi, Y.; Okada, K.; Inaba, M. Bipedal Oriented Whole Body Master-Slave System for Dynamic Secured Locomotion with LIP Safety Constraints. In Proceedings of the IEEE International Conference on Intelligent Robots and Systems (IROS), Vancouver, BC, Canada, 24–28 September 2017; pp. 376–382.

4. Kajita, S.; Kanehiro, F.; Kaneko, K.; Fujiwara, K.; Yokoi, K.; Hirukawa, H. A Realtime Pattern Generator for Biped Walking. In Proceedings of the IEEE International Conference on Robotics and Automation (ICRA), Washington, DC, USA, 1–15 May 2002; pp. 31–37.
5. Kajita, S.; Kanehiro, F.; Kaneko, K.; Fujiwara, K.; Harada, K.; Yokoi, K.; Hirukawa, H. Biped walking pattern generation by using preview control of zero-moment point. In Proceedings of the IEEE International Conference on Robotics and Automation (ICRA), Taipei, Taiwan, 14–19 September 2003; Volume 2, pp. 1620–1626.
6. Naveau, M.; Kudruss, M.; Stasse, O.; Kirches, C.; Mombaur, K.; Souères, P. A reactive walking pattern generator based on nonlinear model predictive control. *IEEE Robot. Autom. Lett.* **2017**, *2*, 10–17.
7. Li, Z.; Tsagarakis, N.G.; Caldwell, D.G. Walking trajectory generation for humanoid robots with compliant joints: Experimentation with COMAN humanoid. In Proceedings of the IEEE International Conference on Robotics and Automation (ICRA), Saint Paul, MN, USA, 14–18 May 2012; pp. 836–841.
8. Hu, Y.; Eljaik, J.; Stein, K.; Nori, F.; Mombaur, K. Walking of the iCub humanoid robot: Implementation and performance analysis. In Proceedings of the IEEE-RAS International Conference on Humanoid Robots (Humanoids), Cancun, Mexico, 15–17 November 2016; pp. 690–696.
9. Mombaur, K.; Clever, D. Inverse optimal control as a tool to understand human movement. In *Geometric and Numerical Foundations of Movements*; Laumond, J.P., Mansard, N., Lasserre, J.B., Eds.; Springer STAR Series: Berlin/Heidelberg, Germany, 2017.
10. Ames, A.D. Human-inspired control of bipedal walking robots. *IEEE Trans. Autom. Control* **2014**, *59*, 1115–1130.
11. Mombaur, K. Humanoid Motion Optimization. In *Humanoid Robotics—A Reference*; Goswami, A., Vadakkepat, P., Eds.; Springer: Berlin/Heidelberg, Germany, 2017.
12. Lengagne, S.; Vaillant, J.; Yoshida, E.; Kheddar, A. Generation of whole-body optimal dynamic multi-contact motions. *Int. J. Robot. Res.* **2013**, *32*, 1104–1119.
13. Miossec, S.; Yokoi, K.; Kheddar, A. Development of a software for motion optimization of robots—Application to the kick motion of the HRP-2 robot. In Proceedings of the IEEE International Conference on Robotics and Automation (ICRA), Kunming, China, 17–20 December 2006; pp. 299–304.
14. Suleiman, W.; Yoshida, E.; Laumond, J.P.; Monin, A. On humanoid motion optimization. In Proceedings of the IEEE-RAS International Conference on Humanoid Robots (Humanoids), Pittsburgh, PA, USA, 29 November–1 December 2007; pp. 180–187.
15. Koch, K.H.; Mombaur, K.; Souères, P. Optimization-based walking generation for humanoids. In *IFAC-SYROCO 2012*; Elsevier: Amsterdam, The Netherlands, 2012; Volume 45, pp. 498–504.
16. Koch, K.H.; Mombaur, K.; Souères, P. Studying the Effect of Different Optimization Criteria on Humanoid Walking Motions. In *Simulation, Modeling, and Programming for Autonomous Robots*; Noda, I., Ando, N., Brugali, D., Kuffner, J.J., Eds.; Springer: Berlin/Heidelberg, Germany, 2012; Volume 7628, pp. 221–236.
17. Koch, K.H.; Mombaur, K.; Souères, P.; Stasse, O. Optimization based exploitation of the ankle elasticity of HRP-2 for overstepping large obstacles. In Proceedings of the IEEE/RAS International Conference on Humanoid Robots (Humanoids 2014), Madrid, Spain, 18–20 November 2014; pp. 733–740.
18. Clever, D.; Harant, M.; Mombaur, K.; Naveau, M.; Stasse, O.; Endres, D. Cocomopl: A novel approach for humanoid walking generation combining optimal control, movement primitives and learning and its transfer to the real robot HRP-2. *IEEE Robot. Autom. Lett.* **2017**, *2*, 977–984.
19. Clever, D.; Mombaur, K.D. An Inverse Optimal Control Approach for the Transfer of Human Walking Motions in Constrained Environment to Humanoid Robots. In Proceedings of the Robotics: Science and Systems, Ann Arbor, MI, USA, 18–22 June 2016.
20. Posa, M.; Kuindersma, S.; Tedrake, R. Optimization and stabilization of trajectories for constrained dynamical systems. In Proceedings of the 2016 IEEE International Conference on Robotics and Automation (ICRA), Stockholm, Sweden, 16–21 May 2016; pp. 1366–1373.
21. Hereid, A.; Cousineau, E.A.; Hubicki, C.M.; Ames, A.D. 3D dynamic walking with underactuated humanoid robots: A direct collocation framework for optimizing hybrid zero dynamics. In Proceedings of the 2016 IEEE International Conference on Robotics and Automation (ICRA), Stockholm, Sweden, 16–21 May 2016; pp. 1447–1454.
22. Metta, G.; Natale, L.; Nori, F.; Sandini, G.; Vernon, D.; Fadiga, L.; Von Hofsten, C.; Rosander, K.; Lopes, M.; Santos-Victor, J.; et al. The iCub humanoid robot: An open-systems platform for research in cognitive development. *Neural Netw.* **2010**, *23*, 1125–1134.

23. Stein, K.; Hu, Y.; Kudruss, M.; Naveau, M.; Mombaur, K. Closed loop control of walking motions with adaptive choice of directions for the iCub humanoid robot. In Proceedings of the IEEE International Conference on Humanoid Robots (Humanoids), Birmingham, UK, 15–17 November 2017.
24. Bock, H.; Plitt, K. *A Multiple Shooting Algorithm for Direct Solution of Optimal Control Problems*; Pergamon Press: Oxford, UK, 1984; pp. 243–247.
25. Mombaur, K.; Longman, R.; Bock, H.; Schlöder, J. Open-loop stable running. *Robotica* **2005**, *23*, doi:0.1017/S026357470400058X.
26. Tikhanoff, V.; Pattacini, U.; Natale, L.; Metta, G. Exploring affordances and tool use on the iCub. In Proceedings of the IEEE-RAS International Conference on Humanoid Robots (Humanoids), Atlanta, GA, USA, 15–17 October 2013; pp. 130–137.
27. Mar, T.; Tikhanoff, V.; Metta, G.; Natale, L. Self-supervised learning of grasp dependent tool affordances on the iCub Humanoid robot. In Proceedings of the IEEE International Conference on Robotics and Automation (ICRA), Seattle, WA, USA, 26–30 May 2015; pp. 3200–3206.
28. Nori, F.; Traversaro, S.; Eljaik, J.; Romano, F.; Del Prete, A.; Pucci, D. iCub whole-body control through force regulation on rigid non-coplanar contacts. *Front. Robot. AI* **2015**, *2*, 6.
29. Parmiggiani, A.; Metta, G.; Tsagarakis, N. The mechatronic design of the new legs of the iCub robot. In Proceedings of the IEEE-RAS International Conference on Humanoid Robots (Humanoids), Osaka, Japan, 29 November–1 December 2012; pp. 481–486.
30. Colasanto, L.; Tsagarakis, N.; Caldwell, D. A Compact Model for the Compliant Humanoid Robot COMAN. In Proceedings of the IEEE International Conference on Biomedical Robotics and Biomechatronics, Rome, Italy, 24–27 June 2012; pp. 688–694.
31. Dallali, H.; Kormushev, P.; Li, Z.; Caldwell, D. On Global Optimization of Walking Gaits for the Compliant Humanoid Robot, COMAN Using Reinforcement Learning. *Cybern. Inf. Technol.* **2012**, *12*, 39–52.
32. Moro, F.L.; Tsagarakis, N.G.; Caldwell, D.G. Walking in the resonance with COMAN robot with trajectories based on human kinematic motion primitives (kMPs). *Auton. Robot.* **2014**, *36*, 331–347.
33. N.G. Tsagarakis, M. Laffranchi, B.V.; Caldwell, D.G. A Compact Soft Actuator Unit for Small Scale Human Friendly Robots. In Proceedings of the International Conference on Robotics and Automation (ICRA), Kobe, Japan, 2–17 May 2009; pp. 4356–4362.
34. Felis, M.; Mombaur, K.D.; Berthoz, A. An Optimal Control Approach to Reconstruct Human Gait Dynamics from Kinematic Data. In Proceedings of the IEEE-RAS 15th International Conference on Humanoid Robots (Humanoids), Seoul, Korea, 3–5 November 2015.
35. Hu, Y.; Mombaur, K. Analysis of human leg joints compliance in different walking scenarios with an optimal control approach. In Proceedings of the 6th IFAC International Workshop on Periodic Control Systems (PSYCO 2016), Eindhoven, The Netherlands, 29 June–1 July 2016; Volume 49, pp. 99–106.
36. Ketchel, J.; Larochelle, P. Collision detection of cylindrical rigid bodies for motion planning. In Proceedings of the IEEE International Conference on Robotics and Automation (ICRA), Orlando, FL, USA, 15–19 May 2006; pp. 1530–1535.
37. Leinweber, D.; Bauer, I.; Bock, H.; Schloeder, J. An efficient multiple shooting based reduced SQP strategy for large-scale dynamic process optimization. Part I: Theoretical aspects. *Comput. Chem. Eng.* **2003**, *27*, 157–166.
38. Felis, M.L. RBDL: An efficient rigid-body dynamics library using recursive algorithms. *Auton. Robot.* **2017**, *41*, 495–511.
39. Alexander, R.M. *Principles of Animal Locomotion*; Princeton University Press: Princeton, NJ, USA 2003.
40. Hu, Y.; Nori, F.; Mombaur, K. Squat Motion Generation for the Humanoid Robot iCub with Series Elastic Actuators. In Proceedings of the IEEE RAS & EMBS International Conference on Biomedical Robotics and Biomechatronics (BioRob), Singapore, 26–29 June 2016; pp. 207–212.
41. Hu, Y.; Mombaur, K. Optimal Control Based Push recovery Strategy for the humanoid robot iCub with Series Elastic Actuators. In Proceedings of the IEEE/RSJ International Conference on Intelligent Robots and Systems (IROS), Vancouver, BC, Canada, 24-28 September 2017; pp. 5842–5852.

Article

Motion Planning for Bipedal Robot to Perform Jump Maneuver

Xinyang Jiang [1], Xuechao Chen [1,2,*], Zhangguo Yu [1,3], Weimin Zhang [1,3], Libo Meng [1] and Qiang Huang [1,2]

1 Intelligent Robotics Institute, School of Mechatronical Engineering, Beijing Institute of Technology, Beijing 100081, China; xinyangjiang@163.com (X.J.); yuzg@bit.edu.cn (Z.Y.); zhwm@bit.edu.cn (W.Z.); menglibo@bit.edu.cn (L.M.); qhuang@bit.edu.cn (Q.H.)
2 Key Laboratory of Biomimetic Robots and Systems, Ministry of Education, Beijing 100081, China
3 Beijing Advanced Innovation Center for Intelligent Robots and Systems, Beijing 100081, China
* Correspondence: chenxuechao@bit.edu.cn; Tel.: +86-010-68-917-658

Received: 14 November 2017; Accepted: 15 January 2018; Published: 19 January 2018

Abstract: The remarkable ability of humans to perform jump maneuvers greatly contributes to the improvements of the obstacle negotiation ability of humans. The paper proposes a jumping control scheme for a bipedal robot to perform a high jump. The half-body of the robot is modeled as three planar links and the motion during the launching phase is taken into account. A geometrically simple motion was first conducted through which the gear reduction ratio that matches the maximum motor output for high jumping was selected. Then, the following strategies to further exploit the motor output performance was examined: (1) to set the maximum torque of each joint as the baseline that is explicitly modeled as a piecewise linear function dependent on the joint angular velocity; (2) to exert it with a correction of the joint angular accelerations in order to satisfy some balancing criteria during the motion. The criteria include the location of ZMP (zero moment point) and the torque limit. Using the technique described above, the jumping pattern is pre-calculated to maximize the jump height. Finally, the effectiveness of the proposed method is evaluated through simulations. In the simulation, the bipedal robot model achieved a 0.477-m high jump.

Keywords: biped robot; jump; electric actuator; motion planning

---

1. Introduction

Recently, the number of disasters, such as earthquakes and tsunamis, is increasing, and humanoid robots have commonly been expected to work in place of humans in such dangerous environments, in which the terrain can be uneven and unpredictable. Improvement of their locomotion abilities by addition of walking, crawling, climbing, fall protection, and jumping capabilities [1–13] is important to allow these robots to navigate across such adverse terrain. Unfortunately, most contemporary humanoid robots only have segmental functions. Some robots are designed simply for walking or crawling [2,4,7], while others are designed for jumping alone [9,10]. The purpose of this paper is to add a jumping pattern to a versatile BHR6 robot, which is actuated using electric motors, and that already has walking, rolling, and fall protection capabilities [11–13].

Quick and harmonized coordination of the jumper's body segments is essential for jumping tasks. Humans take advantage of the elasticity of their tendons to store and use kinetic energy during jumping motions. Jumping using elastic elements has been studied intensively. Raibert and colleagues studied hopping robots [14] that were driven by pneumatic and hydraulic actuators to perform various actions, including somersaults [15]. Other researchers successfully demonstrated jumping motions using robots with artificial muscles [16,17]. Curran [9] and Ugurlu [18] proposed jumping mechanisms that use the potential energy of a spring. These research efforts are focused on development of a jumping mechanism and its control. All of these robots have elastic mechanisms that are used to retrieve kinetic energy during dynamic movement cycles. Use of elastic elements has obvious advantages for storage and use of kinetic energy, but they may also affect the performance of the fundamental functions of the robots, e.g., walking, crawling, climbing, and carrying objects. Because our intention is to add jumping functionality to an already versatile robot BHR6 without any elastic mechanism, the robot is modeled as three rigid planer links.

A jumping robot not only requires a high torque output to accelerate its COM (center of mass) at the beginning of the motion but also must realize a high velocity at the end of the motion. As the joints become increasingly extended, the transfer of the joint angular motion to produce the desired translation of the COM becomes less effective. When the joint is fully extended, the effect of this joint on the COM's translational motion in a specific direction is equal to zero. When these factors are taken into consideration, an appropriate reduction gear is required for the motor [19].

The prime criterion for the robot's obstacle negotiation ability is the required height of the jump, which is dependent on the vertical velocity of the robot's COM at the moment when its feet leave the ground. During the push-off phase, when the mechanical limitations are considered, a high-torque output is required to generate high angular acceleration. Sakka et al. proposed an adaptation of human jumping motion based on the ground force [8]. Using a similar algorithm, Okada et al. designed a nonlinear gear ratio profile to maximize the ground force [19]. Ugurlu et al. used the base resonance frequency to realize high acceleration [18]. By addition of elastic devices, Hondo et al. realized jumping motion in the low-power humanoid robot Nao [20] that would not originally have had sufficient power to jump. All these methods planned jump patterns based on the trajectory of the leg length or the force of the ground. It leads to the lack of consideration of each joint state, e.g., the torque, velocity and acceleration. Consequently, the relevant motors could not work in the ultimate states required to supply higher kinetic energy and thus achieve a higher jump. To produce as much energy as possible, an optimal motion pattern is generated by commanding every joint torque as much as possible to match the profile of the motor torque speed curve.

During the push-off phase of the jump, the robot must keep stable. Okada kept the trunk upright while the shank and thigh have the same length [19]. Urata and his colleagues at the University of Tokyo Jouhou System Kougaku Laboratory, used the same method [21]. Nunez planned COM above the ankle to prevent it from falling [22]. Ugurlu proposed a ZMP-based jumping controller design method [23,24]. The ZMP is defined as the point on the ground about which the sum of all the moments of the active forces equals zero. The ZMP always exists inside the support polygon form by all contact points. The distance from the ZMP to the boundary of the support polygon directly influence the stability of the robot. A larger distance leads to a more stable state. Our strategy is based on fine-tuning of the joint torque on the basis of the maximum torque being limited by the motor characteristics while minimizing the cost function via ZMP constraints.

In this paper, we introduce a motion planning method to allow a bipedal robot to perform a jump maneuver without the use of elastic elements. Using the premise of dynamic balance, we command each joint torque as much as possible to match the motor torque speed curve to ensure that the robot can jump to the greatest possible height. In Section 2, we present the robot model that is used in this paper. The method used to generate the jumping pattern is described in detail in Section 3. In Section 4, using playback of an offline pattern that was generated in the previous section, we present experimental results for a bipedal robot model of jumping about 0.477-m-high. We draw conclusions and address our future plans in Section 5.

2. Robot Model

In this section, we present the physical model to be used in this study, along with the assumptions that are made in our analysis. Throughout the complete jumping pattern, by keeping the robot symmetrical with respect to the sagittal plane [18,22], the movements, torques and forces acting on each leg will be identical. This restriction helps us to study jumping using a simplified model that contains only half of the robot.

2.1. Mechanical Model

The BHR6 robot and its simplified physical model are shown in Figure 1. Referring to the physical properties of BHR6, we built a model of four links (foot, shank, thigh and trunk) and three articulations (ankle, knee and hip). The foot length is d_0, and d_1, d_2 represent the distances from the projection of ankle joint on the ground to the heel and tiptoe, respectively. We therefore find that $d_0 = d_1 + d_2$. The parameter l_0 represents the vertical distance from the ankle joint to the ground and m_0 is the mass of foot. Additionally, $m_i (i = 1, 2, 3)$ and $l_i (i = 1, 2, 3)$ are the masses and lengths of the shank, the thigh and the trunk, respectively, and the mass center of each link is located at its center. Note also that m_3 is only half of the mass of the robot trunk. The model information is listed in detail in Table 1. The total mass and height of the robot are 51 (kg) and 1.53 (m), respectively.

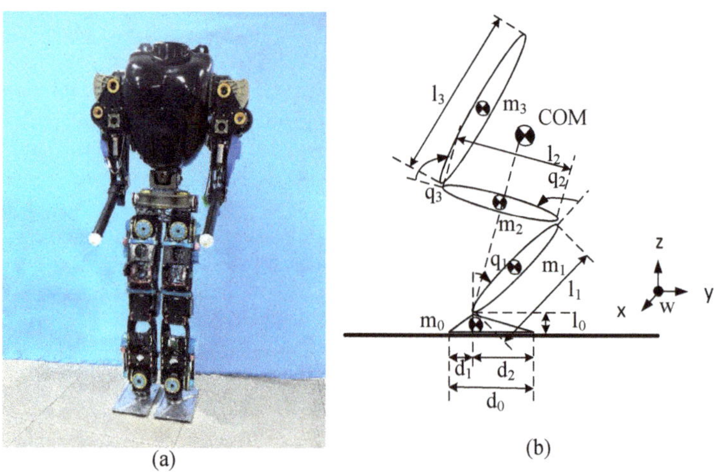

Figure 1. (**a**) BHR6 humanoid robot; (**b**) the simplified model with the similar physical properties of BHR6.

Table 1. Parameters of the simplified model.

m_0	m_1	m_2	m_3	l_0	l_1	l_2	l_3	d_0	d_1	d_2
0.5 (kg)	5.0 (kg)	5.0 (kg)	15 (kg)	0.12 (m)	0.33 (m)	0.33 (m)	0.75 (m)	0.30 (m)	0.10 (m)	0.20 (m)

2.2. Mathematical Model

During the push-off phase, assuming that the friction between the robot foot and the ground is sufficient, the robot will not slide on the ground. Additionally, if the robot foot remain flat on the ground, the model can be represented as a three-part linkage that is fixed to the ground [18,22]. The dynamic equation is given by

$$M(q)\ddot{q} + V(q, \dot{q}) + G(q) = F, \tag{1}$$

where the inertia matrix $M \in R^{3\times3}$, the coriolis and centripetal coupling vector $V \in R^{3\times1}$, and the gravity vector $G \in R^{3\times1}$ are not dependent on the state of the foot. These matrices can be calculated using physical parameters from Table 1 and the instantaneous leg configuration. In this work, we represent all vectors of position, velocity, acceleration, angle, angular velocity and angular acceleration within the world frame $\sum_w$. We define the angle, the angular velocity and the angular acceleration of each joint as q_i, $\dot{q}_i$, and $\ddot{q}_i (i = 1,2,3)$, respectively. $q = (q_1, q_2, q_3)^T$ represents the generalized coordinates and $F = (\tau_1, \tau_2, \tau_3)^T$ represents the joint torque.

3. Method

A jumping robot not only requires a high torque output to accelerate its COM at the beginning of the motion but also must realize a high velocity at the end of the motion [19]. From this viewpoint, (1) we selected the appropriate reduction ratio for a specific motor by using a simplified control system; and (2) using the appropriate reduction ratio, a jumping pattern was generated by commanding each motor produces as much kinetic energy as possible within the limitations of the motor properties and dynamic balance throughout the complete push-off phase.

3.1. Selection of an Appropriate Reduction Ratio

Okada et al. designed a nonlinear gear ratio profile with the aim of performing a high jump using direct current motors [19]. However, this profile led to major difficulties in mechanical design, processing and assembly. Our purpose in this work is to determine an appropriate reduction ratio for a specific motor without consideration of the energy transfer efficiency of the gear reducer.

Because of the nonlinearity of the dynamic equation, i.e., Equation (1), it is difficult to determine three appropriate reduction ratios for three motors simultaneously. With reference to the work of Okada and Urata [19,21], we designed a simple jumping pattern in which the trunk remains upright and above the ankle joint during the push-off phase, as shown in Figure 2. In this simple pattern, we can then estimate that of the three joints, and the knee joint will output the maximum torque and speed.

The BHR6 is equipped with the same motor K044100-6Y-4.98Ams (Parker Hannifin Corp, Clifford, OH, USA) on the ankle, knee and hip joint. In Figure 3, according to the instructions from Parker Hannifin, we showed the torque speed curve and properties of motor K044100-6Y-4.98Ams when the voltage supplied to this motor is 120 V. The blue line is the peak output torque of the motor and the red line is the continue output torque of the motor. Because jumping is a fast and dynamic motion, the peak output torque is used in the push-off phase. By the using of the motor driver, the motor can work at an arbitrary point in the right trapezoid formed by the blue line and axis.

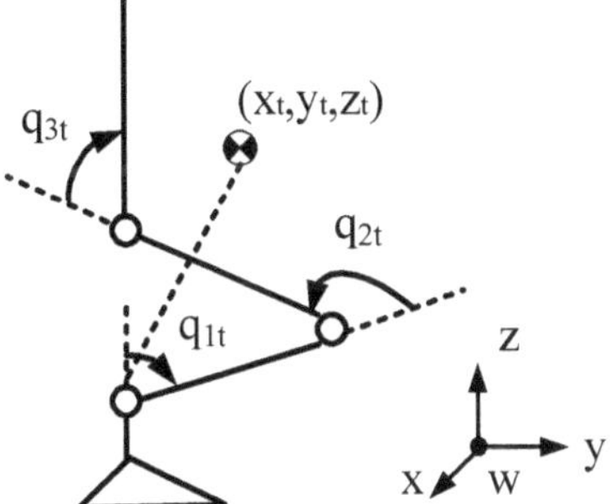

Figure 2. The joint constraint for appropriate reduction ratio selection.

From Figure 3, the peak torque speed curve of this motor can be modeled as a segmented function. When the motor speed $\omega_{motor} \leq \omega_{break} (\omega_{break} = 10634$ (rpm)), the peak motor torque τ_{motor} is $\tau_{peak} = 1.17$ (Nm). When $\omega_{motor} \in [\omega_{break}, \omega_{max}]$ ($\omega_{max} = 18835$ (rpm)), τ_{motor} decreases linearly with increasing ω_{motor}. Given these parameters, we can then obtain the relationship between ω_{motor} and τ_{motor}, as shown below:

$$\tau_{motor} = \begin{cases} \tau_{peak} & \omega_{motor} \leq \omega_{break}, \\ \tau_{peak} - \frac{\tau_{peak} * \omega_{motor}}{(\omega_{max} - \omega_{break})} & \omega_{motor} \in [\omega_{break}, \omega_{max}]. \end{cases} \tag{2}$$

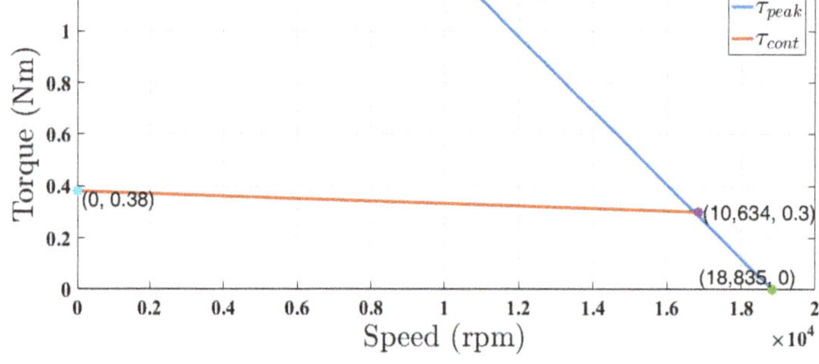

Figure 3. Torque speed curve of the motor K044100-6Y-4.98Ams when the voltage supplied to the motor is 120 V.

We used the harmonic reducer to deliver the motor torque. If the reduction ratio applied to the motor is i_r, then the joint speed ω_{joint} will be $\frac{1}{i_r}$ of the ω_{motor}. If the transfer efficiency of reducer is assumed as 100%, the output power of motor and joint are identical, $\omega_{joint}\tau_{joint} = \omega_{motor}\tau_{motor}$. In addition, the peak joint torque τ_{joint} will be i_r times as τ_{motor}:

$$\tau_{joint} = i_r\tau_{motor} \text{ (Nm)}, \tag{3}$$

$$\omega_{joint} = \frac{\omega_{motor}}{i_r}\text{(rpm)} = \frac{2\pi\omega_{motor}}{60 i_r} \text{ (rad/s)}. \tag{4}$$

It is assumed that the model state at time t is $X_t(q_t, \dot{q}_t)$. $q_t = (q_{1t}, q_{2t}, q_{3t})$ (rad) and $\dot{q}_t = (\dot{q}_{1t}, \dot{q}_{2t}, \dot{q}_{3t})$ (rad/s). Parameter $C_t = (x_t, y_t, z_t)$ (m) represents the coordinate of the COM. The translational velocity of the COM $\dot{C}_t$ can then be calculated using the COM Jacobian matrix $J(q_t)$:

$$\dot{C}_t = J(q_t)\dot{q}_t. \tag{5}$$

Using the parameters given in Table 1, where $l_1 = l_2$, and the simple jumping pattern, we can obtain the joint constraint equations as shown below:

$$\begin{cases} q_{2t} = & -2q_{1t} = -2q_{3t}, \\ \dot{q}_{2t} = & -2\dot{q}_{1t} = -2\dot{q}_{3t}, \\ \ddot{q}_{2t} = & -2\ddot{q}_{1t} = -2\ddot{q}_{3t}. \end{cases} \tag{6}$$

We can estimate that, of the three joints, the knee joint will output the maximum torque and speed. Using the basic idea of maintaining the knee joint torque's compliance with the motor torque speed curve, we substitute $\dot{q}_{2t}$ into Equations (2)–(4) to calculate the peak knee joint torque τ_{2t} (Nm) at time t:

$$\tau_{2t} = \begin{cases} \tau_{peak} i_r & \dot{q}_{2t} \leq \frac{2\pi\omega_{break}}{60 i_r}, \\ \tau_{peak} i_r - \frac{60\tau_{peak} i_r^2 \dot{q}_{2t}}{2\pi(\omega_{max} - \omega_{break})} & \dot{q}_{2t} \in [\frac{2\pi\omega_{break}}{60 i_r}, \frac{2\pi\omega_{max}}{60 i_r}]. \end{cases} \tag{7}$$

With the joint constraint of Equation (6), the second row of Equation (1) then becomes

$$\begin{pmatrix} M_{21} & M_{22} & M_{23} \end{pmatrix} \begin{pmatrix} -0.5 \\ 1 \\ -0.5 \end{pmatrix} \ddot{q}_{2t} + V_2 + G_2 = \tau_{2t}, \tag{8}$$

where V_2, G_2 represent the second rows of V, G in state X_t, respectively. $M_{ij}(i = 2, j = 1,\ 2,\ 3)$ is the i-th row and j-th column element of M. Therefore, the maximum angular acceleration $\ddot{q}_{2t}$ of the knee joint at time t is

$$\ddot{q}_{2t} = \frac{\tau_{2t} - V_2 - G_2}{\begin{pmatrix} M_{21} & M_{22} & M_{23} \end{pmatrix} \begin{pmatrix} -0.5 \\ 1 \\ -0.5 \end{pmatrix}}. \tag{9}$$

Using the joint constraint Equation (6), we can get the q_t, $\dot{q}_t$ and $\ddot{q}_t$ at time t. Thus, the torque of each joint $\tau_t = (\tau_{1t}, \tau_{2t}, \tau_{3t})$ then becomes

$$\tau_t = M(q_t)\ddot{q}_t + V(q_t, \dot{q}_t) + G(q_t). \tag{10}$$

Obviously, the knee joint torque τ_{2t} is equal to the peak torque limited by motor characteristics. By taking the derivative of Equation (5) with respect to t, we can calculate the translational acceleration of COM $\ddot{C}_t$ at time t:

$$\ddot{C}_t = \dot{J}(q_t)\dot{q}_t + J(q_t)\ddot{q}_t. \tag{11}$$

Integration of $\ddot{q}_t$ with respect to t allows the next moment state $X_{t+1}(q_{t+1}, \dot{q}_{t+1})$ to be obtained. Therefore, by providing the initial state $X_0(q_0, \dot{q}_0)$, we can then calculate iteratively the X_t, $\ddot{q}_t$, τ_t, C_t, $\dot{C}_t$, and $\ddot{C}_t$ throughout the entire push-off phase, which guarantees that the COM maintains maximum acceleration. The control system is shown in Figure 4.

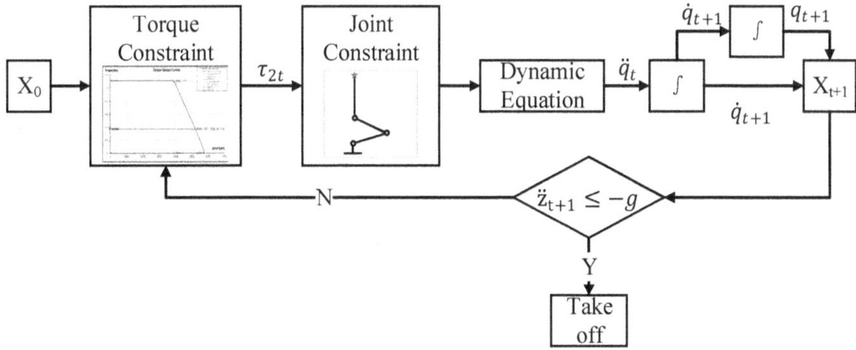

Figure 4. The control system for choosing an appropriate reduction radio for motor K044100-6Y-4.98Ams.

Since our objective is to jump, we have to detect the take-off time to make the robot foot free when the robot is going to take off. At the beginning of the motion, the $\ddot{z}_t$ is positive and the COM is accelerated. When the leg is going to full extension, the $\ddot{z}_t$ will decrease rapidly. Since the computation is iterative and has a certain computation period, the take-off time most likely exists within one of the computation periods. The vertical acceleration of COM calculated is $\leq -g$ before the take-off time, and $\leq -g$ after the take-off time. In addition, $\ddot{z}_{t+1} \leq -g$ is a condition that only happens in mathematical calculations not in the real jump motion. If we get a vertical acceleration of COM, which is $\leq -g$ during the push-off phase, it means that the robot has left the ground. The jump height h is then estimated as

$$h = \frac{\dot{z}_{takeoff}^2}{2g}, \tag{12}$$

where $\dot{z}_{takeoff}$ is the vertical velocity of the COM at the take-off time and g is the acceleration due to gravity.

To select the appropriate reduction ratio, we define the same initial state $X_0 = (-75°, 150°, -75°, 0, 0, 0)$ for different values of the reduction ratio i_r. The results of the relationship between h and i_r can then be presented graphically. Referring to Figure 5, it shows that when $i_r = 97$, the maximum jump height is approximately $h_{max} = 0.343$ (m). However, the gear reduction ratio has only certain specific values, e.g., 30, 50, 80, 100, and 120. Finally, therefore, we choose the most appropriate reduction ratio of $i_r = 100$.

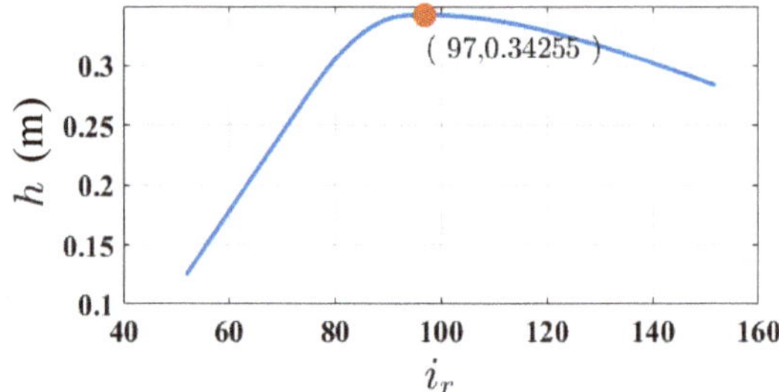

Figure 5. With the same motor and model, different reduction ratio leads to different jump height. When the same initial state $X_0 = (-75°, 150°, -75°, 0, 0, 0)$ is determined, the relationship between h and i_r can be presented.

Most of the robots are equipped with the same or similar motor and reduction ratio on these three joints, e.g., NAO [20], HRP3L [21], DARwin [25] and Honda E2-DR [26] . We equip these three joints with the same K044100-6Y-4.98Ams motor and the same reduction ratio of $i_r = 100$. Using the control system shown in Figure 4, we obtain the torques of the three joints over the entire push-off phase, as shown in Figure 6. In Figure 6b, the purple circle τ_{peak} is the peak output torque of the motor. The other three lines are composed by the working points of three joints when the robot jumps. All the working points are always in the right trapezoid form by τ_{peak} and axis. Furthermore, as expected, the knee joint torque coincides with the maximum torque permitted by the motor's characteristics. However, the ankle and hip motors are ineffective as suppliers of kinetic energy. The hip joint doesn't work and its torque is a constant 0 (Nm), while the ankle joint works at a low efficiency and its torque is much less than the maximal torque allowed. To address this problem, we removed the joint constraint of Equation (6) and take the ZMP constraint into consideration. The details are described in the Section 3.2.

3.2. Motion Planning for Jumping Maneuver

In the original control system in Figure 4, under the joint constraint of Equation (6), we could only ensure that the knee joint torque matched the ultimate torque. To ensure that all three joint torques yield the torque speed curve profile simultaneously as much as possible, we removed the joint constraint and commanded each motor output torque to the torque speed curve. Driven by these torques, the robot will turn over. Thus, we added the ZMP constraint to help the robot to be stable in the push-off phase. If the ZMP constraint is violated, we will fine-tune these torques through optimization. If the ZMP constraint is satisfied, we will directly apply these torques to the joints.

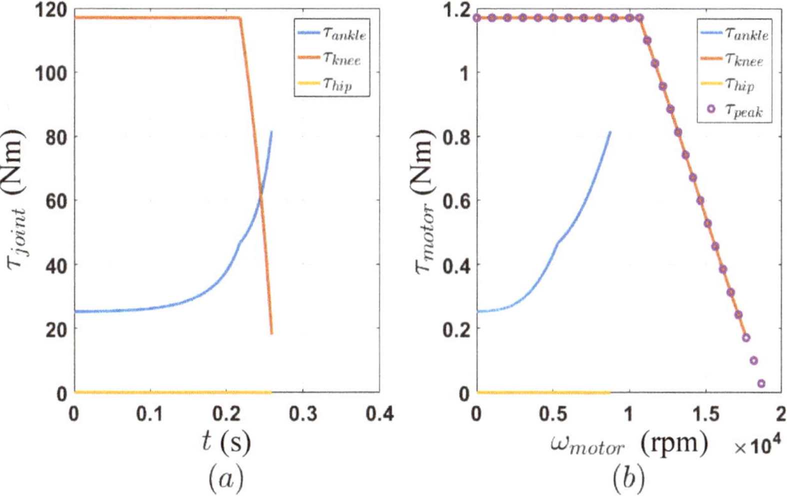

Figure 6. All three joints are equipped with the same K044100-6Y-4.98Ams motor and the same reduction ratio of $i_r = 100$. (**a**) shows the torque of each joint; and (**b**) represents relationships between the motor torque and speed.

We substitute $\dot{q}_{1t}, \dot{q}_{2t}, \dot{q}_{3t}$ into Equations (2)–(4) to calculate the peak ankle, knee and hip joint torques $\tau_{1t}, \tau_{2t}, \tau_{3t}$, respectively, at time t. The maximum angular accelerations of the three joints can then be obtained as follows:

$$\ddot{q}_t = M(q_t)^{-1}(\tau_t - V(q_t, \dot{q}_t) - G(q_t)), \tag{13}$$

where $\tau_t = (\tau_{1t}, \tau_{2t}, \tau_{3t})$ represents the peak torques of the three joints within the limitations of the motor characteristics and M, C, G can be calculated with state $X_t(q_t, \dot{q}_t)$. Using Equations (5) and (11), the ZMP of the robot at time t can be given as below [24]:

$$\begin{pmatrix} p_{xt} \\ p_{yt} \\ p_{zt} \end{pmatrix} = \begin{pmatrix} x_t - \frac{(z_t - p_{zt})\ddot{x}_t}{\ddot{z}_t + g} \\ y_t - \frac{(z_t - p_{zt})\ddot{y}_t}{\ddot{z}_t + g} \\ 0 \end{pmatrix}, \tag{14}$$

where (x_t, y_t, z_t) is the coordinate of the COM that can be calculated with q_t. Generally, if the joints move with maximum angular acceleration calculated by Equation (13), the p_{yt} may move out of the supporting polygon formed by the feet and the ZMP constraint will be violate. Assuming that the limitation to p_{yt} is $p_{ylim} \in (-0.05,\ 0.15)$ (m), which is smaller than the distances from the projection of ankle joint on the ground to the heel d_1 (0.10 (m)) and to the tiptoe d_2 (0.20 (m)) as Figure 7 shows.

At time t, according to the state of the robot $X_t(q_t, \dot{q}_t)$ and Equation (5), (11) and (14), the ZMP can be expressed as a function of $\ddot{q}_t = (\ddot{q}_{1t}, \ddot{q}_{2t}, \ddot{q}_{3t})$. By taking the total derivative, we can get the Δp_{yt} as below:

$$\Delta p_{yt} = \frac{\partial p_{yt}}{\partial \ddot{q}_{1t}} \Delta \ddot{q}_{1t} + \frac{\partial p_{yt}}{\partial \ddot{q}_{2t}} \Delta \ddot{q}_{2t} + \frac{\partial p_{yt}}{\partial \ddot{q}_{3t}} \Delta \ddot{q}_{3t} = \sum_{i=1}^{3} \left(\frac{\partial p_{yt}}{\partial \ddot{q}_{it}} \Delta \ddot{q}_{it}\right). \tag{15}$$

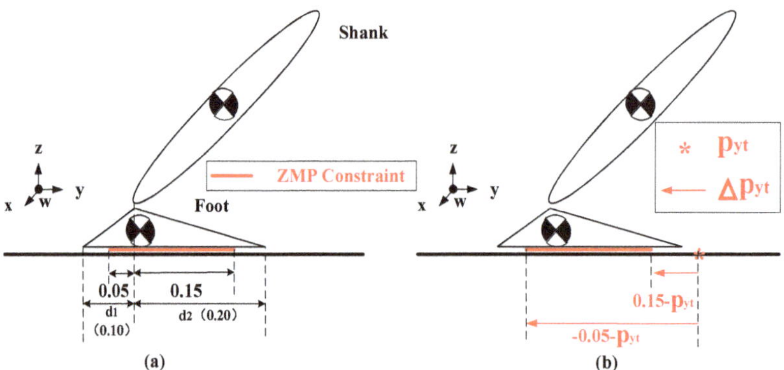

Figure 7. (**a**) the ZMP (zero moment point) limitation in the y-direction $p_{ylim} \in (-0.05,\ 0.15)$ (m); (**b**) the Δp_{yt} added to p_{yt} to maintain the ZMP constraint.

If p_{yt} is located beside p_{ylim}, then the $\Delta\ddot{q}_t$ added to $\ddot{q}_t$ should satisfy the following equation:

$$-0.05 - p_{yt} \leq \sum_{i=1}^{3}\left(\frac{\partial p_{yt}}{\partial \ddot{q}_{it}}\Delta\ddot{q}_{it}\right) \leq 0.15 - p_{yt}. \tag{16}$$

The same method can also be applied to the joint torque constraint. Due to the dynamic equation and $X_t(q_t, \dot{q}_t)$, the joint torque τ_t can be expressed as a function of $\ddot{q}_t = (\ddot{q}_{1t}, \ddot{q}_{2t}, \ddot{q}_{3t})$. By taking the total derivative, we can get the $\Delta\tau_{it}(i = 1, 2, 3)$ as below:

$$\Delta\tau_{it} = \frac{\partial \tau_{it}}{\partial \ddot{q}_{1t}}\Delta\ddot{q}_{1t} + \frac{\partial \tau_{it}}{\partial \ddot{q}_{2t}}\Delta\ddot{q}_{2t} + \frac{\partial \tau_{it}}{\partial \ddot{q}_{3t}}\Delta\ddot{q}_{3t} = \sum_{j=1}^{3}\left(\frac{\partial \tau_{it}}{\partial \ddot{q}_{jt}}\Delta\ddot{q}_{jt}\right); (i = 1, 2, 3). \tag{17}$$

To maintain the torque constraint, the $\Delta\ddot{q}_t$ added to $\ddot{q}_t$ must satisfy the following equation:

$$\left|\sum_{j=1}^{3}\left(\frac{\partial \tau_{it}}{\partial \ddot{q}_{jt}}\Delta\ddot{q}_{jt}\right) + \tau_{it}\right| \leq |\tau_{it}| \quad (i = 1, 2, 3), \tag{18}$$

where $\tau_{it}(i = 1, 2, 3)$ is the maximal torque allowed by torque speed curve at speed $\dot{q}_{it}(i = 1, 2, 3)$. To get the optimal solution for the set of inequalities composed by Equations (16) and (18), we built a cost function

$$Cost(\Delta\ddot{q}_t) = \sum_{i=1}^{3}\sum_{j=1}^{3}\left|\frac{\partial \tau_{it}}{\partial \ddot{q}_{jt}}\Delta\ddot{q}_{jt}\dot{q}_{it}\right|. \tag{19}$$

Since $|\Delta\tau_{it}|$ is the decrease of each joint torque, then $|\Delta\tau_{it}\dot{q}_{it}|$ will be the decrease of each joint power. The sum decreases of the three joint output power can be expressed as the Equation (19). A smaller value of Equation (19) leads to a more output energy that affects the jump height directly. Using the linear programming, we obtain the optimum $\Delta\ddot{q}_t$ required to minimize the cost function of Equation (19) while simultaneously satisfying the the set of inequalities composed by Equation (16) and Equation (18).

Figure 8 shows the control system. In this program process, there is no joint constraint any more. The torque constraint is used to ensure that all three joint torques yield the torque speed curve profile. In addition, the ZMP constraint is used to keep the dynamic balance throughout the entire puss-off phase. From the control system, we can see that the initial state X_0 plays an important role in the jump height. We used the traversal method to obtain the jump height for different $X_0(q_{10} \in [-65°,\ -30°],$

$q_{20} \in [-180°, 180°]$, $q_{30} \in [-180°, 180°]$, $\dot{q}_{10} = 0, \dot{q}_{20} = 0$, $\dot{q}_{30} = 0$). For various initial values of ankle angle $q_{10} \in [-65°, -30°]$, we got the optimal q_{20}, q_{30} and the jump height as shown in Figure 9a–c. With the increasing q_{10}, the jump height h increases while $q_{10} < -57°$ and subsequently decreases. When compared with the original control method, the height h obtained using the proposed method is obviously greater. In particular, at $X_0 = (-57°, 158°, -126°, 0, 0, 0)$, we obtained a maximum jump height of 0.61 (m). The jumping height is increased by nearly 70%. Additionally, we calculated the jumping direction θ with

$$\theta = atan2(\dot{z}_{takeoff}, \dot{y}_{takeoff}). \tag{20}$$

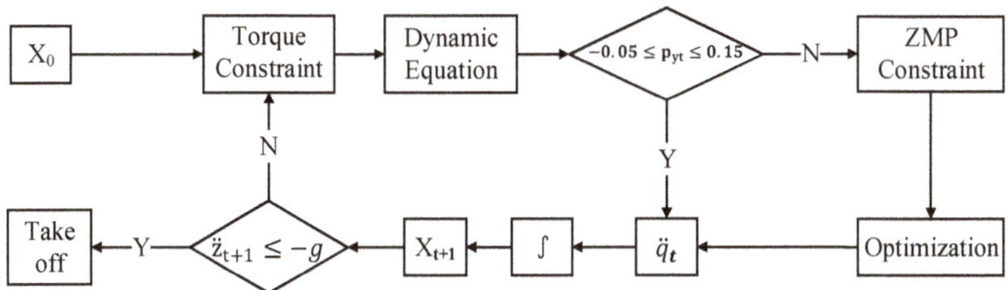

Figure 8. The joint constraint is removed. The torque constraint is used to ensure that all three joint torques yield the torque speed curve profile. Furthermore, the ZMP constraint is used to keep the dynamic balance throughout the entire puss-off phase.

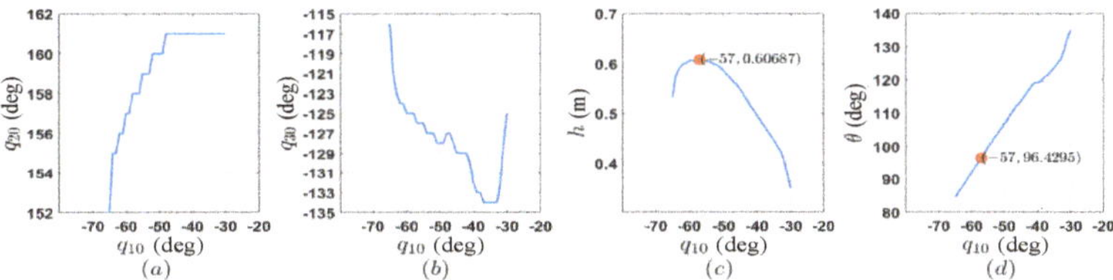

Figure 9. (**a**) optimal knee angle q_{20} for various initial values of ankle angle q_{10}; (**b**) optimal hip angle q_{30} for various initial values of ankle angle q_{10}; (**c**) maximum jump height h for various initial values of ankle angle q_{10}; (**d**) jump direction θ for various initial values of ankle angle q_{10}.

Figure 9d shows that the jumping direction θ and the initial ankle angle q_{10} have an almost linear relationship. Therefore, we can almost control the jumping direction θ via the initial ankle angle q_{10}.

Assuming that $X_0 = (-57°, 158°, -126°, 0, 0, 0)$, Figures 10 and 11 show the torques of the three joints τ_i and the ZMP p_y, respectively, during the entire push-off phase. For most of the motion time, the ankle, knee, and hip joint torques are all equal to the maximal torques allowed by motors, which greatly increases the kinetic energy. At the beginning of the jump motion, we can see the effectiveness of the control system, which reduces the ankle joint torque slightly to satisfy the ZMP constraint. Similar effectiveness can also be observed when the robot is about to take off. During the whole push-off phase, the working points of three joints are always in the right trapezoid form by τ_{peak} and axis.

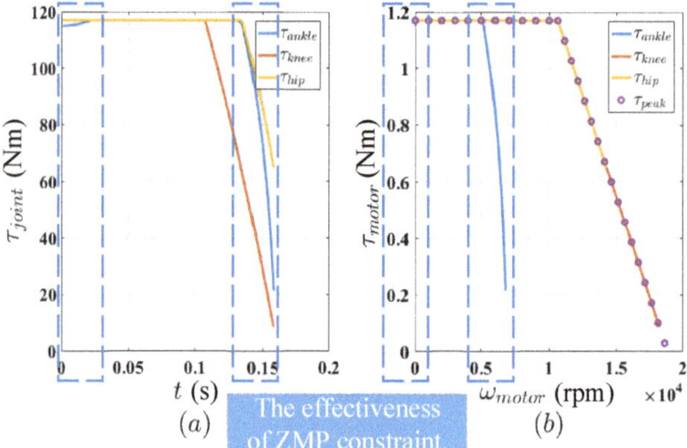

Figure 10. Assuming that $X_0 = (-57°,\ 158°,\ -126°,\ 0,\ 0,\ 0)$, (**a**) shows the torque of each joint; and (**b**) represents relationships between the motor torque and speed. At the beginning of the jump motion, the ankle joint torque is slightly reduced to satisfy the ZMP constraint. Similar effectiveness can also be observed when the robot is about to take off.

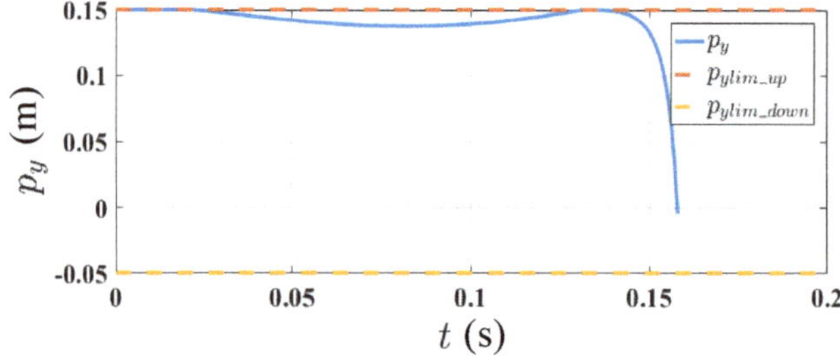

Figure 11. Assuming that $X_0 = (-57°,\ 158°,\ -126°,\ 0,\ 0,\ 0)$, the ZMP is limited by the ZMP constraint to maintain balance throughout the push-off phase.

4. Simulation Results

To verify our control method, we built a simplified model with the parameters in Table 1 in V-rep, a virtual robot experimentation platform. In the simulator, the model is controlled in position mode. In Figure 8, when the robot is going to take off, the jumping controller will not generate data any more, and the angles will be the same as the angles at the take-off time, and each joint will brake suddenly. Since the values of joint angular velocity and angular acceleration are large at the take-off time. If we don't smooth the joint trajectory after take-off, the motor will violate the torque constraint. Thus, we use the cubic interpolation in a joint trajectory to help the joint brake smoothly. The cubic interpolation is also used to smooth the joint trajectory at the beginning of the jump, and the initial state changes from $X_0 = (-57°,\ 158°,\ -126°,\ 0,\ 0,\ 0)$ to $X_0 = (-61.5°,\ 155.2°,\ -123.8°,\ 0,\ 0,\ 0)$. Figure 12 shows the joint trajectory during simulation. By using playback of this offline pattern, we get the velocity of COM as shown in Figure 13. At $t = 0.40$ (s), the translational velocity of COM in vertical direction reaches a maximum 3.42 (m/s). Without consideration of the influence of the added trajectory near the taking off time, the jump height is about 0.596 (m), which can be calculated by Equation (12). The added trajectory leads to the slight decrease in vertical velocity at the take-off time, which is a little delayed. From the simulation, we got the position of the foot and COM in vertical direction and the torque of each joint as shown in Figures 14 and 15. If we take the distance between foot and ground as the jump height, then the maximal jump height is about 0.477 (m) and the COM is increased by 0.83 (m). During the entire motion, the joint torque profiles are within the profile of the motor torque speed curve. Finally, the jumping motion sequence is shown in Figure 16.

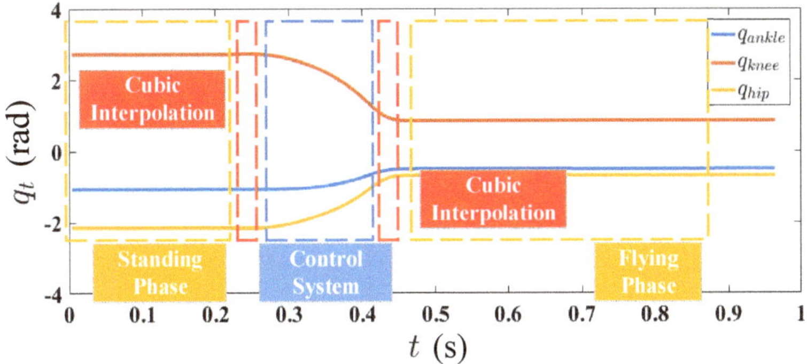

Figure 12. The trajectory of each joint during the simulation.

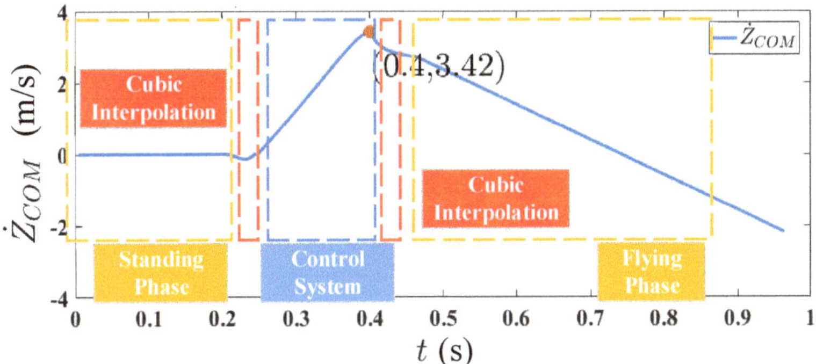

Figure 13. The velocity of COM (center of mass) during the simulation.

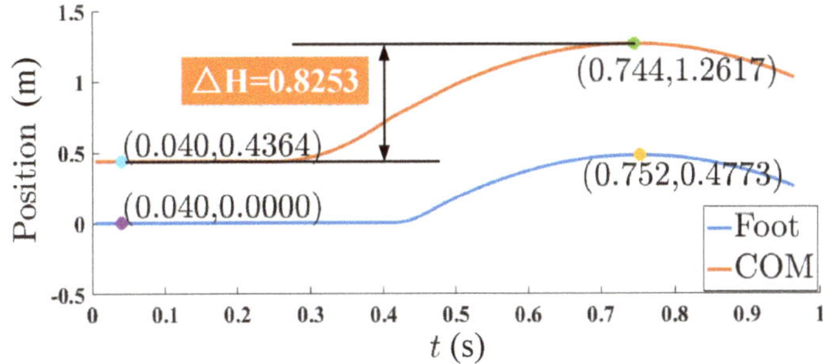

Figure 14. The foot and COM (center of mass) position in vertical direction in the simulation.

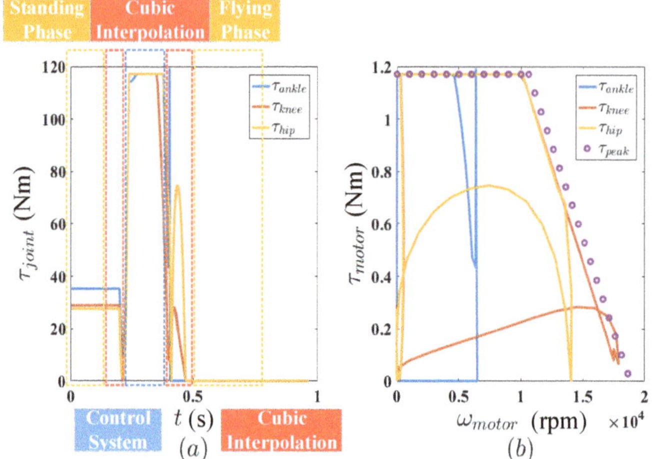

Figure 15. (**a**) shows the torque of each joint in the simulation; and (**b**) represents relationships between the motor torque and speed in the simulation.

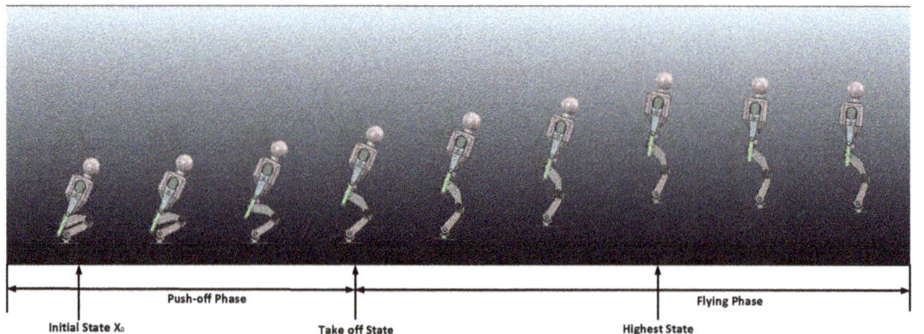

Figure 16. Screenshot of the simulation.

5. Conclusions

This paper proposes a motion planning method to allow a bipedal robot to perform a jumping maneuver. It proves the possibility for bipedal robot driven by electric motors to perform a high jump. Using a simplified model without any elastic elements, we greatly increased the jumping height by commanding every motor as much as possible to yield the profile of the torque speed curve, which reflects the obstacle negotiation capability. During the push-off phase, linear programming is used to fine-tune the joint torques to ensure that the torque and ZMP constraints are satisfied. Finally, in simulations, the robot successfully jumped 0.477 (m) high, and the COM is increased by 0.83 (m). However, some limitations should be noted.

- In this paper, the efficiency of harmonic reducer is not considered. Thus, we have to calibrate the torque speed curve of the motor on a professional motor test platform, with and without reducer.
- While we can obtain an increased jump height based on the premise of the torque and ZMP constraints, how to reduce the impact force and maintain landing equilibrium during the landing phase remains a significant problem.
- We only found that the jump direction θ and the initial ankle angle q_{10} have an almost linear relationship. However, this relationship alone is not sufficiently reliable to specify jumping direction.

In the future, we will take the efficiency of harmonic reducer, flying and landing phase of jump into consideration. Future work should include development of a sequential jumping motion and a running motion to provide further improvements in the obstacle negotiation abilities of the bipedal robot. It is also necessary to design experiments to validate the theory in the future.

Acknowledgments: This work was supported in part by the National Natural Science Foundation of China under Grant 91748202, Grant 61533004, and Grant 61703043, and in part by the Basic Research Program under Grant B132011xx.

Author Contributions: Xinyang Jiang conceived of the presented idea. Xinyang Jiang and Xuechao Chen developed the theoretical formalism. Xinyang Jiang, Weimin Zhang and Qiang Huang designed the simulation experiments and analyzed the data. Xinyang Jiang, Zhangguo Yu and Libo Meng discussed the results and wrote the paper.

Conflicts of Interest: The authors declare no conflict of interest.

References

1. Huang, Q.; Yokoi, K.; Kajita, S.; Kaneko, K.; Arai, H.; Koyachi, N.; Tanie, K. Planning walking patterns for a biped robot. *IEEE Trans. Robot. Autom.* **2001**, *17*, 280–289.
2. Huang, Q.; Kajita, S.; Koyachi, N.; Kaneko, K.; Yokoi, K.; Arai, H.; Komoriya, K.; Tanie, K. A high stability, smooth walking pattern for a biped robot. In Proceedings of the 1999 IEEE International Conference on Robotics and Automation, Detroit, MI, USA, 10–15 May 1999; Volume 1, pp. 65–71.
3. Li, C.; Lowe, R.; Duran, B.; Ziemke, T. Humanoids that crawl: comparing gait performance of iCub and NAO using a CPG architecture. In Proceedings of the 2011 IEEE International Conference on Computer Science and Automation Engineering (CSAE), Shanghai, China, 10–12 June 2011; Volume 4, pp. 577–582.
4. Kuehn, D.; Bernhard, F.; Burchardt, A.; Schilling, M.; Stark, T.; Zenzes, M.; Kirchner, F. Distributed computation in a quadrupedal robotic system. *Int. J. Adv. Robotic. Syst.* **2014**, *11*, 110, doi:10.5772/58733.
5. Fujiwara, K.; Kanehiro, F.; Kajita, S.; Kaneko, K.; Yokoi, K.; Hirukawa, H. Falling motion control to minimize damage to biped humanoid robot. In Proceedings of the Conference on Intelligent Robots and Systems, Lausanne, Switzerland, 30 September–4 October 2002 .
6. Fujiwara, K.; Kajita, S.; Harada, K.; Kaneko, K.; Morisawa, M.; Kanehiro, F.; Nakaoka, S.; Hirukawa, H. Towards an optimal falling motion for a humanoid robot. In Proceedings of the 6th IEEE-RAS International Conference on Humanoid Robots, Genova, Italy, 4–6 December 2006; pp. 524–529.
7. Kajita, S.; Nagasaki, T.; Kaneko, K.; Yokoi, K.; Tanie, K. A hop towards running humanoid biped. In Proceedings of the IEEE International Conference on Robotics and Automation, New Orleans, LA, USA, 26 April–1 May 2004; Volume 1, pp. 629–635.
8. Sakka, S.; Yokoi, K. Humanoid vertical jumping based on force feedback and inertial forces optimization. In Proceedings of the IEEE International Conference on Robotics and Automation, Barcelona, Spain, 18–22 April 2005; pp. 3752–3757.
9. Curran, S.; Orin, D.E. Evolution of a jump in an articulated leg with series-elastic actuation. In Proceedings of the IEEE International Conference on Robotics and Automation, Pasadena, CA, USA, 19–23 May 2008; pp. 352–358.
10. Raibert, M.H.; Brown, H.B., Jr.; Chepponis, M. Experiments in balance with a 3D one-legged hopping machine. *Int. J. Robot. Res.* **1984**, *3*, 75–92.
11. Chen, X.; Zhangguo, Y.; Zhang, W.; Zheng, Y.; Huang, Q.; Ming, A. Bio-inspired Control of Walking with Toe-off, Heel-strike and Disturbance Rejection for a Biped Robot. *IEEE Trans. Ind. Electron.* **2017**, *64*, 7962–7971.
12. Yu, D.; Yu, Z.; Fang, X.; Lei, S.; Chen, X.; Huang, Q.; Meng, L.; Zhou, Q.; Zhang, W.; Han, J. Rolling motion generation of multi-points contact for a humanoid robot. In Proceedings of the International Conference on Advanced Robotics and Mechatronics (ICARM), Macau, China, 18–20 August 2016; pp. 153–158.
13. Li, Q.; Chen, X.; Zhou, Y.; Yu, Z.; Zhang, W.; Huang, Q. A minimized falling damage method for humanoid robots. *Int. J. Adv. Robot. Syst.* **2017**, *14*, 1729881417728016, doi:10.1177/1729881417728016.
14. Raibert, M.H. *Legged Robots That Balance*; MIT Press: Cambridge, MA, USA, 1986.
15. Playter, R.R.; Raibert, M.H. Control of a Biped Somersault in 3D. In Proceedings of the IEEE/RSJ International Conference on Intelligent Robots and Systems, Raleigh, NC, USA, 7–10 July 1992; Volume 1, pp. 582–589.

16. Vanderborght, B.; Van Ham, R.; Verrelst, B.; Van Damme, M.; Lefeber, D. Overview of the lucy project: Dynamic stabilization of a biped powered by pneumatic artificial muscles. *Adv. Robot.* **2008**, *22*, 1027–1051.
17. Hosoda, K.; Takuma, T.; Nakamoto, A.; Hayashi, S. Biped robot design powered by antagonistic pneumatic actuators for multi-modal locomotion. *Robot. Autonom. Syst.* **2008**, *56*, 46–53.
18. Ugurlu, B.; Saglia, J.A.; Tsagarakis, N.G.; Caldwell, D.G. Hopping at the resonance frequency: A trajectory generation technique for bipedal robots with elastic joints. In Proceedings of the IEEE International Conference on Robotics and Automation, Saint Paul, MN, USA, 14–18 May 2012; pp. 1436–1443.
19. Okada, M.; Takeda, Y. Optimal design of nonlinear profile of gear ratio using non-circular gear for jumping robot. In Proceedings of the IEEE International Conference on Robotics and Automation, Saint Paul, MN, USA, 14–18 May 2012; pp. 1958–1963.
20. Hondo, T.; Kinase, Y.; Mizuuchi, I. Jumping motion experiments on a NAO robot with elastic devices. In Proceedings of the 12th IEEE-RAS International Conference on Humanoid Robots (Humanoids), Osaka, Japan, 29 November–1 December 2012; pp. 823–828.
21. Guizzo, E. Japanese Humanoid Robot Can Keep Its Balance After Getting Kicked. Available online: https://spectrum.ieee.org/automaton/robotics/humanoids/japanese-high-power-humanoid-robot-hrp3l-jsk/ (accessed on 8 May 2012).
22. Nunez, V.; Drakunov, S.; Nadjar-Gauthier, N.; Cadiou, J. Control strategy for planar vertical jump. In Proceedings of the 12th International Conference on Advanced Robotics, Seattle, WA, USA, 18–20 July 2005; pp. 849–855.
23. Ugurlu, B.; Kawamura, A. Real-time running and jumping pattern generation for bipedal robots based on ZMP and Euler's equations. In Proceedings of the IEEE/RSJ International Conference on Intelligent Robots and Systems, St. Louis, MO, USA, 10–15 October 2009; pp. 1100–1105.
24. Kajita, S.; Hirukawa, H.; Harada, K.; Yokoi, K. *Introduction to Humanoid Robotics*; Springer: Berlin, Germany, 2014; Volume 101.
25. Hong, Y.D.; Lee, B. Evolutionary Optimization for Optimal Hopping of Humanoid Robots. *IEEE Trans. Ind. Electron.* **2017**, *64*, 1279–1283.
26. Takahide Yoshiike, M.K.; Ujino, R. Development of Experimental Legged Robot for Inspection and Disaster Response in Plants. In Proceedings of the IEEE International Conference on Intelligent Robots and Systems (IROS), Vancouver, BC, Canada, 24–28 September 2017; pp. 4869–4876.

Article

Design and Experimental Development of a Pneumatic Stiffness Adjustable Foot System for Biped Robots Adaptable to Bumps on the Ground

Xizhe Zang [1,*], Yixiang Liu [1,2,3], Wenyuan Li [1], Zhenkun Lin [1] and Jie Zhao [1]

1 State Key Laboratory of Robotics and System, Harbin Institute of Technology, Harbin 150080, China; liuyixiang@163.com (Y.L.); biqing_meng@126.com (W.L.); 15B308008@hit.edu.cn (Z.L.); jzhao@hit.edu.cn (J.Z.)
2 Legs + Walking Lab, Shirley Ryan AbilityLab (Formerly the Rehabilitation Institute of Chicago), Chicago, IL 60611, USA
3 Feinberg School of Medicine, Northwestern University, Chicago, IL 60611, USA
* Correspondence: zangxizhe@hit.edu.cn; Tel.: +86-451-8641-3382

Received: 20 August 2017; Accepted: 26 September 2017; Published: 29 September 2017

Abstract: Walking on rough terrains still remains a challenge that needs to be addressed for biped robots because the unevenness on the ground can easily disrupt the walking stability. This paper proposes a novel foot system with passively adjustable stiffness for biped robots which is adaptable to small-sized bumps on the ground. The robotic foot is developed by attaching eight pneumatic variable stiffness units to the sole separately and symmetrically. Each variable stiffness unit mainly consists of a pneumatic bladder and a mechanical reversing valve. When walking on rough ground, the pneumatic bladders in contact with bumps are compressed, and the corresponding reversing valves are triggered to expel out the air, enabling the pneumatic bladders to adapt to the bumps with low stiffness; while the other pneumatic bladders remain rigid and maintain stable contact with the ground, providing support to the biped robot. The performances of the proposed foot system, including the variable stiffness mechanism, the adaptability on the bumps of different heights, and the application on a biped robot prototype are demonstrated by various experiments.

Keywords: structure design; biped robot; foot mechanism; variable stiffness; rough terrain

---

1. Introduction

A two-legged mobile mechanism that can walk like a human is usually thought as the best suited locomotion method for robots aimed to coexist and collaborate with humans [1]. In the past few decades, bipedal robotic walking has been a hot area in the research of robotics [2–5]. In order to realize stable walking, the majority of existing biped robots are equipped with rigid flat feet and controlled by a trajectory tracking control method based on the zero moment point (ZMP) theory [6–9]. The ZMP theory requires that the location on the ground about which the sum of all the moments of the active forces acting on the robot equals zero is strictly within the support polygon of the foot sole [10]. Therefore, the foot that supports the body weight during walking is always kept flat on the ground to obtain the largest support polygon, especially in the single support phase where there is only one foot in contact with the ground [11,12]. The instantaneous speed of the foot touching the ground is desired to approximate to zero to reduce foot-ground impact. Although effective for walking on flat and structured terrains, this method will cause some problems if the ground is uneven. When a biped robot walks on rough terrain, just a small bump under the sole can significantly reduce the contact area and easily disrupt the walking stability, causing the robot to fall [13]. In addition, if the bump is unforeseen for the control system of the biped robot, the sudden impact with the bump may cause severe shock and vibration to the mechanical system. So several crucial issues involved in walking on rough terrains, such as adaptability on the unevenness, stable foot-ground contact, and shock absorbance still need further investigations.

To handle with the above challenges, some researchers focus on the studies of more advanced walking controllers for biped robots. Wei et al. proposed a landing phase control method based on the non-planar contact model of the flexible foot with the environment, and made the biped robot adaptable to the changes of the ground [14]. To achieve dynamic stable walking on rough ground, Nishiwaki et al. designed a high-frequency pattern generator which considered the current actual motion of the robot as the initial conditions of each generation [15]. In [16], a graph-based footstep planning approach was proposed to generate the whole step sequences in rough terrain scenarios using a black box walking controller. In [17], the preplanned trajectories were modified online to guarantee a smooth landing after the detection of the foot touching the uneven ground. However, the above controllers depended on some sensors, such as inertial measurement units, contact switches, or laser scanners, to obtain the robot motion state and terrain profile information. The sensor signals must be collected and processed correctly and reliably in real-time, and accurate models of robot kinematics, dynamics, foot-ground contact, and terrain shapes were usually required by the online walking pattern generation and modification. Additionally, as the integrations of on-board sensing, pattern generation, and walking control, the controllers had to perform very large computations at a very high speed [18]. These issues cause many difficulties in the implementations of the controllers.

On the other hand, considering that the foot is the only part of the whole biped robot that interacts directly with the environment, it is a feasible approach to improve walking performance on rough terrain through novel foot mechanisms. Some robotic foot systems were designed by adding toe joints and elastic elements to the simple rigid flat foot [19–23]. Segmented to the heel part and toe part, human-like heel contact and toe contact were realized in bipedal robotic walking [20]. These feet were able to dissipate energy from the heel strike with the elasticity of the heel [21,22], and to provide more traction using the toe pad at the push-off phase [23], but they could not adapt to rough terrain. Based on the concept of maintaining multi-point contact on uneven ground, a foot capable of providing stable contact on convex and concave surfaces was developed. The foot sole was equipped with four rigid spikes and corresponding locking mechanisms, each of which had an optical sensors to detect the ground height. According to the sensor values, the foot landing pattern was modified to guarantee a support polygon on uneven terrain with three or four spikes [24]. In [25], a flexible foot with 12 degrees of freedom was designed. Connected by four independently-actuated parts, the foot sole could maintain multi-point contact on complicated terrains. However, the adaptation was also actively controlled, which increased the complexity of the control system. In [26], Piazza et al. developed a completely passive foot by mimicking the longitudinal arch of human foot. The foot was able to vary its shape to comply with uneven terrains as a function of the exerted forces. However, its adaptation capabilities in the coronal plane and push-off movement were limited.

If we observe the human foot closely, we will find that when walking on the ground with trivial obstacles, the human foot sole can adapt to the irregularities passively without adjusting walking patterns or any active control. This is partly due to the intrinsic adaptivity of the soft tissues, such as the plantar fascia underneath the foot sole, which can soften and stiffen accordingly under different contact conditions. In this paper, we aim to design a new stiffness adjustable foot system for biped robots which is adaptable to small-sized bumps and obstacles on the ground on the basis of the inspirations from the human foot. Specifically, the foot can exert high stiffness to provide sufficient support and propulsion to the robot body during walking, and can also exert low stiffness to adapt to the bumps on the ground.

The rest of this paper is organized as follows: Section 2 introduces the working mechanism of the stiffness-adjustable foot system. Section 3 presents the design of the major components of the foot system in detail. Some preliminary experiments are performed to validate the functions of the foot system, and the results are shown in Section 4. Finally, some discussions and conclusions are given in Section 5.

2. Working Mechanism of the Stiffness-Adjustable Foot System

2.1. Overview of the Stiffness-Adjustable Foot System

The central idea of the new foot system is the variable stiffness. This is achieved by attaching some pneumatic variable stiffness units (PVSUs) on the foot sole separately, as shown in Figure 1. Each PVSU mainly consists of a cylindrical pneumatic bladder and a stiffness-adjusting mechanism. In the default state, for example, when the foot is off the ground, the pneumatic bladders are set to high stiffness, being inflated with some compressed air. When the biped robot walks on rough ground with small-sized bumps and obstacles, the stiffness values of the pneumatic bladders may change accordingly with their contact conditions with the ground. Since there are some spaces among the PVSUs, if the bumps happen to be located in the spaces, all pneumatic bladders are just compressed slightly and still remain rigid. If some PVSUs come into contact with the bumps, the corresponding stiffness adjusting mechanisms are triggered to expel the air, making the pneumatic bladders soft, so that more deformation can be generated to comply with the shape of the bumps; meanwhile, the other pneumatic bladders which are in contact with the flat surface remain rigid. It should be noted that in order to reduce the control difficulty, the adjustment from high stiffness to low stiffness is purely passive, depending on the interactions between the pneumatic bladders and the bumps. Figure 2 shows the comparisons between a conventional rigid flat foot and the new foot when there are bumps and obstacles of various sizes and shapes on the ground (the sole of the new foot is set as transparent for clarity). For the flat foot, the posture around the three coordinate axes is greatly changed, which has to be compensated by the movement of at least three joints of other body parts to maintain multi-point contact. However, the new foot has the ability to adapt to the bumps and maintain sufficient contact area by means of its own adaptability.

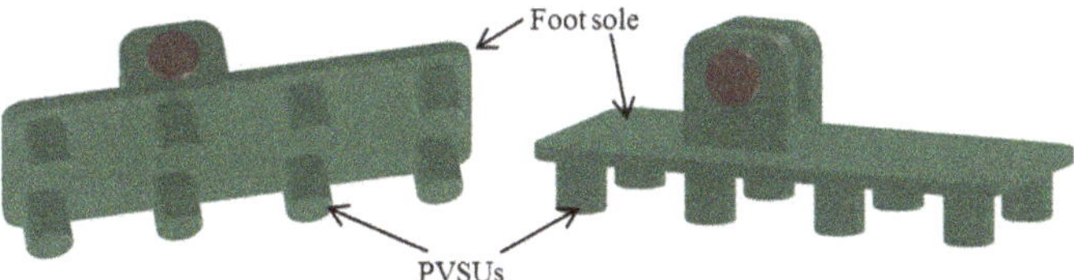

Figure 1. Schematic diagram of the stiffness adjustable foot system.

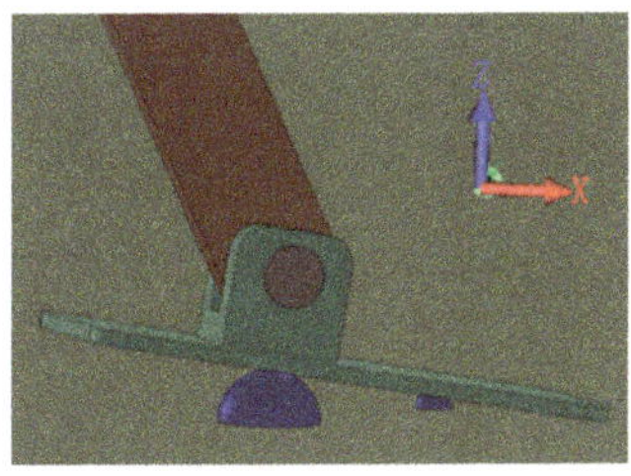

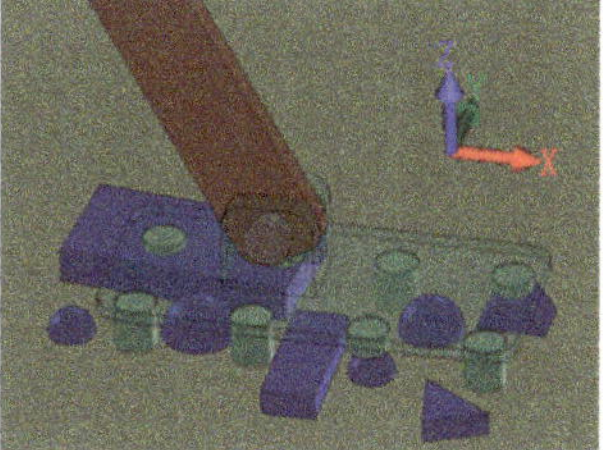

Figure 2. Comparisons between a conventional rigid flat foot and the new foot on the ground with various bumps and obstacles.

2.2. Theoretical Analysis of the Variable Stiffness Mechanism

The PVSU is the most crucial component of the whole foot system because the function of the foot depends on the interactions of each PVSU with the environment. This section introduces the variable stiffness mechanism of a single PVSU. Variable stiffness mechanisms have been more and more widely utilized in vibration isolation devices [27], humanoid robots [28], and rehabilitation devices [29]. Most recent work of variable stiffness mechanisms rely on two techniques. One is to alter the structural or mechanical geometry of an elastic mechanism [30], and the other is to change the elastic modulus of a structure by thermal or electromagnetic stimulation [31]. In this paper, the stiffness-adjusting mechanism of the PVSU is inspired by pneumatic artificial muscles, of which the stiffness can be changed by regulating the air pressure inside the muscles. Figure 3 illustrates the working principle diagram of a single PVSU interacting with the environment. The PVSU includes two major parts: a pneumatic bladder and a stiffness-adjusting mechanism. The stiffness-adjusting mechanism works based on a two-position three-port mechanical reversing valve. The PVSU is mounted underneath the foot sole through the reversing valve. The valve piston is connected with one end of the supporting spring, and the other end of the supporting spring is fixed on the bottom of the pneumatic bladder. The trigger signal of the reversing valve is the external force applied on the pneumatic bladder by the bump. In the normal position, the supply port of the reversing valve is connected with the pneumatic bladder. To trigger the valve, the external force needs to overcome the friction, the restoring spring, as well as the air pressure inside the pneumatic bladder. Once the trigger signal is off, the loaded restoring spring returns the valve to its original position.

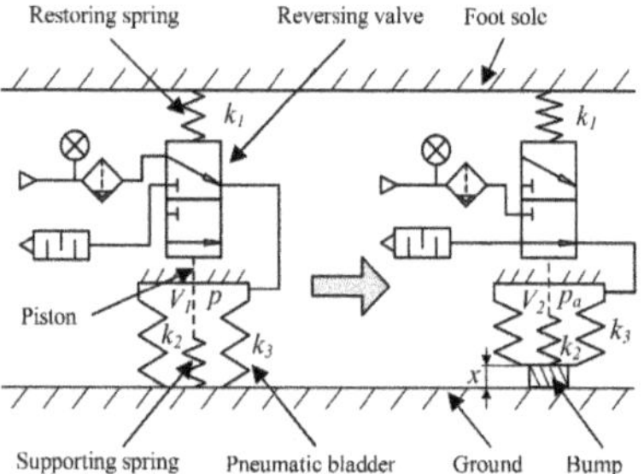

Figure 3. Working principle diagram of the pneumatic variable stiffness unit (PVSU).

The theoretical working process of the PVSU is as follows. In the swing phase of walking when there is no contact between the ground and the foot, the reversing valve is in normal position, supplying compressed air into the pneumatic bladder to keep it stiff. After transferring to the stance phase, the pneumatic bladder is compressed once touching the bump on the ground, and the valve piston is driven to move upward. As soon as the piston reaches the working position, the valve is triggered to expel the air out of the pneumatic bladder. Then, the pneumatic bladder becomes soft and generates more compression to comply with the shape of the bump.

When the PVSU interacts with the bump, compressions can be generated on any one of the restoring spring, supporting spring and pneumatic bladder. According to which elements are compressed, the entire compression process can be divided into three phases. During these phases, the PVSU has different configurations of series and/or parallel elastic elements, and exerts different stiffness values. Prior to analyzing the stiffness in each phase, several assumptions are made to simplify the modeling problem by referring to the mathematical models of air-supported structures and air springs [32–34], specifically, (1) the foot sole, the ground and bumps are perfectly rigid and cannot generate deformation under external forces; (2) the damping effects of the elastic elements are negligible considering that the damping factors are much smaller than stiffness factors [35]; (3) the compression of the PVSU is taken as a quasi-static behavior; (4) the static pressure inside the bladder is assumed to remain stable during each phase; and (5) the inflation and deflation, in other words, the changes of air pressure inside the pneumatic bladder are completed instantaneously.

The first phase starts when the pneumatic bladder touches the bump and ends when the piston arrives at its working position. During this phase, the restoring spring, supporting spring, and the inflated pneumatic bladder are compressed simultaneously. Similar to pneumatic artificial muscles, the stiffness of the pneumatic bladder is dependent upon two factors: one is the inherent stiffness of the bladder in the natural state without inflation, which is determined by its material properties and mechanical structure, and the other is the stiffness that relates to the air pressure inside the bladder [36]. Stiffness of a body can be defined as the infinitesimal force variation with regards to resulting compliant displacement [37]. If we assume F be the external force acted on the pneumatic bladder, and x is the corresponding amount of compression, then the system stiffness of the PVSU in the first phase can be described as:

$$K_1 = \frac{dF}{dx} = \frac{k_1 k_2}{k_1 + k_2} + k_3 + \frac{d(p \cdot A)}{dx} = \frac{k_1 k_2}{k_1 + k_2} + k_3 + p\frac{dA}{dx} \quad (1)$$

where k_1, k_2, and k_3 indicate the stiffness values of the restoring spring, the supporting spring, and the bladder, respectively, A represents the effective circumferential cross-sectional area of the pneumatic bladder, and p represents the air pressure inside the bladder.

When the reversing valve is triggered, the air inside the pneumatic bladder is expelled quickly. Then the compression process enters the second phase and lasts until the piston moves to its limit position. The restoring spring, supporting spring, and pneumatic bladder are all compressed, but the pressure inside the pneumatic bladder is changed to ambient pressure which is represented by p_a. The system stiffness is expressed by:

$$K_2 = \frac{k_1 k_2}{k_1 + k_2} + k_3 \quad (2)$$

The third phase starts as soon as the piston reaches its limit and the restoring spring cannot be compressed anymore. As the external force continues to increase, only the supporting spring and pneumatic bladder undergo further compression. The system stiffness of this phase is given by:

$$K_3 = k_2 + k_3 \quad (3)$$

It should be noted that this section is focused on the introduction of the variable stiffness mechanism in principle, rather than the investigation of the accurate analytical model of the PVSU. Thus, only simple stiffness expressions are provided above. The variable stiffness of the PVSU in the entire compression process is illustrated in Figure 4. The first phase which has the largest stiffness is also termed as the high stiffness phase, and the second and third phases are termed as low stiffness phases. In this figure, h_{ch} and h_{max} represent the trigger displacement and the maximum displacement of the valve piston, respectively.

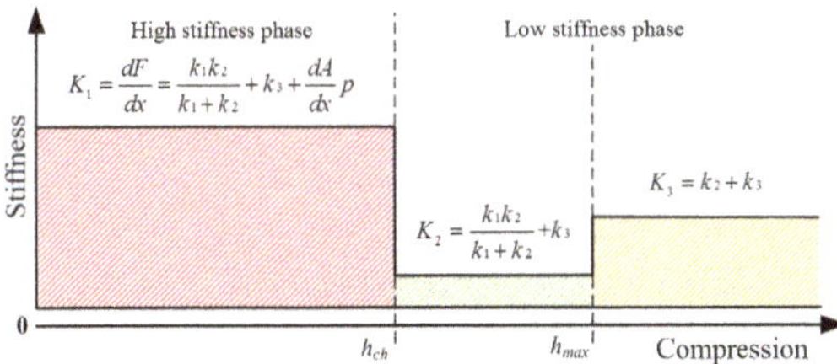

Figure 4. The system stiffness of the PVSU in the entire compression process.

3. Development of the Stiffness Adjustable Foot System

The stiffness-adjustable foot system is designed step by step, i.e., from the pneumatic bladder to the PVSU, then to the whole foot. This section presents the development of the foot system in detail.

3.1. Design of the Pneumatic Bladder

The main structure parameters of the saw-toothed pneumatic bladder include material, diameter, total height, wall thickness, and the angle of the sawtooth. The following points must be taken into account to obtain an appropriate structure: (1) the pneumatic bladder should be stiff enough to hold its shape; (2) the axial deformation after inflation should be homogeneous, while circumferential deformation should be as small as possible; (3) the stiffness of the pneumatic bladder should have good linearity; and (4) the structure of the molds should be considered at the same time.

Finite element analysis is carried out in ANSYS (ANSYS Inc., Canonsburg, PA, USA) for parameter optimization. Applying a pressure load on the virtual model of the pneumatic bladder, deformations under various parameter configurations can be obtained. After trial and error, some key parameters are determined, as shown in Figure 5. The material type is silica gel, the diameter is 30 mm, the total height is 20 mm, and the angle of the sawtooth is 70°.

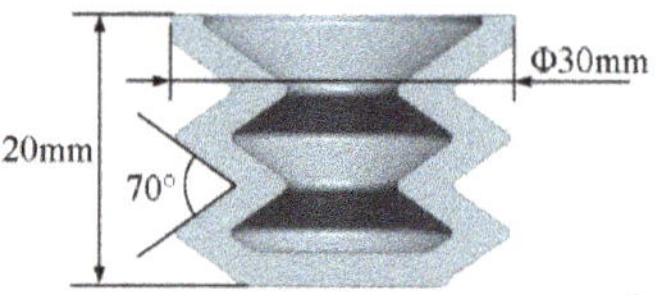

Figure 5. The main structure parameters of the pneumatic bladder.

Furthermore, the hardness of silica gel and the wall thickness, which greatly affect the stiffness of the pneumatic bladder, still need to be determined. The possible values of hardness and wall thickness under consideration are Shore 35A, Shore 40A, 2 mm, and 2.5 mm. The stiffness of the pneumatic bladder under different combinations of hardness and wall thickness is simulated in ANSYS. The simulation results illustrated in Figure 6 show that the stiffness increases with larger hardness and/or wall thickness. In addition, once the hardness and wall thickness are determined, the stiffness is approximately a constant value. On this basis, because the pneumatic bladder is desired to have enough stiffness, Shore 40A and 2.5 mm are selected as the material hardness and wall thickness. The theoretical maximum compression of the pneumatic bladder is 8 mm.

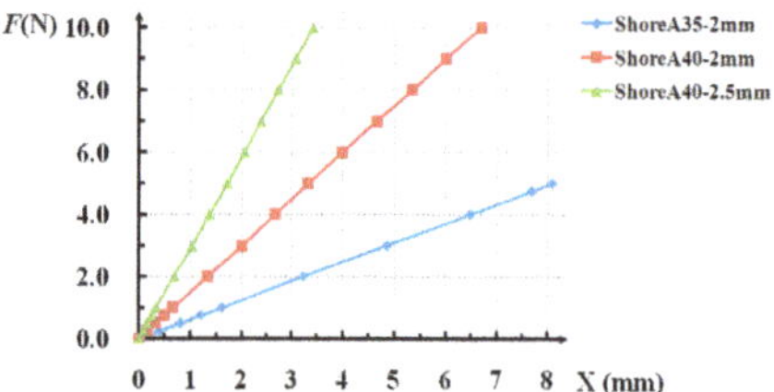

Figure 6. Simulated stiffness of the pneumatic bladder under different combinations of material hardness and wall thickness.

3.2. Design of the PVSU

In the design of the PVSU, there exist the following major structure parameters: the trigger displacement of the piston, the maximum displacement of the piston, and the stiffness of the restoring spring and the supporting spring. The trigger displacement of the piston has important effects on the support ability of the foot, i.e., a higher value means larger external force demanded to trigger the reversing valve, and lower adaptability on rough terrains. In this paper, it is chosen as 2 mm. The maximum displacement of the piston which mainly determines the restoring speed of the piston is set as 4 mm.

Then, the remaining two structural parameters need to be optimized. As before, the finite element analysis method is utilized. The inside pressure load and external force load are applied on the pneumatic bladder which is modelled in ANSYS. Adjusting the stiffness of the restoring spring and the supporting spring, the corresponding displacement of the piston is simulated. According to the analysis, the stiffness values of restoring spring and supporting spring are selected as 0.098 N/mm and 0.975 N/mm, respectively.

After the key parameters are confirmed, the structure of the reversing valve is designed. The overall dimension of the valve is 25 mm × 25 mm × 22 mm. The diameter and height of the piston are both 10 mm. The material of the valve body is 2024 aluminum alloy, and the material of the piston is 304 stainless steel. In addition to the three necessary working ports, an auxiliary port is added inside the valve for quick exhaust. Figure 7a,b show the sectional view and the prototype of the PVSU. In this paper, the pneumatic bladder is manufactured by injecting the liquids of silica gel into the molds. Figure 7c presents the 3D-printed molds developed for the pneumatic bladder.

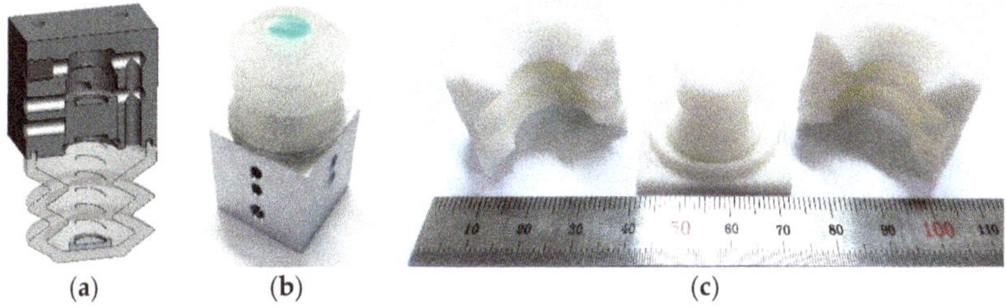

Figure 7. The PVSU. (**a**) The sectional view of the PVSU; (**b**) the prototype of the PVSU; and (**c**) the molds developed for the pneumatic bladder.

3.3. Design of the Foot System

As introduced previously, the foot is designed basically by attaching some PVSUs to the sole. The configuration of these PVSUs on the foot should meet the following requirements, i.e., the foot should have as high adaptability to rough terrain as possible and, at the same time, have the ability to offer stable support to the body weight. The prototype of the foot system is illustrated in Figure 8. In total, there are eight PVSUs symmetrically equipped on the four corners of the foot sole. Each PVSU is fixed on the foot frame through a holder. Between these PVSUs and holders are force-sensitive resistors (FSRs) which are utilized to measure the pressure distribution under the sole. Redundant materials of the frame are cut off to keep weight down, making the total weight as low as about 200 g. Through the flange located on the middle part, the foot system can be conveniently equipped on a biped robot.

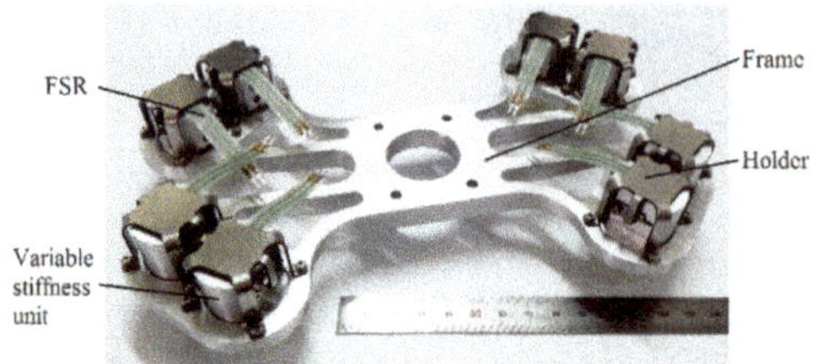

Figure 8. The prototype of the stiffness-adjustable foot system.

4. Experiments

4.1. Experiment on the PVSU

The purpose of this experiment is to verify the variable stiffness mechanism of the PVSU by testing its static stiffness. As mentioned previously, stiffness can be defined as the ratio of steady force acting on a deformable elastic medium to the resulting displacement. Thus, the relation between the forces and resulting displacements when constant pressure is applied to the PVSU is examined in the stiffness identification experiment. Figure 9 presents the schematic diagram of the experimental apparatus. The PVSU is placed on a frame, and its upper end is connected with a slider which can only move along the linear guide in the vertical direction. Below the whole PVSU hangs a weight which causes the PVSU to be compressed. A proportional valve (SMC Corp., Tokyo, Japan) is adopted to regulate the air pressure supplied into the pneumatic bladder. The compression of the PVSU is measured by a linear displacement potentiometer (Novotechnik Messwertaufnehmer OHG, Ostfildern, Germany) installed in parallel with the linear guide, and collected and recorded by data acquisition devices (Advantech Co., Ltd., Taipei, Taiwan) on the computer for further analysis.

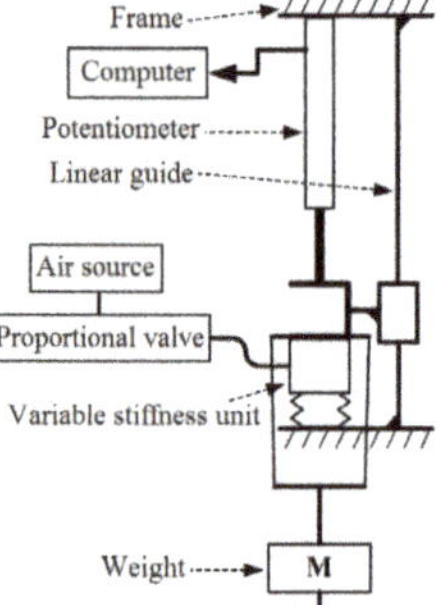

Figure 9. Schematic diagram of the experimental apparatus.

At the start of the experiment, the gauge pressure of the proportional valve was set to zero. Then different weights ranging from 0 to 1.8 kg were applied to the PVSU at a constant increment of 0.1 kg. The weight was changed after the compression of the PVSU stabilized. Corresponding compressions during the process were measured. After that, the above process was repeated in different cases where the gauge pressure was 10 KPa, 20 KPa, 30 KPa, 40 KPa, and 50 KPa, respectively.

After experiments, the relationship between the external loads and the compressions is plotted in Figure 10. From this figure, several findings can be obtained. Firstly, in the point of the response of the PVSU under a certain gauge pressure, it is evident that each curve can be approximately regarded as three segments of straight lines, which signifies that three phases were indeed presented within the entire compression. Secondly, since the slope of the line indicates the system stiffness, the three-segmented curve means that the PVSU exerted different stiffness values during the three phases. Moreover, the stiffness of the second segment is lower than the first one, and the stiffness of the third segment is higher than the second one. This trend coincides exactly with the theoretical analysis in Section 2. Thirdly, the horizontal ordinate of the turning point of each curve is exactly the trigger displacement of the piston. Under different gauge pressures, the trigger displacement of the piston ranges from 2 mm to 3 mm, which is very close to the design value. Synthesizing the above three points, the proposed variable stiffness mechanism, as well as its mechanical structure, can be proved to be feasible.

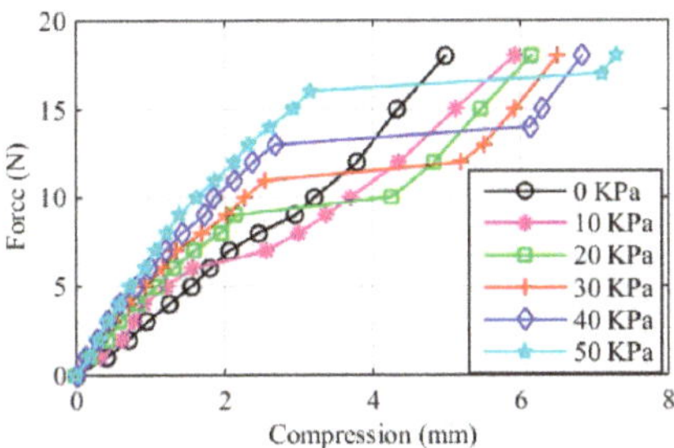

Figure 10. Measured force-compression curves of the PVSU.

In addition, comparing all the measured curves under different gauge pressures, it shows that as the gauge pressure goes up, the piece-wise linearity of the stiffness becomes more obvious, and both the stiffness value in the high stiffness phase and the amount of compression in the low stiffness phase increase. Higher stiffness in the high stiffness phase means more stable support can be provided to the robot body when the foot is in contact with the ground, while a larger compression range in the low stiffness phase means larger unevenness that the pneumatic bladder is able to adapt to, which are exactly what we want. On this basis, taking the stiffness, adaptability, and pressure endurance into consideration together, 50 KPa is adopted as the ideal working pressure of the PVSU.

4.2. Experiment on the Foot System

The most important function of the proposed foot system is to adapt to small-sized bumps on the ground with the help of the PVSUs and, thus, to reduce the change in the posture of the sole and maintain stable foot-ground contact. Several requirements must be met to guarantee the performance of the foot system, specifically, (1) the reversing valves of the PVSUs in contact with the bumps should be triggered reliably; (2) wrong trigger of the reversing valves of the PVSUs in contact with the flat ground should be avoided; and (3) the foot should be able to adapt to bumps of various sizes within a certain range. Therefore, the following experiment is designed to test the function of the foot from the above aspects. First of all, the foot system is placed on a flat surface with all PVSUs inflated with compressed air of 50 KPa, and a 6 kg mass block which is used to imitate the body weight of biped robots is mounted on the top of the foot through a 230 mm long link, as shown in Figure 11a. Then, some small bumps of various sizes are placed under the pneumatic bladder to see if the foot can adapt to them. The adaptability can be evaluated from the posture changes under different experimental conditions. However, the foot posture cannot be directly measured by the FSRs, the only sensors that the foot has. Instead, the projective position of the center of mass (CoM) on the ground is used because they are related with the foot posture and can be calculated from the measurements of FSRs.

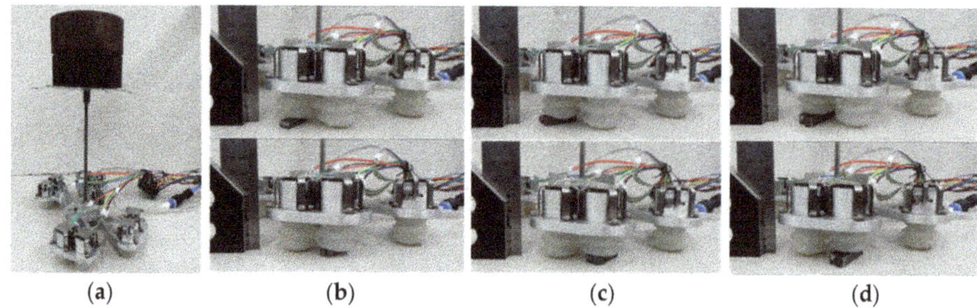

Figure 11. The experiment on the foot system. (**a**) The overall experimental apparatus; (**b**) the foot on a 3-mm high bump; (**c**) the foot on a 5-mm high bump; and (**d**) the foot on an 8-mm high bump.

According to the trigger displacement of the reversing valve and the maximum compression of the pneumatic bladder, the heights of the bumps used in the experiment were selected as 3 mm, 5 mm, and 8 mm. Considering that the foot is symmetric about its center, the bumps were only placed under the two pneumatic bladders on one corner, as illustrated in Figure 11b–d. The experiment showed that under each experimental condition, the reversing valve of the PVSU on the bump was successfully triggered to expel out the air, and the pneumatic bladder became soft and generated more compression under the gravity of the mass block. At the same time, the other pneumatic bladders remained in contact with the flat surface with high stiffness. The foot posture was only slightly changed, enabling the mass block to remain stable without falling down. This result was quite consistent with the theoretical working mechanism of the foot system presented in Section 2. Assuming that the origin of the coordinate system is located at the center of the foot sole, the X axis points to the front, and the Y axis points to the right. Then the stable values of CoM projections under all conditions can be plotted in the same coordinate system, as shown in Figure 12a. From this figure, it can be easily seen that higher bump resulted in larger changes of CoM projections and foot posture. The changes of the CoM projections were within 11.5 mm, and the maximum angle of inclination of the foot sole was measured to be as small as 2.9°. Figure 12b presents the changing curves of the CoM positions in the X and Y axis versus time when the posture change is maximum. In this figure, the CoM position rapidly approached the stable point within 0.1 s, and finally stabilized at the point (11.0, −8.0) after a small-amplitude vibration. Thus, the response time of the pneumatic bladder when it was compressed can be obtained, i.e., less than 0.1 s. The quite short response time indicates that the foot is able to adapt to the bumps on the ground rapidly, and suggests that if the foot is applied to a biped robot, the foot-ground contact time is negligible compared with the gait cycle. All the above results verified the function of the foot system.

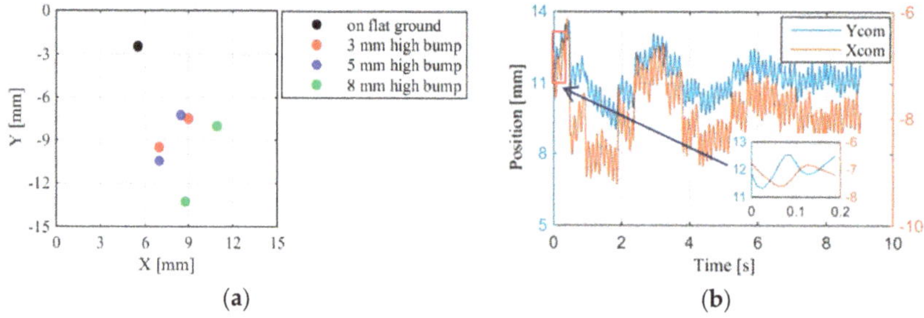

Figure 12. Results of experiment on the foot system. (**a**) The stable values of CoM projections under different experimental conditions; and (**b**) the CoM positions in the X and Y axis versus time.

4.3. Experiment on a Biped Robot

This experiment aims to test the performance of the new foot system on a physical biped robot when the biped robot walks on the ground with small bumps. A lightweight biped robot developed by our research team is utilized in the experiment [38]. The biped robot has two three-segmented legs composed of the thigh, shank, and foot. Each leg has three planar revolute joints, namely, the hip, knee, and ankle joints. The lengths of each link from the thigh to the foot are 0.37 m, 0.36 m and 0.2 m, respectively. Their corresponding masses are 1.8 kg, 1.5 kg and 0.2 kg. The right and left legs are connected by the torso whose mass is 3 kg. The movement of the biped robot in the coronal plane is constrained by a frame presented in Figure 13, without affecting the motion in the sagittal plane [39]. For the sake of comparison, the experiment is performed with two conditions, i.e., (1) the biped robot walks with traditional flat feet; and (2) the biped robot walks with the new feet. The two kinds of feet have the same size, weight, as well as configuration of FSRs. On the ground are some small bumps promiscuously placed, of which the maximum height is 8 mm. The preplanned step length is 0.25 m, and the gait cycle is 1 s. The walking patterns are generated based on the ZMP theory. Under each condition, the biped robot is controlled to walk for ten trials and 5 m each trial. The performances of the two kinds of feet, including the behavior of the foot when contacting with the bumps and the overall walking stability, are compared.

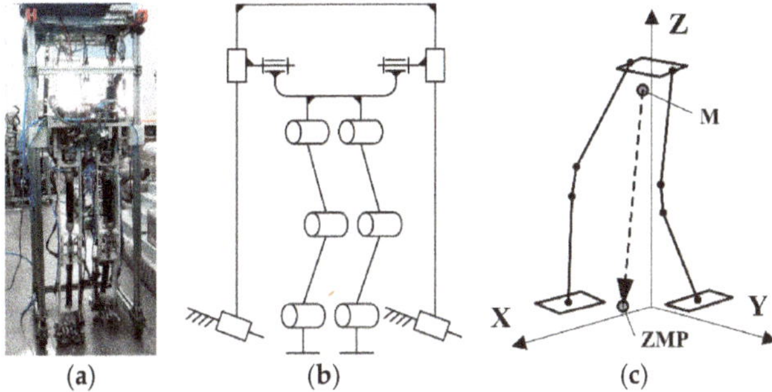

Figure 13. Experimental apparatus of the biped robot. (**a**) Prototype of the biped robot; (**b**) schematic diagram of the biped robot; and (**c**) the coordinate system of the biped robot for ZMP computation.

In the case of walking with the flat feet, the contact area significantly decreased because the foot could not adapt to the bumps. Here we analyze a typical process of the foot contacting with the bumps, which was presented in Figure 14. In this figure, the images in the red panes are partially-enlarged images. During the whole phase from the foot touching the ground to the foot lifting off the ground, three contact states between the foot, ground, and bumps were shown. Firstly, after the foot stepped on the bump, only the trailing edge of the foot, rather than the whole sole, was in contact with the ground (see the upper images in Figure 14). The support polygon was a triangle encircled by the foot's trailing edge and the contact point between the foot and bump. Secondly, as the robot walked forward, the body weight was shifted to this foot which became the new support foot. The forward movement of the CoM of the robot body caused the trailing edge of the foot off the ground. Thus, the whole robot was only supported by the bump (shown in the middle images in Figure 14). Thirdly, the robot continued swinging forward, and the leading edge of the foot touched the ground. The support polygon turned into a triangle encircled by the leading edge of the foot and the contact point between the foot and bump (see the lower images in Figure 14). These three contact states could not provide sufficient and stable support to the biped robot. Consequently, the slippage of the foot on the ground occurred frequently, and the predetermined posture of the robot was easily affected, sometimes causing the robot to fall.

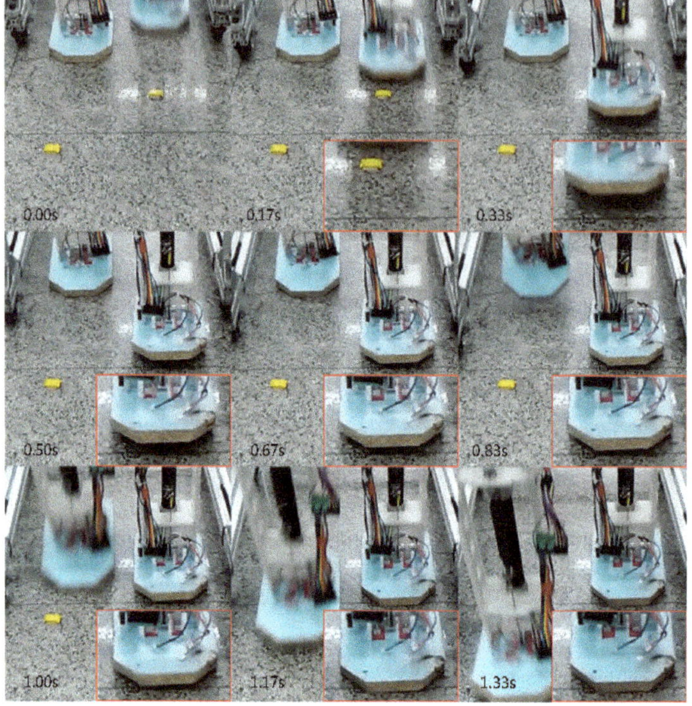

Figure 14. Screenshots of the flat foot contacting with the bumps on the ground during walking.

On the other hand, the performance was much better in the case of walking with the new feet. The behavior of the new feet contacting with the bumps on the ground is presented in Figure 15. When the pneumatic bladders under the foot sole touched the bumps, the corresponding reversing valves were triggered successfully, and the pneumatic bladders became soft and were compressed to follow the shape of the bumps. For the pneumatic bladders which had no contact with the bumps, there were no wrong triggers occurred, and they kept high stiffness and maintained stable contact with the ground. Additionally, the bumps situated in the spaces among the pneumatic bladders did not have any effect on the foot posture. The support polygon was encircled by all the contact areas between the pneumatic bladders and the ground, which was larger than that of the flat feet. The foot slippage was avoided and stable support was provided to the biped robot.

ZMP is the most important and commonly used criterion for evaluating dynamic walking stability of biped robots. During the experiments, the actual ZMP trajectories were calculated from the measurements of FSRs equipped under the feet. The coordinate system for ZMP computation is shown in Figure 13c, of which the origin is centered at the projection of the midpoint between the two hip joints on the ground. The standard coordinate system is such that the positive x-axis points forward along the walking direction, the y-axis is to the left when looking in the x-axis direction, and the z-axis is defined upwards by the right hand rule. Figures 16 and 17 present the ZMP trajectories of the biped robot under the two experimental conditions, as well as the desired support polygons in the single support phase. In the case of rigid flat feet, since the contact between the support foot and the ground was affected by the bumps, the ZMP was shifted outside the support polygon, as marked by the arrows in the figure. However, in the case of the new feet, the changes of ZMP caused by the bumps were smaller, and the ZMP trajectory in the single support phase was located inside the support polygon.

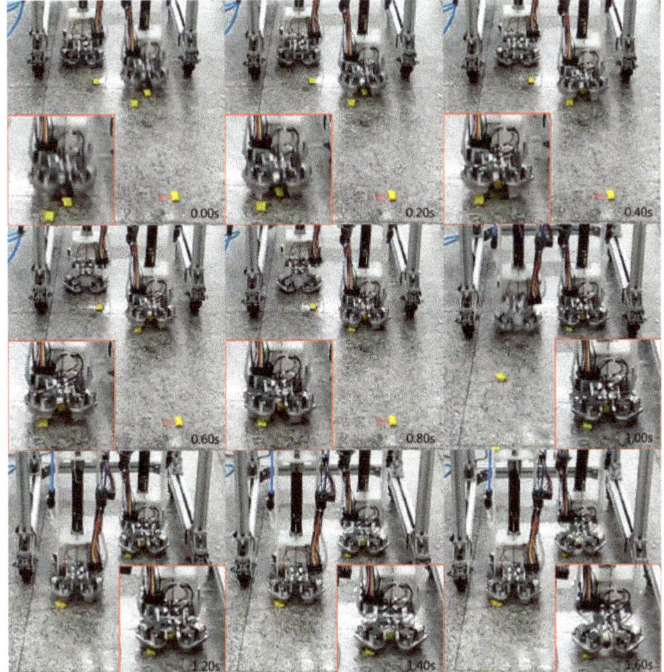

Figure 15. Screenshots of the new foot contacting with the bumps on the ground during walking.

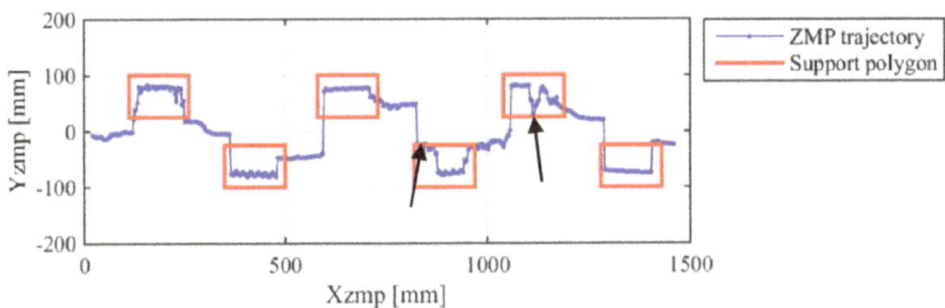

Figure 16. ZMP trajectory of the biped robot in the case of walking with the rigid flat feet.

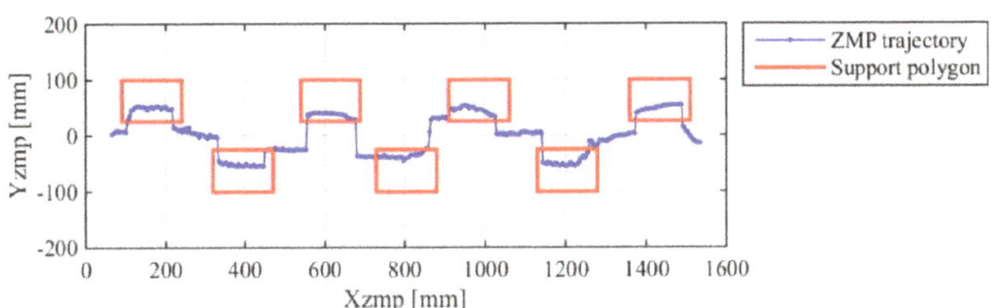

Figure 17. ZMP trajectory of the biped robot in the case of walking with the new feet.

The above comparisons show that the new foot system is able to help to improve the walking performance of the biped robot when walking on the ground with small bumps. The validity of the working mechanism and mechanical structures of the foot system is further confirmed.

5. Discussion and Conclusions

In this paper, a novel variable stiffness mechanism that is able to achieve stiffness adjustment passively under the interactions with the environment is proposed. On this basis, a stiffness adjustable foot system is developed for biped robot walking on the ground with small-sized bumps. The variable stiffness mechanism, the function of the foot system, and the application of the foot system on a prototype of a biped robot are demonstrated to be feasible and effective by the results of various preliminary experiments.

However, there still exist some limitations in our research that need to be further investigated. Firstly, the mathematical model of the variable stiffness mechanism is not studied in detail in this paper, considering that it is not the current focus. However, more assistance can be provided by the accurate analytical model when determining the key design parameters, such as the stiffness of the restoring spring and supporting spring. Secondly, only simple qualitative relationships between the performance of the foot system and the major structural parameters of the PVSU are studied. The mechanical structure of the PVSU is designed by trial and error with the help of the finite element analysis method. Although the variable stiffness mechanism and the desired functions of the foot system are demonstrated to be feasible, more quantitative relationships are necessary for parameter optimization. Thirdly, the adaptability of the foot system on the bumps is somewhat limited. The experiments show that the foot is able to adapt to the bumps with a maximum height of 8 mm with quite a small change in the foot posture. Owing to the small posture change, the biped robot adopted in our research is controlled by a relatively simple controller. We think the foot may be adaptable to bumps as high as 15 mm if a larger posture change can be tolerated by the control system. Therefore, in the future, we will focus on the further improvements of the foot system.

Acknowledgments: The work reported in this paper is supported by the National Natural Science Foundation of China (grant no. 51675116). The second author is funded by the China Scholarship Council (grant no. 201606120094).

Author Contributions: Xizhe Zang, Yixiang Liu, and Wenyuan Li co-organized the work. Wenyuan Li and Zhenkun Lin carried out the experiments. Yixiang Liu wrote the manuscript draft. Jie Zhao supervised the research and commented on the manuscript draft. The manuscript was revised by all authors.

Conflicts of Interest: The authors declare no conflict of interest.

References

1. Hirai, K. Current and future perspective of Honda humanoid robot. In Proceedings of the 1997 IEEE/RSJ International Conference on Intelligent Robots and Systems, Grenoble, France, 11 September 1997; pp. 500–508.
2. Hirai, K.; Hirose, M.; Hikawa, Y.; Takenaka, T. The development of Honda humanoid robot. In Proceedings of the 1998 IEEE International Conference on Robotics and Automation, Leuven, Belgium, 16–20 May 1998; pp. 1321–1326.
3. Kajita, S.; Kanehiro, F.; Kaneko, K.; Yokoi, K.; Hirukawa, H. The 3D linear inverted pendulum mode: A simple modeling for a biped walking pattern generation. In Proceedings of the 2001 IEEE/RSJ International Conference on Intelligent Robots and Systems, Maui, HI, USA, 29 October–3 November 2001; pp. 239–246.
4. Okumura, Y.; Tawara, T.; Endo, K.; Furuta, T.; Shimizu, M. Realtime ZMP compensation for biped walking robot using adaptive inertia force control. In Proceedings of the 2003 IEEE/RSJ International Conference on Intelligent Robots and Systems, Las Vegas, NV, USA, 27–31 October 2003; pp. 335–339.
5. Sugahara, Y.; Hosobata, T.; Mikuriya, Y.; Sunazuka, H.; Lim, H.O.; Takanishi, A. Realization of dynamic human-carrying walking by a biped locomotor. In Proceedings of the 2004 IEEE/RSJ International Conference on Robotics and Automation, New Orleans, LA, USA, 26 April–1 May 2004; pp. 3055–3060.

6. Sakagami, Y.; Watanabe, R.; Aoyama, C.; Matsunaga, S.; Higaki, N.; Fujimura, K. The intelligent Asimo: System overview and integration. In Proceedings of the 2002 IEEE/RSJ International Conference on Intelligent Robots and Systems, Lausanne, Switzerland, 30 September–4 October 2002; pp. 2478–2483.
7. Ogura, Y.; Aikawa, H.; Shimomura, K.; Kondo, H.; Morishima, A.; Lim, H.; Takanishi, A. Development of a new humanoid robot WABIAN-2. In Proceedings of the 2006 IEEE International Conference on Robotics and Automation, Orlando, FL, USA, 15–19 May 2006; pp. 76–81.
8. Nugroho, S.; Prihatmanto, A.; Rohman, A. Design and implementation of kinematics model and trajectory planning for NAO humanoid robot in a tic-tac-toe board game. In Proceedings of the 2014 IEEE 4th International Conference on System Engineering and Technology, Bandung, Indonesia, 24–25 November 2014; pp. 1–7.
9. Tsagarakis, N.; Metta, G.; Sandini, G.; Vernon, D.; Beira, R.; Becchi, F.; Righetti, L.; Santos-Victor, J.; Ijspeert, A.; Carrozza, M.; et al. iCub: The design and realization of an open humanoid platform for cognitive and neuroscience research. *Adv. Robot.* **2007**, *21*, 1151–1175. [CrossRef]
10. Vukobratovic, M. Zero-Moment Point—Thirty five years of its life. *Int. J. Humanoid Robot.* **2001**, *1*, 157–173. [CrossRef]
11. Braun, D.J.; Mitchell, J.E.; Goldfarb, M. Actuated dynamic walking in a seven-link biped robot. *IEEE/ASME Trans. Mechatron.* **2012**, *17*, 147–156. [CrossRef]
12. Shin, H.; Kim, B.K. Energy-efficient gait planning and control for biped robots utilizing the allowable ZMP region. *IEEE Trans. Robot.* **2014**, *30*, 986–993. [CrossRef]
13. Hashimoto, K.; Hosobata, T.; Sugahara, Y.; Mikuriya, Y.; Sunazuka, H.; Kawase, M.; Lim, H.; Takanishi, A. Development of foot system of biped walking robot capable of maintaining four-point contact. In Proceedings of the 2005 IEEE/RSJ International Conference on Intelligent Robots and Systems, Edmonton, AB, Canada, 2–6 August 2005; pp. 1464–1469.
14. Wei, H.; Shuai, M.; Wang, Z. Dynamically adapt to uneven terrain walking control for humanoid robot. *Chin. J. Mech. Eng.* **2012**, *25*, 214–222. [CrossRef]
15. Nishiwaki, K.; Chestnutt, J.; Kagami, S. Autonomous navigation of a humanoid robot over unknown rough terrain using a laser range sensor. *Int. J. Robot. Res.* **2012**, *31*, 1251–1262. [CrossRef]
16. Stumpf, A.; Kohlbrecher, S.; Conner, D.C.; von Stryk, O. Supervised footstep planning for humanoid robots in rough terrain tasks using a black box walking controller. In Proceedings of the 2014 14th IEEE-RAS International Conference on Humanoid Robots, Madrid, Spain, 18–20 November 2014; pp. 287–294.
17. Khadiv, M.; Moosavian, S.; Ali, A.; Yousefi-Koma, A.; Maleki, H.; Sadedel, M. Online adaptation for humanoids walking on uncertain surfaces. *Proc. Inst. Mech. Eng. Part I J. Syst. Control Eng.* **2017**, *231*, 245–258. [CrossRef]
18. Nishiwaki, K.; Chestnutt, J.; Kagami, S. Autonomous navigation of a humanoid robot over unknown rough terrain. In *Robotics Research, Springer Tracts in Advanced Robotics*; Christensen, H., Khatib, O., Eds.; Springer: Cham, Switzerland, 2017; Volume 100, pp. 619–636.
19. Lukac, D.; Siedel, T.; Benckendorff, C. Designing the test feet of the humanoid robot M-Series. In Proceedings of the XXII International Symposium on Information, Communication and Automation Technologies, Bosnia, Serbia, 29–31 October 2009; pp. 1–6.
20. Hashimoto, K.; Sugahar, Y.; Hayash, A.; Kondo, H.; Takashima, T.; Lim, H.; Takanishi, A. Development of new biped foot mechanism mimicking human's foot arch structure. In *ROMANSY 18 Robot Design, Dynamics and Control, Proceedings of the Eighteenth CISM-IFToMM Symposium*; Parenti, C.V., Schiehlen, W., Eds.; Springer: Vienna, Austria, 2010; Volume 524, pp. 249–256.
21. Buschmann, T.; Lobmeier, S.; Ulbrich, H. Humanoid robot LOLA: Design and walking control. *J. Physiol. Paris* **2009**, *103*, 141–148. [CrossRef] [PubMed]
22. Fondahl, K.; Kuehn, D.; Beinersdorf, F.; Bernhardy, F.; Grimminger, F.; Schillingy, M.; Starky, T.; Kirchneret, F. An adaptive sensor foot for a bipedal and quadrupedal robot. In Proceedings of the Fourth IEEE RAS/EMBS International Conference on Biomedical Robotics and Biomechatronics, Rome, Italy, 24–27 June 2012; pp. 270–275.
23. Yamamoto, K.; Sugihara, T.; Nakamura, Y. Toe joint mechanism using parallel four-bar linkage enabling humanlike multiple support at toe pad and toe tip. In Proceedings of the 2007 IEEE International Conference on Humanoid Robots, Pittsburgh, PA, USA, 29 November–1 December 2007; pp. 410–415.

24. Kang, H.; Hashimoto, K.; Kondo, H.; Hattori, K.; Nishikawa, K.; Hama, Y.; Lim, H.; Takanishi, A.; Suga, K.; Kato, K. Realization of biped walking on uneven terrain by new foot mechanism capable of detecting ground surface. In Proceedings of the 2010 IEEE International Conference on Robotics and Automation, Anchorage, AK, USA, 3–7 May 2010; pp. 5167–5172.
25. Yang, H.; Shuai, M.; Qiu, Z.; Wei, H.; Zheng, Q. A novel design of flexible foot system for humanoid robot. In Proceedings of the 2008 IEEE Conference on Robotics, Automation and Mechatronics, Chengdu, China, 21–24 September 2008; pp. 824–828.
26. Piazza, C.; Santina, C.D.; Gasparri, G.M.; Catalano, M.G.; Grioli, G.; Garabini, M.; Bicchi, A. Toward an adaptive foot for natural walking. In Proceedings of the 2016 IEEE-RAS 16th International Conference on Humanoid Robots, Cancun, Mexico, 15–17 November 2016; pp. 1204–1210.
27. Nagarajaiah, S.; Sahasrabudhe, S. Seismic response control of smart sliding isolated buildings using variable stiffness systems: An experimental and numerical study. *Earthq. Eng. Struct. Dyn.* **2006**, *35*, 177–197. [CrossRef]
28. Huang, Y.; Vanderborght, B.; Ham, R.V.; Wang, Q.; Damme, M.V.; Xie, G.; Lefeber, D. Step length and velocity control of a dynamic bipedal walking robot with adaptable compliant joints. *IEEE/ASME Trans. Mechatron.* **2013**, *18*, 598–611. [CrossRef]
29. Wang, R.J.; Huang, H.P. AVSER—Active variable stiffness exoskeleton robot system: Design and application for safe active-passive elbow rehabilitation. In Proceedings of the 2012 IEEE/ASME International Conference on Advanced Intelligent Mechatronics, Kachsiung, Taiwan, 11–14 July 2012; pp. 220–225.
30. Wu, Y.S.; Lan, C.C. Design of a linear variable-stiffness mechanism using preloaded bistable beams. In Proceedings of the 2014 IEEE/ASME International Conference on Advanced Intelligent Mechatronics, Besacon, France, 8–11 July 2014; pp. 605–610.
31. Kuder, I.K.; Arrieta, A.F.; Raither, W.E.; Ermanni, P. Variable stiffness material and structural concepts for morphing applications. *Prog. Aerosp. Sci.* **2013**, *63*, 33–55. [CrossRef]
32. Kind, R.J. Pneumatic stiffness and damping in air-supported structures. *J. Wind Eng. Ind. Aerodyn.* **1984**, *17*, 295–304. [CrossRef]
33. Pan, P.; Liao, G.; Yan, G.; Zhao, Y. Calculation of elastic deformation for the sealing gasket. *J. Harbin Inst. Technol.* **1996**, *28*, 130–134.
34. Liu, H.; Lee, J.C. Model development and experimental research on an air spring with auxiliary reservoir. *Int. J. Automot. Technol.* **2011**, *12*, 839–847. [CrossRef]
35. Wickramatunge, K.C.; Leephakpreeda, T. Empirical modeling of dynamic behaviors of pneumatic artificial muscle actuators. *ISA Trans.* **2013**, *52*, 825–834. [CrossRef] [PubMed]
36. Chou, C.P.; Hannaford, B. Static and dynamic characteristics of McKibben pneumatic artificial muscles. In Proceedings of the 1994 IEEE International Conference on Robotics and Automation, San Diego, CA, USA, 8–13 May 1994; pp. 281–286.
37. Carbone, G. Stiffness analysis and experimental validation of robotic systems. *Front. Mech. Eng.* **2011**, *6*, 182–196.
38. Liu, Y.; Zang, X.; Liu, X.; Wang, L. Design of a biped robot actuated by pneumatic artificial muscles. *Bio-Med. Mater. Eng.* **2015**, *26*, 757–766. [CrossRef] [PubMed]
39. Klein, T.; Lewis, M.A. A neurorobotic model of bipedal locomotion based on principles of human neuromuscular architecture. In Proceedings of the 2012 IEEE International Conference on Robotics and Automation, Saint Paul, MN, USA, 14–18 May 2012; pp. 1450–1455.

Article

Calculation of the Center of Mass Position of Each Link of Multibody Biped Robots

Giovanni Gerardo Muscolo *, Darwin Caldwell and Ferdinando Cannella

Department of Advanced Robotics, Italian Institute of Technology, via Morego, 30, Genova 16163, Italy; darwin.caldwell@iit.it (D.C.); Ferdinando.cannella@iit.it (F.C.)

* Correspondence: giovanni.muscolo@iit.it; Tel.: +39-010-7178-1347

Academic Editors: Toshio Fukuda, Fei Chen and Qing Shi

Received: 26 May 2017; Accepted: 10 July 2017; Published: 14 July 2017

Abstract: In this paper, a novel method to determine the center of mass position of each link of human-like multibody biped robots is proposed. A first formulation to determine the total center of mass position has been tested in other works on a biped platform with human-like dimensions. In this paper, the formulation is optimized and extended, and it is able to give as output the center of mass positions of each link of the platform. The calculation can be applied to different types of robots. The optimized formulation is validated using a simulated biped robot in MATLAB.

Keywords: biped robots; center of mass; balance; biped locomotion; multibody biped robots

---

1. Introduction

One of the pioneers in the field of biped robots was the Waseda University of Tokyo. In 1973, research groups from Waseda University developed WABOT-I, and in 1984 WABOT-2 as to become a professional musician [1]. In 1999, they developed a humanoid with a complete human configuration capable of biped locomotion, WABIAN (Waseda Bipedal humANoid), and in 2011 its Italian version SABIAN [2–4]. In 2017, a more complex version of the WABIAN robot is presented in [5]. In 2013, Google acquired eight advanced robot companies. Boston Dynamics, one of these, is known for its advanced robots including the world's fastest robot, Cheetah [6] (which can travel at 29 mph), Big Dog [7] (the all-rough and tough robot that walks, runs, climbs, and carries heavy loads), and the latest Atlas (a biped humanoid capable of walking in outdoor rough terrain with the upper limbs capable of performing other tasks while walking). Several versions of Atlas [8,9] have been prepared for the DARPA Robotics Challenge program. In 2017, Boston Dynamics presented a very innovative robot with higher locomotion capabilities including wheels in the feet [10].

The general robot design process includes many phases, like the design of every complex machine. A tentative method to define these phases is shown as follows:

- PHASE 1—Determination of the technical specifications which define limits and characteristics that the robot should have.
- PHASE 2—Conceptual design of the robot including analysis of the developed robots in the world; design of novel systems; definition of the whole system including mechanics, electronics, low and high level control.
- PHASE 3—Functional design of the robot including interaction of the robot with the environment; theoretical formulation and optimization; software and hardware design of virtual models (virtual model prototyping using CAD tools, analytical simulations, finite elements analysis, multibody analysis).
- PHASE 4—Development of the robot including rapid prototyping modeling and tests.
- PHASE 5—Realization of the final robot prototype and final tests.

All robots realized in Phase 5 should have, theoretically, the same architecture of the virtual models designed in Phase 3. In particular, the positions of the CoM (Center of Mass) of each link of the robot should have the same position defined in the virtual model. However, the total CoM of the real platform is not in the same calculated position of the total CoM of its respective virtual model. Why are these discrepancies are created? Are these errors (between virtual models and real robots) influences on the robot functionality?

In order to answer to these two questions we could show how the problem was evident in the SABIAN robot (height: 1500 mm; 64 kg). During our tests on the platform, we noted a weight difference of about 5 kg including differences of CoM position between real and virtual SABIAN. These discrepancies are created because during manufacturing, construction, and maintenance of the robot, the tolerances of the joints, cables, drivers, batteries, links, etc. could not be completely respected and errors are created. These errors influence the robot functionality and cannot be eliminated [11]. Hence, the real center of mass (CoM) of the robot is not coincident with the CoM of its virtual model [12]. In our experiments, the robot SABIAN [2,3] had about 5 kg of errors in an unknown position and during locomotion, the controller implemented on the virtual model architecture was not able to control the real platform with this unknown error position. Dynamic balance in locomotion is not simple to control in a biped platform and these errors may disturb biped stability.

Some researchers and specialists in the humanoid robotics field use a posture controller [11,12], in order to reduce the error between the robotic platform and its virtual model. Kwon et al. [13] (2007) proposed a method that uses a closed-loop observer based on a Kalman filter, adopted as estimation framework. Ayusawa et al. (2008) [14] proposed a method based on regression analysis models in order to estimate inertial parameters using a minimal set of sensors. In the work of Sujan and Dubowsky (2003) [15] the dynamic parameters of a mobile robot are calculated using an algorithm based on a mutual-information-based theoretic metric for the excitation of vehicle dynamics. Liu et al. (1998) [16], Khalil et al. (2007) [17], and Swevers et al. (1997) [18] show other methods oriented to improve the balancing performances of mobile biped robots when the center of mass is not precisely known.

In [2,3], the authors proposed a novel approach to determine the correct position of the center of mass in humanoid robots. In order to compensate the errors between the biped platform and its virtual model, an additional mass has been implemented in the virtual model of the humanoid robot. The value of this mass error is the analytical difference between the weight of the robot and the weight of its virtual model. Its position in the space is not known a priori, but it will be approximately calculated with the procedure described in [3]. In order to define its position, the authors of the paper proposed an analytic formula that gives the real position of the CoM of the platform and is based on the application of a procedure that requires only the values of the force-torque sensors, applied on the feet of the humanoid robot, and the values of the motors torque. This procedure standardizes the calibration procedure in order to minimize the errors and it can be applied to every biped platform. The formulation approach has been implemented on the SABIAN robot with dimensions comparable to humans (height: 1500 mm; weight: 64 kg) giving very good results.

The limits underlined in the papers [2,3] were based on the approximation used to put the error mass in the determined CoM with the proposed formula. The approximation has been justified because the real position of the error mass is not known and if an external mass is positioned in the real CoM, its negative influence is reduced. However, the problem remains because an approximation has been used on behalf of exact calculation.

In this paper, the limits underlined in [2,3] are bypassed with the optimization of the formula based on the determination of the CoM position of each link of the robot. With the proposed solution, the error mass is distributed on each link of the biped robot.

Another advantage of the formulation presented in this paper is that if the total CoM position of the platform is known a priori, the first formulation proposed in [2,3] can be bypassed and the CoM positions of each link of the platform can be calculated analytically without using force-torque sensors and the motor's torque.

The paper is structured as follows: Section 2 presents, in synthesis, the first validated theoretical formulation proposed in [2,3]; Section 3 shows results and discussion on the second theoretical formulation to determine the center of mass position of each link of the platform. Section 4 presents validation of the second theoretical formulation. The paper ends with a conclusion and future works.

2. First Validated Theoretical Formulation

2.1. Dynamics of Multibody Biped Robots

Figure 1 shows global and local reference Cartesian system (respectively *G-XYZ* and *P-XpYpZp* and three points in the space (0, 1 and 2). The three points can be considered as belonging to a rigid body in the space; furthermore, the rigid body can be compared to a humanoid platform, or multibody robot, with its center of gravity in the Point 2 and its feet in the Points 0 and 1. A humanoid robot is indeed composed of a trunk and articulated kinematic chains such as legs and arms, connected to the trunk with joints and motors, and with force-torque sensors positioned on the feet and on the hands. The Points 0 and 1 represent the feet, where the force-torque sensors are positioned, and are shown with a light blue colour; the center of mass represented with the Point 2 is shown in red; the other black points indicate the center of mass of the links of the robot.

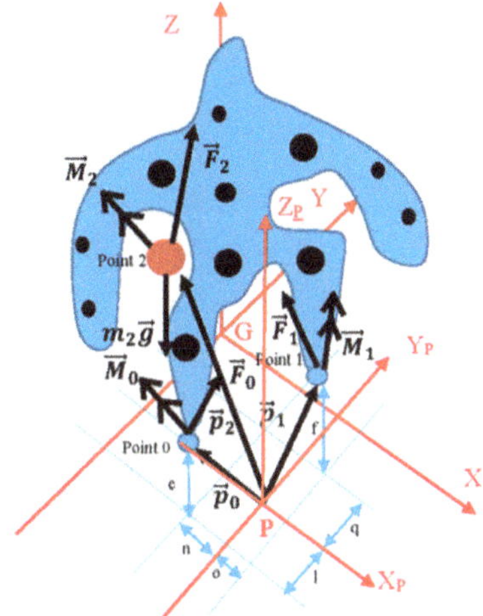

Figure 1. Scheme for determining the total CoM (Center of Mass) position. The direction of the force and torque vectors is only indicative. Points 0 and 1 represent the feet, where the force/torque sensors are positioned; the center of mass is represented with the Point 2; the other black points indicate the mass of the links of the robot.

The dynamics of the system is described by the two equations (see Figure 1):

$$m_2 \cdot \vec{a}_2 = m_2 \cdot \vec{g} + \vec{F}_0 + \vec{F}_1 + \vec{F}_2 \tag{1}$$

$$\vec{M}_P = \vec{M}_0 + \vec{M}_2 + \vec{M}_1 + \vec{p}_0 \times \vec{F}_0 + \vec{p}_1 \times \vec{F}_1 + \vec{p}_2 \times \vec{F}_2 \tag{2}$$

where: $\vec{a}_2 = [a_{X2}, a_{Y2}, a_{Z2}]^T$ is the acceleration of the CoM; m_2 is the total mass of the robot without feet; $\vec{p}_0 = [-n, -l, e]^T$ and $\vec{p}_1 = [o, q, f]^T$ are the position vectors shown in Figure 1. $\vec{p}_2 = [X_2, Y_2, Z_2]^T$ is the CoM position that will be determined with the proposed formula in Section 3. $\vec{M}_0 = [M_{X0}, M_{Y0}, M_{Z0}]^T$, $\vec{M}_1 = [M_{X1}, M_{Y1}, M_{Z1}]^T$, $\vec{M}_2 = [M_{X2}, M_{Y2}, M_{Z2}]^T$, $\vec{F}_0 = [F_{X0}, F_{Y0}, F_{Z0}]^T$, $\vec{F}_1 = [F_{X1}, F_{Y1}, F_{Z1}]^T$, $\vec{F}_2 = [F_{X2}, F_{Y2}, F_{Z2}]^T$ respectively represent the torques $(\vec{M}_0, \vec{M}_1, \vec{M}_2)$ and the forces ($\vec{F}_0, \vec{F}_1, \vec{F}_2$) acting at the Points 0, 1 and 2. $\vec{M}_P = [M_{XP}, M_{YP}, M_{ZP}]^T$ is the resultant moment calculated with respect to the local Cartesian system (see Figure 1). The direction of the force and torque vectors, shown in Figure 1, is only indicative; the positive direction of the force and torque vectors has been considered with the same positive direction of the global reference Cartesian system (*G-XYZ*).

Furthermore, we can say that:

$$\vec{M}_2 = \begin{bmatrix} M_{X2} \\ M_{Y2} \\ M_{Z2} \end{bmatrix} = \begin{bmatrix} \sum M_{j\_roll} \\ \sum M_{j\_pitch} \\ \sum M_{j\_yaw} \end{bmatrix} \quad (3)$$

M_2 includes only the internal torques; M_0 and M_1 include the ground reaction torques. The Equation (3) is based on the assumption that the Jacobian Matrix is equal to the identity Matrix. This assumption is correct if the joints axis of the robot remain parallel, during motion, to the Y axis of the global reference Cartesian system (*G-XYZ*). It means that the motion of the robot for the determination of the formula is performed in a 2D plane. In this paper, the XZ plane is considered. $\sum M_{j\_roll}, \sum M_{j\_pitch}, \sum M_{j\_yaw}$ are the torques of all roll, pitch, and yaw motors of the robot [11] and M_j is obtained from the equation:

$$M_j[Nm] = K[Nm/A] \cdot I[A] \quad (4)$$

The accuracy in the estimation of M_j, calculated in (4) depends on the accuracy of the K value that is a constant parameter set for each motor and on the accuracy of the current I necessary for the motor function. In particular, the resolution of the used A/D converters is a fundamental parameter to define the accuracy of the current I.

2.2. Equilibrium

Considering the robot equilibrium ($m_2\vec{a}_2 = 0$; $\vec{M}_P = 0$) with respect to the Point P as shown in the Figure 1, Equations (1) and (2) can be modified. Moving the point of view from the vector shape to the scalar one, the values of the three components x, y, and z of the force and the torque can be obtained. The new system consists of six equations (five linearly independent) in six unknown values $F_{X2}, F_{Y2}, F_{Z2}, X_2, Y_2, Z_2$. The forces and torques in Points 0 and 1 can be calculated by means of the load cells. The torques M_{X2}, M_{Y2}, M_{Z2} are determined using (3) and (4).

In order to simplify the system, the robot is positioned in two different configurations. The two configurations are chosen in order to have a simplified geometry using $l = q$, $n = o = e = f = 0$ (see Figure 1) obtaining $\vec{p}_0 = [0, -l, 0]^T$ and $\vec{p}_1 = [0, l, 0]^T$.

In the first step the robot is placed on a walking surface and the platform should be kept in a first balance configuration (Scheme A, see Figure 2), allowing a measurement of the forces ($\vec{F}_0, \vec{F}_1$) and the torques ($\vec{M}_0, \vec{M}_1$) by means of the force-torque sensors on the feet, and of the armature currents ($\sum M_{j\_roll}, \sum M_{j\_pitch}, \sum M_{j\_yaw}$). In a second step, the robot is placed in a second balance configuration (Scheme B, see Figure 2), and in the same way, forces, torques, and motor currents associated with this new balance configuration are measured. The two balance configurations can be performed as the reader prefers underlining that the robot should be in a balance position. In particular, the coordinates X_{2A} and Z_{2A} are relative to the position of the center of mass of the platform in the configuration A along the first straight common line, which is chosen (in this paper) orthogonal to the plane of standing, and then parallel to the Z axis. m_u, r_u, l_u, are the mass and the position of the center of mass of the robot ankle link from the floor to the ankle joint; U is the length of the ankle. m_w, r_w, l_w, are the mass and the position of the center of mass of the remaining links of the platform. In the second balance configuration B, the components of the body are aligned according to a second straight common line, inclined to the vertical line with an angle θ_t. While m_u, m_w, r_u, l_u remain constant, r_w and l_w change their values. In this case, X_{2B} and Z_{2B} identify the coordinates of the center of mass of the body in the second balance configuration B. The coordinates of the two feet are the same because we chose this configuration as input. The implementations have been done positioning the robot in this initial position using a leveller and the encoders of the motors.

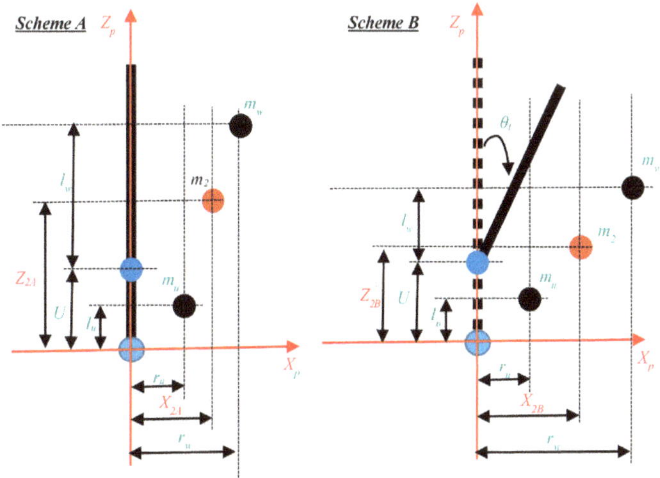

Figure 2. Two examples of balance configurations; scheme A and scheme B. m_2 is the total CoM. m_u and m_w are the CoMs of the two links.

Based on the choice $l = q$, $n = o = e = f = 0$ ($\vec{p}_0 = [0, -l, 0]^T$ and $\vec{p}_1 = [0, l, 0]^T$), the Equations (1) and (2) can be rewritten in a general form that is function of the balance i configuration ($I = A$ or $I = B$). Thus, Equations from (5) to (9) are obtained.

$$F_{X2i} = -F_{X0i} - F_{X1i} \tag{5}$$

$$F_{Y2i} = -F_{Y0i} - F_{Y1i} \tag{6}$$

$$F_{Z2i} = m_2 \cdot g - F_{Z0i} - F_{Z1i} \tag{7}$$

$$\begin{aligned} X_{2i} = &-(M_{Y0i} + M_{Y1i} + M_{Y2i})/(F_{Z0i} + F_{Z1i}) \\ &+[(F_{X0i} + F_{X1i})/(F_{Z0i} + F_{Z1i})] \cdot Z_{2i} \end{aligned} \tag{8}$$

$$\begin{aligned} Y_{2i} = &[(M_{X0i} + M_{X1i} + M_{X2i})/(F_{Z0i} + F_{Z1i})] \\ &+[l \cdot (F_{Z1i} - F_{Z0i})/(F_{Z0i} + F_{Z1i})] + \\ &+[(F_{Y0i} + F_{Y1i})/(F_{Z0i} + F_{Z1i})] \cdot Z_{2i} \end{aligned} \tag{9}$$

2.3. Proposed Coefficients

In following, four novel coefficients (α_i, β_i, γ_i, δ_i) are introduced

$$-(M_{Y0i} + M_{Y1i} + M_{Y2i})/(F_{Z0i} + F_{Z1i}) = \alpha_i \tag{10}$$

$$(F_{X0i} + F_{X1i})/(F_{Z0i} + F_{Z1i}) = \beta_i \tag{11}$$

$$\begin{aligned} &(M_{X0i} + M_{X1i} + M_{X2i})/(F_{Z0i} + F_{Z1i}) \\ &+[l \cdot (F_{Z1i} - F_{Z0i})/(F_{Z0i} + F_{Z1i})] = \gamma_i \end{aligned} \tag{12}$$

$$(F_{Y0i} + F_{Y1i})/(F_{Z0i} + F_{Z1i}) = \delta_i \tag{13}$$

Rewriting Equations (8) and (9) (for $I = A$ or $I = B$)

$$X_{2i} = \alpha_i + \beta_i \cdot Z_{2i} \tag{14}$$

$$Y_{2i} = \gamma_i + \delta_i \cdot Z_{2i} \tag{15}$$

Using the parameters m_u, m_w, r_u, l_u, r_w and l_w, the Equations (16) and (17) can be obtained. Thus, the Equations (14), (16) and (17) can be seen as a system composed of six equations in six unknown variables (for $i = A$ and $i = B$) X_{2A}, Z_{2A}, X_{2B}, Z_{2B}, r_w, l_w; the relation between m_w and m_u is given by the Equation (18). θ_t is fixed by the user ($\theta_t = 0$ in $i = A$), in a way that does not allow to tilt the platform.

$$X_{2i} = [m_u \cdot r_u + m_w \cdot (l_w \cdot \sin\theta_t + r_w \cdot \cos\theta_t)]/m_2 \tag{16}$$

$$Z_{2i} = [m_u \cdot l_u + m_w \cdot (U + l_w \cdot \cos\theta_t - r_w \cdot \sin\theta_t)]/m_2 \tag{17}$$

$$m_w = m_2 - m_u \tag{18}$$

Solving the equations system constituted by (14), (16) and (17), the positions of the center of mass are calculated in both the configurations A and B (for $i = A$ and $i = B$). It must be underlined that only X as a function of Z has been considered, but the same result can be obtained considering Y as a function of Z.

2.4. Determination of the Partial Center of Mass Position

Placing $\theta_t = 0$ (then $i = A$) and substituting (16) and (17) into (14) and placing $\theta_t \neq 0$ (then $i = B$) and substituting (16) and (17) into (14), two different equations will be obtained. Finally, combining these two equations l_w and r_w are obtained.

Placing $\theta_t \neq 0$ (then $i = B$) and rewriting (17) with the latter values given by l_w and r_w, Z_{2B} is obtained.

Z_{2B} represents the general position of the height Z of the center of mass for any value of θ_t. Placing $i = B$ in (14) and (15) and substituting the found value of Z_{2B}, the general formula of the position of the center of mass in (19) is given. Considering the equilibrium configuration, A ($i = A$) and then θ_t equal to zero, the system (20) is obtained.

In particular, the calculation of the center of mass position is strictly related to the proposed coefficients α_A, β_A, γ_A, δ_A, α_B, β_B, γ_B, δ_B, that are numerical values associated with the first and second measurements on the robot. In order to calculate these coefficients, it is necessary to consider the arbitrary θ_t associated with the second balance configuration, in addition to other parameters such as the above mentioned position of the center of mass of the feet (r_u and l_u) and its mass (m_u) calculated using the CAD model. These parameters have a lower weight with respect to other links of the platform and then a lower inertial influence [3]. The two formulations (19) and (20) were tested and validated on the SABIAN platform giving very good results and presented in [2,3].

$$\begin{cases} Z_{2B} = f(\theta_t, m_2, \beta_B, \beta_A, \alpha_A, m_u, l_u, U, r_u, \alpha_B); \\ X_{2B} = \alpha_B + \beta_B \cdot Z_{2B}; \\ Y_{2B} = \gamma_B + \delta_B \cdot Z_{2B}; \end{cases} \tag{19}$$

$$\begin{cases} Z_{2B}(\theta_t = 0) = Z_{2A} \\ = f(m_2, \beta_B, \beta_A, \alpha_A, m_u, l_u, U, r_u, \alpha_B); \\ X_{2B}(\theta_t = 0) = X_{2A} = \alpha_A + \beta_A \cdot Z_{2A}; \\ Y_{2B}(\theta_t = 0) = Y_{2A} = \gamma_A + \delta_A \cdot Z_{2A}; \end{cases} \tag{20}$$

3. Second Theoretical Formulation

3.1. *Procedure for* n *Degrees of Freedom*

In order to find the CoM positions of each link of the platform, the following procedure and formulation must be used. In particular, if n are the degrees of freedom of the platform; i represents the used configurations to calculate the CoM of each link. j represents the initial configuration. The number (k) of the configurations necessary to calculate the CoM position for each link of the system is calculated in the following

$$k = n + 1 \tag{21}$$

The total CoM position (X_{2i}, Y_{2i}, Z_{2i}) of the complete system for each configuration i can be obtained with the following formulas where the coefficients α_i, β_i, γ_i, δ_i, α_j, β_j, γ_j, δ_j, are calculated using respectively (10)–(13).

$$\begin{cases} Z_{2i} = f(\theta_i, m_2, \beta_i, \beta_j, \alpha_j, m_u, l_u, U, r_u, \alpha_i); \\ X_{2i} = \alpha_i + \beta_i \cdot Z_{2i}; \\ Y_{2i} = \gamma_i + \delta_i \cdot Z_{2i}; \end{cases} \tag{22}$$

The following formulas from (23) to (26) allow to determine the CoM positions of each link of the robot. m_w, m_2, m_u, r_u, q_u, l_u, U, are the input of the system as shown in the Section 2. r_{wi}, q_{wi}, l_{wi}, are the components of the vector position $\vec{t}_i = \begin{bmatrix} r_{wi} & q_{wi} & l_{wi} + U \end{bmatrix}^T$ respectively in X_P, Y_P, and Z_P directions (see Figure 1). χ_i and ε_i are the angles of the vector position $\vec{t}_i = \begin{bmatrix} r_{wi} & q_{wi} & l_{wi} + U \end{bmatrix}^T$ as shown in Figure 3.

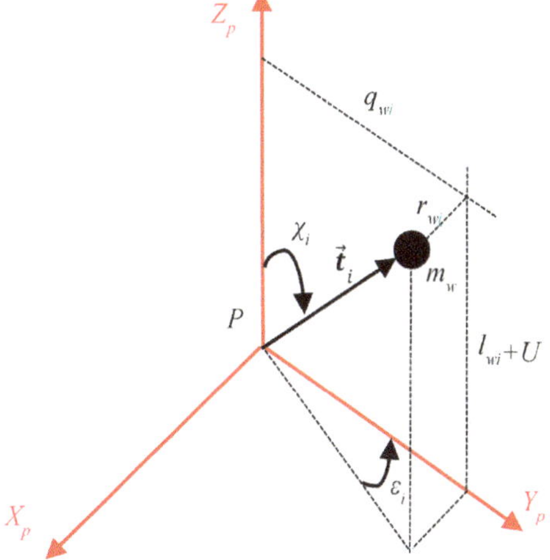

Figure 3. Scheme and vector position of the mass m_w.

$$r_{wi} = \frac{X_{2i} \cdot m_2 - m_u \cdot r_u}{m_w};\ q_{wi} = \frac{Y_{2i} \cdot m_2 - m_u \cdot q_u}{m_w};l_{wi} = \frac{Z_{2i} \cdot m_2 - m_u \cdot l_u}{m_w} - U \tag{23}$$

$$r_{wi} = \frac{X_{2i} \cdot m_2 - m_u \cdot r_u}{m_w};\ q_{wi} = \frac{Y_{2i} \cdot m_2 - m_u \cdot q_u}{m_w};l_{wi} = \frac{Z_{2i} \cdot m_2 - m_u \cdot l_u}{m_w} - U \tag{24}$$

$$\vec{t}_i = \begin{bmatrix} r_{wi} \\ q_{wi} \\ l_{wi} + U \end{bmatrix} = \begin{bmatrix} t_i \sin\chi_i \sin\varepsilon_i \\ t_i \sin\chi_i \cos\varepsilon_i \\ t_i \cos\chi_i \end{bmatrix} \quad (25)$$

$$\vec{t}_i \cdot m_w = \sum_{v=1}^{n} \vec{s}_{iv} \cdot m_{iv} \quad (26)$$

$$\vec{s}_i = \begin{bmatrix} r_i \\ q_i \\ l_i \end{bmatrix} \quad (27)$$

3.2. Procedure for $n = 2$ Degrees of Freedom

Figure 4 shows a sketch of a robot in a plane X_P, Z_P with two degrees of freedom ($n = 2$) and three links (L_a, L_b, L_c). From (21) $k = 3$ is obtained. This result means that two configurations must be used to calculate the coefficients α_i, β_i, γ_i, δ_i, and one configuration must be used to calculate the coefficients α_j, β_j, γ_j, δ_j using respectively (10)–(13) in order to find the CoM position of the links of the robot. In particular, the following iterative procedure should be used:

- The robot is placed on a walking surface and the platform should be kept in a first balance configuration j allowing a measurement of the forces ($\vec{F}_0, \vec{F}_1$) and the torques ($\vec{M}_0, \vec{M}_1$) by means of the force-torque sensors on the feet, and of the armature currents ($\sum M_{j\_roll}$, $\sum M_{j\_pitch}$, $\sum M_{j\_yaw}$). These values are used to calculate the coefficients α_j, β_j, γ_j, δ_j, of the (22) using respectively (10)–(13);
- The robot is placed in a second and third balance configuration i, and in the same way, forces, torques, and motor currents associated with each balance configuration (second and third) are measured. These values are used to calculate the coefficients α_i, β_i, γ_i, δ_i using respectively (10)–(13).
- For each balance configuration i, the total CoM position is calculated using (22);
- Each total CoM position allows to determine r_{wi}, q_{wi}, l_{wi}, using (23) and the CoM position of each link of the robot using (27). For each configuration i, an equation using (26) is created. n linearly independent equations are used to find n vector positions.

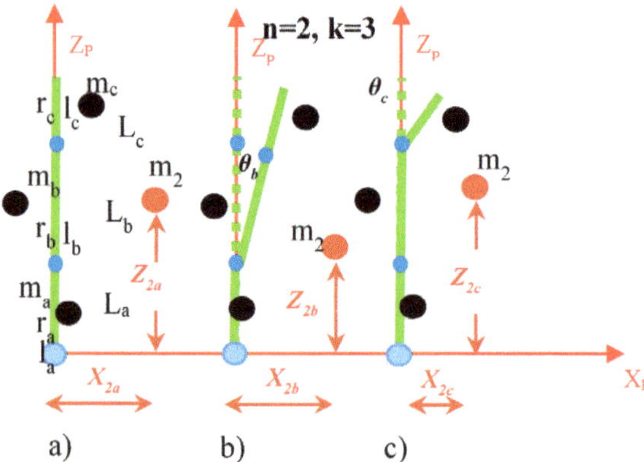

Figure 4. Sketch of a robot in a plane with 2 degrees of freedom. Three Configurations (a)–(c). m_2 is the total mass of the system; $m_a, \ldots, m_c$ are the masses of the links and $r_a, \ldots, r_c$ and $l_a, \ldots, l_c$ are respectively the x and z components of the center of mass of the links respect to the relative revolute joint; $L_a, \ldots, L_c$ are the lengths of the links.

3.3. Implementation of the Analytical Formulation

The example shown in Figure 4 has four unknown parameters (r_b, l_b, r_c, l_c), but four linearly independent equations can be obtained by the formulas shown in Section 3.1. In this case, we suppose to have m_a, m_b, m_c, r_a, l_a, as input. Figure 5 shows in details the configurations b and c shown in Figure 4. In particular, the four unknown parameters (r_b, l_b, r_c, l_c) which should be determined using analytical formulation are found using polar coordinates. $\vec{t}_b$ and $\vec{t}_c$ are the position vectors of the mass m_w respect to the local reference system P-$X_PY_PZ_P$ and respectively of the configuration b and c. $\vec{s}_b$ and $\vec{s}_c$ are respectively the position vectors of the masses, m_b and m_c, respect to the local reference system of each link.

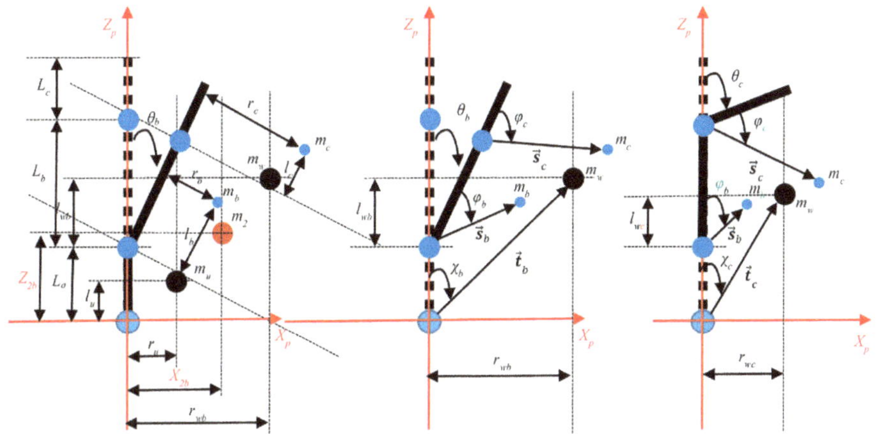

Figure 5. Example for implementing the formula.

Using (22), Z_{2b}, X_{2b}, and Z_{2c}, X_{2c}, can be calculated as shown in following

$$\begin{cases} Z_{2b} = f(\theta_b, m_2, \beta_b, \beta_a, \alpha_a, m_a, l_a, L_a, r_a, \alpha_b) \\ X_{2b} = \alpha_b + \beta_b \cdot Z_{2b} \end{cases} \tag{28}$$

$$\begin{cases} Z_{2c} = f(\theta_c, m_2, \beta_c, \beta_a, \alpha_a, m_a, l_a, L_a, r_a, \alpha_c) \\ X_{2c} = \alpha_c + \beta_c \cdot Z_{2c} \end{cases} \tag{29}$$

Using (23)–(25), r_{wb}, l_{wb}, and r_{wc}, l_{wc}, can be determined (see Figure 5) and the modules and angles of the two position vectors ($\vec{t}_b$ and $\vec{t}_c$) are found

$$r_{wb} = \frac{X_{2b} \cdot m_2 - m_u \cdot r_u}{m_w};\ l_{wb} = \frac{Z_{2b} \cdot m_2 - m_u \cdot l_u}{m_w} - L_a \tag{30}$$

$$r_{wc} = \frac{X_{2c} \cdot m_2 - m_u \cdot r_u}{m_w};\ l_{wc} = \frac{Z_{2c} \cdot m_2 - m_u \cdot l_u}{m_w} - L_a \tag{31}$$

$$t_b = \sqrt{r_{wb}^2 + (l_{wb} + L_a)^2};\ \chi_b = \tan^{-1}\left(\frac{r_{wb}}{(l_{wb} + L_a)}\right) \tag{32}$$

$$t_c = \sqrt{r_{wc}^2 + (l_{wc} + L_a)^2};\ \chi_c = \tan^{-1}\left(\frac{r_{wc}}{(l_{wc} + L_a)}\right) \tag{33}$$

Using (26) and (27), the modules of the position vectors $\vec{s}_b$ and $\vec{s}_c$ (respectively the position vectors of the masses m_b and m_c respect to the local reference system of each link) are determined. In the following, detailed equations are shown which can determine the four unknown parameters (r_b, l_b, r_c, l_c).

$$m_w \cdot r_{wb} = m_b \cdot (l_b \cdot \sin\theta_b + r_b \cdot \cos\theta_b) + m_c \cdot ((L_b + l_c) \cdot \sin\theta_b + r_c \cdot \cos\theta_b) \quad (34)$$

$$\begin{aligned} m_w \cdot (l_{wb} + L_a) = m_b \cdot (L_a + l_b \cdot \cos\theta_b - r_b \cdot \sin\theta_b) + m_c \\ \cdot ((L_b + l_c) \cdot \cos\theta_b - r_c \cdot \sin\theta_b + L_a) \end{aligned} \quad (35)$$

$$m_w \cdot r_{wc} = m_b \cdot r_b + m_c \cdot (l_c \cdot \sin\theta_c - r_c \cdot \cos\theta_c) \quad (36)$$

$$m_w \cdot (l_{wc} + L_a) = m_b \cdot (L_a + l_b) + m_c \cdot (L_a + L_b + l_c \cdot \cos\theta_c - r_c \cdot \sin\theta_c) \quad (37)$$

Solving the system using the Cramer's rule, we obtain the following equations where $\sin\theta_b = s\theta_b$; $\cos\theta_b = c\theta_b$; $\sin\theta_c = s\theta_c$; $\cos\theta_c = c\theta_c$;

$$A = \begin{vmatrix} m_b c\theta_b & m_b s\theta_b & m_c c\theta_b & m_c s\theta_b \\ -m_b s\theta_b & m_b c\theta_b & -m_c s\theta_b & m_c c\theta_b \\ m_b & 0 & -m_c c\theta_c & m_c s\theta_c \\ 0 & m_b & -m_c s\theta_c & m_c s\theta_c \end{vmatrix} \quad (38)$$

$$\begin{vmatrix} a_1 \\ a_2 \\ a_3 \\ a_4 \end{vmatrix} = \begin{vmatrix} m_w r_{wb} - m_c L_b s\theta_b \\ m_w (l_{wb} + L_a) - m_b L_a - m_c L_b c\theta_b - m_c L_a \\ m_w r_{wc} \\ m_w (l_{wc} + L_a) - m_b L_a - m_c (L_a + L_b) \end{vmatrix} \quad (39)$$

$$r_b = \frac{\begin{vmatrix} a_1 & m_b s\theta_b & m_c c\theta_b & m_c s\theta_b \\ a_2 & m_b c\theta_b & -m_c s\theta_b & m_c c\theta_b \\ a_3 & 0 & -m_c c\theta_c & m_c s\theta_c \\ a_4 & m_b & -m_c s\theta_c & m_c s\theta_c \end{vmatrix}}{\det(A)} \quad (40)$$

$$l_b = \frac{\begin{vmatrix} m_b c\theta_b & a_1 & m_c c\theta_b & m_c s\theta_b \\ -m_b s\theta_b & a_2 & -m_c s\theta_b & m_c c\theta_b \\ m_b & a_3 & -m_c c\theta_c & m_c s\theta_c \\ 0 & a_4 & -m_c s\theta_c & m_c s\theta_c \end{vmatrix}}{\det(A)} \quad (41)$$

$$r_c = \frac{\begin{vmatrix} m_b c\theta_b & m_b s\theta_b & a_1 & m_c s\theta_b \\ -m_b s\theta_b & m_b c\theta_b & a_2 & m_c c\theta_b \\ m_b & 0 & a_3 & m_c s\theta_c \\ 0 & m_b & a_4 & m_c s\theta_c \end{vmatrix}}{\det(A)} \quad (42)$$

$$l_c = \frac{\begin{vmatrix} m_b c\theta_b & m_b s\theta_b & m_c c\theta_b & a_1 \\ -m_b s\theta_b & m_b c\theta_b & -m_c s\theta_b & a_2 \\ m_b & 0 & -m_c c\theta_c & a_3 \\ 0 & m_b & -m_c s\theta_c & a_4 \end{vmatrix}}{\det(A)} \quad (43)$$

The proposed representation is general and can be implemented on robots with different types of joints (prismatic, revolute, spherical, helical, etc.). In case of robot conceived in an unconventional way, such as passive or flexible robots or robot with wheels [19–21], the procedure (21) and the formulas from (22) to (27) can be implemented if the two following points are satisfied:

1. Only two force-torque sensors must be the contact elements between the robot and the ground;
2. Joint sensors must give relative position of motion and current values to produce join motion.

Another advantage of the formulation presented in this paper, as underlined in the first section, is that if the total CoM position of the platform is known a priori, the first formulation proposed in [2,3] and shown in Section 2 of this paper can be bypassed and the CoM positions of each link of the platform can be calculated analytically without using force-torque sensors and motor's torque. In this case, only the formulation of the Section 3 can be used.

4. Validation of the Second Theoretical Formulation

4.1. Example

In order to validate the second theoretical formulation shown in Section 3 (the first theoretical formulation was validate in [2,3] as underlined in Sections 1 and 2), a virtual robot with three links and two DoFs in each leg is used. Figure 6 shows the designed robot and Table 1 shows the characteristics of the robot (lengths of the links, weights, CoMs positions, etc.).

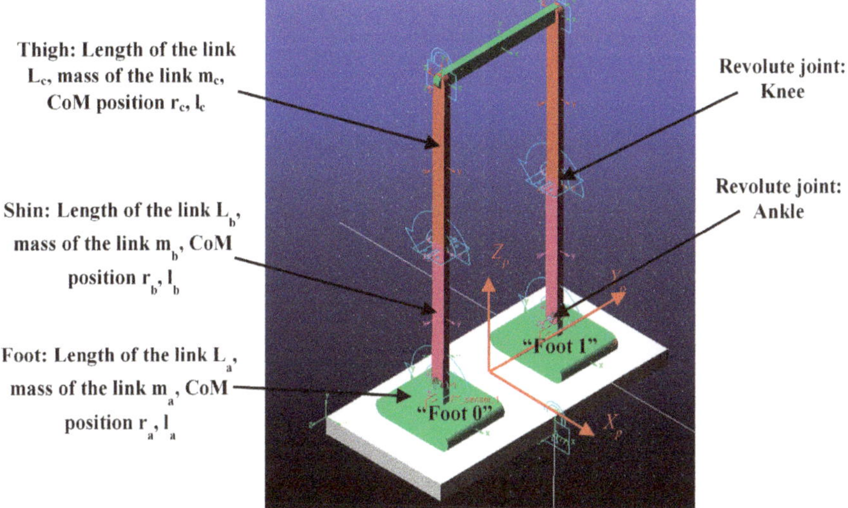

Figure 6. Example for implementing the formula.

Table 1. Characteristics of the robot.

Link	Weight [kg]	Absolute CoM Position Respect to X_p, Y_p, Z_p [mm]	Z_p [mm]	Relative CoM Position in X_p-Z_p Plane [mm]
Foot 0	6.72	0, −200, 26.32	$L_a = 100$	$l_a = 26.32$, $r_a = 0$
Shin 0	0.95	0, −200, 300	$L_b = 400$	$l_b = 200$; $r_b = 0$
Thigh 0	1.16	0, −200, 750	$L_c = 500$	$l_c = 250$; $r_c = 0$
Foot 1	6.72	0, 200, 26.32	$L_a = 100$	$l_a = 26.32$, $r_a = 0$
Shin 1	0.95	0, 200, 300	$L_b = 400$	$l_b = 200$; $r_b = 0$
Thigh 1	1.16	0, 200, 750	$L_c = 500$	$l_c = 250$; $r_c = 0$
Waist	0.94	0, 0, 1000	/	/

4.2. Validation

In following, the example shown in Figure 6 with the characteristics shown in Table 1 is implemented in MATLAB using equations presented in Section 3. The validation consists to give as input the total CoM positions of each configuration and to verify that the CoM position of each link of the platform, calculated with the formulations of the Section 3, has the same value used in input.

```
% MATLAB Example
thb = 10 *pi/180; %radiant angle of the link Lb
thc = 10 *pi/180; %radiant angle of the link Lc
ma = 6.72; %kg weight of the Foot
mb = 0.95; %kg weight of the link b
mc = 1.16 + 0.94; %kg weight of the link c
mw = mb + mc; %kg
m2 = mu + mw; %kg
ra = 0; %mm position of the CoM of the foot
la = 26.32; %mm position of the CoM of the foot
La = 100; %mm length of the foot in Zp direction
Lb = 400; %mm length of the shin in Zp direction
Lc = 500; %mm length of the thigh in Zp direction
lb = 200; %mm INPUT CONDITION FOR VALIDATION
lc = 250; %mm INPUT CONDITION FOR VALIDATION
Z2a = (1/m2) *(ma *la + mb * (La + lb) + mc * (La + Lb + lc)); %mm Total CoM position in the configuration a
X2a = 0; %mm Total CoM position in the configuration a
Z2b = (1/m2) *(ma *la + mb * (La + lb *cos (thb)) + mc *(La + (Lb + lc) *cos (thb))); %mm Total CoM position in the configuration b
X2b = (1/m2) *(mb *lb *sin (thb) + mc *(Lb + lc) *sin (thb)); %mm Total CoM position in the configuration b
Z2c = (1/m2) *(ma *la + mb *(La + lb) + mc * (La + Lb + lc *cos (thc))); %mm Total CoM position in the configuration c
X2c = (1/m2) * (mc *lc *sin (thc)); %mm Total CoM position in the configuration c
rwb = (X2b *m2 – ma *ra)/mw; %mm from Equations (30) and (31)
lwb = (Z2b *m2 – ma *la)/mw – La; %mm from Equations (30) and (31)
rwc = (X2c *m2 – ma *ra)/mw; %mm from Equations (30) and (31)
lwc = (Z2c *m2 – ma *la)/mw – La; %mm from Equations (30) and (31)
B = mw * (lwb + La) – mb *La – mc *Lb *cos (thb) – mc *La; %change of variables
C = mw * (lwc + La) – mb *La – mc * (La + Lb); %change of variables
det_A = (mb *mc *cos (thb) * (cos (thc) – 1)); %from Equation (38)
lb_validation = (mc * (B *cos (thc) – C *cos (thb)))/det_A; %mm from Equation (41)
lc_validation = (mb * (C *cos (thb) – B))/det_A; %mm from Equation (41)
```

5. Conclusions

In this paper, an optimized formulation to determine the center of mass position of each link of a multibody biped robot is presented. The formulation is merged with a procedure that can be applied to each types of robot with two force-torque sensors in contact between the robot and the ground and joint sensors. An advantage of the formulation presented in this paper, as underlined in the paper, is that if the total CoM position of the platform is known a priori, the first formulation proposed in [2,3] (and shown in Section 2 of this paper) can be bypassed and the CoM positions of each link of the platform can be calculated analytically without using force-torque sensors and motors torque. In this case, only the formulation of the Section 3 can be used. The validation confirms the functioning of the proposed formulation.

Acknowledgments: The authors would like to express their sincere gratitude to the Humanot Team.

Author Contributions: This research is completely performed by Giovanni Gerardo Muscolo, who also wrote the paper. Darwin Caldwell and Ferdinando Cannella gave support to the first author.

Conflicts of Interest: The authors declare no conflict of interest.

Nomenclature

$G\text{-}XYZ$	global Cartesian system
$P\text{-}XpYpZp$	local Cartesian system
$\vec{a}_2 = [a_{X2},\ a_{Y2},\ a_{Z2}]^T$	acceleration vector of the total CoM
m_2	total mass of the robot without feet
$\vec{p}_0 = [-n,\ -l,\ e]^T$	position vector of Point 0
$\vec{p}_1 = [o,\ q,\ f]^T$	position vector of Point 1
$\vec{p}_2 = [X_2,\ Y_2,\ Z_2]^T$	position vector of Point 2
$\vec{M}_0 = [M_{X0},\ M_{Y0},\ M_{Z0}]^T$	torque vector of Point 0
$\vec{M}_1 = [M_{X1},\ M_{Y1},\ M_{Z1}]^T$	torque vector of Point 1
$\vec{M}_2 = [M_{X2},\ M_{Y2},\ M_{Z2}]^T$	torque vector of Point 2
$\vec{F}_0 = [F_{X0},\ F_{Y0},\ F_{Z0}]^T$	force vector of Point 0
$\vec{F}_1 = [F_{X1},\ F_{Y1},\ F_{Z1}]^T$	force vector of Point 1
$\vec{F}_2 = [F_{X2},\ F_{Y2},\ F_{Z2}]^T$	force vector of Point 2
$\vec{M}_P = [M_{XP},\ M_{YP},\ M_{ZP}]^T$	torque vector of the resultant moment calculated with respect to the local Cartesian system $P\text{-}XpYpZp$
$\sum M_{j\_roll}, \sum M_{j\_pitch}, \sum M_{j\_yaw}$	torques of all roll, pitch, and yaw motors of the robot
X_{2A}, Y_{2A}, Z_{2A}	position of the centre of mass of the platform in the configuration A
X_{2B}, Y_{2B}, Z_{2B}	position of the centre of mass of the platform in the configuration B
m_u, r_u, q_u, l_u	mass and position of the centre of mass of the robot ankle link from the floor to the ankle joint
U	length of the ankle
m_w, r_w, q_w, l_w	mass and the position of the centre of mass of the remaining links of the platform
θ_t	angle used in the configuration B
K	constant parameter set for each motor
I	current necessary for the motor function
$\alpha_i, \beta_i, \gamma_i, \delta_i$	four novel coefficients for the configuration i
$\alpha_j, \beta_j, \gamma_j, \delta_j$	four novel coefficients for the configuration j
n	degrees of freedom of the platform
k	number of the configurations to calculate positions of the CoM for each link of the system
i, j	used configurations to calculate the CoM of each link
$\vec{t}_i = \begin{bmatrix} r_{wi} & q_{wi} & l_{wi}+U \end{bmatrix}^T$	vector position of m_w
r_{wi}, q_{wi}, l_{wi}	components of the vector position $\vec{t}_i = \begin{bmatrix} r_{wi} & q_{wi} & l_{wi}+U \end{bmatrix}^T$ respectively in X_P, Y_P and Z_P directions
χ_i and ε_i	angles of the vector position $\vec{t}_i = \begin{bmatrix} r_{wi} & q_{wi} & l_{wi}+U \end{bmatrix}^T$ with the the local reference system $P\text{-}X_PY_PZ_P$
$\vec{t}_b$ and $\vec{t}_c$	position vectors of the mass m_w respect to the local reference system $P\text{-}X_PY_PZ_P$ and respectively of the configurations b and c
$\vec{s}_b$ and $\vec{s}_c$	position vectors of the masses m_b and m_c respect to the local reference system of each link

References

1. Sugano, S.; Kato, I. WABOT-2: Autonomous robot with dexterous finger-arm-Finger-arm coordination control in keyboard performance. In Proceedings of the IEEE International Conference on Robotics and Automation (ICRA 1987), Raleigh, NC, USA, 31 March–3 April 1987; pp. 90–97.
2. Muscolo, G.G.; Recchiuto, C.T.; Hashimoto, K.; Laschi, C.; Dario, P.; Takanishi, A. A Method for the calculation of the effective Center of Mass of Humanoid robots. In Proceedings of the 11th IEEE-RAS International Conference on Humanoid Robots (Humanoids 2011), Bled, Slovenia, 26–28 October 2011.
3. Muscolo, G.G.; Recchiuto, C.T.; Molfino, R. Dynamic balance optimization in biped robots: Physical modeling, implementation and tests using an innovative formula. *Robotica* **2015**, *33*, 2083–2099. [CrossRef]
4. Muscolo, G.G.; Hashimoto, K.; Takanishi, A.; Dario, P. A comparison between two force-position controllers with gravity compensation simulated on a humanoid arm. *J. Robot.* **2013**, *2013*, 4. [CrossRef] [PubMed]
5. Otani, T.; Hashimoto, K.; Miyamae, S.; Ueta, H.; Sakaguchi, M.; Kawakami, Y.; Lim, H.O.; Takanishi, A. Angular Momentum Compensation in Yaw Direction using Upper Body based on Human Running. In Proceedings of the IEEE International Conference on Robotics and Automation (ICRA 2017), Singapore, 29 May–3 June 2017.
6. Boston Dynamics 2013: Cheetah—Fastest Legged Robot. Available online: http://bostondynamics.com/robot-cheetah.html (accessed on 15 April 2017).
7. Raibert, M. Dynamic legged robots for rough terrain. In Proceedings of the 10th IEEE-RAS International Conference on Humanoid Robots (Humanoids 2010), Nashville, TN, USA, 13 January 2011; p. 1.
8. Case, S. DARPA Unveils Atlas DRC Robot. July 2013. Available online: http://spectrum.ieee.org/automaton/robotics/humanoids/darpa-unveilsatlas-drc-robot (accessed on 28 March 2017).
9. Boston Dynamics 2013: Atlas—The Agile Anthropomorphic Robot (2013). Available online: http://www.bostondynamics.com/robot_Atlas.html (accessed on 1 April 2017).
10. Boston Dynamics 2017: Introducing Handle. Available online: https://www.youtube.com/watch?v=-7xvqQeoA8c (accessed on 8 May 2017).
11. Kim, J.-H.; Kim, J.-Y.; Oh, J.-H. Adjustment of Home Posture of Biped Humanoid Robot Using an Inertial Sensor and Force Torque Sensors. In Proceedings of the 2007 IEEE/RSJ International Conference on Intelligent Robots and Systems, San Diego, CA, USA, 29 October–2 November 2007.
12. Nunez, V.; Nadjar-Gauthier, N.; Yokoi, K.; Blazevic, P.; Stasse, O. Inertial Forces Posture Control for Humanoid Robots Locomotion. In *Humanoid Robots: Human-like Machines*; Hackel, M., Ed.; Itech: Vienna, Austria, 2007; p. 642.
13. Kwon, S.J.; Oh, Y. Estimation of the Center of Mass of Humanoid Robot. In Proceedings of the 2007 International Conference on Control, Automation and Systems, COEX, Seoul, Korea, 17–20 October 2007.
14. Ayusawa, K.; Venture, G.; Nakamura, Y. Identification of Humanoid Robots Dynamics Using Floating-base Motion Dynamics. In Proceedings of the 2008 IEEE/RSJ International Conference on Intelligent Robots and Systems, Acropolis Convention Center, Nice, France, 22–26 September 2008.
15. Sujan, V.A.; Dubowsky, S. An Optimal Information Method for Mobile Manipulator Dynamic Parameter Identification. *IEEE/ASME Trans. Mechatron.* **2003**, *2*, 215–225. [CrossRef]
16. Liu, G.; Iagnemma, K.; Dubowsky, S.; Morel, G. A Base Force/Torque Sensor Approach to Robot Manipulator Inertial Parameter Estimation. In Proceedings of the 1998 IEEE International Conference on Robotics & Automation, Leuven, Belgium, 20–22 May 1998.
17. Khalil, W.; Gautier, M.; Lemoine, P. Identification of the payload inertial parameters of industrial manipulators. In Proceedings of the 2007 IEEE International Conference on Robotics and Automation, Roma, Italy, 10–14 April 2007.
18. Swevers, J.; Ganseman, C.; Tukel, D.B.; De Schutter, J.; Van Brussel, H. Optimal Robot Excitation and Identification. *IEEE Trans. Robot. Autom.* **1997**, *13*, 730–740. [CrossRef]
19. Muscolo, G.G.; Recchiuto, C.T. Flexible Structure and Wheeled Feet to Simplify Biped Locomotion of Humanoid Robots. *Int. J. Hum. Robot.* **2017**, *14*. [CrossRef]

20. Muscolo, G.G.; Caldwell, D.; Cannella, F. Multibody Dynamics of a Flexible Legged Robot with Wheeled Feet. In Proceedings of the ECCOMAS Thematic Conference on Multibody Dynamics, Prague, Czech Republic, 19–22 June 2017.
21. Muscolo, G.G.; Caldwell, D.; Cannella, F. Biomechanics of Human Locomotion with Constraints to Design Flexible-Wheeled Biped Robots. In Proceedings of the AIM 2017IEEE International Conference on Advanced Intelligent Mechatronics, Munich, Germany, 3–7 July 2017.

Article

Human-Like Walking with Heel Off and Toe Support for Biped Robot

Yixiang Liu [1,2,3], Xizhe Zang [1,*], Shuai Heng [1], Zhenkun Lin [1] and Jie Zhao [1]

1. State Key Laboratory of Robotics and System, Harbin Institute of Technology, Harbin 150080, China; liuyixiang@163.com (Y.L.); heng13514479054@163.com (S.H.); 15B308008@hit.edu.cn (Z.L.); jzhao@hit.edu.cn (J.Z.)
2. Legs + Walking Lab, Shirley Ryan AbilityLab (Formerly the Rehabilitation Institute of Chicago), Chicago, IL 60611, USA
3. Department of Physical Medicine and Rehabilitation, Northwestern University, Chicago, IL 60611, USA

* Correspondence: zangxizhe@hit.edu.cn; Tel.: +86-451-8641-3382

Academic Editors: Toshio Fukuda, Fei Chen and Qing Shi
Received: 31 March 2017; Accepted: 5 May 2017; Published: 18 May 2017

Abstract: The under-actuated foot rotation that the heel of the stance leg lifts off the ground and the body rotates around the stance toe is an important feature in human walking. However, it is absent in the realized walking gait for the majority of biped robots because of the difficulty and complexity in the control it brings about. In this paper, a hybrid control approach aiming to integrate the main characteristics of human walking into a simulated seven-link biped robot is presented and then verified with simulations. The bipedal robotic gait includes a fully actuated single support phase with the stance heel supporting the body, an under-actuated single support phase, with the stance toe supporting the body, and an instantaneous double support phase when the two legs exchange their roles. The walking controller combines virtual force control and foot placement control, which are applied to the stance leg and the swing leg, respectively. The virtual force control assumes that there is a virtual force which can generate the desired torso motion on the center of mass of the torso link, and then the virtual force is applied through the real torques on each actuated joint of the stance leg to create the same effect that the virtual force would have created. The foot placement control uses a path tracking controller to follow the predefined trajectory of the swing foot when walking forward. The trajectories of the torso and the swing foot are generated based on the cart-cable model. Co-simulations in Adams and MATLAB show that the desired gait is achieved with a biped robot under the action of the proposed method.

Keywords: biped robot; human-like walking; foot rotation; virtual force control; foot placement control

1. Introduction

Humans are the most important inspiration for research on biped robots because of their resemblance. The unique neural and morphological mechanisms enable humans to show very versatile and efficient locomotion. Consequently, realizing human-like walking on biped robots has been an attractive research field for decades. Up to now, various advanced biped robots have been developed, from childlike robots NAO [1,2] and iCub [3,4] to human-sized robots Asimo [5,6], HRP [7–9], and Atlas [10,11]. For these biped robots, the most popular control approach is based on the Zero Moment Point (ZMP) theory, i.e. the contact point of the foot with the ground where the total of horizontal inertia and gravity forces equals zero is strictly within the support polygon [12,13]. Following the ZMP criterion ensures that these biped robots walk stably and robustly on a variety of terrains. However, in order to increase the stability of walking, the stance foot of biped robots is always flat on the ground, making the gait unnatural and energy-inefficient.

Actually, closer investigation on human walking reveals that, by the end of the single support phase, the heel of the stance foot rises from the ground and rotates around the toe, with the stance toe touching the ground and supporting the body [14,15]. Even from the gait of persons with lower limb disabilities wearing prostheses can foot rotation be observed [16]. The so-called "heel off and toe support" yields a higher foot clearance, a higher walking speed, and less energy consumption [17–19]. Taking the human as a mechanical locomotion system, it will be under-actuated when foot rotation occurs because there is no actuation between the stance toe and the ground [20]. Although dealing with the changes in dynamic characteristics is seemingly effortless for humans, it is much more difficult for biped robots. Therefore, it is of great important to study human-like walking of biped robots with under-actuated foot rotation.

In order to attain this goal, some researchers have adopted the ZMP-based method [21–23]. The desired walking pattern that includes foot rotation was generated by solving an inverse kinematics problem with a predefined ZMP and foot trajectories and achieved under servo controllers such as preview control. Chevallereau et al. [24] proposed a path-following control strategy to handle more human-like walking. The control strategy was defined in such a way that only the kinematic evolution of the state of the robot was controlled, but not its temporal evolution, allowing simultaneous control of the ZMP and the joint positions. Since the ZMP could be directly regulated, stable bipedal walking that includes prescribed foot rotation was realized. Sinnet and Zhao et al. [25–27] used a human-inspired methodology to translate human locomotion to a biped robot. By analyzing human locomotion data, the human-inspired outputs of the robot were defined based on the human outputs and the representation function. Then, a controller was designed for the biped robot, enabling it to mimic human kinematics behaviors by tracking the functions of the kinematics.

Although bipedal robotic walking with under-actuated foot rotation has been realized in the above research, the methods are usually very complex. Thus, a different and simpler approach is proposed in this paper. The walking control scheme adopts a hybrid controller that combines the ideas of virtual force control and foot placement control. The virtual force control, which is applied on the stance leg, assumes that there is a virtual force on the center of mass of the torso that can generate the desired torso motion including foot rotation, and the real torques are exerted on each actuated joint of the stance leg to create the same effect that the virtual force would have created. The foot placement control, which is applied on the swing leg, uses a path-tracking controller to follow the predefined curve of the swing foot to step forward. Using this hybrid controller, the sophisticated control task of dynamic walking is disintegrated into two simpler ones, making it much easier to achieve.

The rest of this paper is organized as follows. Section 2 presents the proposed hybrid control approach. As the most prevalent reference for biped robots, the human walking pattern is analyzed first. On this basis, a seven-link biped robot model and human-like robotic gait are introduced. Then the method of how to realize human-like walking on the biped robot is introduced in detail. To verify the method, co-simulations are carried out in Adams and MATLAB, and these simulation results are shown in Section 3. Finally, some conclusions and discussion are provided in Section 4.

2. Methods

2.1. Human Walking Pattern Analysis

In order to reproduce anthropomorphic walking on a biped robot, it is important to study and understand the walking pattern of humans. Human walking is the repetition of a basic movement, namely, the step. Generally, a normal step can be roughly divided into two successive phases according to the foot contact with the ground: the double support phase (or the stance phase), when both feet are on the ground, and the single support phase (or the swing phase), when only one foot is in contact with the ground [28]. Figure 1 shows a single gait cycle, from when one foot strikes the ground with its heel to when the other foot strikes the ground. In the initial portion of the single support phase, the sole of the stance foot is flat on the ground, and the center of pressure locates within the stance heel. For the rest of the single support phase, the heel of the stance foot lifts off the ground and rotates around the toe [29]. The center of the pressure moves forward from the stance heel to the stance toe before the heel of the swing foot contacts the ground. The foot rotation around the toe in the single support phase is named "heel off and toe support", which is a significant feature of human walking and has an important effect on walking performance.

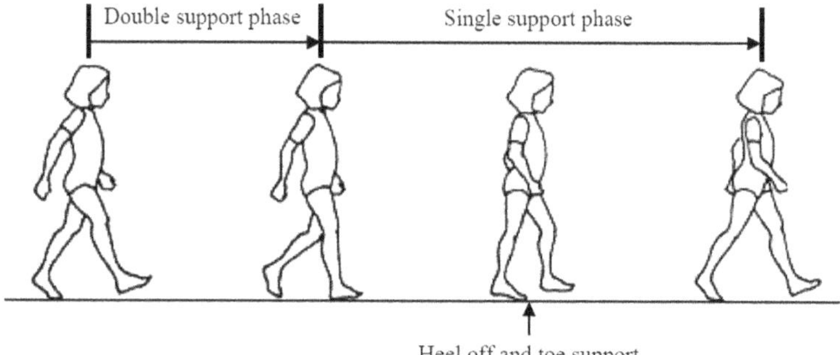

Figure 1. A single gait cycle of the human walking pattern. In the single support phase, the heel of the stance foot lifts off the ground and rotates around the toe, which is named "heel off and toe support".

2.2. The Biped Robot Model

The model considered in this paper is a planar seven-link biped robot of which the movement is constrained in the sagittal plane. The biped robot has two three-segmented legs connected by the right and left hip joints at the torso. Each leg is composed of a thigh, shank, and foot that are interconnected by the knee joint and the ankle joint. Each joint is regarded as an independently actuated ideal or theoretically frictionless revolute joint with only one degree of freedom (DOF). Therefore, the biped robot has a total of six internal DOFs, namely, one in each hip joint, one in each knee joint, and one in each ankle joint.

Figure 2 illustrates the biped robot model. The right and left legs are supposed to be identical and exchange their roles between steps. Thus, the properties of the two legs are pairwise equal. l_i and m_i represent the length and mass of link i, respectively. The distance between the local center of mass (CoM) and the lower end of link i is indicated by $\eta_i l_i$ where $0 < \eta_i < 1$. Having six actuated joints, the biped robot dynamics is driven by six joint driving torques, i.e., $\tau = [\tau_1, \tau_2, \tau_3, \tau_4, \tau_5, \tau_6]^T$. At any time during a step, the configuration of the biped robot can be defined by a set of generalized coordinates, namely, $q = [\theta_1, \theta_2, \theta_3, \theta_4, \theta_5, \theta_6, \theta_7]^T$. The base coordinate frame is attached to the base of the robot, i.e., the tiptoe of the stance foot. According to the kinematics theory of robot, θ_i ($i = 1, 2, 3, 4, 5, 6, 7$) indicates the amount of rotation around the z-axis needed to align the axis of link $i - 1$ with the axis of link i. Specifically, θ_1 represents the angle around the z-axis from the x-axis of the base frame to the axis of the stance foot of the biped robot, and θ_2 to θ_7 represent the angles between each two adjacent links from the stance foot to the swing foot. Positive rotation follows the right hand rule. In this case, the direction of counterclockwise rotation about the z-axis is assumed positive.

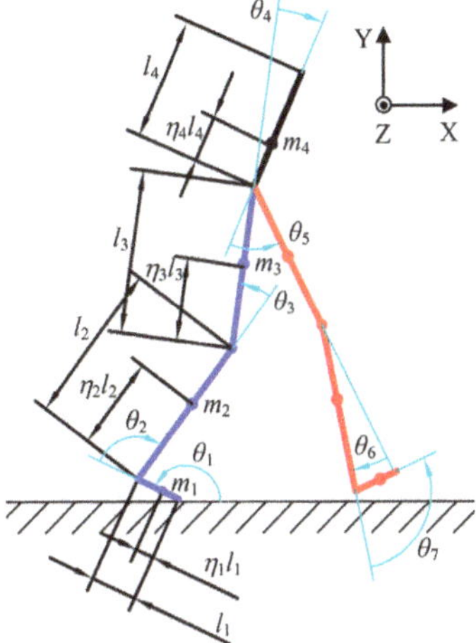

Figure 2. The seven-link biped robot model with associated geometric and inertial properties. The corresponding links on each leg are geometrically and inertially identical.

2.3. The Bipedal Robotic Gait

When the contact point between the foot and ground changes during walking, the biped robot translates into the next walking phase that has quite different dynamical characteristics, resulting in many difficulties to the control of the biped robot. This paper is focused on the realization of the so-called "heel off and toe support" on a biped robot. Therefore, the bipedal robotic gait is simplified from the human walking pattern to some extent. First, the double support phase is assumed to be instantaneous. Second, when the swing foot strikes the ground, the sole is parallel to the ground. The desired robotic walking motion can then be regarded as including three phases: a heel support phase, a toe support phase, and an instantaneous double support phase (or impact phase). Figure 3 shows the bipedal robotic walking pattern with state transition events. Each phase is introduced as follows.

(1) Heel support phase (HSP). This phase starts when the toe of the swing leg lifts off the ground. The whole foot of the stance leg is assumed to remain flat and motionless on the ground without slipping. At the same time, the swing leg swings in the forward direction. In this phase, the biped robot is fully actuated because it has the same numbers of degrees of freedom (DOF) and actuators.

(2) Toe support phase (TSP). This phase starts when the heel of the stance leg lifts off the ground. In this phase, the stance foot rotates around its toe, which is virtually pivoted to the ground, and the swing leg continues to swing forward. The biped robot is under-actuated because there is no actuation between the stance foot and the ground.

(3) Impact Phase (IP). Leg roles exchange takes place in this phase. After the swing foot strikes the ground, the swing leg becomes the new stance leg, while the original stance leg rises from the ground and becomes the swing leg.

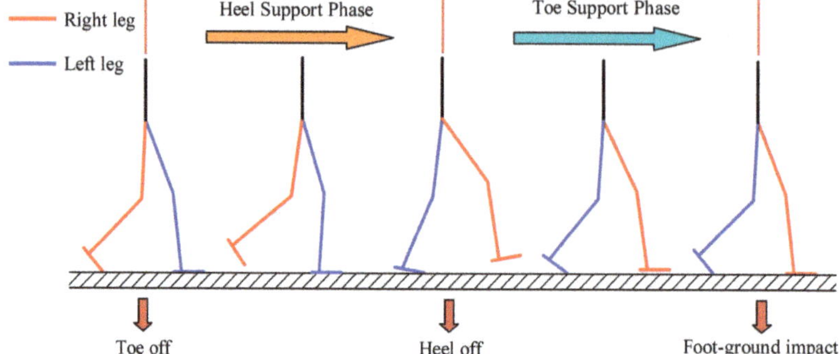

Figure 3. The bipedal robotic walking pattern, which consists of a heel support phase, a toe support phase, and an impact phase. The state transition events are toe off, heel off, and foot–ground impact, respectively.

2.4. The Hybrid Control Scheme

The angle between the stance foot and the ground is determined by the dynamics of the biped robot and cannot be controlled directly. Thus, the greatest difficulty in the control of biped robots is how to integrate the under-actuated toe support phase into the motion. Considering that the high dimensional walking task can be viewed as a collection of several decoupled tasks of lower dimensionality, the control of the biped robot is decoupled into two tasks, i.e., the control of the torso to track the desired motion, and the control of the swing leg to follow the predefined foot placement point. Therefore, a hybrid control scheme is proposed in this paper. The control scheme combines the ideas of virtual force control and foot placement control, which are applied on the stance leg and the swing leg, respectively. The working principle of the hybrid control scheme as illustrated in Figure 4 is detailed below.

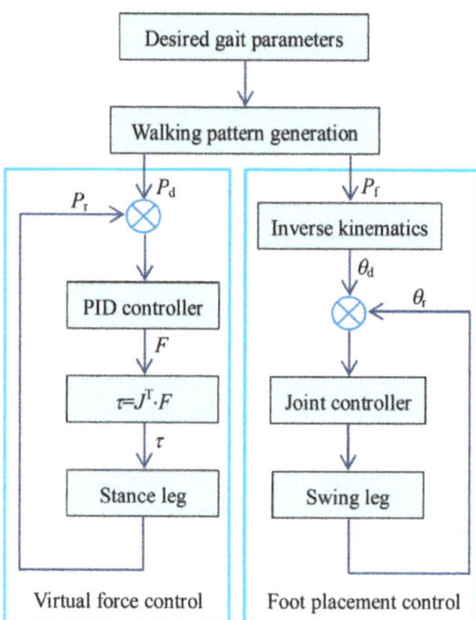

Figure 4. The hybrid walking control scheme. The virtual force control is applied on the stance leg, while the foot placement control is applied on the swing leg.

The main idea of virtual force control is quite straightforward. Taking the CoM of the torso as the end-effector, its motion in the sagittal plane is predefined by the walking pattern generator according to the given gait parameters. In fact, the position and posture of the torso are determined by the motion of the previous joints, i.e., from the toe to the hip of the stance leg. However, it is assumed that a virtual force is directly acted on the center of mass of the torso link, which leads to the desired torso motion. Based on the dynamics theory, the real torques on the previous joints can be derived, which create the same effect that the virtual force would have created had they existed [30].

Suppose that the origin of the global reference frame (X_0, Y_0, Z_0) is set at the tiptoe of the stance foot. The kinematical equations of the biped robot can be easily analyzed by a geometrical method. The position vector of the local CoM of the torso indicated by $P = [P_x, P_y, \theta_z]^{\mathrm{T}}$ can be described as

$$P = \begin{bmatrix} P_x \\ P_y \\ \theta_z \end{bmatrix} = \begin{bmatrix} l_1c_1 + l_2c_{1+2} + l_3c_{1+2+3} + \eta_4 l_4 c_{1+2+3+4} \\ l_1s_1 + l_2s_{1+2} + l_3s_{1+2+3} + \eta_4 l_4 s_{1+2+3+4} \\ \theta_1 + \theta_2 + \theta_3 + \theta_4 \end{bmatrix} \quad (1)$$

where P_x represents the position in the x-direction, P_y represents the position in the y-direction, θ_z represents the posture around the z-axis, and, for simplicity, $c_i = \cos\theta_i$, $s_i = \sin\theta_i$, $c_{i+j} = \cos(\theta_i + \theta_j)$, and $s_{i+j} = \sin(\theta_i + \theta_j)$.

The Jacobian matrix can be obtained from the partial differentiation of the above equation:

$$J = \begin{bmatrix} -l_1s_1 - l_2s_{1+2} - l_3s_{1+2+3} - \eta_4 l_4 s_{1+2+3+4} & -l_2s_{1+2} - l_3s_{1+2+3} - \eta_4 l_4 s_{1+2+3+4} & -l_3s_{1+2+3} - \eta_4 l_4 s_{1+2+3+4} & -\eta_4 l_4 s_{1+2+3+4} \\ l_1c_1 + l_2c_{1+2} + l_3c_{1+2+3} + \eta_4 l_4 c_{1+2+3+4} & l_2c_{1+2} + l_3c_{1+2+3} + \eta_4 l_4 c_{1+2+3+4} & l_3c_{1+2+3} + \eta_4 l_4 c_{1+2+3+4} & \eta_4 l_4 c_{1+2+3+4} \\ 1 & 1 & 1 & 1 \end{bmatrix}. \quad (2)$$

Use $F = [F_x, F_y, T_z]^{\mathrm{T}}$ to indicate the virtual force acting on the torso, with the three components representing the force in the x-direction, the force in the y-direction, and the torque around the z-axis respectively. Then, the real actuator torques $\tau = [\tau_0, \tau_1, \tau_2, \tau_3]^{\mathrm{T}}$ can be calculated using the Jacobian matrix, which maps the force on the end-effector to the joint torques:

$$\tau = J^{\mathrm{T}} \cdot F. \quad (3)$$

It should be noted that the torque τ_0 acting on the stance toe actually equals zero since the stance toe is unactuated. Therefore, the three components of the virtual force are subject to the following constraint:

$$\tau_1 = (-l_1s_1 - l_2s_{1+2} - l_3s_{1+2+3} - \eta_4 l_4 s_{1+2+3+4})F_x + (l_1c_1 + l_2c_{1+2} + l_3c_{1+2+3} + \eta_4 l_4 c_{1+2+3+4})F_y + T_z = 0 \quad (4)$$

which means that, once any two of these three components are determined, the third one can be solved from this equation.

In this paper, the virtual force and torque F_y and T_z are first specified by applying two PID controllers on the torso to track the desired position in the y-direction and the desired posture around the z-axis. F_y and T_z can be obtained from the summation of the output of the PID controller and gravity term, written as

$$\begin{aligned} F_y &= K_{P,y}\left(P_{y,d} - P_{y,r}\right) + K_{I,y}\int_0^t \left(P_{y,d} - P_{y,r}\right)dt + K_{D,y}\frac{d\left(P_{y,d} - P_{y,r}\right)}{dt} + m_4 g \\ T_z &= K_{P,z}(P_{z,d} - P_{z,r}) + K_{I,z}\int_0^t (P_{z,d} - P_{z,r})dt + K_{D,z}\frac{d\left(P_{z,d} - P_{z,r}\right)}{dt} \end{aligned} \quad (5)$$

where $P_{y,d}$ and $P_{z,d}$ are desired positions of the torso, and $P_{y,r}$ and $P_{z,r}$ are real positions of the torso.

Combining Equations (4) and (5), the virtual force F_x can be derived. Plugging F_x, F_y, and T_z back into Equation (3), the joint torques can be calculated.

The foot placement control aims to output the appropriate step location of the biped robot to obtain the desired walking gait [31]. For the swing leg, given the walking parameters such as step length, step height, and cycle time, the trajectory of the swing foot is generated. Once the trajectories of the torso and swing foot are both already known, the inverse kinematics can be resolved to calculate the joint angles of the swing leg. Then, position controllers are applied on the joints to track the desired trajectories.

3. Results

3.1. Walking Pattern Generation

The role of the walking pattern generator is to output the desired trajectories of the CoM of the torso and the swing foot. In this paper, the cart-cable model shown in Figure 5 is adopted, with the cart representing the CoM of the robot, and the foot of the table representing the stance foot of the robot [32,33]. It is assumed that the mass of the biped robot is lumped with the CoM of the robot, and the CoM is kept at a constant height. Based on the model, the relationship between the trajectories of the CoM and the reference ZMP can be described as [34]

$$\ddot{x}_{\mathrm{COM}} = \frac{g}{z_{\mathrm{COM}}}(x_{\mathrm{COM}} - x_{\mathrm{ZMP}}) \tag{6}$$

where x_{COM} is the horizontal position of the CoM, x_{ZMP} is the horizontal position of the ZMP, z_{COM} is the height of the CoM, and g is the gravity acceleration.

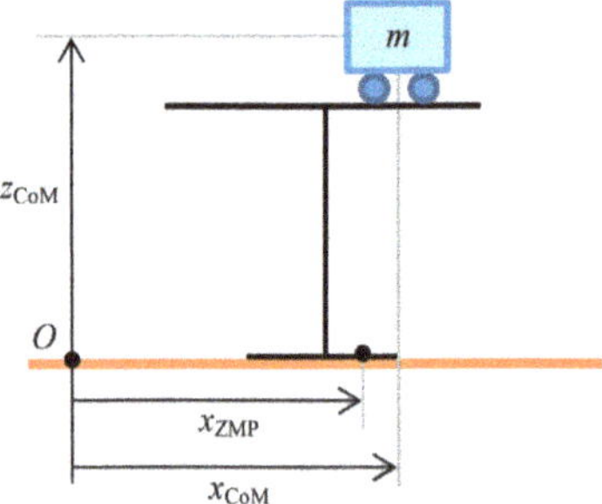

Figure 5. The cart-table model with the cart representing the center of mass (CoM) of the biped robot, and the foot of the table representing the stance foot of the robot.

The reference ZMP trajectory of the biped robot needs to be predefined according to the walking phases described previously. In the heel support phase (from time T_k to $T_k + T_{hs}$), the ZMP moves from the heel to the toe of the stance foot. In the toe support phase (from time $T_k + T_{hs}$ to T_{k+1}), the ZMP is kept constant at the tiptoe. In the impact phase, the ZMP transfers from the tiptoe of the stance foot to the heel of the swing foot. Figure 6 shows the generated reference ZMP trajectory in the horizontal plane. Using Equation (6), the corresponding CoM trajectory is deduced. The trajectory of the CoM of the torso link can then be obtained by adding an offset to the reference CoM trajectory.

The trajectory of the swing foot is planned using polynomial interpolation that satisfies certain constraints in a complete walking cycle. There are four constraints on the curve, including two positional constraints and two constraints on the gradient. Thus, a polynomial of degree four is adopted to describe the trajectory. The trajectory consists of two sections related to the heel support phase and the toe support phase. The coefficients are obtained respecting the continuity of position and velocity. Moreover, in order to preserve a smooth landing of the swing foot, the velocity before the impact moment is considered to be zero.

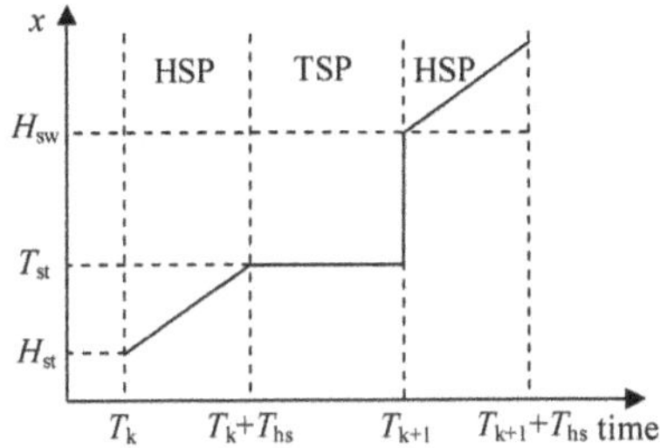

Figure 6. The reference Zero Moment Point (ZMP) trajectory of the biped robot. In the heel support phase, the ZMP moves from the heel to the toe of the stance foot. In the toe support phase, the ZMP is kept constant at the tiptoe. In the impact phase, the ZMP transfers from the tiptoe of the stance foot to the heel of the swing foot.

3.2. Simulation

To verify the effectiveness of the previously proposed control scheme, co-simulations are carried out in Adams and MATLAB. Adams is the most famous and widely used multibody dynamics software, and MATLAB has powerful functions in computing, programming, and control system designing. Co-simulations are able to utilize the advantages of both programs by combining the virtual mechanical system together with the complex control system. The virtual mechanical model of the biped robot is built in Adams, and the control scheme is programmed using the Simulink toolkit in MATLAB. With the Adams/Control plugin, information can be exchanged between Adams and MATLAB, in other words, outputting joint torques to the virtual model in Adams, and feeding joint motion information back to the control system in MATLAB.

The main physical parameters of the biped robot model are shown in Table 1. The lengths of the links from the torso to the foot are 0.4, 0.37, 0.36 and 0.21 m, respectively. Their corresponding masses are 3, 1.8, 1.5 and 0.3 kg respectively. The planned duration of the heel support phase is 0.4 s, and the toe support phase is 0.3 s. In the implementation of the hybrid control scheme, the main work is to tune the parameters of the controllers by trial and error. If the parameters are selected appropriately, the desired walking pattern can be realized. Figure 7 shows the screenshots of the realized bipedal robotic walking in one cycle. Figure 8 presents the joint angles of the biped robot during walking. These angles have the same meanings as defined in Figure 2. From these two figures, it is clear to see that from the start point to 0.4 s, the sole of the stance foot is flat on the ground when the biped robot steps forward. After 0.4 s, the angle between the ground and the stance foot sole indicated by θ_1 begins to decrease, which means the stance heel lifts off the ground and rotates around the tiptoe. Thus, both the heel support phase and the toe support phase are realized, demonstrating that the proposed hybrid walking controller is effective.

Table 1. The main physical parameters of the biped robot model.

Link	Length (m)	Mass (kg)
Torso	0.4	3
Thigh	0.37	1.8
Shank	0.36	1.5
Foot	0.21	0.3

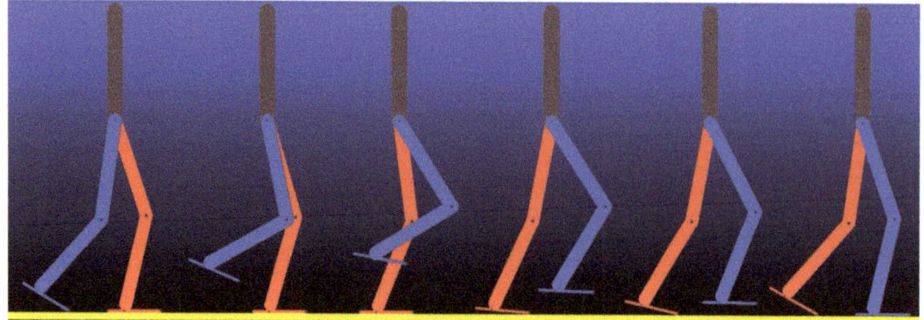

Figure 7. Snapshots of the simulated bipedal robotic walking gait in one cycle.

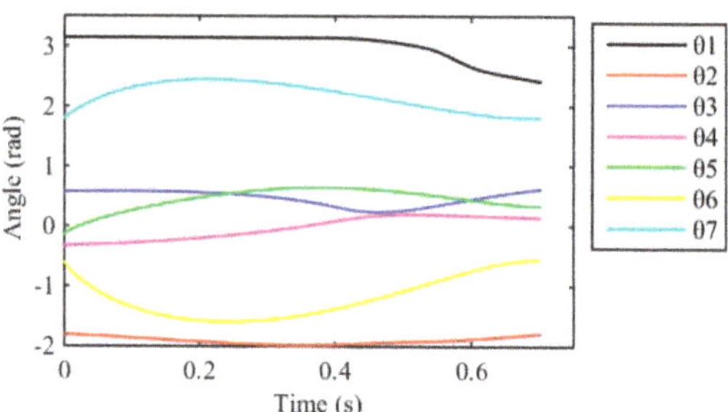

Figure 8. Joint angles of the biped robot during simulated walking. These angles have the same definitions as those in Figure 2.

4. Conclusions and Discussion

This paper introduces a solution to the problem of the realization of human-like bipedal robotic walking with under-actuated foot rotation. The studied biped robot model is a planar seven-like biped robot with feet. The desired gait includes three successive phases, i.e., a fully actuated phase where the heel supports the body, an under-actuated phase where the toe supports the body, and an instantaneous double support phase where the foot–ground impact takes place and the two legs exchange their roles. To achieve this gait, a hybrid walking controller is proposed by combining virtual force control and foot placement control, which are applied on the stance leg and the swing leg, respectively. The controller decouples the high-dimensional dynamic walking into two simpler tasks of lower dimensionality. Therefore, compared with the current methods, fewer control efforts and computation are required to realize human-like walking on a biped robot. The validity of the proposed approach is verified by co-simulations in Adams and MATLAB.

However, there are still limitations in this paper. The double support phase, different from the human walking pattern, is assumed to be instantaneous. In the future, we will add the finite-time double support phase to the bipedal robotic walking gait. We will also study the implementation of the hybrid control approach on a physical biped robot prototype.

Acknowledgments: The work reported in this paper was supported by the National Natural Science Foundation of China (grant no. 51675116). The first author was funded by the China Scholarship Council (grant no. 201606120094).

Author Contributions: Yixiang Liu and Xizhe Zang co-organized the work and wrote the manuscript draft. Shuai Heng and Zhenkun Lin performed the simulations. Jie Zhao supervised the research and commented on the manuscript draft.

Conflicts of Interest: The authors declare no conflict of interest.

References

1. Nugroho, S.; Prihatmanto, A.; Rohman, A. Design and implementation of kinematics model and trajectory planning for NAO humanoid robot in a tic-tac-toe board game. In Proceedings of the 2014 IEEE 4th International Conference on System Engineering and Technology, Bandung, Indonesia, 24–25 November 2014; pp. 1–7.
2. Ghassemi, P.; Masouleh, M.; Kalhor, A. Push recovery for NAO humanoid robot. In Proceeding of the 2nd RSI/ISM International Conference on Robotics and Mechatronics, Tehran, Iran, 15–17 October 2014; pp. 35–40.
3. Tsagarakis, N.; Metta, G.; Sandini, G.; Vernon, D.; Beira, R.; Becchi, F.; Righetti, L.; Santos-Victor, J.; Ijspeert, A.; Carrozza, M.; et al. iCub: The design and realization of an open humanoid platform for cognitive and neuroscience research. *Adv. Robot.* **2007**, *21*, 1151–1175. [CrossRef]
4. Metta, G.; Natale, L.; Nori, F.; Sandini, G.; Vernon, D.; Fadiga, L.; Hofsten, C.; Rosander, K.; Lopes, M.; Santos-Victor, J.; et al. The iCub humanoid robot: An open-systems platform for research in cognitive development. *Neural Netw.* **2010**, *23*, 1125–1134. [CrossRef] [PubMed]
5. Sakagami, Y.; Watanabe, R.; Aoyama, C.; Matsunaga, S.; Higaki, N.; Fujimura, K. The intelligent asimo: System overview and integration. In Proceedings of the 2002 IEEE/RSJ International Conference on Intelligent Robots and Systems, Lausanne, Switzerland, 30 September–4 October 2002; pp. 2478–2483.
6. Kanazawa, M.; Nozawa, S.; Kakiuchi, Y.; Kanemoto, Y.; Kuroda, M.; Okada, K.; Inaba, M.; Yoshiike, T. Robust vertical ladder climbing and transitioning between ladder and catwalk for humanoid robots. In Proceedings of the 2015 IEEE/RSJ International Conference on Intelligent Robots and Systems, Hamburg, Germany, 28 September–2 October 2015; pp. 2202–2209.
7. Kaneko, K.; Harada, K.; Kanehiro, F.; Miyamori, G.; Akachi, K. Humanoid robot HRP-3. In Proceedings of the 2008 IEEE/RSJ International Conference on Intelligent Robots and Systems, Nice, France, 22–26 September 2008; pp. 2471–2478.
8. Kaneko, K.; Kanehiro, F.; Morisawa, M.; Tsuji, T.; Miura, K.; Nakaoka, S.; Kajita, S.; Yokoi, K. Hardware improvement of cybernetic human HRP-4C for entertainment use. In Proceedings of the 2011 IEEE/RSJ International Conference on Intelligent Robots and Systems, San Francisco, CA, USA, 25–30 September 2011; pp. 4392–4399.
9. Kaneko, K.; Kanehiro, F.; Morisawa, M.; Akachi, K.; Miyamori, G.; Hayashi, A.; Kanehira, N. Humanoid robot HRP-4—Humanoid robotics platform with lightweight and slim body. In Proceedings of the 2011 IEEE/RSJ International Conference on Intelligent Robots and Systems, San Francisco, CA, USA, 25–30 September 2011; pp. 4400–4407.
10. Feng, S.; Xinjilefu, X; Atkeson, C.; Kim, J. Optimization based controller design and implementation for the atlas robot in the DARPA robotics challenge finals. In Proceedings of the 2015 IEEE-RAS International Conference on Humanoid Robots, Seoul, Korea, 3–5 November 2015; pp. 1028–1035.
11. Chong, Z.; Hung, R.; Lee, K.; Wang, W.; Ng, T.; Newman, W. Autonomous wall cutting with an Atlas humanoid robot. In Proceedings of the 2015 IEEE International Conference on Technologies for Practical Robot Applications, Woburn, MA, USA, 11–12 May 2015; pp. 1–6.
12. Vukobratovic, M. Zero-moment point—Thirty-five years of its life. *Int. J. Hum. Robot.* **2001**, *1*, 157–173. [CrossRef]
13. Vukobratović, M.; Borovac, B.; Potkonjak, V. ZMP: A review of some basic misunderstandings. *Int. J. Hum. Robot.* **2006**, *3*, 153–175. [CrossRef]
14. Adamczyk, P.G.; Collins, S.H.; Kuo, A.D. The advantages of a rolling foot in human walking. *J. Exp. Biol.* **2006**, *209*, 3953–3963. [CrossRef] [PubMed]
15. Choi, J.H.; Grizzle, J.W. Planar bipedal walking with foot rotation. In Proceedings of the 2005 American Control Conference, Portland, OR, USA, 8–10 June 2005; pp. 4909–4916.
16. Eilenberg, M.F.; Geyer, H.; Herr, H. Control of a powered ankle-foot prosthesis based on a neuromuscular model. *IEEE Trans. Neural Syst. Rehabil. Eng.* **2010**, *18*, 164–173. [CrossRef] [PubMed]
17. Tlalolini, D.; Chevallereau, C.; Aoustin, Y. Comparison of different gaits with rotation of the feet for a planar biped. *Robot. Auton. Syst.* **2009**, *57*, 371–383. [CrossRef]
18. Kouchaki, E.; Sadigh, M.J. Effect of toe-joint bending on biped gait performance. In Proceedings of the 2010 IEEE International Conference on Robotics and Biomimetics, Tianjin, China, 14–18 December 2010; pp. 697–702.

19. Mahdokht, E.; Khadiv, M.; Moosavian, S.A.A. Effects of toe-off and heel-off motions on gait performance of biped robots. In Proceedings of the 3rd RSI International Conference on Robotics and Mechatronics, Tehran, Iran, 7–9 October 2015; pp. 007–012.
20. Tlalolini, D.; Chevallereau, C.; Aoustin, Y. Human-like walking: Optimal motion of a bipedal robot with toe-rotation motion. *IEEE/ASME Trans. Mechatron.* **2011**, *16*, 310–320. [CrossRef]
21. Sellaouti, R.; Stasse, O.; Kajita, S.; Yokoi, K.; Kheddar, A. Faster and smoother walking of humanoid HRP-2 with passive toe joints. In Proceedings of the 2006 IEEE/RSJ International Conference on Intelligent Robots and Systems, Beijing, China, 9–15 October 2006; pp. 4909–4914.
22. Kajita, S.; Kaneko, K.; Morisawa, M.; Nakaoka, S.; Hirukawa, H. ZMP-based biped running enhanced by toe springs. In Proceedings of the 2009 IEEE International Conference on Robotics and Automation, Roma, Italy, 10–14 April 2007; pp. 3963–3969.
23. Miura, K.; Morisawa, M.; Kanehiro, F.; Kajita, S.; Kaneko, K.; Yokoi, K. Human-like walking with toe supporting for humanoids. In Proceedings of the 2011 IEEE/RSJ International Conference on Intelligent Robots and Systems, San Francisco, CA, USA, 25–30 September 2011; pp. 4428–4435.
24. Chevallereau, C.; Djoudi, D.; Grizzle, J.W. Stable bipedal walking with foot rotation through direct regulation of the zero moment point. *IEEE Trans. Robot.* **2008**, *24*, 390–401. [CrossRef]
25. Sinnet, R.W.; Powell, M.J.; Shah, R.P.; Ames, A.D. A Human-inspired hybrid control approach to bipedal robotic walking. In Proceedings of the 18th World Congress/The International Federation of Automatic Control, Milano, Italy, 28 August–2 September 2011; pp. 6904–6910.
26. Zhao, H.; Ma, W.; Zeagler, M.B. Human-inspired multi-contact locomotion with AMBER2. In Proceedings of the 2014 ACM/IEEE International Conference on Cyber-Physical Systems, Berlin, Germany, 14–17 April 2014; pp. 199–210.
27. Zhao, H.; Hereid, A.; Ma, W.; Ames, A.D. Multi-contact bipedal robotic locomotion. *Robotica* **2017**, *35*, 1072–1106. [CrossRef]
28. Torricelli, D.; Gonzalez, J.; Weckx, M.; Jiménez-Fabián, R.; Vanderborght, B.; Sartori, M.; Dosen, S.; Farina, D.; Lefeber, D.; Pons, J.L. Human-like compliant locomotion: State of the art of robotic implementations. *Bioinspir. Biomim.* **2016**, *11*, 051002. [CrossRef] [PubMed]
29. Ames, A.D.; Vasudevan, R.; Bajcsy, R. Human-data based cost of bipedal robotic walking. In Proceedings of the 14th International Conference on Hybrid Systems: Computation and Control, Vienna, Austria, 12–14 April 2011; pp. 153–162.
30. Pratt, J.; Dilworth, P.; Pratt, G. Virtual model control of a bipedal walking robot. In Proceedings of the 1997 IEEE International Conference on Robotics and Automation, Albuquerque, NM, USA, 25 April 1997; pp. 193–198.
31. Li, Z.; Vanderborght, B.; Tsagarakis, N.G.; Caldwell, D.G. Human-like Walking with straightened knees, toe-off and heel-strike for the humanoid robot iCub. In Proceedings of the 2010 UKACC International Conference on Control, Coventry, UK, 7–10 September 2010.
32. Chen, X.; Zhou, Y.; Huang, Q.; Yu, Z.; Ma, G.; Meng, L.; Fu, C. Bipedal walking with toe-off, heel-strike and compliance with external disturbances. In Proceedings of the 2014 14th IEEE-RAS International Conference on Humanoid Robots, Madrid, Spain, 18–20 November 2014; pp. 506–511.
33. Kajita, S.; Kanehiro, F.; Kaneko, K.; Fujiwara, K.; Harada, K.; Yokoi, K.; Hirukawa, H. Biped walking pattern generation by using preview control of zero-moment point. In Proceedings of the 2003 IEEE International Conference on Robotics and Automation, Taipei, Taiwan, 14–19 September 2003; pp. 1620–1626.
34. Kajita, S.; Kanehiro, F.; Kaneko, K.; Yokoi, K.; Hirukawa, H. The 3D linear inverted pendulum model: A simple modeling for a biped walking pattern generation. In Proceedings of the IEEE/RSJ International Conference on Intelligent Robots and Systems, Maui, HI, USA, 29 October–3 November 2001; pp. 239–246.

Article

Upper-Body Control and Mechanism of Humanoids to Compensate for Angular Momentum in the Yaw Direction Based on Human Running

Takuya Otani [1,†,*], Kenji Hashimoto [2,3,†], Shunsuke Miyamae [4,†], Hiroki Ueta [4,†], Akira Natsuhara [4,†], Masanori Sakaguchi [5,†], Yasuo Kawakami [6,†], Hum-Ok Lim [3,7,†] and Atsuo Takanishi [1,3,†]

1. Department of Modern Mechanical Engineering, Waseda University, 2-2 Wakamatsu-cho, Shinjuku-ku, Tokyo 162-8480, Japan; takanisi@waseda.jp
2. Waseda Institute for Advanced Study, Waseda University, No. 41-304, 17 Kikui-cho, Shinjuku-ku, Tokyo 162-0044, Japan; hashimoto@aoni.waseda.jp
3. Humanoid Robotics Institute (HRI), Waseda University, 2-2 Wakamatsu-cho, Shinjuku-ku, Tokyo 162-8480, Japan; holim@kanagawa-u.ac.jp
4. Graduate School of Science and Engineering, Waseda University, No. 41-304, 17 Kikui-cho, Shinjuku-ku, Tokyo 162-0044, Japan; galspa1123@ruri.waseda.jp (S.M.); hirowase2028@ruri.waseda.jp (H.U.); natsu-basuke39@suou.waseda.jp (A.N.)
5. ASICS Corporation, Institute of Sport Science, 6-2-1 Takatsukadai, Nishi-ku, Kobe, 651-2271, Japan; masanori.sakaguchi@asics.com
6. Faculty of Sport Sciences, Waseda University, 2-579-15 Mikajima, Tokorozawa-shi, Tokyo 359-1192, Japan; ykawa@waseda.jp
7. Faculty of Engineering, Kanagawa University, 3-27-1 Rokkakubashi, Kanagawa-ku, Yokohama 221-8686, Japan

* Correspondence: t-otani@takanishi.mech.waseda.ac.jp; Tel.: +81-3-3203-4394

† These authors contributed equally to this work.

Received: 23 October 2017; Accepted: 27 December 2017; Published: 3 January 2018

Abstract: Many extant studies proposed various stabilizing control methods for humanoids during the stance phase while hopping and running. Although these methods contribute to stability during hopping and running, humanoid robots do not swing their legs rapidly during the flight phase to prevent rotation in the yaw direction. Humans utilize their torsos and arms when running to compensate for the angular momentum in the yaw direction generated by leg movement during the flight phase. In this study, we developed an angular momentum control method based on human motion for a humanoid upper body. The method involves calculation of the angular momentum generated by the movement of the humanoid legs and calculation of the torso and arm motions required to compensate for the angular momentum of the legs in the yaw direction. We also developed a humanoid upper-body mechanism having human link length and mass properties, using carbon-fiber-reinforced plastic and a symmetric structure for generating large angular momentum. The humanoid robot developed in this study could generate almost the same angular momentum as that of a human. Furthermore, when suspended in midair, the humanoid robot achieved angular momentum compensation in the yaw direction.

Keywords: humanoid; angular momentum; flight phase; upper body

1. Introduction

Humanoid robots are expected to be useful in various environments where people live. The reason for this is that humanoid robots, which are close to humans in behavior and functionality, are easy to adapt to the living environment designed for human beings. In addition, while other robots have only a few controllable joints, humanoid robots have more than 20 joints and can negotiate various scenarios. Locomotion is necessary to work in various situations, and so far, many studies have been performed on stable walking motion generation techniques. In recent years, running, which is a movement mode including the flight phase with a faster moving speed compared to walking, has also attracted attention, and research is being advanced on running motion generation methods to improve the movement ability of a humanoid robot. Raibert et al. developed a running robot with a single linear leg [1]. The bipedal robot ATRIAS has a four-bar leg mechanism that includes a series of elastic springs [2,3]. Hyon et al. developed a biologically inspired robot based on a dog-leg model [4]. However, these robots do not have a human-like structure. Some studies have shown that bipedal humanoid robots can run [5–9]. For example, the Advanced Step in Innovative MObility (ASIMO) humanoid robot, which was designed and developed by Honda, can run at a speed of 2.5 m/s [10]. Toyota's bipedal humanoid robot can run using a zero-moment-point (ZMP)-based running control system [11]. The athlete robot developed by Niiyama et al. has a human-like musculoskeletal system built to execute dynamic motions, such as running [12]. The bipedal robot MABEL, developed by researchers at the University of Michigan, has leg elasticity that originates from a leaf spring. It is the fastest-running of all currently available bipedal robots, having achieved a speed of 3 m/s with axial constraints on the y-axis [13].

However, present humanoid robots cannot run as fast and stably as human beings, who can run at speeds ranging between 2 m/s and 13 m/s. The reason that ordinary humanoid robots cannot run as fast and stably as humans because they require both a large power output for kicking the ground and various stabilization control methods. In general, to increase the power output, actuators having large power output capacity are required. However, high power actuators are heavy and their use in the humanoid robots renders the robots heavy. Moreover, the actuators require higher power. Therefore, it is difficult to design humanoid robots that can achieve high power output and are light enough to jump. Ordinary humanoid robots can attain a power output of approximately 3.5 W/kg for the joints in the leg; however, humans generate around 16.7 W/kg in leg joints while running [14]. In addition, various stabilization control methods such as considering the center of mass position, landing point, ground reaction force, and linear and angular momentum, are needed for stable running. However, present running stabilization methods do not consider motion during the flight phase. For example, several studies on running control have used the spring-loaded inverted pendulum (SLIP) model [15–17]. This simple model does not consider the mass of the legs, which generates angular momentum by leg swinging [18]. However, a humanoid that can perform human-like fast lower-leg swinging, generates large angular momentum during the flight phase, especially in the yaw direction. Some studies considered the angular momentum of the entire body; however, they focused mainly on only stance-phase movement [19,20]. Thus, in these studies, humanoids could run slowly without leg swing during the flight phase to decrease the angular momentum generated by leg motion. Therefore, the humanoid requires a method to compensate for angular momentum during the flight phase for high speed running.

For fast and stable running, various characteristics of human running have been identified in studies on human sciences and sport sciences, including the following:

- The stance leg acts like a linear spring. A human leg can be modeled as a SLIP model.
- The knee and ankle joints in the stance leg serve as torsion springs that provide leg elasticity and the ability to kick the ground strongly [21,22].
- The leg and joint stiffness change depending on the running speed [23].
- Rapid knee bending occurs in the swing phase to avoid contact of the foot with the ground [24].
- The pelvis rotates in the frontal plane to increase the jumping force [25].
- Moment compensation is accomplished using torso and arm swinging in yaw direction [26,27].

We are working to develop a robot that can run like a human by mimicking the above characteristics. Previously, we developed a lower-body robot that mimics human characteristics, such as leg joint stiffness and pelvic rotations in the frontal plane and can hop with a large joint output of approximately 1000 W by not dissipating the energy at landing but storing the energy in its elastic parts [25,28,29]. In addition, the lower-body robot has the human link length and mass properties, such as the mass, center-of-mass (COM) position in the link, and inertial matrix of each link. The reason for this is that link length and mass properties would result in angular momentum. The currently available humanoid robots do not have human like link lengths or mass properties. These humanoid robots are presumed to have been developed with a focus on walking, which is a movement slower than running. When the angular velocities of joints are not very large, the angular momentum generated is not large and has little influence on the whole-body motion. Some researchers are interested in methods of generating stable motion during the stance phase. Sugihara et al. are developing stable control based on the momentum as a norm during walking exercises [30]. Hyon et al. investigated a back-handspring motion with a multi-link robot and proposed motion planning considering global physical quantities such as the center of mass or angular momentum during the stance phase [31]. On the other hand, during the flight phase, the angular momentum of the whole body of the robot cannot be modified. The upper body and legs are connected to the waist, where the movement of legs and an upper body are generated. It means that when the robot swings only its legs without active moving of the upper body joints during the flight phase, the waist and upper body rotate in yaw direction due to the angular momentum generated by the legs movement. As a result, when the direction of the waist deviates from the traveling direction, the robot cannot perform straight running. To solve it, humans move their upper body actively including arms for generating the angular momentum in opposite direction from that generated by the legs movement [26,27].

In this study, our focus is on the flight phase that occurs during the hopping or running motion of a robot, and we aimed to prevent rotation of the waist by aggressively generating angular momentum with the upper body equivalent to that generated by the legs during the flight phase. To this end, we propose an angular momentum compensation control method that uses the arms and torso inspired by the mechanisms of human running. Moreover, to realize the proposed method with a real humanoid, we developed an upper-body mechanism to mimic human link lengths and mass properties. We performed experiments with the robot that we developed to evaluate the proposed methods. We confirmed that the humanoid robot could compensate for the angular momentum in the yaw direction that is generated by lower-body movement in midair.

The remainder of this paper is organized as follows. In Section 2, we describe the proposed method for compensating for angular momentum in the yaw direction and the design of the upper-body mechanism that mimics the human mass properties. In Section 3, we present the experimental results. In Section 4, we present a discussion. Finally, in Section 5, we present our conclusions.

2. Angular Momentum Compensation

2.1. Requirements for Angular Momentum Compensation

We identified the characteristics of upper-body movement of human running on the basis of previous research in human and sport sciences, and we based our determination of the requirements for a control method and an upper-body mechanism on these characteristics. Table 1 summarizes the characteristics we identified. Humans utilize the torso and arms to compensate for angular momentum in the yaw direction generated by the movement of the lower body during the flight phase, during which the human is not in contact with the ground [26]. Here, the trunk refers to the parts higher than the lumbar vertebrae, and the trunk joints refer to the virtual joint between the trunk and the lumbar vertebrae, which consolidates the complex movements of the spine. Rapid leg swinging during the flight phase is responsible for the large angular momentum produced. To compensate for the large angular momentum, both the torso yaw joint and the shoulder pitch joints swing. The shoulder pitch joints swing widely and rapidly because of the inertial moment of the arms in the yaw direction. Most of that upper-body movement is in the yaw direction. In addition, the shoulder roll and yaw joints can change arm postures. A human extends his/her elbow joints according to his/her running speed [27]. The elbow joints can be used to adjust the inertial moment of the arms when large angular momentum is generated by quick lower-body movements. Therefore, we determined that the requisite joint motions of the upper-body mechanism are the three torso joint motions (pitch, roll, and yaw), the three shoulder joint motions (pitch, roll and yaw), and one elbow joint motion inspired by the human configuration.

Table 1. Characteristics of the upper body during human running.

Joints	Characteristics
Trunk Roll	Bending for angular momentum compensation
Trunk Pitch	Bending for angular momentum compensation
Trunk Yaw	Swing for generating angular momentum
Shoulder Roll	For changing the position of the center of mass
Shoulder Pitch	Wide and rapid swing for generating large angular momentum
Shoulder Yaw	For changing the position of the center of mass
Elbow Pitch	Changing the position of the center of mass and moment of inertia of the arm

Furthermore, we determined the requirements for the link lengths and mass properties of the upper-body mechanism based on human data [32,33] (see Tables 2 and 3; note that in the tables, S.D. means the standard deviation). In Table 3, the moment of inertia is with respect to the center of mass position of each link. These parameters influence the angular momentum; however, ordinary humanoid robots have very light arms so as to decrease the required output power of the leg joint, and thereby cannot generate the large angular momentum to compensate that generated by the leg. Therefore, we assumed that employing human-like parameters would be useful to utilize the human-inspired angular momentum compensation method. The requirements of the total mass and total height are 60 kg and 1600 mm based on human data [33]. In addition, we determined the requirements of the joints. The requirements for the movable angles of the joints were determined based on those of a human [34]. The angular velocity was determined based on human running data obtained in our previous research [25]. The torque was calculated from the angular velocity and mass properties of each link. These requirements are listed in Table 4.

Table 2. Mass, COM (center-of-mass) position, and link length requirements for an upper-body mechanism.

Links	Mass (S.D.) kg	COM Position (S.D.) mm	Link Length (S.D.) mm
Trunk	16	120 (10) [a]	270 (20)
Upper arm	1.6 (0.10)	150 (10) [b]	270 (14)
Forearm	0.90 (0.06)	90 (5.0) [c]	220 (11)

[a] from the waist; [b] from the shoulder; [c] from the elbow.

Table 3. Moment of inertia requirements for an upper-body mechanism.

Links	I_{xx} (Roll) kgm^2	I_{yy} (Pitch) kgm^2	I_{zz} (Yaw) kgm^2
Trunk	1.5×10^{-1}	1.4×10^{-1}	1.0×10^{-1}
Upper arm	6.5×10^{-3}	7.0×10^{-3}	1.7×10^{-3}
Forearm	4.1×10^{-3}	4.2×10^{-3}	7.0×10^{-4}

Table 4. Requirements for upper-body joints.

Joints	Movable Range Deg	Angular Velocity (S.D.) rpm	Torque (S.D.) Nm
Shoulder Pitch	−180–50	50 (15)	21 (8.8)
Shoulder Roll	0–180	25 (5.9)	12 (6.8)
Shoulder Yaw	−80–60	29 (1.9)	4.8 (2.2)
Elbow Pitch	−145–5	69 (12)	17 (1.9)
Trunk Pitch	−45–30	18 (9.5)	77 (27)
Trunk Roll	−50–50	14 (4.7)	64 (25)
Trunk Yaw	−40–40	36 (9.2)	40 (15)

2.2. Upper-Body Control Method Based on Angular Momentum

We developed an angular momentum compensation method inspired by humans. In this method, the upper body, including the torso and arms, is controlled to compensate for the angular momentum generated by the movement of the legs during the flight phase. By using both torso and arms, the upper body can generate large angular momentum. In this paper, we present all vectors of position, angular velocity, linear momentum and angular momentum, and all rotation matrices in the Cartesian frame fixed on the ground. The inertia matrix is basically with respect to the center of mass position of the link presented in the Cartesian frame fixed on the ground. The process of the proposed method consists of four steps:

(1) Selection of an angular momentum reference for a waist of the robot;
(2) Calculation of the angular momentum generated by legs movement;
(3) Calculation of the angular momentum that needs to be generated by movement of the torso and arms;
(4) Generation of movement of each upper-body joint.

The angular momentum of the waist is first determined, based on if rotation in the yaw direction is needed. The angular momentum reference of the waist $\mathbf{L}_{waist\_target}$ should be 0 kgm^2/s to perform straight running motion. Second, the angular momentum generated by leg movement is determined. The legs are controlled with controllers for running. In our previous studies, the controller decided a landing position and swung the legs to achieve the desired landing. When the leg movement is different, the angular momentum generated by the legs also changes. The legs' angular momentum $\mathbf{L}_{legs}$ is described by the following equation, considering the angular momentum of each link.

$$\mathbf{L}_{legs} = \sum_{i=1}^{Legs\_link} \mathbf{L}_i \tag{1}$$

where, $\mathbf{L}_i$ is the angular momentum of the ith link, which is generally calculated as shown in Equation (2) [19].

$$\mathbf{L}_i = \mathbf{c}_i \times \mathbf{P}_i + \mathbf{R}_i{}^T \bar{\mathbf{I}}_i \mathbf{R}_i \boldsymbol{\omega}_i \tag{2}$$

where, $\mathbf{c}_i$ is the COM position of the ith link, $\mathbf{P}_i$ is the linear momentum of the ith link, $\mathbf{R}_i$ is the rotation matrix of the ith link, $\bar{\mathbf{I}}_i$ is the inertia matrix of the ith link with respect to the COM position of the link, and $\boldsymbol{\omega}_i$ is the angular velocity of the ith link. The inertia matrices are known as the designed parameters including the moment of inertia and products of inertia. The rotation matrix and the angular velocity are measured by the joint angle sensors implemented in the robot, and the linear momentum is calculated with mass and the COM position.

In the third step of the process, the angular momentum reference of the upper body $\mathbf{L}_{upperbody\_target}$ is calculated from the angular momentum of the whole body at take-off $\mathbf{L}_{all}$ and the angular momentum $\mathbf{L}_{legs}$ generated by the leg movement. By the same method for the calculation of the angular momentum

$\mathbf{L}_{legs}$ generated by the legs, the angular momentum of the whole body $\mathbf{L}_{all}$ is calculated at take-off because it does not change during the flight phase.

$$\mathbf{L}_{upperbody\_target} = \mathbf{L}_{all} - \mathbf{L}_{legs} - \mathbf{L}_{waist\_target} \tag{3}$$

The angular momentum reference of the upper body is divided into torso and arms components for compensation in the yaw direction. $\mathbf{L}_{trunk\_target}$ and $\mathbf{L}_{arms\_target}$ are the references of angular momentums generated by the movements of the torso and arms, respectively. The arm movement contribution is represented with gain $\mathbf{K}$ as shown in Equations (4) and (5).

$$\mathbf{L}_{arms\_target} = \mathbf{K} \cdot \mathbf{L}_{upperbody\_target} \tag{4}$$

$$\mathbf{L}_{trunk\_target} = (\mathbf{E} - \mathbf{K}) \cdot \mathbf{L}_{upperbody\_target} \tag{5}$$

The gain should be determined by the desired motion and the capacity of generating the angular momentum by each part of the robot. For stable running, both the trunk and arms should be used for generating large angular momentum. When the robot has very light arms and cannot generate angular momentum with arms, the gain should be lower.

Finally, the movements of the joints are determined. To generate $\mathbf{L}_{trunk\_target}$ and $\mathbf{L}_{arms\_target}$, the robot controls its trunk and shoulder joints. The joint angular velocity reference of each joint is calculated using Equations (6)–(8).

$$\boldsymbol{\omega}_{right\_shoulder\_ref} = \mathbf{R}_{right\_arm}{}^{T}\bar{\mathbf{I}}^{-1}_{right\_arm}\mathbf{R}_{right\_arm}\left(\frac{\mathbf{L}_{arms\_target}}{2} - \mathbf{c}_{right\_arm} \times \mathbf{P}_{right\_arm}\right) \tag{6}$$

$$\boldsymbol{\omega}_{left\_shoulder\_ref} = -\boldsymbol{\omega}_{right\_shoulder\_ref} \tag{7}$$

$$\boldsymbol{\omega}_{trunk\_ref} = \mathbf{R}_{upperbody}{}^{T}\bar{\mathbf{I}}^{-1}_{upperbody}\mathbf{R}_{upperbody}(\mathbf{L}_{trunk\_target} - \mathbf{c}_{upperbody} \times \mathbf{P}_{upperbody}) \tag{8}$$

where, $\mathbf{c}_{right\_arm}$ and $\mathbf{c}_{upperbody}$ are the COM positions of the right arm and the upper body, respectively. $\mathbf{P}_{right\_arm}$ and $\mathbf{P}_{upperbody}$ are the linear momentum of the right arm and the upper body, respectively. $\mathbf{R}_{right\_arm}$ and $\mathbf{R}_{upperbody}$ are the rotation matrices of the right arm and the upper body, respectively. $\bar{\mathbf{I}}_{right\_arm}$ and $\bar{\mathbf{I}}_{upperbody}$ are the inertial matrices of the right arm and the upper body with respect to the COM position of each part, respectively. The left and right shoulder joint motions are symmetrical with respect to the center.

Using this method, the joints of the torso and arms are controlled so that the angular momentum of the waist in the yaw direction is kept close to zero. The method will be used for various humanoid robots by changing the arm movement gain according to the capacity of generating the angular momentum by each part of the robot.

2.3. Design of the Upper-Body Mechanism

To fulfill the above requirements of an upper-body robot mechanism, we developed an upper-body mechanism that has human-like mass properties and can perform the motions observed during human running (see Figure 1). Some studies have been conducted using humanoid robots with upper-body mechanisms [7,10,11,35], but it is difficult to achieve human-like size, mass properties and motion. The reason for this is that human-like motion requires a high-power output, but high-power actuators are too heavy to mimic a human mass properties. For example, the mass of the upper arm of a human may be 1600 g. Three actuators are required in the upper arm for the shoulder and elbow joints. However, the mass of each actuator, including the gear, is approximately 400 g. We considered the use of brushless direct-current (DC) motors as actuators, because they are small and light and can output high power. Based on the upper body motion during human running shown in Table 4, the maximum joint power of the shoulder pitch and trunk joints is around 150 W, and that of other joints is even smaller. Implementation of such actuators requires that the mass of the other parts, such as the frames connecting each joint and the electric parts, be less than 400 g. We must save the weight of frames; however, the frames must bear the load derived from large angular momentum for preventing yaw rotation.

Figure 1. Computer-aided design (CAD) of the developed humanoid.

Therefore, we considered the material properties required for the structural members of the humanoid robot to best mimic the size and mass properties of a human. Aluminum is typically used for the structural members of a robot. The structure of the upper body of most humanoid robots are exoskeleton structures, wherein the frame shapes the body, and every component, such as motors and electrical parts, is stored in the frame. It is difficult to save the weight of the upper body because the amount of aluminum parts for shaping the body are increased. For weight saving in the links, we used carbon-fiber-reinforced plastic (CFRP), which is extremely strong yet also light. The density of CFRP (1.5 g/cm^3) is much lower than that of aluminum (2.7 g/cm^3). CFRP pipes were used in the center of each link, similar to the inner skeleton structure based on the human structure, with actuators, motor controllers and cables placed around the pipes (see Figures 2 and 3). In the torso link, a large twisting moment is loaded for generating angular momentum in yaw direction. To stand the twisting moment, two pipes were implemented instead of one big pipe (Figure 3). Consequently, the size of each link was close to that of a human's, and the mass of each link was approximately 11% less than if it were made of aluminum.

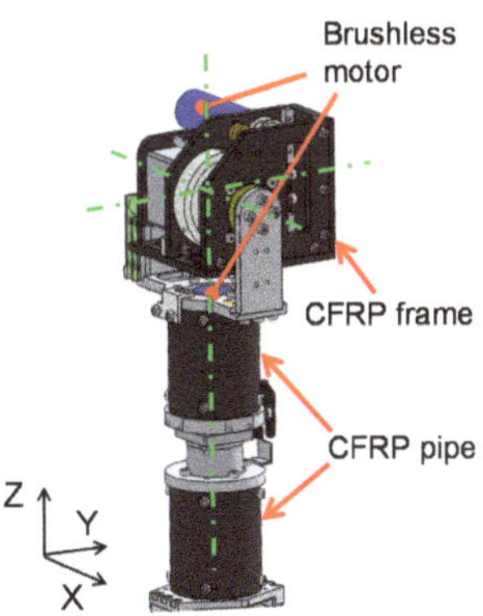

Figure 2. CAD of the humanoid robot arm. The green line is the rotational axis. CFRP: carbon-fiber-reinforced plastic.

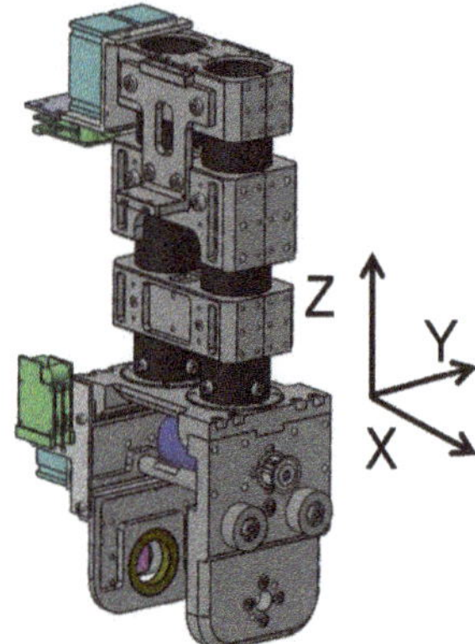

Figure 3. CAD of the humanoid robot trunk. Two CFRP pipes are implemented in the center. Bottom parts are connected to trunk joints.

Achieving a human-like COM position and inertial moment requires axial symmetry of the structure of the torso. Some humanoids have a torso mechanism actuated by a motor and a gear implemented on only one side (see Figure 4a). In this type of mechanism, the COM position of the torso is inclined from the center in the frontal plane and is thus different from that of a human. It therefore produces an angular momentum that is different from that of a human. A humanoid robot developed using this type of mechanism cannot mimic human-like motion by generating the same angular momentum. A counterweight can be used to shift the COM position, but the total mass of the mechanism may still be too large compared with that of a human. To solve this problem, we implemented a symmetrical joint mechanism with two of the same actuators and gears (see Figure 4b). The COM position was thus shifted to the center of the torso, as in a human. With two actuators in parallel, each actuator and gear required less power when only one actuator was used, which made it possible to decrease the size and weight of the actuator. As a result, the mass of the mechanism was decreased by approximately 550 g compared to that associated with using a counterweight. The weight of each part of the upper-body mechanism could therefore be decreased. Thus, the weight could be freely distributed, and the COM position could be regulated more easily.

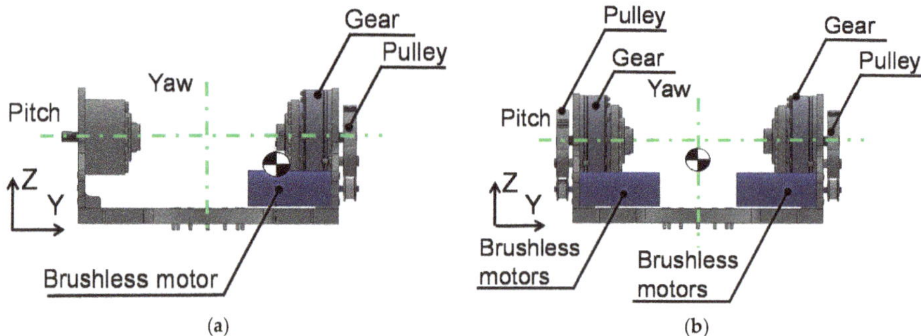

Figure 4. CAD of the trunk joint mechanism of the humanoid robot. The green line is the rotational axis. (**a**) Ordinary mechanism; (**b**) Mechanism developed in this study.

We combined the upper body developed as described above with the lower body to produce a whole-body humanoid (see Figure 5). The humanoid has 22 degrees-of-freedom (DOF) in total.

The brushless DC motors which can output 200 W are implemented in the joints which needs large output such as the shoulder pitch, trunk, hip pitch, and knee joints. On the other hand, the smaller motors which can output 100 W are implemented in other joints. The specifications of the upper body for the link length and mass properties are listed in Tables 5 and 6. The parameter values of the humanoid mimic those of a human in almost every respect. The mass of the humanoid is 60 kg, and the height is 1500 mm. In the case of a human, the COM position of the whole body is located near the pelvis, the height of which is approximately 56% of the standing body height [36]. The robot mimics the whole-body COM position by mimicking the link length, mass, and COM position of each link.

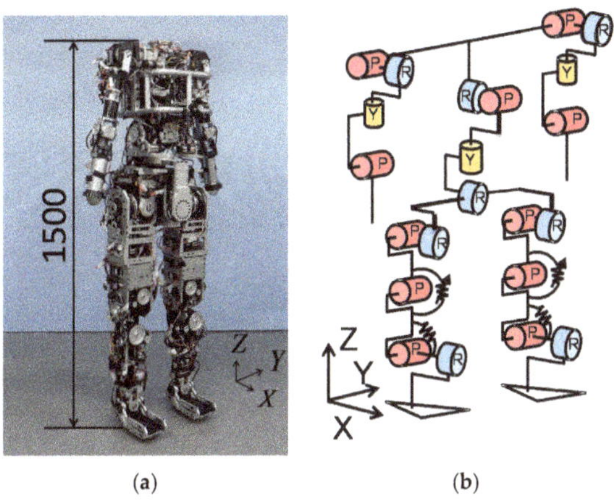

Figure 5. Whole-body running robot. The blue joint is the roll axis, the red joint is the pitch axis, and the yellow joint is the yaw axis. In the knee and ankle joints, a variable joint stiffness mechanism is implemented for storing jumping energy. (**a**) Humanoid robot; (**b**) DOF (degrees-of-freedom) configuration.

Table 5. Upper-body specifications of the humanoid.

Links	Mass (S.D.) kg	COM Position (S.D.) mm	Link Length (S.D.) mm
Trunk	16	130	270
Upper arm	1.6	140	260
Forearm	0.90	90	210

Table 6. Moment of Inertia of the upper body of the humanoid.

Links	I_{xx} (Roll) kgm^2	I_{yy} (Pitch) kgm^2	I_{zz} (Yaw) kgm^2
Trunk	1.4×10^{-1}	1.3×10^{-1}	0.60×10^{-1}
Upper arm	7.3×10^{-3}	7.4×10^{-3}	1.1×10^{-3}
Forearm	4.5×10^{-3}	4.4×10^{-3}	4.4×10^{-4}

3. Experiments and Results

We performed two experiments to verify the capacity of generating angular momentum of the upper body and the effectiveness of the developed control method.

3.1. Verification of Active Generation of Angular Momentum as Large as that of a Human

We conducted an experiment using the humanoid upper body developed in this study to evaluate its capability to generate large angular momentum with human-like motion. We implemented the upper body on a six-axis force sensor for measuring the moment which the upper body applies to the lower part such as the waist of the robot. We calculated the angular momentum generated by the upper body of the robot integrating the moment data measured by the six-axis force sensor. The motion of the upper body was the same as that of human running motion obtained in our previous research [25]. The motion of each joint of human can be expressed as an approximate sine wave as Equation (9).

$$\theta_{JOINT} = \theta_{JOINT\_INITIAL} - A_{JOINT}\sin(\omega t) \tag{9}$$

where, $\theta_{JOINT\_INITIAL}$ is the initial joint angle, A_{JOINT} is the amplitude of the joint motion, ω is the natural frequency of the joint motion, and t is the elapsed time from the start of the experiment. Therefore, we determined the motion of the upper body of the robot as Equation (9). The experimental parameters and their value of each joint are listed in Table 7.

Table 7. Experimental conditions.

Parameters		Value
Trunk pitch amplitude	deg	4.7
Trunk pitch initial angle	deg	12
Trunk roll amplitude	deg	13.8
Trunk roll initial angle	deg	0
Trunk yaw amplitude	deg	15.5
Trunk yaw initial angle	deg	0
Shoulder pitch amplitude	deg	20.5
Shoulder pitch initial angle	deg	37.8
Movement period	s	0.6

Moreover, we calculated the angular momentum generated by the human upper body and the simulation model with a 300 g lighter forearm at the same motion for evaluating the influence of the difference of the weight of the link on the generated angular momentum. After that, we compared the angular momentum generated by the upper body of the developed robot with that of human. The angular momentum of a human's upper body was calculated with the human motion data and body parameters in Tables 2 and 3.

Table 8 summarizes the results for angular momentum generated in the experiment, in simulation, and in human running. The results confirm that the upper-body mechanism developed in this study can generate angular momentum as large as in human motion. In addition, we calculated the angular momentum in the yaw direction with the upper-body model, which has a forearm 300 g lighter than a robot forearm. The maximum angular momentum generated with the upper-body model was 1.7 kgm^2/s. On the other hand, a maximum angular momentum of 1.5 kgm^2/s was generated with the lighter forearm model. The total weight of the lighter forearm model decreased by only 2.3%; however, the angular momentum in the yaw direction decreased by 12%. We confirmed the developed upper body can be used for active angular momentum control as human.

Table 8. Max. angular momentum generated by upper-body of human and the humanoid.

Objects	Max. Generated Angular Momentum kgm^2/s		
	Roll	Pitch	Yaw
Humanoid upper body	5.5	2.3	1.7
Simulation with light forearm	5.3	2.3	1.5
Human upper body	5.5	2.5	1.5

3.2. Angular Momentum Compensation in the Yaw Direction

We conducted an experiment to evaluate how effectively the humanoid could compensate for the angular momentum generated by the leg motion during the flight phase. In this experiment, the humanoid was suspended in midair to perform running motion during the flight phase without any constraint in the yaw direction. The lower-body joints were controlled according to the joint angle references based on the human motion data [25]. We verified the humanoid turning angles in the yaw direction depends on whether control is provided, which depended on the angular momentum in the yaw direction. We measured the turning angle of the waist in the yaw direction using a motion capture system that included infrared cameras used to determine three-dimensional marker positions at 120 Hz. Spherical retro-reflective markers were attached to the surface of the waist parts of the humanoid near the COM of the whole body. The motion capture system calculated the segment attitude using the position data for some of the markers. When angular momentum compensation was not employed, each joint of the upper body maintained a neutral position. When angular momentum compensation was employed, the angular momentum generated by the leg motion was calculated, and the upper body moved to compensate for the angular momentum actively. The ratios of the arms contribution and trunk contribution were determined to be 0.8 and 0.2, respectively, based on the human running data [26] in which the angular momentum generated by the arms was 1.2 kgm^2/s and that generated by the trunk was 0.3 kgm^2/s. The lower-body motion of each joint was determined to follow an approximate sine wave as Equation (10), based on the human running data [15].

$$\theta_{HIP} = \theta_{HIP\_INITIAL} - A_{HIP}\sin(\omega t) \tag{10}$$

where, $\theta_{HIP\_INITIAL}$ is the initial joint angle, A_{HIP} is the amplitude of the joint motion, ω is the natural frequency of the joint motion, and t is the elapsed time from the start of the experiment. The experimental parameters and their values of joint movements are listed in Table 9.

Table 9. Experimental conditions.

Parameters			Value
Hip pitch amplitude	A_{HIP}	deg	15
Hip pitch initial angle	$\theta_{HIP\_INITIAL}$	deg	10
Knee pitch amplitude		deg	20
Knee pitch initial angle		deg	150
Ankle pitch amplitude		deg	0
Ankle pitch initial angle		deg	90
Movement period		s	1.0
Arm gain of angular momentum control		K	0.8
Torso gain of angular momentum control			0.2

Figure 6 illustrates the rotation angle of the humanoid waist in the yaw direction. Figure 7 presents photographs of the experiment. When the angular momentum compensation control was not employed, the humanoid rotated approximately 15 deg in the yaw direction. On the other hand, when the control was used, the rotation decreased to approximately 8 deg. The results confirm the effectiveness of the proposed angular momentum compensation control for stabilization in the yaw direction. It was assumed that the friction force in the yaw direction of the hanger influenced the angular momentum because it was small but not zero.

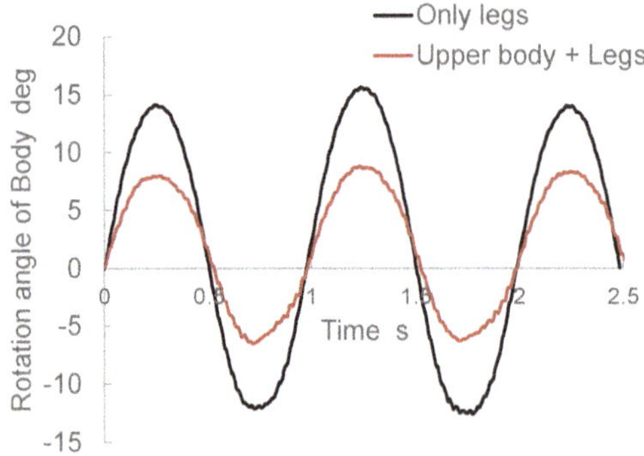

Figure 6. Experimental results for angular momentum compensation.

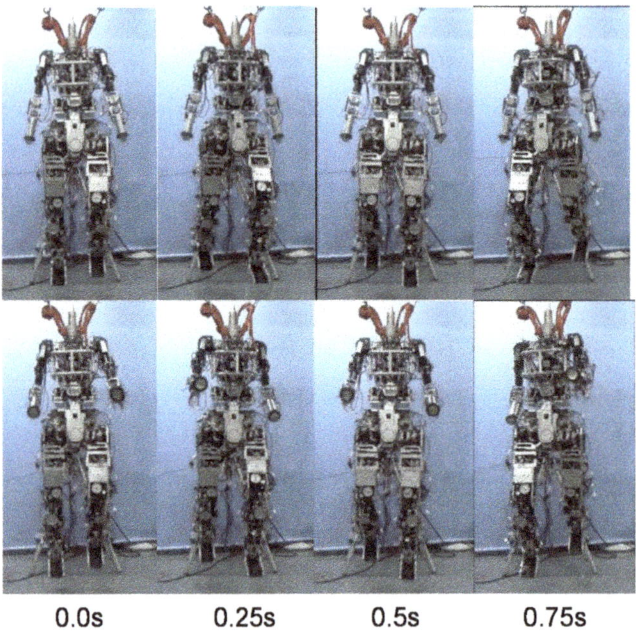

Figure 7. Experiment with a real robot. Upper row shows the experiment without angular momentum control. Lower row shows the experiment with angular momentum control. The humanoid was suspended with orange cables connected to a hanger, which could rotate passively in the yaw direction, and did not contact with the ground.

4. Discussion

In experiment 2, we could not perform a running experiment with the humanoid developed in this study because more work is required to develop methods for stabilization in the pitch and roll directions. During running, the vertical component of the ground reaction force can be up to 1800 N, which is much larger than the horizontal component of approximately 180 N [37]. Therefore, the ground reaction force influences stabilization in the pitch and roll directions but not in the yaw direction. When the ground reaction force does not act in alignment with the COM of the humanoid, angular momentum is generated and the robot falls. To prevent this situation, some researchers have focused on stabilization in the pitch and roll directions by controlling the ground reaction force [38,39]. In contrast, in this study, we focused on yaw stabilization using the upper body, because the ground reaction force does not have a large influence on stabilization in the yaw direction.

In this study, we designed an upper-body mechanism that has human-like link length and mass properties to achieve angular momentum compensation in the yaw direction during running. In experiment 1, we confirmed that the upper body could generate almost the same angular momentum as a human's upper body during running. In addition, we calculated the angular momentum in the yaw direction with the 300 g lighter forearm. These results indicate that mass parameters have a large influence on the angular momentum generated during running. To achieve angular momentum compensation during running, it is important to incorporate the capacity of generating angular momentum in designing a humanoid robot. The upper body developed in this study was found to be able to perform fast movement as well as a human. This upper body will be useful in future research on effective sports movements such as ball throwing.

We propose in this paper an angular momentum compensation method using a humanoid upper body. This method can be used with other humanoid robots that do not have human-like mass properties. The reason for this is that the robot can calculate the angular momentum required using its mass property data and the control method developed. In addition, the control method can be applied to active turning by changing the angular momentum reference of the waist. In experiment 2, which involved running on a straight course, the angular momentum reference of the waist was set to zero.

In addition, we assume that the proposed method for stabilization in the yaw direction can be applied to stabilization in the pitch and roll directions with a change in the ground reaction force. In general, the upper body has a large mass and moment of inertia, and it can thus generate large angular momentum in the pitch, roll and yaw directions. The angular momentum generated by the ground reaction force can be compensated for by the upper-body motion. When the humanoid cannot effectively use its upper body, the humanoid should generate motion within a range that maintains the stabilization with only its legs and torso. By integrating these methods, the humanoid can perform faster and wider-ranging motions. In future work, we intend to apply the proposed method for stabilization to the pitch and roll directions. In addition, we will utilize the robot in the place of human subjects to confirm various running characteristics.

In summary, we found that the angular momentum compensation in the yaw direction using the upper body during the flight phase can improve the capability of humanoid hopping and running performance. Moreover, the upper body design should focus on the capability of generating angular momentum for rapid movement such as hopping and running. For stable running without constraints, the proposed method should be integrated with other control methods; however, humanoid robots will perform stable and faster motions by using the proposed methods, and the developed upper body will be useful in research about finding effective sport movements such as ball throwing.

5. Conclusions

In this paper, we propose an angular momentum compensation method to achieve angular momentum compensation in the yaw direction during the flight phase of running. The method is based on the human-running mechanism. To compensate for the angular momentum generated by lower-body movement during the flight phase, the angular momentum compensation method calculates the angular momentum and generates upper-body motion that activates the torso and arms, as in humans. The humanoid robot can thereby change its upper-body motion according to changes that occur in the lower-body motion, such as shifts in running speed. We also developed an upper-body mechanism that has the link and mass parameters similar to that of a human, and that can generate large angular momentum. We evaluated the developed upper body and noted that it could generate large angular momentum similar to that of humans. Furthermore, the minor differences in the link and mass parameters can significantly influence the capacity of generating angular momentum. Moreover, we confirmed that the humanoid robot could compensate for the angular momentum generated by the lower-body movement when the robot was suspended in midair.

The developed control method can contribute to improving the stability of humanoids that can perform dynamic movements including the flight phase, such as jumping, hopping and running. Because most other control methods for stabilization of the dynamic motion of humanoids focus on the stance phase, the developed method will be integrated without interference. However, using the developed control method requires the generation of large angular momentum by the upper body. To do that, the upper body design, which was not focused on in relevant studies, will also be changed to consider the mass, mass position, inertia matrix and link length for generating large angular momentum. Thanks to improved stability, humanoids will be able to advance into human living spaces and work stably.

Acknowledgments: This study was conducted with the support of the Research Institute for Science and Engineering, Waseda University; the Institute of Advanced Active Aging Research, Waseda University; the Future Robotics Organization, Waseda University, and as part of the humanoid project at the Humanoid Robotics Institute, Waseda University. It was also financially supported in part by JSPS KAKENHI Grant Nos. 25220005, 25709019, 17H00767; a Waseda University Grant for Special Research Projects (Project number 2017K-215); SolidWorks Japan K.K.; the DYDEN Corporation; and Cybernet Systems Co., Ltd. We thank all of these for the financial and technical support provided. We would like to thank Editage (www.editage.jp) for the English language editing.

Author Contributions: Takuya Otani, Kenji Hashimoto, Hiroki Ueta, Akira Natsuhara, Hum-Ok Lim, and Atsuo Takanishi developed the upper-body control; Takuya Otani, Kenji Hashimoto, and Shunsuke Miyamae performed the experiments; Masanori Sakaguchi and Yasuo Kawakami analyzed the human motion data; Hum-Ok Lim and Atsuo Takanishi helped to draft the manuscript, Takuya Otani wrote the paper, and all of the authors read and approved the final manuscript.

Conflicts of Interest: The authors declare no conflict of interest.

References

1. Raibert, M.H. *Legged Robots that Balance*; MIT Press: Cambridge, MA, USA, 1986.
2. Grimes, J.A.; Hurst, J.W. The Design of ATRIAS 1.0 a Unique Monopod, Hopping Robot. In Proceedings of the International Conference on Climbing and Walking Robots (CLAWAR), Baltimore, MD, USA, 23–26 July 2012; pp. 548–554.
3. Martin, W.C.; Wu, A.; Geyer, H. Experimental Evaluation of Deadbeat Running on the ATRIAS Biped. *IEEE Robot. Autom. Lett.* **2017**, *2*, 1085–1092. [CrossRef]
4. Hyon, S.; Emura, T.; Mita, T. Dynamics-based control of one-legged hopping robot. *J. Syst. Control Eng. Proc. Inst. Mech. Eng. Part I* **2003**, *217*, 83–98. [CrossRef]
5. Nagasaka, K.; Kuroki, Y.; Suzuki, S.; Itoh, Y.; Yamaguchi, J. Integrated motion control for walking, jumping and running on a small bipedal entertainment robot. In Proceedings of the 2004 IEEE International Conference on Robotics and Automation, New Orleans, LA, USA, 26 April–1 May 2004; pp. 3189–3194.
6. Kajita, S.; Nagasaki, T.; Kaneko, K.; Yokoi, K.; Tanie, K. A Running Controller of Humanoid Biped HRP-2LR. In Proceedings of the 2005 IEEE International Conference on Robotics and Automation, Barcelona, Spain, 18–22 April 2005; pp. 618–624.
7. Cho, B.K.; Park, S.S.; Oh, J.H. Controllers for running in the humanoid robot, HUBO. In Proceedings of the IEEE-RAS International Conference on Humanoid Robots 2009, Paris, France, 7–10 December 2009; pp. 385–390.
8. Wensing, M.P.; Orin, E.D. High-speed humanoid running through control with a 3D-SLIP model. In Proceedings of the 2013 IEEE/RSJ International Conference on Intelligent Robots and Systems, Tokyo, Japan, 3–7 November 2013; pp. 5134–5140.
9. Tamada, T.; Ikarashi, W.; Yoneyama, D.; Tanaka, K.; Yamakawa, Y.; Senoo, T.; Ishikawa, M. High Speed Bipedal Robot Running Using High Speed Visual Feedback. In Proceedings of the 14th IEEE-RAS International Conference on Humanoid Robots, Madrid, Spain, 18–20 November 2014; pp. 140–145.
10. Takenaka, T.; Matsumoto, T.; Yoshiike, T. Real-time Dynamics Compensation with Considering Ground Reaction Force and Moment Limit for Biped Robot. *J. Robot. Soc. Jpn.* **2014**, *32*, 295–306. [CrossRef]
11. Tajima, R.; Honda, D.; Suga, K. Fast Running Experiments Involving a Humanoid Robot. In Proceedings of the IEEE International Conference on Robotics and Automation, Kobe, Japan, 12–17 May 2009; pp. 1571–1576.

12. Niiyama, R.; Nishikawa, S.; Kuniyoshi, Y. Biomechanical Approach to Open-loop Bipedal Running with a Musculoskeletal Athlete Robot. *Adv. Robot.* **2012**, *26*, 383–398. [CrossRef]
13. Grizzle, J.W.; Hurst, J.; Morris, B.; Park, H.W.; Sreenath, K. MABEL, a new robotic bipedal walker and runner. In Proceedings of the American Control Conference, St. Louis, MO, USA, 10–12 June 2009; pp. 2030–2036.
14. Endo, T.; Miyashita, K.; Ogata, M. Kinetics factors of the lower limb joints decelerating running velocity in the last phase of 100 m race. *Res. Phys. Educ.* **2008**, *53*, 477–490. [CrossRef]
15. Cavagna, G.A.; Franzetti, P.; Heglund, N.C.; Willems, P. The determinants of the step frequency in running, trotting and hoppin in man and other vertebrates. *J. Physiol.* **1988**, *29*, 81–92. [CrossRef]
16. Blickhan, R. The Spring-mass Model for Running and Hopping. *J. Biomech.* **1989**, *22*, 1217–1227. [CrossRef]
17. McMahon, T.; Cheng, G. The Mechanics of Running: How does Stiffness Couple with Speed? *J. Biomech.* **1990**, *23*, 65–78. [CrossRef]
18. Ounpuu, S. The biomechanics of walking and running. *Clin. Sports Med.* **1994**, *13*, 843–863. [PubMed]
19. Kajita, S.; Kanehiro, F.; Kaneko, K.; Fujiwara, K.; Harada, K.; Yokoi, K.; Hirukawa, H. Resolved Momentum Control: Humanoid Motion Planning based on the Linear and Angular Momentum. In Proceedings of the 2003 IEEE/RSJ International Conference on Intelligent Robots and Systems, Las Vegas, NV, USA, 27–31 October 2003; pp. 1644–1650.
20. Orin, D.E.; Goswami, A.; Lee, S.H. Centroidal dynamics of a humanoid robot. *Auton. Robots* **2013**, *35*, 161–176. [CrossRef]
21. Gunther, M.; Blickhan, R. Joint stiffness of the ankle and the knee in running. *J. Biomech.* **2002**, *35*, 1459–1474. [CrossRef]
22. Kuitunen, S.; Komi, P.V.; Kyrolainen, H. Knee and ankle joint stiffness in sprint running. *Med. Sci. Sports Exerc.* **2002**, *34*, 166–173. [CrossRef] [PubMed]
23. Ferber, R.; Davis, I.M.; Williams, D.S., III. Gender Differences in Lower Extremity Mechanics during Running. *Clin. Biomech.* **2003**, *18*, 350–357. [CrossRef]
24. Chapman, A.E.; Caldwell, G.E. Factors determining changes in lower limb energy during swing in treadmill running. *J. Biomech.* **1983**, *16*, 69–77. [CrossRef]
25. Otani, T.; Hashimoto, K.; Yahara, M.; Miyamae, S.; Isomichi, T.; Hanawa, S.; Sakaguchi, M.; Kawakami, Y.; Lim, H.; Takanishi, A. Utilization of Human-Like Pelvic Rotation for Running Robot. *Front. Robot.* **2015**. [CrossRef]
26. Hinrichs, N.R. Upper Extremity Function in Running. II: Angular Momentum Considerations. *Int. J. Sport Biomech.* **1987**, *3*, 242–263. [CrossRef]
27. Hinrichs, N.R. Whole Body Movement: Coordination of Arms and Legs in Walking and Running. In *Multiple Muscle Systems*; Springer: New York, NY, USA, 1990; pp. 694–705.
28. Otani, T.; Hashimoto, K.; Isomichi, T.; Sakaguchi, M.; Kawakami, M.; Lim, H.O.; Takanishi, A. Joint Mechanism That Mimics Elastic Characteristics in Human Running. *Machines* **2016**, *4*, 5. [CrossRef]
29. Otani, T.; Hashimoto, K.; Yahara, M.; Miyamae, S.; Isomichi, T.; Sakaguchi, M.; Kawakami, Y.; Lim, H.O.; Takanishi, A. Running with Lower-Body Robot That Mimics Joint Stiffness of Humans. In Proceedings of the of the IEEE International Conference on Intelligent Robots and Systems 2015, Hamburg, Germany, 28 September–2 October 2015; pp. 3969–3974.
30. Sugihara, T.; Nakamura, Y.; Inoue, H. Realtime Humanoid Motion Generation through ZMP Manipulation based on Inverted Pendulum Control. In Proceedings of the IEEE International Conference on Robotics and Automation, Washington, DC, USA, 11–15 May 2002; pp. 1404–1409.
31. Hyon, S.H.; Yokoyama, N.; Emura, T. Back handspring of a multi-link gymnastic robot—Reference model approach. *Adv. Robot.* **2006**, *20*, 93–113. [CrossRef]
32. Ae, M.; Tang, H.; Yokoi, T. Estimation of inertia properties of the body segments in Japanese athletes. *Soc. Biomech.* **1992**, *11*, 23–33. [CrossRef]
33. Kouchi, M.; Mochimaru, M. *Human Dimension Database*; AIST Digital Human Research Center: Tokyo, Japan, 2005.
34. Nakamura, R.; Saito, H. *Fundamental Kinesiology*, 4th ed.; Kendall Hunt Publishing Company: Dubuque, IA, USA, 1992.
35. Nagasaki, T.; Kajita, S.; Kaneko, K.; Yokoi, K.; Tanie, K. A Running Experiment of Humanoid Biped. In Proceedings of the IEEE/RSJ International Conference on Intelligent Robots and Systems, Sendai, Japan, 28 September–2 October 2004; pp. 136–141.

36. Shima, K. Effect of Masticatory Movement on Head, Trunk and Body sways during Standing Position. Ph.D. Thesis, Hokkaido University, Sapporo, Japan, 2015.
37. Farley, C.T.; Ferris, D.P. Biomechanics of walking and running: Center of mass movements to muscle action. *Exerc. Sport Sci. Rev.* **1998**, *26*, 253–285. [CrossRef] [PubMed]
38. Sreenath, K.; Park, H.W.; Poulakakis, I.; Grizzle, J.W. Embedding active force control within the compliant hybrid zero dynamics to achieve stable, fast running on MABEL. *Int. J. Robot. Res.* **2013**, *32*, 324–345. [CrossRef]
39. Kajita, S.; Kaneko, K.; Morisawa, M.; Nakaoka, S.; Hirukawa, H. ZMP-based Biped Running Enhanced by Toe Springs. In Proceedings of the IEEE International Conference on Robotics and Automation, Roma, Italy, 10–14 April 2007; pp. 3963–3969.

Article

A Bio-Inspired Control Strategy for Locomotion of a Quadruped Robot

Yinquan Zeng [1], Junmin Li [1,*], Simon X. Yang [2,*] and Erwei Ren [1]

1 School of Mechanical Engineering, Xihua University, Chengdu 610039, China; y_q_zeng@163.com (Y.Z.); rew991468085@163.com (E.R.)

2 Advanced Robotics and Intelligent Systems Laboratory, School of Engineering, University of Guelph, Guelph, ON N1G 2W1, Canada

* Correspondence: lijunmin1975@163.com (J.L.); syang@uoguelph.ca (S.X.Y.)

Received: 4 October 2017; Accepted: 20 December 2017; Published: 2 January 2018

Abstract: In order to effectively plan the robot gaits and foot workspace trajectory (WT) synchronously, a novel biologically inspired control strategy for the locomotion of a quadruped robot based on central pattern generator—neural network—workspace trajectory (CPG-NN-WT) is presented in this paper. Firstly, a foot WT is planned via the Denavit-Hartenberg (D-H) notation and the inverse kinematics, which has the advantages of low mechanical shock, smooth movement, and sleek trajectory. Then, an improved central pattern generator (CPG) based on Hopf oscillators is proposed in this paper for smooth gait planning. Finally, a neural network is designed and trained to convert the CPG output to the preplanned WT, which can make full use of the advantages of the CPG-based method in gait planning and the WT-based method in foot trajectory planning simultaneously. Furthermore, virtual prototype simulations and experiments with a real quadruped robot are presented to validate the effectiveness of the proposed control strategy. The results show that the gait of the quadruped robot can be controlled easily and effectively by the CPG with its internal parameters; meanwhile, the foot trajectory meets the preplanned WT well.

Keywords: bio-inspired control; quadruped robot; robot trajectory; CPG; neural network

---

1. Introduction

Compared with wheeled robots, quadruped robots with their good environmental adaptability and motion flexibility have become a research hot spot in recent years [1–7]. Specifically, the bio-heuristic control strategy based on the CPG model is widely used in the locomotion control of legged robots due to its good simulation of the rhythmic movement of animals [8–11]. At present, the typical CPG models for legged robots' locomotion control can be roughly divided into three categories: CPG model based on neuron oscillators, CPG model based on nonlinear oscillators and other kinds of CPG model. The typical CPG models based on neuron oscillators are the Matsuoka model [12], the Kimura model [13], etc. [10]. The parameters of this kind of CPG model generally have a clear biological significance. Since the equations are generally nonlinear, multi-parameter and multi-dimension, its parameter's tuning and dynamic analysis are very complex. The Kuramoto phase oscillator [14–16], Hopf oscillator [17,18] and Van Der Pol (VDP) oscillator [19] are the typical nonlinear oscillators to constitute the second kind of CPG models. Although their parameters are not as clear as the first kind of CPG models in biological significances, this kind of CPG model is mature and fewer parameters need to be tuned. Other kinds of CPG model, such as the cosine function CPG model, hardware CPG model, etc., are seldom used.

The common way to control a legged robots' locomotion using CPG-based method is that the hip joints are directly controlled by the CPG output and the knee joints are controlled by mapping the hip joint control signals in a certain way. However, most of the previous studies focused on the CPG model and its structure and connection topological relations, while only few studies were concerned with the CPG output waveform and the inter-coordination of the joints within a leg to meet the desired foot WT [4]. Generally, for a dynamic system that has stable periodic solutions (limit cycle), the waveform of the solutions is defined by the shape of the limit cycle [11]. Universally, the limit cycle shape of the equations used as a CPG model is extremely difficult to control. Thus, it leads to the situation where it is hard to control the foot WT while using CPG output to control legged robots' locomotion directly, which leads to a series of problems, such as large mechanical shock and movement shake, high energy consumption, etc.

In contrast to the CPG-based method, the WT-based method has great advantages in the foot WT arbitrarily designing and planning [4,20–22], such as excellent capability of over-obstacle, smooth motion, smaller mechanical shock, lower energy consumption, etc. [4,6,23,24]. The general operation mode while using WT-based control strategy is: firstly, planning a foot WT with a specific advantage; secondly, solving the intra-leg joints' coordinated function via inverse kinematics; finally, coupling the legs in an "up layer" through certain phase relationships to obtain the relative gait. Although this method solves the problem of the legged robots' foot WT, generally, gaits lack flexibility and its mutual transitions are stiff.

As we analyzed above, in order to generate the desired WT, the coordination of the two joints (or more) within a leg is very important. However, the shape of periodic solution (limit cycle) of CPG is difficult to control, for cooperating with another joint (or the other joints) to generate the desired WT. Thus, it is very difficult to generate the desired WT with the CPG and inverse kinematics. Fortunately, we found that this complex problem becomes very simple with a "NN" layer, meanwhile, the advantages of CPG-based method in gait planning and WT-based method in foot trajectory planning can be made full use of, synchronously.

In simple terms, the presented biologically inspired control strategy based on CPG-NN-WT is: the CPG is responsible for the gait-related tuning and control, and the NN is responsible for translating the CPG output into the preplanned WT to control the locomotion of a quadruped robot, which simplifies the complex control tasks into relatively simple problems.

2. The Foot WT Planning

In this section, the modeling and kinematics analysis of a quadruped robot are first presented. Then biologically inspired foot WT with both swing phase and stance phase is planned with improvement.

2.1. Modeling and Kinematics Analysis of a Quadruped Robot

A simplified quadruped robot plane model is shown in Figure 1, which consists of nine links: a body, four femurs and four shins. To simplify the analysis and control, the joints configuration of the robot is all-knee.

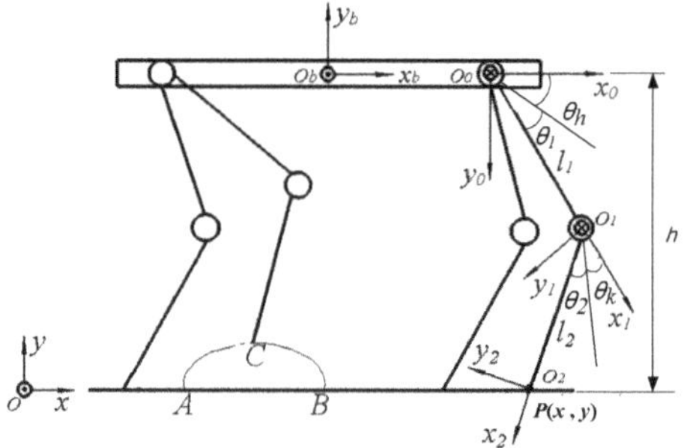

Figure 1. Schematic and parameters of the quadruped robot model.

The coordinate system Oxy and $O_bx_by_b$ are the world coordinate system and the local coordinate system fixed in the geometric center of the robot's body respectively. The orientations are shown in Figure 1, where x_b points to the forward direction of the robot, y_b is perpendicular to the robot's body and pointing to its back from its abdomen, and z_b is determined by the right hand rule of the coordinate system. To make the modeling more general, a leg is chosen arbitrarily to establish the single leg D-H coordinate system, where the right front leg is selected and the coordinate system $O_0x_0y_0$ is located at its hip joint, as shown in Figure 1. The point $P(x,y)$ is the position description of the robot foot in the coordinate system $O_0x_0y_0$. According to the D-H notation, the D-H parameters can be described as shown in Table 1.

Table 1. The D-H notation parameters.

Link No. i	Joint Variable θ_i (Rad)	Offset d_i (mm)	Link Length a_{i-1} (mm)	Link Angle a_{i-1} (Rad)
1	$\theta_1 + \theta_h$	0	l_1	0
2	$\theta_1 + \theta_k$	0	l_2	0

Where θ_h and θ_k are the balance position angles of hip and knee joints, respectively; θ_1 and θ_2 are the rotation angles of the hip and knee joints with respect to their balance position, respectively; l_1 and l_2 are the lengths of the femur and shin, respectively. By the D-H notation, the angles θ_1 and θ_2 corresponding to point $P(x,y)$ can be solved via the inverse kinematics, which is given by

$$\left.\begin{aligned} \theta_1 &= \arcsin\frac{x^2+y^2+{l_1}^2-{l_2}^2}{2l_1\sqrt{x^2+y^2}} - \arctan2(x,y) - \theta_h \\ \theta_2 &= \arccos\frac{x^2+y^2-{l_1}^2-{l_2}^2}{2{l_1}^2{l_2}^2} - \theta_k \end{aligned}\right\} \tag{1}$$

2.2. Foot WT Planning

Foot WT mainly involves swing phase and stance phase. As shown in Figure 1, the arc $\overset{\frown}{ACB}$ is the swing phase trajectory and $\overset{\frown}{BA}$ is the stance phase trajectory. In [10], a trajectory planning method based on compound cycloid is used to design the swing phase trajectory with the advantage of smooth motion and small mechanical shock. In addition, the method is modified and applied to a quadruped robot foot WT planning in [11]. The modified trajectory is defined as

$$\begin{aligned} x &= S\left[\frac{t}{T_m} - \frac{1}{2\pi}\sin\left(2\pi\frac{t}{T_m}\right)\right] \\ y &= \begin{cases} 2H\left[\frac{t}{T_m} - \frac{1}{4\pi}\sin\left(4\pi\frac{t}{T_m}\right)\right] & 0 \le t < \frac{T_m}{2} \\ 2H\left[1 - \frac{t}{T_m} + \frac{1}{4\pi}\sin\left(4\pi\frac{t}{T_m}\right)\right] & \frac{T_m}{2} \le t < T_m \end{cases} \end{aligned} \tag{2}$$

where S is the stride length, H is the maximum height of leg raise, and T_m is the swing phase period. By analyzing the first and second derivatives of the equations, we can know that the velocities and accelerations on x and y of the foot are 0 at the time t = 0 and t = T_m, which reduces the mechanical shock, energy loss and makes the robot's movement more smooth.

Generally, constant motion is used to plan the stance phase trajectory [11,24]. However, huge acceleration jumps will occur at the start and end moment of the stance phase while using the constant motion to plan it directly, which is harmful to the robot and its locomotion. While taking the example of the planning method of swing phase trajectory to plan the stance phase trajectory, the robot will continue to accelerate and decelerate in its stance phase, which affects the stability and locomotion performance of the robot terribly. In this paper, a compromise approach using the constant motion guiding with a short period of sinusoidal velocity at its start and end moment is adopted to plan the stance phase trajectory. Thus, the smooth transition from zero to constant velocity and constant velocity to zero is realized, which greatly reduce the acceleration jumps and speed fluctuations.

Assuming T_d is the period that the velocity starts from 0 to constant velocity v, which is given by

$$T_d = kT_S \qquad (0 < k < 0.5) \tag{3}$$

where k is the scaling factor of the transition period T_d and the support phase period T_S. The velocity v is determined by

$$S = \int_0^{T_d} v \sin\left(\frac{\pi\, t}{2T_d}\right) dt + v(T_S - 2T_d) + \int_{T_S - T_d}^{T_S} v \sin\left(\frac{\pi(T_S - t)}{2T_d}\right) dt \tag{4}$$

Then the improved stance phase trajectory is described as

$$\begin{aligned} x &= \begin{cases} \frac{2vT_d}{\pi}\left[1 - \cos\left(\frac{\pi t}{2T_d}\right)\right] & 0 \le t < T_d \\ \frac{2vT_d}{\pi} + v(t - T_d) & T_d \le t < T_S - T_d \\ \frac{2vT_d}{\pi} + v(T_S - 2T_d) + \frac{2vT_d}{\pi}\cos\left(\frac{\pi(T_S - t)}{2T_d}\right) & T_S - T_d \le t < T_S \end{cases} \\ y &= h \\ v &= \frac{\pi}{\pi + 4k - 2k\pi} \cdot \frac{S}{T_S} \end{aligned} \tag{5}$$

where h is the height from the foot to the corresponding hip joint. We set the maximum height of leg raise $H = 30$ mm, the stride length $S = 200$ mm, the swing and stance phase period $T_m = T_S = 0.4$ s. The single-leg single-period foot WT in the $x_bO_by_b$ plane of coordinate system $O_bx_by_b$ is shown in Figure 2 (for a unified form, the trajectory is shifted to the origin of the coordinate system $O_bx_by_b$).

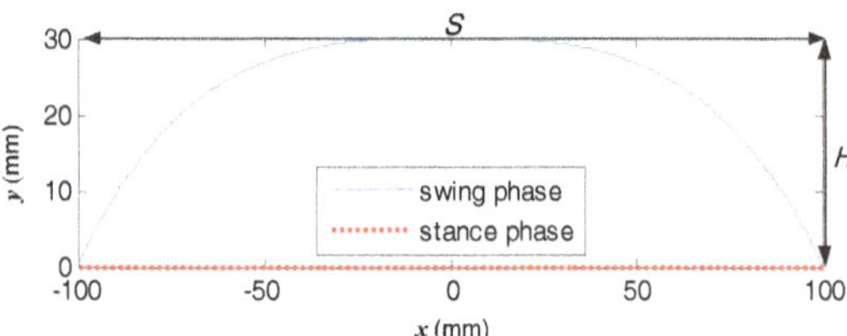

Figure 2. The single-leg single-period foot workspace trajectory (WT).

We set the scaling factor $k = 0.1$, the hip and knee joints' balance position $\theta_h = \theta_k = \frac{\pi}{3}$, and the parameters used in Equation (1) are chosen as $l_1 = l_2 = 150$ mm. In the hip joint coordinate system $O_0x_0y_0$, with the foot WT planned above, the coordinated angular displacement curves of the hip joint (θ_1) and knee joint (θ_2) in single gait period are shown in Figure 3. Besides, from Figure 3 we see that curves in joint space (motor commands) are smooth and fluent too.

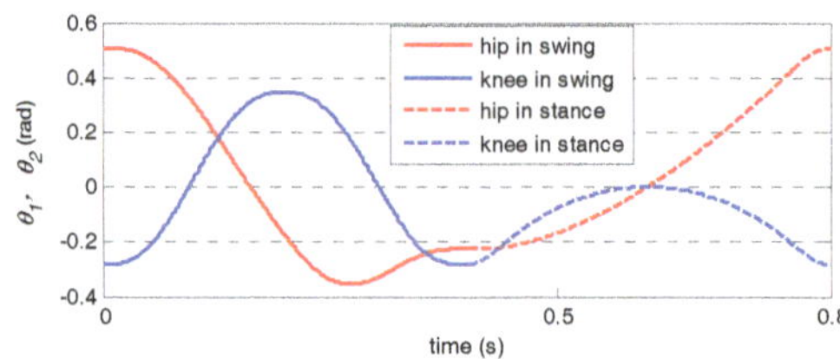

Figure 3. The angular displacement curves of the hip and knee joints.

3. Improved CPG

In this section, a modified CPG oscillator unit model is first presented for CPG parametric control of a quadruped robot. After that, an improved CPG control model is proposed with smoother gait planning.

3.1. Modified CPG Oscillator Unit Model

In order to realize the CPG parametric control of a quadruped robot, Righetti and Ijspeert [25] presented a modified Hopf oscillator and its phase and frequency can be independently controlled. Their modified Hopf oscillator is applied to control the locomotion of a quadruped robot [26], which is given by

$$\begin{cases} \dot{x} = \alpha(\mu - r^2)x - \omega y \\ \dot{y} = \gamma(\mu - r^2)y + \omega x \end{cases} \tag{6}$$

$$\omega = \frac{\omega_{stance}}{e^{-by} + 1} + \frac{\omega_{swing}}{e^{by} + 1} \tag{7}$$

where $r = \sqrt{x^2 + y^2}$; $\sqrt{\mu}$ is the amplitude of oscillator; ω is the frequency of oscillator; ω_{stance} and ω_{swing} are the frequencies of the stance and swing phases, respectively; b is a large positive constant which determines the conversion speed of ω between ω_{stance} and ω_{swing}; and α and γ are positive constants which control the convergence speed of the limit cycle. The greater the value of them, the faster the limit cycle converges. x and y are the two state variables of the oscillator. The oscillator parameters and dynamic characteristics are analyzed in detail in lots of studies [25–29], It should be noted that, whatever the initial value is, except for the $(0, 0)$ singular point, the Hopf limit cycle is stable.

In order to adjust the quadruped robot's gait period and duty factor (the ratio of the stance phase period to the gait period) directly and independently to produce the desired rhythm gaits, Equation (7) is modified to

$$\omega = \frac{\pi}{\beta T(e^{-by} + 1)} + \frac{\pi}{(1 - \beta)T(e^{by} + 1)} \tag{8}$$

where β is the duty factor (in this paper, the rise part of the oscillator output is used as the stance phase), and T is the gait period.

3.2. Improved CPG Control Model

The CPG mechanism of creatures contains the coupling relationship of the time domain and the spatial domain at the same time. Similarly, the CPG control network is a distributed network system composed of multiple oscillator units through a certain topological relationship, which simulate the biological CPG mechanism commendably and is very suitable for multi-legged robot distributed control [10,11]. Aiming at the control of a quadruped robot, four CPG oscillator units based on the modified Hopf oscillators are used to form a fully symmetric CPG control network, as shown in Figure 4.

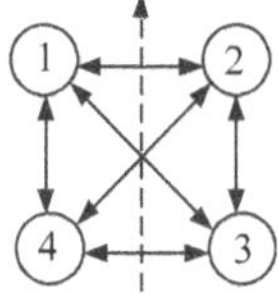

Figure 4. Full symmetric central pattern generator (CPG) control network of a quadruped robot.

In Figure 4, the numbers 1, 2, 3 and 4 are the left front (LF), right front (RF), right hind (RH), and left hind (LH) legs of a quadruped robot, respectively. The dashed arrow points to the head while the solid arrows represent a certain coupling relationship among the CPG oscillator units. Considering the CPG-NN-WT-based control strategy (which will be introduced in detail in Section 4) presented in this paper, the CPG only needs to output 4 smooth control signals with a certain phase coupling relationship instead of controlling any joint, and its duty factor and gait period are adjustable. To realize the smooth gait planning of quadruped, the CPG based on Hopf oscillators presented in [29] is improved as follows:

(1) In order to adjust the gait period and duty factor directly and independently, the frequency ω of the Hopf oscillator is modified as Equation (8).
(2) Parameter δ is introduced as coupling intensity coefficient among the CPG oscillator units to control the gait transition speed and waveform.
(3) Multiple feedbacks with their corresponding reflex coefficients are simultaneously introduced into the two states (x and y) of the CPG oscillator unit. With the reflex information matrix and reflex coefficient vector, a clear way of expression and implementation is provided to realize the biological reflex modeling.

Then the improved CPG is described as follows:

$$\begin{bmatrix} \dot{x}_i \\ \dot{y}_i \end{bmatrix} = \begin{bmatrix} \alpha(\mu - r_i^2) & -\omega_i \\ \omega_i & \gamma(\mu - r_i^2) \end{bmatrix} \begin{bmatrix} x_i \\ y_i \end{bmatrix} + \delta \sum_{j=1}^{4} \boldsymbol{R}\left(\theta_i^j\right) \begin{bmatrix} x_i \\ y_i \end{bmatrix} + \sum_{k=1}^{m} g_k \begin{bmatrix} s_{ik}^x \\ s_{ik}^y \end{bmatrix}, \; i = 1, \ldots, 4 \quad (9)$$

$$\begin{cases} r_i = \sqrt{x_i^2 + y_i^2} \\ \omega_i = \frac{\pi}{\beta T\left(e^{-by_i}+1\right)} + \frac{\pi}{(1-\beta)T\left(e^{by_i}+1\right)} \end{cases} \quad (10)$$

where x_i and y_i are the outputs of oscillator i, and x_i is chosen as the main output of the CPG. The second and third terms on the right side of Equation (9) are the coupling term and feedback term, respectively; δ is the coupling intensity coefficient among the CPG oscillator units; g_k is the reflex coefficient of the k-th feedback; s_{ik}^x and s_{ik}^y are the k-th feedback input of x_i and y_i of oscillator i, respectively; m is the total number of feedback items; θ_i^j is the relative phase between the oscillator i and oscillator j; and $\boldsymbol{R}\left(\theta_i^j\right)$ is the rotation matrix that describes the phase coupling relationship among the CPG oscillator units. Other parameters are the same as described in Equations (6)–(8). Matrix $\boldsymbol{R}\left(\theta_i^j\right)$ is given by

$$\boldsymbol{R}\left(\theta_i^j\right) = \boldsymbol{R}_{ji} = \begin{bmatrix} cos\theta_{ji} & -sin\theta_{ji} \\ sin\theta_{ji} & cos\theta_{ji} \end{bmatrix} \quad (11)$$

where $\theta_i^j = \theta_{ji} = \varphi_i - \varphi_j$, and φ_i is the phase of the oscillator i.

Equation (9) can be rewritten as

$$\dot{\boldsymbol{Q}} = \boldsymbol{F}(\boldsymbol{Q}) + \delta \boldsymbol{R}\boldsymbol{Q} + \boldsymbol{S}\boldsymbol{G} \quad (12)$$

where $\boldsymbol{Q} = \begin{bmatrix} x_1 & y_1 & x_2 & y_2 & x_3 & y_3 & x_4 & y_4 \end{bmatrix}^{\mathrm{T}}$, and $\boldsymbol{R} = \begin{bmatrix} R_{11} & R_{21} & R_{31} & R_{41} \\ R_{12} & R_{22} & R_{32} & R_{42} \\ R_{13} & R_{23} & R_{33} & R_{43} \\ R_{14} & R_{24} & R_{34} & R_{44} \end{bmatrix}$ is the connection weight matrix of the CPG control network, which determines the output of the CPG control network, such as the different gaits and their mutual transition, $\boldsymbol{S} = \begin{bmatrix} s_{11}^x & s_{11}^y & \cdots & s_{41}^x & s_{41}^y \\ \vdots & \vdots & \ddots & \vdots & \vdots \\ s_{1m}^x & s_{1m}^y & \cdots & s_{4m}^x & s_{4m}^y \end{bmatrix}^T$

and $G = \begin{bmatrix} g_1 \\ \vdots \\ g_m \end{bmatrix}$ are the reflex information matrix and reflex coefficient vector respectively. The walk, trot, pace and gallop are the four typical gaits of quadruped robots and their phase relationships is shown in Table 2 respectively.

Table 2. Four typical gaits and their matrixes.

Gait	Gait Matrix (Phase Relationships)
walk	$(0, \pi, \pi/2, 3\pi/2)$
trot	$(0, \pi, 0, \pi)$
pace	$(0, \pi, \pi, 0)$
gallop	$(0, 0, \pi, \pi)$

After introducing the gait matrix into Equation (11), we can obtain the corresponding connection weight matrix $\boldsymbol{R}$. Further, the connection weight matrices of the four typical gaits are given by Figure 5.

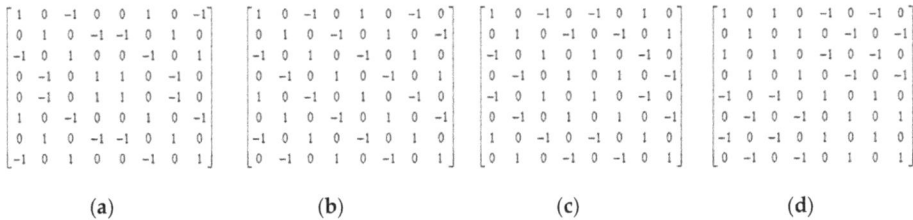

$$\begin{bmatrix} 1 & 0 & -1 & 0 & 0 & 1 & 0 & -1 \\ 0 & 1 & 0 & -1 & -1 & 0 & 1 & 0 \\ -1 & 0 & 1 & 0 & 0 & -1 & 0 & 1 \\ 0 & -1 & 0 & 1 & 1 & 0 & -1 & 0 \\ 0 & -1 & 0 & 1 & 1 & 0 & -1 & 0 \\ 1 & 0 & -1 & 0 & 0 & 1 & 0 & -1 \\ 0 & 1 & 0 & -1 & -1 & 0 & 1 & 0 \\ -1 & 0 & 1 & 0 & 0 & -1 & 0 & 1 \end{bmatrix}$$

(**a**)

$$\begin{bmatrix} 1 & 0 & -1 & 0 & 1 & 0 & -1 & 0 \\ 0 & 1 & 0 & -1 & 0 & 1 & 0 & -1 \\ -1 & 0 & 1 & 0 & -1 & 0 & 1 & 0 \\ 0 & -1 & 0 & 1 & 0 & -1 & 0 & 1 \\ 1 & 0 & -1 & 0 & 1 & 0 & -1 & 0 \\ 0 & 1 & 0 & -1 & 0 & 1 & 0 & -1 \\ -1 & 0 & 1 & 0 & -1 & 0 & 1 & 0 \\ 0 & -1 & 0 & 1 & 0 & -1 & 0 & 1 \end{bmatrix}$$

(**b**)

$$\begin{bmatrix} 1 & 0 & -1 & 0 & -1 & 0 & 1 & 0 \\ 0 & 1 & 0 & -1 & 0 & -1 & 0 & 1 \\ -1 & 0 & 1 & 0 & 1 & 0 & -1 & 0 \\ 0 & -1 & 0 & 1 & 0 & 1 & 0 & -1 \\ -1 & 0 & 1 & 0 & 1 & 0 & -1 & 0 \\ 0 & -1 & 0 & 1 & 0 & 1 & 0 & -1 \\ 1 & 0 & -1 & 0 & -1 & 0 & 1 & 0 \\ 0 & 1 & 0 & -1 & 0 & -1 & 0 & 1 \end{bmatrix}$$

(**c**)

$$\begin{bmatrix} 1 & 0 & 1 & 0 & -1 & 0 & -1 & 0 \\ 0 & 1 & 0 & 1 & 0 & -1 & 0 & -1 \\ 1 & 0 & 1 & 0 & -1 & 0 & -1 & 0 \\ 0 & 1 & 0 & 1 & 0 & -1 & 0 & -1 \\ -1 & 0 & -1 & 0 & 1 & 0 & 1 & 0 \\ 0 & -1 & 0 & -1 & 0 & 1 & 0 & 1 \\ -1 & 0 & -1 & 0 & 1 & 0 & 1 & 0 \\ 0 & -1 & 0 & -1 & 0 & 1 & 0 & 1 \end{bmatrix}$$

(**d**)

Figure 5. Connection weight matrixes of the four typical gaits. (**a**) walk; (**b**) trot; (**c**) pace; (**d**) gallop.

An important feature of CPG control network is the output waveform of gaits and their mutual transitions. The CPG oscillation units can be coupled with any relative phase, since they are coupled to each other with its two output states by the rotation matrix. Besides, a gait can be transformed to another gait by replacing the corresponding target $\boldsymbol{R}$ directly, and the gait transition point can be selected arbitrarily when expecting the gait transition.

As the walk and trot gaits are the most representative, with the coupling intensity coefficient $\delta = 1$ and $\delta = 0.3$ respectively, the output waveform of CPG in the walk and trot gaits and their mutual transitions are shown in Figure 6, where $\mu = 1$, $\alpha = \gamma = b = 50$, $T = 1$ s, the third term on the right side of Equation (9) is 0 since there is no feedback to the CPG, and $\beta = 0.75$ in the walk gait while $\beta = 0.5$ in the trot gait.

As can be seen from Figure 6, the gait transition speed and its output waveform are adjusted by δ independently. Specifically, when δ is relatively large (as shown in the upper part (a) of Figure 6, $\delta = 1$), a gait can be transformed to another gait quickly (the transition time $\leq T$). However, due to the relatively strong coupling, the output waveforms rise slowly in the front part of the ascending phase while rising violently in the rear part in walk gait with the duty factor $\beta > 0.5$, and the greater the duty factor β exceeds 0.5, the more obvious the over-coupling phenomenon is. However, when a smaller coupling intensity coefficient δ is used, the gait transition speed is reduced relatively, but it can effectively avoid the over-coupling phenomenon and the whole transition process is smoother (as shown in the lower part (b) of Figure 6, $\delta = 0.3$). Therefore, δ can be used to control the gait transition speed in different situations.

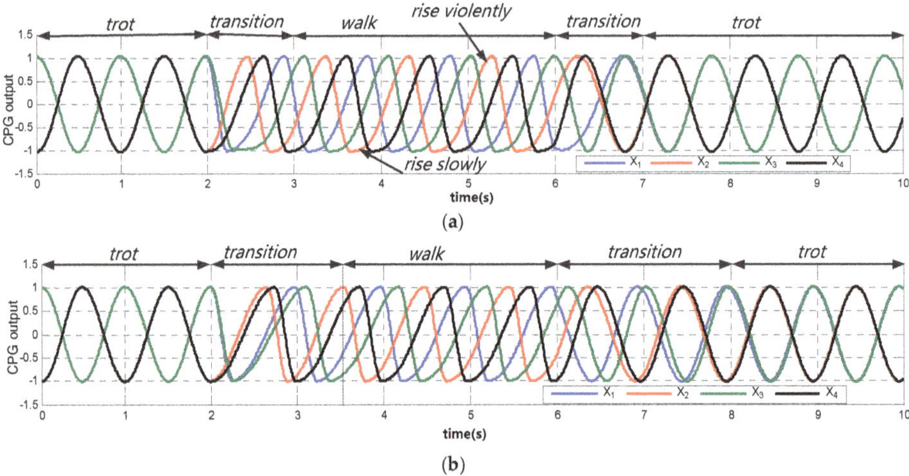

Figure 6. The output waveform of CPG in trot and walk gaits and their mutual conversions. (**a**) $\delta = 1$, the output waveforms of CPG in tort and walk gait and their mutual transitions; (**b**) $\delta = 0.3$, the output waveforms of CPG in tort and walk gait and their mutual transitions. The robot's leg lift sequence is 1-3-2-4 in walk gait.

4. CPG-NN-WT-Based Control Strategy

In this section, the CPG-NN-WT model is first presented for the control of a quadruped robot, which take advantage of both CPG-based method and WT-based method. Then the implementation of the proposed CPG-NN-WT-based control strategy is described in detail.

4.1. CPG-NN-WT Control Model

The fundamental idea of the CPG-NN-WT-based control strategy is to combine the advantages of CPG-based method and WT-based method by a plastic intermediate layer, the NN, which converts the CPG output to the preplanned WT. The relationship among them can be depicted as Figure 7.

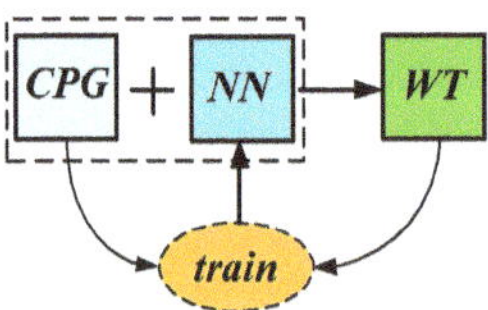

Figure 7. Central pattern generator—neural network—workspace trajectory (CPG-NN-WT) relationship diagram.

As shown in Figure 7, the NN can be obtained by training sample pairs, which are the CPG output and the preplanned WT. Consequently, the preplanned WT can be reappeared when using the CPG-NN-WT control model.

In this paper, the CPG-NN-WT control model for a quadruped robot is shown in Figure 8.

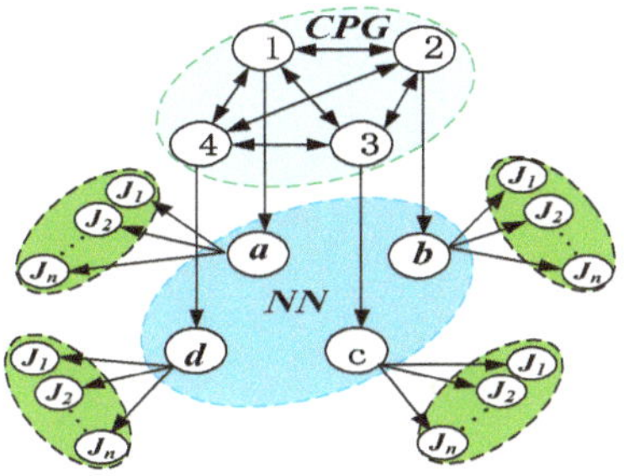

Figure 8. The CPG-NN-WT control model for a quadruped robot.

In Figure 8, the uppermost elliptical dashed box is the CPG layer and the middle elliptical dashed box is the trained NN layer, where a, b, c and d are the four identical NNs, which convert the CPG output into the multi-joint coordinated control signals within a leg (as shown in Figure 8, for example, the NN converts the CPG unit 1 control signal into the coordinated control signals of joint $J_1, J_2, \ldots J_n$ of the left front leg), since the NN has the capability of multi-input and multi-output. What is important is that the more joints in a leg that need to be coordinated controlled, the more obvious the advantages of this control model. In Figure 8, we set the subscript $n = 2$, since the quadruped robot model used in this paper has only hip joint and knee joint within a leg.

4.2. Realization of CPG-NN-WT Control Model

With the preplanned WT and CPG introduced above, the CPG-NN-WT-based control strategy can be achieved after the following steps.

Step 1: the NN choosing and its input and output determining

In this paper, the radial basis function neural network (RBFNN) is chosen for two reasons: (1) It is a local approximation network, which learns fast and can effectively avoid falling into the local minimal. (2) It is a three-layer feed-forward network with a single hidden layer and can approximate any continuous function with arbitrary precision [7,30–34].

Considering the performance and the structure size of the NN as well as computational cost, the NN is defined as the follows: one input to receive the CPG output and outputting two coordinated signals to simultaneously control the hip and knee joints within a leg. However, it can be seen easily form Figure 9 that the CPG output falls in the same range of $(-1,1)$ during both its swing and stance phases. That is, the input of NN is overlapped. Although the NN has the capability to automatically classify if the leg should be in swing or stance phase based on the CPG outputs, it makes the structure size bigger, relatively. Fortunately, each CPG oscillator unit has two outputs, no matter what state it is in, there is a strict correspondence between its two outputs, as shown in Figure 9 (where $i = 1$).

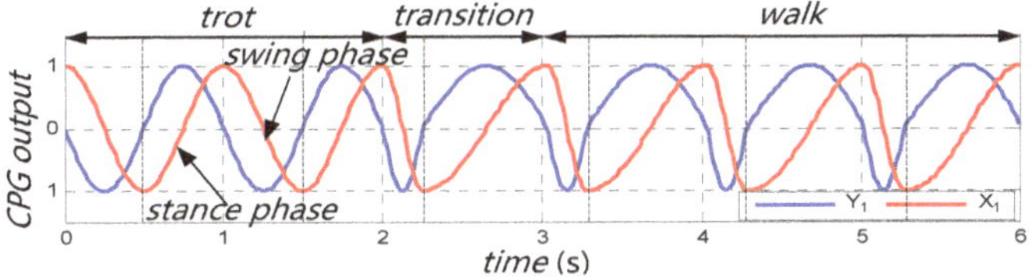

Figure 9. The output of single CPG oscillation unit.

The corresponding relationship is: when the CPG oscillator unit output x_i is in the swing phase (in this paper, the decline part of the curve x_i is the swing phase), the other output $y_i \leq 0$; while x_i is in stance phase, the other output $y_i > 0$.

In order to solve the problem of NN input overlap and make the structure of NN smaller (the smaller the structure of NN, the smaller the computational cost), the corresponding relationship between the two outputs of the CPG oscillator unit is used with the piecewise function method. Two NNs corresponding to the swing and stance phases of a CPG oscillator unit respectively are used to receive the CPG output. Then, the NN is given by

$$N_{network} = \begin{cases} N_{swing} & y_i \leq 0 \\ N_{stance} & y_i > 0 \end{cases} \tag{13}$$

where $N_{network}$ is the RBFNN; and N_{swing} and N_{stance} are the subsections of the RBFNN corresponding to the swing and stance phases, respectively.

Step 2: NN construction and training

The key point of the RBFNN construction is the nodes in its hidden layer, which is related to the network performance and its computation complexity. In this paper, the hip and knee joints' coordinated control signals are used as the output, and the main output x_i of the CPG unit with the duty factor β = 0.5 in a gait period is used as the input to train the RBFNN. The total number of the training sample pairs is 2002 (the swing and stance phase are evenly distributed in 1001 pairs respectively). The distribution density of the radial basis function (RBF) is 0.34 for the swing phase neural network N_{swing} and 0.8 for the stance phase neural network N_{stance}. The MATLAB neural network toolbox is used to analyze its number of nodes in hidden layer and other parameters of the RBFNN with its training performance, as shown in Figure 10.

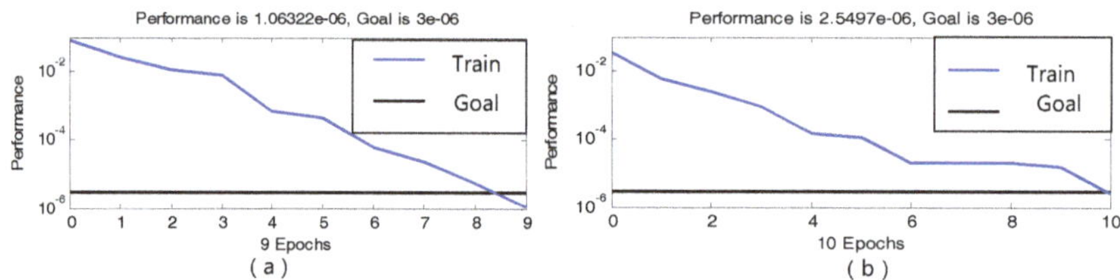

Figure 10. The training performance of the radial basis function neural network (RBFNN). (**a**) the performance of N_{swing} training; (**b**) the performance of N_{stance} training.

In view of the locomotion control of a quadruped robot, the goal of RBFNN training mean square error is chosen to be 3×10^{-6}. It can be seen from Figure 10 that the goal is achieved when the number of the hidden layer nodes of the N_{swing} and N_{stance} reach 9 and 10, respectively. The network structure is shown in Figure 11.

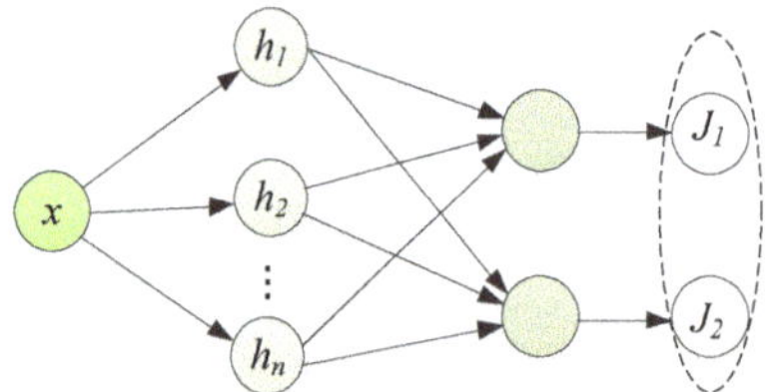

Figure 11. RBFNN network structure.

In this study, 2000 pairs of test samples (the swing phase and stance phase are evenly distributed in 1000 pairs respectively) are used to test the trained RBFNN. The test results are shown in Figure 12.

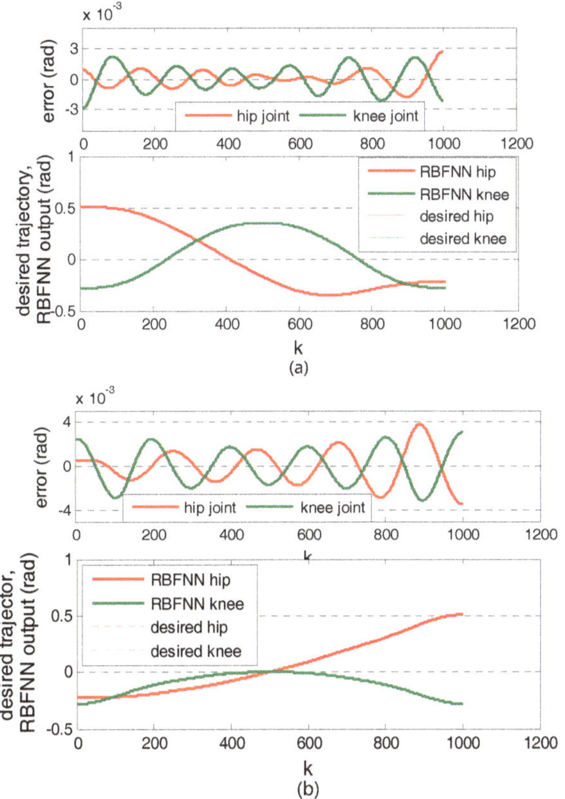

Figure 12. Test performance of the RBFNN. (**a**) Test performance of N_{swing}; (**b**) Test performance of N_{stance}.

It can be seen from Figure 12 that the hip joint and knee joint angular displacement curves of the RBFNN are quite coincident with the curves of the preplanned WT, and their maximum absolute error is under 3×10^{-3} *rad* in swing phase and under 4×10^{-3} *rad* in stance phase respectively.

Step 3: The Output of CPG-NN-WT Control Model

In simple terms, for the output of CPG-NN-WT control model, its essence is the output of CPG + NN. As walk and trot gaits are the most typical gaits of the quadruped robot, Figure 13 shows the CPG-NN-WT output curves of the walk and trot gaits and their mutual transitions, where $\delta = 0.3$, the walk and trot gaits duty factor are $\beta = 0.75$, $\beta = 0.5$ respectively. Other parameters are same as described in Figure 6.

In Figure 13, the first row of curves is the CPG output, the 2–5 rows are the CPG + NN outputs, namely the CPG-NN-WT outputs, which simultaneously controls the hip and knee joints within a leg to adjust the foot positions on the preplanned WT. As shown in Figure 13, no matter what gaits, or the processes of gait transition the robot is in (Figure 13 shows the most typical walk and trot gaits and their mutual transitions), the CPG-NN-WT output curves are in accordance with the preplanned foot WT which is shown in Figure 3 (Figure 12 analyzes its error), and the curves are continuous, smooth and without any sharp point.

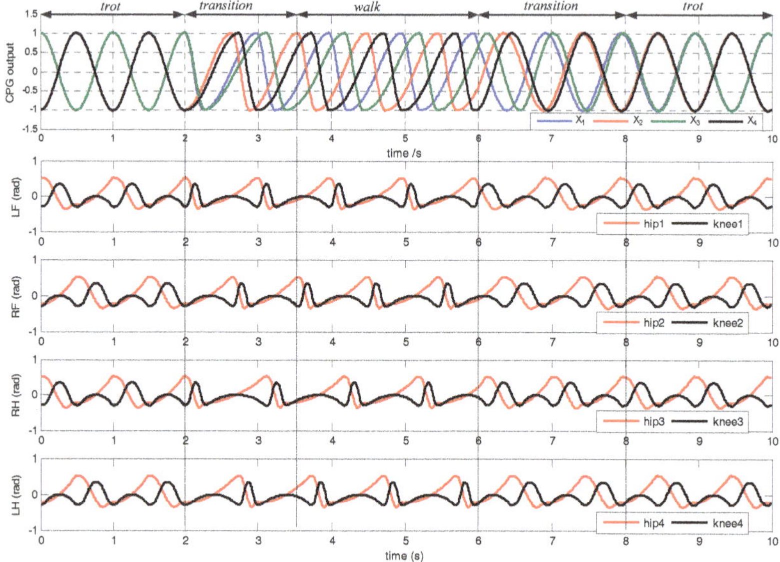

Figure 13. The CPG-NN-WT outputs.

5. Results and Discussion

In order to validate the correctness and validity of the presented CPG-NN-WT-based control strategy, the virtual prototype simulation based on Webots and the experiments with a real quadruped robot are carried out, respectively.

5.1. Virtual Prototype Simulation

Webots is a development environment used to model, program and simulate mobile robots. First of all, a quadruple robot virtual prototype is created in the Webots simulation platform, as shown in Figure 14, where its components and configurations are the same as described in Figure 1. In addition, in order to facilitate the observation of locomotion performance and foot trajectory of the robot, the CPS sensors (used to obtain the location of GPS points in the world coordinate system of the simulation platform) are added in its body geometric center and foot point respectively. Meanwhile, the touch sensors were added at its foot, which were used to measure the force of the foot in the vertical direction.

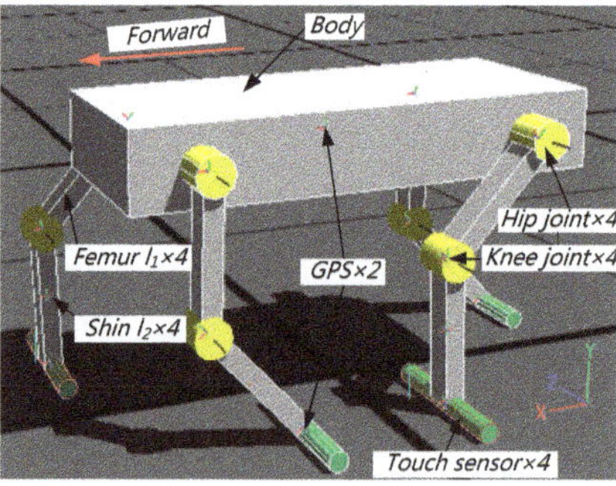

Figure 14. The quadruped robot virtual prototype.

The basic parameters of the robot virtual prototype are shown in Table 3.

Table 3. The robot virtual prototype parameters.

Parameters	Value
size: $L \times W \times H$ (mm)	$500 \times 200 \times 405$
mass (kg)	4
l_1 (mm)	150
l_2 (mm)	150
hip joint range	$(-\pi, \pi)$
knee joint range	$(-\pi, \pi)$

The simulation results with the representative trot and walk gaits and their mutual transitions are shown in Figures 15–18, where the parameters of the CPG-NN-WT control model are the same as in Figure 13.

The upper part of Figure 15 is the robot centroid displacement (geometric center of the robot body), and the lower part is the foot force bearing status of the left front leg. As shown in the upper part of Figure 15, during the trot, walk gaits and their mutual transition process, the centroid displacement curves in the forward direction x_c is smooth and coherent without large fluctuations. In its vertical direction y_c, the curve is smooth and has almost no fluctuation. Further, the slope of the forward direction curve x_c of centroid displacement is large in the trot gait but small in the walk gait due to the different duty factors of the two gaits. In addition, since the yaw control is not taken into account, the lateral displacement curve z_c of centroid is shifted during gait transition. As shown in the lower part of Figure 15, during the trot, walk gaits or their mutual transition process, the force of the foot is balanced and has small mechanical impact without sharp mutation relatively. (In order to more intuitively reflect the stress state, there is no filtering processing).

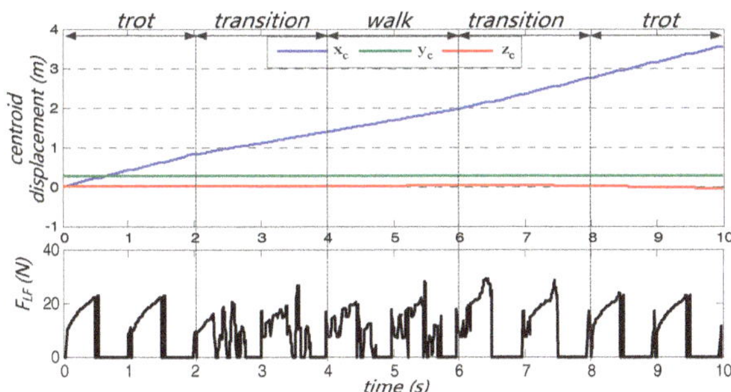

Figure 15. Centroid displacement and the foot force bearing status of the robot virtual prototype.

Figure 16 shows the comparison of the preplanned WT (desired trajectory) with real trajectory of the left front leg when robot is traveling. As shown in Figure 16, there are left or right shifts between the real trajectory curves and the preplanned WT curve due to the yaw angle, which is generated during the gait transition but which is not taken into account while processing data. However, the real trajectories are still in accordance with the preplanned WT if the effect of yaw angle is not taken into account, e.g., the real trajectory of which the yaw angle is 0 during the first gait period is in accordance with the preplanned WT well, and its maximum absolute error in x is under 0.4 mm, which is about 2×10^{-3} S and its maximum absolute error in y is under 0.2 mm, which is about 6.7×10^{-3} H, as shown in Figures 17 and 18.

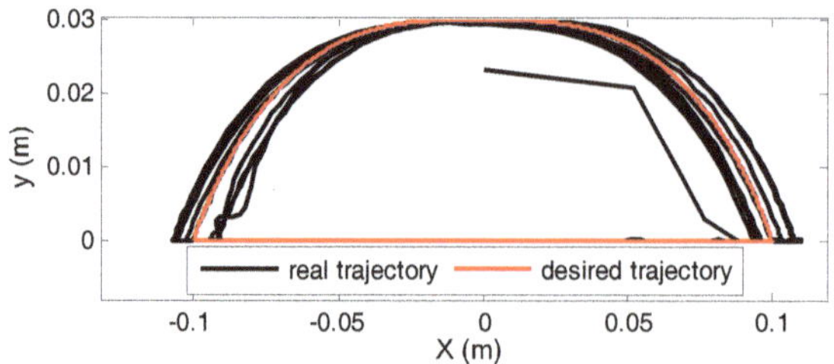

Figure 16. Foot trajectories.

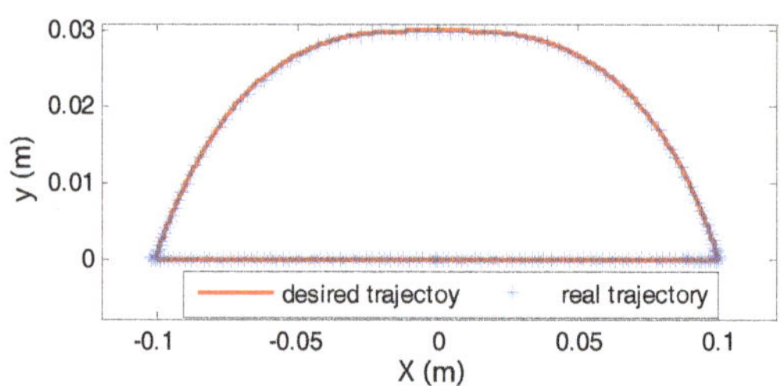

Figure 17. The foot trajectories during the first gait period.

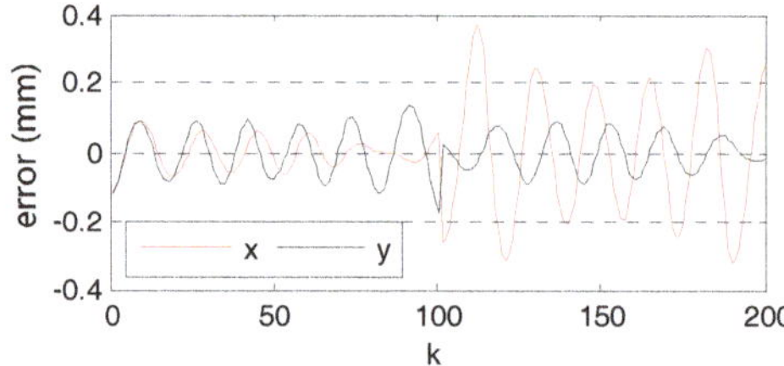

Figure 18. Absolute error of the foot trajectory during the first gait period.

Figure 19 shows the simulation video captures of the virtual prototype in the Webots platform. Throughout the whole simulation process, the locomotion of the quadruped robot is smooth and fluent. Referring to Figure 13, it can be seen that during the whole simulation process, the "CPG part" has played its advantages in gait planning and controlling well, while the "NN part" has the ability to adjust the foot positions according to the preplanned WT. It is verified that the CPG-NN-WT-based control strategy presented in this paper can effectively integrate the advantages of CPG-based method with WT-based method in the locomotion control of a quadruped robot.

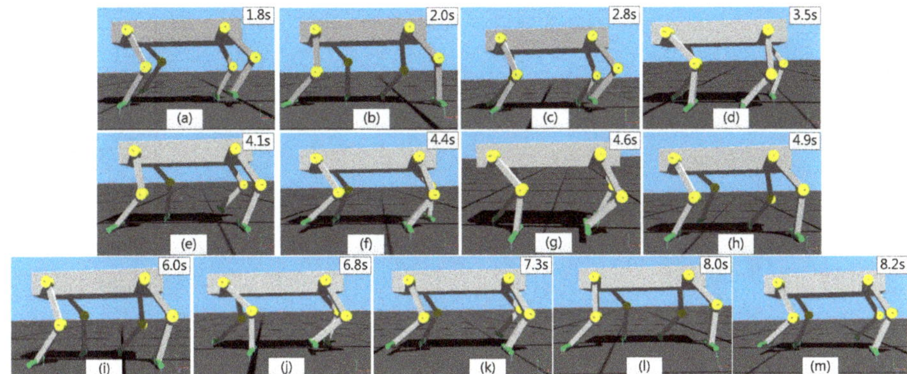

Figure 19. The simulation video capture, where (**a**) the typical trot gait; (**b**) starting the transition from trot to walk gait; (**c**) the gait transition process; (**d**) transition finished; (**e**), (**f**), (**g**) and (**h**) the typical walk gait, robot's leg lift sequence is 1-3-2-4; (**i**) starting the transition from walk to trot gait; (**j**) and (**k**) the gait transition process; (**l**) transition finished; (**m**) the typical trot gait.

5.2. Experiment with a Real Quadruped Robot

After the virtual prototype simulation, the CPG-NN-WT control model with the same parameters is applied to a physical prototype whose basic parameters are shown in Table 4. Besides, an accelerometer is added in its body geometric center to measure the accelerations in three directions, the x-axis (lengthways), the y-axis (vertical) and the z-axis (lateral).

Table 4. Physical prototype parameters.

Parameters	Value
size: $L \times W \times H$ (mm)	$400 \times 220 \times 420$
mass (kg)	3.6
l_1 (mm)	150
l_2 (mm)	150
hip joint range	$(-\pi, \pi)$
knee joint range	$(-\pi, \pi)$

In order to validate the presented control strategy, the experiment results of the trot, walk gaits and their mutual conversions are shown in Figure 20. The robot's actual stride length S is about 0.2 m, the gait period T is about 1 s, and the maximum height of leg raise H is about 0.03 m, which is basically consistent with the parameters set in CPG-NN-WT and preplanned foot WT. Besides, the average move speed in 10 seconds is about 0.4 m/s and the biggest lateral shifting is 0.1 m, which is about 0.5 S.

The acceleration curves of the robot when moving are shown in Figure 21. In the vertical direction (y-axis), the acceleration is stable at around 9.8 m/s^2 (gravitational acceleration), which shows the predesigned WT reduce the mechanical shock well. In the lengthways direction (x-axis), the fluctuation of the acceleration is bigger in trot but smaller in walk due to the walk gait being more stable than trot. In the lateral direction (z-axis), except for the gaits transition period, the acceleration range is very small.

To sum up, from the view of the whole moving process, the locomotion of the physical prototype is fluent and the body shakes are in an acceptable range. In addition, from the hip joint coordinate system, the foot trajectory is still in accordance with the preplanned WT well, which verifies that the presented control strategy is correct and effective.

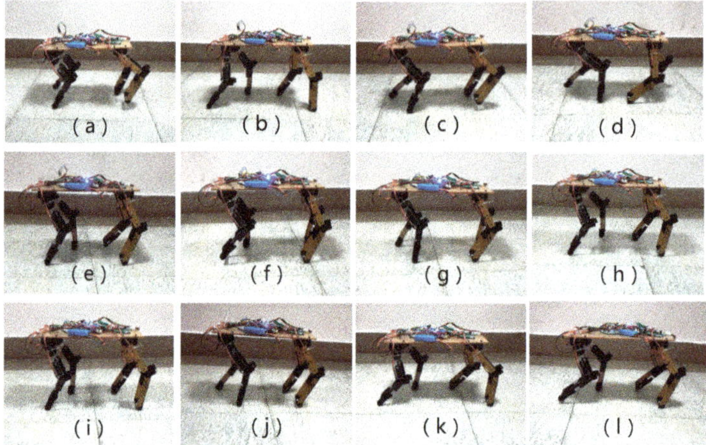

Figure 20. Photos of real robot experiment. where (**a**) the typical trot gait; (**b**) starting the transition from trot to walk gait; (**c**): the gait transition process; (**d**) transition finished; (**e**), (**f**), (**g**) and (**h**) the typical walk gait; (**i**) starting the transition from walk to trot gait; (**j**) the gait transition process; (**k**) transition finished; (**l**) the typical trot gait.

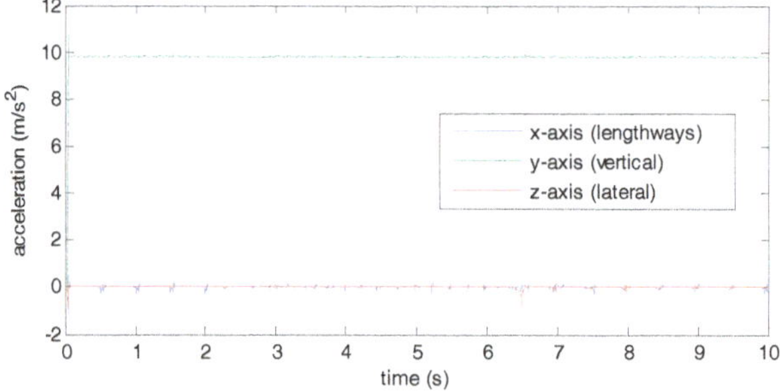

Figure 21. Acceleration curves of the robot.

6. Conclusions

Some concluding remarks would be summarized as follows.

(1) An improved foot WT based on the compound cycloid is planned with advantages of low mechanical impact, smooth movement and sleek trajectory.

(2) An improved CPG based on Hopf oscillators put forward in this paper can effectively realize the smooth gait planning by adjusting its internal parameters.

(3) A biologically inspired control strategy based on CPG-NN-WT is presented for locomotion control of a quadruped robot, which can effectively integrate the advantages of CPG-based method with WT-based method. Besides, the presented control strategy provides an effective way to realize the multi-joint coordination control within a leg, since the NN has the capability of multi-input and multi-output. Furthermore, theoretically, the CPG-NN-WT control model can output any desired periodic WT, which depends only on the complexity of the NN.

By bringing in feedbacks and referring to the neural system of legged animals, the adaptive dynamic walking on irregular terrains using reflexes and online learning of neural networks is being studied.

Acknowledgments: We would like to thank the support from the National Natural Science Foundation of China(No. 51575456), the Open Research Fund of Key Laboratory of Integration and application of solar energy technology (2017-TYN-Y-02), the Key scientific research fund of Xihua University (Grand No. Z1420210, Z1620211), Sichuan International S & T Cooperation and Exchange R & D Project (No. 2017HH0049), the Open Research Fund of Health Management Development Center (Xihua University, No. s2jj2017-023, s2jj2017-039), and the Innovation Fund of Postgraduate, Xihua University (ycjj2017034).

Author Contributions: Yinquan Zeng and Junmin Li co-organized the work and wrote the manuscript draft; Yinquan Zeng and Erwei Ren co-performed the experiments; Simon X. Yang supervised the research and commented on the manuscript writing.

Conflicts of Interest: The authors declare no conflict of interest.

References

1. Ijspeert, A.J. Central pattern generators for locomotion control in animals and robots: A review. *Neural Netw.* **2008**, *21*, 642–653. [CrossRef] [PubMed]
2. Kalakrishnan, M.; Buchli, J.; Pastor, P.; Mistry, M.; Schaal, S. Learning, planning, and control for quadruped locomotion over challenging terrain. *Int. J. Robot. Res.* **2011**, *30*, 236–258. [CrossRef]
3. Kalakrishnan, M.; Buchli, J.; Pastor, P.; Mistry, M.; Schaal, S. Fast, robust quadruped locomotion over challenging terrain. In Proceedings of the 2010 IEEE International Conference on Robotics and Automation, Anchorage, AK, USA, 3–7 May 2010.
4. Lei, J.; Wang, F.; Yu, H.; Wang, T.; Yuan, P. Energy efficiency analysis of quadruped robot with trot gait and combined cycloid foot trajectory. *Chin. J. Mech. Eng.* **2014**, *27*, 138–145. [CrossRef]
5. Xie, H.; Shang, J.; Ahmadi, M. Adaptive control strategies for quadruped robot on unperceived slopedterrain. *Int. J. Robot. Autom.* **2015**, *30*, 90–111.
6. Li, J.; Wang, J.; Yang, S.X.; Zhou, K.; Tang, H. Gait Planning and Stability Control of a Quadruped Robot. *Comput. Intell. Neurosci.* **2016**, *2016*, 9853070. [CrossRef] [PubMed]
7. Qian, X.; Huang, H.; Chen, X.; Huang, T. Generalized Hybrid Constructive Learning Algorithm for Multioutput RBF Networks. *IEEE Trans. Cybern.* **2017**, *47*, 3634–3648. [CrossRef] [PubMed]
8. Matos, V.; Santos, C.P. Omnidirectional locomotion in a quadruped robot: A CPG-based approach. In Proceedings of the 2010 IEEE/RSJ International Conference on Intelligent Robots and Systems, Taipei, Taiwan, 18–22 October 2010; Volume 6219, pp. 3392–3397.
9. Koco, E.; Mutka, A.; Kovacic, Z. New parameterized foot trajectory shape for multi-gait quadruped locomotion with state machine-based approach for executing gait transitions. In Proceedings of the 2014 22nd Mediterranean Conference on Control and Automation, Palermo, Italy, 16–19 June 2014; pp. 1533–1539.
10. Wu, Q.; Liu, C.; Zhang, J.; Chen, Q. Survey of locomotion control of legged robots inspired by biological concept. *Sci. China Ser. F Inf. Sci.* **2009**, *52*, 1715–1729. [CrossRef]
11. Tran, D.T.; Koo, I.M.; Lee, Y.H.; Moon, H.; Park, S.; Koo, J.C.; Choi, H.R. Central pattern generator based reflexive control of quadruped walking robots using a recurrent neural network. *Robot. Auton. Syst.* **2014**, *62*, 1497–1516. [CrossRef]
12. Matsuoka, K. Mechanisms of frequency and pattern control in the neural rhythm generators. *Biol. Cybern.* **1987**, *56*, 345–353. [CrossRef] [PubMed]
13. Kimura, H.; Fukuoka, Y.; Hada, Y.; Takase, K. *Adaptive Dynamic Walking of a Quadruped Robot on Irregular Terrain Using a Neural System Model*; Springer: Berlin/Heidelberg, Germany, 2003.
14. Tsujita, K.; Toui, H.; Tsuchiya, K. Dynamic turning control of a quadruped locomotion robot using oscillators. *Adv. Robot.* **2005**, *19*, 1115–1133. [CrossRef]
15. Sun, X. *Kuramoto Model*; Springer: New York, NY, USA, 2013.
16. Acebron, J.A.; Bonilla, L.L.; Vicente, C.J.P.; Ritort, F.; Spigler, R. The Kuramoto model: A simple paradigm for synchronization phenomena. *Rev. Mod. Phys.* **2005**, *77*, 137–185. [CrossRef]
17. Zhang, J.; Masayoshi, T.; Chen, Q.; Liu, C. Dynamic Walking of AIBO with Hopf Oscillators. *Chin. J. Mech. Eng.* **2011**, *24*, 612–617. [CrossRef]
18. Hu, Y.; Liang, J.; Wang, T. Parameter Synthesis of Coupled Nonlinear Oscillators for CPG-Based Robotic Locomotion. *IEEE Trans. Ind. Electron.* **2014**, *61*, 6183–6191. [CrossRef]
19. Yu, H.; Gao, H.; Ding, L.; Li, M.; Deng, Z.; Liu, G. Gait Generation With Smooth Transition Using CPG-Based Locomotion Control for Hexapod Walking Robot. *IEEE Trans. Ind. Electron.* **2016**, *63*, 5488–5500. [CrossRef]

20. Kwon, O.; Jeon, K.S.; Park, J.H. Optimal trajectory generation for biped robots walking up-and-down stairs. *J. Mech. Sci. Technol.* **2006**, *20*, 612–620. [CrossRef]
21. Ma, S.; Tomiyama, T.; Wada, H. Omnidirectional static walking of a quadruped robot. *IEEE Trans. Robot.* **2005**, *21*, 152–161. [CrossRef]
22. Liu, C.J.; Wang, D.W.; Chen, Q.J. Locomotion Control Of Quadruped Robots Based on Workspace Trajectory Modulations. *Int. J. Robot. Autom.* **2012**, *27*, 345–354. [CrossRef]
23. Sakakibara, Y.; Kan, K.; Hosoda, Y.; Hattori, M. Foot trajectory for a quadruped walking machine. In Proceedings of the IROS '90 IEEE International Workshop on Intelligent Robots and Systems '90 'towards A New Frontier of Applications', Ibaraki, Japan, 3–6 July 1990; Volume 1, pp. 315–322.
24. Wang, L.; Wang, Z.; Wang, S.; He, Y. Strategy of Foot Trajectory Generation for Hydraulic Quadruped Robots Gait Planning. *J. Mech. Eng.* **2013**, *49*, 39–44. [CrossRef]
25. Righetti, L.; Ijspeert, A.J. *Design Methodologies for Central Pattern Generators: An Application to Crawling Humanoids. Robotics: Science and Systems II*; University of Pennsylvania: Philadelphia, PA, USA, 2006.
26. Righetti, L.; Ijspeert, A.J. Pattern generators with sensory feedback for the control of quadruped locomotion. In Proceedings of the IEEE International Conference on Robotics and Automation, Pasadena, CA, USA, 19–23 May 2008; pp. 819–824.
27. Smith, J.A. Galloping in an Underactuated Quadrupedal Robot. *Int. J. Robot. Autom.* **2015**, *30*. [CrossRef]
28. Ren, J.; Xu, H.; Gan, S.; Wang, B. CPG modele design based on hopf oscillator for hexapod robots gait. *CAAI Trans. Intell. Syst.* **2016**, *11*, 627–634.
29. Li, H.; Han, B.; Luo, Q. Inter-limb and intra-limb coordination control of quadruped robots. *J. Beijing Inst. Technol.* **2015**, *4*, 478–486.
30. Lee, N.K.; Wang, D. Realization of Generalized RBF Network. *Appl. Spectrosc.* **2003**, *62*, 341–344.
31. Nabney, I.T. Efficient training of RBF networks for classification. *Int. J. Neural Syst.* **2004**, *14*, 201–208. [CrossRef] [PubMed]
32. He, W.; Chen, Y.; Yin, Z. Adaptive Neural Network Control of an Uncertain Robot With Full-State Constraints. *IEEE Trans. Cybern.* **2015**, *46*, 620–629. [CrossRef] [PubMed]
33. Chen, H.; Yu, G.; Xia, H. Online Modeling With Tunable RBF Network. *IEEE Trans. Cybern.* **2013**, *43*, 935–947. [CrossRef] [PubMed]
34. Fortuna, L.; Arena, P.; Balya, D.; Zarandy, A. Cellular neural networks: A paradigm for nonlinear spatio-temporal processing. *IEEE Circuits Syst. Mag.* **2001**, *1*, 6–21. [CrossRef]

Article

Parameters Sensitivity Analysis of Position-Based Impedance Control for Bionic Legged Robots' HDU

Kaixian Ba [1], Bin Yu [1,2,*], Zhengjie Gao [1], Wenfeng Li [1], Guoliang Ma [1] and Xiangdong Kong [1,2,3]

[1] School of Mechanical Engineering, Yanshan University, Qinhuangdao 066004, China; bkx@ysu.edu.cn (K.B.); gzj@stumail.ysu.edu.cn (Z.G.); xrlzlwf@stumail.ysu.edu.cn (W.L.); mgl@stumail.ysu.edu.cn (G.M.); xdkong@ysu.edu.cn (X.K.)

[2] Hebei Provincial Key Laboratory of Heavy Machinery Fluid Power Transmission and Control, Qinhuangdao, China

[3] National Engineering Research Center for Local Joint of Advanced Manufacturing Technology and Equipment, Yanshan University, Qinhuangdao 066004, China

* Correspondence: yb@ysu.edu.cn; Tel.: +86-0335-807-4618

Received: 2 September 2017; Accepted: 30 September 2017; Published: 10 October 2017

Abstract: For the hydraulic drive unit (HDU) on the joints of bionic legged robots, this paper proposes the position-based impedance control method. Then, the impedance control performance is tested by a HDU performance test platform. Further, the method of first-order sensitivity matrix is proposed to analyze the dynamic sensitivity of four main control parameters under four working conditions. To research the parameter sensitivity quantificationally, two sensitivity indexes are defined, and the sensitivity analysis results are verified by experiments. The results of the experiments show that, when combined with corresponding optimization strategies, the dynamic compliance composition theory and the results from sensitivity analysis can compensate for the control parameters and optimize the control performance in different working conditions.

Keywords: bionic legged robots; hydraulic drive unit (HDU); position-based impedance control; sensitivity analysis

1. Introduction

Bionic legged robots are better at adapting to unknown and unstructured environments. Their unique advantages, such as overcoming obstacles and executing tasks in the wild, have made them a major focus of research in the robotic domain [1–4]. For the hydraulic drive robot, the highly integrated valve-controlled cylinder composes the drive component, which is called the hydraulic drive unit (HDU) [5,6]. During the robotic motion process, the robotic feet interact with the ground frequently. This means that the demand for HDU not only includes characteristics of response ability and high control accuracy, but also dynamic compliance. Thus, the impact on the hydraulic system can be obviously reduced, which can protect the mechanical structure and components, and improve the moving stability of the robots.

The selection of control methods directly affects the compliance of the HDU. As a commonly used control method for compliance, the impedance control method has been widely applied to motor-driving legged robots such as the Tekken [7], Scout [8] and MIT cheetah robot [9]. In recent years, as the hydraulic-driven legged robot became the focus of increased research, impedance control was also applied to this kind of robot, as exemplified by robots such as Bigdog [10], HyQ [11], and Scalf-1 [12]. Force-based and position-based impedance control methods are often used for dynamic compliance control. Their basic principle can be expressed as follows. Firstly, the hydraulic control system is taken as the control inner loop, and a dynamic control outer loop is attached to the system. When an external disturbance acts on the system, the input signal of the control inner loop can be changed through the control outer loop. Thus, the system is able to possess the desired dynamic compliance.

In this paper, position-based impedance control is the focus of research. There are many parameters in hydraulic systems, such as structure parameters, working parameters, and control parameters. Considering the uncertainty of the parameter variation, the system cannot reach the ideal performance, which involves the effects of parameter variation on robot's overall compliance. So, in order to optimize the system more efficiently, it is necessary to know how greatly the parameters, particularly the main control parameters, influence the impedance control performance. The parameters that affect the system more should be compensated and optimized emphatically, while the ones that affect system less can be ignored. Thus, the effect of parameter variation on system dynamic characteristics can be quantificationally analyzed, and the analysis results can be used to optimize the robot's compliance performance. Sensitivity analysis is used to analyze the effect of parameter variation on system characteristics for both linear and nonlinear systems. By computing methods, sensitivity analysis methods can be classified into the trajectory sensitivity method, output sensitivity method, matrix sensitivity method, comparison sensitivity method, characteristic root sensitivity method, etc. By computing accuracy, sensitivity analysis methods can be classified into ties of first-order sensitivity method, second-order sensitivity method, or high-order sensitivity method. These methods have different characteristics in computing; they differ from each other in aspects such as computing accuracy, computing mode, and computing complexity. These differences give each method unique advantages and application ranges. In recent years, sensitivity analysis methods have been commonly used in many fields, but only a few have been applied to hydraulic systems. Vilenus et al. [13] is the first scholar to apply first-order sensitivity analysis to hydraulic systems. For the position control system of valve-controlled cylinders, the sensitivity of 10 main parameters when they changed 1% in a single working condition is researched. Farasat et al. [14] built a fourth-order linear mathematical model for the position control system of valve-controlled cylinders. In the model, the valve's pressure-flow nonlinearity is partly linearized. Based on Vilenus's research, he showed the first-order sensitivity analysis results of an extra seven parameters, and proposed four assessment methods to quantify the sensitivity of state variables when parameters changed by 1%. Kong et al. [15] built a fifth-order linear mathematical model for the position-based control system, and studied the effect on system output when 14 system parameters changed by 1%. Based on the first-order trajectory sensitivity, Kong et al. [16] deduced the method of second-order trajectory sensitivity and analyzed the effect on system output when 14 system parameters changed from 1% to 20%. Moreover, experiments are also conducted to verify the effect.

In the position-based impedance control system, the above achievements adopted first-order and second-order sensitivity analysis methods to study the effect on control characteristics when parameters change. However, the above papers didn't analyze the impedance control methods on HDU. In addition, the trajectory sensitivity analysis methods they used, particularly the second-order trajectory sensitivity analysis method, are very difficult to compute. To solve the two problems, this paper is organized as follows. First, based on the mathematical model of HDU position control system, the method of position-based impedance control is deduced and tested by experiments. Then, to solve the difficulties in sensitivity computing, matrix sensitivity analysis, an easier method, is adopted to analyze the sensitivity of four main control parameters. Further, to obtain the optimal method, the results from matrix sensitivity analysis are compared with the results from trajectory sensitivity analysis. Moreover, the quantificational analysis results of the four main parameters are shown in this paper. Finally, the sensitivity analysis results are verified by experiments.

2. Introduction of the HDU and Its Performance Test Platform

As the driver of the leg joint on bionic legged robots, the HDU is a highly integrated system of servo valve-controlled symmetrical cylinder. The author's institute participates in the design of the hydraulic quadruped robot. The quadruped robot prototype, single leg, and HDU performance test platform are shown in Figure 1a–c, respectively.

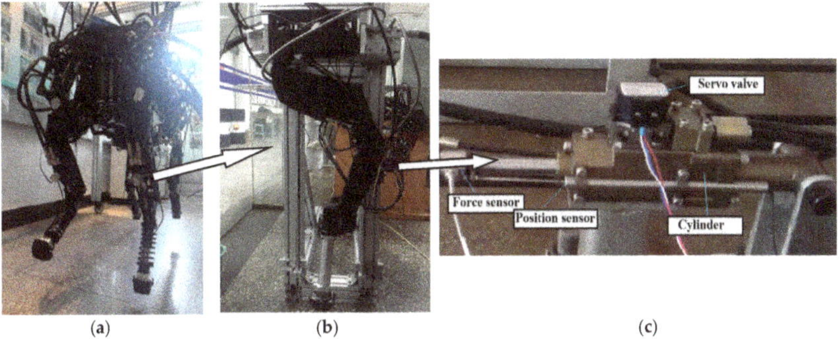

Figure 1. The quadruped robot prototype, single leg and hydraulic drive unit (HDU) performance test platform. (**a**) Quadruped robot prototype; (**b**) Single leg; (**c**) HDU.

The performance of the HDU directly affects the performance of the whole robot. Thus, a special performance test platform is built to study the methods for HDU. The schematic of the test platform is shown in Figure 2.

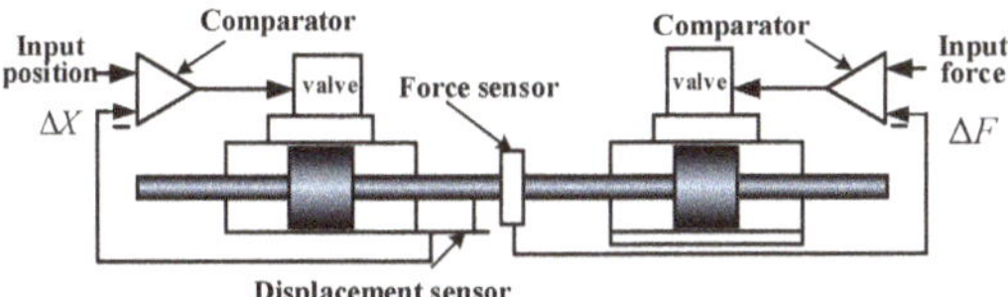

Figure 2. Schematic of HDU performance test platform.

The principle of the electro-hydraulic load simulator in Figure 2 is widely used in many fields such as aviation, aerospace, and vessel and construction machining [5]. In Figure 2, the left part is a HDU-adopted position closed-loop control that contains a small servo valve, servo cylinder, position sensor, and force sensor. The right part is another HDU-adopted force closed-loop control that contains the same type of servo valve and servo cylinder. Two parts' cylinder rods are jointed rigidly by the thread of a force sensor. The HDU performance test platform is showed in Figure 3a. The controller adopted is dSPACE, a semi-physical simulation platform, which is showed in Figure 3b.

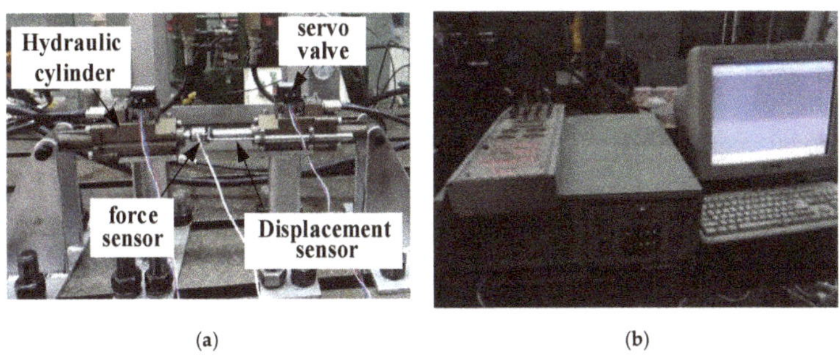

Figure 3. Composition of HDU performance test platform. (**a**) HDU performance test platform; (**b**) dSPACE controller.

3. HDU Position-Based Impedance Control

3.1. Mathematic Model of HDU Position-Based Impedance Control

Impedance control is one type of active compliance control. In particular, it refers to an active compliance control that applies the system equivalent to the second-order mass-spring-damping system. By adopting impedance control, a system can be equipped with the dynamic compliance of a second-order mass-spring-damping system when a disturbance force is applied to the system.

The impedance control is composed by an impedance control inner loop and impedance control outer loop. The impedance control inner loop refers to the closed-loop control, which is realized in the inner loop of the hydraulic position closed-loop control system during the impedance control. In the position-based impedance control of this paper, the impedance control inner loop refers to the position closed-loop control. The impedance control outer loop refers to the open-loop control where the external disturbance signal is transferred to the input signal of impedance control inner loop during the impedance control.

3.1.1. Principles of Impedance Control

The inner loop of the position-based impedance control is a closed-loop control. When the inner loop is affected by a disturbance force, the impedance control outer loop should be added to the system to equip the system with impedance control characteristics. The function of this outer loop is to transform the disturbance force into position error. Then, the desired stiffness K_D can be obtained, which causes an elastic force. In the same way, desired damping C_D and desired mass m_D are obtained, which can cause viscous force and inertia force, respectively. The HDU force schematics with impedance control outer loop are shown respectively in Figures 4 and 5.

During the robot's walking process, the load, such as grounds and steps, provides the disturbance force to the HDU, because the force sensor is mounted on the piston end of the HDU. In this paper, the force control system of the simulated load provides the disturbance force to the performance test platform.

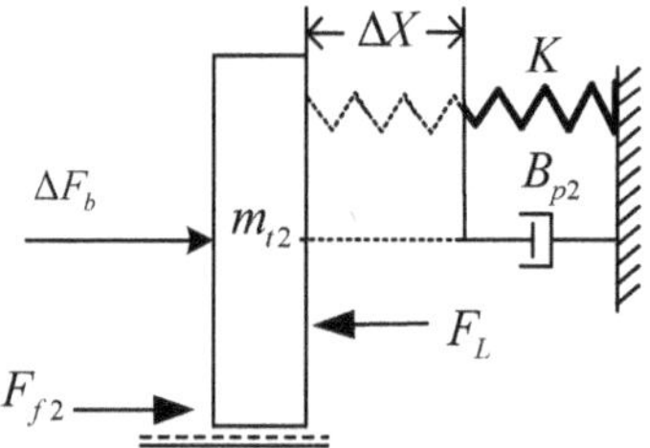

Figure 4. Force schematic of the load.

As it is shown in Figure 4, the load is pressed to position ΔX, B_{p2} refers to damping coefficient at load, m_{t2} refers to equivalent mass at load, and F_{f2} refers to friction at load. The force acting on the piston, which is provided by the sensor, is the disturbance force of the HDU position control system, and is defined as ΔF_a. Further, the force acting on the load, which is provided by the sensor, is defined as ΔF_b. The force balance relation in Figure 4 can be expressed as follows:

$$\Delta F_b = F_L - m_{t2}\Delta \ddot{X} - B_{p2}\Delta \dot{X} - K\Delta X - F_{f2} \tag{1}$$

Third, the force schematic of the HDU is showed in Figure 5.

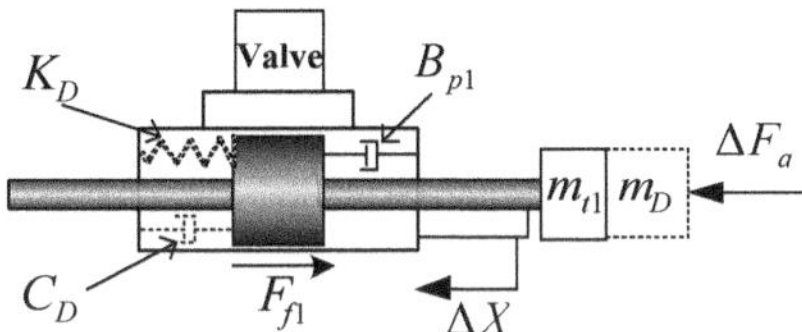

Figure 5. Force schematic of HDU.

Where B_{p1} refers to the viscous damping coefficient, m_{t1} refers to the equivalent mass at piston, and F_{f1} refers to friction in the cylinder. The position error can be expressed as follows:

$$\Delta X = \frac{\Delta F_a - F_{f1}}{(m_{t1} + m_D)s^2 + (B_{p1} + C_D)s + K_D} \tag{2}$$

Based on the theoretical analysis above, the schematic of the position-based impedance control is shown in Figure 6.

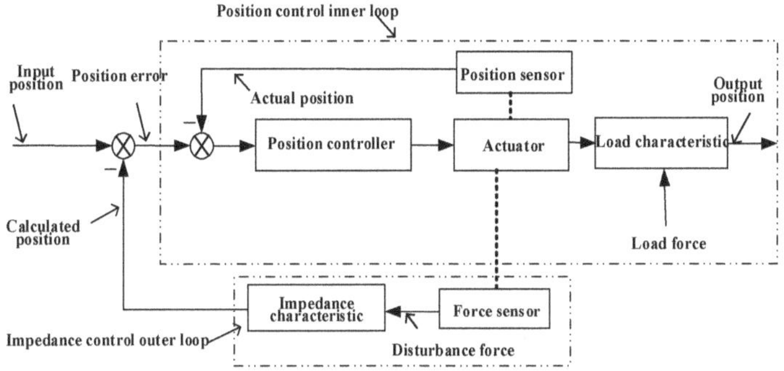

Figure 6. Position-based impedance control schematic.

As can be seen in Figure 6, when the disturbance force tested by the force sensor acts on the HDU, the impedance control outer loop generates a corresponding calculated position that disturbs the input position. Then, the new input to the position control system is formed. Thus, the final input signal enters the position control inner loop, and a new output position is formed to equip the system with impedance characteristics.

3.1.2. The Block Diagram and State Space Presentation of Impedance Control System

The block diagram of HDU position-based impedance control is shown in Figure 7, where the detailed deduction and performance analysis of the inner loop is presented in previous research [5,15].

In Figure 7, m_t is conversion mass (including the piston, the displacement sensor, the force sensor, the connecting pipe, and the oil in the servo cylinder), K_x is input position, is position sensor gain, K_{PID} is proportion-integration-differentiation (PID) controller gain including proportional gain K_P, integral gain K_I and differential gain K_D, K is load stiffness, B_p is load damping, F_L is load force, X_v is servo valve spool displacement, X_p is servo cylinder piston displacement, V_{g1} is volume of input oil pipe, V_{g2} is volume of output oil pipe, F_f is friction, U_r is input voltage, U_g is controller output voltage, Q_1 is inlet oil flow, and Q_2 is outlet oil flow.

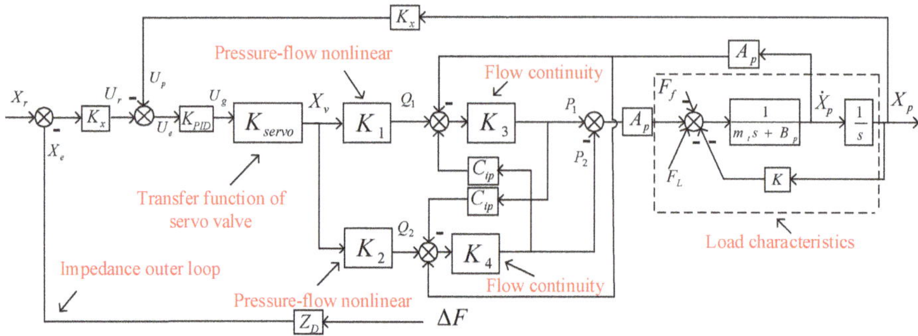

Figure 7. Block diagram of HDU position-based impedance control.

Denote,

$$K_{servo} = \frac{k_{axv}}{(\frac{s^2}{\omega^2} + \frac{2\zeta}{\omega}s + 1)} \tag{3}$$

where K_{servo} is the transfer function of the servo valve, ω is the natural frequency of the servo valve, ζ is the damping ratio of the servo valve, and K_{axv} is the servo valve gain.

$$K_1 = K_d\sqrt{\left\{\frac{[1+\text{sgn}(x_v)]P_s}{2} + \frac{[-1+\text{sgn}(x_v)]P_0}{2}\right\} - \text{sgn}(x_v)P_1} \tag{4}$$

$$K_2 = K_d\sqrt{\left\{\frac{[1-\text{sgn}(x_v)]P_s}{2} + \frac{[-1-\text{sgn}(x_v)]P_0}{2}\right\} + \text{sgn}(x_v)P_2} \tag{5}$$

where, K_1 and K_2 express the transfer function of nonlinear pressure flow, $K_d = C_dW\sqrt{2/\rho}$ (K_d is defined as conversion coefficient in this paper), C_d is the orifice flow coefficient of the spool valve, W is the area gradient of the spool valve, ρ is the density of hydraulic oil, p_s is the system supply's oil pressure, p_1 is the left cavity pressure of the servo cylinder, p_2 is the right cavity pressure of the servo cylinder, and p_0 is the system return oil pressure.

$$K_3 = \frac{1}{C_{ip} + C_{ep} + \frac{V_{g1}+A_pL_0+A_pX_p}{\beta_e}s} \tag{6}$$

$$K_4 = \frac{1}{-C_{ip} - C_{ep} - \frac{V_{g2}+A_p(L-L_0)-A_pX_p}{\beta_e}s} \tag{7}$$

where, K_3 and K_4 express the transfer function of flow continuity, L represents the total piston stroke of the servo cylinder, L_0 is the initial piston position of the servo cylinder, C_{ip} is the internal leakage coefficient of the servo cylinder, C_{ep} is the external leakage coefficient of the servo cylinder, A_p is the effective piston area of the servo cylinder, and β_e is the effective bulk modulus.

The state variables in Figure 7 are expressed as follows:

$$x_1 = x_p,\ x_2 = \dot{x}_p,\ x_3 = x_v,\ x_4 = \dot{x}_v,\ x_5 = \ddot{x}_v - \omega^2K_xK_{axv}K_P\mu_1,\ x_6 = P_1,\ x_7 = p_2,\ x_8 = x_e$$

where the input variables are expressed as follows:

$u_1 = x_r,\ u_2 = F_L + F_f$

Disturbance variables are expressed as follows:

$w_1 = F_L + F_f$

The state space of the system can be expressed as follows:

$$\begin{cases} \dot{x}_1 = x_2 \\ \dot{x}_2 = -\frac{K}{m_t}x_1 - \frac{B_p}{m_t}x_2 + \frac{A_p}{m_t}x_6 - \frac{A_p}{m_t}x_7 - \frac{w_1}{m_t} \\ \dot{x}_3 = x_4 \\ \dot{x}_4 = x_5 + \omega^2 K_x K_{axv} K_P \mu_1 \\ \dot{x}_5 = -\omega^2 K_x K_{axv} K_I x_1 - \omega^2 K_x K_{axv} K_P x_2 - \omega^2 x_4 - 2\zeta\omega x_5 + \omega^2 K_x K_{axv}(\frac{K_D}{C_D}K_P - K_I)x_8 + \omega^2 K_x K_{axv}(K_I - 2\zeta\omega K_P)\mu_1 - \frac{\omega^2 K_x K_{axv} K_P}{C_D}\mu_2 \\ \dot{x}_6 = \beta_e(V_1)^{-1}[-A_p x_2 + K_1 x_3 - C_{ip}x_6 + C_{ip}x_7] \\ \dot{x}_7 = \beta_e(V_2)^{-1}[A_p x_2 - K_2 x_3 + C_{ip}x_6 - (C_{ip} + C_{ep})x_7] \\ \dot{x}_8 = -\frac{K_D}{C_D}x_8 + \frac{1}{C_D}u_2 \end{cases} \quad (8)$$

$$V_1 = V_{01} + A_p x_p = V_{g1} + A_p L_0 + A_p x_p \quad (9)$$

$$V_2 = V_{02} - A_2 x_p = V_{g2} + A_p(L - L_0) - A_p x_p \quad (10)$$

The physical meanings and initial values of the parameters in the control system block diagram are shown in Table 1.

Table 1. Parameters and initial values of the simulation model.

Parameter	Initial Value
Gain of servo valve $K_{axv}/(m/V)$	0.0225
Natural frequency of servo valve $\omega/(\mathrm{rad/s})$	628
Damping ratio of servo valve ζ	0.77
Effective piston area A_p/m^2	3.368×10^{-4}
Volume of inlet chamber V_{g1}/m^3	6.2×10^{-7}
Volume of outlet chamber V_{g2}/m^3	8.6×10^{-7}
Piston stroke L/m	0.05
Initial position of piston L_0/m	0.025
Supply pressure P_s/Pa	1×10^{-7}
Tank pressure P_0/Pa	0.5×10^6
Gain of position sensor $K_x/(\mathrm{V/m})$	54.9×10^{-3}
Outer linkage coefficient of servo valve $C_{ep}/(\mathrm{m}^3/(\mathrm{s{\cdot}Pa}))$	0
Inner linkage coefficient of servo valve $C_{ip}/(\mathrm{m}^3/(\mathrm{s{\cdot}Pa}))$	2.38×10^{-13}
Equivalent mass m_t/kg	1.1315
Effective bulk modulus β_e/Pa	8×10^8
Conversion coefficient K_d	1.248×10^{-4}
Load stiffness $K/(\mathrm{N/m})$	0
Viscous damping coefficient $B_p/(N \cdot s/m)$	54.9×10^{-3}
proportional gain K_p	30
differential gain K_i	10
Desired stiffness $K_D/(N/m)$	1×10^6
Desired damp $C_D/(N \cdot s/m)$	5×10^4
Desired mass M_D/kg	0

3.2. Experiment of Position-Based Impedance Control

As a typical signal, the sinusoidal response is able to evaluate the impedance control performance under the input of different frequencies and amplitudes. In this paper, a sinusoidal signal is adopted to analyze the system performance and the sensitivity of the main parameters. To study their variation patterns in different conditions, four working conditions are tested in this paper. The details of the working conditions are shown in Table 2.

Table 2. Four working conditions researched in this paper.

No.	Working Conditions	
	Frequency f (Hz)	**Bias, Amplitude A (N)**
1	1	1500, 1000
2	2	1500, 1000
3	1	2500, 2000
4	2	2500, 2000

In a position-based impedance control system, the load acts as an external disturbance force that is simulated by the force-based control system. Thus, the desired position can be defined as the ratio of the actual force to the desired impedance characteristic Z_D, i.e., the ratio of the value of the force sensor to Z_D. The actual position is the output of the system to be tested, i.e., the value of the position sensor. The experimental and simulation curves of the sinusoidal response are shown in Figures 8–11 in sequence of working conditions [17].

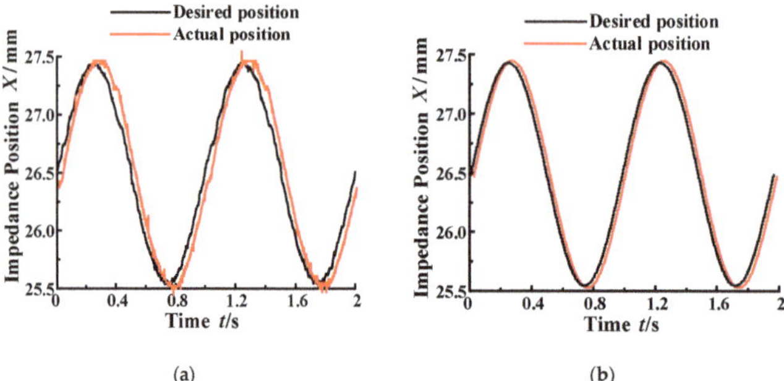

Figure 8. Experimental and simulation curves of sinusoidal response (frequency: 1 Hz, bias: 1500 N, amplitude: 1000 N). (**a**) Experimental curves; (**b**) Simulation curves.

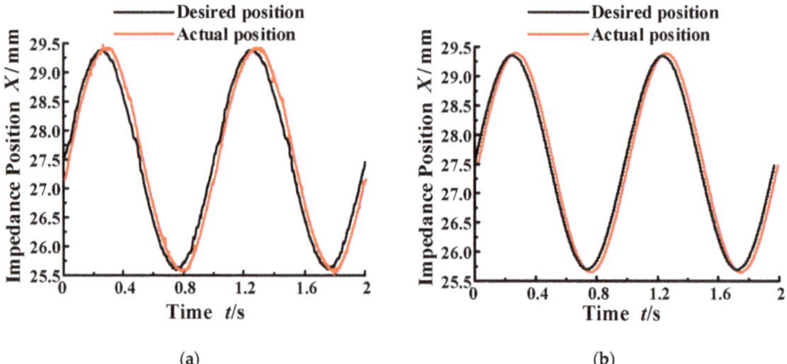

Figure 9. Experimental and simulation curves of sinusoidal response (frequency: 1 Hz, bias: 2500 N, amplitude: 2000 N). (**a**) Experimental curves; (**b**) Simulation curves.

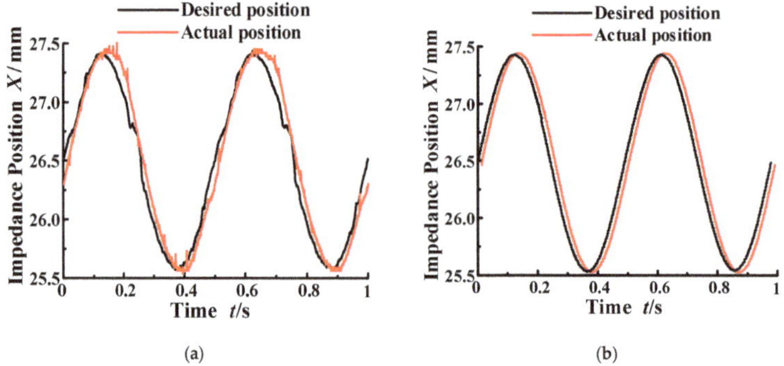

Figure 10. Experimental and simulation curves of sinusoidal response (frequency: 2 Hz, bias: 1500 N, amplitude: 1000 N). (**a**) Experimental curves; (**b**) Simulation curves.

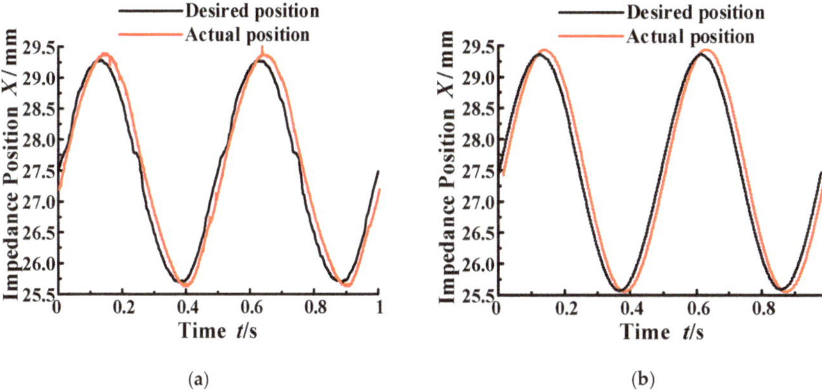

Figure 11. Experimental and simulation curves of sinusoidal response (frequency: 2 Hz, bias: 2500 N, amplitude: 2000 N). (**a**) Experimental curves; (**b**) Simulation curves.

The mean values of performance indexes in different conditions are shown in Table 3.

Table 3. Mean values of performance indexes in different conditions.

Performance Index		No.			
		1	2	3	4
Amplitude reduction (mm)	Experimental	−0.05	−0.09	−0.06	−0.11
	Simulation	−0.03	−0.06	−0.03	−0.06
Phase angle delay (°)	Experimental	7.8	6.7	8.9	9.4
	Simulation	5.2	5.3	7.1	7.5

As it can be seen in Table 3, the values of the two performance indexes are close in experiment and simulation, which indicates that the experimental curves fit the simulation curves well. As the position-based impedance control theory in Figure 8 shows, the actual position is greater than the desired one. Thus, the amplitude attenuation is a negative value. It increases with the increase of the disturbance force's amplitude, but has few relationships with the frequency. In contrast, the phase angle delay increases with the frequency, but has few relationships with the amplitude.

4. Methods of Sensitivity Analysis

4.1. Contrast between First-Order and Second-Order Sensitivity Analysis

Desired stiffness K_D and desired damping C_D are control parameters of the impedance control outer loop. Proportional gain K_P and integral gain K_i are control parameters of the position control inner loop. They affect the impedance control performance in different ways. So, in this paper, the system output position is mainly discussed, which is influenced by the variation of the four parameters. Because of space limitations, only one working condition (bias: 1500 N, amplitude: 1000 N, frequency: 2 Hz) is studied. The first-order and second-order trajectory sensitivity analysis methods researched in our previous works [15,16] are adopted to analyze the position variation when the four parameters increase 10% to 20%. The contrast curves are shown in Figures 12–15.

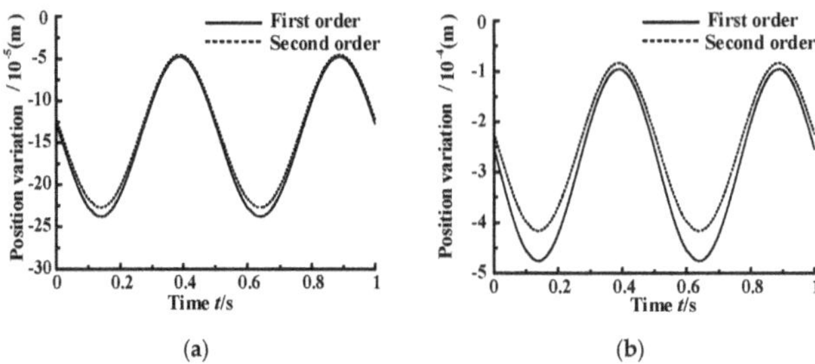

Figure 12. Position variation resulting from K_D variation. (**a**) Four parameters increase 10%; (**b**) Four parameters increase 20%.

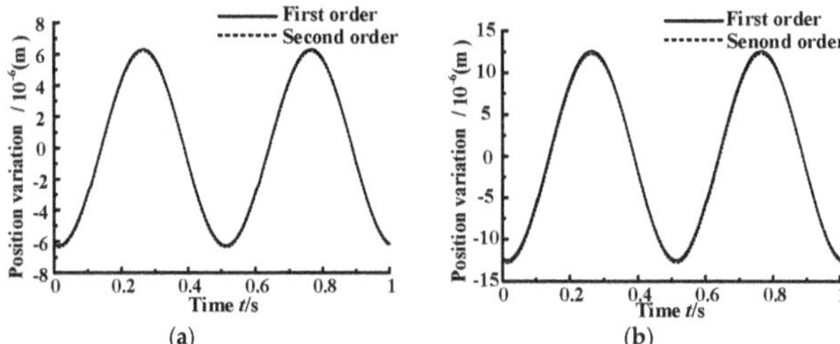

Figure 13. Position variation resulting from C_D variation. (**a**) Four parameters increase 10%; (**b**) Four parameters increase 20%.

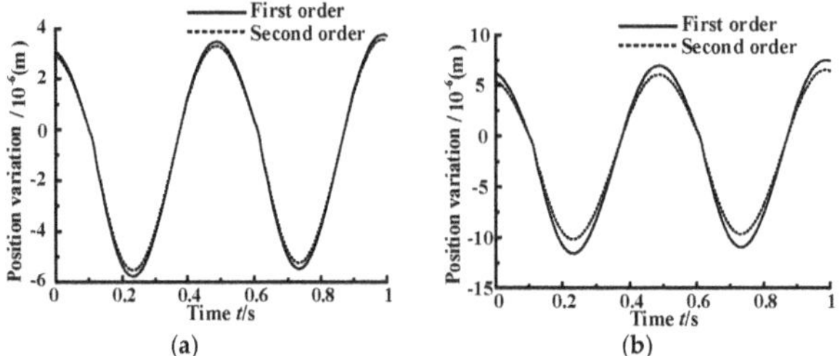

Figure 14. Position variation resulting from K_P variation. (**a**) Four parameters increase 10%; (**b**) Four parameters increase 20%.

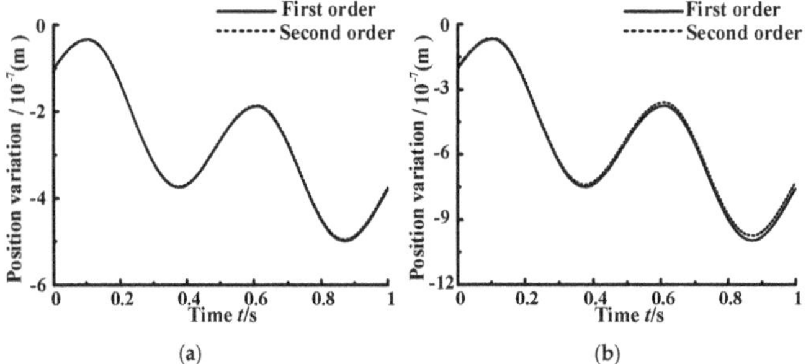

Figure 15. Position variation resulting from K_i variation. (**a**) Four parameters increase 10%; (**b**) Four parameters increase 20%.

The following conclusions can be reached from the four sets of curves. There is little difference between the first-order position variation and the second-order variation when the values of the four parameters increases 10%. Specifically, the curves of desired damping C_D and integral gain K_i almost overlap. In the other two sets, the maximum error is no more than 10% of the amplitude. When the values of the four parameters increase 20%, the variations of first-order and second-order are still close, although there is some increase of their error. Specifically, for desired damping C_D and integral gain K_i, the variation curves of first-order and second-order are still close. As for the proportional gain K_P, its maximum error is no more than 10% of the amplitude. When it comes to the desired stiffness K_D, its maximum error is no more than 20% of the amplitude.

The method of second-order trajectory sensitivity analysis has a very high accuracy, while its calculation is complicated and demands a lot of hard work. Particularly in this paper, when the four parameters increase less than 20%, the corresponding results of the first-order and second order sensitivity analysis method are close. So, in order to ensure the simplicity of calculation and application, the method of first-order sensitivity analysis is adopted to analyze the sensitivity of the four parameters under different working conditions.

The method studied previously can precisely analyze the sensitivity of the parameters. Compared with the second-order method, the calculation has been largely simplified under the method of first-order trajectory sensitivity analysis, but it also requires solving first-order linear non-homogeneous differential equations with time-varying factors, which makes the program complicated. So, in order to ensure high solving accuracy, a new method with an easier calculation is proposed. That is the first-order matrix sensitivity analysis.

4.2. Deduction of First-Order Matrix Sensitivity Analysis Theory

Combined with the position-based impedance control method, the equation of the HDU system can be expressed as follows:

$$\mathbf{g}(\mathbf{x}, \mathbf{u}, \boldsymbol{\alpha}, t) = 0 \tag{11}$$

where $\mathbf{x}$ is m−dimensional state vector, $\mathbf{u}$ is r−dimensional vector unrelated to $\boldsymbol{\alpha}$, $\boldsymbol{\alpha}$ is p−dimensional vector, and t is time.

The initial value of the state vector $\mathbf{x}_0$ can be obtained by giving the initial value of the input vector $\mathbf{u}_0$ and the initial value of parameter vector $\boldsymbol{\alpha}_0$, and the initial state of the equations is:

$$\mathbf{g}(\mathbf{x}_0, \mathbf{u}_0, \boldsymbol{\alpha}_0, t) = \mathbf{0} \tag{12}$$

where the variation of parameter vector $\Delta\boldsymbol{\alpha}$ and input vector $\Delta\mathbf{u}$ can change the value of $\Delta\mathbf{x}$, the error of the state variable $\mathbf{x}$, which is expressed as follows:

$$\mathbf{g}(\mathbf{x}_0 + \Delta\mathbf{x}, \mathbf{u}_0 + \Delta\mathbf{u}, \boldsymbol{\alpha}_0 + \Delta\boldsymbol{\alpha}, t) = \mathbf{0} \tag{13}$$

Expanded in the form of the first-order Taylor Series, Equation (9) can be expressed as follows:

$$\mathbf{g}(\mathbf{x}_0 + \Delta\mathbf{x}, \mathbf{u}_0 + \Delta\mathbf{u}, \boldsymbol{\alpha}_0 + \Delta\boldsymbol{\alpha}, t) = \mathbf{g}(\mathbf{x}_0, \mathbf{u}_0, \boldsymbol{\alpha}_0, t) + \mathbf{g}_x \cdot \Delta\mathbf{x} + \mathbf{g}_u \cdot \Delta\mathbf{u} + \mathbf{g}_\alpha \cdot \Delta\boldsymbol{\alpha} = \mathbf{0} \tag{14}$$

If we bring Equation (12) into Equation (14) and ignore the higher-order terms, then we can get:

$$\mathbf{g}_x \cdot \Delta\mathbf{x} + \mathbf{g}_u \cdot \Delta\mathbf{u} + \mathbf{g}_\alpha \cdot \Delta\boldsymbol{\alpha} = \mathbf{0} \tag{15}$$

Equation (15) can also be expressed as follows:

$$\Delta\mathbf{x} = -\mathbf{g}_x{}^{-1} \cdot \mathbf{g}_u \cdot \Delta\mathbf{u} - \mathbf{g}_x{}^{-1} \cdot \mathbf{g}_\alpha \cdot \Delta\boldsymbol{\alpha} \tag{16}$$

In Equation (17), supposing:

$$\mathbf{S}_u = \mathbf{g}_x{}^{-1} \cdot \mathbf{g}_u \tag{17}$$

where $\mathbf{S}_u$ is $m \times r$ order matrix, and the n-th line indicates the relation among the n-th state variable x_n.

In Equation (17), supposing:

$$\mathbf{S}_\alpha = \mathbf{g}_x{}^{-1} \cdot \mathbf{g}_\alpha \tag{18}$$

where, $\mathbf{S}_\alpha$ is the $m \times p$ order matrix, the n-th line indicates the relation among the n-th state variable and p parameter vectors. Bring Equations (17) and (18) into Equation (16), then:

$$\Delta\mathbf{x} = -\mathbf{S}_u \cdot \Delta\mathbf{u} - \mathbf{S}_\alpha \cdot \Delta\boldsymbol{\alpha} \tag{19}$$

Equation (19) is an approximate expression of $\Delta\mathbf{x}$ resulted from the change of parameter vector $\Delta\alpha$ and input vector $\Delta\mathbf{u}$, in which $\mathbf{S}_\alpha$ indicates an $m \times p$ order parameter sensitivity matrix of parameter vector α with time-varying factors. $\mathbf{S}_u$ indicates an $m \times r$ order input sensitivity matrix of input vector u with time-varying factors.

When taking no account of the variation of input vector, Equation (19) can be simplified as follows:

$$\Delta\mathbf{x} = -\mathbf{S}_\alpha \cdot \Delta\boldsymbol{\alpha} \tag{20}$$

The output equation of the system can be expressed as follows:

$$\Delta\mathbf{Y} = \mathbf{C} \cdot \Delta\mathbf{x} + \mathbf{D} = -\mathbf{C} \cdot \mathbf{S}_\alpha \cdot \Delta\boldsymbol{\alpha} + \mathbf{D} \tag{21}$$

where $\mathbf{C}$ and $\mathbf{D}$ are matrices of output equation factors. The change of output variable $\Delta\mathbf{Y}$ resulting from the parameter variation can be reached after solving parameter sensitivity matrix $\mathbf{S}_\alpha$.

5. Dynamic Sensitivity Analysis

5.1. Contrast between Two Analysis Method of First-Order Sensitivity

The state vectors' initial values of the servo-cylinder's position, velocity, and pressure of two chambers, the servo-valve's position, velocity and acceleration are zero. So, the initial value of parameter sensitivity matrix $\mathbf{S}_\alpha$ can be expressed as follows:

$$\mathbf{S}_{\alpha 0} = \mathbf{0}_{m \times p} \tag{22}$$

Solve the sensitivity matrices in MATLAB, and then compare the results from the trajectory sensitivity in Section 3.1 and $-\mathbf{S}_\alpha$, the inverse value of the control parameter's sensitivity matrix of the system output position. Due to space limitations, only one working condition (1500 N bias, 1000 N amplitude, 2 Hz frequency) is shown. The curves of parameter dynamic sensitivity in this situation is shown in Figure 16.

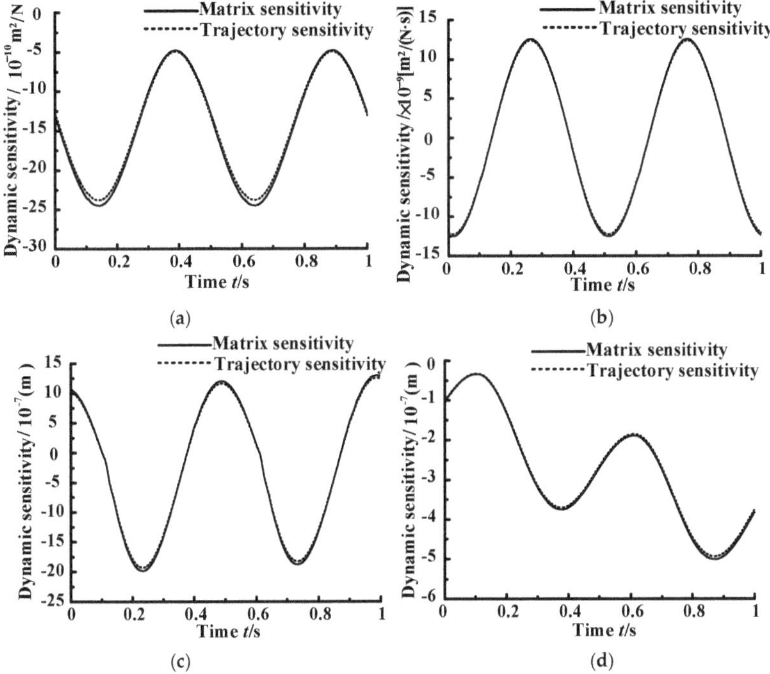

Figure 16. The curves of dynamic sensitivity between two sensitivity methods. (**a**) First-order dynamic sensitivity of K_D; (**b**) First-order dynamic sensitivity of C_D; (**c**) First-order dynamic sensitivity of K_p; (**d**) First-order dynamic sensitivity of K_i.

It can be seen that the dynamic compliance curves of first-order matrix sensitivity deviate little from the curves of first-order trajectory sensitivity. Particularly, by comparing Figures 12–15 in Section 3.1, it can be found that the value calculated by first-order matrix sensitivity is more approximate to the result of second-order trajectory sensitivity and has higher precision than first-order trajectory sensitivity. Moreover, only a two-dimension matrix calculation is needed, which avoids solving complicated differential equations with time-varying factors. So, in the research field of this paper, the first-order matrix sensitivity analysis method is more adapted than the first-order trajectory sensitivity analysis method. Whether calculating a high-order and multi-dimensional matrix is easier than solving differential equations with time-varying factors cannot be determined, so it requires further research to find which is better between the high-order matrix sensitivity analysis method and the high-order trajectory sensitivity analysis method. However, due to space constraints, this will not be discussed in this paper.

5.2. Contrast of Dynamic Sensitivity Analysis in Each Working Condition

For the convenience of contrast, the variation of system position response when each parameter increases 10% is calculated according to Equation (21). The curves of position variation with time is shown in Figure 17.

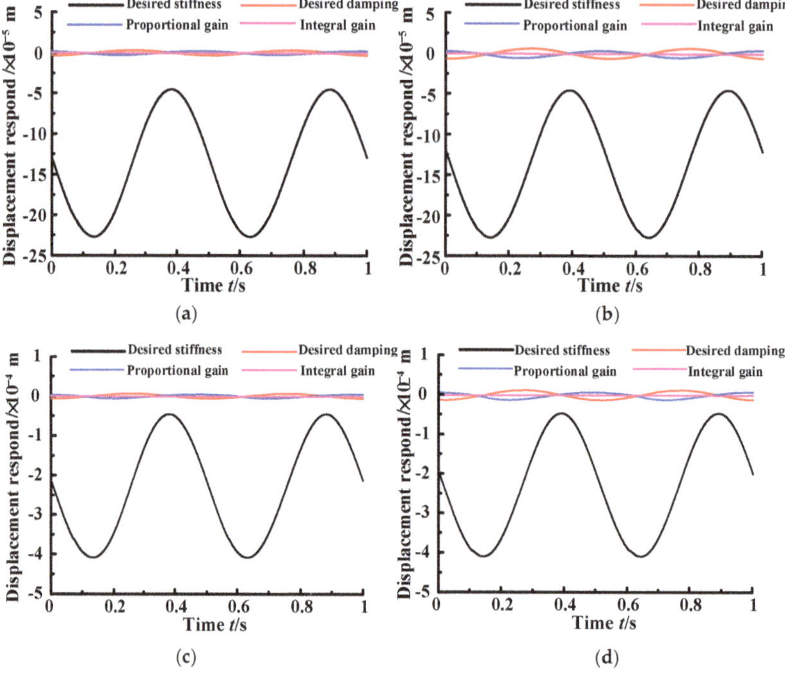

Figure 17. Variation of position response when each parameter increases 10%. (**a**) The first working condition; (**b**) The second working condition; (**c**) The third working condition; (**d**) The fourth working condition.

As it can be seen in Figure 17:

1. The variation of each parameter affects impedance control position output. The position varies periodically with the sinusoidal disturbance force. Among the parameters, the desired stiffness K_D affects the output position much more than the others. The influence from integral gain K_i is the most irrelevant. Desired damping C_D and proportional gain K_p have similar influences on the output position. With the disturbance force in sinusoidal variation, C_D varies by following the curve of minus cosine, and K_p varies by following the curve of cosine.
2. The order of magnitudes of output position increases with the increase of sinusoidal disturbance force. However, there isn't an obvious relationship between the effects on output position and the frequency of the disturbance force.

Using the method of dynamic sensitivity analysis, the qualitative effects on impedance control performance is analyzed. In order to analyze the effects on main system performance indexes and the varying patterns of parameter sensitivity under different working conditions, a quantificational analysis is needed.

6. Sensitivity Quantitative Analysis

6.1. Sensitivity Indexes

Two measurement indexes are introduced to analyze the effects on main performance indexes resulting from the variation of parameters in different working conditions.

For sinusoidal response, in a stable sinusoidal period, the variation of parameters results in the change of the output position amplitude. The mean of amplitude attenuation is defined as the first sensitivity measurement index s_1, which is expressed as follows:

$$s_1 = \text{mean}(\Phi_1 + \Phi_2) \tag{23}$$

where,

$$\begin{aligned}\Phi_1 &= [\max(x_r - x_e) - \max(x_1 - \mathbf{S}_\alpha^1 \cdot \Delta\alpha_i)] - [\max(x_r - x_e) - \max(x_1)] \\ &= \max(x_1) - \max(x_1 - \mathbf{S}_\alpha^1 \cdot \Delta\alpha_i)\end{aligned} \tag{24}$$

$$\begin{aligned}\Phi_2 &= [\min(x_r - x_e) - \min(x_1)] - [\min(x_r - x_e) - \min(x_1 - \mathbf{S}_\alpha^1 \cdot \Delta\alpha_i)] \\ &= \min(x_1 - \mathbf{S}_\alpha^1 \cdot \Delta\alpha_i) - \min(x_1)\end{aligned} \tag{25}$$

Similarly, for phase angle delay, another important index, its mean is defined as the second sensitivity measurement index s_2, which is expressed as follows:

$$s_2 = \text{mean}(\Psi) \tag{26}$$

where,

$$\Psi = [\arcsin(\tfrac{x_r - x_e}{\max(x_r - x_e)}) - \arcsin(\tfrac{x_1 - \mathbf{S}_\alpha^1 \cdot \Delta\alpha_i}{\max(x_1 - \mathbf{S}_\alpha^1 \cdot \Delta\alpha_i)})] - [\arcsin(\tfrac{x_r - x_e}{\max(x_r - x_e)}) - \arcsin(\tfrac{x_1}{\max(x_1)})] \tag{27}$$

Using the above two indexes, s_1 and s_2, the effect on output position resulting from the variation of parameters can be quantificationally analyzed.

6.2. Sensitivity Histograms in Different Working Conditions

According to Equations (23)–(27), the two sensitivity histograms are shown in Figures 18–21, when the four control parameters increase 10% in four working conditions.

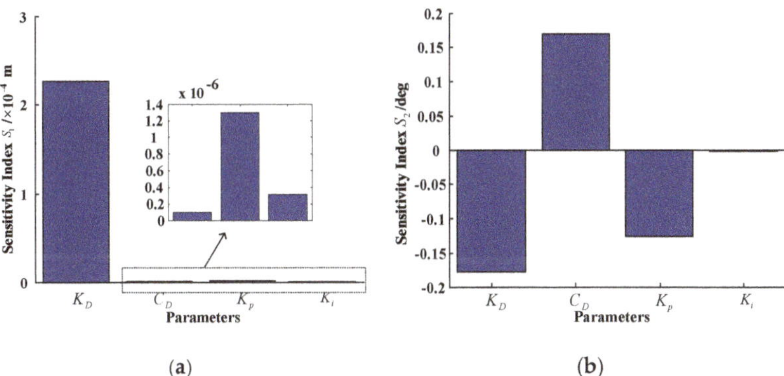

Figure 18. Sensitivity histogram in the first working condition. (**a**) Sensitivity index S_1; (**b**) Sensitivity index S_2.

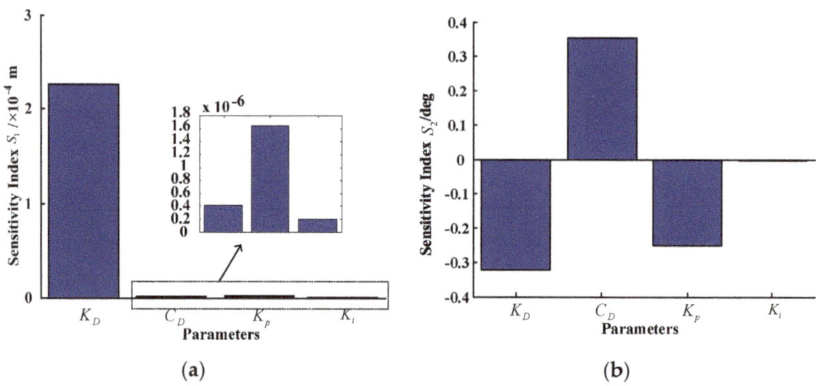

Figure 19. Sensitivity histogram in the second working condition. (**a**) Sensitivity index S_1; (**b**) Sensitivity index S_2.

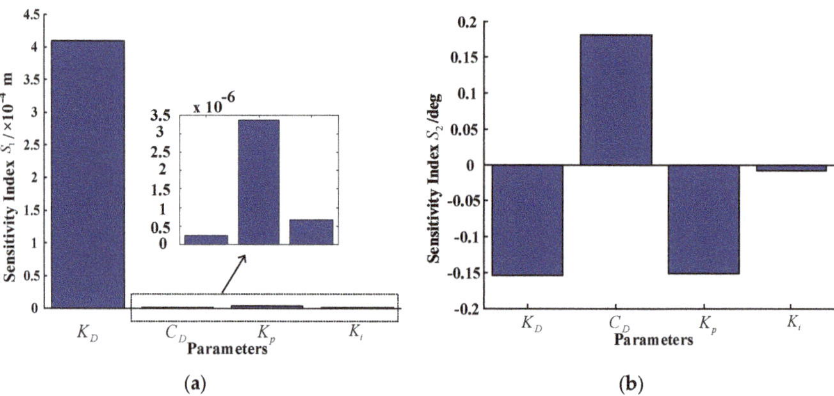

Figure 20. Sensitivity histogram in the third working condition. (**a**) Sensitivity index S_1; (**b**) Sensitivity index S_2.

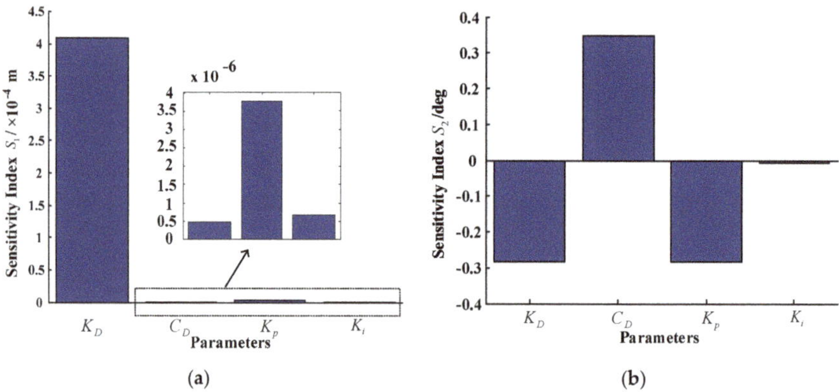

Figure 21. Sensitivity histogram in the four working conditions. (**a**) Sensitivity index S_1; (**b**) Sensitivity index S_2.

As it can be seen in the above Figures 18–21:

1. In all conditions, S_1 is a positive value, so the increase of the four parameters will result in the reduction of the position amplitude attenuation. The S_1 of desired stiffness K_D is much greater than the other parameters, which illustrates that the variation of K_D has a remarkable influence on output position amplitude attenuation, and more relation with disturbance force amplitude than the disturbance force frequency. Under the first working condition, amplitude attenuation reduced about 0.2 mm when K_D increased 10%. The S_1 of proportional gain K_p is greater than that of C_D and K_i. The S_1 of K_D and K_p have some correlation with disturbance force amplitude, but have little correlation with disturbance force frequency.
2. The S_2 of the four control parameters has positive or negative values in different conditions. An increase of the parameters has different influences on the phase angle delay of the output position. Specifically, having more relation with disturbance force frequency than the amplitude, the S_2 of K_D and C_D have nearly the same absolute value, but an opposite sign. Their increase has totally different influences on phase angle delay. An approximate 10% increase of them generates about 0.18° of phase angle delay error. The S_2 of K_p has correlation with both disturbance force amplitude and frequency. The larger the frequency, the more remarkable the influence on phase delay. The S_2 of K_p is close to the S_2 of K_D. The S_2 of K_i is about 0.001° on magnitude order, much less than the other control parameters.

7. Experiment

Compared with structural parameters and some working parameters, control parameters can change and be measured during the working process. The two sensitivity indexes of four control parameters are studied by experiments in this section. The mean of several samples is included to ensure the accuracy of the experiment. By comparing the measured data and the simulation result of the first-order sensitivity matrix, the contrast histograms of experiment value and simulation value for the two sensitivity indexes when the four parameters increase 10% under four working conditions are shown in Figures 22–25.

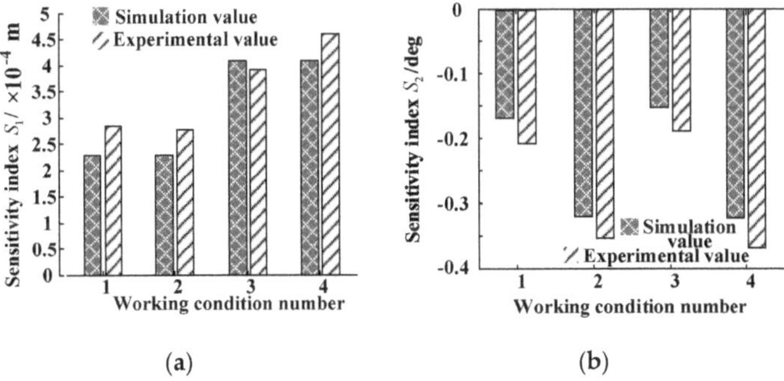

Figure 22. Sensitivity index histograms of K_D. (**a**) Sensitivity index S_1; (**b**) Sensitivity index S_2.

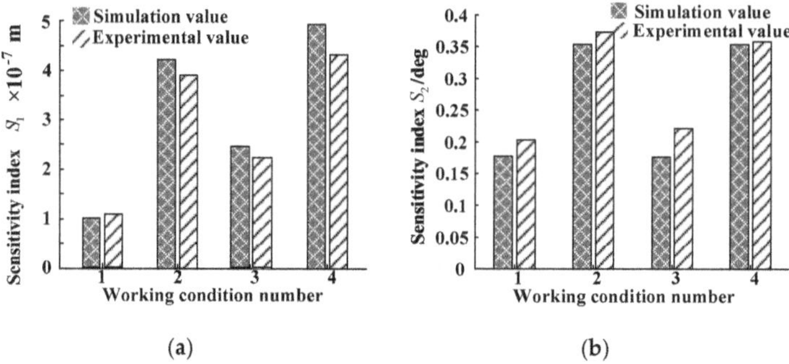

Figure 23. Sensitivity index histograms of C_D. (**a**) Sensitivity index S_1; (**b**) Sensitivity index S_2.

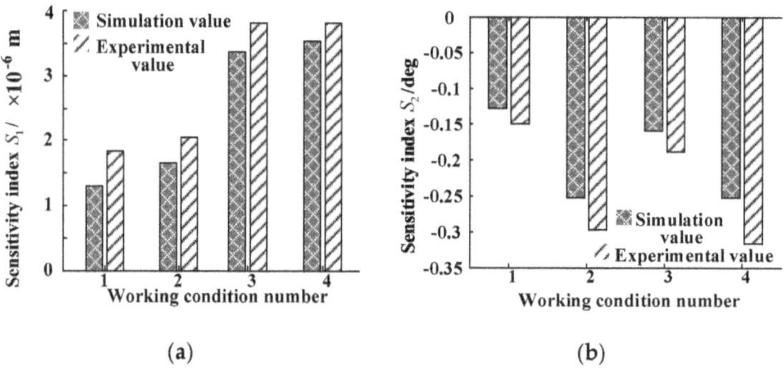

Figure 24. Sensitivity index histograms of K_p. (**a**) Sensitivity index S_1; (**b**) Sensitivity index S_2.

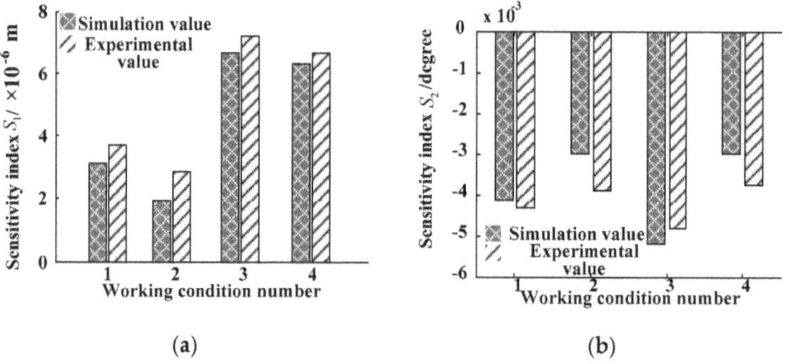

Figure 25. Sensitivity index histograms of K_i. (**a**) Sensitivity index S_1; (**b**) Sensitivity index S_2.

In Figures 22–25, the maximum and mean errors between the experimental value and the simulation value are shown in Table 4.

Table 4. Maximum and mean errors between experimental value and simulation value.

Parameters	S_1 Max Error	S_1 Mean Error	S_2 Max Error	S_2 Mean Error
K_D	19%	12%	20%	16%
C_D	11%	5%	18%	9%
K_p	31%	16%	15%	10%
K_i	35%	15%	21%	11%

According to Figures 22–25 and Table 4, the experimental value and simulation value have the same magnitude and similar variation patterns. Except that the maximum errors of S_1 of K_p and K_i are above 30%, the others are less than 20%. The mean errors of the four parameters are less than 16%. For S_2, the maximum error is about 20%. Except that mean error of S_2 for K_D is 16%, the others' mean errors are about 10%.

8. Conclusions

In this paper, a HDU position-based impedance control method is proposed. The method of sensitivity analysis is selected in research. Dynamic sensitivity analysis and quantificational sensitivity analysis are conducted to study the four main control parameters. The results from the research are verified by experiments. Here are the conclusions summarized in this paper:

1. Sensitivity analysis method is used to analyze the four parameters, which are the proportional gain and integral gain of the inner loop PID controller, and the desired stiffness and desired damping of the impedance outer loop. The first-order and second-order effects on position are analyzed by the trajectory sensitivity analysis, when the four parameters increase no more than 20%. In spite of its higher accuracy, the second-order sensitivity analysis requires a complex solving process with lots of hard work. Considering that the largest error between the first-order and second-order sensitivity analysis methods is below 20% of the amplitude, this paper proposed the easier first-order matrix sensitivity analysis method. Compared with first-order trajectory sensitivity, the first-order matrix sensitivity method is simpler and more accurate in computing.
2. The effects of parameter variation on output position varies periodically with sinusoidal disturbance force. Among the parameters, desired stiffness affects the output position much more than the other parameters. The influence of integral gain is the least. Desired damping and proportional gain have a similar influence on the output position.
3. The S_1 of each parameter is a positive value, which shows that the increase of the four parameters will result in the reduction of the position amplitude attenuation. The S_2 of each parameter has positive or negative values in different conditions. Their increase has different influences on the phase angle delay of output positions.

Acknowledgments: The project is supported by National Natural Science Foundation of China (Grant No. 51605417), Key Project of Hebei Province Natural Science Foundation (Grant No. E2016203264).

Author Contributions: B.Y. designed the experiments; K.B. wrote the paper; X.K. conceived the research idea; Z.G. performed the experiments; G.M. and W.L. analyzed the data.

Conflicts of Interest: The founding sponsors had no role in the design of the study; in the collection, analyses, or interpretation of data; in the writing of the manuscript, and in the decision to publish the results.

References

1. Montes, H.; Armada, M. Force Control Strategies in Hydraulically Actuated Legged Robots. *Int. J. Adv. Robot. Syst.* **2016**, *13*, 50. [CrossRef]
2. Nabulsi, S.; Sarria, J.; Montes, H.; Armada, M. High Resolution Indirect Feet-Ground Interactions Measurement for Hydraulic Legged Robots. *J. IEEE Trans. Instrum. Meas.* **2009**, *58*, 3396–3404. [CrossRef]
3. Huang, Q.; Oka, K. Phased compliance control with virtual force for six-legged walking robot. *Int. J. Innov. Comput. Inf. Control* **2008**, *4*, 3359–3373.
4. Kong, X.D.; Ba, K.X.; Yu, B.; Cao, Y.; Zhu, Q.; Zhao, H. Research on the force control compensation method with variable load stiffness and damping of the hydraulic drive unit force. *Chin. J. Mech. Eng. (English Ed.)* **2016**, *29*, 454–464. [CrossRef]
5. Ba, K.X.; Yu, B.; Kong, X.D.; Zhao, H.-L.; Zhao, J.-S.; Zhu, Q.-X.; Li, C.-H. The dynamic compliance and its compensation control research of the highly integrated valve-controlled cylinder position control system. *Int. J. Control Autom. Syst.* **2017**, *15*, 1814–1825. [CrossRef]

6. Li, M.T.; Jiang, Z.Y.; Wang, P.F.; Sun, L.N.; Ge, S.S. Control of a Quadruped Robot with Bionic Springy Legs in Trotting Gait. *J. Bionic Eng.* **2014**, *11*, 188–198. [CrossRef]
7. Kimura, H.; Fukuoka, Y.; Cohen, A.H. Adaptive dynamic walking of a quadruped robot on natural ground based on biological concepts. *Int. J. Robot. Res.* **2007**, *26*, 475–490. [CrossRef]
8. Poulakakis, I.; Smith, J.A.; Buehler, M. Modeling and experiments of untethered quadrupedal running with a bounding gait: The scout ii robot. *Int. J. Robot. Res.* **2005**, *24*, 239–256. [CrossRef]
9. Seok, S.; Wang, A.; Meng, Y.C.; Otten, D. Design principles for highly efficient quadrupeds and implementation on the MIT Cheetah robot. In Proceedings of the IEEE International Conference on Robotics and Automation, Karlsruhe, Germany, 6–10 May 2013; IEEE: Piscataway, NJ, USA, 2013; pp. 3307–3312.
10. Playter, R.; Buehler, M.; Raibert, M. BigDog. In *Unmanned Systems Technology VIII, Proceedings of SPIE*; SPIE: Bellingham, WA, USA, 2006; Volume 6230, pp. 1–6.
11. Semini, C.; Barasuol, V.; Boaventura, T.; Frigerio, M.; Focchi, M. Towards versatile legged robots through active impedance control. *Int. J. Robot. Res.* **2015**, *34*, 1003–1020. [CrossRef]
12. Rong, X.; Li, Y.; Ruan, J.; Li, B. Design and simulation for a hydraulic actuated quadruped robot. *J. Mech. Sci. Technol.* **2012**, *26*, 1171–1177. [CrossRef]
13. Vilenus, M. The Application of Sensitivity Analysis to Electrohydraulic Position Control Servos. *J. Dyn. Syst. Meas. Control* **1983**, *105*, 77–82. [CrossRef]
14. Farasat, S.; Ajam, H. Sensitivity Analysis of Parameter Changes in Nonlinear Hydraulic Control Systems. *Int. J. Eng.* **2005**, *18*, 239–252.
15. Kong, X.D.; Yu, B.; Quan, L.; Ba, K.; Wu, L. Nonlinear Mathematical Modeling and Sensitivity Analysis of Hydraulic Drive Unit. *Chin. J. Mech. Eng.* **2015**, *28*, 999–1011. [CrossRef]
16. Kong, X.D.; Ba, K.X.; Yu, B.; Quan, L.; Wu, L. Trajectory Sensitivity Analysis of First Order and Second Order on Position Control System of Highly Integrated Valve−controlled Cylinder. *J. Mech. Sci. Technol.* **2015**, *29*, 4445–4464. [CrossRef]
17. Kong, X.; Zhao, H.; Li, B.W.; Ba, K. Analysis of position-based impedance control method and the composition of system dynamic compliance. In Proceedings of the IEEE International Conference on Aircraft Utility Systems (AUS), Beijing, China, 10–12 October 2016; pp. 454–459.

 *applied sciences*

Article

Omnidirectional Jump of a Legged Robot Based on the Behavior Mechanism of a Jumping Spider

Yaguang Zhu [1,2,*], Long Chen [1], Qiong Liu [1], Rui Qin [1] and Bo Jin [2]

[1] Key Laboratory of Road Construction Technology and Equipment of MOE, Chang'an University, Xi'an 710064, China; chenlong@chd.edu.cn (L.C.); liuqiong@chd.edu.cn (Q.L.); qinrui@chd.edu.cn (R.Q.)

[2] State Key Laboratory of Fluid Power and Mechatronic Systems, Zhejiang University, Hangzhou 310028, China; bjin@zju.edu.cn

* Correspondence: zhuyaguang@chd.edu.cn; Tel.: +86-187-9285-2585

Received: 14 November 2017; Accepted: 29 December 2017; Published: 1 January 2018

Abstract: To find a common approach for the development of an efficient system that is able to achieve an omnidirectional jump, a jumping kinematic of a legged robot is proposed based on the behavior mechanism of a jumping spider. To satisfy the diversity of motion forms in robot jumping, a kind of 4 degrees of freedom (4DoFs) mechanical leg is designed. Taking the change of joint angle as inspiration by observing the behavior of the jumping spider during the acceleration phase, a redundant constraint to solve the kinematic is obtained. A series of experiments on three types of jumping—vertical jumping, sideways jumping and forward jumping—is carried out, while the initial attitude and path planning of the robot is studied. The proposed jumping kinematic is verified on the legged robot experimental platform, and the added redundant constraint could be verified as being reasonable. The results indicate that the jumping robot could maintain stability and complete the planned task of jumping, and the proposed spider-inspired jumping strategy could easily achieve an omnidirectional jump, thus enabling the robot to avoid obstacles.

Keywords: jumping spider; jumping robot; omnidirectional jump; redundant DoF

---

1. Introduction

Compared with walking [1] and crawling robots [2], the jumping robot can walk, run, and jump [3]. Jumping locomotion has characteristics of isolated footholds, and powerful and explosive jumping force [4], which contribute to the quick and effective jumping locomotion of bio-inspired robots to stride over obstacles that are several times the size of their bodies, or cross a gully that is several times the length of their own step, and avoid danger in time. As for the jumping robot, the key point is usually planning the trajectory of their center of gravity (CoG) to achieve the jumping process reasonably, and realize the multi-directional jumping moment in various environments. The jumping robot can realize the jumping process by controlling the take-off speed, attitude and landing stability. When the jumping robot takes off, the robot achieves certain acceleration because its foot generates enough reactive force by impacting with the ground. Then, the robot adjusts its posture in the air, and finally lands smoothly. Hence, the study of the jumping process is an important part of the jumping robot. The jumping robot realizes the physical design of the robot and the jump movements by imitating animal body structures or biological movement mechanisms. The bionics [5] include the structure bionics, the motion bionics and the control bionics [6]. Nowadays, research on the bionic robot based on a bionic structure is abundant. There are both robots that imitate mammals, such as the bionic cheetah and bionic kangaroo [7,8], and robots that imitate amphibious creatures, such as bionic frogs and bionic toads [9,10]. Additionally, there are robots that imitate arthropods, such as bionic spiders, bionic locusts, bionic cockroaches and imitation water insects [11–15]. The mammal-like robot has the characteristics of being fast at running and jumping, as well as having smooth motion and a high energy utilization rate. However, there is little research surrounding this because of its large volume and heavy weight. The amphibious robots are almost all bionic frog robots. The robot uses intermittent motion so that it may control its posture, in addition to being able to effectively control the energy accumulation and release. Therefore, this type of robot is characterized by its flexible jumping, powerful explosive force and environmental adaptability [16]. Compared with these types of robots, the arthropod robot has the advantage of fast acceleration, low energy consumption and high energy efficiency in the process of jumping because of its small size, light weight and good bounce ability [17].

The structure model of the jumping robot can be divided into a single-legged model and a multi-legged model. In terms of the single-legged jumping robot, based on the high mobility requirements of jumping, some scholars have proposed a single leg motion mode driven by hydraulics [18,19]. The influence posture and ground impact of the robot on its structure in the vertical jump motion are analyzed, and the overall stability evaluated, to ensure that the robot can complete the take-off task. Ge et al. [20–22] proposed a scheme based on the jumping mechanism of the kangaroo. They studied the jumping movement of the kangaroo and simplified their bodies into single-legged models to study and discuss. Then, they proposed three models: a rigid body jumping model, a compliant jump mechanism model and a rigid flexible hybrid model. The motion mechanisms of the three models were analyzed to see which one was most successful in making the robot jump smoothly. However, the single-legged model is a naturally unstable system, which cannot remain in the stationary state. Therefore, the single-legged robot has braced structures to maintain stability. Furthermore, it is necessary to adjust the initial attitude angle to achieve a smooth jump. In addition, the single-legged robots only realized the bionic jumping function, and were less involved in the overall movement mechanism. Animals in nature predominately have multiple legs. The biped robot, the quadruped robot and the six-legged robot are the main forms of multi-legged robots [23–25]. Fumitaka et al. [26] have designed a quadruped robot, which can jump even in rugged external environments and can achieve the task of crossing obstacles. Some scholars [27–29] have proposed a rigid mode of imitating the motion principle of the cricket by studying the robot's ability to jump and kick; the robot can adjust its own dynamic balance while it is jumping. Thus, compared with the single robot, the multi-legged robot has better overall stability.

Currently, most research focuses on the single direction jump, especially the vertical and forward jump. A vertical jump analysis based on the hydraulic drive has been proposed [16,30]. Surmounting ability and jumping efficiency are analyzed, and then the vertical jumping form is optimized. Thanhtam et al. [31] have proposed a new structure of quadruped robot to accomplish the task of vertical jumps or forward jumps. The joints of the robot are driven by hydraulics to meet the requirements of torque, compactness, speed and impact resistance. Hyunsoo et al. [32] proposed a quadruped jumping robot, which is based on the servo motor drive. The legs are equipped with gears, springs and other components, and the robot can complete a high jump task through two kinds of movement: spring compression and gear drive mechanism.

However, there is no in-depth discussion about the research on omnidirectional jumping. In other words, in the process of jumping, the robot jumps mainly through fixed jumping form, and it only achieves a single direction jump of height and distance. The robot must adjust its posture and jumping direction when it is trying to avoid an obstacle. The initial posture must be adjusted first if the robot is going to change the jumping direction, causing the efficiency to be greatly reduced. Hence, without changing the initial pose of the robot, multi-direction jumps become the key problem to be solved. In this paper, a bionic six-legged robot structure with the ability of omnidirectional jumping is proposed, which is based on the jumping mechanism of jumping spiders. The omnidirectional jumping form has been proposed through observing the jumping form, jumping posture and leg stretch of jumping spiders, which allows it to avoid and cross obstacles in all directions. To verify the rationality of the jumping form proposed in this paper, a series of experiments on a six-legged robot is carried out, and the results show that the proposed multi-direction jumping form has outstanding performance, which provided a good theoretical basis for jumping research.

2. Bio-Inspiration and Materials

Arthropods are the largest group of animals, including over one million species of invertebrates, and accounting for almost 84% of all species. Members of the arthropod family are various; they can be found from the abyssal sea to inland areas. Due to the differentiation between arthropod bodies, and the diversity of physical changes, which give arthropods a highly adaptive capacity, they have adapted to all sorts of surroundings, maintaining themselves even under the most rigorous conditions. After hundreds of millions of years of evolution, arthropods have become very flexible in their ability to move [33]. Arthropods can hunt prey quickly or avoid predators, and when they run into obstacles or ravines, they can quickly run and jump on the terrain to avoid obstacles. Therefore, the agility of arthropods can provide inspiration for the exploration of jumping robots [34,35]. The arthropod is composed of several different structures and functions; the body is symmetrical; its feet are evenly distributed on both sides of the body; and the legs can be coordinated with each other. This not only allows the flexibility of movement of the robot, but also the ability to jump in any direction and remain flexible in various terrains [36]. Jumping spiders are the most common of these arthropods.

Since the jumping spider has multiple joints on each leg, the legs can be stretched long enough to move around or camouflage themselves to prey. Especially due to its jumping ability, the jumping spider can quickly avoid predators and overcome obstacles. In this paper, we mainly focus on the jumping spider and study its motion structure and morphology. When the jumping spider jumps, the feet fall and the legs stretch, and the effective leg length increase rapidly, before the spider accelerates and impacts with the ground. When the jumping spider skips sideways, the front spider legs shrink and the rear legs extend in the jumping direction, and the spider has a certain attitude angle. The spider accelerates highly when its legs begin to extend. When the front legs reach the maximum effective length, the spider reaches the ground velocity across the barrier, and the spider jumps off the ground. When the jumping spider skips forward, the effective length of the front legs is constant in the direction of the jump, and the effective length of the rear legs decreases, and the spider has a certain attitude angle. The spider can achieve high acceleration when its legs extend to full extension. When the hind legs of the spider reach the maximum effective length, the spider reaches the ground velocity across the barrier, and the spider jumps off the ground.

Figure 1 is a schematic diagram of the whole body and a sketch of the spider's leg joint and leg structure. Considering the characteristics of the coxal joints of spiders, the leg structure of the robot is simplified; the blue solid line indicates the leg, and the black circle indicates the joints. The coxal joint can move freely in any direction, so that the spider can jump in any direction. The robot model is designed by observing the schematic diagram of the jumping spider and the spider leg, and the robot experiment platform is shown in Figure 2a. To simplify the design, the spider robot is designed with six legs rather than eight legs. Compared with other hexapod robots that have been studied, in this paper each leg of the robot has 4DoFs, which can increase the flexibility of the robot during activity. The width and length of the body are 134 mm and 228 mm respectively; the length of the patella-tibia is 120 mm; the length of the tibia-metatarsus is 120 mm; the length of the tarsus is 160 mm; and the maximum effective length of the robot leg is 350 mm. The robot's six legs are symmetrically distributed on both sides of the fuselage; the angle between the legs of left foreleg (LF), right foreleg (RF), left hind leg (LH) and right hind leg (RH) (as seen in Figure 2a), and the axis direction of the robot is 60°, while the legs of left middle leg (LM) and right middle leg (RM) (as seen in Figure 2a) are perpendicular to the fuselage. The robot adopts the design principles of having a bionic structure and being light weight. The total weight is only 4 Kg, since the root body, shank, foot and connector of digital motors are all made from aluminum alloy. Each leg has four rotating joints including the coxal joint, complex femur–patella joint, tibia–metatarsus joint and the metatarsus–tarsus joint. The torque of each joint is provided by a digital motor. The rotation of the digital motion of the coxal joint (θ_1) enables the robot to swing back and forth, providing forward power and controlling the step size. According to the robot's mechanical structure, the minimum value of θ_1 is $-30°$, and the maximum value of θ_1 is +30°. The rotation of the digital motor at the femur–patella joint (θ_2), the tibia–metatarsus joint (θ_3) and the metatarsus–tarsus joint (θ_4) can achieve leg extension and control the height of the body. The value of θ_2 ranges from $-90°$ to $+75°$, the value of θ_3 ranges from $0°$ to $+150°$, and the range of θ_4 varies with θ_2 and θ_3. In this paper, the robot is designed to complete tasks that require a maximum jumping height of 100 mm and a maximum jumping distance of 250 mm, while following the form of omnidirectional jumping. Furthermore, the robot can walk quickly, and turn in any direction. The D-H kinematic model was established according to the John method, as shown in Figure 2b, and the kinematic model parameters are shown in Table 1.

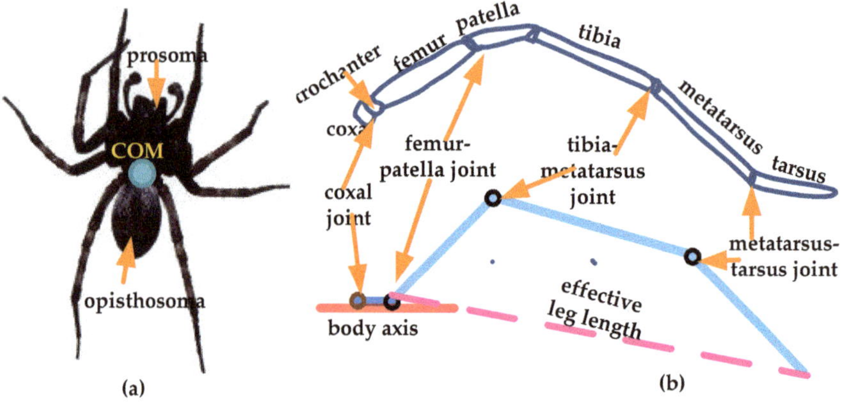

Figure 1. (**a**) The schematic diagram of the whole body; and (**b**) sketch of the spider's leg joint and leg structure.

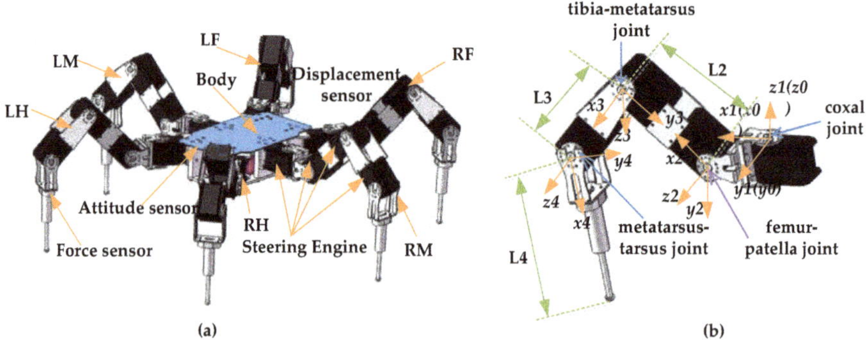

Figure 2. (**a**) The schematic diagram of the robot; and (**b**) sketch of the joint and structure.

Table 1. Kinematic model parameters.

Linkage (i)	$\alpha_{j\text{-}1}$	$d_{j\text{-}1}$	$a_{j\text{-}1}$	θ_j
1	0°	0	0	θ_1
2	0°	0	0	θ_2
3	−90°	0	L_2	θ_3
4	0°	0	L_3	θ_3

The architecture of the control algorithm of the robot is shown in Figure 3. During the jumping process of the robot, the whole system is composed of the time signal, path planning, foot trajectory planning, experimental prototype and the sensor signal. As the input signal of the whole system, time provides the drive for the control system. The foot trajectory planning and path planning are mainly used to plan the jumping path of the robot. When the robot jumps, the body jump trajectory is given and, correspondingly, we can get the trajectory of the robot foot. Then, the kinematic displacement of each joint is calculated by inverse kinematics. The robot can realize the motion form of the jump through the reasonable path planning of the robot. The robot is equipped with various sensors that are mounted inside the body. An attitude sensor is mounted to monitor the motion state of the robot in real time. A displacement sensor and a torque sensor are arranged on each leg joint to detect the real-time position and the torque. A force sensor is used to monitor the load on each leg. Figure 3 is the algorithm framework. We can understand clearly the control process to the jumping robot. At present, we start with the position control, which is consistent with the control mode of the robot platform we are building, and it is easy to quickly implement. The close position loop takes into account the internal torque loop, and the next step will be studied from the dynamics.

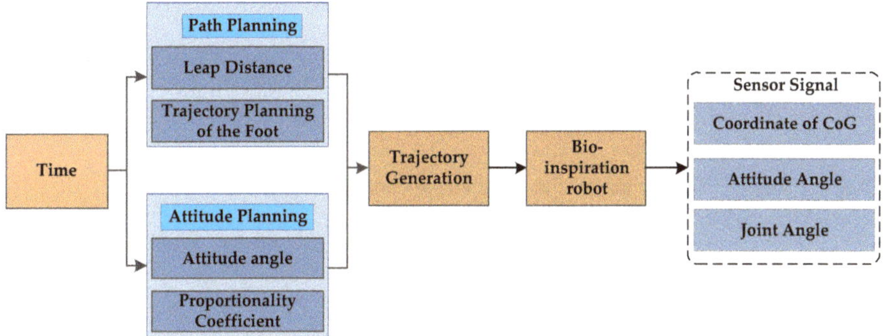

Figure 3. Algorithm framework.

3. Methods

3.1. Kinematics with Redundant DoF

3.1.1. Forward Kinematics Analysis

Based on the D-H model established in Section 2, the length of each link and the rotation angle of each joint are known in the base coordinate system of the robot, and the trajectory equation of the foot is derived as follows:

$$ {}_{\text{tip}}^{0}\mathbf{P} = \begin{bmatrix} {}_{\text{tip}}^{0}P_X \\ {}_{\text{tip}}^{0}P_Y \\ {}_{\text{tip}}^{0}P_Z \end{bmatrix} = \begin{bmatrix} C_1C_{234}L_4 + C_1C_{23}L_3 + C_1C_2L_2 \\ S_1C_{234}L_4 + S_1C_{23}L_3 + S_1C_2L_2 \\ -S_{234}L_4 - S_{23}L_3 - S_2L_2 \end{bmatrix}, \quad (1) $$

is the position vector of the robot's foot relative to the reference coordinate system of the coxal joint. $C_i = \cos(\theta_i)$, $S_i = \sin(\theta_i)$, $C_{ij} = \cos(\theta_i + \theta_j)$, $S_{ij} = \sin(\theta_i + \theta_j)$, $C_{ijk} = \cos(\theta_i + \theta_j + \theta_k)$.

3.1.2. Inverse Kinematics Analysis

According to the position vector of the robot foot relative to the reference coordinate system of the coxal joint, the four joint angles of the leg can be obtained. However, the four joint angles can be driven directly only through being given the position vector of the foot. Therefore, a constraint is added. Firstly, the rotation angle of the coxal joint (θ_1) is solved by the algebraic method

$$\theta_1 = a\tan(\frac{{}^{0}_{tip}P_Y}{{}^{0}_{tip}P_X}). \tag{2}$$

The rotation angles of the femur–patella joint (θ_2), the tibia–metatarsus joint (θ_3) and the metatarsus–tarsus joint (θ_4) are solved by the geometric method. The base coordinate system {O} is assumed to attach to the coordinate system of the coxal joint. The mechanical leg is projected in the X–Z plane coordinate system, and the simplified model of the linkage is shown in Figure 4.

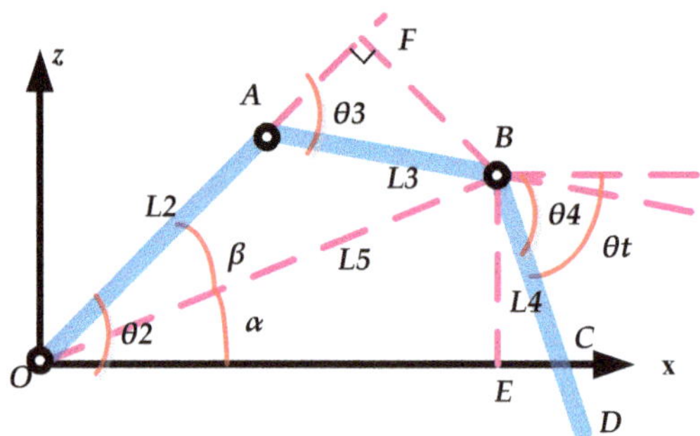

Figure 4. The simplified model of the linkage. O represents the femoral–patella joint, A represents the tibia–metatarsus joint, B represents the metatarsus–tarsus joint, and D represents the foot tip of the robot.

In Figure 4, O represents the femoral–patella joint, A represents the tibia–metatarsus joint and B represents the metatarsus–tarsus joint. We have aligned the coxal joint and the femoral–patella joint at O for convenience in calculating the angles. The angle between the connecting rod BD and the X axis or the ground is θ_t; θ_t is the attitude angle of the foot. The geometric relation between the links of the linkage mechanism can be obtained by θ_t and θ_2, θ_3 and θ_4, which satisfy the constraint relation in Equation (3)

$$\theta_t = \theta_2 + \theta_3 + \theta_4. \tag{3}$$

The plane coordinate of D relative to O is (P_X, P_Z). According to the projection principle, the relation between P_X, P_Z and ${}^{0}_{tip}P_X$, ${}^{0}_{tip}P_Z$ is

$$\begin{cases} P_X = {}^{0}_{tip}P_X / \cos\theta_1 \\ P_Z = {}^{0}_{tip}P_Z \end{cases}. \tag{4}$$

In the ΔOBE, according to cosine theorem, length L_5 between O and B is

$$L_5 = \sqrt{(P_X - L_4\cos(\theta_t))^2 + (P_Z + L_4\sin(\theta_t))^2}. \tag{5}$$

In the ΔOAB, the θ_3 holds

$$\theta_3 = \arccos(\frac{{L_5}^2 - {L_2}^2 - {L_3}^2}{2L_2L_3}). \tag{6}$$

In the ΔOBE

$$\tan\alpha = \frac{(L_4\sin(\theta_t) + P_Z)}{P_X - L_4\cos(\theta_t)}, \alpha = -\arctan(\frac{P_Z + L_4\sin(\theta_t)}{P_X - L_4\cos(\theta_t)}). \quad (7)$$

In ΔOBF and ΔABF, they meet the following formula

$$\beta = -\arcsin(\frac{L_3\sin(\theta_3)}{L_5}). \quad (8)$$

Thus, we can obtain

$$\theta_2 = \alpha + \beta = -\arctan(\frac{P_Z + L_4\sin(\theta_t)}{P_X - L_4\cos(\theta_t)}) - \arcsin(\frac{L_3\sin(\theta_3)}{L_5}). \quad (9)$$

From the constraints relation above, θ_3 solutions can be

$$\theta_4 = \theta_t - \theta_2 - \theta_3. \quad (10)$$

Hence, the inverse kinematics is

$$\begin{cases} \theta_1 = \frac{{}^{0}_{tip}P_Y}{{}^{0}_{tip}P_X} \\ \theta_2 = -\arctan(\frac{P_Z+L_4\sin(\theta_t)}{P_X-L_4\cos(\theta_t)}) - \arcsin(\frac{L_3\sin(\theta_3)}{L_5}) \\ \theta_3 = \arcsin(\frac{L_5{}^2-L_2{}^2-L_3{}^2}{2L_2L_3}) \\ \theta_4 = \theta_t - \theta_2 - \theta_3 \end{cases}. \quad (11)$$

In this paper, we proposed a study based on bionic kinematics with redundant freedom. Figure 5 provides a sketch of the leg flexion and the extension of the spider. The attitude angle of the foot θ_t is required to solve the leg joint angles. Thus, the attitude angle of the foot can be obtained by using the relationships between the changes of joint angles and the foot posture during jumping. In the jumping process of the spider, the rapid extension of the leg makes the effective leg length increase rapidly to complete the fast jumping. Simultaneously, the joint angle of the spider leg changes regularly with the effective length, and the attitude angle of the foot vary regularly with the joint angle. Figure 5 shows a sketch of the leg flexion and the extension of the spider.

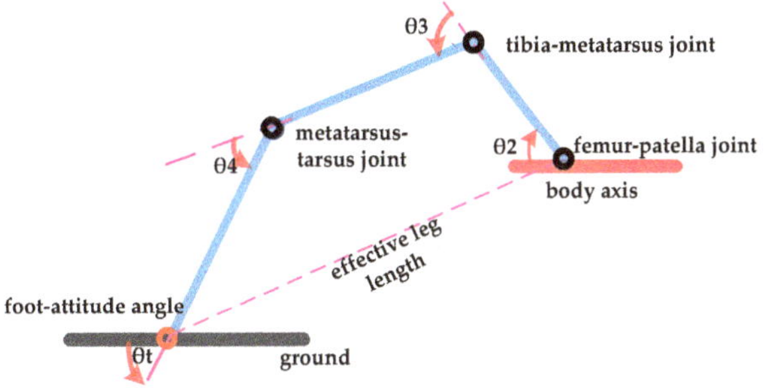

Figure 5. Schematic of the leg at defined instances. Black circles indicate the joint and the red circle indicates the foot. The arrows indicate the movements of the joints. The red line represents the axis of the spider's body, the blue line represents the leg of spider, and black line represents the ground.

Figure 6 shows the angle curve for the duration of the acceleration phase. During the jump process, the spider completes the jump in a fixed pattern. The effective leg length, the body attitude angle, the joint angle and the foot stance angle determine the maximum values of the take-off form, direction, height and distance. The leg extends in the jumping direction during its contact phase, which allows the spider to quickly accumulate speed, while the posture of the spider's body changes depending on the jumping direction, and the attitude angle varies with different jumping forms. When a spider jumps upward, the effective stretch length of each leg is the same, and the body posture hardly changes. When the spider jumps sideways, the body has a certain initial roll angle. At the same time, the length of the hind leg is rapidly stretched along the back of the jumping direction and the effective length of the leg increases rapidly. Then, the forelegs stretch in the jumping direction to achieve the sideways jump. In the process of the sideways jump, the length of the hind legs is longer than that of the forelegs of the spider in the jumping direction and, as a result, the rolling angle of the spider is larger. When the spider jumps forward, it has a certain initial pitch angle in the jump direction. At the same time, the hind legs extend rapidly at first; the effective length of the legs increases rapidly, and then the forelegs rapidly extend to achieve the forward jump. In the whole process of the forward jump, the pitch angle of the body posture becomes bigger. In this paper, we study the kinematic of the jumping spider [7] by observing and studying the relationships between the changes of joint angles and effective leg length during the jumping process. Using the experimental data and the curve of the jumping process of the spider, some groups of jumping motion curves are analyzed, and then several sets of the joint angle curve are fitted out in the interval angle of each joint. According to the D-H kinematic model in the last section and the definition of joint angles, the mathematical relationship between the joint angles of the jumping spider and the joint angles of the robot can be obtained. A suitable angle curve and the attitude angle of the foot are used for the spider-inspired robot. Figure 6 shows the angle curve of the femur–patella joint, femur–patella joint, the metatarsus–tarsus joint and the attitude angle of the foot for the robot during the acceleration phase. Similarly, in the robot's acceleration phase during take-off, the angles of each joint change with the extension of the leg, including the angle of the femoral–patella joint which increases gradually by approximately 90°. The tibia–metatarsal angle decreases gradually by approximately 55°; the metatarsus–tarsus angle decreases gradually and the angle varies by 35°; and the attitude angle of the feet is kept within 70–80°. Through observing the black solid line, the constraint is obtained

$$\theta_t = -0.02\mathrm{t} + 73^\circ. \quad (12)$$

Hence, we propose the method based on bionic kinematics with redundant freedom as it offers an excellent approach to solve the inverse kinematics for a bionic structure with redundant freedom.

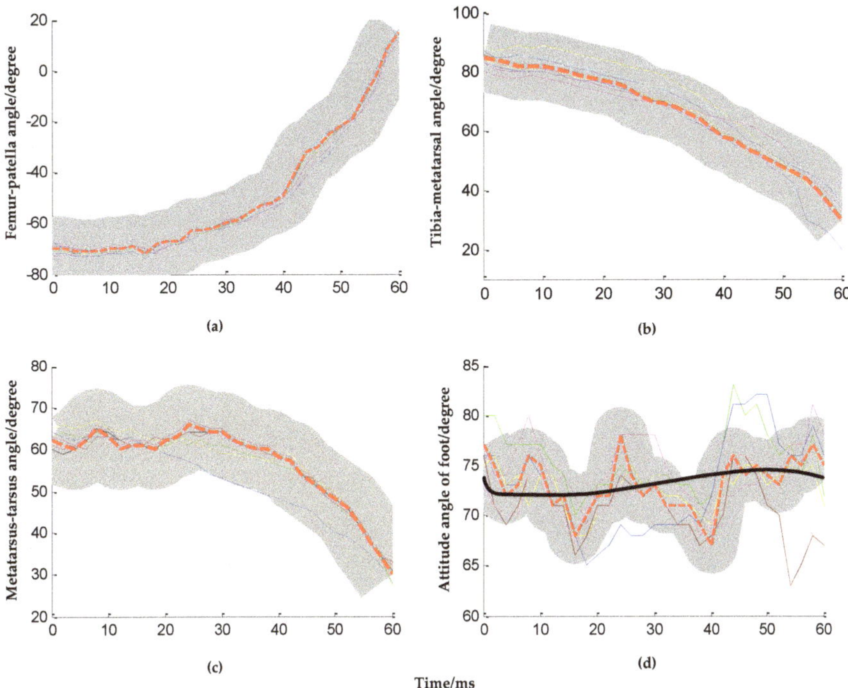

Figure 6. Angle curve for the duration of the acceleration phase: (**a**) femur–patella angle; (**b**) femur–patella angle; (**c**) metatarsus–tarsus angle; and (**d**) attitude angle of the foot. The red dashed line indicates the optimal curve, and the shadow area indicates the range of the angle of the joint indicated by the Quartile Range (IQR) method. The black solid line indicates the optimal curve of the attitude angle of the foot. The other solid line indicates the angle curve for the duration of the acceleration phase.

3.2. Locomotion Planning

3.2.1. Path Planning

There are three phases in the whole jumping process: take-off, flying and touchdown. In this paper, it is mainly the take-off phase of the robot that is studied. The take-off phase refers to the whole movement process of the robot, from a stationary state to jumping away from the ground. In other words, from the moment the robot foot has been subjected to the ground force F_g in the take-off phase until the moment the robot's feet leave the ground. The take-off process of jumping consists of flexing the leg to store energy and extending to release energy. In the first phase, the legs flex and the CoG drops. Then, the initial attitude angle is adjusted with stored energy to take off. In the phase of releasing energy, the robot legs extend, the body mass center rises, and the robot adjusts to reach a proper take-off stance. Then, the legs continue to extend, hit the ground violently and cause impact force, which makes the robot accelerate and realize the jumping movement. Therefore, the trajectory of the mass center decreases slowly and then rises rapidly. When the robot jumps, the speed of the CoG is an important index of the take-off performance of a robot, as it determines the jumping height and the jumping distance when the robot jumps off the ground. Suppose that the moment the robot takes off from the ground is t_f, the velocity of the CoG is $[\dot{X}(t)\dot{Y}(t)\dot{Z}(t)]^{T}$, and the acceleration is $[\ddot{X}(t)\ddot{Y}(t)\ddot{Z}(t)]^{T}$. The foot of the robot generates the impact force via continuous contact with the ground so that the robot will accumulate acceleration and speed and finally realize the jumping task.

In this paper, three different types of jumping—vertical jumping, side jumping and forward jumping—are discussed. The position vector of the robot foot relative to the base coordinate is $[X\ Y\ Z]^T$. When the robot jumps vertically, the whole body has an upward acceleration and velocity relative to the ground. The robot has acceleration and velocity in the Z direction. When the robot jumps sideways, the robot has acceleration and velocity in the X and Z direction. When the robot jumps forward, the robot has acceleration and velocity only in the Y and Z direction. Thus, the constraint conditions of the motion of the CoG in these jumping forms of the robot must be discussed. At first, the reacting force of the ground to the foot of the robot gradually decreases to zero when the foot of the robot leaves the ground. At this instant, the contact force of the foot fulfills the condition

$$F_{gx} = F_{gy} = F_{gz} = 0. \tag{13}$$

The acceleration of CoG fulfills the following constraints:

$$\left.\begin{array}{r} \ddot{X}_f = 0 \\ \ddot{Y}_f = 0 \\ \ddot{Z}_f = -g \end{array}\right\}. \tag{14}$$

At the take-off moment t_f, the robot accumulates sufficient velocity to achieve three types of jumps and the robot leaves the ground. The velocity constraints are as shown in Table 2. Here, V_{xuf}, V_{yuf}, and V_{zuf} denote the velocity of the vertical jump. V_{xsf}, V_{ysf}, and V_{zsf} denote the velocity of the sideways jump. V_{xff}, V_{yff}, and V_{zff} denote the velocity of the forward jump. Therefore,

$$\left.\begin{array}{l} V_{xuf} = \dot{X}_f \\ V_{yuf} = \dot{Y}_f \\ V_{zuf} = \dot{Z}_f \end{array}\right\}, \left.\begin{array}{l} V_{xuf} = \dot{X}_f \\ V_{yuf} = \dot{Y}_f \\ V_{zuf} = \dot{Z}_f \end{array}\right\}, \left.\begin{array}{l} V_{xuf} = \dot{X}_f \\ V_{yuf} = \dot{Y}_f \\ V_{zuf} = \dot{Z}_f \end{array}\right\}. \tag{15}$$

Table 2. Velocity constraints.

Upward Jumping	Sideway Jumping	Forward Jumping
$V_{xuf} = 0$ $V_{yuf} = 0$ $V_{zuf} > 0$	$V_{xsf} > 0$ $V_{xsf} = 0$ $V_{xsf} > 0$	$V_{xff} = 0$ $V_{xff} > 0$ $V_{xff} = 0$

During the whole process of the take-off ($0 \leq t \leq t_f$), it is necessary to ensure that the robot does not leave the ground in advance and that it gains sufficient take-off speed. a_{xuf}, a_{yuf}, and a_{zuf} are used to indicate the acceleration of the robot when the robot jumps vertically. a_{xsf}, a_{ysf}, and a_{zsf} are used to indicate the acceleration of the robot when the robot jumps sideways. a_{xff}, a_{yff}, and a_{zff} are used to indicate the acceleration of the robot when the robot jumps forward. The acceleration constraints are as shown in Table 3. Therefore,

$$\left.\begin{array}{l} a_{xuf} = \ddot{X}_f \\ a_{yuf} = \ddot{Y}_f \\ a_{zuf} = \ddot{Z}_f \end{array}\right\}, \left.\begin{array}{l} a_{xsf} = \ddot{X}_f \\ a_{ysf} = \ddot{Y}_f \\ a_{zsf} = \ddot{Z}_f \end{array}\right\}, \left.\begin{array}{l} a_{xff} = \ddot{X}_f \\ a_{yff} = \ddot{Y}_f \\ a_{zff} = \ddot{Z}_f \end{array}\right\}. \tag{16}$$

Table 3. Acceleration constraints.

Upward Jumping	Sideway Jumping	Forward Jumping
$a_{xuf} = 0$	$a_{xsf} > 0$	$a_{xff} = 0$
$a_{yuf} = 0$	$a_{xsf} = 0$	$a_{xff} > 0$
$a_{zuf} > 0$	$a_{xsf} > 0$	$a_{xff} = 0$

When the robot is ready to take off, the motion of the foot relative to the base coordinate {O} should fulfill the following boundary conditions

$$\begin{cases} X_{foot}{}^{\{o\}}|_{t=0} = X_0, Y_{foot}{}^{\{o\}}|_{t=0} = Y_0, Z_{foot}{}^{\{o\}}|_{t=0} = Z_0 \\ X_{foot}{}^{\{o\}}\Big|_{t=t_f} = X_f, Y_{foot}{}^{\{o\}}|_{t=t_f} = Y_f, Z_{foot}{}^{\{o\}}|_{t=t_f} = Z_f \\ \dot{X}_{foot}{}^{\{o\}}|_{t=t_f} = \dot{X}_f, \dot{Y}_{foot}{}^{\{o\}}|_{t=t_f} = \dot{Y}_f, \dot{Z}_{foot}{}^{\{o\}}|_{t=t_f} = \dot{Z}_f \\ \ddot{X}_{foot}{}^{\{o\}}|_{t=0} = 0, \ddot{Y}_{foot}{}^{\{o\}}|_{t=0} = 0, \ddot{Z}_{foot}{}^{\{o\}}|_{t=0} = 0 \\ \ddot{X}_{foot}{}^{\{o\}}|_{t=t_f} = 0, \ddot{Y}_{foot}{}^{\{o\}}|_{t=t_f} = 0, \ddot{Z}_{foot}{}^{\{o\}}|_{t=t_f} = -g \end{cases}. \tag{17}$$

where $[X_0\ Y_0\ Z_0]^T$ is the position vector of the foot relative to the base coordinate {O} at the initial stage of jumping, and $[X_f\ Y_f\ Z_f]^T$ is the position vector of the foot relative to the base coordinate {O} when the robot leaves the ground.

Then, the robot has the following relation when it is in the flying phase:

$$\begin{cases} L_x = 2\dot{X}_f\dot{Z}_f/g \\ L_y = 2\dot{Y}_f\dot{Z}_f/g \\ H_V = \dot{Z}_f^2/2g \end{cases}. \tag{18}$$

where H_v is the maximum jumping height of the robot, L_y is the maximum distance along the longitudinal body axis, and L_x is the maximum distance along the body lateral axis. During the take-off process, the rational trajectory planning of the foot enables the robot to complete the jump task successfully. The acceleration of the robot while jumping is planned. To make sure that there is no impact between the foot and the ground when the robot makes contact and lifts off, the contact force must be smooth. Here, the quadric curve is used to plan the acceleration of the robot. Then, the acceleration curve equations are

$$\begin{cases} a_x = b_{0x} + b_{1x}t_1 + b_{2x}t_1{}^2 \\ a_y = b_{0x} + b_{1x}t_1 + b_{2x}t_1{}^2 \\ a_z = b_{0z} + b_{1z}t_1 + b_{2z}t_1{}^2 \end{cases}. \tag{19}$$

where the symbols t_1, t_2, and t_3, respectively, indicate the initial time t_0, the ground departure time t_f and the intermediate time t_m of the body. The symbols (a_{x1},a_{y1},a_{z1}), (a_{x2},a_{y2},a_{z2}), and (a_{x3},a_{y3},a_{z3}), respectively, indicate the acceleration vectors corresponding to the three moments. The coefficients of each component are

$$B = [T]^{-1} \cdot [A]. \tag{20}$$

where [T] is time matrix, [A] is coordinate matrix. According to the planning of the acceleration, the instantaneous velocity when the robot foot hits the ground and the trajectory of the foot tip can be obtained as

$$\begin{cases} X(t) = \iint a_x dt \\ Y(t) = \iint a_y dt \\ Z(t) = \iint a_z dt \end{cases}, \begin{cases} v(t) = \int a_x dt \\ v(t) = \int a_y dt \\ v(t) = \int a_z dt \end{cases}. \tag{21}$$

According to the position vector of the robot foot relative to the base coordinate $[X\ Y\ Z]^T$, we can obtain all joint angles of the robot by inverse kinematics. The acceleration of the robot becomes zero at the exact moment of the foot's take-off, and the velocity of the robot reaches its maximum at the same time. During the flying phase of the robot, the maximum jumping height of the robot is 100 mm, the maximum distance along the longitudinal body axis is 250 mm, and the maximum distance along the body lateral axis is 250 mm.

3.2.2. Attitude Planning

In the jump process of the robot, the reasonable initial attitude angle of the robot can affect the jumping height and distance. The initial jumping posture mainly includes the pitch angle, yaw angle and roll angle of the robot when taking off. By controlling the pitch angle of the robot, the maximum height and maximum distance of the robot can be reached when the robot jumps forward. The roll angle affects the sideways jump, while the yaw angle can adjust the jumping direction of the robot, allowing the robot to jump towards the target. During vertical jumps, the yaw angle and pitch angle of the robot are 0°. The initial joint position and angles of each leg of the robot are the same, correspondingly. The robot can jump vertically by reacting to the ground. Before the robot begins jumping sideways, the body leans to the left due to a slight flexion of the left legs, and the robot has a certain roll angle of θ_r in the jumping direction. Meanwhile, the yaw angle and pitch angle are zero. A simple model of the robot's posture during a sideways jump is shown in Figure 7a. When the robot jumps to the right, the effective length of the three left legs of the robot is larger than that of the right legs in the initial state to satisfy a certain proportion. Therefore, we can obtain

$$Z_l = K_1 Z_r, \tag{22}$$

where Z_l is the Z coordination of the three left legs opposite to the base coordinate {O}. Z_r is for the right legs. K_1 is a constant, and it is determined by the initial attitude of the robot.

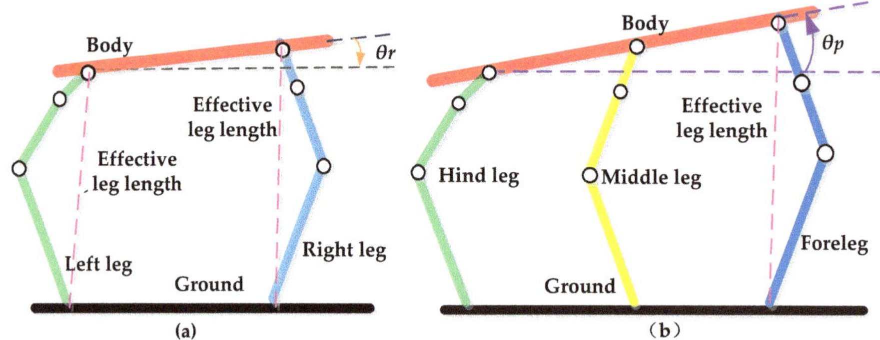

Figure 7. Simplified model of the robot posture. (**a**) Sideways jump: The red solid line indicates the transverse axis of the robot body, the green solid line indicates the left leg of the robot body, the blue solid line indicates the right leg, the purple dashed line indicates the effective length, and the black circles indicate the joints. (**b**) Jump forward: Green solid lines indicate the hind legs of the robot body, the blue solid line indicates the foreleg, and the yellow solid line indicates the middle leg.

When the robot jumps forward, the forelegs extend and the hind legs flex during their contact phase, and the middle legs remain in slight extension. Therefore, the robot has a certain pitch angle in the jump direction; the pitch angle is θ_p. The roll angle and raw angle are 0°. The simple model of the robot posture of the forward jump is shown in Figure 7b. At the beginning of the jump, the effective length of the forelegs of the robot is larger than that of the middle legs in the initial state; meanwhile, the effective length of the middle legs of the robot is larger than that of the hind legs in the initial state, and they satisfy two proportions:

$$\begin{cases} Z_f = K_2 Z_m \\ Z_m = K_2 Z_h \end{cases}, \quad (23)$$

where Z_h is the Z coordination of the hind legs opposite to the base coordinate {O}; Z_f is the forelegs; and Z_m is the middle legs. K_2 is a constant and it is determined by the initial attitude of the robot.

4. Results

To further verify the algorithm of omnidirectional jumping, a series of experiments is conducted using a hexapod robot platform. There are three groups of experiments: the vertical jump, sideways jump, and forward jump. In these experiments, the joint angles as input of the simulation system, and the CoG and the attitude angle as output of the simulation system are used to verify the design requirement and stability of the robot. To avoid the slip phenomenon in the jumping process, and to achieve a certain friction between the ground and the robot foot, we use a wooden floor as the ground. We assume that there is no air resistance and no slipping phenomenon when the robot jumps, since the beginning of the jump is our main concern and the structure is slim with little air resistance. Friction is supposed as a load torque. Robot links are in rigid connections. In this paper, the simulation results are obtained under Matlab-Adams, relevant parameters are set in Adams, and the parameters include static coefficient, dynamic coefficient and stiffness. Experiment parameters are shown in Table 4.

Table 4. Experiment parameters.

Parameters	Symbol	Value
Jumping height	H_v	100 mm
Jumping distance	$L_x(L_y)$	250 mm
Static coefficient	u_S	0.35
Dynamic coefficient	u_D	0.2
Stiffness	T	10000 N/m
Take-off moment	t_f	1 s

4.1. Vertical Jump

In this section, we obtain data on the robot jumping vertically; these data include the joint angle, attitude angle, joint velocity, foot contact force, take-off velocity, trajectory of CoG, jumping height and jumping distance. These data are based on the simulation result under Matlab-Adams, and we can prove the reliability and rationality of vertical jump by analyzing it.

Figure 8a,b shows the simulation platform of the virtual jump prototype. At the beginning of the vertical jump, the body slowly moves downward due to a slight flexion of the robot leg, and the robot begins to store energy. During the acceleration phase of the vertical jump, the effective leg length extends quickly, and the robot starts to take off. Figure 8c shows a sketch of the robot jumping vertically. The initial height of the CoG of the robot is 180 mm and the jumping height is 100 mm. In the vertical jumping experiment, we measure the contact force of the foot and the joint force; we observe the change of joint forces to verify whether the motors meet the requirement.

During the vertical jumping, each joint of the six legs has the same rotation angle. Here, we can see the set of angle curves of the robot leg in Figure 8c. Coxal joint angles of all legs always remain the same during the vertical jump: they are 0 rad. Before the acceleration phase begins, the femur–patella joint angle, tibia–metatarsus joint angle and metatarsus–tarsus joint angle begin to slowly change, and the robot body slows down and adjusts the attitude. Then, the foot moves down quickly, and the robot starts to accelerate by extending its legs at 0.8 s. At 1.0 s, the height of the CoG is 300 mm, and the robot leaves the ground and takes off. At 1.2 s, the robot rises to its maximum height of approximately 400 mm, the robot completes the task of jumping 100 mm in height. From 0.8 s to

1.2 s, the femur–patella joint angle changes by approximately 1.5 rad, the tibia–metatarsus joint angle changes by approximately 1.2 rad and the metatarsus–tarsus joint angle changes by approximately 2.0 rad.

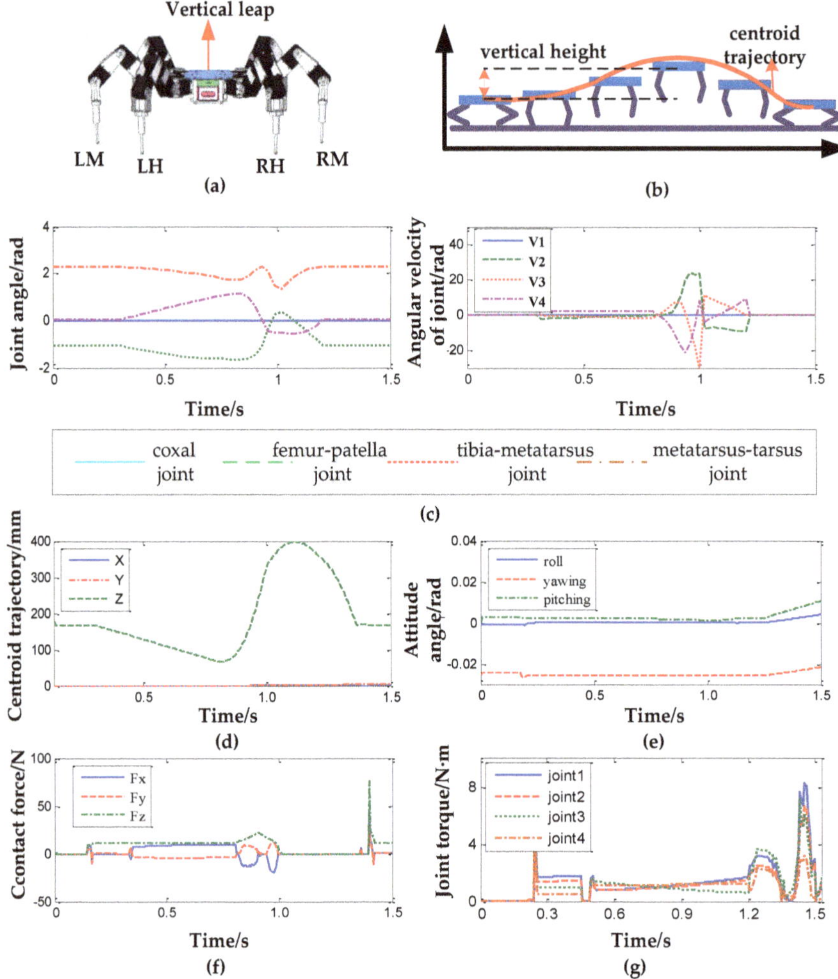

Figure 8. (**a**) Simulation platform of the virtual jump; (**b**) sketch of the robot vertically jumping; (**c**) joint angle curves of the robot leg; (**d**) trajectory of center of gravity (CoG); (**e**) attitude angle of robot; (**f**) contact force; and (**g**) joint torque.

As mentioned, the effective length of the robot's legs always changes during the vertical jump. For the robot to achieve stability, the change of the attitude angle is analyzed. The initial attitude angle of the robot is not 0 rad, which is due to the initial pose of the robot in Figure 8e. The roll angle holds at near 0 rad when the robot jumps. The pitch angle holds at 0.002 rad at approximately 0–1 s, and the roll angle increases gradually to 0.01 rad when the robot lifts off the ground. In the process of jumping, the yaw angle of the robot always holds at approximately −0.025 rad. According to the simulation results, the robot has good stability in the vertical jump process. Figure 8f is the contact force of the foot tip for the robot in the vertical jump process. When the robot jumps vertically, it completes the upward jumping task through the contact force in the Z direction. The forces all cancel each other out which acts on the entire robot in the X and Y directions. As shown in Figure 8f, the initial contact force is 0 N, which is due to the initial state of the robot. At 0.3–0.8 s, the CoG slowly drops and the contact force remains constant; the contact force of X is approximately 10 N, the contact force of Y is 0. The contact force F_Z of Z is approximately 12 N, which is 1/6 of the gravity of the robot. At 0.3–0.8 s, the contact force of the foot tip increased rapidly, and the Z directional force increased to 24 N at 1.0 s, and decreased to 12 N after 1.0 s. Overall, the vertical acceleration of the robot does not reach 0 m/s^2, the speed increases and reaches its maximum at 1 s. After the robot jumps off the ground, the contact force is 0 N until the robot lands. Figure 8g is the joint torque of the robot in the vertical jump process. At the initial stage of simulation, the robot foot does not touch the ground and the joint torque of the robot is 0 Nm. When the robot touches the ground, the joint torques changes slowly between 1 Nm and 3 Nm until the robot begins to adjust its posture at 0.5 s. At 0.8–1.2 s, the joint torque of the robot begins to increase. At 1.4 s, the joint torque of the robot fluctuates wildly, and the maximum valve is approximately 8 Nm, which is caused by the impact of the robot landing.

According to these curves, we can see that the experimental data are consistent with what we expect from our projections; the robot could complete the task of jumping 100 mm in height while maintaining good stability.

4.2. Sideways Jump

The simulation experiment of the robot jumping sideways is carried out to verify whether the joint angle, attitude angle, jumping height and jumping distance are consistent with motion planning or not. Unlike the vertical jump, there is much variability in the joint kinematics and attitude with the sideways jump. At the beginning of the sideways jump, the lateral body axis began to rotate sideways in the jumping direction. The initial attitude is adjusted to prepare to jump. At the same time, the CoG drops and begins to store energy. During the acceleration phase of the sideways jump, the effective leg length of the robot extends quickly and the foot impacts the ground quickly, then the robot starts to take off. Figure 9b shows a sketch of the robot jumping sideways. The initial height of the robot is 180 mm, the jumping height is 100 mm, and the jumping distance is 250 mm.

During the robot's sideways jump, the legs in contact with the ground are first flexed and then extended. Each joint of the six legs has a different rotation angle. Before the acceleration phase begins, the CoG begins to drop and store energy at 0.3 s (Figure 9d); the legs on one side of the robot's body are obviously flexed, but those on the other side remain relatively small in the jumping direction. From 0.3 s to 0.8 s, the robot body slows down and adjusts the attitude. At this instant, the femur–patella joint angle, tibia–metatarsus joint angle and metatarsus–tarsus joint angle begin to slowly change, and the coxal joint angle remain constant until the jump ends (Figure 9c). Then, the foot impacts the ground quickly the robot's legs start to extend and the robot begins to accelerate at approximately 0.8 s. The joint angle begins to change quickly (Figure 9c), while the coxal joint angle of the middle leg remains unchanged; this is due to the robot's structure and path planning of the sideways jump. The displacement of the robot in the jumping direction begins to change, depending on the time at which the robot is in the air. Then, the robot leaves the ground and takes off, with the height of the CoG at 320 mm in this moment. The robot rises to its maximum height of approximately 420 mm and completes the task of jumping 100 mm (Figure 9d). At 1.4 s, the robot will touch down. The robot is in the air for approximately 0.4 s, while the jumping distance is 250 mm (Figure 9d). The robot jumps to the right. From 0.8 s to 1.2 s, the femur–patella joint angle changes by approximately 2 rad, the tibia–metatarsus joint angle changes by approximately 2.5 rad and the metatarsus–tarsus joint angle changes by approximately 0.5 rad in the three left legs (Figure 9c). Concurrently, the femur–patella joint angle changes by approximately 1.5 rad, the tibia–metatarsus joint angle changes by approximately 2.5 rad and the metatarsus–tarsus joint angle changes by approximately 1.0 rad in the three right legs (Figure 9c). The attitude angle directly influences the stability of the robot when it jumps sideways in Figure 9c. The initial attitude angle of the robot depends on attitude planning. The roll angle always holds near 0.02 rad when the robot jumps, and the roll angle changes 0.1 rad until the robot has landed. The pitch angle holds at near 0.2 rad. In the process of jumping, the yaw angle of the robot always holds at 0.02 rad; it increases gradually to 0.03 rad when the robot has landed. According to the simulation results, the attitude angle of the robot is nearly invariable. Hence, the robot has good stability in the sideways jump process, which provides a good theoretical basis for the forward jump of the robot.

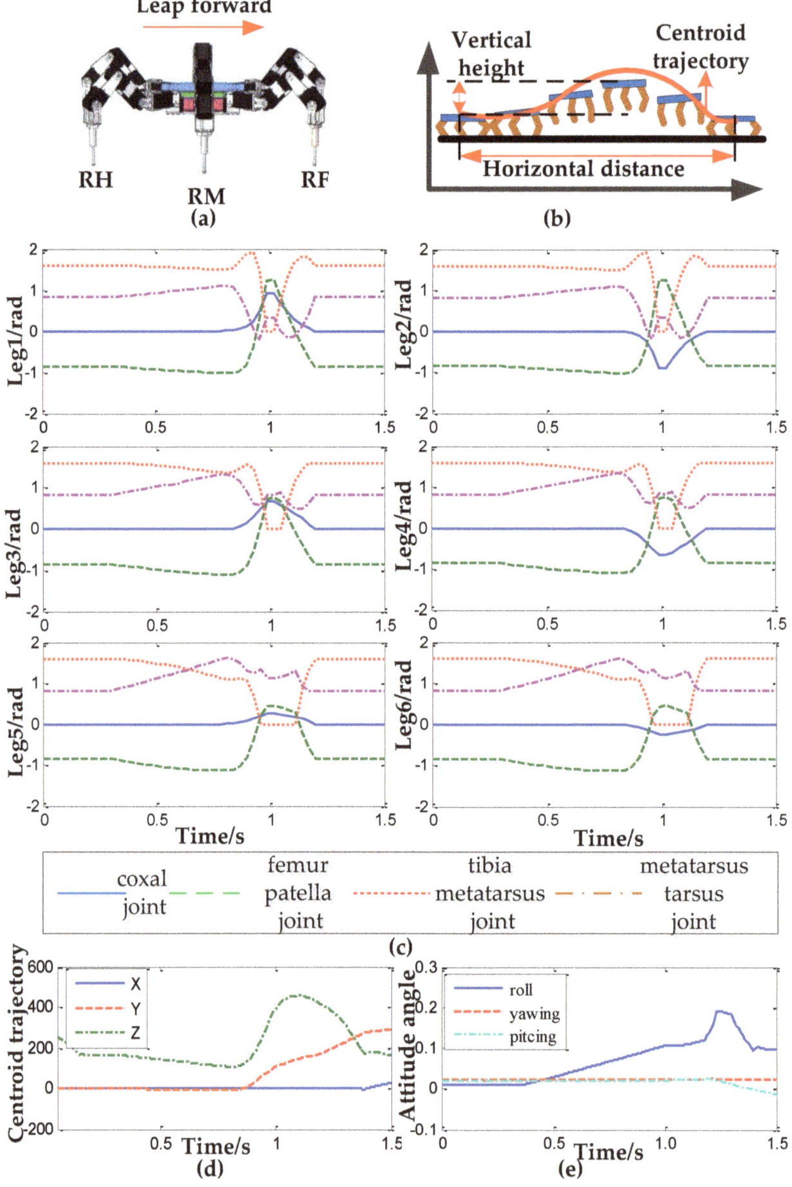

Figure 9. (**a**) Simulation platform of the sideways jump; (**b**) sketch of robot jumping sideways; (**c**) joint angle curves of the robot leg; (**d**) trajectory of the CoG; and (**e**) attitude angle of the robot.

According to the simulation results, we can see that the experimental data are consistent with what we expect from our projections for the sideways jump. Therefore, the robot could complete the task of jumping 100 mm in height and 250 mm in distance, while maintaining good stability.

4.3. Forward Jump

The simulation of the robot jumping forward is carried out to verify whether the joint angle, attitude angle, jumping height and jumping distance is consistent with motion planning or not. The simulation results are seen in Figure 10. In the forward jump, the robot's body slowly moves backwards and downwards due to a slight flexion of the hind legs and the middle legs before the acceleration phase begins. The robot begins to store energy with the CoG dropping. At the same time, the initial attitude is adjusted in preparation to jump. Then, the robot starts to accelerate by extending its legs; the effective leg length increases quickly. Eventually, the robot takes off. Figure 10b shows a sketch of the robot jumping forward. The initial height of the CoG of the robot is 180 mm, the jumping height is 100 mm, and the jumping distance is 250 mm.

While the robot jumps forward, the legs that have contact with the ground are first flexed and then extended. Each joint of the six legs has a different rotation angle (Figure 10c). Before the acceleration phase of the forward jump, the body rotates backwards as the hind legs and the middle of the robot body slightly flex in the jumping direction. Meanwhile, the robot adjusts its attitude by moving backwards and downwards. Then, the foot impacts the ground quickly, and the robot's legs start to extend and, at 0.8 s, the robot begins to accelerate. At the same instant, the joint angle begins to quickly change (Figure 10c). At 1.0 s, the robot loses contact with the ground and takes off; due to the accumulation of velocity, the rotation of the body is reversed. At this instant, the height of the CoG is 320 mm. The robot rises to its maximum height of approximately 420 mm at 1.2 s; it then completes the task of jumping 100 mm in height (Figure 10d). After this instant, none of the joint angles change (Figure 10c).

During the robot's forward jump, and the Coxal joint angles change by approximately 0.8 rad. The femur–patella joint angle changes by approximately 2 rad, the tibia–metatarsus joint angle changes by approximately 1.5 rad and the metatarsus–tarsus joint angle changes by approximately 1 rad. At 1.4 s, the robot lands on the ground. The robot is in the air for approximately 0.4 s, and the jumping distance is 250 mm (Figure 10d). In Figure 10e, the initial attitude angle is not 0, and the initial roll angle is 0.01 rad; the initial yaw angle is 0.025 rad; and the initial pitch angle is 0.02 rad, due to the initial pose of the robot. The roll angle changes by 0.02 rad in the jumping process, and the roll angle decreased by 0.1 rad when the robot landed. The pitch angle holds at 0.01 rad in the jumping process, and it generated a fluctuation of 0.1 rad when the robot landed. The yaw angle was almost constant until the robot landed, maintaining 0.015 rad. According to the simulation results, the robot has good stability in the forward jump process, which provides a good theoretical basis for the forward jump of the robot.

Observing the simulation result, we find that the experimental data are consistent with our projections of the forward jump. The robot could complete the task of jumping 100 mm in height and 250 mm in distance; it maintains good stability in the flying phase.

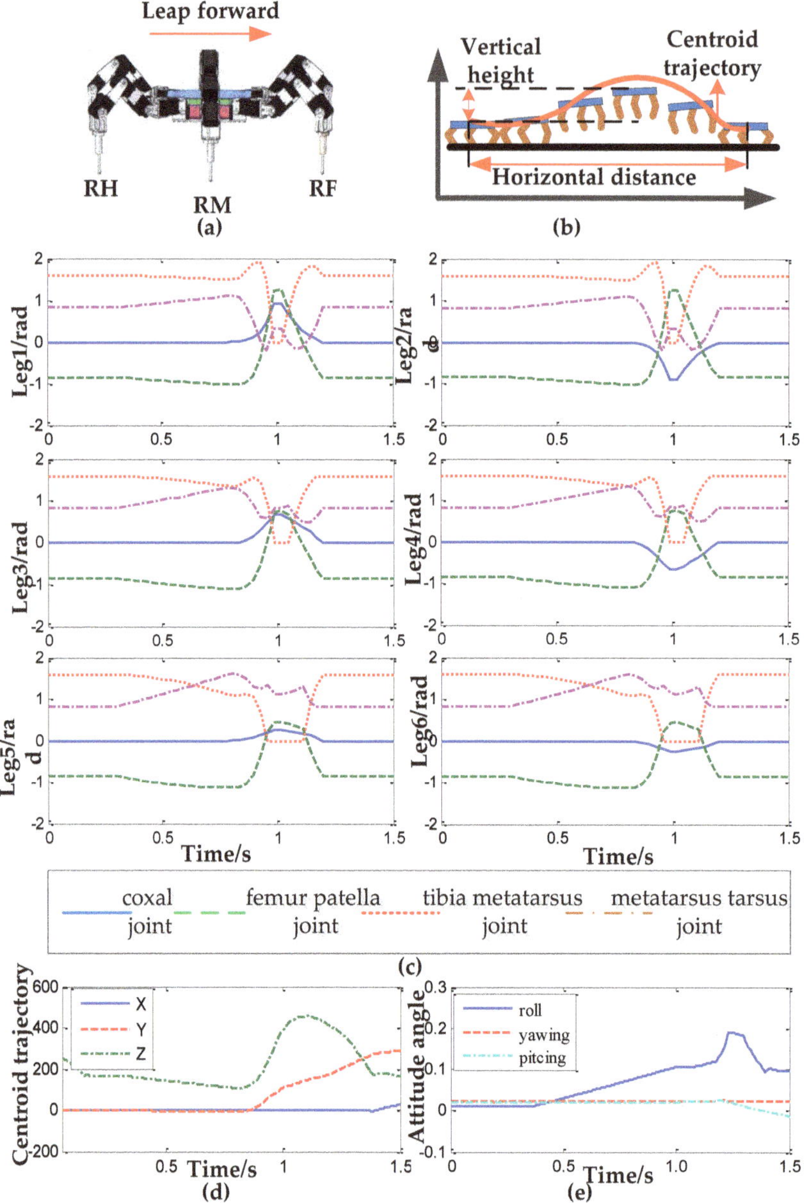

Figure 10. (**a**) Simulation platform of the forward jump; (**b**) sketch of the robot forward jumping; (**c**) joint angle curves of the robot leg; (**d**) trajectory of the CoG; and (**e**) attitude angle of robot.

5. Discussion

In this paper, research on the omnidirectional jump control of the hexapod robot based on the behavior mechanisms of jumping spiders was undertaken. In the first section, the jumping forms of several typical legged robots [16,25,28,34] were described in detail. Normally, the more jumping force a robot has, the more difficult it is to control its jumping movement. However, the robot with more jumping force has a better capacity for avoiding obstacles. In the past decade, research on jumping robots has mainly focused on single-direction jumps, such as the vertical jump and the forward jump. However, this type of jumping robot must first adjust its posture and jumping direction when trying to avoid an obstacle. Therefore, the efficiency of the robot is greatly reduced. We study the change of the effective leg length and joint angle of the jumping spider, before proposing an omnidirectional jump control of the hexapod robot based on the behavior mechanisms of jumping spiders. Through a series of simulation experiments, we verify the possibility of an omnidirectional jump by comparing the simulation and experimental data with the data expected from our projections. Finally, the robot realizes rapid jumping in all directions under any circumstances.

There are several forms of DoF for robot legs, including 1DoF, 3DoFs and 4DoFs. Large DoF will greatly increase the complexity of motion controls, but smaller ones cannot satisfy the amount of motion required. The 1DoF model of a jumping robot leg is usually made by a hydraulic component [37] or an elastic component [38]. However, it is difficult to guarantee the stability of motion with such robots. Currently, 3DoFs for jumping robot legs is the main form, but does not have enough flexibility when the robot jumps omnidirectionally. In this paper, the jumping robot has adapted 4DoFs-mechanical legs, which increases the flexibility of the robot when it jumps omnidirectionally. The attitude angle of the foot tip is obtained based on bionic kinematics, and the constraint of the foot tip's attitude angle is added to calculate the inverse kinematic. Since most studies are in a unidirectional jumping stage to the jumping robot, this type of robot cannot realize omnidirectional jumping, especially avoid obstacles quickly. In this paper, omnidirectional jumping has been proposed as it allows the robot to avoid and cross obstacles in all directions. We proposed a study based on bionic kinematics with redundant freedom, which could potentially solve the inverse kinematics for a bionic structure with redundant freedom and achieve to jump in many different types to robot. Furthermore, we proposed locomotion planning, especially attitude planning, by observing the behavior of the jumping spider.

6. Conclusions

In this paper, the difficulty of omnidirectional jumping for a jumping bio-mimetic spider robot was addressed. The theoretical contribution and novelty of this paper can be summarized as follows:

(1) The path of the robot, the initial attitude and the trajectory of the foot tip must be planned to complete the jumping task. In particular, the reasonable initial attitude angle of the robot can affect jumping height and distance.
(2) To satisfy the diversity of motion forms in robot jumping, each leg has 4DoFs. However, the 4DoFs-mechanical leg is a redundant structure and we must find a constraint condition. According to the change curve of each joint angle in the process of spider jumping, we can obtain the attitude angle curve as the added constraint condition.
(3) Three kinds of jumps are verified on the jumping robot prototype: vertical jumps, sideways jumps and forward jumps. The proposed method is verified by a series of simulation experiments. The results indicate that the jumping robot could maintain stability and complete the task of jumping we planned, and the proposed spider-inspired jumping strategy could easily achieve an omnidirectional jump, and robot able to avoid obstacles quickly.

The results indicate that the robot can perform the omnidirectional jump according to the path planning of the CoG and the initial jumping attitude. The robot also has better stability, as observed by the attitude angle of the jumping robot during its jumps. Therefore, the robot can jump in any direction by providing the trajectory of the CoG and the initial attitude.

Acknowledgments: This study is supported by the National Natural Science Foundation of China (No. 51605039), the China Postdoctoral Science Foundation (No. 2016M592728), Shaanxi Postdoctoral Scientific Research Project (No. 2016BSHYDZZ26) and Open Foundation of the State Key Laboratory of Fluid Power Transmission and Control (No. GZKF-201610).

Author Contributions: Yaguang Zhu designed the algorithm. Yaguang Zhu, Long Chen, and Qiong Liu designed and carried out the Simulation. Yaguang Zhu, Long Chen Rui Qin, Bo Jin and analyzed the data and wrote the paper. Bo Jin gave many meaningful suggestions about the structure of the paper.

Conflicts of Interest: The authors declare no conflict of interest.

References

1. Du, H.; Gao, F. Fault tolerance properties and motion planning of a six-legged robot with multiple faults. *Robotica* **2017**, *35*, 1397–1414. [CrossRef]
2. Lee, C.H.; Kim, S.H.; Kang, S.C. Double-track mobile robot for hazardous environment applications. *Adv. Robot.* **2003**, *17*, 447–459. [CrossRef]
3. Gillis, G.B.; Biewener, A.A. Hind limb Extensor Muscle Function during Jumping and Swimming in the Toad. *J. Exp. Biol.* **2000**, *203*, 3547–3563. [PubMed]
4. Wei, D.W.; Ge, W.J. Research Status and Development Trend of Jumping Robots. *Robot* **2014**, *36*, 502–512. [CrossRef]
5. Xu, H.Y.; Fu, Y.L.; Wang, S.G.; Liu, J.G. Research on biomimetic robotics. *Robot* **2004**, *26*, 283–288.
6. Zak, M.; Rozman, J.; Zboril, F.V. Overview of Bio-Inspired Control Mechanisms for Hexapod Robot. *Int. J. Comput. Inf. Syst. Ind. Manag. Appl.* **2016**, *8*, 125–134.
7. Tom, W.; Michael, K.; Robert, J.F.; Reinhard, B. Jumping kinematics in the wandering spider *Cupiennius salei*. *J. Comp. Physiol. A* **2010**, *196*, 421–438. [CrossRef]
8. Dong, J.H.; Sangok, S.; Jongwoo, L.; Sangbae, K. High speed trot-running: Implementation of a hierarchical controller using proprioceptive impedance control on the MIT Cheetah. *Int. J. Robot. Res.* **2014**, *33*, 1417–1445. [CrossRef]
9. Zhu, X.Y.; Fan, J.Z.; Cai, H.G. Design of Control System for Frog-inspired Jumping Robot. *Mach. Electron.* **2011**, *8*, 63–67.
10. Umberto, S.; Cesare, S.; Paolo, D. Bioinspired Jumping Robot with Elastic Actuators and Passive Forelegs. *Nat. Obs.* **2009**, *54*, 329–338. [CrossRef]
11. Tom, W.; Michael, G.; Reinhard, B. Hydraulic leg extension is not necessarily the main drive in large spiders. *J. Exp. Biol.* **2011**, *215*, 578–583. [CrossRef]
12. Je, S.K.; Kyu, J.C. Development of an Insect Size Micro Jumping Robot. *Living Mach.* **2014**, *8608*, 405–407. [CrossRef]
13. Kosa, G.; Ayali, A.; Hanan, U.B. Design of a Bio-Mimetic Jumping Robot. In Proceedings of the 27th Convention of Electrical and Electronics Engineers in Israel, Eilat, Israel, 14–17 November 2012; IEEE: Piscataway Township, NJ, USA, 2012.
14. Jindrich, D.L.; Full, R.J. Many-legged maneuverability: Dynamics of turning in hexapods. *J. Exp. Biol.* **1999**, *202*, 1603–1623. [PubMed]
15. Huang, X.G.; Hang, Z.Q. Kinematics analysis of frog-jumping robot. *J. Mach. Des.* **2012**, *29*, 23–25.
16. Zhang, Q.; Chen, A.J. Study on the jumping kinematics of the bionic cricket robot. *J. Mach. Des.* **2010**, *20*, 38–41.
17. Zhang, X.F.; Qin, X.S.; Feng, H.S.; Tan, X.Q. Motion Analysis and Control of a Single Leg of Hydraulically Actuated Quadruped Robots during Vertical Jumping. *Robot* **2013**, *35*, 135–141. [CrossRef]
18. Lu, Q.Q.; Liu, C.H.; Liu, J.S.; Tang, K.Q. Simulation Analysis of Jumping Hexapod Robot Drived with Hydraulic. *Mach. Des. Manuf.* **2015**, *8*, 200–205.
19. Zhou, Y.; Cheng, L.; Chen, K. Energy supply strategy for one-legged jumping robot. *J. Tsinghua Univ.* **2015**, *55*, 273–278.
20. Li, Y.; Ge, W.J.; Fan, C.Q. Research and design on the foot of the jumping kangaroo robot based on adaptive structure. *Mach. Des. Manuf.* **2012**, *5*, 187–189.
21. Ge, W.J.; Shen, Y.W.; Yang, F. Research on the Driving Characteristics of Bionic Kangaroo jumping Robot. *Mech. Eng. China* **2006**, *17*, 856–861.
22. Zhang, W.T.; Ge, W.J.; Li, J.H.; Jiang, M.; Shen, P. Design and Research on the Power System of Kangaroo Jumping Robot. *Robot* **2008**, *30*, 359–363.

23. Incaini, R.; Sestini, L.; Garabini, M.; Catalano, M.; Grioli, G.; Bicchi, A. Optimal Control and Design Guidelines for Soft Jumping Robots: Series Elastic Actuation and Parallel Elastic Actuation in Comparison. In Proceedings of the 2013 IEEE International Conference on Robotics and Automation (ICRA), Karlsruhe, Germany, 6–10 May 2013.
24. Tedeschi, F.; Carbone, G. Design Issues for Hexapod Walking Robots. *Robotics* **2014**, *3*, 181–206. [CrossRef]
25. Haynes, G.C.; Rizzi, A.A. Gaits and Gait Transitions for Legged Robots. In Proceedings of the 2006 IEEE International Conference on Robotics and Automation, Orlando, FL, USA, 15–19 May 2006.
26. Kikuchi, F.; Ota, Y.; Hirose, S. Basic Performance Experiments for Jumping Quadruped. In Proceedings of the 2003 IEEE International Conference on Intelligent Robots and Systems, Las Vegas, NV, USA, 27–31 October 2003.
27. Laksanacharoen, S.; Pollack, A.J.; Nelson, G.M.; Quinn, R.D.; Ritzmann, R.E. Biomechanics and Simulation of Cricket for Microrobot Design. In Proceedings of the 2000 IEEE International Conference on Robotics and Automation, San Francisco, CA, USA, 24–28 April 2000; IEEE: Piscataway Township, NJ, USA, 2002.
28. Sakakibara, Y.; Kan, K.; Hosoda, Y.; Hattori, M.; Fujie, M. Foot Trajectory for a Quadruped Walking Machine. In Proceedings of the IEEE International Workshop on Intelligent Robots and Systems, Ibaraki, Japan, 3–6 July 1990.
29. Birch, M.C.; Quinn, R.D.; Hahm, G.; Philips, S.M.; Drennan, B.; Fife, A.; Verma, H.; Beer, R.D. Design of a Cricket Microrobot. In Proceedings of the IEEE International Conference on Robotics and Automation, San Francisco, CA, USA, 24–28 April 2000; IEEE: Piscataway Township, NJ, USA, 2002.
30. Yang, Y.S.; Semini, C.; Guglielmino, E. Water vs. Oil Hydraulic Actuation for a Robot Leg. In Proceedings of the International Conference on Mechatronics and Automation, Changchun, China, 9–12 August 2009; IEEE: Piscataway Township, NJ, USA, 2009.
31. Thanhtam, H.; Sangyoon, L. Design of a Shape Memory Alloy-Actuated Biomimetic Mobile Robot with the Jumping Gait. *Int. J. Control Autom. Syst.* **2013**, *11*, 991–1000. [CrossRef]
32. Hyunsoo, S.; Sangyoon, L. Simulation and Experiments of a Four-Legged Robot That Can Locomote by Crawling and Jumping. In Proceedings of the 2014 IEEE International Conference on Robotics and Biomimetics, Bali, Indonesia, 5–10 December 2014.
33. Li, F.; Liu, W.; Fu, X.; Bonsignori, G.; Scarfogliero, U.; Stefanini, C.; Dario, P. Jumping like an insect: Design and dynamic optimization of a jumping mini robot based on bio-mimetic inspiration. *Mechatronics* **2012**, *22*, 167–176. [CrossRef]
34. Hoover, A.M.; Burden, S.; Fu, X.Y.; Shankar Sastry, S.; Fearing, R.S. Bio-Inspired Design and Dynamic Maneuverability of a Minimally Actuated Six-Legged Robot. In Proceedings of the IEEE RAS and EMBS International Conference on Biomedical Robotics and Biomechatronics, Tokyo, Japan, 26–29 September 2010; IEEE: Piscataway Township, NJ, USA, 11 November 2010.
35. Ackerman, J.; Seipel, J. Energetics of Bio-Inspired Legged Robot Locomotion with Elastically-Suspended Loads. In Proceedings of the IEEE International Conference on Intelligent Robots and Systems, San Francisco, CA, USA, 25–30 September 2011; IEEE: Piscataway Township, NJ, USA, 5 December 2011.
36. Scarfogliero, U.; Stefanini, C.; Dario, P. The use of compliant joints and elastic energy storage in bio-inspired legged robots. *Mech. Mach. Theory* **2009**, *44*, 580–590. [CrossRef]
37. Zhao, M.G.; Qiu, Y.; Chen, X. Control algorithm Based on Time Event for a Pneumatic Single-legged Jumping Robot. *Robot* **2012**, *32*, 525–530. [CrossRef]
38. Bjelonic, M.; Kottege, N.; Beckerle, P. Proprioceptive Control of an Over-Actuated Hexapod Robot in Unstructured Terrain. In Proceedings of the 2016 IEEE International Conference on Intelligent Robots and Systems, Daejeon, South Korea, 9–14 October 2016; IEEE: Piscataway Township, NJ, USA, 2016.

Article

A Synthetic Nervous System Controls a Simulated Cockroach

Scott Rubeo *, Nicholas Szczecinski and Roger Quinn

Department of Mechanical and Aerospace Engineering, Case Western Reserve University, Cleveland, OH 44106, USA; nicholas.szczecinski@case.edu (N.S.); roger.quinn@case.edu (R.Q.)

* Correspondence: scott.rubeo@case.edu; Tel.: +1-513-706-2320

Received: 14 November 2017; Accepted: 19 December 2017; Published: 22 December 2017

Abstract: The purpose of this work is to better understand how animals control locomotion. This knowledge can then be applied to neuromechanical design to produce more capable and adaptable robot locomotion. To test hypotheses about animal motor control, we model animals and their nervous systems with dynamical simulations, which we call synthetic nervous systems (SNS). However, one major challenge is picking parameter values that produce the intended dynamics. This paper presents a design process that solves this problem without the need for global optimization. We test this method by selecting parameter values for SimRoach2, a dynamical model of a cockroach. Each leg joint is actuated by an antagonistic pair of Hill muscles. A distributed SNS was designed based on pathways known to exist in insects, as well as hypothetical pathways that produced insect-like motion. Each joint's controller was designed to function as a proportional-integral (PI) feedback loop and tuned with numerical optimization. Once tuned, SimRoach2 walks through a simulated environment, with several cockroach-like features. A model with such reliable low-level performance is necessary to investigate more sophisticated locomotion patterns in the future.

Keywords: synthetic nervous systems; muscle control; joint control; inter-leg coordination; cockroach; animat; neuromechanical model; insect locomotion

---

1. Introduction

Insects are excellent models for walking robots, for two reasons. First, insects are highly successful animals that can walk through a variety of terrains. Second, decades of behavioral and neurobiological research have revealed many of the neuromechanical mechanisms that underlie this adaptive behavior.

The legs of walking animals propel the body by alternating between stance and swing phases. A leg is in contact with the ground in stance phase, during which it supports and propels the body. A leg is not in contact with the ground in swing phase, during which it is returned from its position at the end of stance phase (posterior extreme position, or PEP) to its position at the beginning of stance phase (anterior extreme position, or AEP). This alternating activity is supported by central pattern generators (CPGs), networks that are capable of producing endogenous oscillatory output. Each leg joint appears to be controlled by its own CPG [1,2]. Although periods of coordinated behavior can be observed [3], each leg joint's activity is generally decoupled from that of the others [2]. CPGs' phases are adjusted by inter-joint signals from load [4–6] and proprioceptive [7,8] sensors, giving rise to the coordination seen during walking. When walking in different directions, some of these inter-joint reflexes are modified [9,10], giving rise to different coordination patterns that better serve locomotion in a particular direction. Indeed, stimulating populations in the insect central complex (CX) that encode locomotion direction can change the sign and phase of inter-joint reflexes as observed in intact animals [11].

There is an interdependent relationship between neuroscience, mathematical modeling, and robotics. The neuroscience of insects promises to inspire new and adaptive robot control systems (for a review, see [12]). However, mathematical modeling of the nervous system and the body's mechanics is often a necessary intermediate step between neuroscience and robotics. First, despite decades of wonderful work, there still exist many gaps in our understanding of how insects walk. Thus, a mathematical model can be constructed that incorporates what is known, and suggests how to bridge gaps in understanding [13,14]. Second, it is valuable to test the control method first in a simulated agent before controlling a robot. This enables a more thorough study of the function and stability of the system in a simplified environment before deploying it on a hardware system in the real world.

Many models exist to address specific questions about motor control in insects. Walknet, for example, has developed into a highly successful model. It uses a combination of finite state machine (FSM) and artificial neural network (ANN) elements to coordinate the walking of kinematic models [15] and hardware robots [16]. While Walknet is a capable walking controller, it is based on behavioral, not neurobiological data, making it difficult to test neurobiological hypotheses. Another model used similar FSM elements to demonstrate how decoupled CPGs could be coordinated into stepping motions via sensory feedback [17]. This model was even used to control a single robotic leg [18]. These works were important proofs of concept for the control of a leg via sensory-coupled CPGs. However, the simplicity of the model made it difficult to produce fine-tuned motions necessary for walking. In addition, the FSM implementation leaves out the details of neural dynamics, which can produce more complicated and potentially more capable motor patterns. More recently, Hodgkin–Huxley neural models of networks in insect nervous systems have been constructed to explore the effect of such neural dynamics on locomotion [13,19–21]. These models have enabled scientists to suggest neural and synaptic mechanisms underlying animal behavior. However, due to their complexity, the full mechanics of the body and environment are rarely simulated, leaving questions about the precise control of muscle activation, and the role of sensory feedback, unaddressed.

Over the years, there has been much successful work building and controlling six-legged robots [15,16,22–24]. However, despite all of this good work, these robots are not as good at locomotion as insects. We believe that an SNS will result in better performance.

To be directly applicable to robotic control, the mechanical model must possess realistic dynamics, and the neural system must use feedback from interactions with the environment to shape motor output.

Our recent work has explored many components of legged locomotion, although always limited to one leg. We have shown how a decoupled system of central pattern generators (CPGs) can give rise to directional stepping with only sparse descending commands [25]. This control approach can be applied to different legs, resulting in unique motions for each leg, as observed in walking insects [14,26–28]. However, adding more legs to the body makes walking a redundant problem, with several paths from body to ground. Successful locomotion requires controlling both the timing and amplitude of motor output to ensure that the body remains supported at all times [29].

This work expands upon our previous model of insect locomotion [14] and presents three novel results. The first is the development of a generalized design method for setting parameters in neuromechanical models. The control method uses the dynamics of the body to establish a resting posture, set passive viscoelastic forces based on the scale of the model, and set parameters in a negative feedback position controller for the antagonistic pair of muscles that actuate each leg joint. The result is a model with realistic dynamics, and accurate control of the leg joints. The second result is a neural implementation of "Holk Cruse's rules" [29], which precisely and stably controls the timing of each leg's swing phase. The third result is the comparison between biological and simulated walking data. The resulting model can walk freely at cockroach-like speed and stepping frequency. Such performance supports the usefulness of the presented design method. This model serves as a baseline for studying more complicated control problems in the future, such as walking in a curve, walking at different speeds, and walking over rough terrain.

2. Methods

2.1. Modeling Overview

All modeling was done in AnimatLab 2, an environment for neuromechanical simulations [30]. The software can simulate the closed-loop interactions between an animat's body, its nervous system, and and its environment. For example, motor neuron voltage affects the force in the muscles, which exert forces on the body to cause motion, which is detected by proprioceptors and force sensors, which affect motor neuron voltage.

Previously, we developed design tools in MATLAB (The MathWorks, Nattick, MA) to perform analysis to aid in the parameter value design process. Versions of these tools existed previously [26,31], but these tools had to be advanced significantly to incorporate additional dynamics due to muscles and exoskeletal forces. This work thus extends tools previously only useful for robots controlled by servomotors to animal models like SimRoach2, which was developed in this project.

2.2. System Dynamics

2.2.1. Neuron Dynamics

A number of different mathematical models exist that attempt to capture the behavior of neurons. Our neural controller uses non-spiking Hodgkin–Huxley (HH) compartments [30], which are based on the original formulation of the HH model [32], but do not include ion channels for spiking. Thus the neurons function as non-spiking leaky integrators, with optional persistent sodium channels which enable more complicated responses. For details on the specific equations of motion, see the Supplementary Materials (Section: Methods Explained).

2.2.2. Muscle and Joint Dynamics

Each joint of SimRoach2 is actuated by two antagonistic Hill muscles [33]. This model contains an active contractile element, and three passive components: a series spring, a parallel spring, and a parallel damper (Figure 1). The primary dynamical variable for the muscle is its tension, T, with the dynamics:

$$\dot{T} = \frac{k_{se}}{b}\left(k_{pe}\Delta x(t) + b\dot{x}(t) - \left(1 + \frac{k_{pe}}{k_{se}}\right)T(t) + A(t)\right), \tag{1}$$

where b is the linear damping coefficient, k_{se} is the stiffness of the series elastic component, and k_{pe} is the stiffness of the parallel elastic component. Δx is the change in muscle length relative to its resting length. $\dot{x}$ is the rate of change of the muscle length. A is the muscle activation and is a function of both the muscle's motor-neuron's voltage and the length of the muscle. A is given by:

$$A(t) = \left(1 - \frac{(\Delta x(t))^2}{l_{width}^2}\right)T_{ce}(t), \tag{2}$$

where l_{width} is the width of the inverted parabola used for the tension-length curve and T_{ce} is a function of the motor-neurons voltage given by the sigmoid:

$$T_{ce} = \frac{T_{max}}{1 + \exp(S_m(x_{off} - V))} + y_{off}, \tag{3}$$

where T_{max}, S_m, x_{off}, and y_{off} are all parameters that define the stimulus-tension sigmoidal curve.

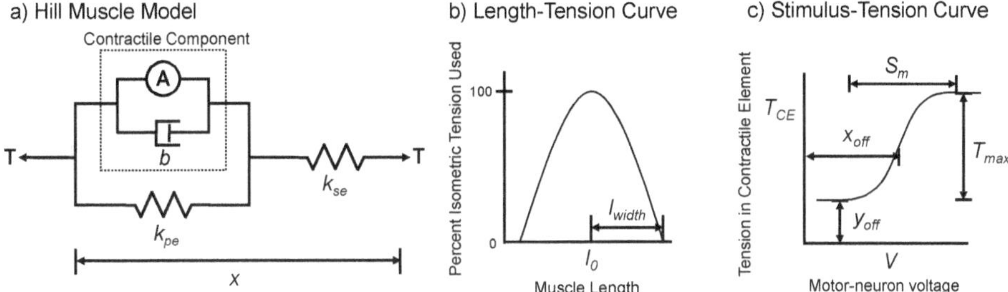

Figure 1. Hill Muscle Model used in AnimatLab 2 (adapted from [34]). (**a**) The model contains a series spring, a parallel spring, a parallel damper, and an active component; (**b**) the length-tension curve adjusts the active tension based on the length of the muscle; (**c**) the stimulus-tension curve adjusts the active tension based on the voltage of the motor-neuron.

In addition to the two muscles acting on each joint, there is also a passive spring with damping attached to each joint to mimic the viscoelasticity of insects' joints [35,36]. This stabilizes the motion of each joint, but makes the relationship between muscle activation and joint rotation more complicated. It shares attachments with the flexor muscle, and produces the tension:

$$T_{pass} = k_{pass} \cdot (x_{rest} - x) + c_{pass} \cdot \dot{x}, \quad (4)$$

where T_{pass} is the tension in the passive element, k_{pass} and c_{pass} are the stiffness and damping constants, respectively, x_{rest} is the resting length of the passive element, and x is the current length of the passive element.

When the leg is not in contact with the ground, there are four external moments acting on the joint: tensions from the flexor and extensor muscles, and spring stiffness and spring damping forces from the passive spring. The equation for the net torque on the joint can be developed as:

$$\sum_{i=1}^{4} r_i \times F_i, \quad (5)$$

where r_i is the moment arm of the force and $i = 1$ corresponds to the flexor muscle force, $i = 2$ corresponds to the extensor muscle force, $i = 3$ corresponds to the linear spring stiffness force, and $i = 4$ corresponds to the linear spring damping force.

2.3. Physical Model

SimRoach2 (Figure 2) was adapted from our previous model [14]. Each leg has five actuated hinge joints (Table 1). Each foot is modeled as a passive ball and socket joint that "sticks" to the ground in stance phase but allows the tibia to rotate over the foot. This functionality mimics insects' tarsal claws, which grip the substrate when external forces are applied [37], and can improve the stability of walking [38]. Furthermore, it has been shown that the adhesive features on an insect's leg plays an important role in walking [39]. Thus, the animat's body has 54 degrees of freedom ($(5+3)$ DOF $\times$ 6 legs + 6 DOF $\times$ 1 body)), 30 of which are actuated, and 6 of which are constrained for each leg when it is in contact with the ground. See Figure 4 in [14] for a description of the different body segments and joint angles in a cockroach.

Table 1. Actuated joints in each leg starting from the joint closest to the body and moving distally outward. Thorax-coxa (ThC) joints act between the thorax and the coxa. Coxa-trochanter (CTr) joints act between the coxa and the trochanter. Trochanter-feumr (TrF) joints act between the trochanter and the femur. Femur-tibia (FTi) joints act between the femur and the tibia.

Actuated Joints		
Front Leg	Middle Leg	Hind Leg
ThC2	ThC2	ThC2
ThC1	ThC1	ThC1
ThC3	CTr	CTr
CTr	TrF	TrF
FTi	FTi	FTi

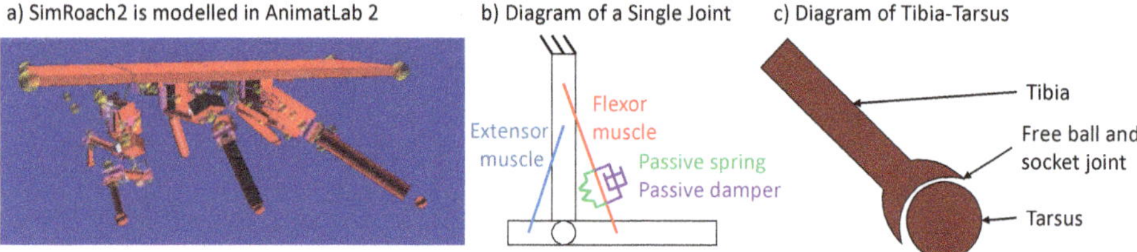

Figure 2. Physical model of the cockroach *Blaberus discoidalis*. (**a**) Picture of SimRoach2 in AnimatLab2; (**b**) each joint consists of a flexor muscle, extensor muscle and a passive spring and damper attached at the same points as the flexor muscle; (**c**) each tarsus (foot) is attached to the tibia with a free ball and socket joint that "sticks" when in contact with the ground.

2.4. Neural Network as Control System

2.4.1. Overview of Synthetic Nervous System

The SNS used to control SimRoach2 is a distributed network made up of functional subnetworks [26,40] (Figure 3). In short, the network functions as follows. A high-level network (gray, Figure 3) sends commands to the thoracic ganglia to stand or walk, and in which direction to walk. Each leg joint control network in each ganglion interprets this signal to decode the direction and amplitude of joint rotation in stance phase (brown, Figure 3). Each joint in the leg has a central pattern generator (CPG) that generates alternating neural activity (red, Figure 3). These have no direct neural connections between them [2]. The descending commands and CPG together establish a commanded position for each joint. Each low level joint controller (green, Figure 3) receives the commanded position signal, moves the joint to that position, and signals its position back to the intermediate (i.e., inter-joint and intra-leg) network. This intermediate network controls the transitions between stance and swing phase, by causing CTr (joints act between the coxa and the trochanter) levation when the other leg joints have reached their posterior extreme position (PEP), and causing CTr depression when the other leg joints have reached their anterior extreme position (AEP).

Additional networks are included to coordinate the walking of multiple legs. Inter-leg reflexes (purple, Figure 3) coordinate the legs for stable walking. Height control networks (blue, Figure 3) are included to control the height of the thorax by adjusting CTr depression while in stance phase. When the leg is in stance and loaded, the tarsus is "stuck" to the ground (yellow, Figure 3).

Each leg in SimRoach2 is controlled by a network like that in Figure 3. The neural implementation of joint position control and inter-leg reflexes are provided in this article. The neural implementation used for the upper level, height control, and for freezing the tarsus in stance phase are provided in the Supplementary Materials (Section: Methods Explained). The neural implementation used in the intermediate level is published in [26], and the CPG network is published in [31]. One can download the AnimatLab 2 file to see the specifics on how the sub-networks are connected to each other (Supplementary Materials File S2). Specific parameters values of the neurons and synapses are available in the Supplementary Materials (Section: Parameter Values) and in the AnimatLab 2 source file.

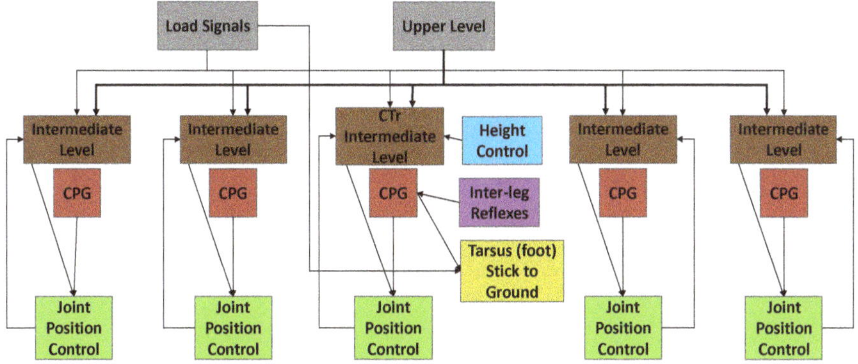

Figure 3. Functional diagram of the synthetic nervous system used for one leg. Each leg consists of five joints. Descending commands from the higher level brain and load signals (gray) affect the intermediate network (brown). Central pattern generators (CPGs, red) generate the rhythmic movement and are present in each joint. The intermediate network and the CPGs establish a commanded position for each joint. Joint controllers (green) take the commanded position and move the joint to that position. Signals from other legs coordinate walking via inter-leg reflexes (purple). Height control networks (blue) are present in each leg. When the leg is loaded and in stance phase, the foot is frozen to the ground (yellow). Note that this is a simple functional diagram; each functional unit consists of dynamics neurons connected to each other via synapses.

2.4.2. Joint Position Control

The force and length of muscles are not easy to control, and the details of how animals control them are not known, although it is known that negative feedback reflexes play a role [41]. The primary challenge is that the muscles and exoskeleton possess many elastic elements that cancel out one another's tension. To overcome this, we designed a neural proportional-integral (PI) network to control a limb's position (Figure 4). We believe this is justified because arthropods have disproportionately slow muscle dynamics for their size [42], which means that their muscle membranes integrate individual spikes from excitatory motor neurons, functionally behaving as integrators [43].

The perceived position (PP) neuron receives an input current proportional to the joint's rotation (e.g., $I_{app} = A \times \theta + B$). The commanded position (CP) neuron uses the same relationship to specify the intended rotation of the joint. The CP and PP neuron voltages are subtracted from each other to create two "error" neurons [40] (green in Figure 4). One corresponds to the CP being greater than the PP (positive error); the other corresponds to the PP being greater than the CP (negative error). Only one of these neurons will be active at a time. Each error neuron excites the appropriate muscle's motor neuron via a synapse tuned using optimization explained in Section 2.5.6. Exciting the motor neuron increases the tension in the muscle, correcting the error in the limb's position and functioning as proportional negative feedback. Proportional feedback alone is not enough to guarantee that the joint will reach the CP, except in very specific cases [44]. Thus, we add an integral subnetwork (described in [31]) to eliminate the steady-state error of proportional feedback alone (orange, Figure 4). The purpose of the "Zero" neuron is to subtract out the steady state activation of the integrator. Just as with the proportional feedback, there are two integral feedback pathways (integral positive and integral negative), only one of which will be active at a time.

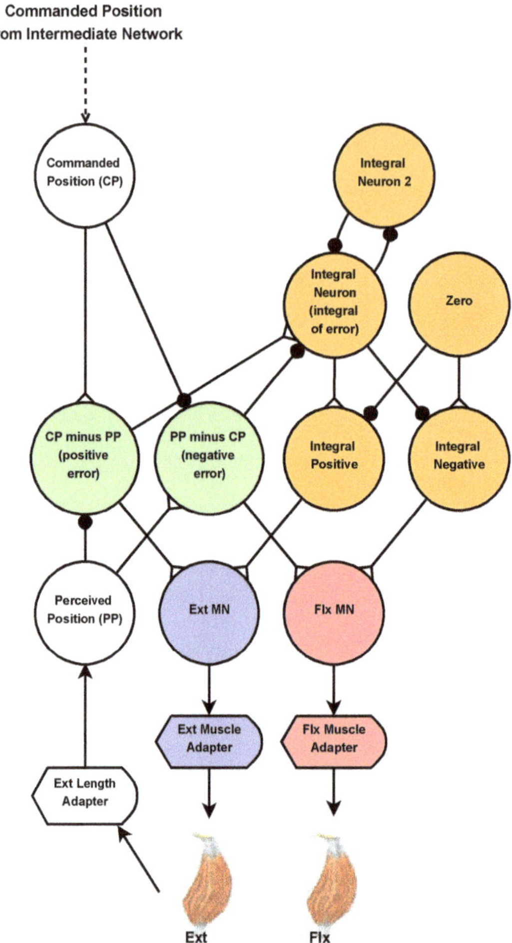

Figure 4. Diagram of the joint control network. Joint angles are linearly mapped to neuron voltage. The commanded position (CP) neuron input signal comes from the intermediate network. The perceived position (PP) neuron input signal is mapped from the length of the extensor. The CP and PP neuron voltage's are subtracted from each other to create two "error" neurons (green). Only one of these neurons can be active at a time. The error neurons excite the appropriate muscle motor-neuron, acting in a very similar way to the proportional part of a standard PID controller. The error neurons are also used as an input to the integral part (orange). The integral part continues to adjust the appropriate motor-neuron (MN) until the voltage of the CP and PP neurons match. Synapses with black circles and white triangles represent inhibitory synapses and excitatory synapses, respectively.

The importance of the integral part of the joint controller is shown in Figure 5. With only the proportional part, the joint does not move to its commanded position; there is always a steady state error. By adding the integral part, the steady state error is removed and the joint moves to its commanded position.

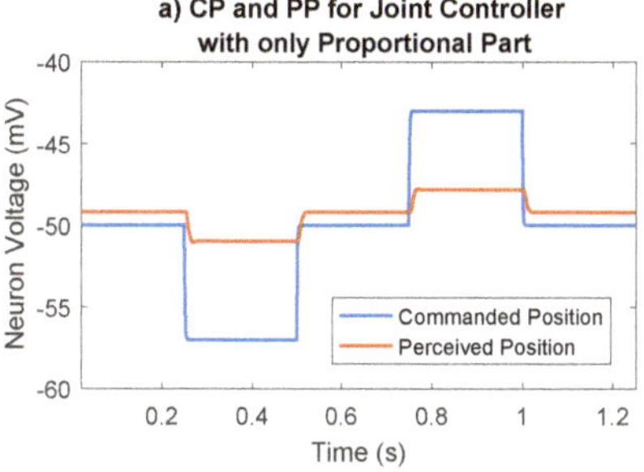

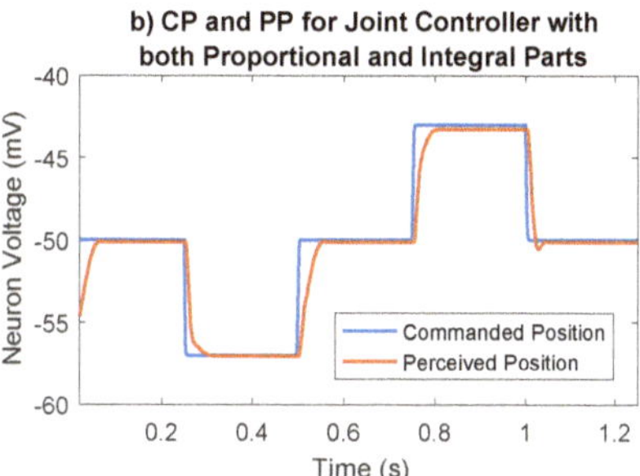

Figure 5. Importance of the integral part in the joint control network. (**a**) Joint controller with integral part disabled. The proportional part alone is not enough to move the joint to the desired position. There is always a steady state error. (**b**) Joint controller with both proportional and integral parts. The integral part removes the steady state error.

2.4.3. Inter-Leg Coordination

The legs of both insects and legged robots must communicate when they are in swing phase to ensure that the body always stays supported. Behavioral experiments have revealed correlations between the positions of different legs in arthropods [29,45], but the underlying networks remain largely unknown. Work in locusts suggests that some direct neural connections between the legs may contribute to coordination during walking [3,46]. Therefore, in SimRoach2, the levator-depressor (CTr) CPGs between legs are connected and neural pathways were engineered to produce the observed behavior.

"Holk Cruse's rules" describe inter-leg influences that control the timing of stepping [29]. Three of these influences have been shown to be the most important to coordinate walking, and were implemented into SimRoach2. Influence 1 inhibits the start of swing phase in the rostral leg while the caudal leg is in its swing phase. Influence 2 excites the start of swing phase in the rostral leg when the caudal leg enters its stance phase. Influence 3 excites the start of swing phase in the caudal leg when the rostral leg is in stance phase, increasing in strength over time as the caudal leg is in stance phase [29]. Influences 2 and 3 are also present between contralateral legs. Table 2 summarizes these three influences. These influences are redundant, such that any perturbation to a walking gait will result in nearly immediate re-coordination of legs. The specific neural networks that were designed to replicate these three influences are shown in Figure 6. Verification of each of these networks is presented in the Supplementary Materials (Section: Methods Explained).

Table 2. Summary of inter-leg influences.

Influence Number	Direction of Influence	Effect of Influence
Influence 1	Rostral	Inhibits the start of swing phase when the posterior leg is in its swing phase
Influence 2	Rostral and contralateral	Excites the start of swing phase when the posterior leg enters its stance phase
Influence 3	Caudal and contralateral	Excites the start of swing phase while the anterior leg is in its stance phase, increasing in strength over time

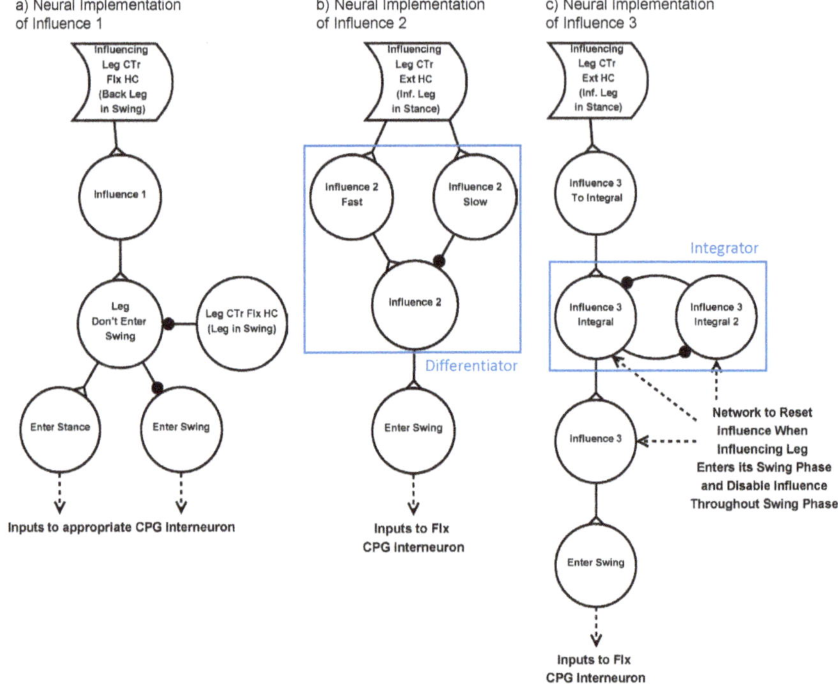

Figure 6. Neural Implementation of three "Holk Cruse's rules" [29]. (**a**) Influence 1 inhibits the start of swing phase when the posterior leg is in its swing phase; (**b**) Influence 2 momentarily excites the start of swing phase in the rostral leg when the caudal leg enters its stance phase. To capture this, we use a differentiator network [40]; (**c**) Excites the start of swing phase while the anterior leg is in its stance phase, increasing in strength over time. To capture this, we use an integrator network [40]. Synapses with black circles and white triangles represent inhibitory synapses and excitatory synapses, respectively.

2.4.4. Height Control

Mechanisms to control the height of insects have been observed to exist while both walking and standing still [47,48]. However, little is known about the underlying neural mechanisms that produce these observed behaviors. Height control is important because the multiple legs form a parallel manipulator, potentially resulting in internal torques that waste energy. SimRoach2 uses a negative feedback loop to adjust CTr depression based on the length of the leg while in contact with the ground. Please see the Supplementary Materials (Section: Methods Explained) for more information.

2.5. Design Process for a Single Joint

This section describes how the parameter values of the system's dynamical components components (e.g., muscles, joints, etc.) are selected to enable our simulated cockroach to walk. Our previous work has explained how to select parameter values for CPGs [31] and descending connectives from higher command centers that control joint motion amplitude [26]. These works, however, assumed that each joint's control system could control the position of the joint. To apply this work to a muscle-actuated (rather than servomotor actuated) model, we first needed to select parameters for the viscoelasticity of each joint, the active and passive dynamics of each joint's muscles, and the synaptic strengths in each joint's controller. The name, equation number, and quantity of each are listed in Table 3.

Table 3. Parameter values to select.

Object	Parameters	Equation Number	Number of Parameters (No. × Joints)
Joint	$k_{pass}, c_{pass}, x_{rest}$	(4)	$3 \times 30 = 90$
Muscle	$k_{se}, k_{pe}, b, l_{width}, T_{max}, S_m, x_{off}, y_{off}$	(1), (2), (3)	$8 \times 2 \times 30 = 480$
Controller	$G_{syn,P}, G_{syn,I}$		$2 \times 30 = 60$
Total			630

The following sections will describe a serial design process for systematically selecting these parameter values. First, we find a rest posture for the animat. This enables us to calculate the torque on each joint due to gravity. Next, we use this gravitational torque to estimate the viscoelastic torques of the exoskeleton acting on each joint. Then, we select the muscle parameter values such that the muscles can apply enough tension to move the limb. Finally, a numerical optimization is performed to find maximum muscle activations and feedback gains that result in stable, rapid motion.

2.5.1. Resting Posture

We seek a resting posture that holds the body at a desired height above the ground, h_{des}, while keeping each joint as far away from its limits as possible. This ensures that when the animat walks, it can move every joint in both directions. This process can be formulated as a constrained optimization problem. Let the vector $\boldsymbol{\theta} \in \mathbb{R}^{n \times 1}$ be the configuration (i.e., joint angles) of a leg with n joints. We then define $\mathbf{x}$ as $\boldsymbol{\theta}$ normalized to the lower and upper bounds of motion. The i-th element of $\mathbf{x}$ is defined as:

$$x_i = \frac{\theta_i - \theta_i^{lb}}{\theta_i^{ub} - \theta_i^{lb}}, \tag{6}$$

where θ_i^{lb} and θ_i^{ub} are the lower and upper bounds of joint i, respectively. This simplifies the definition of the objective function f,

$$f(\mathbf{x}) = -\frac{4}{n}\mathbf{x}^T(\mathbf{x} - \mathbf{p}), \tag{7}$$

where $\mathbf{p} \in \mathbb{R}^{n \times 1}$ is a vector of ones. The factor of $4/n$ ensures that $f = -1$ when $\mathbf{x} = \mathbf{p}/2$, i.e., when every joint is in the center of its range of motion.

Let us also define $g(\mathbf{x})$ as the vertical distance from the body to the foot, as computed with the forward kinematic map of the leg. We can then define the resting posture as the solution to the optimization problem:

$$\begin{aligned} \underset{x}{\text{minimize}} \quad & f(\mathbf{x}) \\ \text{subject to} \quad & g(\mathbf{x}) = h_{des} \end{aligned}$$

We set $h_{des} = 1.2$ cm, a value used for oiled plate walking experiments with cockroaches [14]. The resulting posture, shown in Figure 7, is remarkably cockroach-like. Figure 7c graphically shows the final values of $\mathbf{x}$ on "sliders" between 0 and 1, showing that most joints are near to the center of allowed rotation. This analysis is now automated by our design toolbox KinematicOrganism [31].

a) SimRoach2 Standing

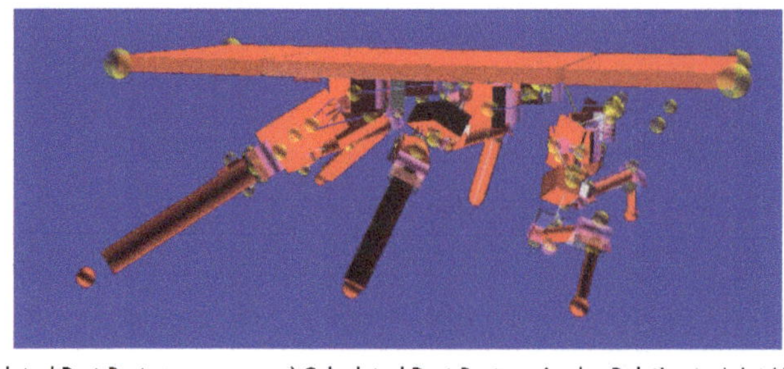

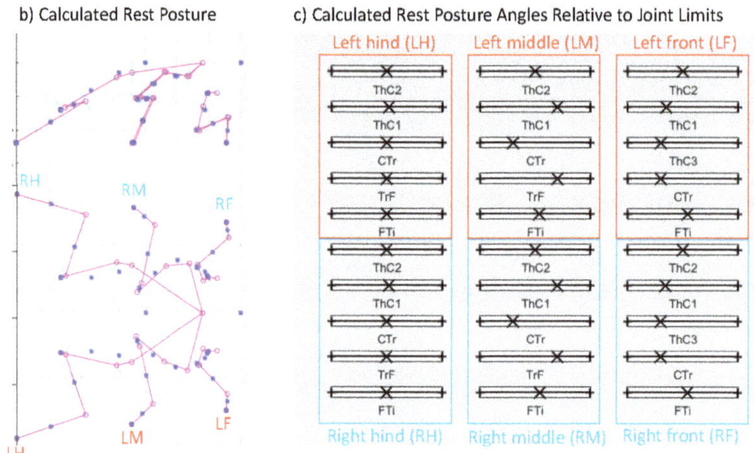

Figure 7. Resting posture of SimRoach2. (**a**) Screenshot of SimRoach2 standing in AnimatLab 2; (**b**) rest posture of SimRoach2 calculated in MATLAB; (**c**) graphical representation of **x**, the rest posture angle of each joint relative to its joint limits. No joint's rest posture angle is near its joint limits.

2.5.2. Calculation of Passive Spring Stiffness and Damping Coefficients

The rest lengths of the passive spring and both muscles in each joint were set to the length that matched the joint angle based on the muscle attachment points. To determine the stiffness of every spring, we assumed that the animat could support its weight using the passive spring alone. This is consistent with the observation that insects' exoskeletons are stiff even when their muscles are cut from them [35,36]. Each leg is assumed to hold an equal amount of the weight. We use the manipulator Jacobian from [49] to calculate the torques required for each joint to support its weight. Then, to calculate the stiffness of the spring, we deflect the joint a small amount (0.1 radian) and increase the stiffness of the springs until the spring produces the required torque. Detailed calculations can be found in the Supplementary Materials (Section: Methods Explained).

The damping of the passive element was calculated in the following way. For each joint, the distal joints were assumed to be held rigidly in place. This was used to compute the effective moment of inertia about the joint in question. Then, together with the stiffness value already calculated, the damping was computed to ensure that motion was overdamped, consistent with observations of insect exoskeletons [35]. Detailed calculations can be found in the Supplementary Materials (Section: Methods Explained).

2.5.3. Passive Muscle Force Parameters

After tuning the spring coefficients, the next step was to add passive muscle components. The parameters of interest for this section are muscle parameters k_{se}, k_{pe}, and b in Equation (1). Values for these parameters vary widely from muscle-to-muscle and even from organism-to-organism [50]. Guschlbauer cited k_{se} and k_{pe} values of approximately 6.32 N/m and 3.156 N/m (Figure 6 in [51]) while Blümel cited k_{se} and k_{pe} values of approximately 45 N/m and 11.24 N/m (Figure 3 in [50]). Testing different values with the entire design process did not noticeably change the resulting performance of the control system, so we simply chose k_{se} and k_{pe} values of 45 N/m and 11.24 N/m. Because the joints are already damped, we chose the small value $b = 0.1$ Ns/m for every muscle.

2.5.4. Active Muscle Parameters

Tuning the muscle activation parameters concerns both the muscle activation curve and the length-tension curve. Each muscle activation curve has four parameters that must be chosen: the amplitude, steepness, x-offset, and y-offset (see Equation (3)). We used data from insect muscles to pick the steepness and x-offset of the muscle activation curve such that it looked like the biologically observed data [51–53] (for an example, see Supplementary Materials (Section: Methods Explained)). The y-offset is constrained such that the muscle produces 0 N of active tension when the motor-neuron is at rest. The only unconstrained parameter is the maximum active tension, which is selected as part of the optimization process (see Section 2.5.6).

A muscle' length-tension relationship reduces its ability to apply active tension as the muscle lengthens or shortens (Figure 1c). The parameter l_{width} in Equation (2) defines the change in length at which the muscle can no longer contract. This parameter is chosen such that the muscle cannot apply tension at 133% and 67% of its resting length [54].

2.5.5. Synaptic Strength to Position Control Integral Neuron

The synaptic strength between the difference neurons and the integral neuron in Figure 4 was an additional parameter that must be chosen. This parameter controls the speed of integration, and therefore the strength of the integral portion of the joint control network. The integrator could saturate if this synaptic strength was too high (for example, Figure 8), so it was set to 10 nS, a value that prevented saturation in most cases. The synaptic output of this network was later numerically optimized, enabling us to fine-tune the integral feedback in the controller.

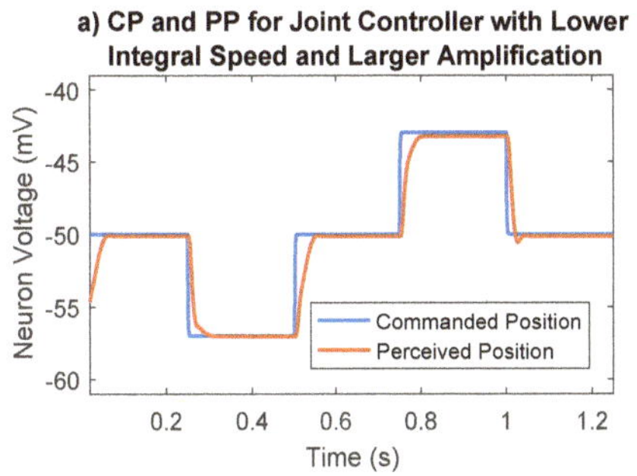

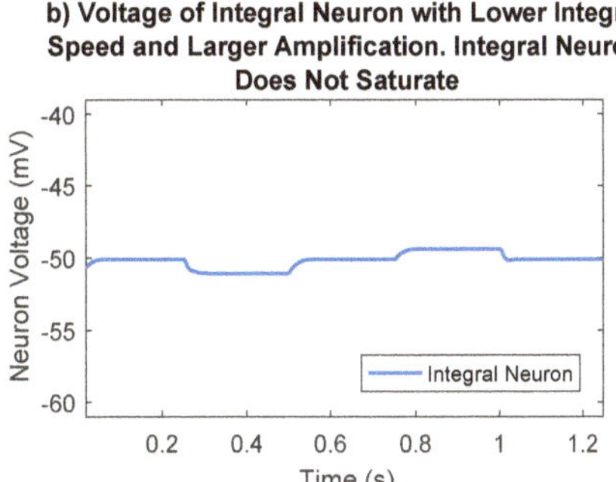

Figure 8. *Cont.*

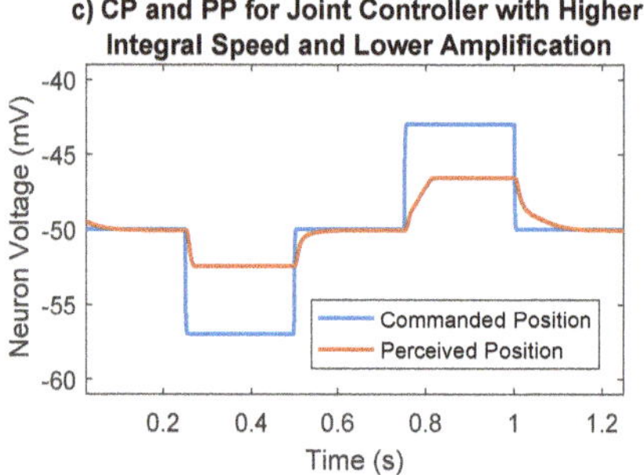

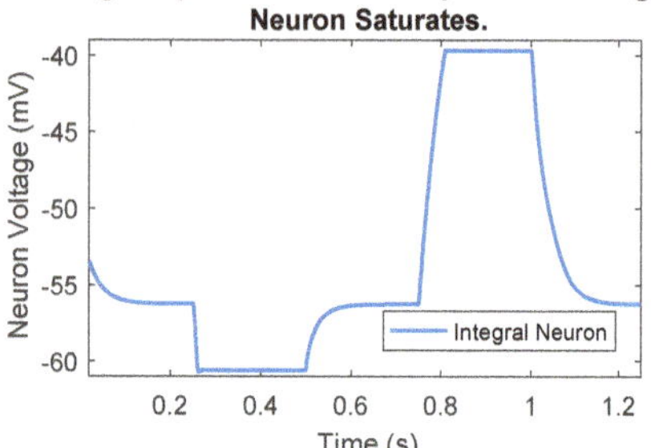

Figure 8. Importance of synaptic strength to the integral neuron in joint position control. (**a**) Commanded position and perceived position for the selected case. Lower integration speed and larger amplification tuned during the optimization process results in good reference tracking; (**b**) Voltage of integral neuron for the selected case. Lower integral speed results in voltages far away from saturation points; (**c**) Commanded position and perceived position for the undesirable case. The joint never gets to its commanded position because the integral neuron saturates; (**d**) Voltage of integral neuron for the undesirable case. Higher integration speeds cause the integral neuron to reach the synaptic thresholds, losing its ability to send information.

2.5.6. Optimization Process

The last part of our design method is the optimization of four parameters per joint: $G_{syn,P}$, $G_{syn,I}$, $T_{max,ext}$, $T_{max,flx}$. $G_{syn,P}$ and $G_{syn,I}$ are the synapse strengths from the difference and integral positive and negative neurons in the joint position control network, respectively (Effectively the proportional and integral gain of the feedback controller, respectively; Figure 4). $T_{max,ext}$ and $T_{max,flx}$ are the amplitudes of the muscle stimulus-tension curves (Equation (3)). These parameters together determine the stiffness and torque output of each joint. The optimization was automated for each joint in our SimScan tool (Section 2.1). For each joint, an identical, four-step reference trajectory was provided, and the following objective function, f_{obj}, was minimized using MATLAB's built in `fmincon` function, an interior-point, gradient-based nonlinear program solver:

$$f_{obj} = (\text{maxOvershoot} - \text{desiredOvershoot})^2 + (1000 \times (\text{maxRiseTime} - \text{desiredRiseTime}))^2 + (1000 \times (\text{settlingTimeMax} - \text{desiredSettlingTime}))^2 \quad (8)$$

This function calculates the squared deviation of the tracking performance from the desired values. This value is computed separately for each step in the reference trajectory, and added for all four steps to evaluate one parameter value combination. In our trials, desiredOvershoot, desiredRiseTime, and desiredSettlingTime where chosen to be 0, 0.02, and 0.03, respectively. The factors of 1000 were included to weight the different terms in f_{obj} approximately equally.

Every joint's parameters were optimized to track the commanded position very well (Figure 9). One example of a joint controller's reference tracking before and after optimization is included. This chosen case, while still good, was the worst joint in terms of reference tracking; the other 29 joints that were optimized produced even better reference tracking results.

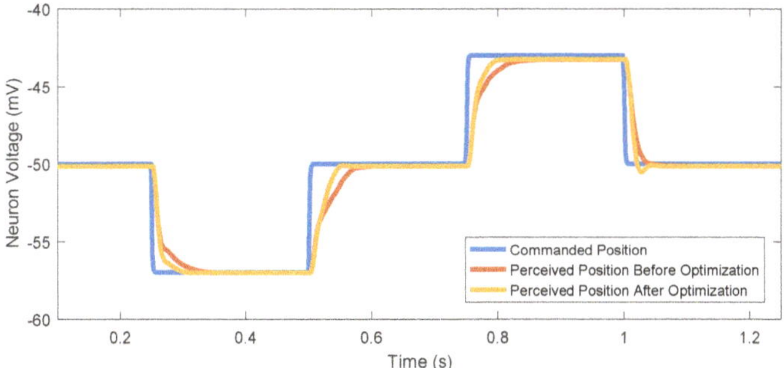

Figure 9. Optimizing parameters $G_{syn,P}$, $G_{syn,I}$, $T_{max,ext}$, and $T_{max,flx}$ improves reference tracking. The commanded position trajectory is chosen arbitrarily.

3. Results

To show that the design method works, and that SimRoach2 will enable us to explore the control of insect locomotion, we need to show that the resulting model can indeed walk. In this section, we present data and video of SimRoach2, and compare it to data from cockroaches. All data were collected in simulation using AnimatLab 2.

3.1. Forward Walking

A video showing the walking of SimRoach2 is attached with this paper in the Supplementary Materials (Video S1). Figure 10 shows the walking path and walking speed of SimRoach2. The walking speed is calculated by smoothing and differentiating the total distance walked. After a transient startup, SimRoach2 walks in a straight path, with insect-like side-to-side motion (Figure 10). The speed of SimRoach2 in steady state is just below 20 cm/s (Figure 10).

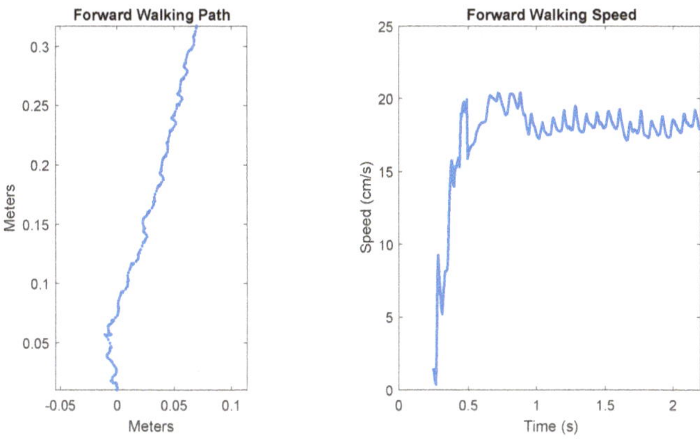

Figure 10. Forward walking path and speed.

The pitch and roll of the body during steady state walking is presented in Figure 11. Actual cockroach data presented in [55] measured the pitch to be ±4 degrees and the roll to be ±7 degrees. These reference lines are plotted in Figure 11 to allow for comparison.

Individual joint kinematic data from both SimRoach2 and a cockroach are presented in Figures 12–14. The cockroach data used for comparison was taken from [14]. For each joint in each leg during simulated walking, the joint angle at the AEP and PEP were measured for a number of steps. $\Delta\theta$ was calculated by subtracting the AEP from the PEP. Details on the specific method for calculated the AEP and PEP from raw data are provided in the Supplementary Materials (Section: Methods Explained).

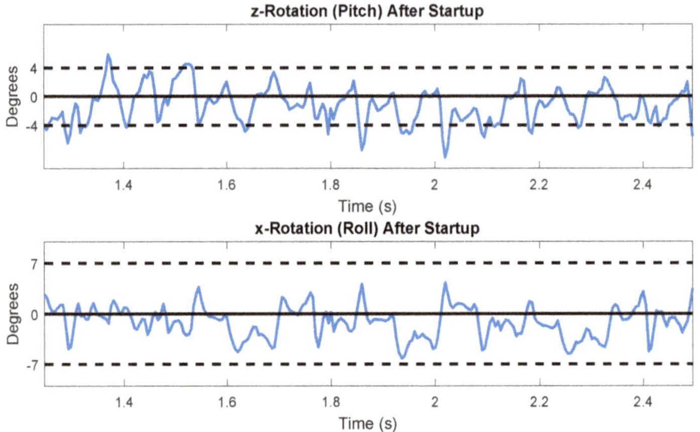

Figure 11. Pitch and roll of the body during steady state walking.

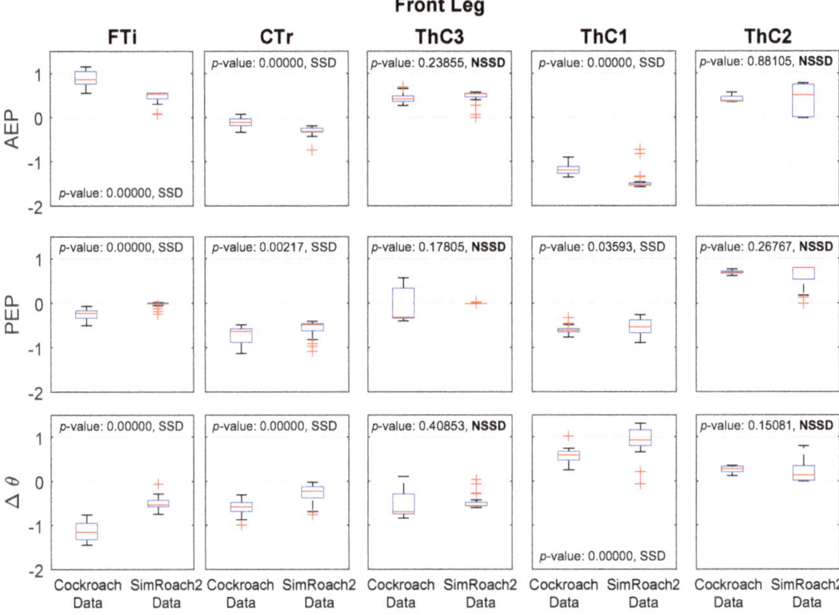

Figure 12. Front leg anterior extreme position (AEP), posterior extreme position (PEP), and $\Delta\theta$ joint angle data in a cockroach and in SimRoach2. Thirty six cockroach data points were used. Thirty three SimRoach2 data points were used. p-values from the two-sample t-tests and whether there is a statistically-significant difference (SSD, bad match) or no statistically-significant difference (NSSD, good match) between the samples are provided.

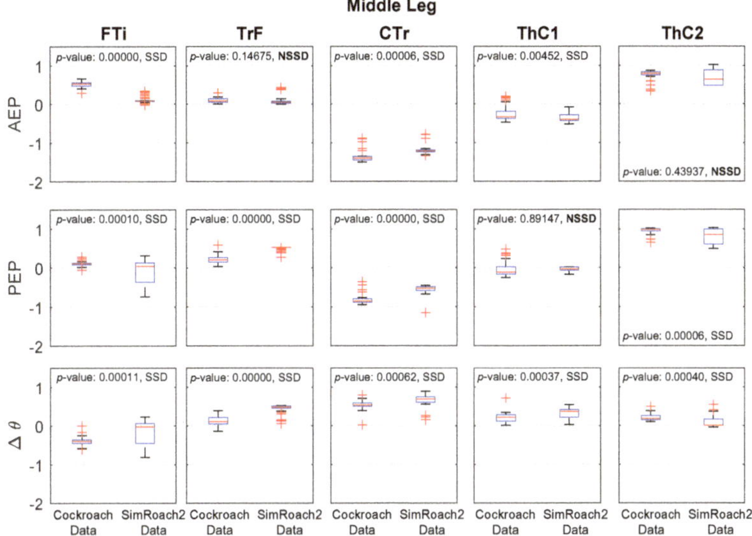

Figure 13. Middle leg anterior extreme position (AEP), posterior extreme position (PEP), and $\Delta\theta$ joint angle data in a cockroach and in SimRoach2. Thirty one cockroach data points were used. Forty six SimRoach2 data points were used. p-values from the two-sample t-tests and whether there is a statistically-significant difference (SSD, bad match) or no statistically-significant difference (NSSD, good match) between the samples are provided.

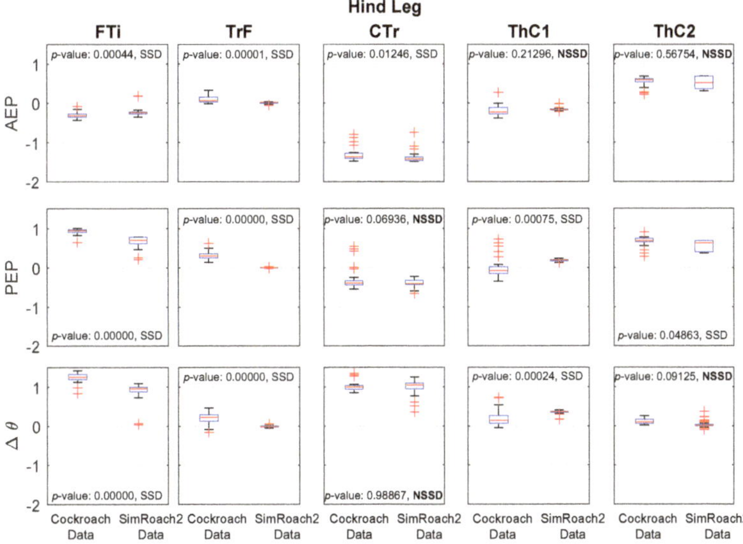

Figure 14. Hind leg anterior extreme position (AEP), posterior extreme position (PEP), and $\Delta\theta$ joint angle data in a cockroach and in SimRoach2. Twenty nine cockroach data points were used. Forty five SimRoach2 data points were used. p-values from the two-sample t-tests and whether there is a statistically-significant difference (SSD, bad match) or no statistically-significant difference (NSSD, good match) between the samples are provided.

For each set of joint angle data, a two-sample t-test was run to see if the mean of the data sets were statistically significantly different. The statistical data is presented in Figures 12–14. For the t-tests, the variances were assumed to be unequal.

3.2. Inter-Leg Coordination

The inter-leg connections of SimRoach2 were based on biological observations and designed to enforce a tripod gait. Cockroaches walk and run almost exclusively with a tripod gait [56]. The three different inter-leg influences (Table 2) were redundant in that any perturbation to the gait would result in the near immediate re-coordination of the legs. The result of these inter-leg influences show that SimRoach2 walks with a tripod gait (Figure 15).

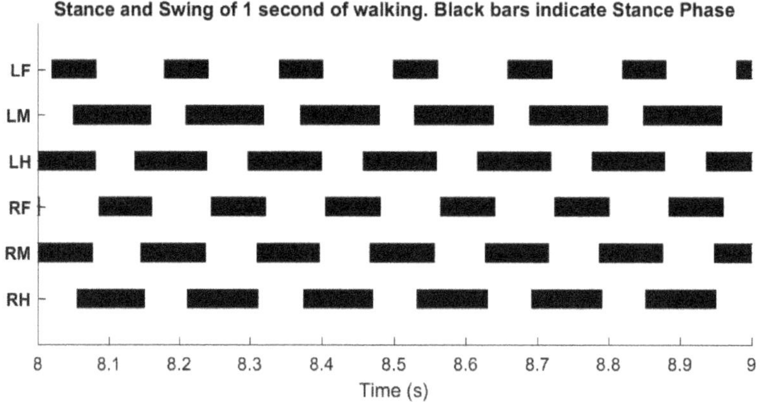

Figure 15. Inter-leg influences result in a stable tripod gait. Data presented over one second of walking. Black bars indicate stance phase.

4. Discussion and Future Work

A cockroach model actuated by an antagonistic pair of muscles and controlled by a synthetic nervous system, informed by the current state of knowledge of insect nervous systems, successfully walked in a simulated environment. The basic parameters of locomotion are consistent with cockroaches, and this model will enable us to study more sophisticated control questions in the future. The primary challenge in this study is tuning the large number of parameter values for proper locomotion. A parameter selection process was developed that heavily depends on biological observations, and sets the few remaining parameters based on desired engineering performance.

This work improves upon the performance of a previously published neuromechanical model [57]. Since then, we have developed several methods for tuning neuron dynamics [31,40] and muscle dynamics. Most of this research has been applied to our robot MantisBot. The presented work used what has been learned in the past few years to improve upon our original simulated cockroach model such that it better reflects the biomechanics and nervous system in the animal [58]. Studying the control of more complex locomotion, such as walking in a curve or changing speed, require reliable control of the muscles actuating each joint, a functionality not incorporated in our original model. Other networks that improved the reliability of walking, such as height control and inter-leg influences, were also designed and implemented.

Many parameters required tuning for proper operation, so a design process was developed to choose these parameters based on biological data and observations, and engineering usefulness. Most of the system's parameters are constrained by the geometry of the body, and a few are selected via numerical optimization. We believe that the presented design process is generic enough to be applied to any neuromechanical model of an insect.

In addition to enabling SimRoach2 to walk stably, two of our previously developed MATLAB design tools (KinematicOrganism [26], FeedbackDesign [31]) were improved. Before this work, all of the tools were built around working with the servomotors that MantisBot uses. Understanding the nervous system, however, requires understanding muscle dynamics, because they ultimately transform neural activity into behavior. Integrating muscle compatibility to the tools will enable us to explore how the nervous system uses muscles to produce motion. If one was interested in making their own hexapod model using our method, one could create a new hexapod model in AnimatLab 2 and tune parameters using our design tool KinematicOrganism (see Supplementary Materials File S3). To that note, KinematicOrgranism does assume a certain structure to the network, but one could potentially modify the program to tune different network structures.

4.1. Importance of Simulated Animats for Biologically-Inspired Robots

Biologists and roboticists study animal locomotion in different ways with different goals in mind. Although their goals are different, the research and findings of each side advances the other. At the intersection between biology and robots are animats such as SimRoach2. Animats facilitate the interaction between robotics and biology, enabling biologists to test hypotheses quantitatively [15,17], which leads to new understanding that can eventually be incorporated into robot designs.

While SimRoach2 can be improved in many ways, it has served as a useful model of insect locomotion. Much of the SNS used to control SimRoach2 is the same as the controller for our robot MantisBot. This work has demonstrated the potential of SNS for controlling locomotion at different scales and of differing geometry. Additionally, exploring and testing inter-leg connections on a robot's hardware is challenging and there is a risk of breaking parts; so, testing inter-leg connections with simulated models such as SimRoach2 is useful. Lastly, much of this work involved learning about and improving muscle-actuated locomotion, and we know that muscle-like controllers are useful for robotic locomotion [59].

4.2. Walking Comparison between SimRoach2 and a Cockroach

The main goal of this work was to develop a design method for an insect model actuated with an antagonistic pair of muscles. To show that the design method works, walking data was collected and presented. In this section, the data from SimRoach2 is compared to a cockroach.

It is important to note that the cockroach kinematic data was not used during the design process. We did not design our model to "match" this data. We believe that the walking data of our model and a cockroach are similar is because of the detailed work that went into the many different parts of this model and because the model was informed by the biomechanics and nervous system of the animal. These results are a consequence of our process, and the fact that there are some similarities suggest that this model captures some of the underlying principles.

When commanded to walk forward, SimRoach2 walks straight after reaching steady state (Figure 10). The exact path of the body does include some side to side wobble, but this is also seen in cockroaches [55]. It is interesting to note that in SimRoach2, there is no correction mechanism to alter the path against error. The straight walking path is a result of the completeness of the design method in this paper. In the future, we will apply our work on directional stepping with MantisBot [60] to SimRoach2, giving us a complete-body model with which we can develop a model of the central complex, a region of the brain implicated in the control of walking speed and direction [11].

The walking speed of SimRoach2 is just below 20 cm/s (Figure 10). This is similar to cockroach data presented in [61] for free walking. The pitch and roll of a walking cockroach were measured to be between ± 4 degrees and ± 7 degrees, respectively [55]. SimRoach2 walks with pitch rotations just slightly outside these limits and with roll rotations well within these limits (Figure 11). Again, we believe that our results are similar to that of a cockroach because of the detailed work that went into many different parts of SimRoach2, and this similarity will serve as a launching point for future studies on locomotion.

Joint Angle Comparison

For most joints in SimRoach2, the positions of the AEP and PEP appear to match fairly well, especially compared to the range of motion of these joints (Figures 12–14), suggesting that the change between stance and swing phase happens at similar positions in each leg. The means of the data sets were also compared using statistical *t*-tests. Of the 30 tests on the AEP and PEP for the 15 joints, 20 of the tests showed a statistically significant difference between the animal and animat. It is difficult to know if the model needs to capture these small details to inform neurobiology or robotics. Past studies of insect locomotion have focused on the changes in individual joints' range of motion between different conditions [14,27]. As we increase the complexity of SimRoach2's locomotion, we plan to take a similar approach; not focusing as much on how the joint kinematics match the animal, as much as how they change in different scenarios. We believe such changes help reveal what parameters the nervous system is controlling during locomotion.

4.3. Redundant Parameters

Throughout the duration of this work, it was found that many parameter values were redundant in that their overall effect on the system was very similar. For example, the angular velocity of a joint could be modulated by (1) changing the amplitude of the muscle activation curve, (2) changing the synapse strengths from the proportional or (3) integral part to the motor-neurons; or (4) changing the synapse strength to the integral neuron. Because of these parameter redundancies, it is possible that a model with very different parameter values could produce very similar locomotion. This is an interesting observation, but also not that surprising. Animal experiments [52] and modeling studies alike [62] suggest that parameter values in the nervous system vary largely, even between individuals in the same species, despite indistinguishable performance. This suggests that laboring over the decision of how to tune redundant parameters is not important, as long as the resulting motion is useful.

4.4. Velocity Control for Muscles

Insects control the velocity of their legs and joints when propelling their bodies [63,64]. Modeling studies have shown that the complete continuum of gaits can be produced simply by reducing the speed of the leg in stance phase relative to swing phase, and allowing "Holk Cruse's rules" to coordinate the legs [15]. Therefore, to study how SimRoach2 might change walking speed, our single-joint controller must be enhanced to limit joint velocity. Currently, the joints rotate to their commanded positions very quickly, so we do not observe that "fluid" motion we like to see in walking gaits. Previous work with MantisBot has shown that limiting the velocity of the servomotors, and incorporating reflexes seen in insects, together contribute to asymmetrical stance and swing phase durations, which should lead to a continuum of gaits [65]. We have already begun such work, and find that increased biological detail in fact improves the performance of the controller (Naris, Szczecinski, and Quinn, in preparation). Models of other, slower walking insects such as stick insects may be constructed in the future to more thoroughly explore how "gait" emerges in insects.

Supplementary Materials: The following are available online at http://www.mdpi.com/2076-3417/8/1/6/s1. Video S1: SimRoach2 forward walking; File S2: Animatlab 2 source files; File S3: KinematicOrganism Toolbox; Table S4: Rest lengths of spring, flexor and extensor; Table S5: Spring stiffness and damping coefficients; Table S6: Muscle amplitude and Y-offset values; Table S7: proportional and integral synapse parameter values; Table S8: Upper level synapse parameter values; Section S9: Methods explained.

Author Contributions: S.R. led the research efforts and led the preparation of this paper. N.S. aided in the research efforts, aided in preparation of this paper and provided critical oversight. R.Q. aided in the preparation of this paper and provided critical oversight.

Conflicts of Interest: The authors declare no conflict of interest.

Abbreviations

The following abbreviations are used in this manuscript:

AEP	Anterior extreme position
ANN	Artificial neural network
CP	Commanded position
CPG	Central pattern generator
CTr	Coxa-trochanter
CX	Central complex
DOF	Degree of freedom
FSM	Finite state machine
FTi	Femur-tibia
HH	Hodgkin–Huxley
MN	Motor-neuron
PEP	Posterior extreme position
PI	Proportional-integral
PP	Perceived position
SNS	Synthetic nervous system
ThC	Thorax-coxa
TrF	Trochanter-femur

References

1. Ryckebusch, S.; Laurent, G. Rhythmic patterns evoked in locust leg motor neurons by the muscarinic agonist pilocarpine. *J. Neurophysiol.* **1993**, *69*, 1583–1595.
2. Büschges, A.; Schmitz, J.; Bässler, U. Rhythmic patterns in the thoracic nerve cord of the stick insect induced by pilocarpine. *J. Exp. Biol.* **1995**, *198*, 435–456.
3. Ryckebusch, S.; Laurent, G. Interactions between segmental leg central pattern generators during fictive rhythms in the locust. *J. Neurophysiol.* **1994**, *72*, 2771–2785.
4. Noah, J.A.; Quimby, L.; Frazier, S.F.; Zill, S.N. Sensing the effect of body load in legs: Responses of tibial campaniform sensilla to forces applied to the thorax in freely standing cockroaches. *J. Comp. Physiol. A* **2004**, *190*, 201–215.
5. Zill, S.; Schmitz, J.; Büschges, A. Load sensing and control of posture and locomotion. *Arthropod Struct. Dev.* **2004**, *33*, 273–286.
6. Akay, T.; Haehn, S.; Schmitz, J.; Büschges, A. Signals From Load Sensors Underlie Interjoint Coordination During Stepping Movements of the Stick Insect Leg. *J. Neurophysiol.* **2004**, *92*, 42–51.
7. Bucher, D.; Akay, T.; DiCaprio, R.a.; Buschges, A. Interjoint coordination in the stick insect leg-control system: The role of positional signaling. *J. Neurophysiol.* **2003**, *89*, 1245–1255.
8. Hess, D.; Büschges, A. Role of Proprioceptive Signals From an Insect Femur-Tibia Joint in Patterning Motoneuronal Activity of an Adjacent Leg Joint. *J. Neurophysiol.* **1999**, *81*, 1856–1865.
9. Mu, L.; Ritzmann, R.E. Kinematics and motor activity during tethered walking and turning in the cockroach, Blaberus discoidalis. *J. Comp. Physiol. A* **2005**, *191*, 1037–1054.
10. Hellekes, K.; Blincow, E.; Hoffmann, J.; Buschges, A. Control of reflex reversal in stick insect walking: Effects of intersegmental signals, changes in direction, and optomotor-induced turning. *J. Neurophysiol.* **2012**, *107*, 239–249.
11. Martin, J.P.; Guo, P.; Mu, L.; Harley, C.M.; Ritzmann, R.E. Central-Complex Control of Movement in the Freely Walking Cockroach. *Curr. Biol.* **2015**, *25*, 2795–2803.
12. Buschmann, T.; Ewald, A.; von Twickel, A.; Büschges, A. Controlling legs for locomotion-insights from robotics and neurobiology. *Bioinspir. Biomim.* **2015**, *10*, 041001.
13. Daun-Gruhn, S. A mathematical modeling study of inter-segmental coordination during stick insect walking. *J. Comput. Neurosci.* **2011**, *30*, 255–278.

14. Szczecinski, N.S.; Brown, A.E.; Bender, J.A.; Quinn, R.D.; Ritzmann, R.E. A neuromechanical simulation of insect walking and transition to turning of the cockroach Blaberus discoidalis. *Biol. Cybern.* **2014**, *108*, 1–21.
15. Cruse, H.; Kindermann, T.; Schumm, M.; Dean, J.; Schmitz, J. Walknet - A biologically inspired network to control six-legged walking. *Neural Netw.* **1998**, *11*, 1435–1447.
16. Schilling, M.; Paskarbeit, J.; Hoinville, T.; Hüffmeier, A.; Schneider, A.; Schmitz, J.; Cruse, H. A hexapod walker using a heterarchical architecture for action selection. *Front. Comput. Neurosci.* **2013**, *7*, 126.
17. Ekeberg, Ö.; Blümel, M.; Büschges, A. Dynamic simulation of insect walking. *Arthropod Struct. Dev.* **2004**, *33*, 287–300.
18. Rutter, B.L.; Taylor, B.K.; Bender, J.A.; Blümel, M.; Lewinger, W.A.; Ritzmann, R.E.; Quinn, R.D. Descending commands to an insect leg controller network cause smooth behavioral transitions. In Proceedins of the IEEE International Conference on Intelligent Robots and Systems, San Francisco, CA, USA, 25–30 September 2011; pp. 215–220.
19. Daun-Gruhn, S.; Tóth, T.I. An inter-segmental network model and its use in elucidating gait-switches in the stick insect. *J. Comput. Neurosci.* **2011**, *31*, 43–60.
20. Toth, T.I.; Schmidt, J.; Büschges, A.; Daun-Gruhn, S. A Neuro-Mechanical Model of a Single Leg Joint Highlighting the Basic Physiological Role of Fast and Slow Muscle Fibres of an Insect Muscle System. *PLoS ONE* **2013**, *8*, e78247.
21. Tóth, T.I.; Grabowska, M.; Rosjat, N.; Hellekes, K.; Borgmann, A.; Daun-Gruhn, S. Investigating inter-segmental connections between thoracic ganglia in the stick insect by means of experimental and simulated phase response curves. *Biol. Cybern.* **2015**, *109*, 349–362.
22. Schilling, M.; Hoinville, T.; Schmitz, J.; Cruse, H. Walknet, a bio-inspired controller for hexapod walking. *Biol. Cybern.* **2013**, *107*, 397–419.
23. Schmitz, J.; Schneider, A.; Schilling, M.; Cruse, H. No need for a body model: Positive velocity feedback for the control of an 18-DOF robot walker. *Appl. Bionics Biomech.* **2008**, *5*, 135–147.
24. Dupeyroux, J.; Passault, G.; Ruffier, F.; Viollet, S.; Serres, J. Hexabot: A small 3D-printed six-legged walking robot designed for desert ant-like navigation tasks. In Proceedings of the IFAC Word Congress, Toulouse, France, 9–14 July 2017.
25. Szczecinski, N.S.; Getsy, A.P.; Martin, J.P.; Ritzmann, R.E.; Quinn, R.D. Mantisbot is a robotic model of visually guided motion in the praying mantis. *Arthropod Struct. Dev.* **2017**, *46*, 736–751.
26. Szczecinski, N.S.; Quinn, R.D. Template for the neural control of directed stepping generalized to all legs of MantisBot. *Bioinspir. Biomim.* **2017**, *12*, 045001.
27. Gruhn, M.; Hoffmann, O.; Dübbert, M.; Scharstein, H.; Büschges, A. Tethered stick insect walking: A modified slippery surface setup with optomotor stimulation and electrical monitoring of tarsal contact. *J. Neurosci. Methods* **2006**, *158*, 195–206.
28. Durr, V. The behavioural transition from straight to curve walking: Kinetics of leg movement parameters and the initiation of turning. *J. Exp. Biol.* **2005**, *208*, 2237–2252.
29. Cruse, H. What mechanisms coordinate leg movement in walking arthropods? *Trends Neurosci.* **1990**, *13*, 15–21.
30. Cofer, D.; Cymbalyuk, G.; Reid, J.; Zhu, Y.; Heitler, W.J.; Edwards, D.H. AnimatLab: A 3D graphics environment for neuromechanical simulations. *J. Neurosci. Methods* **2010**, *187*, 280–288.
31. Szczecinski, N.S.; Hunt, A.J.; Quinn, R.D. Design process and tools for dynamic neuromechanical models and robot controllers. *Biol. Cybern.* **2017**, *111*, 105–127.
32. Hodgkin, A.L.; Huxley, A.F. A quantitative description of membrane current and its application to conduction and excitation in nerve. *Bull. Math. Biol.* **1990**, *52*, 25–71.
33. Hill, A.V. The Heat of Shortening and the Dynamic Constants of Muscle. *Proc. R. Soc. B Biol. Sci.* **1938**, *126*, 136–195.
34. Shadmehr, R.; Arbib, M.A. A mathematical analysis of the force-stiffness characteristics of muscles in control of a single joint system. *Biol. Cybern.* **1992**, *66*, 463–477.
35. Hooper, S.L.; Guschlbauer, C.; Blumel, M.; Rosenbaum, P.; Gruhn, M.; Akay, T.; Buschges, A. Neural Control of Unloaded Leg Posture and of Leg Swing in Stick Insect, Cockroach, and Mouse Differs from That in Larger Animals. *J. Neurosci.* **2009**, *29*, 4109–4119.
36. Ache, J.M.; Matheson, T. Passive Joint Forces Are Tuned to Limb Use in Insects and Drive Movements without Motor Activity. *Curr. Biol.* **2013**, *23*, 1418–1426.

37. Zill, S.N.; Chaudhry, S.; Exter, A.; Büschges, A.; Schmitz, J. Positive force feedback in development of substrate grip in the stick insect tarsus. *Arthropod Struct. Dev.* **2014**, *43*, 441–455.
38. Paskarbeit, J.; Otto, M.; Schilling, M.; Schneider, A. Stick(y) Insects—Evaluation of Static Stability for Bio-inspired Leg Coordination in Robotics. In Proceedings of the Conference on Biomimetic and Biohybrid Systems, Edinburgh, UK, 19–22 July 2016; Volume 1, pp. 239–250.
39. Ramdya, P.; Thandiackal, R.; Cherney, R.; Asselborn, T.; Benton, R.; Ijspeert, A.J.; Floreano, D. Climbing favours the tripod gait over alternative faster insect gaits. *Nat. Commun.* **2017**, *8*, 14494.
40. Szczecinski, N.S.; Hunt, A.J.; Quinn, R.D. A functional subnetwork approach to designing synthetic nervous systems that control legged robot locomotion. *Front. Neurorobot.* **2017**, *11*, 1–19.
41. Bässler, D.; Büschges, A.; Meditz, S.; Bässler, U. Correlation between muscle structure and filter characteristics of the muscle-joint system in three orthopteran insect species. *J. Exp. Biol.* **1996**, *199*, 2169–2183.
42. Wolf, H. Inhibitory motoneurons in arthropod motor control: Organisation, function, evolution. *J. Comp. Physiol. A* **2014**, *200*, 693–710.
43. Hooper, S.L.; Guschlbauer, C.; von Uckermann, G.; Buschges, A. Different Motor Neuron Spike Patterns Produce Contractions With Very Similar Rises in Graded Slow Muscles. *J. Neurophysiol.* **2007**, *97*, 1428–1444.
44. Garcia-Sanz, M. Chapter 3 P.I.D. control: Structure. In *EECS 475 Applied Control*, Case Western Reserve University: Cleveland, OH, USA, 2016.
45. Dürr, V.; Schmitz, J.; Cruse, H. Behaviour-based modelling of hexapod locomotion: Linking biology and technical application. *Arthropod Struct. Dev.* **2004**, *33*, 237–250.
46. Mantziaris, C.; Bockemühl, T.; Holmes, P.; Borgmann, A.; Daun, S.; Büschges, A. Intra- and intersegmental influences among central pattern generating networks in the walking system of the stick insect. *J. Neurophysiol.* **2017**, *118*, 2296–2310.
47. Cruse, H.; Riemenschneider, D.; Stammer, W. Control of body position of a stick insect standing on uneven surfaces. *Biol. Cybern.* **1989**, *61*, 71–77.
48. Cruse, H.; Schmitz, J.; Braun, U.; Schweins, A. Control of body height in a stick insect walking on a treadwheel. *J. Exp. Biol.* **1993**, *181*, 141–155.
49. Murray, R.M.; Li, Z.; Sastry, S.S. *A Mathematical Introduction to Robotic Manipulation;* CRC Press: Boca Raton, FL, USA, 1994.
50. Blumel, M.; Hooper, S.L.; Guschlbauer, C.; White, W.E.; Buschges, A. Determining all parameters necessary to build Hill-type muscle models from experiments on single muscles. *Biol. Cybern.* **2012**, *106*, 543–558.
51. Guschlbauer, C.; Scharstein, H.; Büschges, A. The extensor tibiae muscle of the stick insect: Biomechanical properties of an insect walking leg muscle. *J. Exp. Biol.* **2007**, *210*, 1092–1108.
52. Blümel, M.; Guschlbauer, C.; Daun-Gruhn, S.; Hooper, S.L.; Büschges, A. Hill-type muscle model parameters determined from experiments on single muscles show large animal-to-animal variation. *Biol. Cybern.* **2012**, *106*, 559–571.
53. Hooper, S.L.; Guschlbauer, C.; Blümel, M.; von Twickel, A.; Hobbs, K.H.; Thuma, J.B.; Büschges, A. *Muscles: Non-Linear Transformers of Motor Neuron Activity*; Springer Series in Computational Neuroscience; Springer: New York, NY, USA, 2016; pp. 163–194.
54. Rassier, D.; MacIntosh, B.; Herzog, W. Length dependence of active force production in skeletal muscle. *J. Appl. Physiol.* **1999**, *86*, 1445–1457
55. Schroer, R.; Boggess, M.; Bachmann, R.; Quinn, R.; Ritzmann, R. Comparing cockroach and Whegs robot body motions. In Proceedings of the IEEE International Conference on Robotics and Automation, New Orleans, LA, USA, 26 April–1 May 2004; Volume 4, pp. 3288–3293.
56. Hughes, G.M. The Co-ordination of Insect Movements. *J. Exp. Biol.* **1952**, *29*, 267–285.
57. Szczecinski, N.S. Massively Distributed Neuromorphic Control for Legged Robots Modeled after Insect Stepping. Master's Thesis, Case Western Reserve University, Cleveland, OH, USA, 2013.
58. Rubeo, S.E. Control of Simulated Cockroach Using Synthethic Nervous Systems. Master's Thesis, Case Western Reserve University, Cleveland, OH, USA, 2017.
59. Rutter, B.L.; Lewinger, W.A.; Blumel, M.; Buschges, A.; Quinn, R.D. Simple Muscle Models Regularize Motion in a Robotic Leg with Neurally-Based Step Generation. In Proceedings of the 2007 IEEE International Conference on Robotics and Automation, Roma, Italy, 10–14 April 2007; pp. 630–635.

60. Szczecinski, N.S.; Getsy, A.P.; Bosse, J.W.; Martin, J.P.; Ritzmann, R.E.; Quinn, R.D. MantisBot Uses Minimal Descending Commands to Pursue Prey as Observed in Tenodera Sinensis. In *Biomimetic and Biohybrid Systems*; Springer International Publishing: Cham, Switzerland, 2016; pp. 329–340.
61. Bender, J.A.; Simpson, E.M.; Tietz, B.R.; Daltorio, K.A.; Quinn, R.D.; Ritzmann, R.E. Kinematic and behavioral evidence for a distinction between trotting and ambling gaits in the cockroach Blaberus discoidalis. *J. Exp. Biol.* **2011**, *214*, 2057–2064.
62. Golowasch, J.; Goldman, M.S.; Abbott, L.F.; Marder, E. Failure of Averaging in the Construction of a Conductance-Based Neuron Model. *J. Neurophysiol.* **2002**, *87*, 1129–1131.
63. Cruse, H. Which Parameters Control the Leg Movement of a Walking Insect? I. Velocity Control during the Stance Phase. *J. Exp. Biol.* **1985**, *116*, 343–355.
64. Gruhn, M.; von Uckermann, G.; Westmark, S.; Wosnitza, A.; Buschges, A.; Borgmann, A. Control of Stepping Velocity in the Stick Insect Carausius morosus. *J. Neurophysiol.* **2009**, *102*, 1180–1192.
65. Szczecinski, N.S.; Quinn, R.D. MantisBot Changes Stepping Speed by Entraining CPGs to Positive Velocity Feedback. In *Biomimetic and Biohybrid Systems*; Springer: Cham, Switzerland, 2017; pp. 440–452.

Article

Germinal Center Optimization Applied to Neural Inverse Optimal Control for an All-Terrain Tracked Robot

Carlos Villaseñor [1], Jorge D. Rios [1], Nancy Arana-Daniel [1,*], Alma Y. Alanis [1], Carlos Lopez-Franco [1] and Esteban A. Hernandez-Vargas [2]

1 Centro Universitario de Ciencias Exactas e Ingenierías, Universidad de Guadalajara, Blvd. Marcelino García Barragán 1421, 44430 Guadalajara, Mexico; cavp@outlook.com (C.V.); jorge_rios.1xyz@yahoo.com (J.D.R.); almayalanis@gmail.com (A.Y.A.); clzfranco@gmail.com (C.L.-F.)

2 Frankfurt Institute For Advanced Studies, Ruth-Moufang-Straße 1, 60438 Frankfurt am Main, Germany; vargas@fias.uni-frankfurt.de

* Correspondence: nancy.arana@cucei.udg.mx; Tel.: +52-331-547-3877

Received: 16 November 2017; Accepted: 21 December 2017; Published: 27 December 2017

Abstract: Nowadays, there are several meta-heuristics algorithms which offer solutions for multi-variate optimization problems. These algorithms use a population of candidate solutions which explore the search space, where the leadership plays a big role in the exploration-exploitation equilibrium. In this work, we propose to use a Germinal Center Optimization algorithm (GCO) which implements temporal leadership through modeling a non-uniform competitive-based distribution for particle selection. GCO is used to find an optimal set of parameters for a neural inverse optimal control applied to all-terrain tracked robot. In the Neural Inverse Optimal Control (NIOC) scheme, a neural identifier, based on Recurrent High Orden Neural Network (RHONN) trained with an extended kalman filter algorithm, is used to obtain a model of the system, then, a control law is design using such model with the inverse optimal control approach. The RHONN identifier is developed without knowledge of the plant model or its parameters, on the other hand, the inverse optimal control is designed for tracking velocity references. Applicability of the proposed scheme is illustrated using simulations results as well as real-time experimental results with an all-terrain tracked robot.

Keywords: Germinal Center Optimization; Artificial Immune Systems; Evolutionary Computing; neural identification; inverse optimal control; extended kalman filter

---

1. Introduction

Nowadays, in computer science research is important to offer optimal techniques for a variety of problems, nevertheless, for most of these problems are difficult to formalize a mathematical model to optimize. Soft-computing optimization techniques, such as Evolutionary Computing (EC) [1], Artificial Neural Networks (ANN) [2] and Artificial Immune Systems (AIS) [3–5], approach these kinds of problems by offering good approximate solutions in an affordable time. EC algorithms offer an analogy of the competitive process in natural selection applied to multi-agent search for multi-variate problems, in the same way, AIS are based on the adaptive properties of the vertebrates immune system.

The vertebrates immune system has been developed through time by natural selection to overcome many diseases, although some of this protection mechanisms are inheritable, the immune system is capable of adapting to a new variety of Antigens (AGs) (foreign particles) in order to acquire specific protection [6]. This specific protection is given by Antibodies (ABs) that attach to AGs with certain affinity in the so-called humoral immunity. ABs are produced by the differentiation of the lymphocyte

B (B-cell). In the case that the body does not have a specific AB for an AG, the B-cells compete for producing a better affinity AB with the help of lymphocyte T CD4^{+} (Th-cell), this competition is the inspiration of the Clonal Selection algorithm [7], which has many variants and improvements [8,9].

When an infection prevails, the innate immune response is not capable of managing it. In this case, the adaptive immune response starts a process called clonal expansion of B-cells, looking for a B-cell with high-affinity ABs [6]. The highest affinity of ABs is achieved by a biological process called Germinal Center reaction. The Germinal Centers are temporal sites in the secondary lymph nodes histologically recognizable, where inactive B-cell enclose active B-cells, Follicular Dendritic Cells (FDC) and Th-cells with the objective of maturating the affinity through a competitive process. For a better understanding of the biological phenomenon, we refer the interested readers to [10].

In this paper, we use Germinal Center Optimization (GCO), a new multi-variate optimization algorithm, inspired by the germinal center reaction, that hybridizes some concepts of EC and AIS, for optimization of an inverse optimal controller applied to an all-terrain tracked robot. The principal feature of GCO is that the particle selection for crossover is guided by a competitive-based non-uniform distribution, this embedded the idea of temporal leadership, as we explain in Section 2.3.

On the other hand, most of the modern control techniques need the knowledge of a mathematical model of the system to be controlled. This model can be obtained using system identification in which the model is obtained using a set of data obtained from practical experiments with the system. Even when the system identification technique does not obtain an exact model, satisfactory models can be obtained with reasonable effort. There is a number of system identification techniques, to name a few: neural networks, fuzzy logic, auxiliary model, hierarchical identification. Among these system identification techniques, system identification using neural networks stands out, especially using recurrent neural networks which have a dynamic behavior [2,11,12].

The Recurrent High Order Neural Networks (RHONNs) are a generalization of the first order Hopfield network [11,12]. The presence of recurrent and high order connections gives the RHONN compared to a first order feedforward neural networks [11–13]: strong approximation capabilities, a faster convergence, greater storage capacity, a Higher fault tolerance, robustness against noise and dynamic behavior. Also, the RHONNs have the following characteristics [11,12,14]:

- They allow an efficient modeling o complex dynamic systems, even those with time-delays.
- They are good candidates for identification, state estimation, and control.
- Easy implementation.
- A priori information of the system to be identified can be added to the RHONN model.
- On-line or off-line training is possible.

The goal of the inverse optimal control is to determine a control law which forces the system to satisfy some restrictions and at the same time to minimize a cost functional. The difference with the optimal control methodology is that the inverse optimal control avoids the need of solving the associated Hamilton-Jacobi-Bellman (HJB) equation which is not an easy task and it has not been solved for general nonlinear systems. Furthermore, for the inverse approach, a stabilizing feedback control law, based on a priori knowledge of a Control Lyapunov Function (CLF), is designed first and then it is established that this control law optimizes a cost functional [15].

The control scheme consisting of a neural identifier and an inverse optimal control technique is named neural inverse optimal control (NIOC), this control scheme has shown good results in the literature for trajectory tracking [15–17]. However, the designer has to tune the appropriate value of some parameters of the controller discuss later in this work, the quality of the controller depends directly on this selection.

In this work, the main contribution is the introduction of an optimization process using GCO, in order to find the appropriate values for the controller parameters which minimize the tracking error of the system to be controlled. Performance of the optimization is shown presenting simulation and experimental tests comparing the results of the trajectory tracking using the NIOC with the parameters selected by the designer and the results using the parameters given by the GCO algorithm.

This work is organized as follows: In Section 2 the Germinal Center Optimization algorithm is described. In Section 2.1 the vertebrates adaptive immune system is briefly explained, in Section 2.2 we detail the germinal center reaction and in Section 2.3 the computation analogy of the germinal center reaction is presented along with the algorithm description. Section 3 introduces the Neural Inverse Optimal Control (NIOC) scheme for this work, where Section 3.1 presents the RHONN identifier and the extended kalman filter (EKF) training, and in Section 3.1.2 the design of the inverse optimal control law is discussed. Section 4 unveils comparative simulations (Section 4.2) and experimental (Section 4.3) results between the selection of the parameter of the controller using the GCO algorithm and the classic way which is let completely to the designer for an application of the NIOC to an All-Terrain Tracked Robot. Conclusions of this work are included in Section 5.

2. Germinal Center Optimization

In this section, we briefly overview the principal processes in the vertebrates immune system, and we detail the germinal center reaction. After that, we propose the computational analogy for multi-variate optimization, with the proper algorithm description.

2.1. Adaptive Immune System

The vertebrates immune system (VIS) is the biological mechanism for protecting the body from AG. There are two types of immunities, the innate immunity, and the adaptive one. The innate immunity is conformed by epithelial barriers that prevent the entrance of AG, phagocytes that swallow AG, FDCs that capture antigen and lymphocytes NKs (Natural Killers) that destroy any non-self cell.

The innate immunity is an inheritable protection that has been developed through natural selection, but if a new type of AG gets inside the body could overcome this basic protections, in this case, the adaptive immunity takes place. The adaptive immunity is conformed by B-cells whose main functions are to internalize AG for presentation and generate ABs, Th-cells that give a life signal to high-affinity B-cells and cytotoxic T-cells that kills own cells that are already infected or kills carcinogenic cells [6].

The affinity of the innate immunity is not diverse because it is coded in the germinal line, in the other hand the adaptive immunity has a high-affinity diversity because the receptors are produced for somatic recombination and variate with somatic hyper-mutation [6,18].

There are two types of adaptive immune response, the humoral immunity, and the cellular immunity. The first one is based on ABs that travel in the bloodstream, attaching to every compatible particle. The B-cells compete for antigen internalization and presentation to the Th-cell, whose reward them with a life signal, then the B-cell proliferates by clonal expansion and differentiates into plasmatic cells, releasing higher affinity ABs in the bloodstream.

There are some AGs that infect the owner cells and hide inside them, these cells are destroyed by the cellular immunity with the cytotoxic T-cell. The adaptive immune response has the following features, as is shown in [6]:

- Specificity: Ensure to produce a specific AB for a specific AG
- Diversity: The immune system is capable of responding to a great variety of AGs
- Memory: Using memory B-cells, the immune system is capable of fighting repeated infections
- Clonal expansion: Increase the number of lymphocytes with high affinity of certain AGs
- Homeostasis: The immune system recover from an infection by itself
- No self-reactivity: The immune system does not attack the host body in the present of AGs

2.2. Germinal Center

When the body does not have a specific AB for an infection, the body starts the process of affinity maturation with the germinal center reaction. Germinal centers are micro-anatomical regions in the secondary lymph nodes, that form in the present of antigen [10]. The AGs that survive the other immunity mechanism arrives at the secondary lymph nodes, where are capture for the FDCs. The FDCs activate near B-cells. The active B-cells end up being enclosed by the inactive ones forming a natural barrier that allow the active cells to proliferate, mutate and be selected.

The B-cells start to proliferate inside the GC, and compete for the antigen, this competition polarizes the GC in two distinct zones, the dark zone and the light zone. The dark zone is where B-cells proliferate through clonal expansion and somatic hyper-mutation, this process ensures the diversity of ABs. On the other hand, the light zone is where the B-cells are selected in accordance with their affinity.

On the light zone, the B-cells must find AG and internalize it, with the final purpose of digest the AG and expose their peptides to the Th-cell. The Th-cell gives a life signal allowing B-cells high affinity to live more time and therefore, proliferate and mutate with higher success.

In Figure 1, we show a schematic summary of the process. The GC reaction ensures the diversity through clonal expansion and somatic hyper-mutation in the dark zone, while in the light zone is a competitive process that reward the more adapted B-cells. The B-cells reentry to dark zone making this process an iterative refinement of the affinity [19,20].

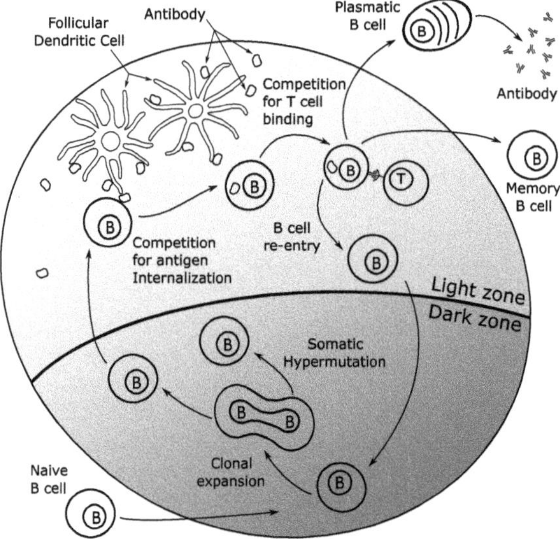

Figure 1. Germinal Center reaction.

Finally, when the GC generates high affinity B-cells for certain AG, some B-cells differentiate into plasmatic cells and release their ABs. A few B-cells become Memory B-cells that could live a long time and keeps information about this particular AG. Then, the GCs are capable of generating specific AB for a specific AG, and keep this information in Memory B-cells for future infections [21].

2.3. Algorithm Description

In this section, we explain the GCO algorithm. In Table 1, we present the computational analogy between germinal center and the optimization problem.

Table 1. Computational analogy between Germinal Center (GC) and Optimization.

Germinal Center	Optimization Problem
Antigen	Objective function
Antibody and B-cell	Candidate solution
Affinity	Objective function evaluation
T-help cell binding	Incrementation of life-signal

The GC reaction has multiple competitive processes, the GCO algorithm does not try to simulate GC reaction per se but to use some of its competitive mechanisms. A key factor in the GC reaction is the distinction between the dark zone and the light zone. The dark zone represents the diversification of the solutions that could be understood like a mutation process, many algorithms, such as Differential evolution and Genetic algorithms [1], already have the idea of mutation, but the dark zone includes not only a mutation process (somatic hyper-mutation), but also the clonal expansion. This clonal expansion is guided by the life signal of the B-cell, denoted by $\mathbb{L} \in [0, 100]$. The B-cells with greater life signal are more likely to clone, increasing the B-cell multiplicity.

In the light zone, B-cells with the best affinity are rewarded and the other cells age (lower their life signal). Then, the affinity-based selection in the light zone changes the probability of clone or death of a B-cell. In Figure 2, we present the GCO algorithm flowchart, where "For each B-cell" indicates that the process is applied in every candidate solution.

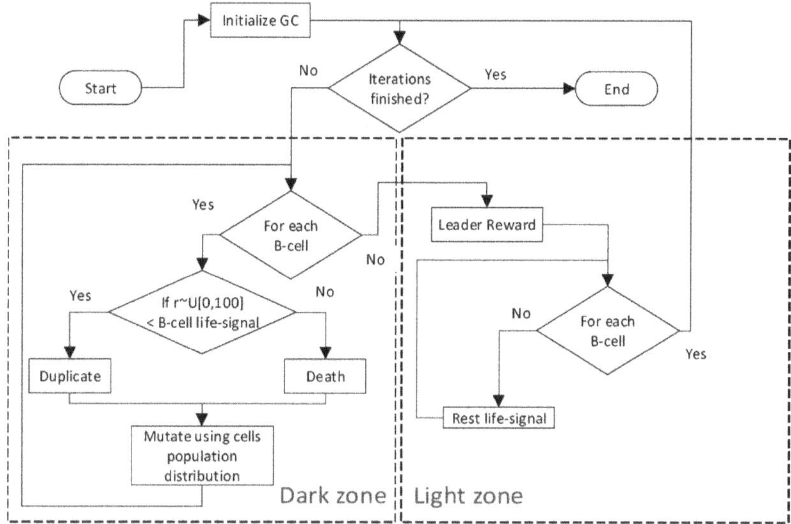

Figure 2. Germinal Center Optimization (GCO) algorithm flowchart.

As we are dealing with a population-based algorithm [1], there is a particular interest in the cells mutation and which information we use for crossover particles. In the GCO algorithm, we use the distribution of the cells multiplicity, denoted with C, to select three individuals for crossover. It is important to note that initially, all the cells have a multiplicity of one, then the individuals are uniformly selected, but this distribution changes through iterations modeled by the competitive process. This kind of distribution offers different types of leadership behaviors in the collective intelligence, for example, initially the GCO algorithm behaves like Differential Evolution in the DE/rand/1 strategy [1], and when a particle wins for many iterations, the algorithm behaves like Particle Swarm optimization [1]. However, the leadership in GCO is not only dynamic, but it also includes temporal leadership, this is implemented when a particle mutates to a better solution, this new candidate substitutes the actual particle resetting the cell multiplicity to one.

Then, GCO algorithm offers a bio-inspired technique of adaptive leadership in collective intelligence algorithms for multivariate optimization problems. In Algorithm 1, we include an explicit pseudocode, where B_i is the i-esim B-cell, M is a mutant B-cell and a new candidate solution, $F \in [0, 2]$ is the mutation factor and $CR \in [0, 1]$ is the cross-ratio, C is the distribution of cells multiplicity, and $\mathbb{L}$ is the life signal.

Algorithm 1: GCO algorithm

```
Initialize B-cells (B_i)
foreach k ∈ {1, ··· , Iterations} do
    /* Dark-zone process                                                  */
    foreach i ∈ {1, ··· , N} do
        if r_l ~ U[0, 100] < 𝕃 of B_i then
            /* B-cell Duplication                                         */
            Add one to GC cells counter
            Add one to B_i cells counter
        else
            /* B-cell Death                                               */
            if Cells in B_i > 1 then
                Rest one to GC cells counter
                Rest one to B_i cells counter
            end
        end
        /* B-cell mutation                                                */
        Calculate distribution C
        Using r_1, r_2, r_3 ~ C choose 3 different B-cells: B_{r_1}, B_{r_2} and B_{r_3}
        Create new Mutant M
        foreach j ∈ {1, ··· , D} do
            if r ~ U[0, 1] < CR then
                M(j) ← B_{r_1}(j) + F(B_{r_2}(j) − B_{r_3}(j))
            else
                M(j) ← B_i(j)
            end
        end
        Evaluate objective function for M
        if M is better than B_i then
            Substitute B_i for M
            if M is the best then
                Save i index
            end
        end
    end
    /* Light-zone process                                                 */
    Add 10 units to 𝕃 of the Best B-cell
    foreach i ∈ {1, ··· , N} do
        Rest 10 units to 𝕃 of B_i
    end
end
```

3. Neural Inverse Optimal Control

In this section, we introduced the neural identifier based on a RHONN and its EKF training, and the inverse optimal control law designed. Consider the following affine discrete-time nonlinear system

$$\mathbf{x}(k) = f(\mathbf{x}(k) + g(\mathbf{x}(k))\mathbf{u}(k) \tag{1}$$

where $\mathbf{x} \in \Re^n$ is the state vector of the system, $\mathbf{u} \in \Re^m$ is the control input vector, $f \in \Re^n \rightarrow \Re^n$ and $g \in \Re^n \rightarrow \Re^{n \times m}$ are smooth maps.

3.1. Neural Identification with Recurrent High Order Neural Networks (RHONN)

To identify the system (1), we used the following RHONN identifier based on a RHONN in series-parallel model:

$$\begin{aligned} \widehat{\chi}_i(k+1) &= \omega_i^\top(k)\mathbf{z}_i(\mathbf{x}(k-l),\mathbf{u}(k)) \\ i &= 1,2,\cdots,n \end{aligned} \quad (2)$$

with

$$\mathbf{z}_i(\mathbf{x}(k),\mathbf{u}(k)) = \begin{bmatrix} z_{i_1} \\ z_{i_2} \\ \vdots \\ z_{i_{L_i}} \end{bmatrix} = \begin{bmatrix} \Pi_{j\in I_1}\xi_{i_j}^{d_{ij}(1)} \\ \Pi_{j\in I_2}\xi_{i_j}^{d_{ij}(2)} \\ \vdots \\ \Pi_{j\in I_{L_i}}\xi_{i_j}^{d_{ij}(L_i)} \end{bmatrix}, \quad \xi_i = \begin{bmatrix} \xi_{i_1} \\ \vdots \\ \xi_{i_n} \\ \xi_{i_{n+1}} \\ \vdots \\ \xi_{i_{n+m}} \end{bmatrix} = \begin{bmatrix} S(x_1(k-l)) \\ \vdots \\ S(x_n(k-l)) \\ u_1(k) \\ \vdots \\ u_m(k) \end{bmatrix} \quad (3)$$

where $S(v) = 1/(1+\exp(-\beta v))$, $\beta > 0$, n is the state dimension, $\widehat{\chi}$ is state vector of the neural network, ω is the weight vector $\mathbf{x}$ is the plant state vector, and $\mathbf{u} = [u_1, u_2, \ldots, u_m]^\top$ is the input vector to the neural network.

The neural identifier (2) is presented in [14]. This neural identifier does not need previous knowledge of the model of the system, also, it does not need information of the disturbances and delays. Moreover, this model is semi-globally uniformly ultimately bounded (SGUUB) and the proof can be found in [14].

3.1.1. Training of RHONN with Extended Kalman Filter

The extended kalman filter estimates the state of a system with additive white noise in the input and in the output using a recursive solution in which each update of the state is estimated from the previous estimated state and the new input data [11,22].

For the case of neural networks the extended kalman filter training goal is to find the optimal weight vector which minimizes the prediction error. Due to the fact that the neural network mapping is non-lineal the extended kalman filter (EKF) is required. The EKF-based training algorithm [11] is (4):

$$\omega_i(k+1) = \omega_i(k) + \eta_i\mathbf{K}_i(k)e_i(k) \quad (4)$$

$$\mathbf{K}_i(k) = \mathbf{P}_i(k)\mathbf{H}_i(k)[R_i(k) + \mathbf{H}_i^\top(k)\mathbf{P}_i(k)\mathbf{H}_i(k)]^{-1} \quad (5)$$

$$\mathbf{P}_i(k+1) = \mathbf{P}_i(k) - \mathbf{K}_i(k)\mathbf{H}_i^\top(k)\mathbf{P}_i(k) + \mathbf{Q}_i(k) \quad (6)$$

with

$$e_i(k) = x_i(k) - \widehat{\chi}_i(k) \quad (7)$$

$$H_{ij} = \left[\frac{\partial\widehat{\chi}_i(k)}{\partial\omega_{ij}(k)}\right]^\top \quad (8)$$

where $i = 1\cdots n$, $\omega_i \in \Re^{L_i}$ is the on-line adapted weight vector, $\mathbf{K}_i \in \Re^{L_i}$ is the Kalman gain vector, $e_i \in \Re$ is the identification error, $\mathbf{P}_i \in \Re^{L_i\times L_i}$ is the weight estimation error covariance matrix, χ_i is the i-th state variable of the neural network, $\mathbf{Q}_i \in \Re^{L_i\times L_i}$ is the estimation noise covariance matrix, $R_i \in \Re$ is the error noise covariance matrix and $\mathbf{H}_i \in \Re^{L_i}$ is a vector in which each entry H_{ij} is the derivative of the neural network state ($\widehat{\chi}_i$) with respect to one neural network weight (ω_{ij}) and it is given by (8). $\mathbf{P}_i$ and $\mathbf{Q}_i$ are initialized as diagonal matrices with entries $\mathbf{P}_i(0)$ and $\mathbf{Q}_i(0)$, respectively. It is important to remark that $\mathbf{H}_i(k)$, $\mathbf{K}_i(k)$ and $\mathbf{P}_i(k)$ for the EKF are bounded.

3.1.2. Inverse Optimal Control

Optimal control finds a control law for a system such that a performance criterion is minimized. The criterion is a cost functional based on the state and control variables. The solution of the optimal control leads to the HJB equation which solution is not an easy task. Inverse optimal control is an alternative to optimal control, avoiding the HJB equation solution. For the inverse optimal control approach a stabilizing feedback control law based on a priori knowledge of a control Lyapunov function (CLF), is designed first, and then it is established that this control law optimizes a cost functional, then, the CLF is modified in order to achieve asymptotic tracking for given trajectory references [15]. The existence of a CLF implies stability and every CLF can be considered as a cost functional. The CLF approach for control synthesis has been applied successfully to systems for which a CLF can be established, such as feedback linearizable, strict feedback and feed-forward ones [15].

The system (1) is supposed to have an equilibrium point $\mathbf{x}(0) = \mathbf{0}$. Moreover, the full state $\mathbf{x}(k)$ is assumed to be available. In order to ensure stability of the system (1) the following control Lyapunov fuction is proposed:

$$V(\mathbf{x}(k)) = \frac{1}{2}\mathbf{x}(k)^\top \mathbf{P}\mathbf{x}(k), \quad \mathbf{P} = \mathbf{P}^\top > 0 \tag{9}$$

The inverse optimal control law for the system (1) with (9) is:

$$\begin{aligned} \mathbf{u}(k) &= -\frac{1}{2}\mathbf{R}^{-1}(\mathbf{x}(k))g^\top(\mathbf{x}(k))\frac{\partial V(\mathbf{x}(k))}{\partial \mathbf{x}(k+1)} \\ &= -\frac{1}{2}(\mathbf{R}(\mathbf{x}(k)) + \frac{1}{2}g^\top(\mathbf{x}(k))\mathbf{P}g(\mathbf{x}(k)))^{-1}g^\top(\mathbf{x}(k))\mathbf{P}f(\mathbf{x}(k)) \end{aligned} \tag{10}$$

where $\mathbf{R}(\mathbf{x}(k)) = \mathbf{R}(\mathbf{x}(k))^\top > \mathbf{0}$ is a matrix whose elements can be functions of the system state or can be fixed. P is a matrix such that the inequality (11) holds.

$$V_f(\mathbf{x}(k)) - \frac{1}{4}\mathbf{P}_1^\top(\mathbf{x}(k))(\mathbf{RP}(\mathbf{x}(k)))^{-1}\mathbf{P}_1(\mathbf{x}(k)) \leq -\mathbf{x}^\top(k)\mathbf{Q}\mathbf{x}(k) \tag{11}$$

with

$$\mathbf{RP}(\mathbf{x}(k)) = \mathbf{R}(\mathbf{x}(k)) + \mathbf{P}_2(\mathbf{x}(k)) \tag{12}$$

$$V_f(\mathbf{x}(k)) = \frac{1}{2}f^\top(\mathbf{x}(k))\mathbf{P}f(\mathbf{x}(k)) - V(\mathbf{x}(k)) \tag{13}$$

$$\mathbf{P}_1(\mathbf{x}(k)) = g^\top(\mathbf{x}(k))\mathbf{P}f(\mathbf{x}(k)) \tag{14}$$

$$\mathbf{P}_2(\mathbf{x}(k)) = \frac{1}{2}g^\top(\mathbf{x}(k))\mathbf{P}g(\mathbf{x}(k)) \tag{15}$$

$$\mathbf{Q} = \mathbf{Q}^\top > 0 \tag{16}$$

In [15], it is demonstrated that control law (10) is globally asymptotically stable. Moreover, (10) is inverse optimal in the sense that minimizes a cost functional [15].

4. Results

In this section, we briefly describe how GCO is applied to improve NIOC performance, we also show simulation results and real-time experimental results. The all-terrain tracked robot is a modified HD2® (HD2 is a registered trademark of SuperDroid Robots), shown in Figure 3. The changes of the modified HD2® are (Figure 4): the replacement of the original board for a system based on Arduino® (Arduino is a registered trademark of Arduino LLC), and an attachment of a wireless router. Chassis, tracks, batteries, and motors remained without modifications.

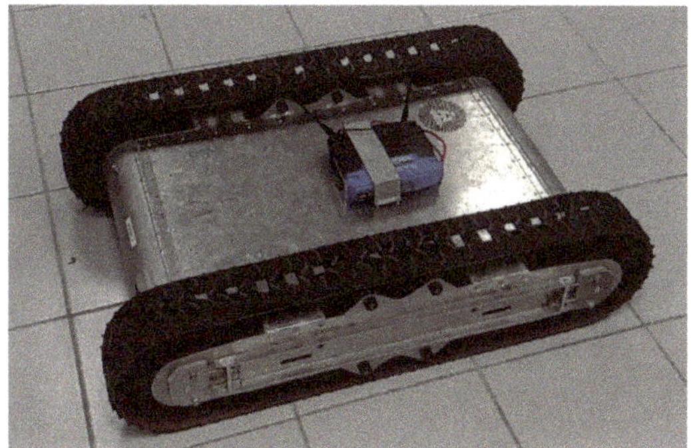

Figure 3. Modified HD2® Treaded All-Terrain Tracked Robot (ATR) Tank Robot Platform with wireless communication.

Figure 4. The interior of the Modified HD2® Treaded ATR Tank Robot.

4.1. Application to All-Terrain Tracked Robot Control

Considered as the most important type of mobile robots, a tracked robot runs on continuous tracks instead of wheels which develop a thrust higher than a wheeled robot. This kind of robots is ideal for working in tasks under rough terrains. Among the applications tracked robots can achieve are urban reconnaissance, forestry, mining, agriculture, rescue mission scenarios, autonomous planetary explorations [23–25].

A tracked robot consists of the following state variables [17,26,27] position x, position y, position θ, velocity 1, velocity 2, current 1 and current 2. In this work, we focus on the controller tracking performance for x, y and θ (Figure 5) for given references x_r, y_r and θ_r. The objective is to improve the NIOC results presented in [17] by using GCO to find the optimal parameters of the controller. These parameters are included in the matrices $\mathbf{P}_1$ and $\mathbf{P}_2$ defined in (11) and (12) respectively. The $\mathbf{P}_1$ and $\mathbf{P}_2$ are symmetric positive definite matrices, therefore we can define the set of variables $\{\psi_1, \psi_2, \psi_3, \psi_4, \psi_5\}$ for the optimization problem with the definition in (17).

$$\mathbf{P}_1 = \begin{bmatrix} \psi_1 & \psi_2 & \psi_3 \\ \psi_2 & \psi_1 & \psi_4 \\ \psi_3 & \psi_4 & \psi_1 \end{bmatrix} \quad \mathbf{P}_2 = \begin{bmatrix} \psi_5 & 0 \\ 0 & \psi_5 \end{bmatrix} \tag{17}$$

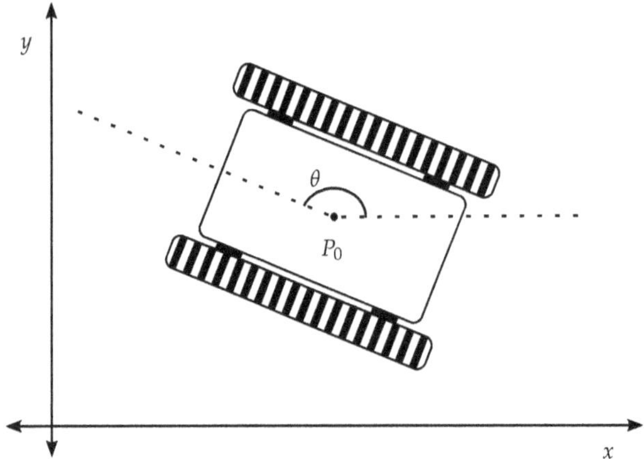

Figure 5. Schematic representation of a tracked robot, where x and y are the coordinates of P_0 and θ is the heading angle.

The lower bound of the search space is given by $\{1, 1, 1, 1, 1\}$ and the upper bound is given by $\{4 \times 10^7, 4 \times 10^7, 4 \times 10^7, 4 \times 10^7, 4 \times 10^4\}$. Next, the objective of the optimization is to get a better control tracking, in order to achieve this, we minimize the sum of the Root Mean Square Error (RMSE) in every state. This idea is described by (18), where n is the number of samples of the reference and estimated functions, we are using $n = 3335$ for real-time experiments and $n = 5000$ for simulation experiments.

$$f = \min_{\psi_1,\psi_2,\psi_3,\psi_4,\psi_5} \sqrt{\frac{\sum_{i=1}^{n}(\hat{x} - x)^2}{n}} + \sqrt{\frac{\sum_{i=1}^{n}(\hat{y} - y)^2}{n}} + \sqrt{\frac{\sum_{i=1}^{n}(\hat{\theta} - \theta)^2}{n}} \tag{18}$$

Then, the GCO algorithm will find optimal values that minimize (18). For the following experiments we use a GCO algorithm in five dimension running 150 iterations using 30 B-cells (4500 executions) for a test of 10 seconds; we set the parameters $F = 1.25$ and $CR = 0.7$. We include graphics for one simulation test using the all-terrain robot model presented in [17], and the results of one experimental test using the modified HD2® Treaded ATR Tank Robot Platform with wireless communication (Figure 3) presented in [17].

4.2. Simulation Results

In [17], $\mathbf{P}_1$ and $\mathbf{P}_2$ are defined as shown in (19) for simulation. We show in (20) $\mathbf{P}_1^*$ and $\mathbf{P}_2^*$, which contain the parameters found by the GCO algorithm.

$$\mathbf{P}_1 = 14400 \begin{bmatrix} 162 & 1 & 2 \\ 1 & 162 & 3 \\ 2 & 3 & 162 \end{bmatrix} \quad \mathbf{P}_2 = 20 \begin{bmatrix} 20 & 0 \\ 0 & 20 \end{bmatrix} \tag{19}$$

$$\mathbf{P}_1^* = 1 \times 10^6 \begin{bmatrix} 5.7564 & 0 & 0 \\ 0 & 5.7564 & 0 \\ 0 & 0 & 5.7564 \end{bmatrix} \quad \mathbf{P}_2^* = \begin{bmatrix} 40000 & 0 \\ 0 & 40000 \end{bmatrix} \tag{20}$$

Figures 6–8 show the tracking performance and the error comparison for the position x, position y and position θ of the simulation test. For each figure, the graph on the left side shows the obtained signals for reference, the identified signal using NIOC [17], and the identified signal using the optimized

NIOC for its respective state variable. The graph on the right side shows the obtained the error signals for the NIOC [17] and the optimized NIOC for its respective state variable. This arrangement is maintained for all the following figures in this section.

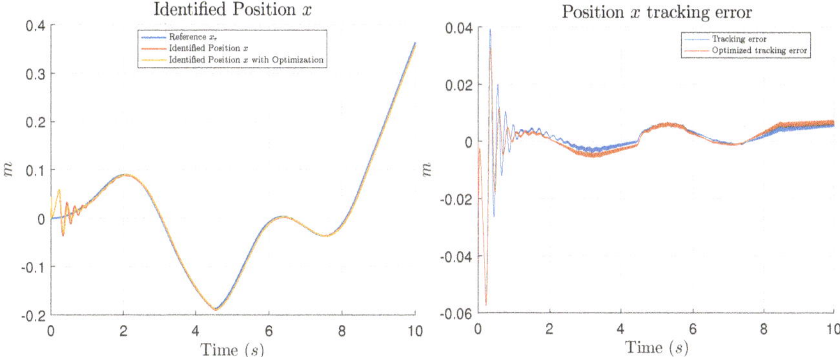

Figure 6. Tracking of x position (**left**) and error comparison (**right**) for the simulation test.

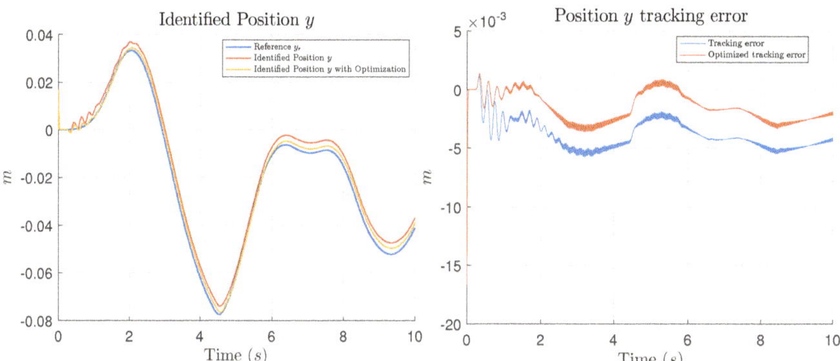

Figure 7. Tracking of y position (**left**) and error comparison (**right**) for the simulation test.

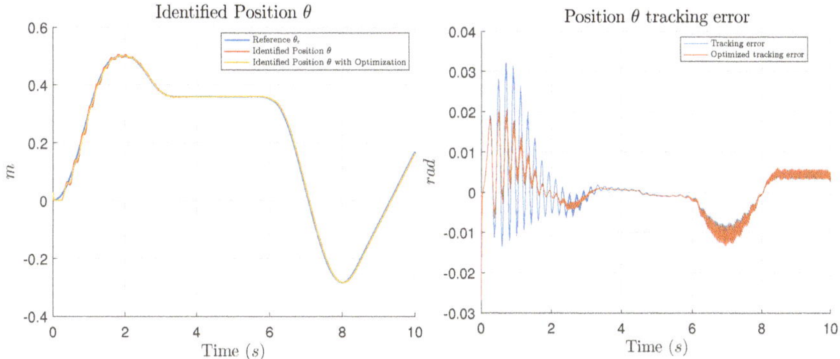

Figure 8. Tracking of θ (**left**) and error comparison (**right**) for the simulation test.

Table 2 shows the RMSE of each state variable and their total which is the evaluation of Equation (18) for this simulation test. Total does not have a physical meaning, it is the minimum found in the objective function in (18).

Table 2. Root Mean Squared Error (RMSE) in states for second simulation. Bold values highlight the best result.

Control	RMSE x (m)	RMSE y (m)	RMSE θ (rad)	Total
NIOC [17]	**0.0073**	0.0043	0.0064	0.0180
NIOC with GCO	**0.0073**	**0.0025**	**0.0049**	**0.0147**

A second simulation test was made resulting in the tracking errors shown in Table 3.

Table 3. Root Mean Squared Error (RMSE) in states for first simulation. Bold values highlight the best result.

Control	RMSE x (m)	RMSE y (m)	RMSE θ (rad)	Total
NIOC [17]	**0.0077**	0.0055	0.0064	0.0195
NIOC with GCO	0.0079	**0.0041**	**0.0049**	**0.0169**

Additionally to the presented results, in [17] it was demonstrated via simulations that the NIOC has a better performance than a super twisting scheme for a tracked robot model. The tracking comparison results shown in Table 4 are reported in [17].

Table 4. Tracking error comparison of NIOC [17] and a Super Twisting controller. Bold values highlight the best result.

RMSE	x (m)	y (m)	θ (rad)
NIOC [17]	**0.0260**	0.0362	**0.0158**
Super Twisting	0.0317	**0.0036**	0.0652

4.3. Real-Time Results

The work [17] presents a NOIC for the Modified HD2® Treaded ATR Tank Robot Platform with wireless communication (Figure 3), this implementation uses an RHONN identifier as (2) to identify the unknown model of the HD2®. The obtained model is then used as the based to synthesize the control law using the inverse optimal control approach. In [17] the values of $\mathbf{P}_1$ and $\mathbf{P}_2$ are defined in (21) for real-time operation. The following results show a comparison obtained with the values from [17] and $\mathbf{P}_1^*$ and $\mathbf{P}_2^*$ in (22) found by the GCO algorithm. In this section, there are presented two experiments named as "test 1" and "test 2", respectively.

$$\mathbf{P}_1 = 72000 \begin{bmatrix} 162 & 1 & 2 \\ 1 & 162 & 3 \\ 2 & 3 & 162 \end{bmatrix} \quad \mathbf{P}_2 = 10000 \begin{bmatrix} 20 & 0 \\ 0 & 20 \end{bmatrix} \tag{21}$$

$$\mathbf{P}_1^* = 1 \times 10^7 \begin{bmatrix} 3.2039 & 0 & 4 \\ 0 & 3.2039 & 0 \\ 4 & 0 & 3.2039 \end{bmatrix} \quad \mathbf{P}_2^* = 1 \times 10^4 \begin{bmatrix} 2.4128 & 0 \\ 0 & 2.4128 \end{bmatrix} \tag{22}$$

Figures 9–11 show the tracking performance and the error comparison for the position x, position y and position θ of the experimental test 1.

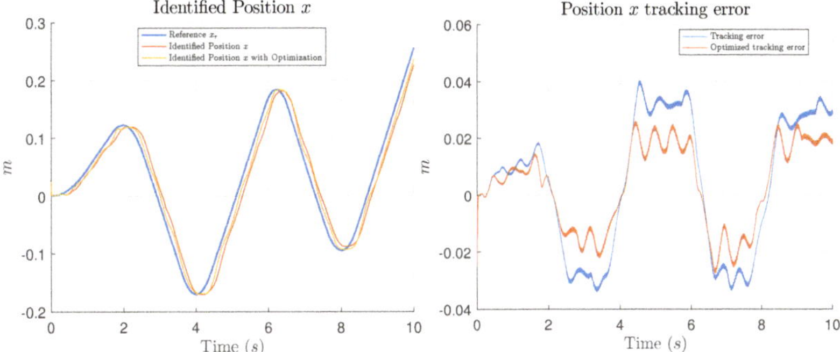

Figure 9. Tracking of x position (**left**) and error comparison (**right**) for the experimental test 1.

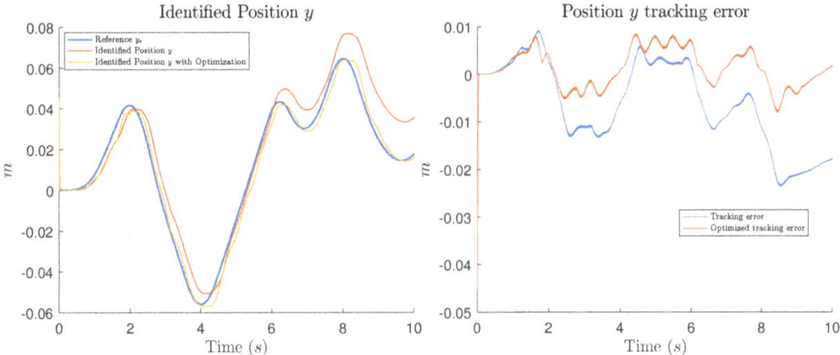

Figure 10. Tracking of y position (**left**) and error comparison (**right**) for the experimental test 1.

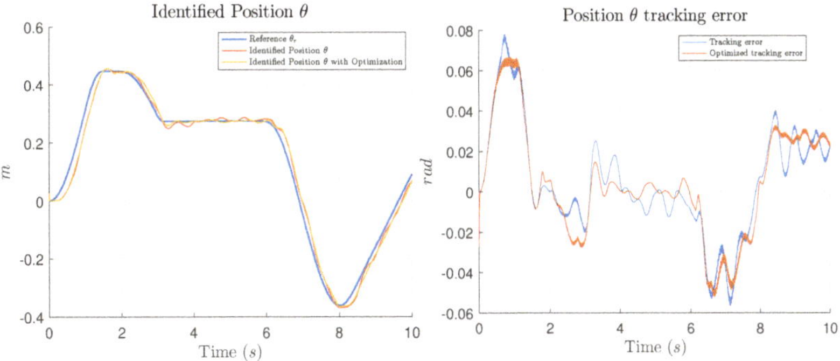

Figure 11. Tracking of θ (**left**) and error comparison (**right**) for the experimental test 1.

Table 5 shows the RMSE of each state variable and their total which is the evaluation of Equation (18) for the experimental test 1.

Table 5. Root Mean Squared Error in states for Real-Time experimental test 1. Bold values highlight the best result.

Control	RMSE x (m)	RMSE y (m)	RMSE θ (rad)	Total
NIOC [17]	0.0238	0.0109	**0.0267**	0.0614
NIOC with GCO	**0.0152**	**0.0043**	0.0270	**0.0465**

Figures 12–14 show the tracking performance and the error comparison for the position x, position y and position θ of the experimental test 2.

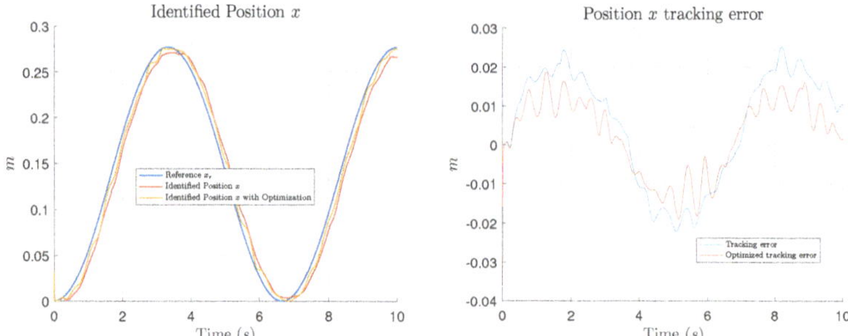

Figure 12. Tracking of x position (**left**) and error comparison (**right**) for the experimental test 2.

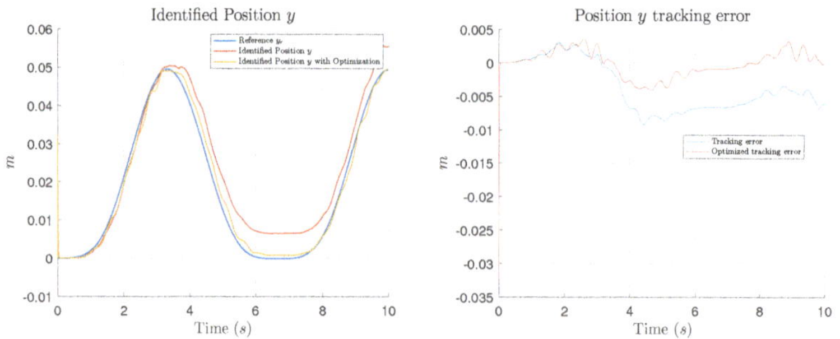

Figure 13. Tracking of y position (**left**) and error comparison (**right**) for the experimental test 2.

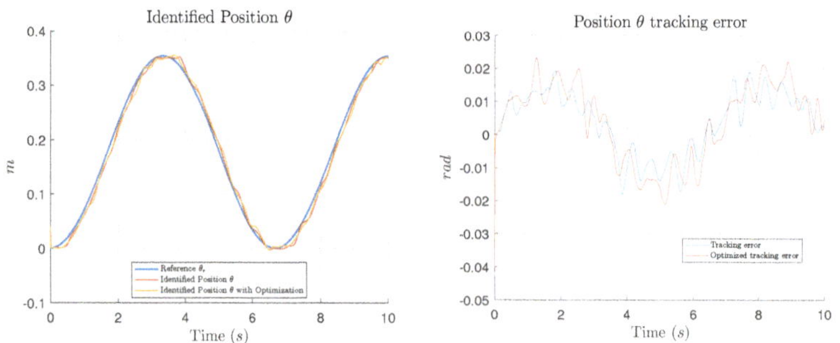

Figure 14. Tracking of θ (**left**) and error comparison (**right**) for the experimental test 2.

Table 6 shows the RMSE of each state variable and their total which is the evaluation of Equation (18) for the experimental test 2.

Table 6. Root Mean Squared Error (RMSE) in states for Real-Time experimental test 2. Bold values highlight the best result.

Control	RMSE x (m)	RMSE y (m)	RMSE θ (rad)	Total
NIOC [17]	0.0153	0.0052	**0.0101**	0.0305
NIOC with GCO	**0.0100**	**0.0021**	0.0123	**0.0244**

5. Conclusions

GCO is a hybridization between Evolutionary Computing and Artificial Immune System based on the Germinal Center reaction which is a biological process in vertebrates immune system that maturates affinity of antibodies. GCO models a population of B-cells and the competitive process in their proliferation, then, a dynamic distribution of the cells multiplicity is constructed over the performance of the candidate solutions. This distribution allows to select B-cells for crossover with an adaptive leadership. The adaptive leadership takes the advantage of both high leadership and none leadership algorithms allowing it to find a better solution.

In this work, it is shown how GCO can help with the overall performance of a control technique like inverse optimal control, which depends on a number of designed parameters. Our results reveal a better performance of the controller version with the parameters obtained with GCO. It is also important to mention that the found parameters by GCO for the NIOC are not unique for a reference, those parameters can work for a number of references.

Acknowledgments: The authors thank the support of CONACYT Mexico, through Projects CB256769 and CB258068 ("Project supported by *Fondo Sectorial de Investigación para la Educación*").

Author Contributions: Jorge D. Rios, Alma Y. Alanis and Carlos Lopez-Franco contribute to the NIOC design for an All-Terrain Tracked Robot. Carlos Villaseñor, Nancy Arana-Daniel and Esteban A. Hernandez-Vargas designed the GCO algorithm, the objective function and perform the optimization.

Conflicts of Interest: The authors declare no conflict of interest.

References

1. Simon, D. *Evolutionary Optimization Algorithms*; John Wiley & Sons: Hoboken, NJ, USA, 2013.
2. Haykin, S. *Neural Networks and Learning Machines*; Prentice Hall: Upper Saddle River, NJ, USA, 2009.
3. Brabazon, A.; O'Neill, M.; McGarraghy, S. Artificial Immune Systems. In *Natural Computing Algorithms*; Springer: Berlin, Germany, 2015; pp. 301–332.
4. Dasgupta, D.; Yu, S.; Nino, F. Recent advances in artificial immune systems: Models and applications. *Appl. Soft Comput.* **2011**, *11*, 1574–1587.
5. Timmis, J.; Hone, A.; Stibor, T.; Clark, E. Theoretical advances in artificial immune systems. *Theor. Comput. Sci.* **2008**, *403*, 11–32.
6. Abbas, A.K.; Lichtman, A.H.; Pillai, S. *Cellular and Molecular Immunology*; Elsevier Health Sciences: Amsterdam, The Netherlands, 2014.
7. Ulutas, B.H.; Kulturel-Konak, S. A review of clonal selection algorithm and its applications. *Artif. Intell. Rev.* **2011**, *36*, 117–138.
8. Dai, H.; Yang, Y.; Li, H.; Li, C. Bi-direction quantum crossover-based clonal selection algorithm and its applications. *Expert Syst. Appl.* **2014**, *41*, 7248–7258.
9. Gao, S.; Dai, H.; Zhang, J.; Tang, Z. An expanded lateral interactive clonal selection algorithm and its application. *IEICE Trans. Fundam. Electron. Commun. Comput. Sci.* **2008**, *91*, 2223–2231.
10. Victora, G.D.; Nussenzweig, M.C. Germinal centers. *Ann. Rev. Immunol.* **2012**, *30*, 429–457.
11. Sanchez, E.N.; Alanis, A.Y.; Loukianov, A.G. *Discrete-Time High Order Neural Control: Trained with Kalman Filtering*; Springer: Berlin, Germany, 2008.
12. Alanis, A.; Sanchez, E. *Discrete-Time Neural Observers: Analysis and Applications*; Elsevier Science: Amsterdam, The Netherlands, 2017.

13. Zhang, M. *Artificial Higher Order Neural Networks for Economics and Business*; IGI Global Research Collection, Information Science Reference: Hershey, PA, USA, 2008.
14. Alanis, A.Y.; Rios, J.D.; Arana-Daniel, N.; Lopez-Franco, C. Neural identifier for unknown discrete-time nonlinear delayed systems. *Neural Comput. Appl.* **2016**, *27*, 2453–2464.
15. Sanchez, E.; Ornelas-Tellez, F. *Discrete-Time Inverse Optimal Control for Nonlinear Systems*; EBL-Schweitzer, CRC Press: Boca Raton, FL, USA, 2016.
16. Lopez, V.G.; Sanchez, E.N.; Alanis, A.Y.; Rios, J.D. Real-time neural inverse optimal control for a linear induction motor. *Int. J. Control* **2017**, *90*, 800–812.
17. Rios, J.D.; Alanis, A.Y.; Lopez-Franco, M.; Lopez-Franco, C.; Arana-Daniel, N. Real-time neural identification and inverse optimal control for a tracked robot. *Adv. Mech. Eng.* **2017**, *9*, 1687814017692970.
18. Di Noia, J.M.; Neuberger, M.S. Molecular mechanisms of antibody somatic hypermutation. *Annu. Rev. Biochem.* **2007**, *76*, 1–22.
19. Meyer-Hermann, M.; Mohr, E.; Pelletier, N.; Zhang, Y.; Victora, G.D.; Toellner, K.M. A theory of germinal center B cell selection, division, and exit. *Cell Rep.* **2012**, *2*, 162–174.
20. Zhang, Y.; Meyer-Hermann, M.; George, L.A.; Figge, M.T.; Khan, M.; Goodall, M.; Young, S.P.; Reynolds, A.; Falciani, F.; Waisman, A.; et al. Germinal center B cells govern their own fate via antibody feedback. *J. Exp. Med.* **2013**, *210*, 457–464.
21. Adachi, Y.; Onodera, T.; Yamada, Y.; Daio, R.; Tsuiji, M.; Inoue, T.; Kobayashi, K.; Kurosaki, T.; Ato, M.; Takahashi, Y. Distinct germinal center selection at local sites shapes memory B cell response to viral escape. *J. Exp. Med.* **2015**, *212*, 1709–1723.
22. Haykin, S. *Kalman Filtering and Neural Networks*; Adaptive and Cognitive Dynamic Systems: Signal Processing, Learning, Communications and Control; Wiley: New York, NY, USA, 2004.
23. Wong, J.; Huang, W. "Wheels vs. tracks"—A fundamental evaluation from the traction perspective. *J. Terramech.* **2006**, *43*, 27–42.
24. Siegwart, R.; Nourbakhsh, I.; Scaramuzza, D. *Introduction to Autonomous Mobile Robots*; Intelligent Robotics and Autonomous Agents; MIT Press: Cambridge, MA, USA, 2011.
25. González, R.; Rodríguez, F.; Guzmán, J. *Autonomous Tracked Robots in Planar Off-Road Conditions: Modelling, Localization, and Motion Control*; Springer: Berlin, Germany, 2014.
26. Das, T.; Kar, I.N. Design and implementation of an adaptive fuzzy logic-based controller for wheeled mobile robots. *IEEE Trans. Control Syst. Technol.* **2006**, *14*, 501–510.
27. Moosavian, S.A.A.; Kalantari, A. Experimental slip estimation for exact kinematics modeling and control of a Tracked Mobile Robot. In Proceedings of the 2008 IEEE/RSJ International Conference on Intelligent Robots and Systems, Nice, France, 22–26 September 2008; pp. 95–100.

Article

Trajectory Tracking of an Omni-Directional Wheeled Mobile Robot Using a Model Predictive Control Strategy

Chengcheng Wang [1,2,3], Xiaofeng Liu [1,2,3,*], Xianqiang Yang [4], Fang Hu [5], Aimin Jiang [1,2,3] and Chenguang Yang [6]

1 College of IoT Engineering, Hohai University, Changzhou 213022, China; wangcc_1992@hhu.edu.cn (C.W.); jiangam@hhuc.edu.cn (A.J.)
2 Changzhou Key Laboratory of Robotics and Intelligent Technology, Hohai University, Changzhou 213022, China
3 Jiangsu Key Laboratory of Special Robots, Hohai University, Changzhou 213022, China
4 Research Institute of Intelligent Control and Systems, Harbin Institute of Technology, Harbin 150080, China; xianqiangyang@hit.edu.cn
5 School of Electromechanical and Automobile Engineerng, Changzhou Vacational Institue of Engineering, Changzhou 213164, China; 8000000244@email.czie.net
6 Key Laboratory of Autonomous Systems and Networked Control, College of Automation Science and Engineering, South China University of Technology, Guangzhou 510640, China; cyang@ieee.org
* Correspondence: xfliu@hhu.edu.cn, Tel.: +086-519-85191725

Received: 13 November 2017; Accepted: 30 January 2018; Published: 2 February 2018

Abstract: This paper addresses trajectory tracking of an omni-directional mobile robot (OMR) with three mecanum wheels and a fully symmetrical configuration. The omni-directional wheeled robot outperforms the non-holonomic wheeled robot due to its ability to rotate and translate independently and simultaneously. A kinematics model of the OMR is established and a model predictive control (MPC) algorithm with control and system constraints is designed to achieve point stabilization and trajectory tracking. Simulation results validate the accuracy of the established kinematics model and the effectiveness of the proposed MPC controller.

Keywords: omni-directional mobile robot (OMR); kinematics model; model predictive control; point stabilization; trajectory tracking

---

1. Introduction

Industrial automation is a prerequisite for intelligent manufacturing, and mobile robots represent a core component of industrial automation systems. In recent decades, due to the broad applications for mobile robots, the core issues in motion control of mobile robots, including point stabilization, path planning, trajectory tracking, and real-time avoidance, have attracted considerable attention from both academic scholars and practitioners [1–5].

Although large numbers of mobile robots have begun to enter homes, hospitals, and many other industrial areas, mobile robots with new structures (as opposed to the traditional dual-drive wheeled mobile robots), are required to meet continuously increasing demands with respect to high flexibility, efficiency, and safety of complex and diverse applications in reality. Omni-directional mobile robots (OMRs) with mecanum wheels provide an alternative and have drawn much attention from researchers. In [6,7], two different omni-directional mobile structures with mecanum wheels were proposed. In [7,8], the kinematics of OMRs were analyzed and the motion control of robots was achieved. In [9], a comprehensive omni-directional soccer player robot with no head direction was proposed and it was capable of more sophisticated behaviors, such as ball passing and goal keeping. Compared with the traditional nonholonomic dual-drive wheeled robot, the omni-directional mobile robot is able to synchronize steering and linear motion in any direction. This advantage not only improves the flexibility of the robot greatly in order to achieve fast target tracking and obstacle avoidance, but also provides more references for robot motion control methods.

Motion control is a core issue in the mobile robot system and guarantees the smooth movement of the robot. Due to the diversity of motion and increasing performance requirements, motion control algorithms are becoming more and more complex. In [10], a receding horizon (RH) controller was developed for the tracking control of a nonholonomic mobile robot. In [11], a novel biologically inspired tracking control approach for real-time navigation of a nonholonomic mobile robot was proposed by integrating a backstepping technique and a neurodynamics model. In [12], an approach was proposed for complete characterization of the shortest paths to a goal position and it was used for path planning of the robot. A novel method referred to as the visibility binary tree algorithm for robot navigation with global information was introduced in [13]. The set of all complete paths between the robot and target in the global environment with some circle obstacles was obtained by using the tangent visibility graph. Then, an algorithm based on the visibility binary tree created for the shortest paths was run to obtain the optimal path for the robot. In [14], a real-time obstacle avoidance approach for mobile robots was presented based on the artificial potential field concept. The motion control of omni-directional mobile robot has also been investigated and many results have been reported [15–18]. Among them, the model predictive control (MPC) algorithm is the most common control method used for trajectory tracking of the OMR.

The MPC algorithm is an efficient predictive control technique and is composed of model prediction, rolling optimization, and feedback correction. The MPC algorithm has been widely used in industrial control applications and is able to take various constraints into consideration. The core of MPC control problem finding the optimal solution of a cost function constructed according to the error between system output at the sampling time and the predicted output of the established prediction system model. The optimal solution is taken as the optimal control input in the future prediction time domain and the first control vector of the optimal solution is used as the real control input. The advantages of the MPC algorithm are: (1) It is easy to model; (2) It has a rolling optimization strategy with good dynamic control effect; (3) It can correct the output by feedback which improves the robustness of the control system; (4) As a computer optimization control algorithm, it is easy to realize on a computer. In recent years, the MPC algorithm has also been widely used to achieve optimized motion control of mobile robots. In [19], a novel visual servo-based model predictive control method was proposed to steer a wheeled mobile robot moving in a polar coordinate. In [20], a model-predictive trajectory-tracking controller, which used linearized tracking-error dynamics to predict future system behavior, was presented. In [21], a state feedback MPC method was proposed and applied for trajectory tracking of a three-wheeled OMR. The cost function in the proposed MPC method is constructed over finite horizon and is optimized in the linear matrix inequalities (LMI) framework. In [22], a virtual-vehicle concept and an MPC strategy were combined to handle robot motion constraints and the path-following problem.

In this paper, the kinematic analysis of the OMR is conducted and then an MPC algorithm for point stabilization and trajectory tracking is proposed. In the proposed MPC controller, the rotation speeds of the three wheels and the state outputs of the robot are taken as manipulated variables and controlled variables, respectively. In stabilization control of the robot, a target point and an arbitrary initial direction, which lie within a given constraint range of the coordinate, are set and then the robot tracks the target point and stabilizes itself automatically. In the trajectory tracking of the robot, a pulsed route is planned and different target angles at each node point are set. The robot can move from the start point of the trajectory to the end point along the given route with target angles.

This paper is organized as follows: A brief description and kinematics analysis of the considered OMR are given in Section 2. Section 3 presents the MPC strategy including MPC elements and the optimization method. Simulation results of the proposed MPC algorithm are presented in Section 4. Finally, Section 5 concludes the work in this paper.

2. Omni-Directional Wheeled Mobile Robot Model

2.1. Omni-Directional Mobile Structure

The concept of the mecanum wheel was proposed by a Swedish mecanum company. The mecanum wheel is mainly composed of a motor-controlled wheel hub and some passive rollers, which are evenly distributed at a certain angle along the outer edge of the wheel.

The mecanum wheels used in this paper are light-load 90° omni-directional CL-10 wheels produced by Chengdu Hangfa Hydraulic Engineering Co., Ltd. (Chengdu, China). As shown in Figure 1, the outer diameter of the mecanum wheel is 101.6 mm and it has 16 rollers distributed in both of outer sides. The rollers can rotate in a perpendicular direction to the hub rotation and the envelopes on the outside of these rollers form a cylindrical surface, which enables smooth movement of the wheel. Each roller is equipped with two nylon sliding bearings, which gives the wheels good wear resistance and flexibility. Each wheel can withstand loads of up to 50 kg. Therefore, this series of mecanum wheels is a good choice for lightly-loaded omni-directional wheeled mobile platforms.

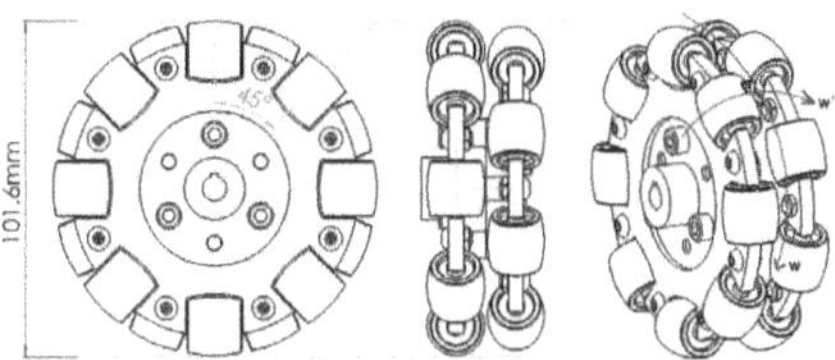

Figure 1. The structure of the mecanum wheel.

As shown in the bottom view of the OMR in Figure 2b, the platform is equipped with three motors and each motor is used to control one mecanum wheel. The axes of these three motors join at the center of the chassis and the angles between the neighboring two motors are the same. In Figure 2a,c, the front view and the overall model of the platform diagram are shown, respectively. Based on the characteristics of mecanum wheel and the analysis of system model, the platform can conduct omnidirectional motion with the advantages of strong adaptability, high sensitivity, good stability, and flexible rotation.

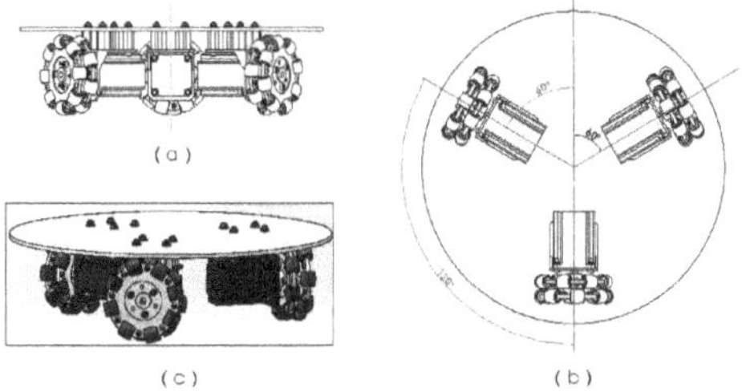

Figure 2. The omni-directional mobile robot (OMR) structure diagram.

2.2. Kinematics Model

For the OMR, each wheel has two position parameters: the position relative to the center of the platform, and the respective attitude angle. The coordinates of the mecanum wheel are shown in Figure 3. In this figure, xoy is the coordinate frame attached to the center of the platform, $x'o'y'$ is a coordinate frame attached to the axis center of one mecanum wheel, i represents the order of the mecanum wheel, α_i is the deflection angle of each roller, β_i is the angle between lines $\overline{oo'}$, and $\overline{ox}$, $(l_{ix}, l_{iy}, \theta_i)$ are the wheel's location and pose, respectively, in the coordinate xoy. l_{ix} and l_{iy} can be expressed as

$$l_{ix} = l_i \cos \beta_i, \ l_{iy} = l_i \sin \beta_i \tag{1}$$

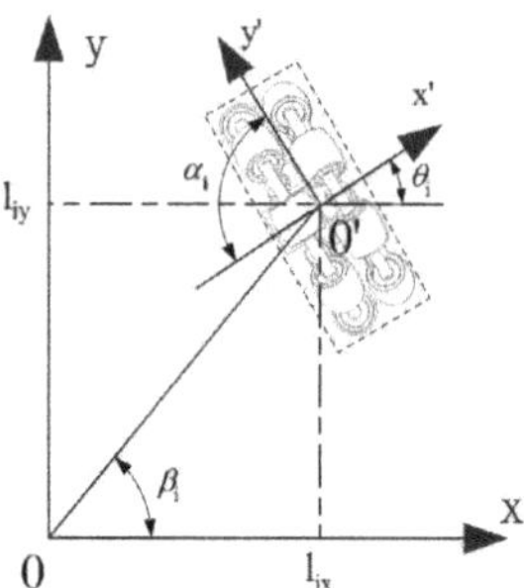

Figure 3. The wheel location and pose in the platform center.

The structure of mecanum wheel is shown in Figure 4. In this figure, r denotes radius of the mecanum wheel, V_{ir} denotes velocity vector of the roller center, and ω_i denotes rotation velocity of the motor. There are three speed vectors. $\left[v_x \ v_y \ \omega\right]^T$ describes the center speed of the platform relative to the ground, $\left[v_{ix} \ v_{iy} \ \omega_i\right]^T$ presents the center speed of the mecanums wheels relative to the coordinate xoy, and $\left[v'_{ix} \ v'_{iy} \ \omega'_i\right]^T$ denotes the center speed of the mecanums relative to the coordinate $x'o'y'$. The kinematic models of the OMR are described as

$$\begin{bmatrix} v'_{ix} \\ v'_{iy} \end{bmatrix} = K_{i1} \begin{bmatrix} \omega_i \\ v_{ir} \end{bmatrix}, \ K_{i1} = \begin{bmatrix} 0 & \sin \alpha_i \\ r & \cos \alpha_i \end{bmatrix} \tag{2}$$

$$\begin{bmatrix} v_{ix} \\ v_{iy} \end{bmatrix} = K_{i2} \begin{bmatrix} v'_{ix} \\ v'_{iy} \end{bmatrix} = K_{i2}K_{i1} \begin{bmatrix} \omega_i \\ v_{ir} \end{bmatrix} \ K_{i2} = \begin{bmatrix} \cos \theta_i & -\sin \theta_i \\ \sin \theta_i & \cos \theta_i \end{bmatrix} \tag{3}$$

$$\begin{bmatrix} v_{ix} \\ v_{iy} \end{bmatrix} = K_{i3} \begin{bmatrix} v_x \\ v_x \\ \omega \end{bmatrix}, \ K_{i3} = \begin{bmatrix} 1 & 0 & -l_{iy} \\ 0 & 1 & l_{ix} \end{bmatrix} \tag{4}$$

Based on Equations (3) and (4) , the inverse kinematics equation of the system can be obtained as

$$K_{i2}K_{i1} \begin{bmatrix} \omega_i \\ v_{ir} \end{bmatrix} = K_{i3} \begin{bmatrix} v_x \\ v_x \\ \omega \end{bmatrix}, i = 1, 2, 3, \tag{5}$$

and $\det(K_{i2}K_{i1}) \neq 0$.

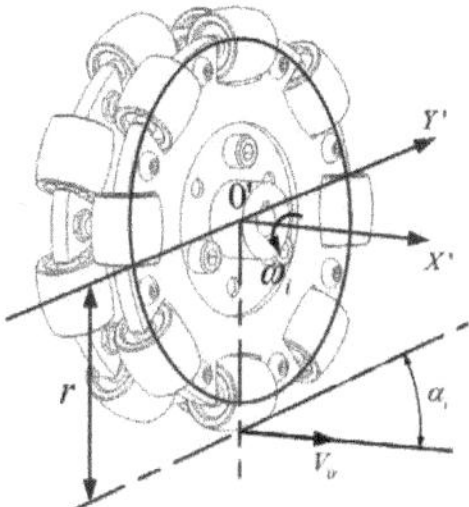

Figure 4. The wheel location and pose in the platform center.

Assuming that

$$K_i = [K_{i1}]^{-1}[K_{i2}]^{-1} * K_{i3},\ i = 1,2,3,\ \gamma_i = \theta_i - \alpha_i$$

then, the inverse kinematics equation of each mecanum wheel is derived as

$$\begin{bmatrix} \omega_i \\ v_{ir} \end{bmatrix} = [K_{i1}]^{-1}[K_{i2}]^{-1}K_{i3} \begin{bmatrix} v_x \\ v_y \\ \omega \end{bmatrix} = K_i \begin{bmatrix} v_x \\ v_y \\ \omega \end{bmatrix} \quad (6)$$

where

$$K_i = \frac{1}{-r\sin\alpha_i} \begin{bmatrix} \cos(\gamma_i) & \sin(\gamma_i) & l_{ix}\sin(\gamma_i) - l_{iy}\cos(\gamma_i) \\ r\cos\theta_i & -r\sin\theta_i & l_{iy}r\cos\theta_i - l_{ix}r\sin\theta_i \end{bmatrix}$$

Assuming that

$$\mathrm{E} = \begin{bmatrix} \frac{\cos(\gamma_1)}{\sin\alpha_1} & \frac{\sin(\gamma_1)}{\sin\alpha_1} & \frac{l_1\sin(\gamma_1-\beta_1)}{\sin\alpha_1} \\ \frac{\cos(\gamma_2)}{\sin\alpha_2} & \frac{\sin(\gamma_2)}{\sin\alpha_2} & \frac{l_2\sin(\gamma_2-\beta_2)}{\sin\alpha_2} \\ \frac{\cos(\gamma_3)}{\sin\alpha_3} & \frac{\sin(\gamma_3)}{\sin\alpha_3} & \frac{l_3\sin(\gamma_3-\beta_3)}{\sin\alpha_3} \end{bmatrix}$$

the rotating speed of mecanum wheels is expressed as

$$\begin{bmatrix} \omega_1 \\ \omega_2 \\ \omega_3 \end{bmatrix} = \frac{1}{-r} E \begin{bmatrix} v_x \\ v_y \\ \omega \end{bmatrix} \quad (7)$$

Based on the relationship $v = r\omega$, Equation (7) can be transformed into

$$v = \begin{bmatrix} v_1 \\ v_2 \\ v_3 \end{bmatrix} = -E \begin{bmatrix} \dot{x} \\ \dot{y} \\ \omega \end{bmatrix} \quad (8)$$

Actually, the Cartesian motion control problems are the main issues in the research on mobile robots. The state of the OMR in Cartesian coordinate $x_c o_c y_c$ is denoted as (x_c, y_c, θ_c) and is shown in Figure 5.

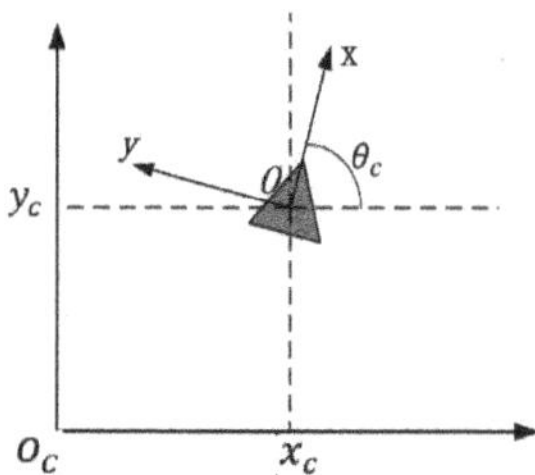

Figure 5. The state of the OMR in Cartesian coordinates.

According to Figure 5, it is derived that

$$\begin{bmatrix} \dot{x}_c \\ \dot{y}_c \\ \omega_c \end{bmatrix} = \begin{bmatrix} \cos\theta_c & -\sin\theta_c & 0 \\ \sin\theta_c & \cos\theta_c & 0 \\ 0 & 0 & 1 \end{bmatrix} \begin{bmatrix} \dot{x} \\ \dot{y} \\ \omega \end{bmatrix} \tag{9}$$

Assuming that

$$H = \begin{bmatrix} \cos\theta_c & -\sin\theta_c & 0 \\ \sin\theta_c & \cos\theta_c & 0 \\ 0 & 0 & 1 \end{bmatrix},$$

the inverse kinematics equation of the OMR in Cartesian coordinate can be transformed into

$$\begin{bmatrix} v_1 \\ v_2 \\ v_3 \end{bmatrix} = -EH^{-1} \begin{bmatrix} \dot{x}_c \\ \dot{y}_c \\ \omega_c \end{bmatrix}. \tag{10}$$

The Jacobian matrix of the inverse kinematics equation of the OMR is

$$R = -EH^{-1} \tag{11}$$

Actually, the Jacobi matrix of the inverse kinematics equation determines the OMR's omni-directionality. If the Jacobi matrix is in the state of dissatisfied rank, there are some nonholonomic constraints in the structure and the degree of freedom movement is reduced, which means that the system cannot achieve all-round movement [23]. It can be derived that the necessary condition to achieve the all-round movement of OMR is $rank\,(R) = 3$.

Through tests, the parameters of the OMR are obtained. They are shown in Figure 6 and Table 1.

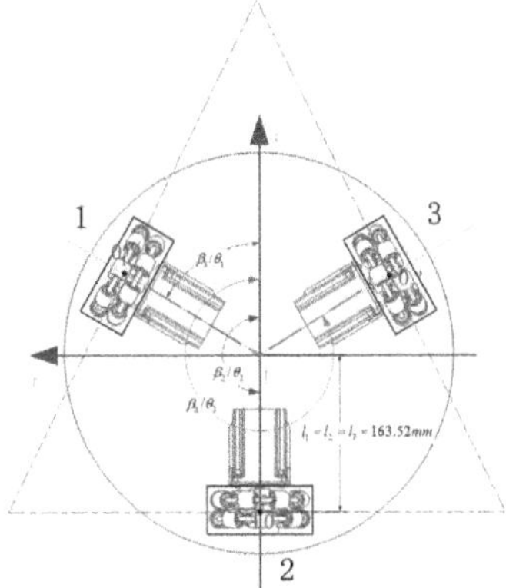

Figure 6. Three-wheeled structure layout of the platform.

Table 1. The parameters of system structure.

Para	β_1	β_2	β_3	θ_1	θ_2	θ_3	α_1	α_2	α_3
Value	60°	180°	300°	60°	180°	300°	90°	90°	90°

Substituting these parameters into Equation (10) yields

$$\begin{bmatrix} v_1 \\ v_2 \\ v_3 \end{bmatrix} = \begin{bmatrix} -0.866\cos\theta_c - 0.5\sin\theta_c & 0.5\cos\theta_c - 0.866\sin\theta_c & 0.16352 \\ \sin\theta_c & -\mathrm{con}\,\theta_c & 0.16352 \\ 0.866\cos\theta_c - 0.5\sin\theta_c & 0.5\cos\theta_c + 0.866\sin\theta_c & 0.16352 \end{bmatrix} \begin{bmatrix} \dot{x}_c \\ \dot{y}_c \\ \omega_c \end{bmatrix} \tag{12}$$

It can be derived from the result that $rank\,(R) = 3$, which means the OMR can achieve all-round movement.

3. MPC Strategy

The MPC algorithm has been widely used in motion control of mobile robots and it has three basic characteristics: model prediction, rolling optimization, and feedback correction. It has unique advantages for solving motion control problems, such as point stabilization and trajectory tracking of mobile robots. Traditional methods often neglect or simplify the robot kinematics and dynamics constraints which have a significant impact on the robot's control performance. Meanwhile, the MPC algorithm takes the constraints into consideration to improve the system robustness. The existing controllers have high requirements with respect to the accuracy of the system model and tend to obtain effective control parameters based on the model and multiple experimental tests;. However, the MPC algorithm has low requirements with respect to the accuracy of the system model and can acquire the control parameters through simple experiments. Through applying the characteristics of rolling optimization and feedback correction, the influence of time-delay in the closed-loop system can be effectively reduced or even eliminated, and the motion control is optimized to achieve improved control performance. Particularly, the MPC algorithm can handle multivariate and constrained problems effectively.

The MPC problem can be described as solving the optimal solution of the cost function. At each sampling time, the outputs of the future N sampling instants are predicted according to the system model, and the cost function is constructed based on the errors between the predicted outputs and the true state outputs of the system. The optimal control inputs of the next N sampling instants are obtained by minimizing the cost function, however, only the first control vector is used as the system input. At the next sampling time, this process is repeated and rolling optimization is performed.

In this section, an MPC-based kinematic controller is designed to ensure that the mobile robot can be driven to the desired position accurately, smoothly, and stably.

Define $q = [x_c \; y_c \; \theta_c]^T$ as the state of mobile robot in coordinate $x_c o_c y_c$ and $S = [S_1 \; S_2 \; S_3]^T = R^{-1}$. Then, it can be derived that

$$\dot{q}(t) = S(t)\, v(t) \tag{13}$$

where

$$S_1(t) = \begin{bmatrix} -0.5774\cos\theta_c(t) - 0.3333\sin\theta_c(t) \\ 0.6667\sin\theta_c(t) \\ 0.5774\cos\theta_c(t) - 0.3333\sin\theta_c(t) \end{bmatrix}^T$$

$$S_2(t) = \begin{bmatrix} 0.3333\cos\theta_c(t) - 0.5774\sin\theta_c(t) \\ -0.6667\cos\theta_c(t) \\ 0.3333\cos\theta_c(t) + 0.5774\sin\theta_c(t) \end{bmatrix}^T$$

$$S_3(t) = \begin{bmatrix} 0.002038 & 0.002038 & 0.002038 \end{bmatrix}$$

Based on zero-order hold (ZOH), a continuous-time system can be transformed into a discrete-time form with a sampling period T.

$$q(k+1) = q(k) + \dot{q}(k)\, T \tag{14}$$

The discrete-time system (14) can be transformed into a discrete state space system as

$$q(k+1) = Aq(k) + Bu(k) \tag{15}$$

where

$$A = I \in \mathbb{R}^{n\times n},\; B = S(k)T \in \mathbb{R}^{n\times m}$$

The parameter n is the number of state variables and m is the number of input variables. The matrix A is a constant identity matrix, and matrix B is determined by orientation of the system dynamically.

Based on the state space model (15), the MPC algorithm is designed to control the system. Here, a cost function is required and optimized through solving a quadratic program (QP) problem to obtain the optimal control sequence in predictive horizon N. The cost function for the MPC can be defined as

$$J(k) = \sum_{j=1}^{N} \left[\|q(k+j|k)\|_Q^2 + \|u(k+j-1|k)\|_R^2 \right] \tag{16}$$

where $Q \in \mathbb{R}^{n\times n}$ and $R \in \mathbb{R}^{m\times m}$ are appropriate weighting matrices. $q(k+j|k)$ and $u(k+j-1|k)$ denote the predicted state and control vector in predictive horizon at time k.

Through solving the following finite-horizon optimal control problem online, the optimal control sequence can be obtained.

$$u^* = \arg\min_u \{J(k)\} \tag{17}$$

Since the torques generated by motors are limited by the performance of the motors, the $u(k)$ has an upper and lower bound, and the range of state variable $q(k)$ is also constrained. Thus, the following constraints should be imposed on the system:

$$u_{min} \leq u\,(k) \leq u_{max} \tag{18}$$

$$q_{min} \leq q\,(k) \leq q_{max} \tag{19}$$

In order to solve the optimal control problem, Equation (17) is transformed into a standard quadratic form which a QP solver can solve. The following prediction vector is defined:

$$\bar{q}(k) = [q\,(k+1|k)\,, \cdots, q(k+N|k)]^T \in \mathbb{R}^{n\times N} \tag{20}$$

$$\bar{u}\,(k) = [u\,(k|k)\,, \cdots, u(k+N-1|k)]^T \in \mathbb{R}^{m\times N} \tag{21}$$

Based on Equation (15), the relationship between vectors $\bar{q}(k)$ and $\bar{u}(k)$ can be described as

$$\bar{q}(k) = Gq(k|k) + L(k)\bar{u}(k) \tag{22}$$

where

$$G = [A, A^2, \cdots, A^N]^T \in \mathbb{R}^{n\times nN}$$

$$L(k) = \begin{bmatrix} \alpha_{11}(k) & 0 & \ldots & 0 \\ \alpha_{21}(k) & \alpha_{22}(k) & \cdots & 0 \\ \vdots & \vdots & \ddots & \vdots \\ \alpha_{N1}(k) & \alpha_{N2}(k) & \ldots & \alpha_{NN}(k) \end{bmatrix} \in \mathbb{R}^{nN\times mN}$$

with $\alpha_{ij}(k)$ defined as

$$\alpha_{ij}(k) = A^{i-j}B(k+j-1),\ i \geq j$$

Two block diagonal matrices are defined:

$$\bar{Q} = \begin{bmatrix} Q & 0 & \ldots & 0 \\ 0 & Q & \ldots & 0 \\ \vdots & \vdots & \ddots & \vdots \\ 0 & 0 & \ldots & Q \end{bmatrix} \in \mathbb{R}^{nN\times nN},\ \bar{R} = \begin{bmatrix} R & 0 & \ldots & 0 \\ 0 & R & \ldots & 0 \\ \vdots & \vdots & \ddots & \vdots \\ 0 & 0 & \ldots & R \end{bmatrix} \in \mathbb{R}^{mN\times mN}$$

Thus, the optimal control problem (17) can be rewritten as

$$\begin{aligned} \arg\min_u J(k) &= min\,\|\bar{q}\,(k)\|_{\bar{Q}}^2 + \|\bar{u}\,(k)\|_{\bar{R}}^2 \\ &= min\,\|Gq(k|k) + L(k)\bar{u}(k)\|_{\bar{Q}}^2 + \|\bar{u}\,(k)\|_{\bar{R}}^2 \end{aligned} \tag{23}$$

subjected to

$$\bar{u}_{min} \leq \bar{u}\,(k) \leq \bar{u}_{max} \tag{24}$$

$$\bar{q}_{min} \leq Gq(k|k) + L(k)\bar{u}(k) \leq \bar{q}_{max} \tag{25}$$

where $\bar{u}_{min}$ and $\bar{u}_{max}$ are the lower and upper bounds of the control vector in predictive horizon, respectively. $\bar{q}_{min}$ and $\bar{q}_{max}$ are the lower and upper bounds of the predictive state vector, respectively.

After some algebraic manipulations, the optimization problem (23) is transformed into a standard quadratic form:

$$min\frac{1}{2}\bar{u}^T(k)W(k)\bar{u}(k) + C^T(k)\bar{u}(k) \tag{26}$$

with

$$W(k) = 2(L^T(k)\bar{Q}L(k) + \bar{R}) \in \mathbb{R}^{mN\times mN}$$
$$C(k) = 2L^T(k)\bar{Q}Gq(k|k) \in \mathbb{R}^{m\times N}$$

where matrix $W(k)$ denotes the quadratic part of the objective function and the vector $C(k)$ describes the linear part.

The QP problem (26) can be solved online to obtain an optimal prediction control sequence. The first control vector of the sequence is applied to the system at time k. In the next sampling instant, this process is repeated to achieve rolling optimization.

4. Simulation Verification

In this section, the simulation is carried out to verify the accuracy of the derived kinematic model of the OMR and the performance of the MPC algorithm. In this simulation, the range of motion for the OMR is set to (-10 m, 10 m) and it serves as the state constraint; The rotation velocity of each mecanum wheel is set to (-2 m/s, 2 m/s), which is treated as the input constraint. Two tests, point stabilization and trajectory tracking, are performed. In the point stabilization test, the OMR moves to a target point and changes the orientation in the moving process. In the trajectory tracking test, the OMR moves along a known pulse trajectory with different target orientations at different nodes.

4.1. Point Stabilization

In this test, the goal and initial states are predefined. Here, the goal and initial position are set to (3 m, 2 m, $\pi/3$ rad) and (0 m, 0 m, 0 rad), respectively. In order to simulate the real movement of the OMR, the noise is added to the output state.

In each prediction period, a control matrix $\bar{u}$ is obtained through solving the QP problem (26). The $\bar{u}$ is the velocity input sequence to control the movement of the OMR. Actually, only the first control vector in $u\,(j)$ is used as the effective velocity input. The velocity input throughout the whole motion process is shown in Figure 7. In this figure, there are two red dotted horizontal lines, which present the velocity constraints. It can be seen from this figure, with the application of MPC algorithm, that the control velocities of three wheels converge to zero gradually and smoothly. Meanwhile, velocity u_3 is affected by the maximum control input constraint at the beginning.

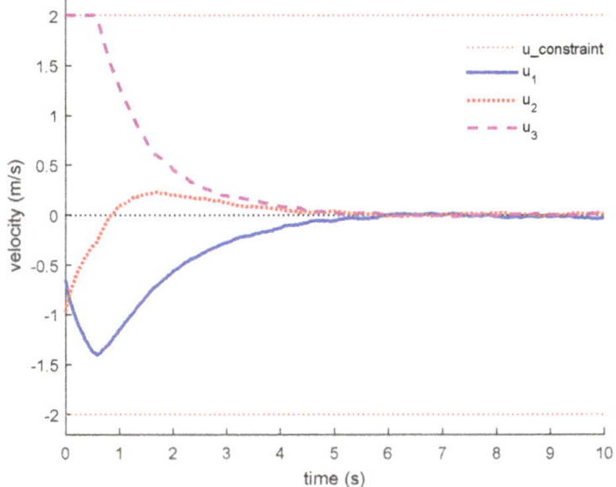

Figure 7. The velocity inputs of three mecanum wheels.

The state trajectories of the OMR in Cartesian coordinates are shown in Figure 8. Two position parameters, x_c and y_c, are given in the upper subfigure of Figure 8. The initial position of (x_c, y_c) is set to (0 m, 0 m) and the target position is set to (3 m, 2 m). The orientation angle θ_c of the OMR in the moving process is presented in the lower subfigure of Figure 8. The initial orientation angle is 0 rad and the target orientation angle is set to $\pi/3$ rad. As shown in Figure 8, the MPC controller can steer the OMR to the target quickly and steadily.

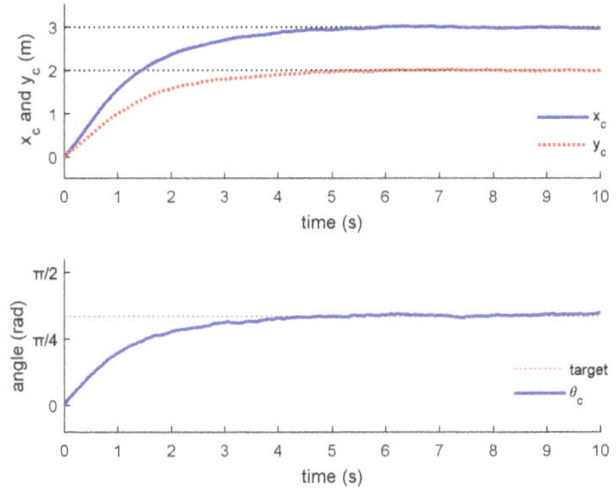

Figure 8. The state trajectories of the OMR in Cartesian coordinates.

The coordinates, (x_c, y_c), of the OMR in moving process are shown in Figure 9. It can be seen from this figure that the movement trajectory is very close to the target trajectory. The tracking error between the movement trajectory and the target trajectory is shown in Figure 10. The maximal tracking error is about 4 cm. From Figures 8 and 10, it can be seen that the OMR moves along the predefined target trajectory to the target point, which demonstrates the effectiveness of the developed MPC algorithm.

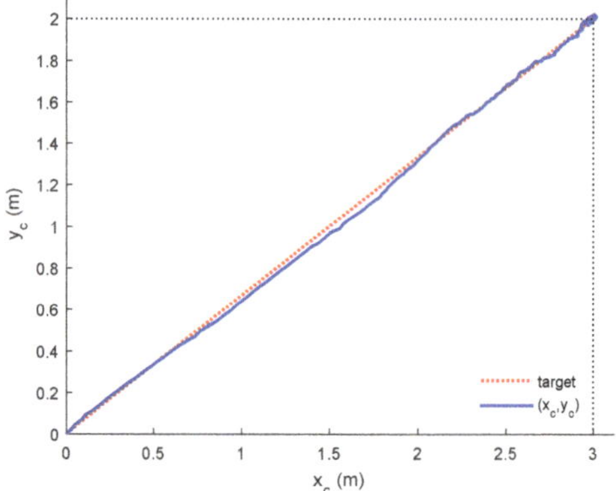

Figure 9. The trajectory coordinates of the OMR.

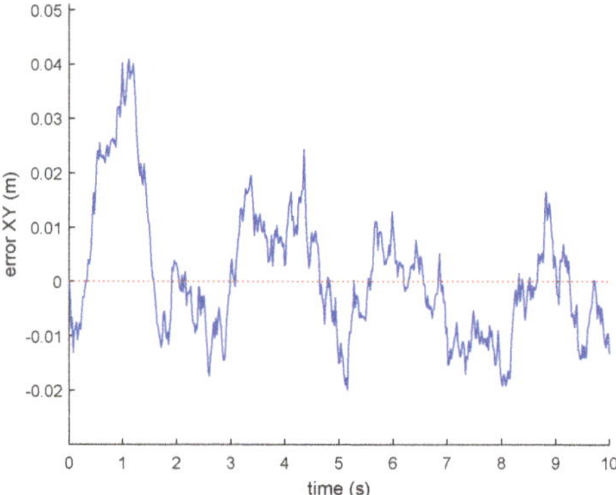

Figure 10. The tracking error.

4.2. Trajectory Tracking

In this test, the trajectory tracking of the OMR along a pulse trajectory is conducted to evaluate the performance of the proposed MPC strategy. The node parameters of the trajectory are shown in Table 2.

Table 2. The node parameters of the tracking trajectory.

Index i	x_c (m)	y_c (m)	θ_c (rad)
1	0	1	0
2	2	1	0
3	2	3	1.57
4	4	3	0
5	4	1	1.57
6	6	1	0

There are six nodes, including the starting point and end point, in the pulse trajectory. The trajectory tracking is achieved by tracking each node according to the position and target direction given in Table 2. In order to simulate the real state of the OMR, the noise is added to the output state.

The linear velocity and angular velocity curves of the OMR in the trajectory-tracking process are shown in Figure 11. It can be seen from this figure that at each node, the axis component of linear velocity fluctuates due to steering action, but converges to zero soon.

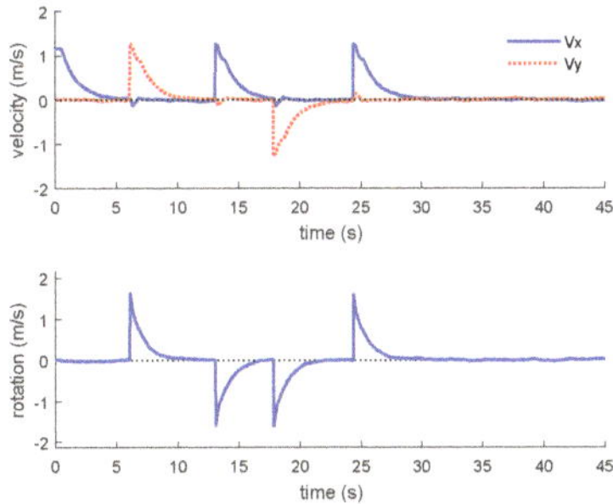

Figure 11. The line velocity and angular velocity of the OMR.

The comparisons of true and tracked trajectory and direction are shown in Figures 12 and 13, respectively. As shown in these figures, the position and angle of the OMR change at the same time and the angle reaches the target faster than the position. The trajectory tracking error is presented in Figure 14. The maximal tracking error of the OMR is about 0.08 m.

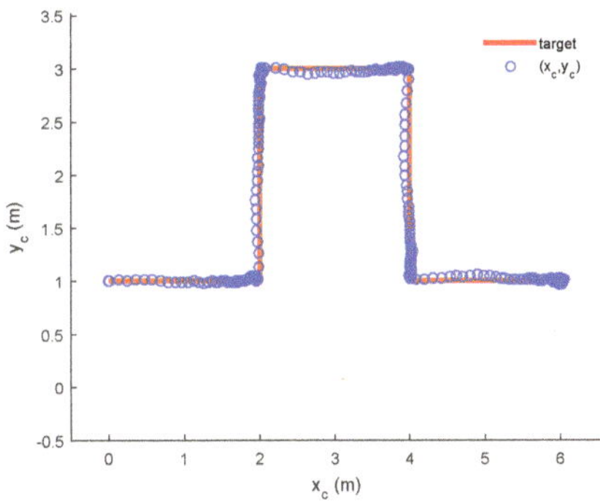

Figure 12. The real trajectory of the OMR.

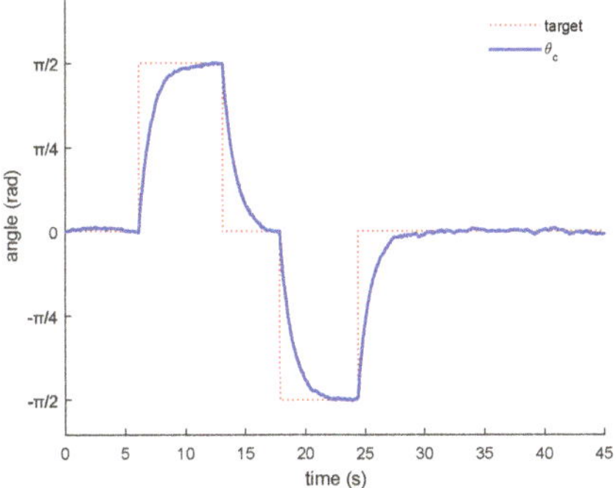

Figure 13. The orientation of the OMR.

The omni-directivity and flexibility of the kinematic structure of the OMR are verified through these two tests. The OMR can achieve movement of any direction without turning, which solves the problem that nonholonomic wheeled mobile robots are not flexible in real-time avoidance. In addition, the results of the first test show that the MPC algorithm can be applied in stabilization control of the robot and the target position and angle can be reached quickly and steadily.

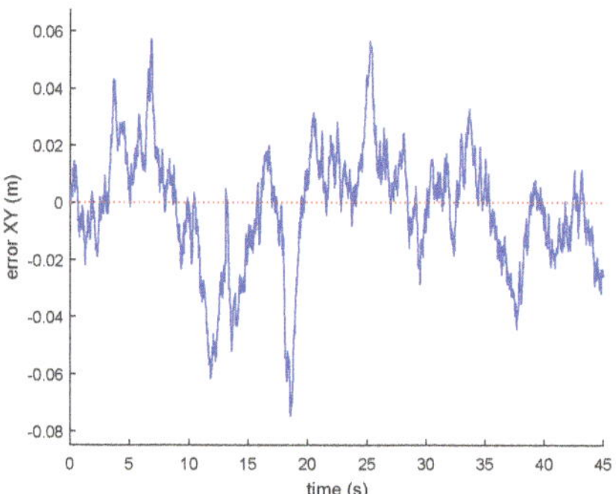

Figure 14. The trajectory offset of the OMR.

5. Conclusions

This paper considers the trajectory tracking of an omni-directional mobile robot with three fully symmetric distribution mecanum wheels. The kinematic analysis of the OMR is conducted and the kinematic model is established. An MPC-based target-tracking motion controller with input and output state constraints is designed and guarantees the optimal solution is effective for the OMR. In addition, the advantages of the omni-directional structure compared with the traditional dual-wheeled structure with nonholonomic constraints are also discussed. The effectiveness of point stabilization and trajectory tracking of the OMR by using the designed MPC controller is demonstrated through two tests.

Acknowledgments: This research was supported in part by the key research and development program of Jiangsu (BE2017071 and BE2017647), projects of international cooperation and exchanges of Changzhou (CZ20170018), and the National Nature Science Foundation (NSFC) under Grant 61473120.

Author Contributions: C.W. and X.L. conceived and designed the simulations and experiments; C.W., F.H. and X.L. performed the numerical simulations and experiments; X.Y., A.J., C.Y. and X.L. wrote the paper.

Conflicts of Interest: The authors declare no conflict of interest.

References

1. Zhang, X.; Fang, Y.; Sun, N. Visual servoing of mobile robots for posture stabilization: From theory to experiments. *Int. J. Robust Nonlinear Control* **2015**, *25*, 1–15.
2. Li, W.; Yang, C.; Jiang, Y.; Liu, X.; Su, C. Motion Planning for Omnidirectional Wheeled Mobile Robot by Potential Field Method. *J. Adv. Transport.* **2017**, 4961383.
3. Micaelli, A.; Samson, C. Trajectory tracking for two-steering-wheels mobile robots. *IFAC Proc. Vol.* **1994**, *14*, 249–256.
4. De Wit, C.C.; Sørdalen, O.J. Exponential stabilization of mobile robots with nonholonomic constraints. *IEEE Trans. Autom. Control* **1992**, *37*, 1791–1797.
5. Serkies, P.J.; Szabat, K. Application of the MPC to the Position Control of the Two-Mass Drive System. *IEEE Trans. Ind. Electron.* **2013**, *60*, 3679–3688.
6. Mohd Salih, J.E.; Rizon, M.; Yaacob, S.; Adom, A.H.; Mamat, M.R. Designing omni-directional mobile robot with mecanum wheel. *Am. J. Appl. Sci.* **2006**, *3*, 1831–1835.

7. Asama, H.; Sato, M.; Bogoni, L.; Kaetsu, H. Development of an omni-directional mobile robot with 3 DOF decoupling drive mechanism. In Proceedings of the IEEE International Conference on Robotics and Automation, Nagoya, Japan, 21–27 May 1995; Volume 2, pp. 1925–1930.
8. Zhao, D.B.; Qiang, Y.J.; Yue, D.X. Structure and Kinematic Analysis of Omni-Directional Mobile Robots. *Robot* **2003**, *25*, 394–398.
9. Samani, H.A.; Abdollahi, A.; Ostadi, H.; Rad, S.Z. Design and Development of a Comprehensive Omni Directional Soccer Player Robot. *Int. J. Adv. Robot. Syst.* **2004**, *1*, 191–200.
10. Gu, D.; Hu, H. Receding horizon tracking control of wheeled mobile robots. *IEEE Trans. Control Syst. Technol.* **2006**, *14*, 743–749.
11. Yang, S.X.; Zhu, A.; Yuan, G.; Meng, Q.H. A Bioinspired Neurodynamics-Based Approach to Tracking Control of Mobile Robots. *IEEE Trans. Ind. Electron.* **2012**, *59*, 3211–3220.
12. Salaris, P.; Fontanelli, D.; Pallottino, L.; Bicchi, A. Shortest paths for a robot with nonholonomic and field-of-view constraints. *IEEE Trans. Robot.* **2010**, *26*, 269–281.
13. Rashid, A.T.; Ali, A.A.; Frasca, M.; Fortuna, L. Path planning with obstacle avoidance based on visibility binary tree algorithm. *Robot. Auton. Syst.* **2013**, *61*, 1440–1449.
14. Khatib, O. Real-Time Obstacle Avoidance for Manipulators and Mobile Robots. *Int. J. Rob. Res.* **1986**, *5*, 90–98.
15. Kalmár-Nagy, T.; D'Andrea, R.; Ganguly, P. Near-optimal dynamic trajectory generation and control of an omnidirectional vehicle. *Robot. Auton. Syst.* **2004**, *46*, 47–64.
16. Kanjanawanishkul, K.; Zell, A. Path following for an omnidirectional mobile robot based on model predictive control. In Proceedings of the IEEE International Conference on Robotics and Automation, Kobe, Japan, 12–17 May 2009; pp. 3341–3346.
17. Huang, H.C.; Tsai, C.C. Adaptive Trajectory Tracking and Stabilization for Omnidirectional Mobile Robot with Dynamic Effect and Uncertainties. *IFAC Proc. Vol.* **2008**, *41*, 5383–5388.
18. Ortíz, J.M.; Olivares, M. Trajectory Tracking Control of an Omnidirectional Mobile Robot Based on MPC. In Proceedings of the IEEE Fourth Latin American Robotics Symposium (Lars 2007), Monterrey, Mexico, 8–9 November 2007.
19. Li, Z.; Yang, C.; Su, C.Y.; Deng, J.; Zhang, W. Vision-Based Model Predictive Control for Steering of a Nonholonomic Mobile Robot. *IEEE Trans. Control Syst. Technol.* **2016**, *24*, 553–564.
20. Klančar, G.; Škrjanc, I. Tracking-error model-based predictive control for mobile robots in real time. *Robot. Auton. Syst.* **2007**, *55*, 460–469.
21. Araújo, H.X.; Conceição, A.G.S.; Oliveira, G.H.C.; Pitanga, J. Model Predictive Control based on LMIs Applied to an Omni-Directional Mobile Robot. *IFAC Proc. Vol.* **2011**, *44*, 8171–8176.
22. Kanjanawanishkul, K. MPC-Based Path Following Control of an Omnidirectional Mobile Robot with Consideration of Robot Constraints. *Adv. Electr. Electron. Eng.* **2015**, *13*, 54–63.
23. Holmberg, R.; Khatib, O. Development and Control of a Holonomic Mobile Robot for Mobile Manipulation Tasks. *Int. J. Robot. Res.* **1999**, *19*, 1066–1074.

Article

Transition Analysis and Its Application to Global Path Determination for a Biped Climbing Robot

Haifei Zhu [1], Shichao Gu [1], Li He [1], Yisheng Guan [1,*] and Hong Zhang [1,2]

1 School of Electromechanical Engineering, Guangdong University of Technology, Guangzhou 510006, China; hfzhu@gdut.edu.cn (H.Z.); xuguqian9@163.com (S.G.); heli@gdut.edu.cn (L.H.); hzhang@ualberta.ca (H.Z.)
2 Department of Computing Science, University of Alberta, Edmonton, AB T6G2H1, Canada
* Correspondence: ysguan@gdut.edu.cn; Tel.: +86-20-3932-2212

Received: 26 December 2017; Accepted: 11 January 2018; Published: 16 January 2018

Abstract: Biped climbing robots are considered good assistants and (or) substitutes for human workers carrying out high-rise truss-associated routine tasks. Flexible locomotion on three-dimensional complex trusses is a fundamental skill for these robots. In particular, the capability to transit from one structural member to another is paramount for switching objects to be climbed upon. In this paper, we study member-to-member transition and its utility in global path searching for biped climbing robots. To compute operational regions for transition, hierarchical inspection of safety, reachability, and accessibility of grips is taken into account. A novel global path rapid determination approach is subsequently proposed based on the transition analysis. This scheme is efficient for finding feasible routes with respect to the overall structural environment, which also benefits the subsequent grip and motion planning. Simulations are conducted with Climbot, our self-developed biped climbing robot, to verify the efficiency of the presented method. Results show that our proposed method is able to accurately determine the operational region for transition within tens of milliseconds and can obtain global paths within seconds in general.

Keywords: transition analysis; global path determination; path planning; biped climbing robot; truss-climbing robot

---

1. Introduction

Trusses typically comprise a number of triangular units constructed with straight rigid members whose ends are connected through joints. In modern architecture, spatial trusses are widely used in the construction of roofs, towers, bridges, and the like. Celebrated buildings include, for example, the Bird's Nest Stadium in Beijing, the Eiffel Tower in Paris, and the Auckland Harbor Bridge. Besides, scaffolds on which workers process the exterior of buildings are also typical spatial trusses. At present, truss-associated routine tasks such as construction, painting, inspection, maintenance, and so on, rely highly on manual labor, signifying a great risk to workers' safety.

Robots are ideal assistants or substitutes for human workers carrying out these high-rise and high-intensity tasks. In the past decades, a number of robots have been developed for climbing on trusses or truss members, including, SM^2 [1], ROMA [2], the brachiating robot [3], TREPA [4,5], WOODY [6,7], Shady3D [8], RiSE [9,10], UT-PCR [11], 3DCLIMBER [12], Treebot [13], the tendril-based bio-inspired robot [14], Climbot [15,16], and the Snake Robot [17]. Configuring different locomotion and attaching mechanisms, these robots differ significantly in mobility and flexibility. In-depth discussions on this topic could be found in [18,19]. Among these robots, SM^2, ROMA, Shady3D, 3DCLIMBER, Treebot and Climbot, having the characteristic of bipedal climbing, are considered to be dominant. These biped climbing robots (BiCRs) generally comprise of an arm-like serial body for locomotion and grippers at both ends for attachment. Benefiting from the bipedal climbing patterns inspired by arboreal primates [20], BiCRs have the flexibility to imitate transition between branches, as illustrated in Figure 1. This transition capability is paramount for robots switching objects to be climbed upon, especially for executing tasks on large-scale spatial complex structures. Another distinct advantage of BiCRs is the combination of manipulation and mobility. Hence, BiCRs are also known as mobile manipulators [1,15]. As a featured example, SM^2 was designed to work on the truss and other exterior surfaces of the Space Station Freedom for performing routine tasks. Besides, the two underactuated miniature climbing robots in [21], MATS [22], Frambot [23] and W-Climbot [24] are also typical BiCRs. Although coupled climbing robots have been developed in the past, most research focused on the system prototype and experimental verification of climbing patterns on simple structures [25,26]. Few attention in the literature has been paid to the perception and planning of this type of robots when traveling in truss environments. In this paper, we study the BiCR climbing path planning problem.

Figure 1. The inspiration for biped climbing robots.

Routine truss-associated tasks usually involve the transition from one member to another. Therefore, the transition issue must be handled well before looking into the climbing path planning problem. BiCR transit was qualitatively described in [2,27,28], but neither quantitative analysis nor executable output can be found in detail. In [29], a BiCR with five degrees of freedom (DoF) was proven capable of transiting between two cylindrical members at any relative orientation. Unfortunately, the distance of the members was not considered in this paper, and it was assumed that the robot could always reach its target member at any time. This is not the case in real situations. Moreover, the issue of where and how to transit was not dealt with, preventing the transition from practice. In [30,31], approaches for determining the graspable region in a climbing cycle were presented. The resulting graspable region, however, corresponds to a fixed grip only, i.e., one end attached at a known point. Theoretically, the entire operational region for transition could be numerically obtained by sampling the gripping point on both members alternatively, solving the graspable region for each and then merging all regions. Despite probabilistic completeness, this is intensive in terms of time and computational resources.

Lacking a close-formed solution for transition analysis, concrete transiting motion is always ignored in studies of path planning on trusses. Hierarchical control structure and multi-phase control strategy were discussed in [32], where only the truss junctions were considered attachable by SM^2. Additionally, the initial and final states of the robot's two feet for each climbing cycle were artificially designated. In other words, neither path planning nor member-to-member transition were required by SM^2. In [2], the path planning of ROMA, a robot designed for inspection applications in 3D environments, was modeled as a classical traveling salesman problem (TSP), taking a climbing step as the smallest unit for the consideration of energy cost. The TSP was finally solved by optimizing the energy consumption. However, only the middle and both ends of a member are allowed to attach for the robot. Similar research was done in [28], with a stricter calculation of the energy consumed in terms of a specific climbing gait. One obvious but important issue in both [2,28] is that the ability of a gripper to grip a member on the expected point was neglected. In the real world, the spatial relationships of gripper fingers and members must be properly considered. Otherwise, the robot may fail in attachment, and thereby fall down. This issue was well handled in [27] by defining a node with its position, direction, and face on a truss. Truss members were dispersed into a number of limited nodes, each representing a discrete gripping point (The term "gripping point" used in this paper consists of the position vector and the orientation matrix to locate a grip in 3D space.). The best path to a destination on the truss was then determined by optimizing the path length and the cost of difficult maneuvers. However, owing to the limitation in locomotion flexibility, Shady3D has only one type of climbing gait, and performs simple transition within a plane cooperating with another unit. Consequently, the path planning of Shady3D robots is simpler than that of a more flexible BiCR with greater DoF. To assemble a truss with multiple Shady3D robots, the truss navigation problem was investigated in [33], where each Shady3D was treated as a single movable point. The trunks and branches of trees form natural truss structures upon which the path planning problem was studied in [34]. A grid map with bottom-up rings comprised by discrete points was used to model the surfaces of the trunk and branch. A sequence of gripping points was afterwards arranged by considering the motion cost, gravity, robot orientation, and its reachability. This method is applicable to climbing robots with nonenclosure gripping, i.e., Treebot, as stated in the paper, but not to BiCRs, such as SM^2 and Climbot, which use enclosure gripping and can flip over the head and tail in climbing. In a word, to the best of our knowledge, few efforts have been made in the field of BiCR path planning on three-dimensional (3D) complex trusses, while complete transition analysis has not yet been reported.

Our Climbot, originally developed in 2007 [15], is a robot designed to carry out high-rise routine tasks. Figure 2 shows the latest version of Climbot and its kinematic diagram. Compared with other BiCRs, Climbot has more climbing gaits and a stronger ability to negotiate obstacles and transit between structural members owing to its agile body and control, transplanted from powerful and light-weight industrial robots.

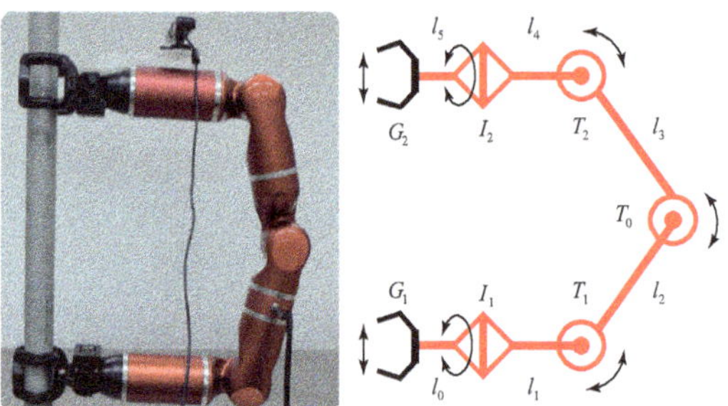

Figure 2. Climbot and its kinematic diagram.

To address the path planning problem of BiCRs in 3D complex truss environments, firstly, we propose a theoretical analysis of the transition between any two given members. The principle of transition analysis was previously presented in [35], and will be further improved by considering the grip accessibility in this paper. As far as we know, our proposed algorithm is the only one capable of determining both the possibility and operational regions for transition in tens of microseconds, considering the safety, reachability, and accessibility of grips. Secondly, we propose a novel path planning algorithm based on transition analysis, for a rapid generation of all feasible global paths on trusses from a start point to the given destination.

The remainder of this paper is organized as follows. We firstly introduce global path planning and transition problems of BiCRs in Section 2. We then analyze the transition requirements and constraints to compute the operational region in Section 3. We apply the transition analysis to fast determination of global climbing paths, with algorithm implementations presented in Section 4. In Section 5, we conduct simulations with Climbot to verify the proposed analysis and algorithms. Finally, we conclude our work in Section 6.

2. Problem Statement

2.1. Global Path for Traveling on Trusses

For BiCRs moving on trusses, path planning determines a sequence of discrete gripping points along the way from the original point to the destination, and the corresponding collision-free motion wags the moving end from one gripping point to another. On a universal 3D complex truss, step-by-step blind exploration is very time- and resource-consuming, resulting in extremely low searching efficiency. In addition, the increase in the number of members will greatly reduce the efficiency of path searching. Global guidance, as a promising solution, is hence important and necessary to avoid inefficient searching. Unlike the path planning problems on 2D or 2.5D terrain, where providing the gradient from the original point to the destination may be sufficient, efficient global guidance for BiCRs traveling on trusses must indicate the entire path, member by member, as shown in Figure 3. Operational regions for performing transition between each pair of members must be also provided, in order to facilitate the subsequent grip planning.

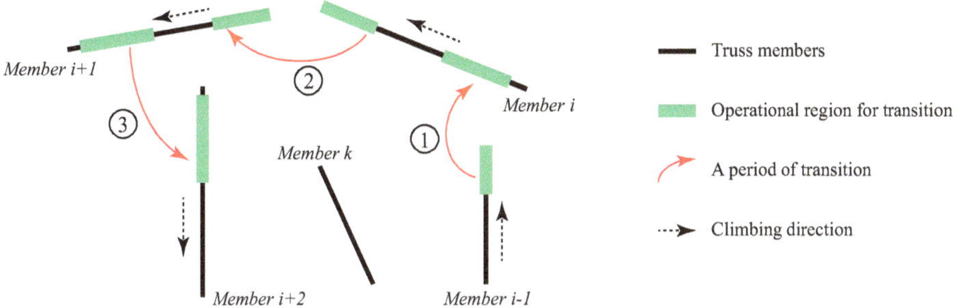

Figure 3. An illustration of global path for BiCRs traveling on a truss.

Since the crucial point of global path planning is to provide global guidance for the subsequent processes, it does not require details such as where to grip and how to move during each climbing cycle, but concentrates on the fast determination of feasible routes based on the overall structural environment. It should be underlined that these feasible routes will be discarded if later processing fails. However, the distinguishing merit is that the searching space is largely narrowed down to a limited number of members with step-by-step guidance by the so-called global path planner. As a result, the searching efficiency is expected to be largely increased.

2.2. Transition from One Member to Another

Operational regions for transition are the junction of a route. Therefore, transition analysis plays a fundamental role in global path planning on trusses. However, transition analysis for BiCRs is challenging due to many constraints, such as robot kinematics, geometries of members, and gripper features.

Basically, a BiPCR performing a transition from *Member 1* to *Member 2* is illustrated in Figure 4 and is described below, supposing the robot originally grips on *Member 1* at the beginning.

(a) The robot starts to move its end effector, i.e., the swinging gripper, towards *Member 2*, as indicated in Figure 4a.
(b) After aligning well with *Member 2*, the robot shifts its swinging gripper to the desired gripping point, as shown in Figure 4b.

(c) The robot grips on *Member 2* with its swinging gripper and holds the two members at the same time, as shown in Figure 4c.
(d) The robot releases the base gripper from *Member 1*. Afterwards the two grippers alternate their roles.
(e) The new swinging gripper moves away from *Member 1*, then towards its new target gripping point. Procedures (d) and (e) correspond to Figure 4d.

Accordingly, a valid transition has the following requirements.

- Safety. The robot must be able to support itself reliably with only one grip, as required in phases (a)–(e), respectively.
- Reachability. The robot must be able to simultaneously grip both members with its two grippers, satisfying kinematic constraints, as illustrated in phase (c).
- Accessibility. Grips on both members must be accessible by corresponding grippers, as required in phases (b) and (e). Each grip's accessibility must be considered in two aspects: when the gripper moves towards the grip, and when the gripper moves away from the grip after alternating its role. Considering the 3/4 envelope pattern used, possibilities for the gripper moving forwards and backwards with respect to a grip in the gripping direction for a specified safe distance, are accounted the corresponding grip's accessibility.

These three criteria form the basis for transition analysis. Regarding safety checks, strict computation considering dynamics is neither necessary nor practical at the global path planning stage, which will be discussed later on. For reachability and accessibility inspection we propose a mathematical model for further analysis.

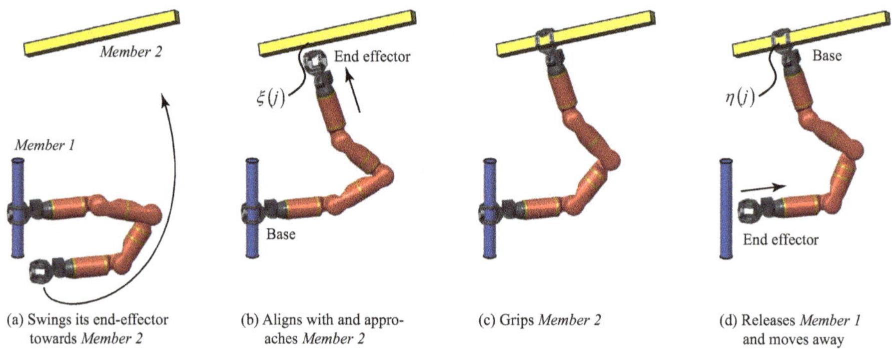

(a) Swings its end-effector towards *Member 2* (b) Aligns with and approaches *Member 2* (c) Grips *Member 2* (d) Releases *Member 1* and moves away

Figure 4. Procedures in the transition from *Member 1* to *Member 2*, illustrated with Climbot.

Without loss of generality, the transition problem can be depicted as a geometric model, as shown in Figure 5. Denote $\{W\}$ as the world coordination frame, and arrange $\{U\}$, $\{V\}$, $\{B\}$ and $\{E\}$ to the reference points of two members, the current and target gripping points, respectively. We use parametric equations to express the gripping positions and transformation matrices to represent grips for BiCRs. Considering reachability and accessibility, we have the following problem statement:

Given:	${}^W_B\boldsymbol{P} = {}^W_U\boldsymbol{P} + t_1{}^W\boldsymbol{d}_1$	Points on *Member 1*
	${}^W_E\boldsymbol{P} = {}^W_V\boldsymbol{P} + t_2{}^W\boldsymbol{d}_2$	Points on *Member 2*
	${}^W_U\mathbf{R}$	Current gripping orientation
	${}^W_V\mathbf{R}$	Goal gripping orientation
	$\boldsymbol{q} = IK({}^B_E\mathbf{T})$	Robot inverse kinematics
To solve:	t_1 and t_2	Operational regions for transition satisfying Equations (1) to (4)

$$ {}^W_B\boldsymbol{T} = \begin{bmatrix} {}^W_U\boldsymbol{R} & {}^W_B\boldsymbol{P} \\ \mathbf{0} & 1 \end{bmatrix}, \quad {}^W_E\boldsymbol{T} = \begin{bmatrix} {}^W_V\boldsymbol{R} & {}^W_E\boldsymbol{P} \\ \mathbf{0} & 1 \end{bmatrix}, \tag{1} $$

$$\exists\, \boldsymbol{q} = IK(\, {}^{W}_{B}\boldsymbol{T}^{-1}\, {}^{W}_{E}\boldsymbol{T}\,), \tag{2}$$

$$\exists\, \boldsymbol{q} = IK(\, {}^{W}_{B}\boldsymbol{T}^{-1}\, {}^{W}_{E}\boldsymbol{T}\, {}^{E}_{E'}\boldsymbol{T}\,), \tag{3}$$

$$\exists\, \boldsymbol{q} = IK(\, \boldsymbol{T}_{Rx}\, {}^{W}_{E}\boldsymbol{T}^{-1}\, {}^{W}_{B}\boldsymbol{T}\, {}^{B}_{B'}\boldsymbol{T}\boldsymbol{T}_{Rx}\,), \tag{4}$$

where ${}^{E}_{E'}\boldsymbol{T}$ and ${}^{B}_{B'}\boldsymbol{T}$ refer to the safe distances for grippers approaching *Member 2* and moving away from *Member 1* linearly. $\boldsymbol{T}_{Rx}\,{}^{E}_{B}\boldsymbol{T}\boldsymbol{T}_{Rx}$ is the operation in order to use the same inverse kinematics as in the case of ${}^{B}_{E}\boldsymbol{T}$, while the robot interchanges grippers for attaching. $\boldsymbol{T}_{Rx}$ is the matrix indicating a pure rotation around the x-axis by π, as

$$\boldsymbol{T}_{Rx} = \begin{bmatrix} \boldsymbol{R}_x(\pi) & \boldsymbol{0} \\ \boldsymbol{0} & 1 \end{bmatrix} = \begin{bmatrix} 1 & 0 & 0 & 0 \\ 0 & -1 & 0 & 0 \\ 0 & 0 & -1 & 0 \\ 0 & 0 & 0 & 1 \end{bmatrix}. \tag{5}$$

Equation (2) reflects the constraints on reachability, while Equations (3) and (4) reflect accessibility.

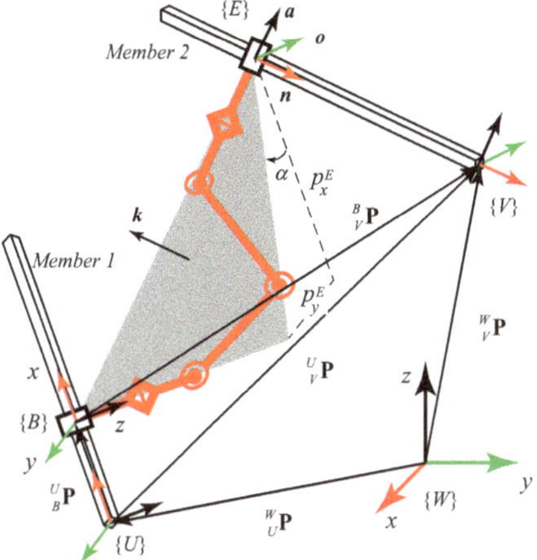

Figure 5. The diagram for transition analysis.

3. Transition Analysis

Transition analysis will be the core component called frequently by the global path planner. In order to rapidly determine whether a member-to-member transition is feasible, hierarchical inspections are conducted in transition analysis. Preliminary requirements for reachability and safety are firstly applied to distinguish apparently infeasible transitions. Strict constraints are then considered to compute the operational region for transition. In this section, we first present the preliminary requirements, and then move on to the strict constraints. The implementation of corresponding algorithms will be presented in Section 4.

3.1. Preliminary Requirements

A transition is definitely infeasible if the target gripping position is out of reach by a BiCR. Denoting D as the distance between two members, the first preliminary requirement is,

$$D \le \sum\nolimits_{i=0}^{n} l_i, \tag{6}$$

where n stands for the degrees of freedom of the robot, and l_i for the lengths of the robot linkages.

Normally, dynamics and gripping force should be analyzed and controlled to ensure the safety of each grip. However, in the global path planning stage, the robot's trajectory is not yet determined and is thereby unknown, regardless of strict verification of grip safety. Observing the geometrical constraints between grippers and members, we use prior knowledge from statics for preliminary safety checks in this paper.

Some BiCRs, for instance, 3DCLIMBER and Climbot mentioned in Section 1, are designed with grippers configuring perpendicular V-shaped grooves as palms. When enclosing members of square or circular cross sections, resulting grips can be classified into four categories according to the circumferential torque required and difficulties for balance, as shown in Figure 6. Suppose the biggest gripping force is always acting on thereby the gripper will never open. Among the above grips, gripping an upright cylindrical or a square member is always safe, no matter how the robot moves, as illustrated in Figure 6a,b. However, when gripping a slanted cylindrical member from an upright direction, as indicated in Figure 6c, the robot has no guarantee of safety, which depends on its motion. More exactly, the robot movement will be limited in an upright plane, in which case there is no circumferential torque generated by gravity. Finally, gripping slanted cylindrical members in other orientations, as illustrated in Figure 6d, is unsafe because the gripper cannot generate sufficient friction force to resist the huge circumferential torque caused by gravity. Taking the Climbot for example to estimate, the gripper must be able to generate a clamping force of 27,000 N to resist the gravity in the worst case. Denoting R as the spatial relationship between *Member 1* and *Member 2*, and $\boldsymbol{d}_G \in \mathbb{R}^3$ as the unit direction vector of the gravity, another preliminary requirement can be expressed as,

$$\neg\{[(\parallel \boldsymbol{d}_1 \times \boldsymbol{d}_G \parallel = 0 \wedge \zeta_1 = 0) \vee (\parallel \boldsymbol{d}_2 \times \boldsymbol{d}_G \parallel = 0 \wedge \zeta_2 = 0)] \wedge R = \textit{staggered} \text{ or } \textit{parallel})\}, \tag{7}$$

where ζ_1 and ζ_2 are flags indicating the cross section type of members. For example, $\zeta_1 = 0$ if the cross section of *Member 1* is cylindrical. Otherwise $\zeta_1 = 1$. In addition, movement constraints could be recorded for future usage in the cases of $R = \textit{intersecting}$ or *collinear*.

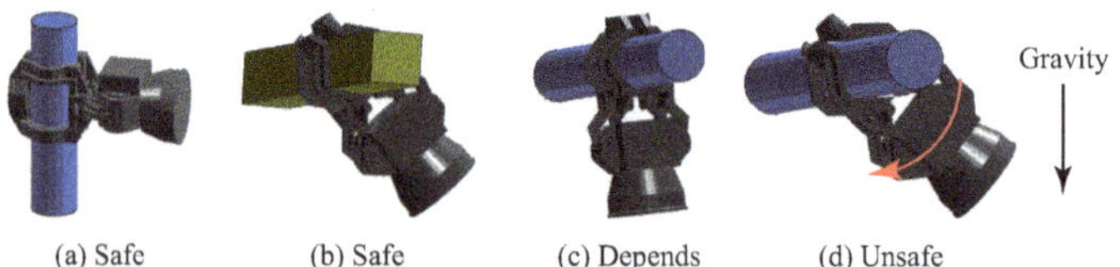

Figure 6. Four categories of grips.

3.2. Strict Constraints

To reach an arbitrary configuration in 3D space, a manipulator needs at least 6 DoF. Hence, a 6DoF BiCR is capable of transiting between any two members with any target orientation. However, with respect to 5DoF BiCRs, such as Climbot, the reachability of its end effector, particularly in terms of orientation, is obviously limited. It has been verified in [31] that 5 DoF in planar configuration is a reasonable trade-off between dexterity and physical limitations such as power of actuators. Accordingly, transitions with 5DoF BiCRs and given members should be analyzed strictly.

From the transition model stated in Section 2.2, for each pair of gripping points, one on *Member 1* and another on *Member 2*, the reachability is equivalent to the existence of solutions when solving the robot inverse kinematics. Motivated by this observation, we first describe *Member 2* with respect to the coordinate frame $\{B\}$. From Figure 5, we can directly write down

$${}^{B}_{V}\boldsymbol{P} = {}^{U}_{V}\boldsymbol{P} - {}^{U}_{B}\boldsymbol{P}, \tag{8}$$

where ${}^{U}_{B}\boldsymbol{P} = \begin{bmatrix} t_1 & 0 & 0 \end{bmatrix}^{\mathrm{T}}$. Then, expressing the origin of $\{V\}$ with respect to $\{U\}$ yields,

$$\begin{bmatrix} {}^{U}_{V}\boldsymbol{P} \\ 1 \end{bmatrix} = {}^{U}_{W}\mathbf{T} \begin{bmatrix} {}^{W}_{V}\boldsymbol{P} \\ 1 \end{bmatrix} = \begin{bmatrix} {}^{W}_{U}\mathbf{R}^{\mathrm{T}} & -{}^{W}_{U}\mathbf{R}^{\mathrm{T}}\,{}^{W}_{U}\boldsymbol{P} \\ \mathbf{0} & 1 \end{bmatrix} \begin{bmatrix} {}^{W}_{V}\boldsymbol{P} \\ 1 \end{bmatrix} = \begin{bmatrix} {}^{W}_{U}\mathbf{R}^{\mathrm{T}}({}^{W}_{V}\boldsymbol{P} - {}^{W}_{U}\boldsymbol{P}) \\ 1 \end{bmatrix}. \tag{9}$$

Substituting Equation (9) into Equation (8), we obtain the reference origin of *Member 2* with respect to $\{B\}$ as,

$${}^{B}_{V}\boldsymbol{P} = {}^{W}_{U}\mathbf{R}^{\mathrm{T}}({}^{W}_{V}\boldsymbol{P} - {}^{W}_{U}\boldsymbol{P}) - {}^{U}_{B}\boldsymbol{P}. \tag{10}$$

The direction unit vector of *Member 2* can be easily transformed to $\{B\}$ as ${}^{B}\boldsymbol{d}_2 = {}^{U}_{B}\mathbf{R}^{-1}\,{}^{W}_{U}\mathbf{R}^{-1}\,{}^{W}\boldsymbol{d}_2$. Thus, we have

$${}^{B}_{E}\boldsymbol{P} = {}^{B}_{V}\boldsymbol{P} + t_2\,{}^{B}\boldsymbol{d}_2. \tag{11}$$

3.2.1. Orientation Constraints

Considering orientation constraints, we propose and prove two propositions which are important for analyzing the member-to-member transition problem with BiCRs, such as Climbot. Note that variable symbols without superscript are used with respect to frame $\{B\}$ hereafter.

Proposition 1. *Given the current base at* $\{B\}$ *and the goal gripping orientation* ${}^{B}_{E}\mathbf{R} = \begin{bmatrix} \boldsymbol{n} & \boldsymbol{o} & \boldsymbol{a} \end{bmatrix}$, *there exists only one feasible configuration for BiCRs, such as Climbot, to transit between two staggered members.*

Proof. Due to the structural configuration of Climbot, its links are always in a plane ("robot plane" for short, highlighted in gray in Figure 5). Therefore, the following orientation constraints must be satisfied,

$$\begin{cases} \boldsymbol{k} \cdot \boldsymbol{a} = 0 \\ \tan\alpha = p^{E}_{y} / p^{E}_{x} \end{cases}, \tag{12}$$

where $\boldsymbol{k} = \begin{bmatrix} -\sin\alpha & \cos\alpha & 0 \end{bmatrix}^{\mathrm{T}}$ represents the norm vector of the robot plane. Substituting Equation (11) and $\boldsymbol{n} = \boldsymbol{d}_2$ into Equation (12) yields

$$t_2 = \frac{p^{V}_{x} a_y - p^{V}_{y} a_x}{n_y a_x - n_x a_y}. \tag{13}$$

On the right side of Equation (13), the values are all constants. Hence, we obtain a unique gripping position for feasible transition by substituting t_2 into Equation (11), when $n_y a_x - n_x a_y \neq 0$. □

It should be underlined that *Member 2* is on the robot plane when $n_y a_x - n_x a_y = 0$. Therefore, all points on *Member 2* satisfy the orientation constraints. In this case, a gripping point on *Member 1* may correspond to unlimited gripping points on *Member 2* for feasible transition. The graspable region determination algorithm presented in [31] could be applied to this case by sampling attaching positions on both members in order to quickly obtain some operational regions for transition.

Proposition 2. *For BiCRs, such as Climbot, to transit between two staggered members, given the base and the goal gripping orientations as* ${}^{W}_{U}\mathbf{R}$ *and* ${}^{W}_{V}\mathbf{R}$, *respectively, the operational region for transition on Member 2 is linear with that on Member 1.*

Proof. Recall Equation (10), and set $\boldsymbol{\lambda} = {}^{W}_{U}\mathbf{R}^{\mathrm{T}}({}^{W}_{V}\boldsymbol{P} - {}^{W}_{U}\boldsymbol{P})$. ${}^{B}_{V}\boldsymbol{P}$ can be written as

$${}^{B}_{V}\boldsymbol{P} = \begin{bmatrix} p^{V}_{x} & p^{V}_{y} & p^{V}_{z} \end{bmatrix}^{\mathrm{T}} = \begin{bmatrix} -t_1 + \lambda_x \\ \lambda_y \\ \lambda_z \end{bmatrix}. \tag{14}$$

Note that since ${}^W_U\mathbf{R}$, ${}^W_V\boldsymbol{P}$, and ${}^W_U\boldsymbol{P}$ are all known invariants, $\boldsymbol{\lambda}$ is actually a constant vector. Substituting Equation (14) into Equation (13) results in

$$t_2 = \frac{(-t_1 + \lambda_x)a_y - \lambda_y a_x}{n_y a_x - n_x a_y} = \sigma t_1 + \delta, \tag{15}$$

where $\sigma = -a_y/(n_y a_x - n_x a_y)$, and $\delta = (\lambda_x a_y - \lambda_y a_x)/(n_y a_x - n_x a_y)$. Both σ and δ are constants. Therefore, t_2 and t_1 are linear, and thereby have one-to-one mapping with each other. Once t_1 is determined, t_2 can be calculated by Equation (15), and vice versa. □

3.2.2. Position Constraints

To confirm whether the transition is feasible, we need to further investigate if ${}^B_E\boldsymbol{P}$ is within the reachable workspace of the robot. Reflected in solving the inverse kinematics, the following inequality must be satisfied,

$$c_3 = \left| \frac{(p'_x c_1 + p'_y s_1)^2 + (p'_z - l_{01})^2 - (l_2^2 + l_3^2)}{2 l_2 l_3} \right| \le 1, \tag{16}$$

where $c_1 \triangleq \cos\theta_1$, $s_1 \triangleq \sin\theta_1$, and $l_{01} \triangleq l_0 + l_1$ (similar expressions are used for shorthand hereafter); $\tan\theta_1 = a_y/a_x$, and $\boldsymbol{P}' = \boldsymbol{P}^V + t_2\boldsymbol{n} - l_{45}\boldsymbol{a}$ represents the position vector of the wrist joint of the robot, e.g., T_2 in Figure 2 for Climbot. Let $\omega_1 = p'_x c_1 + p'_y s_1$, and $\omega_2 = p'_z - l_{01}$, and denote θ_{lim} as the rotation limitation of joint T_0 (see Figure 2) . Since $l_2 = l_3$ for Climbot, Equation (16) can be simplified as

$$2 l_2 \cos\frac{\theta_{lim}}{2} \le \omega_1^2 + \omega_2^2 \le 4 l_2^2. \tag{17}$$

Utilizing t_1 (Equation (15)) to express $\omega_1^2 + \omega_2^2$, we have

$$\omega_1^2 + \omega_2^2 = (A^2 + C^2)t_1^2 + 2(AB + CD)t_1 + B^2 + D^2, \tag{18}$$

where $A = \sigma(n_x c_1 + n_y s_1) - c_1$, $B = (\delta n_x - a_x l_{45} + \lambda_x)c_1 + (\delta n_y - a_y l_{45} + \lambda_y)s_1$, $C = \sigma n_z$, $D = \delta n_z - a_z l_{45} + \lambda_z - l_{01}$.

Let $f(t_1) = \omega_1^2 + \omega_2^2 - 4l_2^2$, and $g(t_1) = \omega_1^2 + \omega_2^2 - 2l_2\cos\frac{\theta_{lim}}{2}$. Equation (17) can be converted to two quadratic functions, as

$$f(t_1) = Et_1^2 + Ft_1 + G \le 0, \tag{19}$$

$$g(t_1) = Et_1^2 + Ft_1 + H \ge 0, \tag{20}$$

where $E = (A^2 + C^2)$, $F = 2(AB + CD)$, $G = B^2 + D^2 - 4l_2^2$, and $H = B^2 + D^2 - 2l_2\cos\frac{\theta_{lim}}{2}$.

In fact, Equation (19) accounts the reachability of the robot, while Equation (20) is the constraint from the rotation limitation of joint T_0. Therefore, discriminants of $f(t_1)$ and $g(t_1)$ determine the feasibility and distribution of operational regions for transition, with the correspondences listed in Table 1.

Table 1. Discriminants corresponding to the distribution of operational regions for transition.

Discriminants [a]	Distribution [b]	Operational Regions
$\Delta_1 < 0$	Out of the workspace	No
$\Delta_1 = 0, \Delta_2 > 0$	Out of the rotation limitation	No
$\Delta_1 = 0, \Delta_2 \le 0$	On the boundary of the workspace	Unique point
$\Delta_1 > 0, \Delta_2 \le 0$	Within the workspace and rotation limitation	One segment
$\Delta_1 > 0, \Delta_2 > 0$	Within the workspace but some points beyond rotation limitation of T_0	Two segments [c]

[a] $\Delta_1 = F^2 - 4EG$ and $\Delta_2 = F^2 - 4EH$ are discriminants of Equations (19) and (20), respectively. [b] The gripping points on *Member 1* satisfying the orientation constraints are described. [c] The two segments have equal lengths and are symmetric about $-F/(2E)$.

Solving Equations (19) and (20), the operational regions on *Member 1* and *Member 2*, respectively, satisfying Equation (2) could be obtained. Similar processes should be conducted for Equations (3) and (4), to discard those reachable but not accessible gripping points from the obtained operational regions.

3.2.3. Length Constraints

Since a real member has limited length, we need to verify whether the obtained operational regions are on the given members or not. In other words, t_1 and t_2 have to satisfy

$$\begin{cases} 0 \le t_1 \le L_1 \\ 0 \le t_2 \le L_2 \end{cases}, \tag{21}$$

where L_1 and L_2 stand for the lengths of the two members, respectively.

An effective way to constrain the operational regions on members is by mapping the length limitation of *Member 2* to t_1 using Equation (15),

$$\begin{cases} \underline{t_1'} = \min\left(-\frac{\delta}{\sigma}, \frac{L_2-\delta}{\sigma}\right) \\ \overline{t_1'} = \max\left(-\frac{\delta}{\sigma}, \frac{L_2-\delta}{\sigma}\right) \end{cases}. \tag{22}$$

Supposing the operational regions on *Member 1* from Section 3.2.2 are $\left[\underline{t_1}, \overline{t_1}\right]$, the real operational regions for the base gripper satisfying constraints including the reachability, accessibility, rotation limitation of T_0, and length limitations become

$$t_1 = \left[0, \ L_1\right] \bigcap \left[\underline{t_1}, \ \overline{t_1}\right] \bigcap \left[\underline{t_1'}, \ \overline{t_1'}\right]. \tag{23}$$

The corresponding operational regions for the swinging gripper on *Member 2* can be updated by Equation (15).

3.2.4. Other Constraints

Besides the above constraints, if the rotation limitations of joints T_1 and T_2 are also taken into account, we additionally need to check whether their rotation is in a valid range. This could be done through solving the inverse kinematics when gripping endpoints of t_1 from Equation (23). Those parts of t_1 and t_2 corresponding to invalid movements should be discarded. An effective method to accomplish this test can be found in [31].

Finally, we obtain accurate and complete operational regions. Holding points involved in these regions, the robot is able to perform the transition from a member to another. Naturally, if the operational region is empty, there is no feasible transition. Note that the presented transition analysis is mainly based on the existence of solutions to a set of quadratic functions. Therefore, it is capable of obtaining the compete solution if the operational region exists for BiCRs, such as Climbot, or with fewer degrees of freedom.

4. Fast Determination of Feasible Global Paths

Making use of the proposed transition analysis, we present a novel effective and efficient path planning approach in this section to address the global route fast determination problem for BiCRs moving on spatial trusses.

4.1. Principle and Flowchart

The basic idea is to find out all the feasible routes from the current state based on our transition analysis in Section 3, and then reserve only those terminated at the expected destination. Tree data structure is a good choice to organize the route exploration and record all the feasible routes. Each node of a tree stores the information of a member and associated operational regions for transiting to from its parent node (the former member). Let the member the robot initially grips on be the root node and ensure the leaf nodes represent the destination member. In this way, each route from the root to the leaf forms a global path from the original point to the destination, containing the necessary information for transition, i.e., members and operational regions. Figure 7 shows us the flowchart.

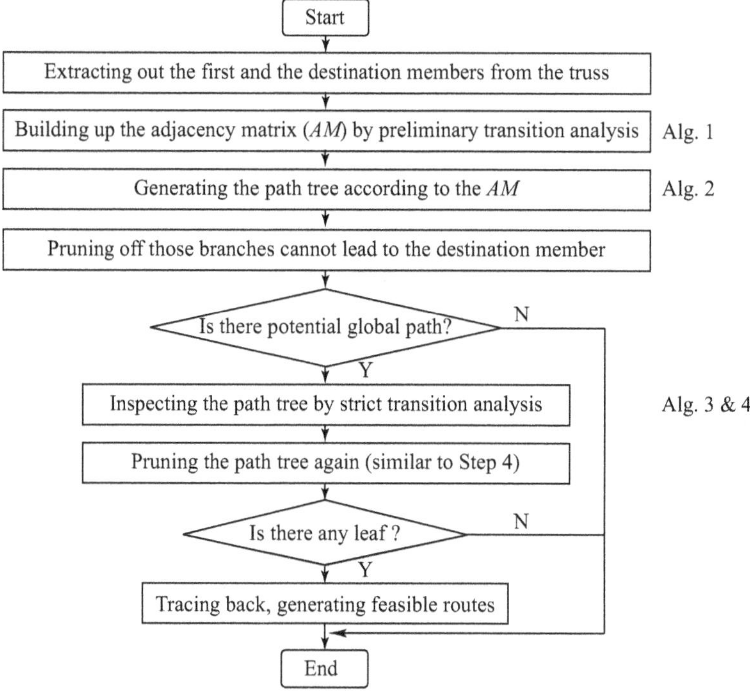

Figure 7. The flowchart of global path planning.

First of all, the original and destination members are extracted by checking which members the points ${}^{W}_{B}\boldsymbol{P}_{init}$ and ${}^{W}_{B}\boldsymbol{P}_{goal}$ are located on.

Secondly, all the members are checked in pairs with preliminary requirements in Equations (6) and (7). The results are stored in an adjacency matrix, as shown in Algorithm 1. Elements of this matrix refer to the preliminary possibility of transition between two members indicated by the row and the column indexes.

Thirdly, a tree data structure (path tree for short) rooted at the original member is then built up according to the adjacency matrix and the following rules, as in Algorithm 2.

- Never go backwards.
- Stop going forwards only when either the destination member is reached or there is no member that it has never been to.

Those branches not terminated at the destination member are pruned off promptly in order to keep the tree data structure on a small scale, which is helpful for improving the solving efficiency.

Fourthly, the above path tree will be traversed and inspected by analyzing strict constraints for transition presented in Section 3.2 and shown in Algorithm 3. This process starts from the root and spreads to the leaves. In other words, each route from the root to any of leaves will be inspected.

Branches will be grafted onto the tree if more than one possible gripping orientations result in operational regions for transition. Conversely, a branch will be pruned off once its corresponding transition is verified to be impossible. It should be noted that the pruning operation must ensure that the tree leaves can only be the destination member at this moment. After traversing the entire tree, the remaining routes if any, are feasible global paths.

Finally, feasible global paths are obtained by tracing back from the leaves to the root. Each route consists of a sequence of members with specific gripping orientations and the corresponding operational regions for transition.

Algorithm 1: Building up the adjacency matrix

Input: ${}^{W}\mathbf{S}$: the truss;
l_i: lengths of robot linkages.
Output: AM: the adjacency matrix.
$N \Leftarrow$ NUMBEROFMEMBERS$({}^{W}\mathbf{S})$
for $i = 1$ **to** $N - 1$ **do**
  for $j = i + 1$ **to** N **do**
    $M_i \Leftarrow$ GETMEMBER$({}^{W}\mathbf{S},\ i)$;
    $M_j \Leftarrow$ GETMEMBER$({}^{W}\mathbf{S},\ j)$
    $D \Leftarrow$ DISTANCE$(M_i,\ M_j)$;
    $R \Leftarrow$ RELATIONSHIP$(M_i,\ M_j,\ {}^{W}\mathbf{G})$;
    if D and R satisfy Equations (6) and (7) **then** // Transition may be possible
      $AM(i,\ j) \Leftarrow$ **true**, $AM(j,\ i) \Leftarrow$ **true**;
    else // Transition is definitely impossible
      $AM(i,\ j) \Leftarrow$ **false**, $AM(j,\ i) \Leftarrow$ **false**;
    end if
  end for
  $AM(i,\ i) \Leftarrow$ **false**;
end for

Algorithm 2: Generating the path tree

Input: AM: the adjacency matrix;
$\chi(i)$: the member the robot moving on in current iteration;
$\chi(m)$: the destination member;
T: the path tree from last iteration.
Output: T: the path tree updated in each iteration.
$M^{potential} \Leftarrow$ GETPOTENTIALMEMBERS$(AM,\ \chi(i))$;
$M^{passed} \Leftarrow$ GETPASTMEMBERS$(T,\ \chi(i))$;
for $j = 1$ **to** NUMBEROFMEMBERS$(M^{potential})$ **do**
  if $M_j^{potential} = \chi(m)$ **then** // Reach the destination member
    return T.ADDNODE$(T,\ \chi(m))$;
  else // Continue to explore
    if $M_j^{potential} \notin M^{passed}$ **then** // New potential member never visited
      T.ADDNODE$(T,\ M_j^{potential})$;
      call Algorithm 2$(AM,\ M_j^{potential},\ \chi(m),\ T)$; // Recursively call the algorithm itself
    end if
  end if
end for

Algorithm 3: Inspecting the path tree

Input: ${}^{W}\boldsymbol{S}$: the truss;
${}^{W}_{B}\boldsymbol{R}_{init}$: initial gripping orientation for base gripper;
${}^{W}_{B}\boldsymbol{R}_{goal}$: final gripping orientation for base gripper;
T: the pruned path tree from Algorithm 2.
Output: T^{ori}: path tree with operational regions and corresponding gripping orientations.

$iter \Leftarrow T$. BreadthFirstIterator( ); // Generate iterators to traverse nodes
T^{ori}. Initialize(${}^{W}_{B}\boldsymbol{R}_{init}$, T. GetRoot( ));
$i \Leftarrow 2$;
while $i <$ Number($iter$) **do**
  $N^{Parent} \Leftarrow T$. GetParentNode($iter(i)$);
  $M1 \Leftarrow$ GetMember(${}^{W}\boldsymbol{S}, N^{Parent}$);
  $M2 \Leftarrow$ GetMember(${}^{W}\boldsymbol{S}$, T. GetNode($iter(i)$));
  for $j = 1$ **to** T. NumberOfParents($iter(i)$) **do**
    ${}^{W}_{U}\mathbf{R} \Leftarrow T$. GetGripOrient($iter(i)$, j); // Extract corresponding gripping orientation
    if T. IsLeaf($iter(i)$) **then** // Reach the destination member
      ${}^{W}_{V}\mathbf{R} \Leftarrow {}^{W}_{B}\boldsymbol{R}_{goal}$, $K = [1]$;
    else if $\zeta_{M2} = 0$ **then** // Transition to a cylindrical member
      ${}^{W}_{V}\mathbf{R} \Leftarrow$ SelectOrientation($M2$), $K = [1]$;
    else // Transition to a squared member
      ${}^{W}_{V}\mathbf{R} \Leftarrow T$. InitGripOrientation($M2$);
      $K = [1, 2, 3, 4]$;
    end if
    for each k in K **do** // Check each potential gripping oirentation
      $(t_1, t_2) \Leftarrow$ Algorithm 4($M1$, $M2$, ${}^{W}_{U}\mathbf{R}$, ${}^{W}_{V}\mathbf{R}$);
      if $t_1 \neq \emptyset$ **then** // Trasisition is feasible
        T^{ori}. AddNode($iter(i)$, ${}^{W}_{V}\mathbf{R}$, t_1, t_2);
      end if
      ${}^{W}_{V}\mathbf{R} \Leftarrow \boldsymbol{R}_n(\pi/2){}^{W}_{V}\mathbf{R}$; // Prepare for another gripping orientation
    end for
  end for
  $i++$;
end while

4.2. Algorithms

Algorithm 1 shows us the implementation of building up the $N \times N$ symmetric adjacency matrix. Each element of the matrix stores a binary value, indicating whether a pair of members is transitable or not, i.e., the transition between *Member i* and *Member j* is possible if $AM(i,j) = true$ and impossible if $AM(i,j) = false$. The probabilities for transition between each pair of members are inspected successively, according to the preliminary requirements concerning about the reachability and safety, i.e., Equations (6) and (7). Binary values are then assigned to corresponding elements.

Algorithm 2 implements the generation of the path tree according to the adjacency matrix from Algorithm 1, and the original and destination members. Algorithm 2 is actually a recursive function that calls itself iteratively, so that the path tree "grows up" from the root to leaves step by step. During each iteration, it extracts all transitable members from the current one, removes those visited, then adds the remaining members to the path tree. Such an iteration repeatedly goes on until the robot arrives at the destination member or has no new members to visit. The obtained path tree should be "scanned" and pruned if necessary, keeping only those branches with the destination member as leaves.

Algorithm 3 traverses the obtained path tree in a breadth-first sequence, inspecting all transitions from a parent node to the present one with strict constraints. For squared members, all four gripping directions will be sent to Algorithm 4 for inspection. This is because a squared member has four potential gripping directions for a V-shaped palm gripper, which are around its center axis by an angular interval of $\pi/2$. As a result, new branches may be grafted onto the path tree if more than one gripping directions result in operational regions for transition. Regarding the transit to an upright cylindrical member, the possible gripping directions are unlimited, as demonstrated in Equation (13). In this case, we can select an optimal gripping direction, for example to balance the difficulties of two consecutive transitions, i.e., from the parent node to the present one, and from the present node to the child one.

Algorithm 4: Solving operational regions for transition

Input: ${}^{W}_{U}\mathbf{P}$: reference point of *Member 1*;
${}^{W}_{V}\mathbf{P}$: reference point of *Member 2*;
${}^{W}_{U}\mathbf{R}$: gripping orientation on *Member 1*;
${}^{W}_{V}\mathbf{R}$: gripping orientation on *Member 2*.
Output: t_1: operational region for transition on *Member 1*;
t_2: operational region for transition *Member 2*;
1: calculate intermediate variables: $\sigma, \delta, A \sim H$;
2: $\Delta_1 \Leftarrow F^2 - 4EG, \Delta_2 \Leftarrow F^2 - 4EH$;
3: **if** $\Delta_1 < 0$ or ($\Delta_1 = 0$ and $\Delta_2 > 0$) **then** // No operational region for transition
4: **return** $t_1 \Leftarrow \varnothing, t_2 \Leftarrow \varnothing$;
5: **else if** $\Delta_1 = 0$ **then** // Unique configuration
6: $t_1 \Leftarrow -F/(2E)$;
7: **else if** $\Delta_1 > 0$ and $\Delta_2 \leq 0$ **then** // A segment of operational region
8: $t_1 \Leftarrow \left[\frac{-F-\sqrt{\Delta_1}}{2E}, -\frac{F+\sqrt{\Delta_1}}{2E}\right]$;
9: **else if** $\Delta_1 > 0$ and $\Delta_2 > 0$ **then** // Two segments of operational regions
10: $t_1 \Leftarrow \left[\frac{-F-\sqrt{\Delta_1}}{2E}, \frac{-F-\sqrt{\Delta_2}}{2E}\right] \cup \left[\frac{-F+\sqrt{\Delta_2}}{2E}, \frac{-F+\sqrt{\Delta_1}}{2E}\right]$;
11: **end if**
12: check accessibility according to Equations (3) and (4);// Similar to Line 1 to 11
13: restrict t_1 according to Equation (23);
14: check rotation limitations of other joints and modify t_1;
15: **if** $t_1 \neq \varnothing$ **then** // Transition feasible
16: $t_2 \Leftarrow \sigma t_1 + \delta$;
17: **else** // Transition infeasible
18: $t_2 \Leftarrow \varnothing$;
19: **end if**

Algorithm 4 details the procedure to compute operational regions for each to-be-inspected transition, dispatched from Algorithm 3. The reachability is firstly taken into account to generate one or two segments of regions on *Member 1* for further inspection, according to Table 1. After that, the resulting operational regions will be shortened if (a) points are not accessible; (b) points are not on *Member 2*; or (c) corresponding movement is beyond the robot joints' rotation limitation. Operational regions on *Member 2* will be finally computed according to Equation (15).

5. Simulations

To verify the proposed analysis and algorithms, simulations are conducted with Climbot. A simulation environment is developed and algorithms are implemented on the platform of MATLAB R2015b. All the simulations are launched on a desktop with Intel Core i7-7700K CPU and 16 GB RAM, running with the 64-bit operating system Windows 10 Pro.

5.1. Result of Transition Analysis

The first part of the simulations is to verify the effectiveness of transition analysis. In this simulation, given two arbitrary squared members in spatial environments, the aim is to solve the operational regions for transition. The two members used for illustration in this paper are

$$^{W}S = \begin{cases} 1187\ 372\ 692\ 428\ 150\ 878\ 1\ 0.5 \\ 225\ 1182\ 1331\ 550\ 807\ 603\ 1\ 9 \end{cases}, \tag{24}$$

where each row represents a member, with the parameters $\boldsymbol{P}_1 \in \mathbb{R}^3$, and $\boldsymbol{P}_2 \in \mathbb{R}^3$ successively standing for positions of end points, and $\zeta \in \mathbb{R}^1$ and $\psi \in \mathbb{R}^1$ indicating the cross-section type and the rotating angle around the member's own axis, respectively.

Figure 8 shows the simulation result of transition analysis. Without consideration of the accessibility of grips, there are four possible cases for transiting from *Member 1* to *Member 2*, with different gripping orientations on the two members, respectively, as shown in Figure 8a. The operational regions are highlighted in green to distinguish them from the other parts of the member. Robots are displayed in specific configurations corresponding to the boundaries of operational regions. Taking the accessibility of grips into account, there remain only two possible cases for transition, as shown in Figure 8b. Moreover, the obtained operational regions in these two possible cases are narrowed down. Shortened parts, painted in red in Figure 8b, must be discarded, owing to corresponding grips being reachable but not accessible.

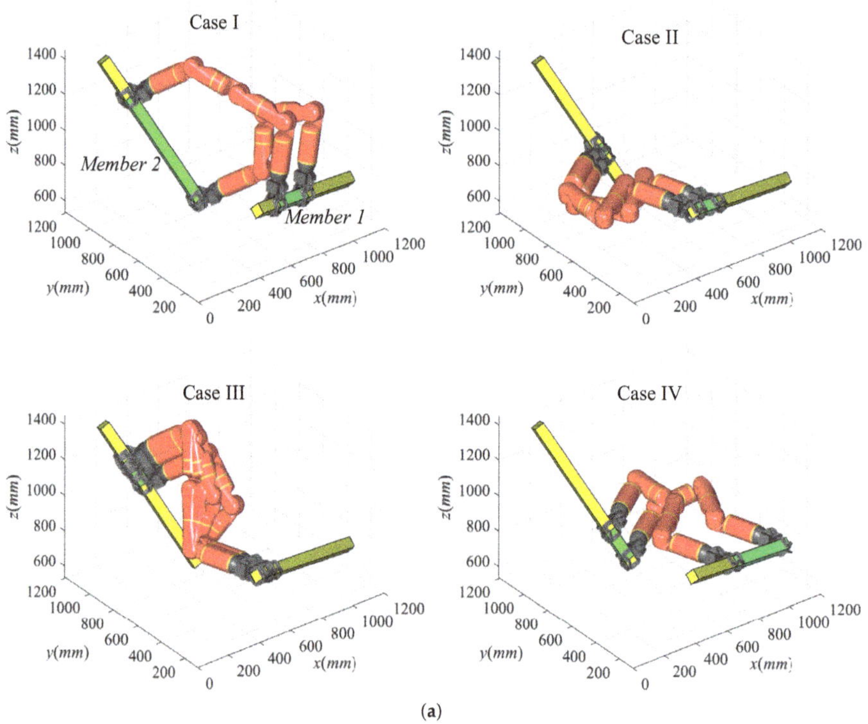

Figure 8. *Cont.*

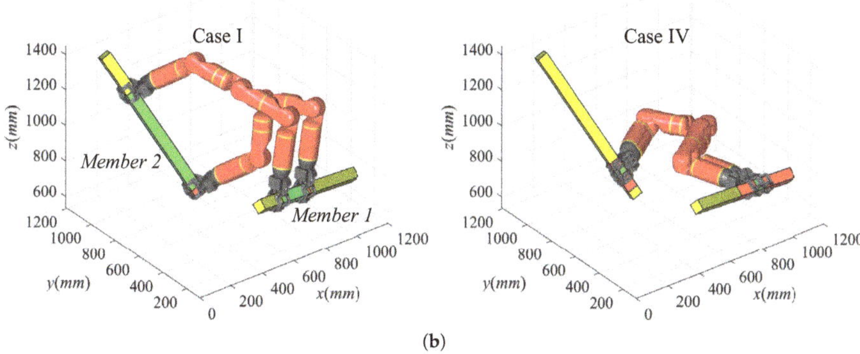

(b)

Figure 8. Results of member-to-member transition analysis: (**a**) without the consideration of grip accessibility; (**b**) considering grip accessibility (no operational region for the cases II and III shown in Sub-figure (**a**)).

Table 2 quantitatively compares the operational regions for transition, computed with and without a consideration of the robot's accessibility. From the table, if the accessibility is taken into account, all the operational regions in four cases are shortened, with a percentage from 7.3% to 100%. To further investigate the importance of considering accessibility, we conducted 1000 rounds of comparative transition analysis with randomly-generated members. In total 94.6% of operational regions were narrowed down, with an average shortening percentage of 59.9% at a time cost of 32.7 ms (over 19.6 ms if we do not account for accessibility). In fact, configurations corresponding to the boundaries always suffer from singularity or joint rotation limitation. When planning specific grips subsequently, optimal grips should keep a distance from the boundaries of operational regions. Therefore, it is necessary to consider accessibility in addition to reachability for each grip.

Table 2. Comparison of transition analysis results with and without the consideration of accessibility.

Characteristics	Operational Regions [a]		Shortening Percentage [b]
	On *Member 1*	On *Member 2*	
	Case I		
Without	[408, 653.7]	[881.0, 229.35]	7.3%
With	[426.7, 653.7]	[831.4, 229.3]	
	Case II		
Without	[628.6, 801.0]	[600.0, 535.0]	100.0%
With	\	\	
	Case III		
Without	[694.0, 742.6] ∪ [760.9, 770.1]	[760.9, 770.1] ∪ [230.5, 212.2]	100.0%
With	\	\	
	Case IV		
Without	[0, 470.6]	[837.0, 660.0]	84.5%
With	[252.0, 325.0]	[742.0, 714.5]	

[a] Represented with parameters of t_1 and t_2, respectively. [b] Computed as $1 - \Delta t_1' / \Delta t_1$, where $\Delta t_1'$ and Δt_1 refer to operational regions with and without the consideration of accessibility.

5.2. Result of Global Path Determination

The second part of the simulation is to verify the proposed algorithms used to determine global paths. Two virtual truss scenes, simple and complicated, consisting of 7 and 25 members, respectively, were deployed for validation. Figures 9 and 10 show us the results. In the simulation, the starting point and the destination are specified manually but arbitrarily, highlighted with a green and a red sphere, respectively. The feasible global routes are illustrated with member-to-member transitions. Corresponding operational regions for transition are also highlighted with green. Nearby numbers indicate the sequence of transitions to be performed along the way. Sub-figures on the right side are simplified diagrams of routes for easy understanding. Arrows indicate the directions Climbot goes forwards from the starting point to the destination.

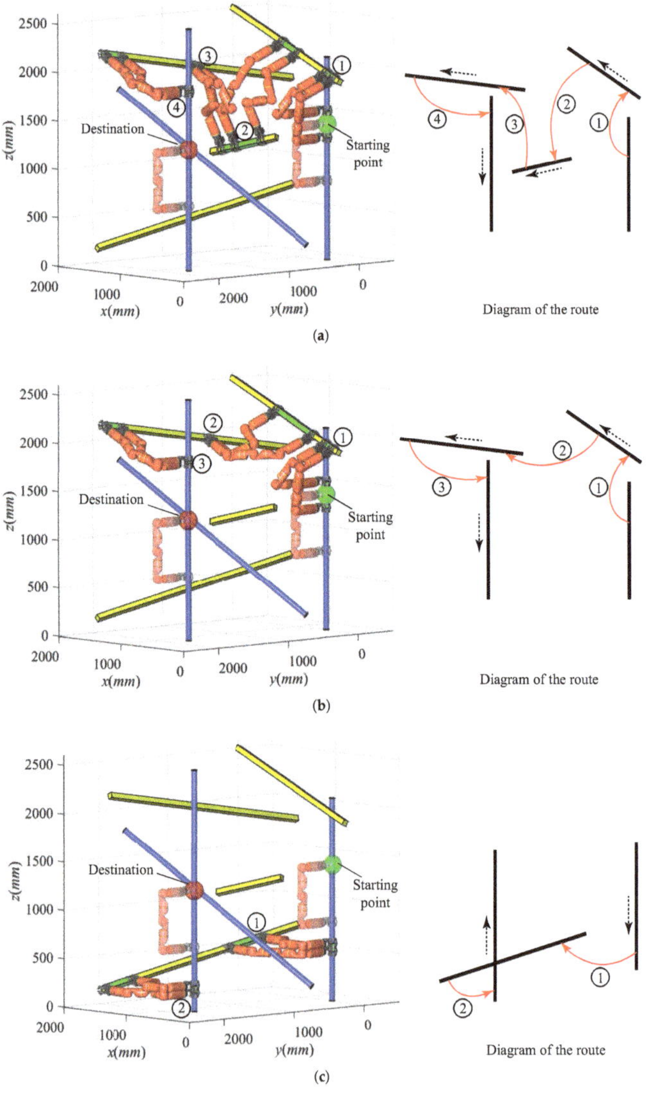

Figure 9. Results of global path determination in Scene I consisting of 7 members. (**a**) Route I; (**b**) Route II; (**c**) Route III.

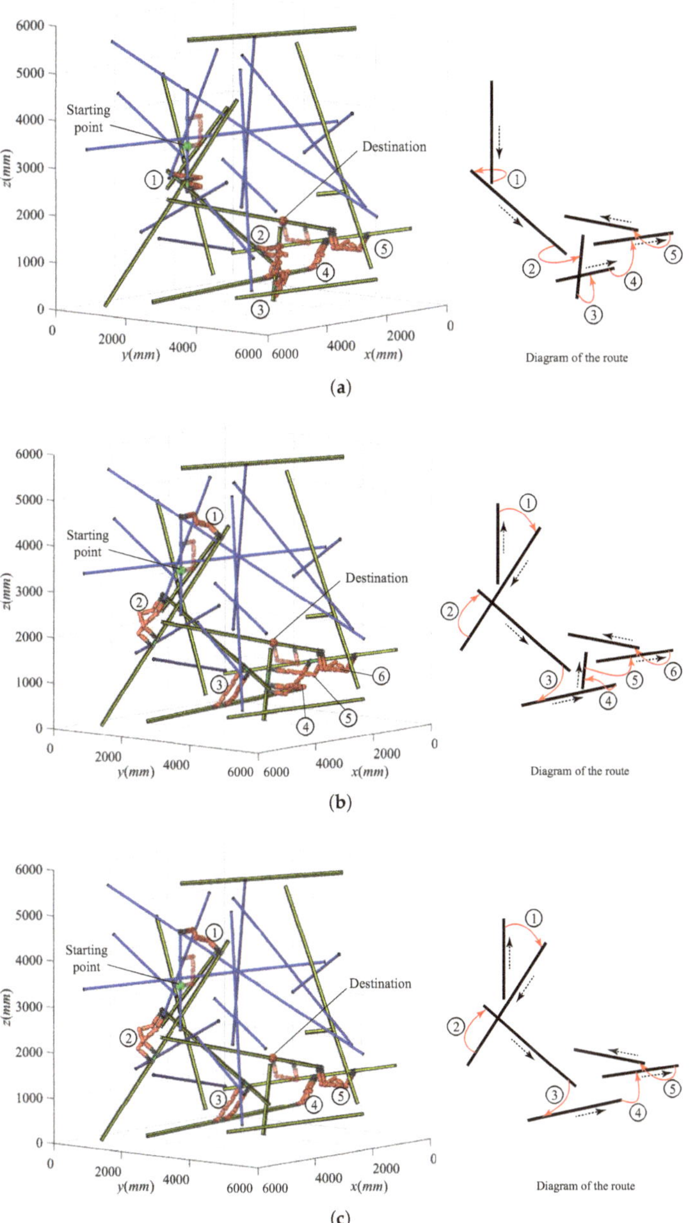

Figure 10. Results of global path determination in Scene II consisting of 25 members. (**a**) Route I; (**b**) Route II; (**c**) Route III.

From Figure 9, there are in total three possible routes globally, passing through 3, 2, and 1 in-between members, respectively. Correspondingly, the robot needs to perform 4, 3, and 2 periods of transition. Among the possible routes, Route III shown in Figure 9c is the best solution, regardless of whether the point of transition numbers or the total path length are evaluated. With regard to the complicated truss scene shown in Figure 10, whose members are randomly generated, the proposed algorithms also determine three possible global routes. From the results, we can see that the robot needs to perform 5 to 6 periods of transition in order to reach the target member. Routes I and III have a superiority in terms of transition number and path length over Route II. However, the routes are not so different from each other. They mainly branch off at the lower right local region where several squared members are concentrated, providing more potential gripping orientations. As a comparison of time consumed, the determination process requires 207.6 ms for the simple case, as compared to 1274.6 ms for the complicated one.

6. Conclusions and Future Work

Biped climbing robots represent an ideal automation solution so that human workers do not need to perform high-rise truss-related routine tasks. To freely move in the truss environments, member-to-member transition is a basic and important ability for biped climbing robots.

In this paper, we presented a complete approach to compute the feasibility and corresponding operational regions for transiting from one member to another with biped climbing robots, such as Climbot. The transition analysis takes the safety, reachability, and accessibility of grips in to account. This achievement was then applied to the rapid determination of global paths as a core evaluation for the first time. Simulations successfully verified the effectiveness and efficiency of the presented analysis and algorithms. A novel contribution of this paper is the presentation of a systematic scheme to quantitatively analyze the feasibility of member-to-member transition performed with biped climbing robots. This scheme solves the operational region completely. However, the proposed transition analysis has its own limitation. Owing to the introduction of planar configuration constraints (not usually the case for robots with more than 5 DoF), the transition analysis is only applicable to robots with up to 5 DoF. Another contribution of this paper is a novel path planning algorithm to rapidly determinate feasible global routes for biped climbing robots to move in 3D complex truss environments. This path planning algorithm is greatly beneficial to all the biped climbing robots if their transition capabilities can be modeled and calculated, by outputting global guidance for subsequent planning procedures.

In the near future, we will explore a more general idea for performing transition analysis, which will be applicable to all biped climbing robots. We also plan to propose criteria for the evaluation of the global routes' qualities, and then develop subsequent planners to generate the entire climbing path, i.e., optimal grip sequencer and single-step motion planner. Extensive climbing experiments on various trusses will be conducted to further verify our planning algorithms.

Acknowledgments: This research is supported in part by the National Natural Science Foundation of China (Grant Nos. 51605096, 51705086), the NSFC-Guangdong Joint Fund (Grant No. U1401240), the Frontier and Key Technology Innovation Special Funds of Guangdong Province (Grant Nos. 2014B090919002, 2015B010917003, 2015B090922003, 2017B050506008), and the Program of Foshan Innovation Team of Science and Technology (Grant No. 2015IT100072).

Author Contributions: Haifei Zhu and Yisheng Guan co-organized the work. Haifei Zhu and Shichao Gu conceived and designed the planner and performed the simulation work. Haifei Zhu wrote the manuscript. Li He analyzed the simulation data, and co-worked with Hong Zhang to prepare the final manuscript. Yisheng Guan and Hong Zhang co-supervised the research.

Conflicts of Interest: The authors declare no conflict of interest.

References

1. Xu, Y.; Brown, B.; Aoki, S.; Kanade, T. Mobility and Manipulation of a Light-weight Space Robot. *Robot. Auton. Syst.* **1994**, *13*, 1–12.
2. Balaguer, C.; Gimenez, A.; Pastor, J.; Padron, V.M. A Climbing Autonomous Robot for Inspection Application in 3D Complex Environments. *Robotica* **2000**, *18*, 287–297.
3. Nakanishi, J.; Fukuda, T.; Koditschek, D.E. A Brachiating Robot Controller. *IEEE Trans. Robot. Autom.* **2000**, *16*, 109–123.

4. Saltaren, R.; Aracil, R.; Reinoso, O.; Scarano, M. Climbing Parallel Robot: A Computational and Experimental Study of its Performance around Structural Nodes. *IEEE Trans. Robot.* **2005**, *21*, 16–22.
5. Aracil, R.; Saltaren, R.; Reinoso, O. A Climbing Parallel Robot: A Robot to Climb along Tubular and Metallic Structures. *IEEE Robot. Autom. Mag.* **2006**, *13*, 16–22.
6. Kushihashi, Y.; Koji, Y.; Yoshikawa, K. Development of Tree-climbing and Pruning Robot WOODY-1: Simplification of Control Using Adjust Function of Grasping Power. In Proceedings of the JSME Conference on Robotics and Mechatronics, Tokyo, Japan, 26–28 May 2006; pp. 1A1-E08_1–1A1-E08_2. (In Japanese)
7. Takeuchi, M.; Namba, H.; Suga, Y.; Shirai, Y.; Sugano, S. Development of Street Tree Climbing Robot WOODY-2. In Proceedings of the JSME Conference on Robotics and Mechatronics, Fukuoka, Japan, 24–26 May 2009; pp. 1A2-D07_1–1A2-D07_2. (In Japanese)
8. Detweiler, C.; Vona, M.; Yoon, Y.; Yun, S.; Rus, D. Self-assembling Mobile Linkages. *IEEE Robot. Autom. Mag.* **2007**, *14*, 45–55.
9. Spenko, M.J.; Haynes, G.C.; Saunders, J.A.; Cutkosky, M.R.; Rizzi, A.A. Biologically Inspired Climbing with a Hexapedal Robot. *J. Field Robot.* **2008**, *24*, 223–242.
10. Haynes, G.C.; Khripin, A.; Lynch, G.; Amory, J.; Saunders, A.; Rizzi, A.A.; Koditschek, D.E. Rapid Pole Climbing with a Quadrupedal Robot. In Proceedings of the IEEE International Conference on Robotics and Automation, Kobe, Japan, 12–17 May 2009; pp. 2767–2772.
11. Noohi, E.; Mahdavi, S.; Baghani, A.; Nili-Ahmadabadi, M. Wheel-Based Climbing Robot: Modeling and Control. *Adv. Robot.* **2010**, *24*, 1313–1343.
12. Tavakoli, M.; Marques, L.; Almeida, A. Development of an Industrial Pipeline Inspection Robot. *Ind. Robot Int. J.* **2010**, *37*, 309–322.
13. Lam, T.L.; Xu, Y. A Flexible Tree Climbing Robot: Treebot—Design and Implementation. In Proceedings of the IEEE International Conference on Robotics and Automation, Shanghai, China, 9–13 May 2011; pp. 5849–5854.
14. Vidoni, R.; Mimmo, T.; Pandolfi, C. Tendril-Based Climbing Plants to Model, Simulate and Create Bio-Inspired Robotic Systems. *J. Bionic Eng.* **2015**, *12*, 250–262.
15. Guan, Y.; Jiang, L.; Zhang, X. Mechanical Design and Basic Analysis of a Modular Robot with Special Climbing and Manipulation Functions. In Proceedings of the IEEE International Conference on Robotics and Biomimetics, Sanya, China, 15–18 December 2007; pp. 502–507.
16. Guan, Y.; Jiang, L.; Zhu, H.; Zhou, X.; Cai, C.; Wu, W.; Xiao, Z.; Chen, X.; Zhang, H. Climbot: A Bio-inspired Modular Biped Climbing Robot—System Development, Climbing Gaits and Experiments. *ASME J. Mech. Robot.* **2016**, *8*, 021026.
17. Rollinson, D.; Choset, H. Pipe Network Locomotion with a Snake Robot. *J. Field Robot.* **2016**, *33*, 322–336.
18. Balaguer, C.; Gimenez, A.; Jardon, A. Climbing Robots' Mobility for Inspection and Maintenance of 3D Complex Environments. *Auton. Robot.* **2005**, *18*, 157–169.
19. Chu, B.; Jung, K.; Han, C.S.; Hong, D. A Survey of Climbing Robots: Locomotion and Adhesion. *Int. J. Precis. Eng. Manuf.* **2010**, *11*, 633–647.
20. Muscolo, G.G.; Recchiuto, C.T.; Sellers, W.; Molfino, R. Towards a Novel Embodied Robot Bio-inspired by Non-human Primates. In Proceedings of the 41st International Symposium on Robotics/Robotik 2014, Munich, Germany, 2–3 June 2014; pp. 1–7.
21. Tummala, R.L.; Mukherjee, R.; Xi, N.; Aslam, D.; Dulimarta, H.; Xiao, J.; Minor, M.; Dang, G. Climbing the Walls–Presenting Two Underactuated Kinematic Designs for Miniature Climbing Robots. *IEEE Robot. Autom. Mag.* **2002**, *9*, 10–19.
22. Balaguer, C.; Gimenez, A.; Huete, A.J.; Sabatini, A.M.; Topping, M.; Bolmsjo, G. The MATS Robot: Service Climbing Robot for Personal Assistance. *IEEE Robot. Autom. Mag.* **2006**, *13*, 51–58.
23. Chung, W.; Xu, Y. A Novel Frame Climbing robot: Frambot. In Proceedings of the IEEE International Conference on Robotics and Biomimetics, Phuket, Thailand, 7–11 December 2011; pp. 2559–2566.
24. Guan, Y.; Zhu, H.; Wu, W.; Zhou, X.; Jiang, L.; Cai, C.; Zhang, L.; Zhang, H. A Modular Biped Wall-Climbing Robot with High Mobility and Manipulating Function. *IEEE/ASME Trans. Mechatron.* **2013**, *18*, 1787–1798.
25. Silva, M.F.; Machado, J.A.T. A Survey of Technologies and Applications for Climbing Robots Locomotion and Adhesion. In *Climbing and Walking Robots*; InTech: London, UK, 2010; pp. 1–22.
26. Schmidt, D.; Berns, K. Climbing Robots for Maintenance and Inspections of Vertical Structures—A Survey of Design Aspects and Technologies. *Robot. Auton. Syst.* **2013**, *61*, 1288–1305.

27. Yoon, Y.; Rus, D. Shady3D: A Robot that Climbs 3D Trusses. In Proceedings of the IEEE International Conference on Robotics and Automation, Roma, Italy, 10–14 April 2007; pp. 4071–4076.
28. Chung, W.; Xu, Y. Minimum Energy Demand Locomotion on Space Station. *J. Robot.* **2013**, *2013*, 723535.
29. Cai, C.; Zhu, H.; Guan, Y.; Zhang, X.; Zhang, H. A Biologically Inspired Miniature Biped Climbing Robot. In Proceedings of the IEEE International Conference on Mechatronics and Automation, Changchun, China, 9–12 August 2009; pp. 2653–2658.
30. Cai, C.; Guan, Y.; Zhou, X.; Jiang, L.; Zhu, H.; Wu, W.; Zhang, X.; Zhang, H. Joystick-based Control for a Biomimetic Biped Climbing Robot. *Robot* **2012**, *34*, 363–368. (In Chinese)
31. Zhu, H.; Guan, Y.; Wu, W.; Chen, X.; Zhou, X.; Zhang, H. A Binary Approximating Method for Graspable Region Determination of Biped Climbing Robots. *Adv. Robot.* **2014**, *28*, 1405–1418.
32. Xu, Y.; Brown, B.; Kanade, T. Control Systems of the Self-Mobile Space Manipulator. *IEEE Trans. Control Syst. Technol.* **1994**, *2*, 207–219.
33. Yun, S.; Rus, D. Optimal Self Assembling of Modular Manipulators with Active and Passive Modules. *Auton. Robot.* **2011**, *31*, 183–207.
34. Lam, T.; Xu, Y. Motion Planning for Tree Climbing with Inchworm-like Robots. *J. Field Robot.* **2013**, *30*, 87–101.
35. Zhu, H.; Guan, Y.; Wu, W.; Zhou, X.; Zhang, H. Transition Analysis of a Biped Pole-Climbing Robot—Climbot. In Proceedings of the 16th International Conference on Climbing and Walking Robots, Sydney, Australia, 14–17 July 2012; pp. 685–692.

Article

Model-Based Design and Evaluation of a Brachiating Monkey Robot with an Active Waist

Alex Kai-Yuan Lo †, Yu-Huan Yang †, Tsen-Chang Lin, Chen-Wen Chu and Pei-Chun Lin *

Department of Mechanical Engineering, National Taiwan University, Taipei 10617, Taiwan; b00502005@ntu.edu.tw (A.K.-Y.L.); b00502053@ntu.edu.tw (Y.-H.Y.); b00502024@ntu.edu.tw (T.-C.L.); b00502002@ntu.edu.tw (C.-W.C.)

* Correspondence: peichunlin@ntu.edu.tw; Tel.: +886-2-3366-9747

† A.K.-Y.L. and Y.-H.Y. contributed equally to this work.

Received: 4 July 2017; Accepted: 7 September 2017; Published: 14 September 2017

Abstract: We report on the model-based development of a monkey robot that is capable of performing continuous brachiation locomotion on swingable rod, as the intermediate step toward studying brachiation on the soft rope or on horizontal ropes with both ends fixed. The work is different from other previous works where the model or the robot swings on fixed bars. The model, which is composed of two rigid links, was inspired by the dynamic motion of primates. The model further served as the design guideline for a robot that has five degree of freedoms: two on each arm for rod changing and one on the waist to initiate a swing motion. The model was quantitatively formulated, and its dynamic behavior was analyzed in simulation. Further, a two-stage controller was developed within the simulation environment, where the first stage used the natural dynamics of a two-link pendulum-like model, and the second stage used the angular velocity feedback to regulate the waist motion. Finally, the robot was empirically built and evaluated. The experimental results confirm that the robot can perform model-like swing behavior and continuous brachiation locomotion on rods.

Keywords: brachiation; robot; dynamics; swingable rod; rope

---

1. Introduction

While most legged animals move on the ground by using their legs when walking [1], running [2,3], leaping [4], or crawling [5], a special subset of legged animals includes primates who have a different kind of locomotion mechanism owing to the characteristics of the habitat. Monkeys live in trees, which can be regarded as a special kind of terrain that is scattered and randomly distributed in all directions. Thus, using arms for locomotion, or so called brachiation, has naturally evolved in primates such as gibbons and siamangs. Their brachiation mechanism has been studied [6,7], and the influence of their size and proportion on biomechanic characteristics have been reported as well [8]. Brachiation locomotion acts as a unique class of limbed locomotion that can negotiate some types of special terrain, where other methods may not be functional.

Continuous brachiation locomotion in general is composed of a swing motion with respect to a fixed point and the change of fixed points for forward motion. Because the fixed point is usually above the system, the swing motion is similar to that of a pendulum [9,10]. The dynamics of a multi-link pendulum with respect to a fixed position have been well studied, especially for the two-link [11–15] and the three-link [16,17] systems, as acrobatic motion performed by gymnasts can also be modeled in this manner [18]. The research topics have included the pendulum focus on energy transformation mechanism, control strategy, and parameter identification. Most studies have been conducted in a simulation environment, and some have produced experimental results [13,16]. In addition to the study of pendulum dynamics, some research has focused on the change in fixed positions in the performance of complete brachiation locomotion.

Similarly, two-link [19–26] and three-link [27–30] systems have been widely adopted. In these studies, different controllers were applied and simulated to stabilize the behavior and to reject disturbances, and some studies included comparisons between simulations and experimental results [21,24,29]. The empirical robots in this category needed to have extra degrees-of-freedom (DOFs) to grasp/release or switch the fixed positions, and a distance sensor was used to detect the distance and height of the bars for brachiating adjustment [31]. Later, the simulation studies used models with link numbers up to seven to simulate the complex dynamics of primates [32–36].

In our study, we report on the simulation and experimental results of a monkey robot in brachiation locomotion. Unlike other studies where the models/robots grasp the fixed bar, we intend to investigate the behavior while the robot grasps a swingable rod, as the intermediate step toward studying brachiation on the soft rope. The system, which is composed of the robot and the rope/rod, has a more concentrated mass/size distribution and a larger distance to the fixed end (i.e., the top of the rope/rod) than the system composed of the robot and a fixed bar, as reported in other literature. Thus, the mechanical property of the rope/rod changes the overall system dynamics, and the passive rope/rod enlarges the swing amplitude of the robot as well as increases the possibility of catching. Following this setup, the two-step control strategy is developed, so the system can initiate the swing motion from rest as well as maintain the swing with a large amplitude. The system is modeled and the empirical robot is built and experimentally evaluated.

The remainder of the paper is organized as follows. Note that the motivation of this work is neither the development of a sophisticated control strategy of the two-link pendulum nor the development of a complex robot to mimic the monkey itself. Instead, we focus on how to design a simple robot and to simultaneously find a corresponding simple model (i.e., template vs. anchor [37]), so the complex brachiating behavior of the monkey can be abstractly reconstructed. Following this logic, we intend to firstly describe the bio-inspiration process in Section 2, starting from our interest in understanding how a monkey (or Tarzan, a famous fictional character in Disney movies) brachiates between ropes and ending at the conclusion that a two-link model is sufficient to abstract this behavior. Then, in Section 3 we try to design a robot that has limited DOFs but is capable of reconstructing the brachiating motion (i.e., including switching ropes/rods). Next, the dynamic behaviors of the model and the robot are analyzed and reported in Section 4, and Section 5 describes the strategy and behavior of the robot brachiating between the ropes/rods. Finally, Section 6 concludes the work.

2. The Reduced-Order Two-Link Dynamic Model of the Monkey Robot

A monkey swinging in a tree, as shown in Figure 1a, is a spatial and complex dynamic motion. To effectively understand the essential dynamic characteristics of the system, we construct a reduced-order model of the system by considering the following issues: (i) The motion of the monkey is only considered in the sagittal plane, and any dynamics outside this plane are ignored; (ii) The "simplest actuation scheme" is adopted, where a fore/aft swing motion is generated by only one active and rotational degree-of-freedom (DOF) with torque τ. This DOF is located around the geometric center of the body, so the relative configuration of the "upper body" and the "lower body" can be actively controlled. This joint is referred to as "joint 2"; (iii) The upper and lower bodies are assumed to be rigid bodies with mass (m_1 and m_2) and inertia (I_1 and I_2); (iv) The mass of the rope is ignored. In this case, during the monkey swing motion, the rope can be treated as a massless rigid rod that connects to the fixed end by a passive revolute joint (referred to as "joint 1"), as shown in Figure 1b. The viscous damper with the damping coefficient c is added at this joint to represent the energy loss of the model; (v) During the monkey swing motion, the relative configuration between the hand of the monkey and the rope is fixed. Together with the issue described in (iv), the "rope" and the upper body of the monkey can be modeled together as a rigid-body link with length l_1 and its center-of-mass (COM) is located close to the active joint (l_{1c} from the revolute joint); (vi) The motion and dynamic effects of the other arm are not considered because we intend to keep the reduced-order model (or template) simple to extract the essential dynamics of the system. On the robot side, the robot is designed to have less inertia in the arms to fulfill the assumption of the template. After taking these considerations into account, the original complex system can be simplified as a two-link model, as shown in Figure 1b.

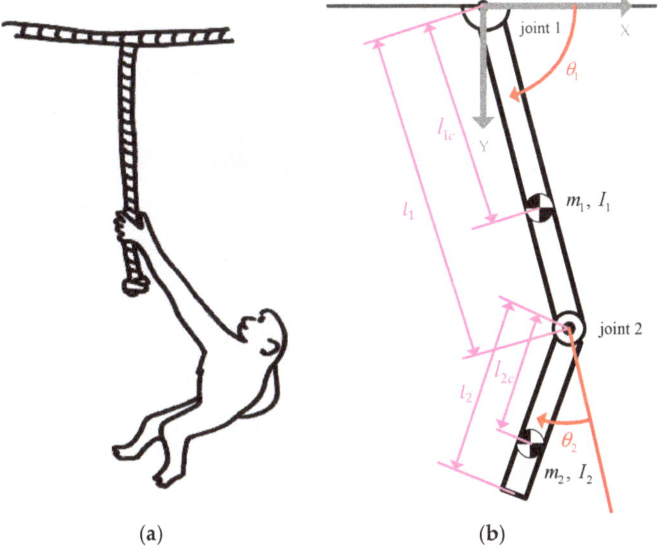

Figure 1. (**a**) A monkey swings in a tree and (**b**) the simplified two-link model.

The equations of motion (EOMs) of the reduced-order two-link model shown in Figure 1b was developed based on Lagrangian mechanics. The origin of the Cartesian coordinate system is located at joint 1. The angles θ_1 and θ_2 represent the configuration of the upper and lower links, respectively. The model can be regarded as a compound pendulum, and its behavior can be parameterized by two variables, θ_1 and θ_2.

The quantitative formulation of the model is described as follows. First, the positions ($\vec{P_1}$ and $\vec{P_2}$) of the masses (m_1 and m_2) can be expressed as

$$\begin{aligned}\vec{P_1} &= l_{1c}cos\theta_1\hat{i} + l_{1c}sin\theta_1\hat{j} \\ \vec{P_2} &= [l_1cos\theta_1 + l_{2c}cos(\theta_1+\theta_2)]\hat{i} + [l_1sin\theta_1 + l_{2c}sin(\theta_1+\theta_2)]\hat{j},\end{aligned} \tag{1}$$

and their velocities ($\vec{V_1}$ and $\vec{V_2}$) can be derived as

$$\begin{aligned}\vec{V_1} &= l_{1c}\dot{\theta}_1(-sin\theta_1\hat{i} + cos\theta_1\hat{j}) \\ \vec{V_2} &= l_1\dot{\theta}_1(-sin\theta_1\hat{i} + cos\theta_1\hat{j}) + l_{2c}(\dot{\theta}_1+\dot{\theta}_2)[(-sin(\theta_1+\theta_2)\hat{i} + cos(\theta_1+\theta_2)\hat{j}),\end{aligned} \tag{2}$$

The kinetic energy (T) and potential energy (V) of the model can be expressed as

$$\begin{aligned}T &= \frac{1}{2}m_1\vec{V_1}\cdot\vec{V_1} + \frac{1}{2}I_1\dot{\theta_1}^2 + \frac{1}{2}m_2\vec{V_2}\cdot\vec{V_2} + \frac{1}{2}I_2(\dot{\theta}_1+\dot{\theta}_2)^2 \\ V &= -m_1gl_{1c}sin\theta_1 - m_2g[l_1sin\theta_1 + l_{2c}sin(\theta_1+\theta_2)],\end{aligned} \tag{3}$$

where g represents the gravity constant. Following this definition, the equations of motion can be expressed as

$$\frac{\mathrm{d}}{\mathrm{dt}}\left(\frac{\partial \mathrm{L}}{\partial \dot{\theta_1}}\right) - \frac{\partial \mathrm{L}}{\partial \theta_1} = -c\dot{\theta_1}\frac{d}{dt}\left(\frac{\partial L}{\partial \dot{\theta_2}}\right) - \frac{\partial L}{\partial \theta_2} = \tau, \tag{4}$$

where $L = T - V$ is the Lagrangian of the model and c is the damping coefficient. Together with Equations (1)–(3), the EOMs of the model can be derived as

$$\begin{aligned} & m_1 l_{1c}^2 \ddot{\theta}_1 + I_1 \ddot{\theta}_1 + I_2\left(\ddot{\theta}_1 + \ddot{\theta}_2\right) + m_2\left\{l_{2c}^2\left(\ddot{\theta}_1 + \ddot{\theta}_2\right) + l_1^2 \ddot{\theta}_1 + l_1 l_{2c}\left[\left(2\ddot{\theta}_1 + \ddot{\theta}_2\right) cos\theta_2 - \right.\right. \\ & \quad \left.\left.\dot{\theta}_2\left(2\dot{\theta}_1 + \dot{\theta}_2\right) sin\theta_2\right]\right\} - m_1 g l_{1c} cos\theta_1 - m_2 g[l_1 cos\theta_1 + l_{2c} cos(\theta_1 + \theta_2)] = -c\dot{\theta}_1 \\ & I_2(\ddot{\theta}_1 + \ddot{\theta}_2) + m_2 l_{2c}^2(\ddot{\theta}_1 + \ddot{\theta}_2) + m_2 l_1 l_{2c}(\dot{\theta}_1^2 sin\theta_2 + \ddot{\theta}_1 cos\theta_2) - m_2 l_{2c} g cos(\theta_1 + \theta_2) = \tau, \end{aligned} \tag{5}$$

The derived EOMs are programmed in Matlab® to study the dynamic behavior of the model. The equations are rewritten in state-space with variables ($\theta_1, \dot{\theta}_1, \theta_2, \dot{\theta}_2$) and solved by function ode45. When the active joint 2 (θ_2) is not force controlled but position controlled, the state θ_2 and $\dot{\theta}_2$ are regarded as priori. In this case, only the states of θ_1 and $\dot{\theta}_1$ need to be solved.

3. The Design of the MonkeyBot

The design of the monkey robot (hereafter referred to as the MonkeyBot) basically aligns with the configuration of the two-link model described in Section 2, so the latter (the "template") can correctly serve as the reduced-order model of the former (the "anchor") [37]. In addition to this constraint, the design of the robot is also based on several considerations that will be described separately as follows.

The computer-aided drafting (CAD) drawing of the robot is shown in Figure 2a. The robot has two arms, which are designed to grab/release the rope (i.e., the swing rods). By programming the arms to grab/release the rope/rod alternatively, the robot is capable of swinging among different ropes/rods as a monkey does. Thus, the upper body of the model described in Section 2, in reality, is composed of the main upper robot body, two arms, and the rope/rod. In contrast, though a monkey has two legs and a tail, the lower robot body is designed as a rigid box, which allows us to directly map that to the lower body of the model. The rigid box is designed to have several mounting holes and a small basket, as shown in the figure, so the length and mass of the lower body can be easily adjusted. Because of these adjustable designs, we can alter the variable l_2 and m_2 in Section 2, which can enhance the robot's moment of inertia and make the movement optimal.

The MonkeyBot has five actuators in total, as shown in Figure 2a. One of them is to control the relative configuration of the upper and lower robot bodies and is equivalent to the control of joint 2 (θ_2) of the model. Each arm has two actuators: One of them controls the open/close of the hand, so the hand can grasp/release the rope/rod. The other actuator changes the rotational configuration of the arm relative to the upper robot body (i.e., the shoulder joint). When the MonkeyBot swings, the arm that grasps the rope/rod poses vertically. In the meantime, the other arm poses horizontally toward the front side, in preparation for grasping the next rope/rod.

Each arm/hand is composed of dual 4-bar linkage structures, as shown in Figure 2b. The robot hand, or the simple 2-finger gripper, is composed of two linkages, one from each 4-bar linkage structure. By using one actuator to drive these two 4-bar linkages simultaneously, the hand can perform the open/close motion. The use of the 4-bar linkages to drive the hand motion has several advantages: First, the actuator, which is relatively heavier than other components on the arm/hand, can be installed close to the shoulder joint, so the arm/hand inertia can be reduced. This helps to increase the dynamic response of the arm motion. Second, when the hand grasps the rope/rod, the arm and body weight can be supported by the linkage structure, but not by the actuator torque. The shoulder joint also has to carry the body weight. Thus, instead of relying on the actuator motor shaft and aluminum horn, a ring bearing is installed to support the force in the axial and radial directions of the joint. Third, the motion range can be amplified. The open/close angle of the gripper is increased to 80° from its original 40° on the actuator side. The actuator (55 g) is the heaviest component on the arm, so it is placed close to the shoulder joint to reduce its inertia effect as shown in Figure 2b. Thus, motion of the arm that is not grasping the rope/rod does not have a significant effect on the dynamics of the robot's overall motion.

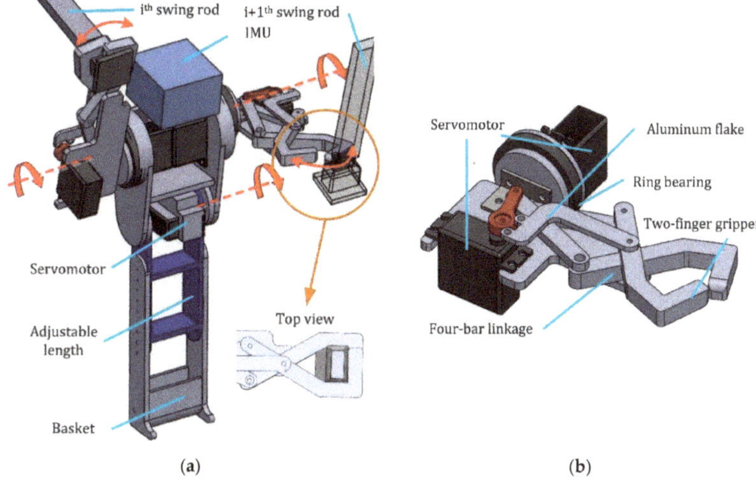

Figure 2. (**a**) A computer-aided drafting (CAD) drawing of the whole MonkeyBot and (**b**) the design of the right hand-arm.

When the hand closes, the enclosed inner area is in a rectangular shape, which matches the cross section of the rigid rod, as shown in the blow-up subfigure in Figure 2a. Thus, when the hand of the robot grasps the rod, the yaw and roll disturbance of the robot can be constrained, so the robot can have motion only in the sagittal plane as the model requires. In addition, a small piece of rubber is mounted on the inner surface of the hand to increase the friction force between the hand and the rigid rod.

The mechatronic system of the MonkeyBot is briefly described as follows. It has an embedded system (MyRIO-1900, National Instruments, Austin, TX, USA), which has a real-time processor operating at a 50 Hz loop rate and an integrated field programmable gate array (FPGA). The former is utilized to deploy the main control algorithm, and the latter is for high-speed computation and input/output (I/O) signal exchange. The servomotors (MG995, Tower Pro, Taipei, Taiwan) are used to control the arm motion and hand motion. The high-speed servomotor (PL-8509, PowerStar, New Taipei City, Taiwan) is used to control the configuration between the upper and lower bodies, and it works at a speed of 2.32 rad/s, which is much faster than the natural frequency of the system. Because it can achieve a swing frequency that is higher than the natural frequency of the system, the motion control is feasible. The inertia measurement unit (IMU) (ADIS-16364, Analog Device, Norwood, MA, USA) is installed close to the COM of the upper body. A limit switch is mounted on each hand to detect the grasp condition between the hand and the rigid rod. Figure 3 shows a photo of the robot. The physical parameters of the robot are listed in Table 1.

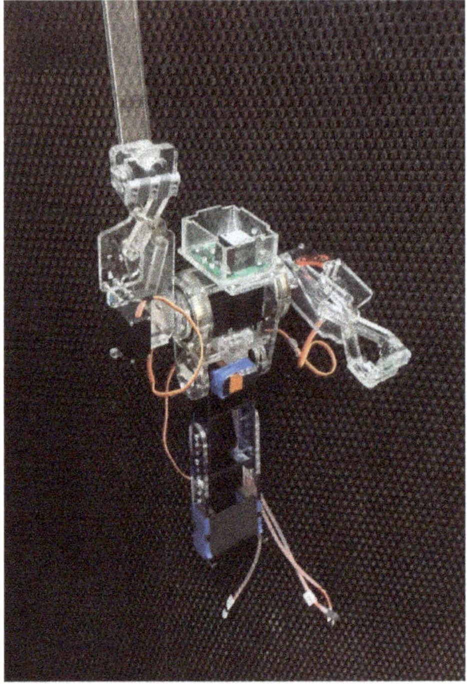

Figure 3. A photo of the MonkeyBot.

Table 1. Physical parameters of the robot.

m_1	0.995 kg
m_2	0.19 kg
I_1	1.26×10^{-3} kg·m^2
I_2	0 kg·m^2
l_1	0.495 m
l_{1c}	0.410 m
l_2	0.217 m
l_{2c}	0.217 m
c	0.0185

The swing angle of joint 1 (θ_1) is an important index to evaluate the swing performance of the robot. Because the relative configuration of the rod and the arm, the upper robot body remains the same during the swing motion, θ_1 equals the body pitch, and this information can be estimated by utilizing the inertia sensor mounted on the body. The inertia sensor provides the 3-axis linear acceleration (a_x, a_y, a_z) and 3-axis angular velocity $(\omega_x, \omega_y, \omega_z)$ of the body. The body pitch can be estimated by the following method. The aim is to integrate the angular velocity to yield body pitch. The computed body pitch would suffer from the notorious drifting error if the integration time were large. Because the robot swings periodically as the angular velocity alternatively and periodically exhibits positive and negative values, the integration can be "reset" to zero when the robot reaches its highest position ($\dot{\theta}_1 = 0$). Assuming the swing motion is symmetric, the integrated angle represents an angle that is twice as large as the swing amplitude ($\Delta\theta_1$).

4. Swing Dynamics of the Model and the MonkeyBot

The reduced-order two-link model described in Section 2 is designed to simulate and predict the dynamic behavior of the MonkeyBot described in Section 3, which helps to ease the process of developing the control law and investigating the effect of the parameters of the model/robot. The behavior of the empirical robot is much more complex than that of the model; for example, energy loss is one of the important issues. We do understand a discrepancy between the model and the robot definitely exists, but this is a typical synergy of "template vs. anchor" [37] and it is tolerable as long as the model (i.e., template) is good enough to provide the correct behavioral trend of the robot (i.e., anchor). On the model side, in addition to the use of the active joint θ_2 to represent the actuator effect of the robot, the model has to include the energy loss term to simulate the energy loss of the robot. Here, to make the reduced-order model as simple as possible, we intend to use just one viscous damping dashpot mounted at the swing joint (i.e., at joint θ_1) to represent the overall energy loss of the robot. While the physical specifications of the MonkeyBot, such as the mass, inertia, and lengths, can be directly mapped to the model parameters, the only parameter that needs to be determined is the damping coefficient c of the "resultant" viscous damping at the swing joint.

The free swing test of the model/robot was conducted to determine the damping coefficient c of the model. The experimental environment was set up to record the robot's motion for quantitative analysis. Five markers were mounted in the setup: two on the top of two rods (i.e., two joint 1s), one on the COM, one on the arm, and one on the lower body. When the robot swung or brachiated on the rod, its sagittal-plane motion was recorded by a stationary high-definition (HD) camcorder (HDR-SR11, Sony, Tokyo, Japan). While the qualitative motion of the robot can be easily observed by the video, the quantitative motion of the robot can be computed based on the markers' positions in the sequential snapshots taken by the camcorder. In the experiments, the MonkeyBot was posed at two different initial heights ($\theta_1 = 100°, 115°$, corresponding to $90° \pm 10°$ and $90° \pm 25°$ swings) and released. The robot started to swing, and the swing amplitude gradually decreased, owing to the energy loss. The profile θ_1 versus time t was recorded for data analysis. On the model side, models with various damping coefficients c were simulated to generate the corresponding θ_1 versus t profiles. Next, the root mean squared (RMS) errors between the peak value of trajectories in the first five periods of the robot's swing motion, and the models with various c values were computed. The damping coefficient $c = 0.0185$ was selected because this model has the smallest RMS error. Figure 4 plots the trajectories θ_1 versus t of the robot and the model with this damping coefficient, and the subplots (a) and (b) correspond to the initial condition $\theta_1 = 100°$ and $\theta_1 = 115°$, respectively. Both plots show a decent trajectory match between the robot and the model, and the RMS errors are 0.42° and 0.44°, accordingly.

The frequencies shown in Figure 4 represent the natural frequency of the model and the robot. According to a paper [38], the stride frequencies of different-sized animals have a certain relation to their body mass, and those frequencies also match the natural frequencies if the animals are modeled as the spring-mass systems (i.e., resonance region) because the brachiating motion of the robot is basically a pendulum motion that has a certain natural frequency. To actuate the waist of the robot/model at a frequency similar to the natural frequency is similar to driving the legs of the animals at a frequency similar to the natural frequency. The model has a natural swing frequency of around $\omega_n = 0.714$ Hz when $\theta_1 = 90°$. If the model starts with the at rest configuration and the joint 2 is actuated with a swing signal at the same frequency, the model can gradually increase its swing magnitude (θ_1), as shown in Figure 5a. This control scheme is hereafter referred to as the "open-loop" method. Figure 5a also reveals that the open-loop method can quickly build up energy to increase the swing amplitude θ_1 to $90° \pm 50°$ within 15 s. Because the double pendulum is a nonlinear system and the natural frequency changes when the configuration changes, its swing motion does not gradually converge to a certain amplitude, but exhibits amplitude variations. In addition, the model has less capability for disturbance rejection.

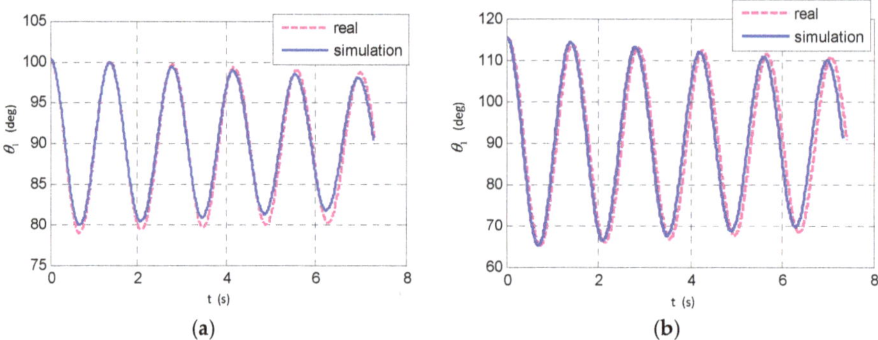

Figure 4. A simulation (blue solid curve) of the two-link model and the experimental result (red dashed curve) of the MonkeyBot with same damping coefficients and two different initial heights: (**a**) $c = 0.0185$, $\theta_{1\_0} = 100^{\circ}$ and (**b**) $c = 0.0185$, $\theta_{1\_0} = 115^{\circ}$.

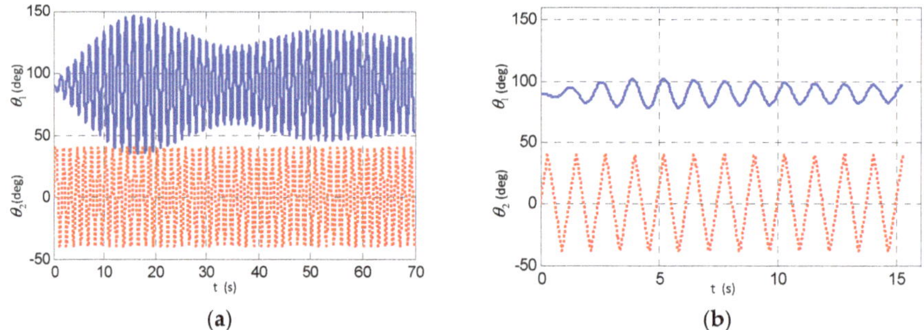

Figure 5. A simulation of the two-link model with (**a**) open-loop and (**b**) closed-loop control methods.

The closed-loop strategy is developed to compensate for the drawbacks of the open-loop strategy. When joint 2 is actuated with the same rotation direction as joint 1 ($\dot{\theta}_1$), the energy generated from joint 2 can be injected into the model. Figure 6 demonstrates this phenomenon. It can be assumed that the model initially swings toward the left ($\dot{\theta}_1 > 0$) and joint 2 is fixed at a positive angle ($\theta_2 > 0$), as shown in Figure 6a. After the model reaches its highest left position, as shown in Figure 6b, and starts to change motion direction ($\dot{\theta}_1 < 0$), if joint 2 is actuated to have a negative angular velocity ($\dot{\theta}_2 < 0$), as shown in Figure 6c, the swing motion of joint 1 ($\dot{\theta}_1 < 0$) can be maintained to overcome the damping loss or even increased to enlarge the swing amplitude. When the model reaches its highest right position, as shown in Figure 6d, joint 2 has a negative value ($\theta_2 < 0$) and is ready for the next actuation ($\dot{\theta}_2 > 0$) after the model starts to change motion direction ($\dot{\theta}_1 > 0$). This configuration is identical to the one shown in Figure 6a; thus, the swing motion can be periodically generated. Figure 5b plots the trajectories of the joint 1 angles of the model versus time when the model is actuated with this control scheme. This scheme is hereafter referred to as the "closed-loop" method. The model can successfully build up the swing amplitude θ_1 to $90^{\circ} \pm 10^{\circ}$ within 4 s; however, the amplitude seems not to increase until the end of simulation (15 s). Moreover, this method was not functional on the robot when it started at a rest condition $\theta_1 = 90^{\circ}$ for two reasons: First, the joint angle measurement (θ_1) of the robot has noise, and this may lead to a wrong actuation initiation, especially when θ_1 is close to 90°. Second, empirically, the actuation of joint 2 (θ_2) needs time to finish its configuration change. When θ_1 is close to 90°, the control algorithm may quickly switch the motion direction of joint 2, and in this case, the robot cannot correctly initiate its periodic motion pattern. Therefore, the hybrid control scheme is utilized: The model/robot firstly uses an open-loop control strategy. After the swing amplitude approaches the maximum value, the control algorithm switches to a closed-loop to increase the motion control stability. The overall control flow chart is shown in Figure 7.

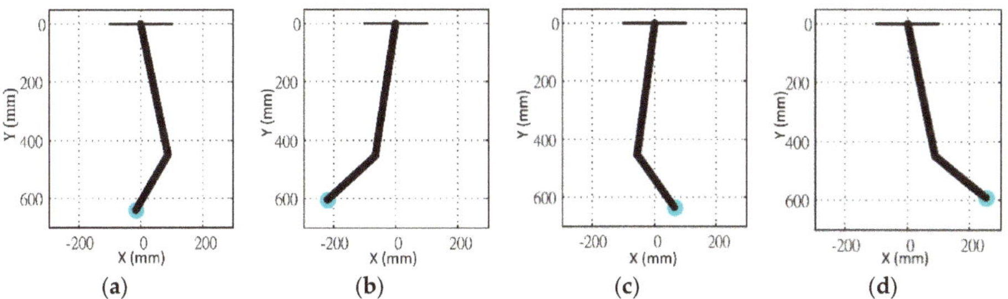

Figure 6. Illustrative motion sequence (**a–d**) of the two-link model with the closed-loop control method.

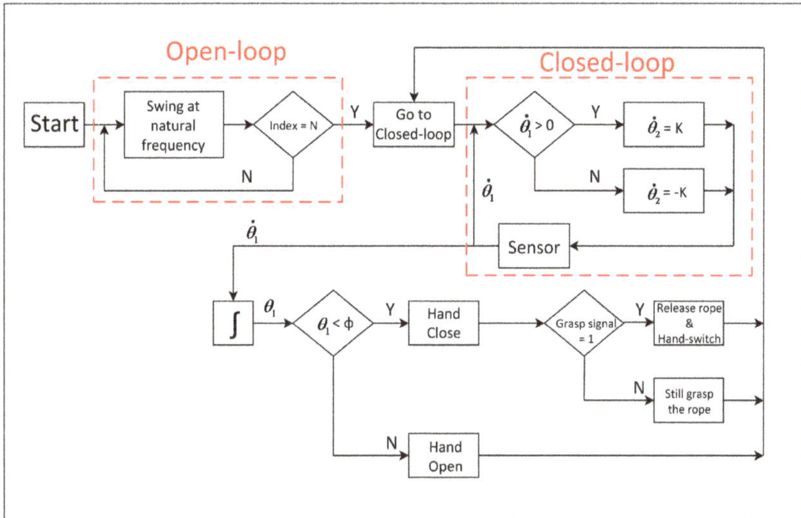

Figure 7. The overall control flowchart.

While the size and mass of the upper body is, in general, fixed and determined by the characteristics of the components, the size and mass of the lower body has some freedom to be adjusted for better dynamic performance of the robot itself. The evaluation process was done in simulation, where the link length l_2 and mass m_2 of the model were varied, and the model was set to swing six periods from its rest position ($\theta_1 = 90°$ and $\theta_2 = 0°$) with the open-loop control strategy. Because the model with a different size and mass has different natural frequencies, different models use different frequency settings. The performance was judged by the final swing amplitude of the model. Figure 8 shows the simulation result. The figure reveals that as the mass and length of the lower body increases, the swing amplitude increases. Though the longer length and larger weight are preferred, the actuator power is constrained to a certain value because the size and mass of the upper body are given. Thus, the link length l_2 and mass m_2 are maximized to the value that the motor can sustain. The selected length and mass of the lower body are marked in the figure.

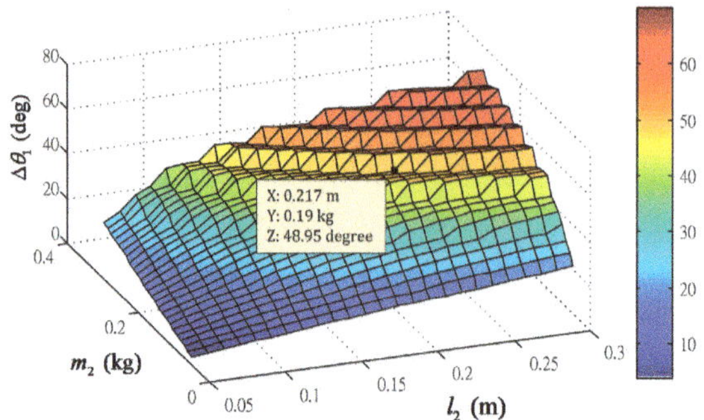

Figure 8. A simulation of the two-link model with different lengths and mass values.

The robot brachiating on the swingable rope/rod has a more concentrated mass/size distribution and a larger distance than that brachiating on a fixed point. The added rope/rod changes the overall system dynamics, and the passive rope/rod enlarges the swing amplitude of the robot, which increases the possibility of grasping the next rope/rod. Figure 9 compares the difference in behavior between the two systems, where the lengths of l_1 have different values. One includes the length of rope/rod, and the other does not. As shown in the figure, though the latter has a faster response, the former has a larger swing amplitude.

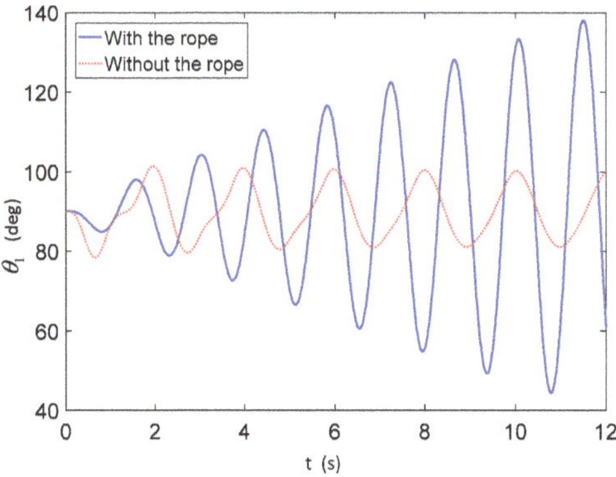

Figure 9. The swing amplitudes of the robot grasping the fixed point and the rod.

After the development of the control strategy and optimization of the size and mass, the model and robot are ready for performance evaluation. Figure 10 shows several snapshots of the robot during the experiments, and it confirms that the robot can successfully swing by using the proposed control strategy. Figure 11a shows the simulation results of the model with a hybrid control strategy, and Figure 11b shows the corresponding experimental result of the robot with this hybrid control strategy. We found that the swing frequency of the robot is similar to that of the simulation, but the swing amplitude of the robot is about 2/3 of the simulation. The root mean squared error of the peak amplitude between the model and the robot is 25.7°. We believe that the discrepancy mainly comes from the empirically tested non-rigid grasp of the hand to the rod. Though the hand grasps the rod with a certain contact area, as shown in Figure 2a, the grasping cannot generate enough resistance moment to hold the rod tightly. As a result, when joint 2 of the robot rotates, the arm and the rod are not able to keep their relative configuration as a rigid link, as the model assumes. This phenomenon can also be observed in Figure 10. This less than ideal situation results in two behaviors: (i) The swing energy generated by joint 2 has a certain loss, so the achievable amplitude decreases; (ii) Because the robot COM lies in different sagittal planes than the arms/rods, when the robot swings, the non-rigid grasping causes the robot to have other dynamic behaviors, such as a yaw swing motion. Thus, some of the swing energy generated by joint 2 is consumed by these unwanted dynamics. Except for the discrepancy in swing amplitude, the swing frequency of the robot can be predicted by the simple two-link model. The dynamic behavior of the robot with both open-loop and closed-loop controllers can also be predicted by the model. Therefore, though the model is quite simple, it can serve as a template of the robot.

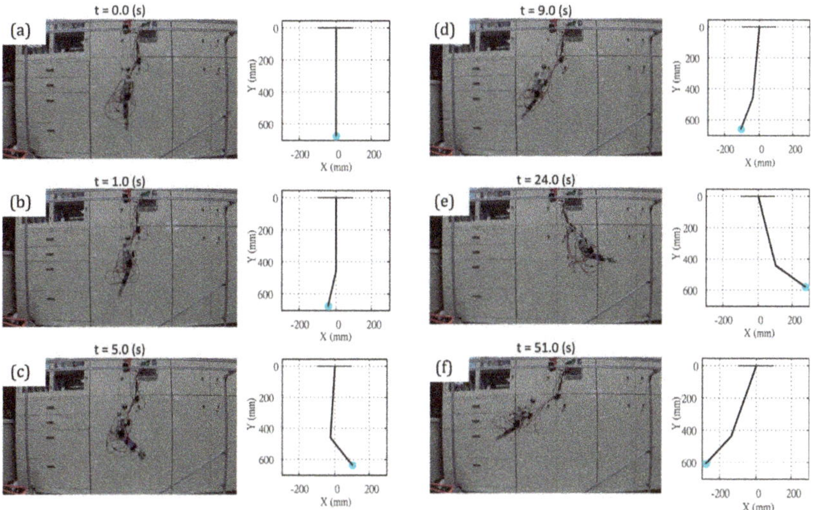

Figure 10. Snapshots of the robot (left) and the two-link model (right) starting from the at rest condition. The robot/model uses the open-loop control method (**a**–**d**) and then the closed-loop control method (**e**,**f**).

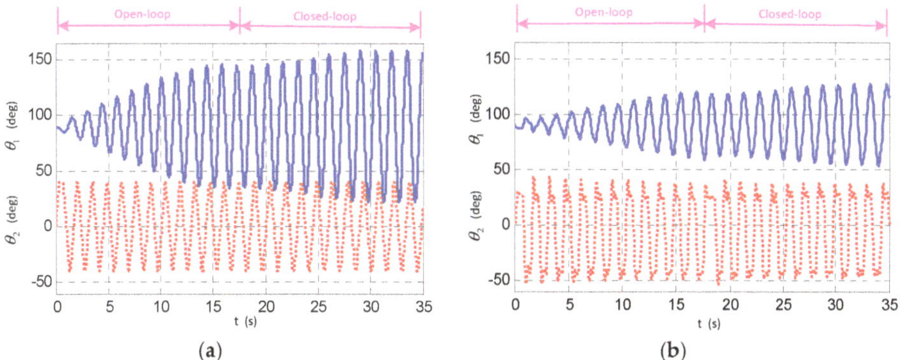

Figure 11. A simulation of the (**a**) two-link model and the experimental result of the (**b**) MonkeyBot with the hybrid control strategy.

5. Switching Rods of the MonkeyBot

Following the capability of the robot to swing, the next step is to develop the rope/rod-switching behavior of the MonkeyBot. The overall control flow chart is shown in Figure 7. To simplify the problem, the relative displacement between the ropes/rods is assumed a priori. As described in Section 3, the robot can estimate the joint angle 1 (θ_1, equal to the body pitch). By using the trigonometric relation, the robot can compute its relative displacement to the next rope/rod. When the robot hand can reach the next rope/rod, the hand opens and grasps it. A limit switch mounted in the hand is utilized to detect the grasping condition. When the grasping signal of the new rod is positive, the other hand of the robot releases the previous rod, so the robot can start its new swing motion on the new rod. Because the robot on the new rod swings from a certain initial height (i.e., not from an at rest position), the robot does not need to use the open-loop method to swing itself up to gain potential and kinetic energy. Instead, the closed-loop strategy is directly adopted for continuous motion, and the robot can grasp/release the front rod and continuously swing forward. Figure 12h shows the plane trajectories of the robot while it swings from the left rod to the right rod. The experimental setup is described in Section 4, and the locations of the markers are depicted in the CAD drawing of the robot and the swing rods, as shown in Figure 12g. The image shown in the upper right corner of Figure 12a–d is the snapshot captured by the camcorder after the intensity filter, so only the positions of the reflective markers are preserved. By collecting the positions of these markers versus time, the motion of the robot can be quantitatively analyzed. As shown in Figure 12h, the robot uses one of its arms to grasp the rod and swing forward (i.e., the left sections of the curves). In the meantime, the other arm of the robot (i.e., with the marker) poses forward and prepares to grasp the right rod. After the other arm grasps the right rod, the first arm releases its grasp to the left rod, and the robot keeps swinging forward (i.e., the right section of the curves). The short trajectories between the left and right trajectories represent the transition when the robot switches rods, and this exists because, empirically, the switching is not instant but takes a short period of time for the hands to grasp or release the bar. In short, the robot can successfully brachiate between the rods.

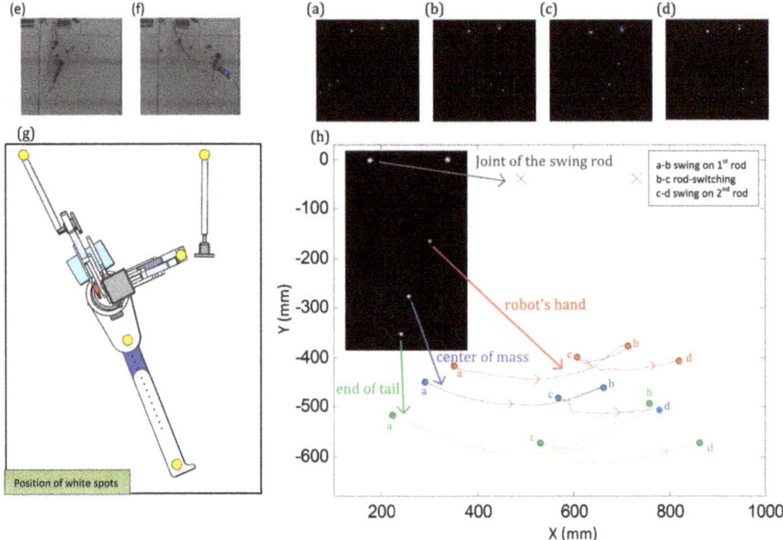

Figure 12. The MonkeyBot in brachiating motion. (**a–f**) The snapshots of the system captured by the camcorder before (**e**,**f**) and after (**a–d**) image processing; (**g**) The markers, whose locations are depicted in CAD drawing, were mounted on the MonkeyBot when it swung and switched rods; (**h**) The overall motion trajectories of the markers/system. The instants marked with a-d correspond with the snapshots (**a–d**).

6. Conclusions

We report on the model-based development of a monkey robot that is capable of performing continuous brachiation locomotion on swingable rods. The two-link model was inspired by the brachiation of primates, and its quantitative formulation was derived based on the Lagrangian method. The dynamic behaviors of the model were simulated in MATLAB®, and the model was also utilized for controller design. Following this, the robot was empirically built, and its dimensions were determined based on the simulation results. In addition, the controller developed during simulation was also implemented on the robot.

The simulation and experiment results reveal several facts and characteristics of this study: (i) The open-loop control strategy is effective in initiating the swing motion of the robot from its at rest configuration. However, it is hard to maintain a steady swing motion owing to the nonlinearity of the model and the disturbance; (ii) In contrast, the closed-loop control strategy can stabilize the model with a certain amplitude to perform a steady swing motion. However, when the initial amplitude is small, this strategy in simulation cannot amplify the swing magnitude. Furthermore, this strategy in the empirical robot is not functional because of the noise and latency of the sensor information and actuation; (iii) As a result, the hybrid control strategy was adopted, and it was confirmed to be effective, both in simulation and in experiments; (iv) The robot with the same control strategy as the model can successfully initiate its swing motion. The swing frequency maintains a similar value as that of the model, but the swing magnitude is about 2/3 of that of the model because of the non-rigid contact between the hand and the rod. This causes the empirical energy loss and the initiation of unwanted yaw dynamics in the robot. Except for the amplitude discrepancy, the robot exhibits similar dynamics to the model. Thus, the model can serve as a template of the original complex robot; (v) The robot can also successfully swing between the rods, performing brachiation locomotion.

We are in the process of revising the grasping mechanism of the robot hand, so the grasping can be more precise and rigid than the current method. In addition, we are exploring the energy flow of the robot during its swing and rod changing motions, so the control can be better fused with the natural dynamics of the system.

Acknowledgments: This work is supported by the National Science Council (NSC), Taiwan, under the contract 103-2815-C-002-094-E.

Author Contributions: Alex Kai-Yuan Lo and Yu-Huan Yang co-organized the work and wrote the manuscript draft. Alex Kai-Yuan Lo, Tsen-Chang Lin, and Chen-Wen Chu built the robot and performed the robot experiments. Yu-Huan Yang performed the simulation work. Pei-Chun Lin supervised the research and prepared the inal manuscript.

Conflicts of Interest: The authors declare no conflict of interest.

Nomenclature

m_1, m_2	Mass of the two-link model
I_1, I_2	Inertia of the two-link model
l_1, l_2	Link lengths of the two-link model
l_{1c}, l_{2c}	COM positions of the links
θ_1, θ_2	Configurations of the links 1 and 2 in the model
$\vec{P_1}, \vec{P_2}$	Positions of two masses in the model
$\vec{V_1}, \vec{V_2}$	Velocities of two masses in the model
T, V	Kinetic energy and potential energy of the model
L	Lagrangian of the model
g	Gravity constant
τ	Actuation torque at joint 2
c	Damping coefficient at joint 1

References

1. Lin, P.C.; Komsuoglu, H.; Koditschek, D.E. A leg configuration measurement system for full-body pose estimates in a hexapod robot. *IEEE Trans. Robot.* **2005**, *21*, 411–422.
2. Huang, K.J.; Huang, C.K.; Lin, P.C. A Simple Running Model with Rolling Contact and Its Role as a Template for Dynamic Locomotion on a Hexapod Robot. *Bioinspir. Biomim.* **2014**, *9*, 046004. [CrossRef] [PubMed]
3. Huang, K.J.; Chen, S.C.; Komsuoglu, H.; Lopes, G.; Clark, J.; Lin, P.C. Design and performance evaluation of a bio-inspired and single-motor-driven hexapod robot with dynamical gaits. *ASME J. Mech. Robot.* **2015**, *7*, 031017. [CrossRef]
4. Chou, Y.C.; Huang, K.J.; Yu, W.S.; Lin, P.C. Model-based development of leaping in a hexapod robot. *IEEE Trans. Robot.* **2015**, *31*, 40–54. [CrossRef]
5. Chou, Y.C.; Yu, W.S.; Huang, K.J.; Lin, P.C. Bio-inspired step-climbing in a hexapod robot. *Bioinspir. Biomim.* **2012**, *7*, 036008. [CrossRef] [PubMed]
6. Fleagle, J. Dynamics of a brachiating siamang [Hylobates (Symphalangus) syndactylus]. *Nature* **1974**, *248*, 259–260. [CrossRef] [PubMed]
7. Michilsens, F.; Vereecke, E.E.; D'août, K.; Aerts, P. Functional anatomy of the gibbon forelimb: Adaptations to a brachiating lifestyle. *J. Anat.* **2009**, *215*, 335–354. [CrossRef] [PubMed]
8. Preuschoft, H.; Demes, B. Influence of size and proportions on the biomechanics of brachiation. In *Size and Scaling in Primate Biology*; Jungers, W., Ed.; Springer: New York, NY, USA, 1985; pp. 383–399.
9. Swartz, S. Pendular mechanics and the kinematics and energetics of brachiating locomotion. *Int. J. Primatol.* **1989**, *10*, 387–418. [CrossRef]
10. Mahindrakar, A.D.; Banavar, R.N. A swing-up of the acrobot based on a simple pendulum strategy. *Int. J. Control* **2005**, *78*, 424–429. [CrossRef]
11. Xin, X.; Kaneda, M. The swing up control for the Acrobot based on energy control approach. In Proceedings of the 41st IEEE Conference on Decision and Control, Las Vegas, NV, USA, 10–13 December 2002.
12. Araki, N.; Okada, M.; Konishi, Y.; Ishigaki, H. Parameter identification and swing-up control of an Acrobot system. In Proceedings of the IEEE International Conference on Industrial Technology, Hong Kong, China, 14–17 December 2005.
13. Xin, X.; Yamasaki, T. Energy-based swing-up control for a remotely driven Acrobot: Theoretical and experimental results. *IEEE Trans. Control Syst. Technol.* **2012**, *20*, 1048–1056. [CrossRef]
14. Wan, D.; Cheng, H.; Ji, G.; Wang, S. Non-horizontal ricochetal brachiation motion planning and control for two-link Bio-primate robot. In Proceedings of the 2015 IEEE International Conference on Robotics and Biomimetics (ROBIO), Zhuhai, China, 6–9 December 2015.
15. Ebrahimi, K.; Namvar, M. Port-Hamiltonian control of a brachiating robot via generalized canonical transformations. In Proceedings of the 2016 American Control Conference (ACC), Boston, MA, USA, 6–8 July 2016.
16. Michitsuji, Y.; Sato, H.; Yamakita, M. Giant swing via forward upward circling of the Acrobat-robot. In Proceedings of the American Control Conference, Arlington, VA, USA, 25–27 June 2001.
17. Xin, X.; Kaneda, M. Swing-up control for a 3-DOF gymnastic robot with passive first joint: Design and analysis. *IEEE Trans. Robot.* **2007**, *23*, 1277–1285. [CrossRef]
18. Henmi, T.; Wada, T.; Deng, M.; Inoue, A.; Ueki, N.; Hirashima, Y. Swing-up control of an Acrobot having a limited range of joint angle of two links. In Proceedings of the 5th Asian Control Conference, Melbourne, Victoria, Australia, 20–23 July 2004.
19. Zhao, Y.; Cheng, H.; Zhang, X. Swing control for two-link brathiation robot based on SMC. In Proceedings of the Chinese Control and Decision Conference, Yantai, Shandong, China, 2–4 July 2008.
20. Fukuda, T.; Saito, F.; Arai, F. A study on the brachiation type of mobile robot (heuristic creation of driving input and control using CMAC). In Proceedings of the IEEE/RSJ International Workshop on Intelligent Robots and Systems, Osaka, Japan, 3–5 November 1991.
21. Yamafuji, K.; Fukushima, D.; Maekawa, K. Study of a mobile robot which can shift from one horizontal bar to another using vibratory excitation. *JSME Int. J.* **1992**, *35*, 456–461. [CrossRef]
22. Saito, F.; Fukuda, T.; Arai, F. Swing and locomotion control for a two-link brachiation robot. *IEEE Control Syst.* **1994**, *14*, 5–12. [CrossRef]

23. Hasegawa, Y.; Fukuda, T.; Shimojima, K. Self-scaling reinforcement learning for fuzzy logic controller-applications to motion control of two-link brachiation robot. *IEEE Trans. Ind. Electron.* **1999**, *46*, 1123–1131. [CrossRef]
24. Nakanishi, J.; Fukuda, T.; Koditschek, D.E. A brachiating robot controller. *IEEE Trans. Robot. Autom.* **2000**, *16*, 109–123. [CrossRef]
25. Xin, X.; Liu, Y. A set-point control for a two-link underactuated robot with a flexible elbow joint. *J. Dyn. Syst. Meas. Control* **2013**, *135*, 051016. [CrossRef]
26. Ordaz, P.; Espinoza, E.S.; Muñoz, F. Research on swing up control based on energy for the pendubot system. *J. Dyn. Syst. Meas. Control* **2014**, *136*, 041018. [CrossRef]
27. Nishimura, H.; Funaki, K. Motion control of brachiation robot by using final-state control for parameter-varying systems. In Proceedings of the 35th IEEE Conference on Decision and Control, Kobe, Japan, 13 December 1996.
28. Odagaki, H.; Moran, A.; Hayase, M. Analysis of the dynamics and nonlinear control of under-actuated brachiation robots. In Proceedings of the 36th SICE Annual Conference, Tokushima, Japan, 29–31 July 1997.
29. Nishimura, H.; Funaki, K. Motion control of three-link brachiation robot by using final-state control with error learning. *IEEE/ASME Trans. Mechatron.* **1998**, *3*, 120–128. [CrossRef]
30. Lai, X.Z.; Pan, C.Z.; Wu, M.; Yang, S.X.; Cao, W.H. Control of an underactuated three-link passive–active–active manipulator based on three stages and stability analysis. *J. Dyn. Syst. Meas. Control* **2014**, *137*, 021007. [CrossRef]
31. Toshio, F.; Keiichi, I.; Sekiyama, K. Vision-based real time trajectory adjustment for brachiation robot. In Proceedings of the International Symposium on Micro-NanoMechatronics and Human Science, Nagoya, Japan, 5–8 November 2006.
32. Hasegawa, Y.; Fukuda, T. Motion coordination of behavior-based controller for brachiation robot. In Proceedings of the IEEE International Conference on Systems, Man, and Cybernetics, Tokyo, Japan, 12–15 October 1999.
33. Hasegawa, Y.; Tanahashi, H.; Fukuda, T. Continuous locomotion of brachiation robot by behavior phase shift. In Proceedings of the Joint 9th IFSA World Congress and 20th NAFIPS International Conference, Vancouver, BC, Canada, 25–28 July 2001.
34. Hasegawa, Y.; Tanahashi, H.; Fukuda, T. Behavior coordination of brachiation robot based on behavior phase shift. In Proceedings of the IEEE/RSJ International Conference on Intelligent Robots and Systems, Maui, HI, USA, 29 October–3 November 2001.
35. Kajima, H.; Hasegawa, Y.; Fukuda, T. Learning algorithm for a brachiating robot. *Appl. Bionics Biomech.* **2003**, *1*, 57–66. [CrossRef]
36. Gomes, M.W.; Ruina, A.L. A five-link 2D brachiating ape model with life-like zero-energy-cost motions. *J. Theor. Biol.* **2005**, *237*, 265–278. [CrossRef] [PubMed]
37. Full, R.J.; Koditschek, D.E. Templates and anchors: Neuromechanical hypotheses of legged locomotion on land. *J. Exp. Biol.* **1999**, *202*, 3325–3332. [PubMed]
38. Heglund, N.C.; Taylor, C.R.; McMahon, T.A. Scaling stride frequency and gait to animal size: Mice to horses. *Science* **1974**, *186*, 1112–1113. [CrossRef] [PubMed]

Review

Perception-Driven Obstacle-Aided Locomotion for Snake Robots: The State of the Art, Challenges and Possibilities †

Filippo Sanfilippo [1,*], Jon Azpiazu [2], Giancarlo Marafioti [2], Aksel A. Transeth [2], Øyvind Stavdahl [1] and Pål Liljebäck [1]

1 Department of Engineering Cybernetics, NTNU – Norwegian University of Science and Technology, 7491 Trondheim, Norway; oyvind.stavdahl@ntnu.no (Ø.S.); Pal.Liljeback@itk.ntnu.no (P.L.)
2 Mathematics and Cybernetics, SINTEF Digital, 7465 Trondheim, Norway; jon.azpiazu@gmail.com (J.A.); Giancarlo.Marafioti@sintef.no (G.M.); Aksel.A.Transeth@sintef.no (A.A.T.)
* Correspondence: filippo.sanfilippo@ntnu.no; Tel.: +47-942-58-929

† This paper is an extended version of our paper published in Sanfilippo, F.; Azpiazu, J.; Marafioti, G.; Transeth, A.A.; Stavdahl, Ø.; Liljebäck, P. A review on perception-driven obstacle-aided locomotion for snake robots. In Proceedings of the 14th International Conference on Control, Automation, Robotics and Vision (ICARCV), Phuket, Thailand, November 13–15, 2016; pp. 1–7.

Academic Editors: Toshio Fukuda, Fei Chen and Qing Shi
Received: 8 February 2017; Accepted: 25 March 2017; Published: 29 March 2017

Abstract: In nature, snakes can gracefully traverse a wide range of different and complex environments. Snake robots that can mimic this behaviour could be fitted with sensors and transport tools to hazardous or confined areas that other robots and humans are unable to access. In order to carry out such tasks, snake robots must have a high degree of awareness of their surroundings (i.e., *perception-driven* locomotion) and be capable of efficient obstacle exploitation (i.e., *obstacle-aided locomotion*) to gain propulsion. These aspects are pivotal in order to realise the large variety of possible snake robot applications in real-life operations such as fire-fighting, industrial inspection, search-and-rescue, and more. In this paper, we survey and discuss the state of the art, challenges, and possibilities of perception-driven obstacle-aided locomotion for snake robots. To this end, different levels of autonomy are identified for snake robots and categorised into environmental complexity, mission complexity, and external system independence. From this perspective, we present a step-wise approach on how to increment snake robot abilities within guidance, navigation, and control in order to target the different levels of autonomy. Pertinent to snake robots, we focus on current strategies for snake robot locomotion in the presence of obstacles. Moreover, we put obstacle-aided locomotion into the context of perception and mapping. Finally, we present an overview of relevant key technologies and methods within environment perception, mapping, and representation that constitute important aspects of perception-driven obstacle-aided locomotion.

Keywords: obstacle-aided locomotion; environment perception; snake robots

---

1. Introduction

Bio-inspired robots have developed rapidly in recent years. Despite the great success of bio-robotics in mimicking biological snakes, there is still a large gap between the performance of bio-mimetic robot snakes and biological snakes. In nature, snakes are capable of performing an astounding variety of tasks. They can locomote, swim, climb, and even glide through the air in some species [1]. However, one of the most interesting features of biological snakes is their ability to exploit roughness in the terrain for locomotion [2], which allows them to be remarkably adaptable to different types of environments. To achieve this adaptability and to locomote more efficiently, biological snakes may push against rocks, stones, branches, obstacles, or other environment irregularities. They can also exploit the walls and surfaces of narrow passages or pipes for locomotion.

Building a robotic snake with such agility is one of the most attractive steps to fully mimic the movement of biological snakes. The development of such a robot is motivated by the fact that different applications may be realised for use in challenging real-life operations, in earthquake-hit areas, pipe inspection for the oil and gas industry, fire-fighting operations, and search-and-rescue. Snake robot locomotion in a cluttered environment where the snake robot utilises walls or external objects other than the flat ground as means of propulsion can be defined as *obstacle-aided locomotion* [3,4]. To achieve such a challenging control scheme, a mathematical model that includes the interaction between the snake robot and the surrounding operational environment is beneficial. This model can take into account the external objects that the snake robot uses in the environment as push-points to propel itself forwards. From this perspective, the environment perception, mapping, and representation is of fundamental importance for the model. To highlight this concept even further, we adopt the term *perception-driven obstacle-aided locomotion* as locomotion where the snake robot utilises a sensory-perceptual system to exploit the surrounding operational space and identifies walls, obstacles, or other external objects for means of propulsion [5]. Consequently, we can provide a more comprehensive characterization of the whole scientific problem considered in this work. The underlying idea is shown in Figure 1. The snake robot exploits the environment for locomotion by using augmented information: obstacles are recognised, potential push-points are chosen (shown as cylinders), while achievable normal contact forces are illustrated by arrows.

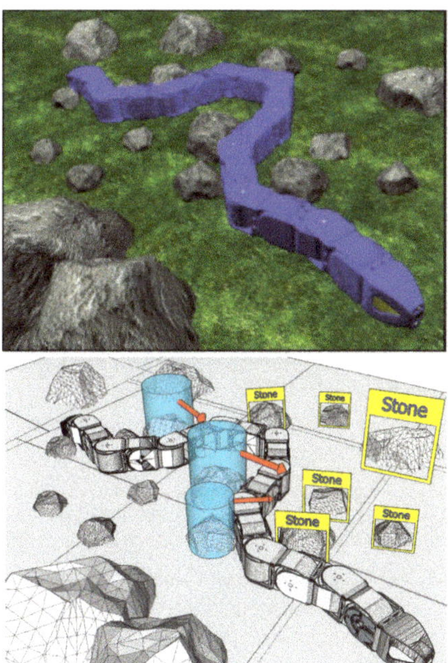

Figure 1. The underlying idea of snake robot perception-driven obstacle-aided locomotion: a snake robot perceives and understands its environment in order to utilise it optimally for locomotion.

The goal of this paper is to further raise awareness of the possibilities with perception-driven obstacle-aided locomotion for snake robots and provide an up-to-date stepping stone for continued research and development within this field.

In this paper, we survey the state of the art within perception-driven obstacle-aided locomotion for snake robots, and provide an overview of challenges and possibilities within this area of snake robotics. We propose the identification of different levels of autonomy, such as environmental complexity, mission complexity, and external system dependence. We provide a step-wise approach on how to increment snake robot abilities within guidance, navigation, and control in order to target the different levels of autonomy. We review current strategies for snake robot locomotion in the presence of obstacles. Moreover, we discuss and present an overview of relevant key technologies and methods within environment perception, mapping, and representation which constitute important aspects of perception-driven obstacle-aided locomotion. The contribution of this work is summarised by three fundamental remarks:

1. necessary conditions for lateral undulation locomotion in the presence of obstacles;
2. lateral undulation is highly dependent on the actuator torque output and environmental friction;
3. knowledge about the environment and its properties, in addition to its geometric representation, can be successfully exploited for improving locomotion performance for obstacle-aided locomotion.

The paper is organised as follows. A survey and classification of snake robots is given in Section 2. Successively, the state-of-the-art concerning control strategies for obstacle-aided locomotion is described in Section 3. Challenges related to the environment perception, mapping, and representation are described in Section 4. Finally, conclusions and remarks are outlined in Section 5.

2. Challenges and Possibilities in the Context of Autonomy Levels for Unmanned Systems

In this section, we survey snake robots as unmanned vehicle systems to highlight challenges and possibilities in the context of autonomy levels for unmanned systems. Moreover, we focus on the ability of snake robots to perform a large variety of tasks in different operational environments. When designing such systems, different levels of autonomy can be identified from an operational point of view. Based on this idea, the so-called framework of autonomy levels for unmanned systems (ALFUS) [6] is successively adopted and applied to provide a more in-depth overview for the design of snake robot perception-driven obstacle-aided locomotion. Following this, the autonomy and technology readiness assessment (ATRA) framework [7,8] is adopted and presented to better understand the design of these systems.

2.1. Classification of Snake Robots as Unmanned Vehicle Systems

An uncrewed or unmanned vehicle (UV) is a mobile system not having or needing a person, a crew, or staff operator on board [9]. UV systems can either be remote-controlled or remote-guided vehicles, or they can be autonomous vehicles capable of sensing their environment and navigating on their own. UV systems can be categorised according to their operational environment as follows: unmanned ground vehicle (UGV; e.g., autonomous cars or legged robots); unmanned surface vehicle (USV; i.e., unmanned systems used for operation on the surface of water); autonomous underwater vehicle (AUV) or unmanned undersea/underwater vehicle (UUV) for the operation underwater; unmanned aerial vehicle (UAV), such as unmanned aircraft generally known as "drones"; unmanned spacecraft, both remote-controlled ("unmanned space mission") and autonomous ("robotic spacecraft" or "space probe").

According to this terminology, snake robots can be classified as uncrewed vehicle (UV) systems. In particular, a snake robot constitutes a highly adaptable UV system due to its potential ability to perform a large variety of tasks in different operational environments. Following the standard nomenclature for UV systems, snake robots can be classified as shown in Figure 2, where some of the most significant systems are reported based on the literature for the sake of illustration. Several snake robots are implemented as unmanned ground vehicle (UGV) systems that are capable of performing the following tasks:

- locomotion on flat or slightly rough surfaces, such as the *ACM III* snake robot [10], which was the world's first snake robot, or the *toroidal skin drive* (TSD) snake robot [11], which is equipped with a skin drive propulsion system;
- climbing slopes, pipes, or trees, such as the *Creeping snake Robot* [12], which is capable of obtaining an environmentally-adaptable body shape to climb slopes, or the *PIKo* snake robot [13], which is equipped with a mechanism for navigating complex pipe structures, or the *Uncle Sam* snake robot [14], which is provided with a strong and compact joint mechanism for climbing trees;
- locomoting in the presence of obstacles, such as the *Aiko* snake robot [3], which is capable of pushing against external obstacles apart from a flat ground, or the *Kulko* snake robot [15], which is provided with a contact force measurement system for obstacle-aided locomotion.

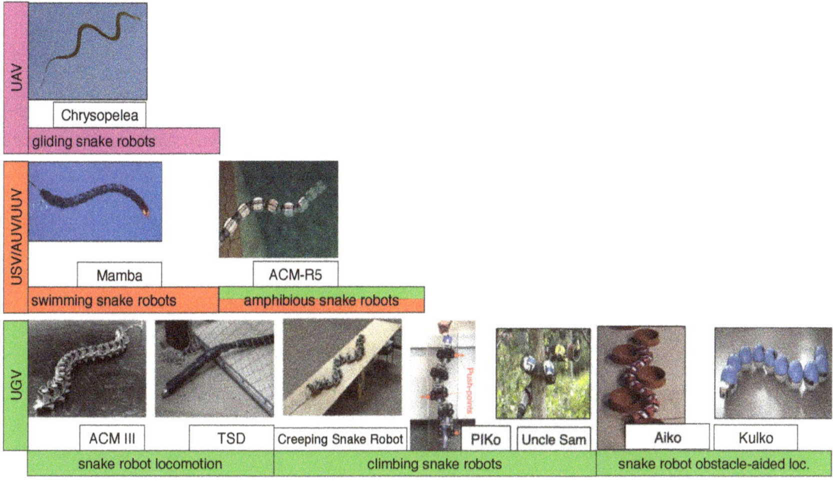

Figure 2. The variety of possible application scenarios for snake robots. AUV: autonomous underwater vehicle; TSD: toroidal skin drive snake robot; UAV: unmanned aerial vehicle; UGV: unmanned ground vehicle; USV: unmanned surface vehicle; UUV: unmanned undersea/underwater vehicle.

To better assess the different working environments in which these robotic systems operate, we may consider various metrics. For instance, the environmental complexity (EC)—a measure of entropy and the compressibility of the environment as seen by the robot's sensors [16]—may be used. Another useful metric is the mission complexity (MC), which is an estimation of the complicatedness of the environment as seen by the robot's perception system [6]. Both environmental complexity EC and MC gradually increase when moving from the former examples to the last ones. With a slight additional increase in terms of EC and MC, other snake robot systems are designed as unmanned surface vehicle (USV), autonomous underwater vehicle (AUV), or unmanned undersea/underwater vehicle (UUV) systems. An example of such systems is the *Mamba* snake robot [17], which is capable of performing underwater locomotion. Another notable example of such systems is the *ACM-R5* snake robot [18], which is an amphibious snake-like robot characterised by its hermetic dust and waterproof body structure.

With an additional increase in terms of both EC and MC, flying or gliding snake robots may be designed as unmanned aerial vehicle (UAV) systems. These systems can be inspired by the study and analysis of the gliding capabilities in flying snakes, such as the *Chrysopelea*, more commonly known as the flying snake or gliding snake [19]. This category of snake robot still does not exist, but perhaps one day it will be realised when the necessary technological advances become a reality. However, other designs may be more practical for flying.

2.2. Similarities and Differences between Traditional Snake Robots and Snake Robots for Perception-Driven Obstacle-Aided Locomotion

Some similarities can be identified between traditional snake robots and snake robots specifically designed to achieve perception-driven obstacle-aided locomotion. Both have a number of links serially attached to each other by means of joints that can be moved by some type of actuator. Therefore, they share similar structures from a kinematic point of view. However, snake robots specifically designed for obstacle-aided locomotion must be capable to exploit roughness in the terrain for locomotion. To achieve this, both their sensory system and their control system fundamentally differ with respect to the systems of traditional snake robots, which instead aim at avoiding or at most accommodating obstacles. The design guidelines for snake robot perception-driven obstacle-aided locomotion are described in the following sections of the paper.

2.3. The ALFUS Framework for Snake Robot Perception-Driven Obstacle-Aided Locomotion

In this work, the focus is on snake robots designed as unmanned ground vehicle (UGV) systems with the aim of achieving perception-driven obstacle-aided locomotion. To the best of our knowledge, little research has been done in the past concerning this topic. When designing such systems, different levels of autonomy can be identified from an operational point of view. In Section 2.1, we have already briefly touched upon the concepts of environmental complexity (EC) and mission complexity (MC) to better categorise existing snake robots. By additionally considering the external system independence (ESI) metric, which represents the independence of snake robots from other external systems or from human operators, the so-called ALFUS framework [6] can be adopted and applied to provide a more in-depth overview for the design of snake robot perception-driven obstacle-aided locomotion, as shown in Figure 3.

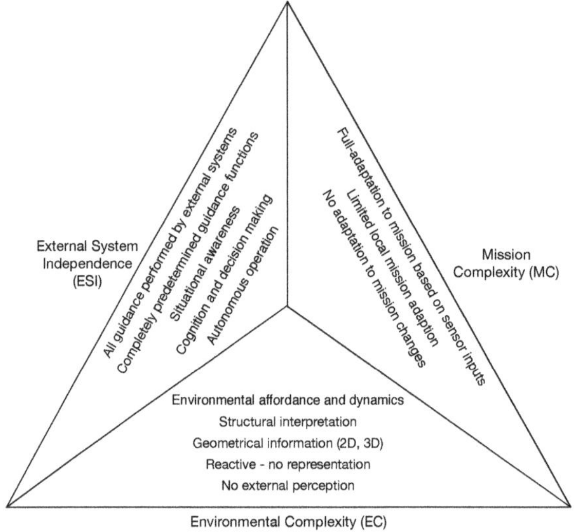

Figure 3. The autonomy levels for unmanned systems (ALFUS) framework [6] applied to snake robot perception-driven obstacle-aided locomotion.

The levels of autonomy for a system can be defined for various aspects of the system and categorised according to several taxonomies [20]. Regarding the external system independence for snake robots, we introduce the following autonomy levels (AL) according to gradually increased complexity:

- All guidance performed by external systems. In order to successfully accomplish the assigned mission within a defined scope, the snake robot requires full guidance and interaction with either a human operator or other external systems;
- Completely predetermined guidance functions. All planning, guidance, and navigation actions are predetermined in advance based on perception. The snake robot is capable of very low adaptation to environmental changes;
- Situational awareness [21]. The snake robot has a higher level of perception and autonomy with high adaptation to environmental changes. The system is not only capable of comprehending and understanding the current situation, but it can also make an extrapolation or projection of the actual information forward in time to determine how it will affect future states of the operational environment;
- Cognition and decision making. The snake robot has higher levels of prehension, intrinsically safe cognition, and decision-making capacity for reacting to unknown environmental changes;
- Autonomous operation. The snake robot is capable of fully autonomous capabilities. The system can achieve its assigned mission successfully without any intervention from human or any other external system while adapting to different environmental conditions.

Concerning the environmental complexity (EC), the following autonomy levels (AL) can be identified according to gradually increased complexity:

- No external perception. The snake robot executes a set of preprogrammed or planned actions in an open loop manner;
- Reactive—no representation. The snake robot does not generate an explicit environment representation, but the motion planner is able to react to sensor input feedback;
- Geometrical information (2D, 3D). Starting from sensor data, the snake robot can generate a geometric representation of the environment which is used for planning—typically for obstacle avoidance;
- Structural interpretation. The environment representation includes structural relationships between objects in the environment;
- Environmental affordance and dynamics. Higher-level entities and properties can be derived from the environment perception, including separate treatment for static and dynamic elements; different properties from the objects which the snake robot is interacting with might be of interest according to the specific task being performed.

With reference to mission complexity (MC), the following autonomy levels (AL) can be identified according to gradually increased complexity:

- No adaptation to mission changes. The mission plan is predetermined, the snake robot is not capable of any adaptation to mission changes;
- Limited local mission adaptation. The snake robot has low adaptation capabilities to small, externally-commanded mission changes;
- Full-adaptation to mission based on sensor inputs. The snake robot has high and independent adaptation capabilities.

Clearly, in all three cases, an increase of the complexity generates more challenging problems to solve; however, it opens new possibilities for snake robot applications as autonomous systems.

2.4. A Framework for Autonomy and Technology Readiness Assessment

To better understand the design of snake robot perception-driven obstacle-aided locomotion, some design examples of similarly demanding systems can be considered as sources of inspiration and prototyping purposes. For instance, the design of autonomous unmanned aircraft systems (UAS) may provide solid directions for establishing a flexible design and prototyping framework. In particular, to systematically evaluate the autonomy levels (AL) of a UAS and to correctly measure the maturity of their autonomy-enabling technologies, the autonomy and technology readiness assessment (ATRA) framework may be adopted [7,8]. The ATRA framework combines both autonomy levels (AL) and technology readiness level (TRL) metrics. Borrowing this idea from UAS, the same framework concept can be used to provide a comprehensive picture of how snake robot perception-driven obstacle-aided locomotion may be realised in a realistic operational environment, as shown in Figure 4.

Level	Description	Guidance	Navigation	Control	ESI	EC	MC
9	Fully autonomous	Human-level decision-making that outperform biological snakes, accomplishment of most missions without any intervention from ES (100% ESI).	Human-like navigation capabilities that outperform biological snakes for most missions. Situational awareness in extremely complex environments and situations.	Same or better control performance as for biological snakes in the same situation and conditions.	100% ESI	Extreme EC	Highest Level MC
8	Full mission planning	High-level decision making. Evaluation and optimisation of mission performance.	Higher level entities and properties are derived from the environment perception according to the desired task to be performed.	Same as previous levels.	High Level ESI	Difficult EC	High Level tasks
7	Dynamic global planning	Same as Level 6 but planning in a dynamic environment.	Same as Level 6 but mapping in a dynamic environment.	Same as previous levels.			
6	Global planning	Goal waypoint provided by ES. Global path planning determines optimal path to goal.	Global map includes structural properties of the environment. Localisation of snake robot relative to map.	Same as previous levels.			
5	Local planning with environment awareness	Same as Level 4 but local motion planner takes also into account structural properties of the environment (stiffness, friction, etc.).	Same as Level 4 but local map also includes structural properties of the environment (stiffness, friction, etc.).	Same as previous levels.	Mid Level ESI	Moderate EC	Mid Level MC
4	Local planning	Local motion planner optimises locomotion for given immediate surroundings. ES commands direction of locomotion.	Local mapping with geometrical representation of immediate surroundings. Localisation of snake robot relative to map.	Local adaptive control to compensate for possible deviations between map and actual environment.			
3	Reactive control	Motion planner reacts to sensor input feedback and detects if snake robot body is jammed in environment.	Snake robot can detect contacts between its own body and the environment. No mapping.	Local adaptation to resolve jammed joint(s) and/or local terrain adaptation.	Low Level ESI	Simple EC	Low Level tasks
2	Pre-planned motion	Pre-programmed motion patterns.	Same as Level 1.	Automatic control to follow specified motion pattern. No adaptation.			
1	Remote control	All guidance functions are performed by external systems (mainly human operator).	Sensors may be adopted, but all data is processed and analysed by a remote ES. No mapping. No localisation.	All control commands are given by a remote ES (mainly human operator) on joint level.	0% ESI	Lowest EC	Lowest MC

Figure 4. The ATRA framework [7,8] applied to snake robot perception-driven obstacle-aided locomotion with the different levels of external system independence (ESI), of environmental complexity (EC) and of mission complexity (MC). ES refers to "External System".

When considering snake robot perception-driven obstacle-aided locomotion, identifying and differentiating between consecutive autonomy levels is very challenging from a design point of view. Nevertheless, it is crucial to clearly distinguish autonomy levels during the design process in order to provide the research community with a useful evaluation and comparison tool. Inspired by similarly demanding systems [7], a nine-level scale is proposed based on gradual increase (autonomy as a gradual property) of guidance, navigation, and control (GNC) functions and capabilities. Referring to Figure 4, the key GNC functions that enable each autonomy level are verbally described along with their correspondences with mission complexity (MC), environmental complexity (EC), and external system independence (ESI) metrics (illustrated with a colour gradient). It should be noted that the motion planner is assumed to take the environment representation as input (i.e., a map of the environment built as a fusion between different sensor data and previous stored knowledge). Then, the control function is supposed to compensate with adjustments for any possible deviations between map and actual environment.

To a certain degree, the three key GNC functions can be independently considered with respect to the information flow, as shown in Figure 5a. The snake robot's sensory-perceptual data and external system commands are used to provide an input for the guidance system, which is responsible for decision-making, path-planning, and mission planning activities. The navigation system is responsible for achieving all the functions of perception, mapping, and localisation. The processed information is then adopted by the control system, which is responsible for low-level adaptation and control tasks.

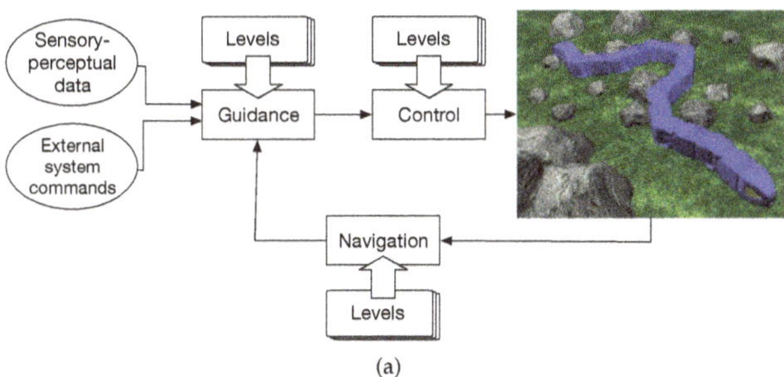

(a)

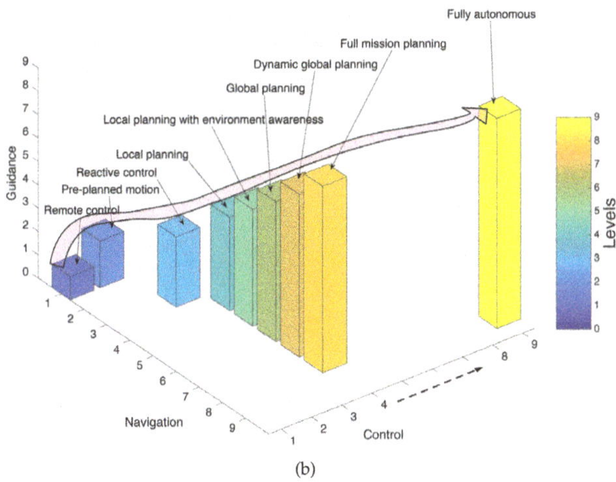

(b)

Figure 5. (**a**) Information flow through the different functions and capabilities of guidance, navigation, and control (GNC); (**b**) the possible design levels depicted in Figure 4 are represented in a three-dimensional space.

This design approach allows for the achievement a good level of modularity. For instance, concerning level 6 (global planning), the expected guidance function could first be carried out for a map which only includes a geometrical representation. At a later stage of the prototyping process, some structural properties of the environment may be gradually considered, such as stiffness, friction, and other parameters. Different combinations are possible, and they depend upon the particular prototyping approach that designers can adopt. The possible design levels depicted in Figure 4 are only provided for the sake of illustration. Our goal is to provide an overview of some of the main combinations of the three key GNC functions which we think are the most important steps in order to achieve fully autonomous snake robot operations.

Regarding our choice of describing a nine-level scale, it should be noted that the amount of effort needed in order to transit from one level to the next is far from equal for all levels. In particular, a significant and more challenging leap in functionality will be required to transit from level 7 to level 8, and from level 8 to level 9. We have chosen not to further refine the potentially existing sub-levels between these main levels (i.e., levels 7–9) in this paper. Instead, we have chosen an agile development-based and rapid-prototyping approach where we currently focus on refining the steps necessary in order to transit to the next immediate levels (compared to the state-of-the-art), and then leave the details for the higher levels until more information is available on the exact requirements and functionalities needed for these levels.

To provide a more intuitive overview of the design guidelines depicted in Figure 4, the same design levels can be represented in a three-dimensional space as shown in Figure 5b. It should be noted that control level 4 enables levels 4–8 concerning both guidance and navigation. The possible path that can be followed to achieve the desired level of independence is highlighted in the same figure.

It should be noted that the framework discussed in this work provides possible guidelines on how to design a system capable of adapting to more difficult terrains as the autonomy level of the robot increases. In particular, in Figure 4, the environmental complexity (EC) increases with the rise of complexity of the different design levels in the ATRA framework. Note that these design levels are also visualised in 3D in Figure 5b.

Referring to Figures 4 and 5b, the current cutting edge technology for snake robots can exhibit level 4 or at most level 5 characteristics to the best of the authors' knowledge. In fact, the authors believe that the current technology for snakes falls a little behind with respect to the advances of the current cutting-edge technology of non-snake-type robots.

3. Control Strategies for Obstacle-Aided Locomotion

The greater part of the existing literature on the control of snake robots considers motion across smooth—usually flat—surfaces. Different research groups have extensively investigated this particular operational scenario. Various approaches to mathematical modelling of the snake robot kinematics and dynamics have been presented as a means to simulate and analyse different control strategies [22]. In particular, many of the models presented in the early literature focus purely on kinematic aspects of locomotion [23,24], while more recent studies also include the dynamics of motion [25,26].

Among the different locomotion patterns inspired by biological snakes, lateral undulation is the fastest and most commonly implemented locomotion gait for robotic snakes in the literature [27]. This particular pattern can be realised through phase-shifted sinusoidal motion of each joint of the robotic snake [28]. This approach has been investigated for planar snake robots with metallic ventral scales [29] placed on the outer body of the robot, passive wheels [30], or for snake robots with anisotropic ground friction properties [31].

Even though these previous studies have provided researchers with a better understanding of snake robot dynamics, most of the past works on snake robot locomotion have almost exclusively considered motion across smooth surfaces. However, many real-life environments are not smooth, but cluttered with obstacles and irregularities. When the operational scenario is characterised by a surface that is no longer assumed to be flat and which has obstacles present, snake robots can move by sensing the surrounding environment. In the existing literature, not much work has been done to develop control tools specifically designed for this particular operational scenario. Next, we analyse and group relevant literature for snake robot locomotion in environments with obstacles, as shown in Table 1.

3.1. Obstacle Avoidance

A traditional approach to dealing with obstacles consists of trying to avoid them. Collisions may make the robot unable to progress and cause mechanical stress or damage to equipment. Therefore, different studies have focused on obstacle avoidance locomotion. For instance, principles of artificial potential field (APF) theory [32] have been adopted to effectively model imaginary force fields around objects that are either repulsive or attractive on the robot. The target position emits an attractive force field while obstacles, other robots, or the robot itself emits repulsive force fields. The strength of these forces may increase as the robot gets closer. Based on these principles, a controller capable of obstacle avoidance was presented in [33]. However, the standard APF approach may cause the robot to end up trapped in a local minima. In this case, the repulsive forces from nearby obstacles may leave the robot unable to move. To escape local minima, a hybrid control methodology using APF integrated with a modified simulated annealing (SA) optimization algorithm for motion planning of a team of multi-link snake robots was proposed in [34]. An alternative methodology was developed in [35]. Central pattern generators (CPGs) were employed to allow the robot to avoid obstacles or barriers by turning the robot's body from its trajectory. A phase transition method was also presented in the same work, utilising the phase difference control parameter to realise the turning motion. This methodology also provides a way to incorporate sensory feedback into the CPG model, allowing for the detection of possible collisions.

3.2. Obstacle Accommodation

By using sensory feedback, a more relaxed approach to obstacle avoidance can be considered. Rather than absolutely avoiding collisions, the snake robot may be allowed to collide with obstacles, but collisions must be controlled so that no damage to the robot occurs. This approach was first investigated in [36], where a motion planning system was implemented to provide a snake-like robot with the possibility of accommodating environmental obstructions by continuing the motion towards the target while in contact with the obstacles. In [37], a general formulation of the motion constraints due to contact with obstacles was presented. Based on this formulation, a new inverse kinematics model was developed that provides joint motion for snake robots under contact constraints. By using this model, a motion planning algorithm for snake robot motion in a cluttered environment was also proposed.

3.3. Obstacle-Aided Locomotion

Even though obstacle avoidance and obstacle accommodation are useful features for snake robot locomotion in unstructured environments, these control approaches are not sufficient to fully exploit obstacles as means of propulsion. As observed in nature, biological snakes exploit the terrain irregularities and push against them so that a more efficient locomotion gait can be achieved. In particular, the entire snake's body bends itself, and all sections consistently follow the path taken by the head and neck [2]. Snake robots may adopt a similar strategy. A key aspect of practical snake robots is therefore obstacle-aided locomotion [3,4]. To understand the mechanism underlying the functionalities of biological snakes on the basis of a synthetic approach, a model of a serpentine robot with viscoelastic properties was presented in [38,39]. It should be noted that the authors adopted the term *scaffold-assisted serpentine locomotion*, which is conceptually similar to the idea of obstacle-aided locomotion. The authors also designed an autonomous decentralised control scheme that employs local sensory feedback based on the muscle length and strain of the snake body, the latter of which is generated by the body's softness. Through modelling and simulations, the authors demonstrated that only two local reflexive mechanisms which exploit sensory information about the stretching of muscles and the pressure on the body wall, are crucial for realising locomotion. This finding may help develop robots that work in undefined environments and shed light on the understanding of the fundamental principles underlying adaptive locomotion in animals.

Table 1. Snake locomotion in unstructured environments.

Obstacle Avoidance	Obstacle Accommodation	Obstacle-Aided Locomotion
[32–35]	[36,37]	[3,4,40–48]

However, to the best of our knowledge, little research has been done concerning the possibility of applying this locomotion approach to snake robots. For instance, a preliminary study aimed at understanding snake-like locomotion through a novel push-point approach was presented in [40].

Remark 1. *In [40], an overview of the lateral undulation as it occurs in nature was first formalised according to the following conditions:*

- *it occurs over irregular ground with vertical projections;*
- *propulsive forces are generated from the lateral interaction between the mobile body and the vertical projections of the irregular ground, called push-points;*
- *at least three simultaneous push-points are necessary for this type of motion to take place;*
- *during the motion, the mobile body slides along its contacted push-points.*

Based on the conditions described in Remark 1, the authors of the same work considered a generic planar mechanism and a related environment that suit to satisfy the fundamental mechanical phenomenon observed in the locomotion of terrestrial snakes. A simple control law was applied and tested via dynamic simulations with the purpose of calculating the contact forces required to propel the snake robot model in a desired direction. Successively, these findings were tested with practical experiments in [41], where closed-loop control of a snake-like locomotion through lateral undulation was presented and applied to a wheel-less snake-like mobile mechanism. To sense the environment and to implement this closed-loop control approach, simple switch sensors located on the side of each module were adopted. A more accurate sensing approach was introduced in [42], where a design process for the electrical, sensing, and mechanical systems needed to build a functional robotic snake capable of tactile and force sensing was presented. Through manipulation of the body shape, the robot was able to move in the horizontal plane by pushing off of obstacles to create propulsive forces. Instead of using additional hardware, an alternative and low-cost sensing approach was examined in [43], where robot actuators were used as sensors to allow the system to traverse an elastically deformable channel with no need for external tactile sensors.

Some researchers have focused on asymmetric pushing against obstacles. For instance, a control method with a predetermined and fixed pushing pattern was presented in [44]. In this method, the information of contact affects not only adjacent joints but also a couple of neighboring joints away from a contacting link. Furthermore, the distribution of the joint torques is empirically set asymmetrically in order to propel the snake robot forward. Later on, a more general and randomised control method that prevents the snake robot to get stuck in crowded obstacles was proposed by the same research group in [45].

When locomoting through environments with obstacles, it is also important to achieve body shape compliance for the snake robot. Some researchers have focused on shape-based control approaches, where a simple motion pattern is propagated along the snake's body and dynamically adjusted according to the surrounding obstacles. For instance, a general motion planning framework for body shape control of snake robots was presented in [46]. The applicability of this framework was demonstrated for straight line path following control, and for implementing body shape compliance in environments with obstacles. Compliance is achieved by assigning mass-spring-damper dynamics to the shape curve defining the motion of the robot.

The idea of adopting a shape-based control approach to achieve compliance is particularly challenging when considering snake robots with many degrees of freedom (DOF). In fact, having many DOF is both a potential benefit and a possible disadvantage. A snake robot with a large number of DOF can better comply to and therefore better move in complex environments. Yet, possessing many degrees of freedom is only an advantage if the system is capable of coordinating them to achieve desired goals in real-time. In practice, there is a trade-off between the capability of highly articulated robots to navigate complex unstructured environments and the high computational cost of coordinating their many DOF. To facilitate the coordination of many DOF, the robot shape may be used as an important component in creating a middle layer that links high-level motion planning to low-level control for the robust locomotion of articulated systems in complex terrains. This idea was presented in [47], where shape functions were introduced as the layer of abstraction that formed the basis for creating this middle layer. In particular, these shape functions were used to capture joint-to-joint coupling and provide an intuitive set of controllable parameters that adapt the system to the environment in real-time. This approach was later extended to the definition of spatial frequency and temporal phase serpenoid shape parameters [48]. This approach provides a way to intuitively adapt the shape of highly articulated robots using joint-level torque feedback control, allowing a robot to navigate its way autonomously through unknown, irregular environment scenarios.

Remark 2. *Most of the previous studies highlight the fact that lateral undulation is highly dependent on the actuator torque output and environmental friction.*

Based on Remark 2, interesting approaches were discussed in [4,49,50]. The authors focused on how to optimally use the motor torque inputs, which result in obstacle forces suitable to achieve a user-defined desired path for a snake robot. Less importance was given to the forces generated by the environmental friction. In detail, assuming a desired snake robot trajectory and the desired link angles at the obstacles in contact, the contribution was on how to map the given trajectory to obstacle contact forces, and these forces to control inputs. Based on a 2D-model of the snake robot, a convex quadratic programming problem was solved, minimizing the power consumption and satisfying the maximum torque constraint and the obstacle constraints. To implement this strategy in practice, the optimization problem needs to be solved on-line recursively. In addition, when the configuration of the contact points changes, some of the parameters need to be updated accordingly. Finally, as the authors pointed out, there are two main issues to practically using their method for obstacle-aided locomotion. The first is the definition of an automatic method for finding the desired link angles at the obstacles. The second is the automatic calculation of the desired path. However, an interesting result is that one could use the approach in [4] to check the quality of a given path by verifying if useful forces can be generated by the interaction with a number of obstacles for that path, and this could be done off-line.

4. Environment Perception, Mapping, and Representation for Locomotion

In order for robots to be able to operate autonomously and interact with the environment in any of the ways mentioned in Section 3 (obstacle avoidance, obstacle accommodation, or obstacle-aided locomotion), they need to acquire information about the environment that can be used to plan their actions accordingly. This task can be divided into three different challenges that need to be solved:

1. *sensing*, using the adequate sensor or sensor combinations to capture information about the environment;
2. *mapping*, which combines and organises the sensing output in order to create a representation that can be exploited for the specific task to be performed by the robot;
3. *localisation*, which estimates the robot's pose in the environment representation according to the sensor inputs.

These topics are well studied for different types of robots and environments, and are tackled by the Simultaneous Localisation And Mapping (SLAM) community which has been the foremost research area for the last years in robotics [51]. However, comparatively little work has been done in this field for snake robots, as research has been focused on understanding the fundamentals of snake locomotion, and on the development of the control techniques.

Table 2 summarises the sensors most commonly found in the robotics literature for environment perception aimed at navigation. The table also contains some basic evaluation of the suitability of the specific sensor or sensing technology for the requirements and limitations of snake robots. For the sake of completeness, we have included references to representative robots different from snake robots for those technologies where we were not able to find examples of applications involving snake robots.

Table 2. Sensors for environment perception

Sensor/Sensing Technology	Pros	Cons	References
Proprioceptive	No need for additional payload.	Depends on accuracy of the robot's model. Low level of detail. Does not allow to plan in advance.	[52,53]
Contact/Force	Bioinspired. Suitable for simple obstacle-aided locomotion.	Low level of detail. Reactive, does not allow to plan in advance.	[10,15,54–58]
Proximity (US and IR)	Suitable for simple obstacle-aided locomotion. Allows for some lookahead planning.	Low level of detail. Additional payload.	[59–62]
LiDAR	Well-known sensor in robotics community. Provides dense information about environment.	Usually requires sweeping and/or rotating movement for full 3D perception.	[62–67]
Laser triangulation	Provides high accuracy measurements.	Very limited measurement range. Requires sweeping movement. Limitations in dynamic environments.	[68]
ToF camera	Provides direct 3D measurements.	Low resolution, low accuracy. Not suitable for outdoor operation.	[69,70]
Structured light—Temporal coding	Provides direct 3D measurements. High accuracy, high resolution.	Limited measurement range. Not suitable for outdoor operation. Limitations in dynamic environments. Sensor size.	Non-snake: [71]
Structured light—Light coding	Provides direct 3D measurements. Small sensor size.	Noisy measurements. Not suitable for outdoor operation.	Non-snake: [72]
Stereovision	High accuracy, high resolution, wide range.	Measuring range limited by available baseline. Computationally demanding. Dependent on texture.	Non-snake: [73]
Monocular—SfM	Small sensor, lightweight, low power. Wide measurement range. No active lighting.	Computationally demanding. Dependent on texture. Scale ambiguity.	Non-snake: [74]
Radar (e.g., UWB)	Sense through obstacles.	Mechanical or electronic sweeping required. Computationally demanding.	Non-snake: [75]

4.1. Sensor Technologies for Environment Perception for Navigation in Robotics

The classification of sensors in Table 2 considers both the measuring principles and sensing devices involved. For more clarity, Figure 6 summarises the sensor technologies that have been taken into account as the most representative ones found in literature regarding environment perception for navigation.

The qualifier *proprioceptive* applied to sensors is used in robotics to distinguish between those measures of values internal to the robot (e.g., motor speed) from other sensors that obtain information from the robot's environment (e.g., distance to objects)—that is, exteroceptive. Contact sensors can inform whether the robot is touching elements from the environment in some specific locations, or even measure the pressure being applied. Range sensors provide distance measurements to nearby objects.

Proximity sensors [76] are intended for short-range distance measurements. They are active sensors that emit either an ultrasound (US) or infrared (IR) pulse and calculate the distance to the closest obstacle based on the time-of-flight principle; i.e., measuring the time it takes for the pulse to travel from the emitter to the obstacle, and then back to the receiver. In the case of infrared-based sensors, in order to simplify the electronics, in some cases instead of measuring the actual travelling time, the emitted pulse is modulated, and the basics for estimating the distance is the shift in the phase of the received signal with respect to the emitted one. Light detection and ranging (LiDAR) sensors are also based on the time-of-flight principle, emitting pulsed laser which is deflected by a rotating mirror. In this way, instead of obtaining a single measurement, the LiDAR system is able to scan the surroundings and provide measurements in a plane at specified angle intervals. Often in literature, LiDARs are mounted on an additional rotating element [77] that allows a full 3D reconstruction to be produced by aligning the subsequent scans in a single point cloud. Radars actuate under similar principles to LiDARs, but use electromagnetic waves instead of relying on optical signals.

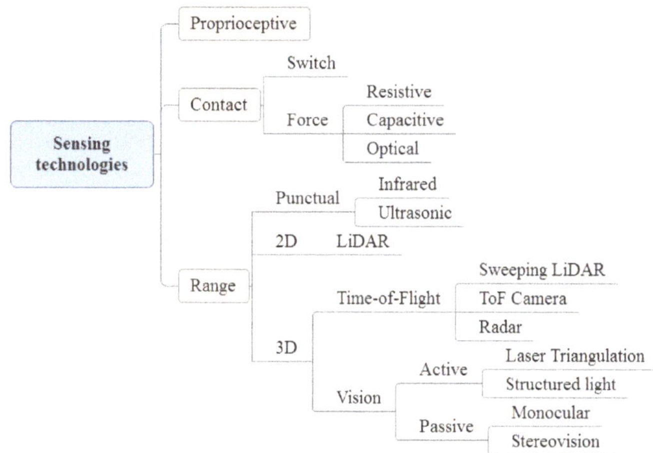

Figure 6. A taxonomy of sensors for environment perception for navigation in robotics.

Active 3D imaging techniques are also referred to as *structured light* [78]; the term "active" is used here in opposition to "passive" to denote that a controlled light source is projected onto the scene. In Table 2, we distinguish between laser triangulation, structured light, and light coding. *Laser triangulation* sensors consist of a laser source calibrated with respect to a camera or imaging sensor (position and orientation). The laser generates a single-point or narrow stripe that is projected onto the scene or object being inspected. The projected element is deformed according to the shape of the scene; this deformation can be observed by the imaging sensor, and then the depth profile is estimated by simple triangulation principle. By providing an additional scanning movement, the different scans can be stuck together to generate a point cloud representation. In order to avoid the scanning movement, bi-dimensional patterns of non-coherent light are used. In literature there are multiple proposals for ways of designing the projected patterns to achieve better reconstructions (e.g., binary codes, gray codes, phase shifting). A clear distinction can be made between methods that perform a temporal coding (which require a sequence of patters that are successively projected onto the scene) versus spatial coding methods which concentrate all the coding scheme in a unique pattern. Following some rather extended nomenclature, we denote temporal coding methods just by *structured light*, and spatial methods by *light coding* [79]. Light coding methods have been popularized in the last years by the introduction of low-cost devices in the consumer market, such as the Microsoft Kinect. Those devices are also often referred to as RGB-D cameras. Despite the name, most commercially available time-of-flight cameras (*ToF camera*) [80] operate by measuring the phase shift of the modulated infrared signal bounced back. The main difference with respect to LiDAR systems is that in ToF cameras the whole scene is captured with each pulse, as opposed to one-point measurements in the case of LiDARs. ToF cameras can provide 3D measurements at high rates, but at the cost of low resolutions and accuracies in the range of a few centimetres.

Referring to passive 3D imaging techniques, *stereovision* [81] consists of two calibrated cameras that acquire images of a scene from two different viewpoints. Calibration involves both the optical characteristics of the cameras, but also the relative position and orientation between them. Points in the scene need to be recognized or matched across the images from the two cameras, and then their position is triangulated to calculate the depth. Stereovision methods can be classified as sparse or dense, according to the nature of the point cloud generated: sparse methods select certain interest points from the images to be triangulated, while dense methods operate on a per-pixel basis. A single camera can also be used to recover the 3D structure of an environment: structure from motion (*SfM*) [82] or visual SLAM (*vSLAM*) [83] methods use different images from a moving camera (or images taken from different viewpoints) to simultaneously estimate the positions from where the images were taken and the 3D structure of the environment.

4.2. Survey of Environment Perception for Locomotion in Snake Robots

Many of the snake robots found in literature are equipped with a camera located in the head of the robot and pointing in the same direction as the navigation direction. However, in most cases, the images captured by the camera are not used for visual feedback for the robot's control, but sent to the operator for her/him to plan the robot's trajectory or to solve the high level applications, which in the case of snake robots are usually inspection or search-and-rescue operations.

One way a snake robot can have a very simple perception of the surrounding environments is by using *proprioception*; i.e., using the robot's own internal state (e.g., joint angles, motor current levels, etc.) to derive some properties of the environment. As an example, [53] showed that the tilt angle of a slope can be inferred by only using the snake robot's state estimated from the joint angles, and used this information to adapt the behaviour of sidewinding and maximise the travelling speed. In [52], a heavily bioinspired method was developed based on the mechanics and neural control locomotion of a worm which also uses a serpentine motion. The objective in this case is to react and adapt to obstacles in the environment. The work in [58] implements strain gauges measuring the deformation in the joint actuators; from these measurements, the contact forces applied to a waterproof snake robot are calculated.

The sensing modality that is most commonly explored in snake robots for environment perception is *force* or *contact* sensors. The first snake robot in 1972 [10] already used contact switches along the body of the snake robot, and demonstrated lateral inhibition with respect to external obstacles. Apart from reacting to contact with obstacles, the main use of force sensors is to adapt the body of the snake robot to the irregularities of the terrain [56,57]. The authors in [54] claim to feature the first full-body 3D-sensing snake. The robot is equipped with 3-DOF force sensors integrated in the wheels (which are also actuated), and uses this data to equally distribute the weight of all the segments, apart from moving away from obstacles at the sides. In [55], a different application of contact sensors is shown, in which lateral switches distributed along the body of the snake are used as touch sensors to guarantee that there are enough push-point contacts so that propulsion can be performed. In case the last contact point was lost, the robot performs an exploratory movement. The snake robot in [15] is already designed aimed at obstacle-aided locomotion, by including two main features: a smooth exterior surface that allows the robot to glide, and contact force sensing by using four force-sensing resistors on each side of the joint module, and assuming that in locomotion on horizontal surfaces it is enough to know on which side of the robot the contact happens.

Robots that base their environment perception on contact-based sensors exclusively allow for limited motion planning. For example, when aiming at obstacle avoidance, it is impossible to achieve full avoidance, as the robot needs to make contact with the obstacle in order to realise its presence. However, in [84,85], it is demonstrated that environment representation can be achieved purely by contact sensors. In this case, whisker-like contact sensors were used in a SLAM framework to produce an environment representation that could potentially be used for planning and obstacle-aided locomotion purposes.

The use of range or *proximity* sensors allows snake robots to not rely on contact in order to perceive the environment, and thus perform obstacle avoidance. The works in [60,61] feature active infrared sensors used to implement reactive behaviours for avoiding obstacles, either by selecting an obstacle-free trajectory in the first case, or by adapting the undulatory motion for narrow corridor-like passages in the second one. It is worth noting that the snake robot in [60] is equipped with actively-driven tracks for propulsion in slippery terrains. Ultrasound sensors are used in a similar fashion, and their data can also be used to estimate the snake robot's speed in case it is approaching an obstacle [59]. A combination of ultrasound sensors for mapping and obstacle avoidance and passive infrared sensors for the detection of human life in urban search-and-rescue applications is proposed in [63], though details on how the sensor information is exploited is not provided.

A more detailed and accurate representation of the environment can be achieved by the use of LiDAR sensors, sometimes combined with ultrasound sensors as in [62]. The use of LiDAR even allows for the generation of richer and more complete maps. The use of such sensors in a SLAM framework is demonstrated in [64]: the snake robot is equipped with a LiDAR sensor in the head, and a camera and infrared sensors in the sides. The camera is used to provide position information to the remote operator who controls the desired velocity of the snake robot using a joystick. The snake then uses the LiDAR to perform SLAM, producing a map and an estimated position of the robot itself in that map. The robot then uses the output from the SLAM to navigate the environment, while using the infrared sensor information in a reactive way to avoid obstacles not detected by the LiDAR and overcome the errors in the SLAM. The work in [66] studies the use of SLAM in snake robots. It emphasizes two important challenges of SLAM in this type of robot in comparison with existing approaches of SLAM: on one hand, the use of odometry-less models, and on the other hand, the lack of features or landmarks that characterizes navigation environments such as the inside of tunnels and pipes, which is a typical focus for snake robots. A rotating LiDAR is used in [65] to scan the environment and generate a 2.5 dimensional map that then can be used to perform motion planning in 3D. The main objective of this system is to overcome challenging obstacles such as stairs, for which the robot also relies on active wheels. The point clouds generated by the LiDAR are also matched across time, estimating the relative localisation that is then used to correct the robot odometry. The work in [67] also approaches step climbing using a LiDAR sensor, which is used to calculate the relative position of the snake's head to the step and its height. Several measurements of the LiDAR are fused through an Extended Kalman Filter in order to reduce the measurement's uncertainty and detect the line segments that correspond to the different planes of the step. This information is then fed into a model predictive control (MPC) algorithm to generate a collision-free trajectory. In [66], the influence of two different SLAM algorithms using serpentine locomotion in a featureless environment is presented.

The use of onboard *vision* systems to perceive the environment and influence the snake robot's motion is limited in the literature. In a simplified example [86], a camera mounted in the head of the snake robot is adopted to detect a black tape attached on the ground and then use that information as the desired trajectory for the snake. Time-of-flight (ToF) cameras provide 3D information of the environment in the form of depth images or point clouds, without requiring any additional scanning movement. A modified version of the iterative closest point (ICP) algorithm is used in [69] to combine the information of an inertial measurement unit (IMU) with a ToF camera. The ICP [87] is a well-known algorithm that calculates the transformation (translation and rotation) to align two point clouds that minimises the mean squared error between the point pairs of the point clouds. The modifications proposed for the ICP are intended to speed up and increase the robustness. The objective of this process is to perform localisation and mapping. Localisation is demonstrated at four frames per second, while map construction was done offline. The concept is demonstrated in a very challenging scenario, which is the Collapsed House Simulation Facility, and adopting the *IRS Souryu* snake robot which uses actively driven tracks for propulsion. The use of ToF cameras is also demonstrated in [70] in a pipe inspection snake robot. The camera is used to detect key aspects of the pipe geometry, such as bends, junctions, and pipe radius. The snake robot's shape can then be adapted to the pipe's features, and navigate that way efficiently, even through vertical pipes.

Laser triangulation is a well-known sensing technology in industry, providing very high resolution and high accuracy measurements. The work in [68] focuses on increasing the snake robot's autonomy, which is demonstrated by autonomous pole climbing. This is a complex behaviour to be achieved by teleoperation. The authors have custom-designed a laser triangulation sensor to fit into the size and power constraints of the snake robot. The robot adopts a stable position with the head raised, and rotates the head to perform environment scanning. The resulting point cloud is filtered and processed to detect pole-like elements in the environment that the snake robot can climb. Once the pole and relative positioning is calculated, the robot pose is estimated by using the forward kinematics and IMU data.

For the sake of completeness, we can refer to vision systems *offboard*, which might not be applicable to more realistic applications such as exploration, search-and-rescue, or inspection. In [88], two cameras with a top-down view of the operating area are used to detect the obstacles and calculate the snake robot's pose. The pose estimation is simplified by placing fiducials in the snake robot (10 orange blocks along the snake robot's length). In a similar setup, [89] uses a stereovision system to measure the head's position of the robot and target coordinates.

4.3. Other Relevant Sensor Technologies for Navigation in Non-Snake Robots

Multiple examples of robots with shapes and propulsion mechanisms others than snake robots, and which use different sensor modalities from the ones introduced above, are common in literature. Structured light sensors based on temporal coding are seldom used for navigation because of their limitations to cope with movement, either due to dynamic elements in the environment or with a moving sensor. However, an example of a moving robot for search-and-rescue applications is introduced in [71]. The work in [72] focuses on improving the performance of SLAM using RGB-D sensors in large-scale environments. The use of stereovision for mapping and navigating in very challenging outdoor scenarios is shown in [73]. A single camera together with an IMU is demonstrated to be enough to produce a map and calculate a pose estimate at 40 Hz in a micro aerial vehicle (MAV) in [74]. One of the most relevant limitations in MAVs is the constraints to the size, power, and weight of the payload. These constraints are addressed by [75] by designing and building a radar sensor based on ultra-wide band (UWB), used for mapping and obstacle detection.

Remark 3. *Knowledge about the environment and its properties, in addition to its geometric representation, can be successfully exploited for improving locomotion performance for obstacle-aided locomotion.*

While knowledge about the environment's geometry might seem an obvious requirement for obstacle avoidance, other kinds of interaction with the environment—including obstacle-aided locomotion—require some further task-relevant knowledge about the environment. From a cognitive perspective, this has been acknowledged by the robotics community by the creation of *semantic maps* [90], which capture higher-level information about the environment, usually linked or grounded to knowledge from other sources. For the T^2 Snake-2s robot [64], the authors also claim that for planning the trajectory, they require the nature of the surrounding obstacles to be considered, as contact with some elements (e.g., fragile, high heat, electrically-charged, or sticky obstacles) might pose a safety risk to the robot and must then be completely avoided. However, safety is not the only reason. The biologically-inspired hexapod robot in [91] represents a good example of how knowledge about the environment is exploited for enhanced navigation. Information about certain terrain characteristics is captured in the environment model, and later adopted as part of the cost function used by the RRT* planner [92]. This way, the planned trajectory considers factors such as terrain roughness, terrain inclination, or mapping uncertainty.

5. Concluding Remarks

In this paper we have surveyed and discussed the state-of-the-art, challenges, and possibilities with perception-driven obstacle-aided locomotion. We have proposed a division of levels of autonomy for snake robots along three main axes: environmental complexity, mission complexity, and external system independence. Moreover, we have further expanded the description to suggest a step-wise approach to increasing the level of autonomy within three main robot technology areas: guidance, navigation, and control. We have reviewed existing literature relevant for perception-driven obstacle-aided locomotion. This includes snake robot obstacle avoidance, obstacle accommodation, and obstacle-aided locomotion, as well as methods and technologies for environment perception, mapping, and representation.

Perception-driven obstacle-aided locomotion is still in its infancy. However, there are strong results within both the snake robot community in particular, and the robotics community in general, which can be used to build further upon. One of the fundamental targets of this paper is to further increase global efforts to realise the large variety of application possibilities offered by snake robots and to provide an up-to-date reference as a stepping-stone for new research and development within this field.

Acknowledgments: This work is supported by the Research Council of Norway through the *Young research talents* funding scheme, project title "SNAKE—Control Strategies for Snake Robot Locomotion in Challenging Outdoor Environments", project number 240072.

Conflicts of Interest: The authors declare no conflict of interest.

References

1. Perrow, M.R.; Davy, A.J. *Handbook of Ecological Restoration: Volume 1, Principles of Restoration*; Cambridge University Press: Cambridge, UK, 2008.
2. Gray, J. The Mechanism of Locomotion in Snakes. *J. Exp. Biol.* **1946**, *23*, 101–120.
3. Transeth, A.; Leine, R.; Glocker, C.; Pettersen, K.; Liljebäck, P. Snake Robot Obstacle-Aided Locomotion: Modeling, Simulations, and Experiments. *IEEE Trans. Robot.* **2008**, *24*, 88–104.
4. Holden, C.; Stavdahl, Ø.; Gravdahl, J.T. Optimal dynamic force mapping for obstacle-aided locomotion in 2D snake robots. In Proceedings of the IEEE/RSJ International Conference on Intelligent Robots and Systems (IROS), Chicago, IL, USA, 14–18 September 2014; pp. 321–328.
5. Sanfilippo, F.; Azpiazu, J.; Marafioti, G.; Transeth, A.A.; Stavdahl, Ø.; Liljebäck, P. A review on perception-driven obstacle-aided locomotion for snake robots. In Proceedings of the 14th International Conference on Control, Automation, Robotics and Vision (ICARCV), Phuket, Thailand, 13–15 November 2016; pp. 1–7.
6. Huang, H.M.; Pavek, K.; Albus, J.; Messina, E. Autonomy levels for unmanned systems (ALFUS) framework: An update. In Proceedings of the 2005 SPIE Defense and Security Symposium, Orlando, FL, USA, 28 March–1 April 2005; International Society for Optics and Photonics: Orlando, FL, USA, 2005; pp. 439–448.
7. Kendoul, F. Towards a Unified Framework for UAS Autonomy and Technology Readiness Assessment (ATRA). In *Autonomous Control Systems and Vehicles*; Nonami, K., Kartidjo, M., Yoon, K.J., Budiyono, A., Eds.; Springer: Tokyo, Japan, 2013; pp. 55–71.
8. Nonami, K.; Kartidjo, M.; Yoon, K.J.; Budiyono, A. *Autonomous Control Systems and Vehicles: Intelligent Unmanned Systems*; Springer Science & Business Media: Tokyo, Japan, 2013.
9. Blasch, E.P.; Lakhotia, A.; Seetharaman, G. Unmanned vehicles come of age: The DARPA grand challenge. *Computer* **2006**, *39*, 26–29.
10. Hirose, S. *Biologically Inspired Robots: Snake-Like Locomotors and Manipulators*; Oxford University Press: Oxford, UK, 1993.
11. McKenna, J.C.; Anhalt, D.J.; Bronson, F.M.; Brown, H.B.; Schwerin, M.; Shammas, E.; Choset, H. Toroidal skin drive for snake robot locomotion. In Proceedings of the IEEE International Conference on Robotics and Automation (ICRA 2008), Pasadena, CA, USA, 19–23 May 2008; pp. 1150–1155.
12. Ma, S.; Tadokoro, N. Analysis of Creeping Locomotion of a Snake-like Robot on a Slope. *Auton. Robots* **2006**, *20*, 15–23.
13. Fjerdingen, S.A.; Liljebäck, P.; Transeth, A.A. A snake-like robot for internal inspection of complex pipe structures (PIKo). In Proceedings of the IEEE/RSJ International Conference on Intelligent Robots and Systems (IROS 2009), Saint Louis, MO, USA, 11–15 October 2009; pp. 5665–5671.
14. Wright, C.; Buchan, A.; Brown, B.; Geist, J.; Schwerin, M.; Rollinson, D.; Tesch, M.; Choset, H. Design and architecture of the unified modular snake robot. In Proceedings of the 2012 IEEE International Conference on Robotics and Automation (ICRA), Saint Paul, MN, USA, 14–18 May 2012; pp. 4347–4354.

15. Liljebäck, P.; Pettersen, K.; Stavdahl, Ø. A snake robot with a contact force measurement system for obstacle-aided locomotion. In Proceedings of the 2010 IEEE International Conference on Robotics and Automation (ICRA), Anchorage, AK, USA, 3–8 May 2010; pp. 683–690.
16. Anderson, G.T.; Yang, G. A proposed measure of environmental complexity for robotic applications. In Proceedings of the 2007 IEEE International Conference on Systems, Man and Cybernetics, Montreal, QC, Canada, 7–10 October 2007; pp. 2461–2466.
17. Kelasidi, E.; Liljebäck, P.; Pettersen, K.Y.; Gravdahl, J.T. Innovation in Underwater Robots: Biologically Inspired Swimming Snake Robots. *IEEE Robot. Autom. Mag.* **2016**, *23*, 44–62.
18. Chigisaki, S.; Mori, M.; Yamada, H.; Hirose, S. Design and control of amphibious Snake-like Robot ACM-R5. In Proceedings of the 2005 JSME Conference on Robotics and Mechatronics, Kobe, Japan, 9–11 June 2005.
19. Socha, J.J.; O'Dempsey, T.; LaBarbera, M. A 3-D kinematic analysis of gliding in a flying snake, Chrysopelea paradisi. *J. Exp. Biol.* **2005**, *208*, 1817–1833.
20. Vagia, M.; Transeth, A.A.; Fjerdingen, S.A. A literature review on the levels of automation during the years. What are the different taxonomies that have been proposed? *Appl. Ergon.* **2016**, *53 Pt A*, 190–202.
21. Yanco, H.A.; Drury, J. "Where am I?" Acquiring situation awareness using a remote robot platform. In Proceedings of the 2004 IEEE International Conference on Systems, Man and Cybernetics, The Hague, The Netherlands, 10–13 October 2004; Volume 3, pp. 2835–2840.
22. Liljebäck, P.; Pettersen, K.Y.; Stavdahl, Ø.; Gravdahl, J.T. *Snake Robots: Modelling, Mechatronics, and Control*; Springer Science & Business Media: London, UK, 2012.
23. Chirikjian, G.S.; Burdick, J.W. The kinematics of hyper-redundant robot locomotion. *IEEE Trans. Robot. Autom.* **1995**, *11*, 781–793.
24. Ostrowski, J.; Burdick, J. The Geometric Mechanics of Undulatory Robotic Locomotion. *Int. J. Robot. Res.* **1998**, *17*, 683–701.
25. Prautsch, P.; Mita, T.; Iwasaki, T. Analysis and Control of a Gait of Snake Robot. *IEEJ Trans. Ind. Appl.* **2000**, *120*, 372–381.
26. Liljebäck, P.; Pettersen, K.Y.; Stavdahl, Ø.; Gravdahl, J.T. Controllability and Stability Analysis of Planar Snake Robot Locomotion. *IEEE Trans. Autom. Control* **2011**, *56*, 1365–1380.
27. Liljebäck, P.; Pettersen, K.Y.; Stavdahl, Ø.; Gravdahl, J.T. A review on modelling, implementation, and control of snake robots. *Robot. Auton. Syst.* **2012**, *60*, 29–40.
28. Hu, D.L.; Nirody, J.; Scott, T.; Shelley, M.J. The mechanics of slithering locomotion. *Proc. Natl. Acad. Sci. USA* **2009**, *106*, 10081–10085.
29. Saito, M.; Fukaya, M.; Iwasaki, T. Serpentine locomotion with robotic snakes. *IEEE Control Syst.* **2002**, *22*, 64–81.
30. Tang, W.; Reyes, F.; Ma, S. Study on rectilinear locomotion based on a snake robot with passive anchor. In Proceedings of the 2015 IEEE/RSJ International Conference on Intelligent Robots and Systems (IROS), Hamburg, Germany, 28 September–3 October 2015; pp. 950–955.
31. Transeth, A.A.; Pettersen, K.Y.; Liljebäck, P. A survey on snake robot modeling and locomotion. *Robotica* **2009**, *27*, 999–1015.
32. Lee, M.C.; Park, M.G. Artificial potential field based path planning for mobile robots using a virtual obstacle concept. In Proceedings of the 2003 IEEE/ASME International Conference on Advanced Intelligent Mechatronics (AIM 2003), Port Island, Japan, 20 July–24 July 2003; Volume 2, pp. 735–740.
33. Ye, C.; Hu, D.; Ma, S.; Li, H. Motion planning of a snake-like robot based on artificial potential method. In Proceedings of the 2010 IEEE International Conference on Robotics and Biomimetics (ROBIO), Tianjin, China, 14–18 December 2010; pp. 1496–1501.
34. Yagnik, D.; Ren, J.; Liscano, R. Motion planning for multi-link robots using Artificial Potential Fields and modified Simulated Annealing. In Proceedings of the 2010 IEEE/ASME International Conference on Mechatronics and Embedded Systems and Applications (MESA), Qingdao, China, 5–17 July 2010; pp. 421–427.
35. Nor, N.M.; Ma, S. CPG-based locomotion control of a snake-like robot for obstacle avoidance. In Proceedings of the 2014 IEEE International Conference on Robotics and Automation (ICRA), Hong Kong, China, 31 May–7 June 2014; pp. 347–352.
36. Shan, Y.; Koren, Y. Design and motion planning of a mechanical snake. *IEEE Trans. Syst. Man Cybern.* **1993**, *23*, 1091–1100.

37. Shan, Y.; Koren, Y. Obstacle accommodation motion planning. *IEEE Trans. Robot. Autom.* **1995**, *11*, 36–49.
38. Kano, T.; Sato, T.; Kobayashi, R.; Ishiguro, A. Decentralized control of scaffold-assisted serpentine locomotion that exploits body softness. In Proceedings of the IEEE International Conference on Robotics and Automation (ICRA), Shanghai, China, 9–13 May 2011; pp. 5129–5134.
39. Kano, T.; Sato, T.; Kobayashi, R.; Ishiguro, A. Local reflexive mechanisms essential for snakes' scaffold-based locomotion. *Bioinspir. Biomim.* **2012**, *7*, 046008, doi:10.1088/1748-3182/7/4/046008.
40. Bayraktaroglu, Z.Y.; Blazevic, P. Understanding snakelike locomotion through a novel push-point approach. *J. Dyn. Syst. Meas. Control* **2005**, *127*, 146–152.
41. Bayraktaroglu, Z.Y.; Kilicarslan, A.; Kuzucu, A. Design and control of biologically inspired wheel-less snake-like robot. In Proceedings of the First IEEE/RAS-EMBS International Conference on Biomedical Robotics and Biomechatronics (BioRob 2006), Pisa, Italy, 20–22 February 2006; pp. 1001–1006.
42. Gupta, A. Lateral undulation of a snake-like robot. Master's Thesis, Massachusetts Institute of Technology, Cambridge, MA, USA, 2007.
43. Andruska, A.M.; Peterson, K.S. Control of a Snake-Like Robot in an Elastically Deformable Channel. *IEEE/ASME Trans. Mechatron.* **2008**, *13*, 219–227.
44. Kamegawa, T.; Kuroki, R.; Travers, M.; Choset, H. Proposal of EARLI for the snake robot's obstacle aided locomotion. In Proceedings of the 2012 IEEE International Symposium on Safety, Security, and Rescue Robotics (SSRR), College Station, TX, USA, 5–8 November 2012; pp. 1–6.
45. Kamegawa, T.; Kuroki, R.; Gofuku, A. Evaluation of snake robot's behavior using randomized EARLI in crowded obstacles. In Proceedings of the 2014 IEEE International Symposium on Safety, Security, and Rescue Robotics (SSRR), Toyako-cho, Japan, 27–30 October 2014; pp. 1–6.
46. Liljebäck, P.; Pettersen, K.Y.; Stavdahl, Ø.; Gravdahl, J.T. Compliant control of the body shape of snake robots. In Proceedings of the IEEE International Conference on Robotics and Automation (ICRA), Hong Kong, China, 31 May–7 June 2014; pp. 4548–4555.
47. Travers, M.; Whitman, J.; Schiebel, P.; Goldman, D.; Choset, H. Shape-Based Compliance in Locomotion. In Proceedings of the Robotics: Science and Systems (RSS), Cambridge, MA, USA, 12–16 July 2016; Number 5, p. 10.
48. Whitman, J.; Ruscelli, F.; Travers, M.; Choset, H. Shape-based compliant control with variable coordination centralization on a snake robot. In Proceedings of the IEEE 55th Conference on Decision and Control (CDC). IEEE, Las Vegas, NV, USA, 12–14 December 2016; pp. 5165–5170.
49. Ma, S.; Ohmameuda, Y.; Inoue, K. Dynamic analysis of 3-dimensional snake robots. In Proceedings of the IEEE/RSJ International Conference on Intelligent Robots and Systems (IROS), Sendai, Japan, 28 September–2 October 2004; Volume 1, pp. 767–772.
50. Sanfilippo, F.; Stavdahl, Ø.; Marafioti, G.; Transeth, A.A.; Liljebäck, P. Virtual functional segmentation of snake robots for perception-driven obstacle-aided locomotion. In Proceedings of the IEEE Conference on Robotics and Biomimetics (ROBIO), Qingdao, China, 3–7 December 2016; pp. 1845–1851.
51. Thrun, S.; Burgard, W.; Fox, D. *Probabilistic Robotics*; MIT Press: Cambridge, MA, USA, 2005.
52. Boyle, J.H.; Johnson, S.; Dehghani-Sanij, A.A. Adaptive Undulatory Locomotion of a *C. elegans* Inspired Robot. *IEEE/ASME Trans. Mechatron.* **2013**, *18*, 439–448.
53. Gong, C.; Tesch, M.; Rollinson, D.; Choset, H. Snakes on an inclined plane: Learning an adaptive sidewinding motion for changing slopes. In Proceedings of the 2014 IEEE/RSJ International Conference on Intelligent Robots and Systems (IROS 2014), Chicago, IL, USA, 14–18 September 2014; pp. 1114–1119.
54. Taal, S.; Yamada, H.; Hirose, S. 3 axial force sensor for a semi-autonomous snake robot. In Proceedings of the IEEE International Conference on Robotics and Automation, (ICRA '09), Kobe, Japan, 12–17 May 2009; pp. 4057–4062.
55. Bayraktaroglu, Z.Y. Snake-like locomotion: Experimentations with a biologically inspired wheel-less snake robot. *Mech. Mach. Theory* **2009**, *44*, 591–602.
56. Gonzalez-Gomez, J.; Gonzalez-Quijano, J.; Zhang, H.; Abderrahim, M. Toward the sense of touch in snake modular robots for search and rescue operations. In Proceedings of the ICRA 2010 Workshop on Modular Robots: State of the Art, Anchorage, AK, USA, 3–8 May 2010; pp. 63–68.
57. Wu, X.; Ma, S. Development of a sensor-driven snake-like robot SR-I. In Proceedings of the 2011 IEEE International Conference on Information and Automation (ICIA), Shenzhen, China, 6–8 June 2011; pp. 157–162.

58. Liljebäck, P.; Stavdahl, Ø.; Pettersen, K.; Gravdahl, J. Mamba—A waterproof snake robot with tactile sensing. In Proceedings of the 2014 IEEE/RSJ International Conference on Intelligent Robots and Systems (IROS 2014), Chicago, IL, USA, 14–18 September 2014; pp. 294–301.
59. Paap, K.; Christaller, T.; Kirchner, F. A robot snake to inspect broken buildings. In Proceedings of the 2000 IEEE/RSJ International Conference on Intelligent Robots and Systems (IROS 2000), Takamatsu, Japan, 30 October–5 November 2000; Volume 3, pp. 2079–2082.
60. Caglav, E.; Erkmen, A.M.; Erkmen, I. A snake-like robot for variable friction unstructured terrains, pushing aside debris in clearing passages. In Proceedings of the IEEE/RSJ International Conference on Intelligent Robots and Systems (IROS 2007), San Diego, CA, USA, 29 October–2 November 2007; pp. 3685–3690.
61. Sfakiotakis, M.; Tsakiris, D.P.; Vlaikidis, A. Biomimetic Centering for Undulatory Robots. In Proceedings of the First IEEE/RAS-EMBS International Conference on Biomedical Robotics and Biomechatronics (BioRob 2006), Pisa, Italy, 20–22 February 2006; pp. 744–749.
62. Wu, Q.; Gao, J.; Huang, C.; Zhao, Z.; Wang, C.; Su, X.; Liu, H.; Li, X.; Liu, Y.; Xu, Z. Obstacle avoidance research of snake-like robot based on multi-sensor information fusion. In Proceedings of the 2012 IEEE International Conference on Robotics and Biomimetics (ROBIO), Guangzhou, China, 11–14 December 2012; pp. 1040–1044.
63. Chavan, P.; Murugan, M.; Vikas Unnikkannan, E.; Singh, A.; Phadatare, P. Modular Snake Robot with Mapping and Navigation: Urban Search and Rescue (USAR) Robot. In Proceedings of the 2015 International Conference on Computing Communication Control and Automation (ICCUBEA), Pune, India, 26–27 February 2015; pp. 537–541.
64. Tanaka, M.; Kon, K.; Tanaka, K. Range-Sensor-Based Semiautonomous Whole-Body Collision Avoidance of a Snake Robot. *IEEE Trans. Control Syst. Technol.* **2015**, *23*, 1927–1934.
65. Pfotzer, L.; Staehler, M.; Hermann, A.; Roennau, A.; Dillmann, R. KAIRO 3: Moving over stairs & unknown obstacles with reconfigurable snake-like robots. In Proceedings of the 2015 European Conference on Mobile Robots (ECMR), Lincoln, UK, 2–4 September 2015; pp. 1–6.
66. Tian, Y.; Gomez, V.; Ma, S. Influence of two SLAM algorithms using serpentine locomotion in a featureless environment. In Proceedings of the IEEE International Conference on Robotics and Biomimetics (ROBIO), Zhuhai, China, 6–9 December 2015; pp. 182–187.
67. Kon, K.; Tanaka, M.; Tanaka, K. Mixed Integer Programming-Based Semiautonomous Step Climbing of a Snake Robot Considering Sensing Strategy. *IEEE Trans. Control Syst. Technol.* **2016**, *24*, 252–264.
68. Ponte, H.; Queenan, M.; Gong, C.; Mertz, C.; Travers, M.; Enner, F.; Hebert, M.; Choset, H. Visual sensing for developing autonomous behavior in snake robots. In Proceedings of the 2014 IEEE International Conference on Robotics and Automation (ICRA), Hong Kong, China, 31 May–7 June 2014; pp. 2779–2784.
69. Ohno, K.; Nomura, T.; Tadokoro, S. Real-Time Robot Trajectory Estimation and 3D Map Construction using 3D Camera. In Proceedings of the 2006 IEEE/RSJ International Conference on Intelligent Robots and Systems, Beijing, China, 9–15 October 2006; pp. 5279–5285.
70. Fjerdingen, S.A.; Mathiassen, J.R.; Schumann-Olsen, H.; Kyrkjebø, E. Adaptive Snake Robot Locomotion: A Benchmarking Facility for Experiments. In *European Robotics Symposium 2008*; Bruyninckx, H., Přeučil, L., Kulich, M., Eds.; Number 44 in Springer Tracts in Advanced Robotics; Springer: Berlin/Heidelberg, Germany, 2008; pp. 13–22.
71. Mobedi, B.; Nejat, G. 3-D Active Sensing in Time-Critical Urban Search and Rescue Missions. *IEEE/ASME Trans. Mechatron.* **2012**, *17*, 1111–1119.
72. Labbé, M.; Michaud, F. Online global loop closure detection for large-scale multi-session graph-based SLAM. In Proceedings of the 2014 IEEE/RSJ International Conference on Intelligent Robots and Systems, Chicago, IL, USA, 14–18 September 2014; pp. 2661–2666.
73. Agrawal, M.; Konolige, K.; Bolles, R.C. Localization and Mapping for Autonomous Navigation in Outdoor Terrains: A Stereo Vision Approach. In Proceedings of the IEEE Workshop on Applications of Computer Vision (WACV '07), Austin, TX, USA, 21–22 February 2007.
74. Weiss, S.; Achtelik, M.W.; Lynen, S.; Chli, M.; Siegwart, R. Real-time onboard visual-inertial state estimation and self-calibration of MAVs in unknown environments. In Proceedings of the 2012 IEEE International Conference on Robotics and Automation (ICRA), Saint Paul, MN, USA, 14–18 May 2012; pp. 957–964.

75. Fontana, R.J.; Richley, E.A.; Marzullo, A.J.; Beard, L.C.; Mulloy, R.W.T.; Knight, E.J. An ultra wideband radar for micro air vehicle applications. In Proceedings of the 2002 IEEE Conference on Ultra Wideband Systems and Technologies, Digest of Papers, Baltimore, MD, USA, 21–23 May 2002; pp. 187–191.
76. Dudek, G.; Jenkin, M. *Computational Principles of Mobile Robotics*; Cambridge University Press: Cambridge, UK, 2010.
77. Surmann, H.; Nüchter, A.; Hertzberg, J. An autonomous mobile robot with a 3D laser range finder for 3D exploration and digitalization of indoor environments. *Robot. Auton. Syst.* **2003**, *45*, 181–198.
78. Sansoni, G.; Trebeschi, M.; Docchio, F. State-of-the-art and applications of 3D imaging sensors in industry, cultural heritage, medicine, and criminal investigation. *Sensors* **2009**, *9*, 568–601.
79. Henry, P.; Krainin, M.; Herbst, E.; Ren, X.; Fox, D. RGB-D mapping: Using Kinect-style depth cameras for dense 3D modeling of indoor environments. *Int. J. Robot. Res.* **2012**, *31*, 647–663.
80. Foix, S.; Alenya, G.; Torras, C. Lock-in Time-of-Flight (ToF) Cameras: A Survey. *IEEE Sens. J.* **2011**, *11*, 1917–1926.
81. Scharstein, D.; Szeliski, R. A taxonomy and evaluation of dense two-frame stereo correspondence algorithms. *Int. J. Comput. Vis.* **2002**, *47*, 7–42.
82. Hartley, R.; Zisserman, A. *Multiple View Geometry in Computer Vision*; Cambridge University Press: Cambridge, UK, 2003.
83. Davison, A.J.; Reid, I.D.; Molton, N.D.; Stasse, O. MonoSLAM: Real-Time Single Camera SLAM. *IEEE Trans. Pattern Anal. Mach. Intell.* **2007**, *29*, 1052–1067.
84. Fox, C.; Evans, M.; Pearson, M.; Prescott, T. Tactile SLAM with a biomimetic whiskered robot. In Proceedings of the 2012 IEEE International Conference on Robotics and Automation (ICRA), Saint Paul, MN, USA, 14–18 May 2012; pp. 4925–4930.
85. Pearson, M.; Fox, C.; Sullivan, J.; Prescott, T.; Pipe, T.; Mitchinson, B. Simultaneous localisation and mapping on a multi-degree of freedom biomimetic whiskered robot. In Proceedings of the 2013 IEEE International Conference on Robotics and Automation (ICRA), Karlsruhe, Germany, 6–10 May 2013; pp. 586–592.
86. Xinyu, L.; Matsuno, F. Control of snake-like robot based on kinematic model with image sensor. In Proceedings of the 2003 IEEE International Conference on Robotics, Intelligent Systems and Signal Processing, Changsha, Hunan, China, 8–13 October 2003; Volume 1, pp. 347–352.
87. Zhang, Z. Iterative point matching for registration of free-form curves and surfaces. *Int. J. Comput. Vis.* **1994**, *13*, 119–152.
88. Xiao, X.; Cappo, E.; Zhen, W.; Dai, J.; Sun, K.; Gong, C.; Travers, M.; Choset, H. Locomotive reduction for snake robots. In Proceedings of the 2015 IEEE International Conference on Robotics and Automation (ICRA), Seattle, WA, USA, 26–30 May 2015; pp. 3735–3740.
89. Yamakita, M.; Hashimoto, M.; Yamada, T. Control of locomotion and head configuration of 3D snake robot (SMA). In Proceedings of the IEEE International Conference on Robotics and Automation (ICRA '03), Taipei, Taiwan, 14–19 September 2003; Volume 2, pp. 2055–2060.
90. Galindo, C.; Fernández-Madrigal, J.A.; González, J.; Saffiotti, A. Robot task planning using semantic maps. *Robot. Auton. Syst.* **2008**, *56*, 955–966.
91. Oberländer, J.; Klemm, S.; Heppner, G.; Roennau, A.; Dillmann, R. A multi-resolution 3-D environment model for autonomous planetary exploration. In Proceedings of the 2014 IEEE International Conference on Automation Science and Engineering (CASE), New Taipei, Taiwan, 18–22 August 2014; pp. 229–235.
92. Karaman, S.; Frazzoli, E. Sampling-based algorithms for optimal motion planning. *Int. J. Robot. Res.* **2011**, *30*, 846–894.

Article

Locomotion Efficiency Optimization of Biologically Inspired Snake Robots

Eleni Kelasidi [1,*], Mansoureh Jesmani [2], Kristin Y. Pettersen [1] and Jan Tommy Gravdahl [2]

1 Centre for Autonomous Marine Operations and Systems, Department of Engineering Cybernetics at NTNU, NO-7491 Trondheim, Norway; kristin.y.pettersen@ntnu.no

2 Department of Engineering Cybernetics at NTNU, NO-7491 Trondheim, Norway; jesmani.mansoureh@gmail.com (M.J.); jan.tommy.gravdahl@ntnu.no (J.T.G.)

* Correspondence: eleni.kelasidi@ntnu.no; Tel.: +47-4518-5796

Received: 6 November 2017; Accepted: 21 December 2017; Published: 9 January 2018

Abstract: Snake robots constitute bio-inspired solutions that have been studied due to their ability to move in challenging environments where other types of robots, such as wheeled or legged robots, usually fail. In this paper, we consider both land-based and swimming snake robots. One of the principal concerns of the bio-inspired snake robots is to increase the motion efficiency in terms of the forward speed by improving the locomotion methods. Furthermore, energy efficiency becomes a crucial challenge for this type of robots due to the importance of long-term autonomy of these systems. In this paper, we take into account both the minimization of the power consumption and the maximization of the achieved forward velocity in order to investigate the optimal gait parameters for bio-inspired snake robots using lateral undulation and eel-like motion patterns. We furthermore consider possible negative work effects in the calculation of average power consumption of underwater snake robots. To solve the multi-objective optimization problem, we propose transforming the two objective functions into a single one using a weighted-sum method. For different set of weight factors, Particle Swarm Optimization is applied and a set of optimal points is consequently obtained. Pareto fronts or trade-off curves are illustrated for both land-based and swimming snake robots with different numbers of links. Pareto fronts represent trade-offs between the objective functions. For example, how increasing the forward velocity results in increasing power consumption. Therefore, these curves are a very useful tool for the control and design of snake robots. The trade-off curve thus constitutes a very useful tool for both the control and design of bio-inspired snake robots. In particular, the operators or designers of bio-inspired snake robots can choose a Pareto optimal point based on the trade-off curve, given the preferred number of links on the robot. The optimal gait parameters for the robot control system design are then directly given both for land-based and underwater snake robots. Moreover, we are able to obtain some observations about the optimal values of the gait parameters, which provide very important insights for future control design of bio-inspired snake robots.

Keywords: bio-inspired snake robots; multi-objective optimization; particle swarm optimization (PSO); energy efficiency

---

1. Introduction

Several robotic systems have been developed over the last few decades with the overall goal to be used for different applications and take the place of humans in dull, distant, or dangerous environments, including space applications, subsea applications and manufacturing processes. Recently, bio-inspired robots have received significant attention in research. The overall idea of bio-inspired robots is to make systems learn the concepts of the locomotion from nature and apply them to the design and the locomotion of the robotic systems. In many instances, to address scientific problems, engineers seek for solutions by getting inspired from the nature. In particular, they study the robust motion capabilities of biological creatures that have emerged through millions of years of evolution, and this process is termed biomimetics. Nowadays, there has been increasing interest on using robotic systems for exploration, monitoring, and surveillance tasks. Bio-inspired robotic systems that mimic the motion of biological snakes or fish can be considered good candidates for these kinds of applications. Inspired by the robustness and the stability of biological snake locomotion, snake robots carry the potential to meet the growing need for new technology that can be used in challenging environments [1]. Several models have been proposed for bio-inspired snake robots moving on land [1]. In [2], the locomotion capabilities of biological snakes has been studied, while the world's first snake-like robot was developed in Japan in 1972 [3]. Bio-inspired snake robots have the potential to be used in several applications such as inspection and intervention operations in hazardous environments in industrial plants, manipulator tasks in tight spaces for instance pipe inspection. In addition, the robotic community signified the need for new technology that can be employed to locate survivors inside a collapsed building. Hence, a very relevant application of bio-inspired snake robots can involve search and rescue missions in earthquake areas in the future [1]. Furthermore, several results have been published regarding the design, modeling and control of bio-inspired underwater robots [4]. One of the main characteristics of snake robots is their adaptability to different motion demands. This attribute is extremely important for snake robots, since they have the ability to move over land as well underwater, while the physiology remains the same.

Nowadays, different types of underwater vehicles and underwater robotic systems have been used for several challenging subsea inspection and interventions tasks [5]. In addition to the existing underwater robotic solutions, recently, the potential of using bio-inspired underwater snake robots has been studied. Due to their long, flexible and slender body the snake robots can be considered as an interesting alternative solution to improve the efficiency and maneuverability of conventional underwater vehicles [6–9]. The swimming snake robot as a mobile manipulator arm can be used for several applications such as inspection and maintenance of subsea oil and gas installations accessing tight structures, in biological community and marine archeology. However, several control related challenges must be addressed before we have fully functional swimming snake robots suitable for challenging tasks in highly uncertain subsea environment. An essential problem to be solved is related to the locomotion efficiency of the bio-inspired swimming snake robots. Hence, in this paper, the energy efficiency is studied since it constitutes the main important factor in order to obtain long term autonomy of these type of robots.

In [10], gait parameters of a land-based snake robot are obtained to optimize head stability and the speed of the robot simultaneously. However, the energy efficiency is not considered in the optimization framework. In [11], the genetic algorithm (GA) is used to find forward head serpentine gait parameters for a land-based snake robot. However, only the speed of the robot is considered in the optimization in order to find the optimal gait parameters of the head link. Moreover, in [12] gradient-free method is implemented by using Powell's method in order to optimize the forward velocity of the land-based and swimming snake robots. However, the frequency of the motion pattern has not been included in the optimized parameters. In [13], empirical rules have been derived for underwater snake robots based on extensive simulation studies describing the relationship between the different gait parameters, the power consumption and the achieved velocity for lateral undulation and eel-like motion patterns. Furthermore, in [14], the energy consumption of underwater snake robots (USRs) and remotely operated vehicles (ROVs) have been studied. The obtained comparison results in [14] between USRs and ROVs showed that the USRs are more energy efficient. Experimental results regarding the locomotion efficiency of USRs have been presented in [15], showing the relationship between the gait parameters, the achieved forward velocity and the energy consumption for the lateral undulation and eel-like motion patterns. An optimization framework is proposed in [16] for underwater snake robots. In particular, in [16] the energy efficiency of underwater snake robots was investigated solving a multi-objective optimization problem for a robot with 10 links. To the best of our knowledge, however, no results have been presented formulating an effective and general optimization framework for bio-inspired snake robots with an arbitrary number of links. In this paper, we consider both the minimization of the power consumption and maximization of the forward velocity, and we propose a framework for solving a multi-objective optimization problem in order to obtain optimal gait parameters. Based on the proposed method, in this paper, we investigate the motion efficiency of both land-based and underwater snake robots using undulatory motion patterns. Furthermore, we here take into account the negative work effect of underwater snake robots, which was not considered in [16]. The optimization framework proposed in this paper is thus an extension of the method presented in [16] and may be used for bio-inspired snake robots locomotion both on land and in water.

In this paper, we solve a motion optimization problem to obtain the optimal gait parameters based on a dynamic model of USR. The optimization problem is challenging as the USR model includes highly nonlinear hydrodynamic effects. Several models have been proposed for USRs in previous studies[7,8,17–20]. A closed form model for USRs was presented in [6,21,22], where both the resistive and reactive fluid forces were considered. In particular, the proposed model in [6,21,22], considers several hydrodynamic effects such as the linear and the nonlinear drag forces, the added mass effect, the fluid moments and current effects. This model is general in the sense that it can be used to simulate the behavior of snake robots moving on land or swimming in the water by considering either ground friction [1] or fluid friction model [6]. In this paper, the optimization results are obtained based on this model.

To optimize both the power consumption and the forward velocity, a weighted sum of both objective functions is proposed as a single goal. Optimization algorithms can be divided into two main classes: gradient-based and derivative free algorithms. Gradient-based algorithms are generally faster than the derivative free ones. However, the optimization solution in the gradient-based methods often depends on the initial points, since these algorithms are more prone to get trapped in local optima [23]. Therefore, derivative-free and more specifically stochastic derivative-free methods are reported to be used more often in problems with highly non-smooth objective functions containing multiple optima, because these types of algorithms can avoid local solutions by their stochastic nature. Consequently, as also mentioned in [8], we find that stochastic derivative-free methods constitute a proper optimization algorithm for the motion optimization problem. In [24], a genetic algorithm (GA) is applied to a three-link swimmer to solve the swimming gait optimization problem. Moreover, pattern optimization of anguilliform swimming is studied in [25]. In particular, a combination of an evolutionary optimization algorithm and a three-dimensional numerical solution are implemented. Two objective functions of efficiency and velocity are considered in [25], and the optimization problem is addressed by solving Navier–Stokes equations. Because of the computational cost of solving the Navier–Stokes equations, only one optimization run is considered for each fitness goal. In [8], optimization is used to generate efficient hyper-redundant mechanism swimming gaits. In particular, the gait parameters were optimized in order to minimize the total required energy over a given distance for a desired swimming velocity. Therefore, a penalty term was added to the objective function to set the velocity to the desired value. Since the penalty factor is chosen relatively large, the velocity tends to stay at the desired value, and thus it is difficult to observe the trade-off between the velocity and the required energy. However, we will show in this paper that a slight reduction of velocity may result in a significant saving in power consumption. In particular, we propose to optimize both the energy and the velocity, simultaneously. Therefore, there is no need to tune the penalty factor or define the desired velocity.

We apply the particle swarm optimization (PSO) algorithm to optimize the power consumption and the forward velocity. PSO iterates on the gait parameters in order to improve a candidate solution. Kennedy and Eberhart [26] developed the PSO algorithm inspired by the social behavior of animals. A random velocity is assigned to each particle according to simple mathematical formulae including the experience of the particle itself and its neighborhoods. The position of the particle is updated using the obtained new velocity. More specifically, the movement of the particle is influenced by the best experienced position of the particle itself and also the best known position of the swarm. The later helps the swarm to move toward the best solutions. In the PSO algorithm, there is few or no assumption about the optimization problem. Moreover, the gradient information is not required in the optimization procedure. Therefore, there is no limitation about differentiability of the optimization problem. In contrast to GA, PSO acquires memory and saves the experience of good solutions during generations. More importantly, the particles in PSO share the information among the neighborhood. Therefore, the mechanism of constructive cooperation results in a quicker convergence rate. In addition, quite recently, the Reinforcement Learning (RL) method has been applied to address optimization problems in robotics field [27]. To converge to the solution, many objective function evaluations are required in both the PSO and RL methods. However, PSO can use parallel computing to evaluate the objective functions of the particles of each generation. Therefore, it is possible to reduce the computational time significantly. [28] compares the PSO and RL methods applied to solve the obstacle avoidance problem for multi-robotic systems. The results in [28] shows that PSO gives a higher fitness than RL.

In this paper, a multi-objective optimization framework is proposed with the overall goal to maximize the achieved forward velocity and simultaneously minimize the average power consumption for bio-inspired snake robots. The optimization problem is solved based on the dynamic model of snake robots proposed in [6,21,22]. In particular, multi-objective optimization is used in order to obtain results for lateral undulation and eel-like motion patterns for swimming snake robots with different numbers of links. Furthermore, results are obtained also considering the negative work effect [8] of underwater snake robots. In addition, for land-based snake robots, optimization results are presented for the lateral undulation motion pattern, which is the most common motion pattern for this type of robot. In [29], a multi-objective optimization method for objectives with input-dependent noise is proposed and applied to optimize the speed and head stability of the sidewinding gait of a snake robot. The energy efficiency of land-based snake robot with 16 links is studied in [30] by comparing the energy efficiency of sidewinding locomotion with that of lateral undulation and sinus-lifting motion. The multi-objective optimization approach is presented in [10] for land-based snake robots to optimize the head stability and speed, simultaneously. The dynamic model of the robot was considered as a black-box model. The speed was simply defined by the center of mass displacement after a given time. In order to consider the head stability, the estimated area through which the desired focal point swept was minimized. In [10], surrogate models are obtained by running a few experiments, while the objective functions are also optimized, simultaneously. Therefore, adding more optimization variables to the problem may lead to an increase in the number of experiments. In this paper, however, we propose to use a dynamic model of snake robots instead of a black-box model. Consequently, we reduce the cost of experiments. Moreover, we also consider minimizing the power consumption in our proposed optimization framework. Furthermore, the proposed optimization method obtains the optimal parameters giving the most efficient motion pattern for both land-based and swimming snake robots. The obtained results thus provide interesting insights regarding the control and the design of bio-inspired snake robots.

The rest of this paper is structured as follows. The dynamic models of underwater and land-based snake robots are presented in Section 2. Section 3 presents the proposed multi-objective optimization framework and Section 4 the results obtained for both land-based and swimming snake robots. Finally, conclusions and suggestions for further research are given in Section 5.

2. Modeling of Snake Robots

In this section, we briefly present the equation of motion of an USR since it will be used in the following sections to study the energy efficiency of snake robots based on the proposed multi-objective optimization framework in this paper. For more details, see [6,21,22].

2.1. Dynamic Model of Underwater Snake Robot

The robot is assumed to consist of n rigid links with each of them having length $2l$, mass m and moment of inertia $J = \frac{1}{3}ml^2$. The links are interconnected by $n-1$ joints. The center of mass (CM) of the link is defined as the middle point, since it is assumed that the mass is uniformly distributed, with the total mass of the USR given by nm. The kinematics and dynamics of the robot will be described in terms of the mathematical symbols described in Table 1 and illustrated in Figures 1 and 2. The following vectors and matrices are used in the subsequent sections:

$$\mathbf{A} = \begin{bmatrix} 1 & 1 & & \\ & \ddots & \ddots & \\ & & 1 & 1 \end{bmatrix}, \mathbf{D} = \begin{bmatrix} 1 & -1 & & \\ & \ddots & \ddots & \\ & & 1 & -1 \end{bmatrix},$$

where $\mathbf{A}, \mathbf{D} \in \mathbb{R}^{(n-1)\times n}$. Furthermore,

$$\mathbf{e} = \begin{bmatrix} 1, & \ldots, & 1 \end{bmatrix}^T \in \mathbb{R}^n, \mathbf{E} = \begin{bmatrix} \mathbf{e} & \mathbf{0}_{n\times 1} \\ \mathbf{0}_{n\times 1} & \mathbf{e} \end{bmatrix} \in \mathbb{R}^{2n\times 2},$$

$$\begin{aligned} \mathbf{S}_\theta &= \text{diag}(\sin\boldsymbol{\theta}) \in \mathbb{R}^{n\times n}, \quad \mathbf{C}_\theta = \text{diag}(\cos\boldsymbol{\theta}) \in \mathbb{R}^{n\times n}, \\ \dot{\boldsymbol{\theta}}^2 &= \begin{bmatrix} \dot{\theta}_1^2, & \ldots, & \dot{\theta}_n^2 \end{bmatrix}^T \in \mathbb{R}^n, \mathbf{K} = \mathbf{A}^T\left(\mathbf{D}\mathbf{D}^T\right)^{-1}\mathbf{D}. \end{aligned}$$

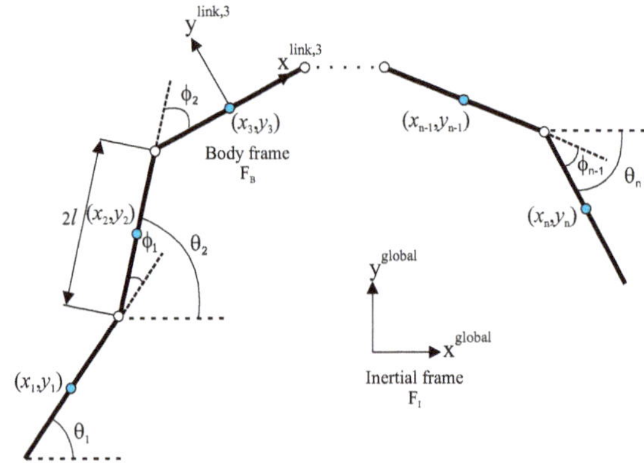

Figure 1. Kinematic parameters of the underwater snake robot.

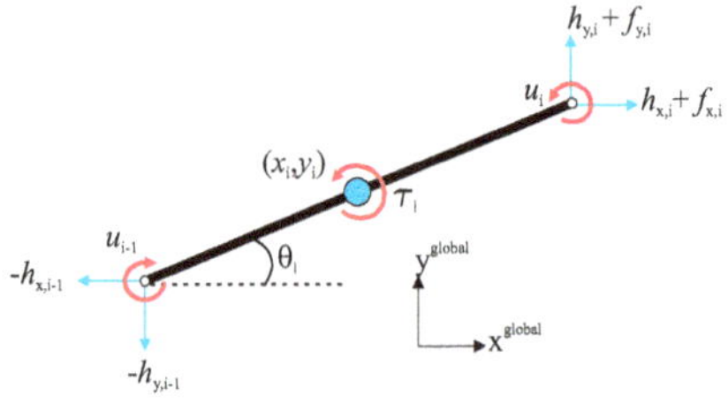

Figure 2. Forces and torques acting on each link of the underwater snake robot.

Table 1. Definition of mathematical terms.

Symbol	Description	Vector
n	The number of links	-
l	The half length of a link	-
m	Mass of each link	-
J	Moment of inertia of each link	-
θ_i	Angle between link i and the global $x-$axis	$\boldsymbol{\theta} \in \mathbb{R}^n$
ϕ_i	Angle of joint i	$\boldsymbol{\phi} \in \mathbb{R}^{n-1}$
(x_i, y_i)	Global coordinates of the CM of link i	$\mathbf{X}, \mathbf{Y} \in \mathbb{R}^n$
(p_x, p_y)	Global coordinates of the CM of the robot	$\mathbf{p}_{\text{CM}} \in \mathbb{R}^2$
u_i	Actuator torque of joint between link i and link $i+1$	$\mathbf{u} \in \mathbb{R}^{n-1}$
u_{i-1}	Actuator torque of joint between link i and link $i-1$	$\mathbf{u} \in \mathbb{R}^{n-1}$
$f_{x,i}$	Fluid force on link i in the $x-$direction	$\mathbf{f_x} \in \mathbb{R}^n$
$f_{y,i}$	Fluid force on link i in the $y-$direction	$\mathbf{f_y} \in \mathbb{R}^n$
τ_i	Fluid torque on link i	$\boldsymbol{\tau} \in \mathbb{R}^n$
$f_{Rx,i}$	Ground friction force on link i in the $x-$direction	$\mathbf{f_{Rx}} \in \mathbb{R}^n$
$f_{Ry,i}$	Ground friction force on link i in the $y-$direction	$\mathbf{f_{Ry}} \in \mathbb{R}^n$
$h_{x,i}$	Joint constraint force in the $x-$direction on link i from link $i+1$	$\mathbf{h_x} \in \mathbb{R}^{n-1}$
$h_{y,i}$	Joint constraint force in the $y-$direction on link i from link $i+1$	$\mathbf{h_y} \in \mathbb{R}^{n-1}$
$h_{x,i-1}$	Joint constraint force in the $x-$direction on link i from link $i-1$	$\mathbf{h_x} \in \mathbb{R}^{n-1}$
$h_{y,i-1}$	Joint constraint force in the $y-$direction on link i from link $i-1$	$\mathbf{h_y} \in \mathbb{R}^{n-1}$

The kinematics of the robot is derived for a motion in horizontal plane. We assume that the robot is fully immersed in water and has in total $n + 2$ degrees of freedom (DOF). The parameters θ_i and $\phi_i = \theta_i - \theta_{i-1}$ represent the link angle and the joint angle of each link, respectively. The link angles and the joint angles are grouped in $\boldsymbol{\theta} = [\theta_1, \ldots, \theta_n]^T \in \mathbb{R}^n$ and $\boldsymbol{\phi} = [\phi_1, \ldots, \phi_{n-1}]^T \in \mathbb{R}^{n-1}$. The position of the CM of the USR $\mathbf{p}_{\text{CM}} \in \mathbb{R}^2$ can be calculated as:

$$\mathbf{p}_{\text{CM}} = \begin{bmatrix} p_x \\ p_y \end{bmatrix} = \begin{bmatrix} \frac{1}{nm} \sum_{i=1}^{n} m x_i \\ \frac{1}{nm} \sum_{i=1}^{n} m y_i \end{bmatrix} = \frac{1}{n} \begin{bmatrix} \mathbf{e}^T \mathbf{X} \\ \mathbf{e}^T \mathbf{Y} \end{bmatrix}, \tag{1}$$

where (x_i, y_i) are the global frame coordinates of the CM of link i, $\mathbf{X} = [x_1, \ldots, x_n]^T \in \mathbb{R}^n$ and $\mathbf{Y} = [y_1, \ldots, y_n]^T \in \mathbb{R}^n$.

Regarding the hydrodynamic modeling, in [6], it is shown that the global frame forces due to interaction of the links with the water can be assembled in vector form as follows:

$$\mathbf{f} = \begin{bmatrix} \mathbf{f_x} \\ \mathbf{f_y} \end{bmatrix} = \begin{bmatrix} \mathbf{f_{A_x}} \\ \mathbf{f_{A_y}} \end{bmatrix} + \begin{bmatrix} \mathbf{f}^{\text{I}}_{\mathbf{D_x}} \\ \mathbf{f}^{\text{I}}_{\mathbf{D_y}} \end{bmatrix} + \begin{bmatrix} \mathbf{f}^{\text{II}}_{\mathbf{D_x}} \\ \mathbf{f}^{\text{II}}_{\mathbf{D_y}} \end{bmatrix}, \tag{2}$$

where the added mass effects $\mathbf{f_{A_x}}$ and $\mathbf{f_{A_y}}$ in x and y direction, respectively, are given by

$$\begin{aligned} \begin{bmatrix} \mathbf{f_{A_x}} \\ \mathbf{f_{A_y}} \end{bmatrix} = &- \begin{bmatrix} \mu_n (\mathbf{S}_\theta)^2 & -\mu_n \mathbf{S}_\theta \mathbf{C}_\theta \\ -\mu_n \mathbf{S}_\theta \mathbf{C}_\theta & \mu_n (\mathbf{C}_\theta)^2 \end{bmatrix} \begin{bmatrix} \ddot{\mathbf{X}} \\ \ddot{\mathbf{Y}} \end{bmatrix} \\ &- \begin{bmatrix} -\mu_n \mathbf{S}_\theta \mathbf{C}_\theta & -\mu_n (\mathbf{S}_\theta)^2 \\ \mu_n (\mathbf{C}_\theta)^2 & \mu_n \mathbf{S}_\theta \mathbf{C}_\theta \end{bmatrix} \begin{bmatrix} \mathbf{V}^a_x \\ \mathbf{V}^a_y \end{bmatrix} \dot{\boldsymbol{\theta}}, \end{aligned} \tag{3}$$

where $\mathbf{V}^a_x = \text{diag}(V_{x,1}, \ldots, V_{x,n}) \in \mathbb{R}^{n \times n}$, $\mathbf{V}^a_y = \text{diag}(V_{y,1}, \ldots, V_{y,n}) \in \mathbb{R}^{n \times n}$ and $[V_{x,i}, V_{y,i}]^T$ represents the constant and irrotational ocean current effects. Furthermore, the linear and nonlinear drag forces on the robot are given by

$$\begin{bmatrix} \mathbf{f}^{\text{I}}_{\mathbf{D_x}} \\ \mathbf{f}^{\text{I}}_{\mathbf{D_y}} \end{bmatrix} = - \begin{bmatrix} c_t \mathbf{C}_\theta & -c_n \mathbf{S}_\theta \\ c_t \mathbf{S}_\theta & c_n \mathbf{C}_\theta \end{bmatrix} \begin{bmatrix} \mathbf{V_{r_x}} \\ \mathbf{V_{r_y}} \end{bmatrix}, \tag{4}$$

$$\begin{bmatrix} \mathbf{f}^{II}_{\mathbf{D_x}} \\ \mathbf{f}^{II}_{\mathbf{D_y}} \end{bmatrix} = - \begin{bmatrix} c_t\mathbf{C}_\theta & -c_n\mathbf{S}_\theta \\ c_t\mathbf{S}_\theta & c_n\mathbf{C}_\theta \end{bmatrix} \operatorname{sgn}\left(\begin{bmatrix} \mathbf{V_{r_x}} \\ \mathbf{V_{r_y}} \end{bmatrix} \right) \begin{bmatrix} \mathbf{V_{r_x}}^2 \\ \mathbf{V_{r_y}}^2 \end{bmatrix}, \tag{5}$$

where the relative velocities are given by

$$\begin{bmatrix} \mathbf{V_{r_x}} \\ \mathbf{V_{r_y}} \end{bmatrix} = \begin{bmatrix} \mathbf{C}_\theta & \mathbf{S}_\theta \\ -\mathbf{S}_\theta & \mathbf{C}_\theta \end{bmatrix} \begin{bmatrix} \dot{\mathbf{X}} - \mathbf{V}_x \\ \dot{\mathbf{Y}} - \mathbf{V}_y \end{bmatrix}. \tag{6}$$

In addition, the fluid torques on all links are

$$\boldsymbol{\tau} = -\boldsymbol{\Lambda}_1\ddot{\theta} - \boldsymbol{\Lambda}_2\dot{\theta} - \boldsymbol{\Lambda}_3\dot{\theta}|\dot{\theta}|, \tag{7}$$

where $\boldsymbol{\Lambda}_1 = \lambda_1\mathbf{I}_n$, $\boldsymbol{\Lambda}_2 = \lambda_2\mathbf{I}_n$ and $\boldsymbol{\Lambda}_3 = \lambda_3\mathbf{I}_n$. The fluid parameters due to the drag and added mass effects are denoted by c_t, c_n, λ_2, λ_3 and μ_n, λ_1, respectively. For more details see [6].

Combining the kinematics and the hydrodynamic model, in [6,21,22], it is shown that the equations of motion of an USR can be expressed as follows:

$$\begin{aligned} \begin{bmatrix} \ddot{p}_x \\ \ddot{p}_y \end{bmatrix} &= -\mathbf{M}_p \begin{bmatrix} \mathbf{k}_{11} & \mathbf{k}_{12} \\ \mathbf{k}_{21} & \mathbf{k}_{22} \end{bmatrix} \begin{bmatrix} l\mathbf{K}^T(\mathbf{C}_\theta\dot{\theta}^2 + \mathbf{S}_\theta\ddot{\theta}) \\ l\mathbf{K}^T(\mathbf{S}_\theta\dot{\theta}^2 - \mathbf{C}_\theta\ddot{\theta}) \end{bmatrix} \\ &- \mathbf{M}_p \begin{bmatrix} \mathbf{k}_{12} & -\mathbf{k}_{11} \\ \mathbf{k}_{22} & -\mathbf{k}_{21} \end{bmatrix} \begin{bmatrix} \mathbf{V}^a_x \\ \mathbf{V}^a_y \end{bmatrix} \dot{\theta} + \mathbf{M}_p \begin{bmatrix} \mathbf{e}^T\mathbf{f_{Dx}} \\ \mathbf{e}^T\mathbf{f_{Dy}} \end{bmatrix}, \end{aligned} \tag{8}$$

$$\mathbf{M}_\theta\ddot{\theta} + \mathbf{W}_\theta\dot{\theta}^2 + \mathbf{V}_\theta\dot{\theta} + \boldsymbol{\Lambda}_3|\dot{\theta}|\dot{\theta} + \mathbf{K_{Dx}}\mathbf{f_{Dx}} + \mathbf{K_{Dy}}\mathbf{f_{Dy}} = \mathbf{D}^T\mathbf{u}, \tag{9}$$

where $\mathbf{f_{Dx}} = \mathbf{f}^{I}_{\mathbf{D_x}} + \mathbf{f}^{II}_{\mathbf{D_x}}$ and $\mathbf{f_{Dy}} = \mathbf{f}^{I}_{\mathbf{D_y}} + \mathbf{f}^{II}_{\mathbf{D_y}}$ represent the drag forces in the x and y directions and $\mathbf{u} \in \mathbb{R}^{n-1}$ the control input. For more details regarding the derivation of the vectors $\mathbf{k}_{11}$, $\mathbf{k}_{12}$, $\mathbf{k}_{21}$ and $\mathbf{k}_{22}$ and the matrices $\mathbf{M}_p$, $\mathbf{M}_\theta$, $\mathbf{W}_\theta$, $\mathbf{V}_\theta$, $\mathbf{K_{Dx}}$ and $\mathbf{K_{Dy}}$, see [21,22].

2.2. Dynamic Model of Land-Based Snake Robot

Note that the dynamic model of an USR given by Equations (8) and (9) is general in the sense that it can be used to model the motion of amphibious snake robots moving both on land and in water. In particular, by setting the the fluid parameters to zero (i.e., setting c_t, c_n, μ_n, λ_1, λ_2 and λ_3 to zero) and replacing the drag forces Equations (4) and (5) with the following viscous friction model proposed in [1]:

$$\begin{bmatrix} \mathbf{f_{R_x}} \\ \mathbf{f_{R_y}} \end{bmatrix} = - \begin{bmatrix} \tilde{c}_t(\mathbf{C}_\theta)^2 + \tilde{c}_n(\mathbf{S}_\theta)^2 & (\tilde{c}_t - \tilde{c}_n)\mathbf{S}_\theta\mathbf{C}_\theta \\ (\tilde{c}_t - \tilde{c}_n)\mathbf{S}_\theta\mathbf{C}_\theta & \tilde{c}_t(\mathbf{S}_\theta)^2 + \tilde{c}_n(\mathbf{C}_\theta)^2 \end{bmatrix} \begin{bmatrix} \dot{\mathbf{X}} \\ \dot{\mathbf{Y}} \end{bmatrix}, \tag{10}$$

the equation of motion of a land-based snake robot can be expressed as

$$\mathbf{M}_\theta\ddot{\theta} + \mathbf{W}_\theta\dot{\theta}^2 - l\mathbf{S}_\theta\mathbf{K}\mathbf{f_{R_x}} + l\mathbf{C}_\theta\mathbf{K}\mathbf{f_{R_y}} = \mathbf{D}^T\mathbf{u}, \tag{11}$$

$$\begin{bmatrix} \ddot{p}_x \\ \ddot{p}_y \end{bmatrix} = \frac{1}{nm} \begin{bmatrix} \mathbf{e}^T\mathbf{f_{R_x}} \\ \mathbf{e}^T\mathbf{f_{R_y}} \end{bmatrix}. \tag{12}$$

The parameters $\tilde{c}_t$ and $\tilde{c}_n$ represent the viscous friction coefficients, while detailed derivation of the matrices $\mathbf{M}_\theta$ and $\mathbf{W}_\theta$ can be found in [1].

3. Optimization of Motion

In this section, we propose an optimization framework to investigate the efficient motion of snake robots. As mentioned in the Introduction, the optimization approach presented in [16] for underwater snake robots is extended in this paper to consider both land-based and swimming snake robots, an arbitrary number of links, and also to take into account the negative work effect of swimming snake robots. Therefore, the proposed optimization framework of this paper is general since it can be widely applied to study locomotion efficiency of robotic systems using undulatory motion patterns to move forward. As shown in Figure 3, the optimization framework consists of three parts: (a) the plant, which in this paper is the model of a snake robot; (b) the system input, which combines the motion pattern generator with a joint actuation controller; and (c) an optimizer. Regarding the optimizer, the objective functions need to be evaluated for different values of the gait parameters for each simulation of the model in order to investigate the locomotion efficiency. The feasible region of the solution is also determined by defining the constraints, which are also inputs of the optimizer. More details regarding the system input and the optimizer parts used in this paper are given in the following sections.

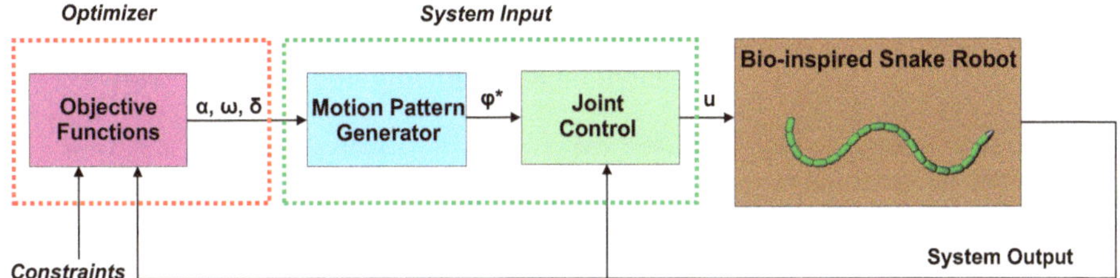

Figure 3. Illustration of the optimization framework.

3.1. System Input

As shown in Figure 3, the system input part consists of the motion pattern generator and the joint controller. In the following, both parts are discussed in detail.

3.1.1. Motion Pattern

Both land-based and swimming snake robots commonly use undulatory motion patterns in order to move forward. Previous studies on snake robots moving on land mostly adopt the lateral undulation motion pattern, which is considered as the fastest form of land-based snake robot locomotion [1]. Furthermore, studies on swimming snake robots often consider both lateral undulation and eel-like motion patterns as common motion modes to achieve propulsion. In [31], a general sinusoidal (undulatory) motion pattern was proposed, which makes it possible to describe different undulatory motion patterns including the lateral undulation and eel-like motion. To achieve an undulatory motion pattern, each joint $i \in \{1, \cdots, n-1\}$ of the snake robot is commanded to follow the reference signal

$$\phi_i^*(t) = \alpha g(i,n) \sin(\omega t + (i-1)\delta) + \gamma, \tag{13}$$

where the proper choice of the scaling faction $g(i,n)$ makes it possible to describe different undulatory motion patterns. For instance, it is possible to achieve lateral undulation by simply choosing $g(i,n) = 1$, while $g(i,n) = (n-i)/(n+1)$ results in an eel-like motion pattern. The gait parameters α, ω and δ give the amplitude, frequency and the phase shift of the sinusoidal motion pattern, respectively. Furthermore, the joint offset parameter γ is commonly used to control the heading of the robot [1].

3.1.2. Joint Actuation Controller

For low-level joint actuation, the following proportional–derivative (PD) controller is used:

$$u_i = k_p(\phi_i^* - \phi_i) + k_d(\dot{\phi}_i^* - \dot{\phi}_i), \quad i = 1, \ldots, n-1, \tag{14}$$

where $k_p > 0$ and $k_d > 0$ denote the gains of the controller. For more details, see [1].

3.2. Formulation of the Optimization Problem

Solving an optimization problem entails finding a solution from all feasible solutions to maximize (or minimize) an objective function. In this paper, we define the motion optimization problem to optimize two objective functions, simultaneously: energy consumption, and the forward velocity. This type of optimization problem is called multi-objective optimization, since more than one objective function is involved [32]. In the following, these two objective functions (the energy consumption and the forward velocity) are formulated. To solve the bi-objective optimization problem, the well-known weighted-sum approach is implemented to combine these two objective functions.

First, we formulate the total energy consumption as a function of the actuation torque and the angular velocity of the joints:

$$E_s = \int_0^T \left(\sum_{i=1}^{n-1} |u_i(t)\dot{\phi}_i(t)| \right) dt, \tag{15}$$

where T represents the time of a complete motion cycle. The actuation torque u_i and the angular velocity $\dot{\phi}_i$ for joint i are calculated by using (14) and $\dot{\phi}_i = \dot{\theta}_i - \dot{\theta}_{i-1}$, respectively, [8,15].

Note that in this paper, we assume the joint to be ideal. Therefore, the total energy of the system is equal to the summation of kinetic energy and the energy which is dissipated due to either the ground friction for the land-based snake robot or the surrounding fluid for the underwater snake robot. The average power consumption is calculated by using the following expression:

$$P_{\text{avg}} = \frac{1}{T} \int_0^T \left(\sum_{i=1}^{n-1} |u_i(t)\dot{\phi}_i(t)| \right) dt. \tag{16}$$

Note that Equation (16) gives the average power consumption taking into account only the absolute value of the power consumed for the joint motion. However, in order to take into account also the effects of negative work [8], the net joint power can be considered instead. In this case, the average power consumption can be calculated using the following expression:

$$P_{\text{avg}} = \frac{1}{T} \int_0^T \left(\sum_{i=1}^{n-1} u_i(t)\dot{\phi}_i(t) \right) dt. \tag{17}$$

For underwater snake robots, both Equations (16) and (17) are examined. In the latter case which is presented in [8], it is shown that by considering the negative work effect the simulated robot is able to recover energy. However, for the land-based snake robot only the absolute values of the theoretical joint power is considered because the ground friction consumes energy when the robot moves on land, i.e., Equation (16) is used.

In addition, the following expression is used to calculate the forward velocity:

$$\bar{v} = \frac{\sqrt{(p_x(T) - p_x(0))^2 + (p_y(T) - p_y(0))^2}}{T}, \tag{18}$$

where $(p_x(0), p_y(0))$ and $(p_x(T), p_y(T))$ denote the initial and final positions of CM of the robot.

It is shown in [16] that the motion optimization problem can be formulated by the following objective functions and bound constraints:

$$\min_{\alpha,\omega,\delta} J_{\text{opt}} = [P_{\text{avg}}, -\bar{v}], \tag{19a}$$

$$\text{s.t:} \mid \phi_i^* \mid \leq \phi_i^{max}, \mid \dot{\phi}_i^* \mid \leq \dot{\phi}_i^{max}, \mid u_i \mid \leq u_i^{max}, \tag{19b}$$

$$0 \leq \alpha \leq \alpha^{max}, 0 \leq \omega \leq \omega^{max}, 0 \leq \delta \leq \delta^{max}, \tag{19c}$$

where Equation (19b) represents the physical limitations of the joints because of the servo motors and special design of the snake robot [33], and Equation (19c) limits the possible range of the parameters of the sinusoidal motion pattern Equation (13).

As stated in [16], for the bi-objective optimization problem Equation (19), instead of a single global solution, there are a set of optimal solutions. Moreover, in this problem, the objective functions are in conflict such that minimizing the power consumption clearly results in decreasing the velocity and vice versa. In [16], we suggested using a Pareto optimality concept for the bi-objective optimization problem. In the optimization problem Equation (19), a solution is called Pareto optimal [32], if any improvement in power consumption cannot be achieved without a penalty of decreasing the forward velocity, or the forward velocity cannot have any higher value without requiring more power. The collection of Pareto optimal solutions is called the Pareto frontier or efficient frontier. Since the considered problem in this paper has two objective functions, it is possible to show the frontier in Cartesian coordinates. As stated before, one of the well-known ways to formulate the multi-objective optimization problem into a standard single objective function problem, is the weighted-sum approach:

$$J_{\text{bal}} = w_p (P_{\text{avg}})_{sc} - w_v (\bar{v})_{sc}, \tag{20}$$

$$w_p = 1 - w_v, \tag{21}$$

where $(P_{\text{avg}})_{sc}$ and $(\bar{v})_{sc}$ represent scaled values of the power consumption and forward velocity, respectively, and w_p and w_v are the weighting coefficients. The solution is always the Pareto frontier, if the weighting factors are positive.

By sweeping the values of w_p in the range of 0 and 1 (and correspondingly varying w_v) and solving the problem for each set of weights, Pareto optimal solutions and consequently Pareto frontier plots can be obtained. The single-objective function optimization problem is formulated by replacing the objective function J_{opt} to J_{bal} in Equation (20). In order to solve the motion optimization problem, the PSO algorithm was proposed in [16]. The PSO algorithm was first introduced by Kennedy and Eberhart in 1995 [26]. The PSO algorithm is a population-based stochastic algorithm and searches the search space using a set of potential solutions at each iteration (or generation). Each potential solution is named a particle, and the set of particles is called a population. The location of each particle in the new generation is updated by some equations. These equations are inspired by the social behavior of animals such as bird flocks or fish schools. For details regarding the implementation of the PSO algorithm, see [16].

Remark 1. *As stated before, in the optimization algorithm the plant is only required to evaluate the measure of efficiency (the objective function). This means that the multi-objective optimization approach proposed in this paper is general and thus can used to study locomotion efficiency of robotic systems using undulatory (sinusoidal) motion patterns such as Equation (13) to move forward.*

Remark 2. *Note that including other design alternatives such as the number of links obviously leads to increasing the size of the optimization problem. Therefore, it might require more function evaluations to converge to the solution. Moreover, the designers might not be able to give us an initial guess for some of the parameters, which in turn leads to more computational time. In this paper we only include bound constraints. Nonlinear constraints, if any, can be handled by some constraint handling methods. In [34], two different constraint handling methods are coupled with the PSO algorithm. In addition, note that the optimality of the Pareto solutions can be checked by perturbing the optimization parameters and comparing the obtained objective function.*

4. Optimization Study

The optimization results presented in this section consider snake robots moving on land and swimming underwater. The models of underwater and land-based snake robots were simulated using Matlab R2013b (MathWorks, Natick, MA, U.S.A, 2013). In particular, the ode23tb solver with both the relative and absolute errors being set to 10^{-4} was used to simulate the dynamic equations presented in Section 2. The PSO is implemented using GenOpt, 3.1.1 (Lawrence Berkeley National Laboratory, Berkeley, CA, U.S.A, 2016), which is developed by Lawrence Berkeley National Laboratory and has parallel simulation run options for computation time reduction and allows using any simulation software to evaluate the cost function [35].

4.1. Parameters of the Robot

For the obtained results, the parameters of the robot are chosen to be identical to the underwater snake robot Mamba [33], i.e., each link having the length $2l = 0.18$ m and mass $m = 0.8$ kg. Different configurations of the robot are investigated. In particular, we present results for $n = 5$, $n = 10$ and $n = 20$ number of links. The hydrodynamic parameters c_t, c_n, μ_n, λ_1, λ_2 and λ_3 were calculated for $2a = 2 \times 0.055$ m and $2b = 2 \times 0.05$ m, $\rho = 1000$ kg/m^3, $C_f = 0.03$, $C_D = 2$, $C_A = 1$ and $C_M = 1$. For more details, see [6]. Furthermore, based on the discussion in [1], the viscous friction parameters are set to $\tilde{c}_t = 0.1$ and $\tilde{c}_n = 10$. The gains of the PD joint controller given by Equation (14) are set to $k_p = 20$, $k_d = 5$ and the sinusoidal motion pattern is achieved using Equation (13). Note that for the optimization results presented in this paper, we have not taken into account the effects from the ocean current, which remains topic of future work.

4.2. Optimization Parameters

The optimization problem formulated in Equation (19) includes 3 decision variables. Following the suggestion in [36], to use a population size equivalent to approximately 5 times the decision vector size, we set the number of particles equal to 16. Like other stochastic approximation algorithms, PSO has a few tunable parameters. There are extensive discussion about tuning of the PSO parameters in the literature. We here follow the recommendations in [36,37], which are based on the trials on a selection of different benchmark problems, to initialize the tuning parameters in the PSO algorithm. The values of the constraints in Equation (19b) are chosen in accordance with the limitations of the servo motors used in the snake robot presented in [33]. Therefore, we set the physical constraints of the joints to $u_i^{max} = 2300$ Nm, $\phi_i^{max} = 90°$, $\dot{\phi}_i^{max} = 429\,°/\text{s}$. Regarding the range of the gait parameters in Equation (19c), Table 2 gives the values for both the underwater snake robot and the land-based snake robot. As can be seen in Table 2 for a snake robot with $n = 20$ links, the values of the parameters $\alpha^{max} = 70°$ and $\delta^{min} = 20°$. These upper and lower limits are set to avoid collisions between the links during the motion of the robot with a large number of links. In addition, note that in this paper we have chosen to use $\gamma = 0$ since we are not aiming to control the heading of the robot.

In this paper, we sweep the range from 0 to 1 for $w_p = 0$ by the step size of $\Delta w_p = 0.05$. Therefore, the first optimization problem is to only optimize the velocity ($w_p = 0$, $w_v = 1$ in Equation (21)). The next optimization problems, the weight of the velocity w_v is reduced gradually, while w_p is increased. In the next step, we start to reduce the weight of the velocity w_v, while increasing the power consumption weight w_p. In order to reduce the computation cost of the PSO algorithm, the initial value of the optimization parameter for the first optimization problem is tuned based on expert knowledge. Moreover, the optimal solution of each problem is used as an initial value for the next optimization problem.

To obtain a Pareto optimal point, 320 simulation runs are required (16 particles and 20 generations). Using a distributed computing framework, the computation time to obtain a Pareto optimal point for the optimization problem considered in this paper is approximately 65 minutes.

Table 2. Range of the parameters of the sinusoidal motion pattern.

n	α^{min}	α^{max}	ω^{min}	ω^{max} (°/s)	δ^{min}	δ^{max}
5	0	90°	0	210	0	90°
10	0	90°	0	210	0	90°
20	0	70°	0	210	20°	90°

4.3. Results for Underwater Snake Robots

In this section, the results of motion efficiency optimization of underwater snake robots are presented for both lateral undulation and eel-like motion patterns. The parameters of the robot are specified in Section 4.1. Pareto fronts are presented in Figures 4 and 5 for a different number of links for lateral undulation and eel-like motion pattern, respectively. In these figures, we also investigate the negative work effect in the calculation of average power consumption (see Equations (16) and (17)). Figures 6–9 show Pareto optimal gait parameters of the sinusoidal motion pattern vs. Pareto optimal average power consumption and achieved forward velocity. As it was expected, the robot consumed the maximum power when moving with the highest forward velocity, while zero power is consumed when the robot is not moving (i.e., for zero forward velocity). To discuss some of the observations, selected Pareto optimal points are compared in Tables 3–6. The maximum achieved forward velocities and corresponding optimal parameters for different numbers of links are presented in these tables. The maximum forward velocity for the underwater snake robot with lateral undulation motion pattern is equal to $\bar{v} = 0.46$ m/s, $\bar{v} = 0.84$ m/s, and $\bar{v} = 1.06$ m/s, respectively, for a number of links equal to 5, 10, and 20. The maximum average power consumption considering the negative work effect is equal to $P_{avg} = 4.83$ W, $P_{avg} = 34.25$ W, and $P_{avg} = 74.01$ W for a number of links equal to $n = 5, 10$, and 20, respectively. If we neglect negative work effect the maximum average power consumption is increased to $P_{avg} = 5.33$ W, $P_{avg} = 34.92$ W, and $P_{avg} = 83.45$ W, respectively, for a number of links equal to 5, 10, and 20. The increment of average power consumption as a result of disregarding negative work effect is equal to 10.35%, 1.97% and 12.76%. As expected, the maximum forward velocity for eel-like motion pattern is lower than that of snake robot with a lateral undulation pattern: $\bar{v} = 0.38$ m/s for $n = 5$, $\bar{v} = 0.60$ m/s for $n = 10$, $\bar{v} = 0.77$ m/s for $n = 20$. The maximum average power consumption for eel-like motion pattern considering negative work effect is equal to $P_{avg} = 3.26$ W, $P_{avg} = 13.44$ W, $P_{avg} = 36.08$ W for $n = 5, 10$ and 20, respectively. In the case of disregarding negative work effect, the maximum power consumption is changed to $P_{avg} = 3.45$ W, $P_{avg} = 14.96$ W and $P_{avg} = 36.78$ W, for a number of links equal to 5, 10, and 20, respectively. This means neglecting negative work effect results in 5.83%, 11.31% and 1.94% increase of power consumption for $n = 5, 10$ and 20, respectively. One can observe from all Pareto optimal points of both motion patterns, the robot consumes more energy when the negative work effect is not considered. It is important to consider this observation in the development of underwater snake robots.

We can also compare the maximum values of objective functions in terms of the number of links. For the lateral undulation motion pattern, 82.61% and 26.19% increases in the forward velocity are achieved by doubling the number of links from 5 to 10 and from 10 to 20, respectively. The penalty of these velocity increments is an increase of power consumption of approximately 7 and 2.2 times, respectively. Similar results are observed for the eel-like motion pattern. We have approximately 4.1 and 2.7 times increment in the average power consumption for 57.89% and 28.33% increment in the achieved forward velocity, respectively, by multiplying the number of links by a factor of 2. In the case of disregarding negative work effect, one can observe that the robot with 10 links consumes approximately 6.55 and 2.4 times more power than a robot with 5 links for lateral undulation and eel-like motion pattern, respectively. Moreover, a robot with 20 links consumes 4.4 and 2.5 times more power than a robot with 10 links, for lateral undulation and eel-like motion patterns, respectively. Hence, the number of links is directly connected to the efficiency of the underwater snake robots. Therefore, it is important to also consider the number of links as a new optimization variable in the problem. The optimum value of the number of links gives useful inputs in the design stage and the implementation of the swimming snake robots.

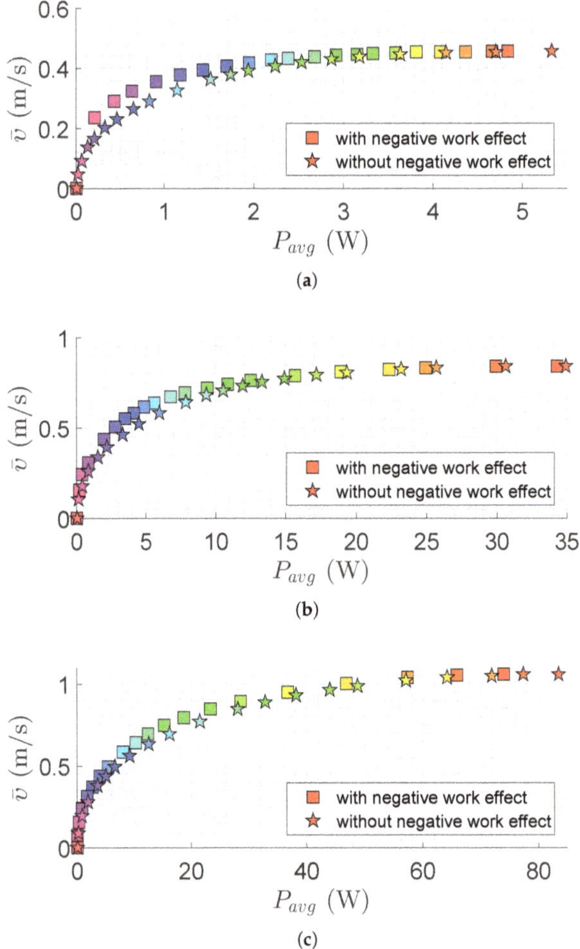

Figure 4. Pareto front for the USR using the lateral undulation motion pattern: (**a**) Pareto front for robot with $n = 5$ links; (**b**) Pareto front for robot with $n = 10$ links; and (**c**) Pareto front for robot with $n = 20$ links. Colors illustrate different weights of the objective functions.

Figures 4 and 5 show that it is possible to obtain a significant decrease in the consumed power by a slight decrease in the achieved forward velocity. This observation is also clearly presented in Tables 3–6. For the lateral undulation motion pattern of the snake robot with five links, 44.51% (or 54.45%) decrease in the average power consumption leads to a slight decrease of 4.35% (or 6.52%) in the forward velocity. Similarly for the eel-like motion pattern, by decreasing 42.02% (or 57.36%) in the power consumption, we have 2.63% (or 5.26%) decrease in the forward velocity. Tables 5 and 6 give similar results for the case that the negative work effect is not considered. In this case, we see that it is possible to decrease the consumed power about 40.34% (or 52.35%) and 45.22% (or 55.94%) for lateral undulation and eel-like motion pattern, respectively. Similar results are also obtained for an underwater snake robot with 10 and 20 links. For the robot with 10 links, we can observe that the robot can consume 44.76% (or 54.19%) and 41.40% (or 52.60%) less power by reducing the forward velocity by 3.57% (or 5.95%) and 3.33% (or 5%), for lateral undulation and eel-like motion, respectively. If we neglect negative work effect, the robot consumes 44.70% (or 57.27%) less power for lateral undulation and 42.98% (or 52.27%) less power for eel-like motion pattern. For the robot with 20 links with lateral undulation pattern, we can reduce the power consumption by 36.85% and 50.34% by reducing the achieved forward velocity by 5.66% and 10.38%, respectively. For the eel-like motion pattern, Figure 5c shows that the average power consumption can be reduced by 42.32% and 59.23% while the corresponding forward velocity is decreased only by 5.19% and 11.69%. If we calculate the power consumption without considering negative work effect, for the mentioned decrease of the forward velocity we can save 31.59% and 47.40% power for lateral undulation, and 39.61% and 49.37% power for eel-like motion pattern. In conclusion, this observation may attract the attention of designers to study Pareto fronts in order to reduce power consumption significantly while decreasing the corresponding forward velocity slightly.

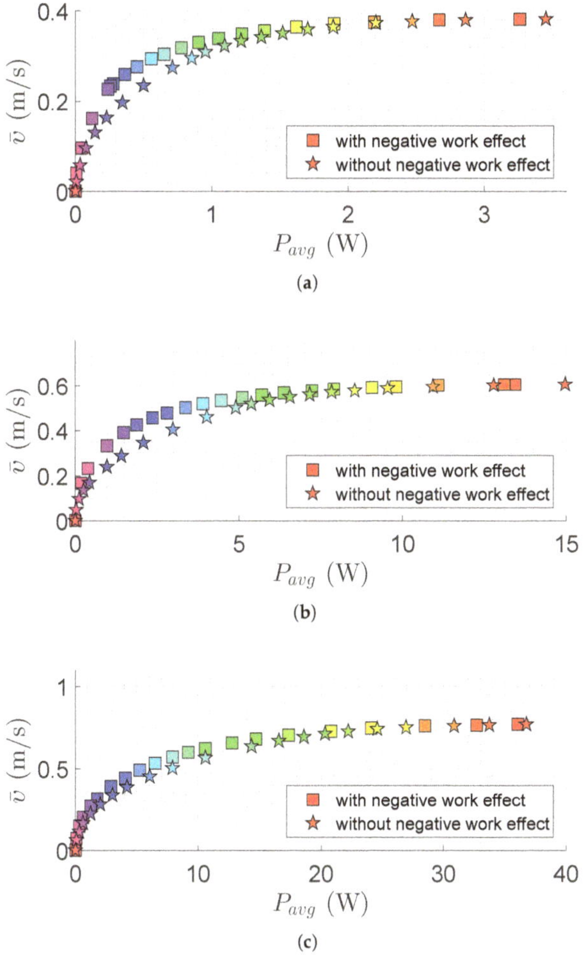

Figure 5. Pareto front for the USR using the eel-like motion pattern: (**a**) Pareto front for robot with $n = 5$ links; (**b**) Pareto front for robot with $n = 10$ links; and (**c**) Pareto front for robot with $n = 20$ links. Colors illustrate different weights of the objective functions.

Figures 6–9 present the inverse correlation of the parameter δ and the forward velocity or the power consumption. Furthermore, we can also observe in Figure 6 that, for the case of considering the negative work effect, the gait parameter α takes values in the range of $[50°, 66°]$, $[31°, 69°]$ and $[25°, 59°]$ for all optimization trials for the underwater snake robot studied in this paper with $n = 5$, $n = 10$ and $n = 20$ links, respectively, for lateral undulation. Figure 7 shows that the range of change of the gait parameter α is equal to $[31°, 80°]$, $[51°, 90°]$ and $[42°, 70°]$ for a robot with $n = 5$, $n = 10$ and $n = 20$ links, for an eel-like motion pattern. The values of the phase shift parameter δ are in the range of $[26°, 90°]$, $[15°, 90°]$, $[20°, 90°]$ and $[34°, 90°]$, $[25°, 90°]$, $[23°, 90°]$ for the robot with $n = 5$, $n = 10$, $n = 20$ links for lateral undulation and an eel-like motion pattern, respectively. In the case that the negative work effect is not considered, the maximum amplitude α for lateral undulation pattern is in the range of $[22°, 49°]$, $[19°, 45°]$, and $[14°, 41°]$ when the number of links is, respectively, equal to 5, 10, and 20. For the eel-like motion pattern, the range of α is equal to $[23°, 80°]$, $[36°, 60°]$, and $[28°, 56°]$, respectively, for $n = 5, 10$, and 20. For lateral undulation pattern, the phase shift δ changes in the range of $[26°, 80°]$, $[15°, 62°]$, and $[20°, 64°]$ for $n = 5, 10$, and 20, respectively. For the eel-like motion pattern, the range of phase shift is equal to $[34°, 81°]$, $[26°, 78°]$, and $[23°, 64°]$ for $n = 5, 10$, and 20, respectively.

We can also observe that the maximum forward velocity is achieved using the maximum values of $\omega = 210\,°/\mathrm{s}$ for all the cases, while the obtained values of the amplitude, α, and the phase-shift, δ, are different for the investigated motion patterns. These observations are useful for both control design purposes and defining the range of the gait parameters for solving the optimization problem more precisely in the future.

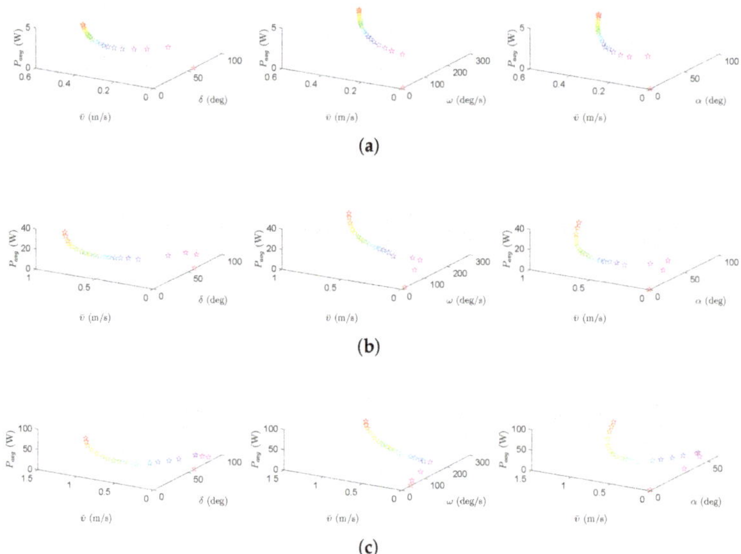

Figure 6. Optimal gait parameters for the lateral undulation motion pattern for the USR, considering the negative work effect: (**a**) 3D plots of optimal gait parameters for robot with $n = 5$ links; (**b**) 3D plots of optimal gait parameters for robot with $n = 10$ links; and (**c**) 3D plots of optimal gait parameters for robot with $n = 20$ links. Colors illustrate different weights of the objective functions.

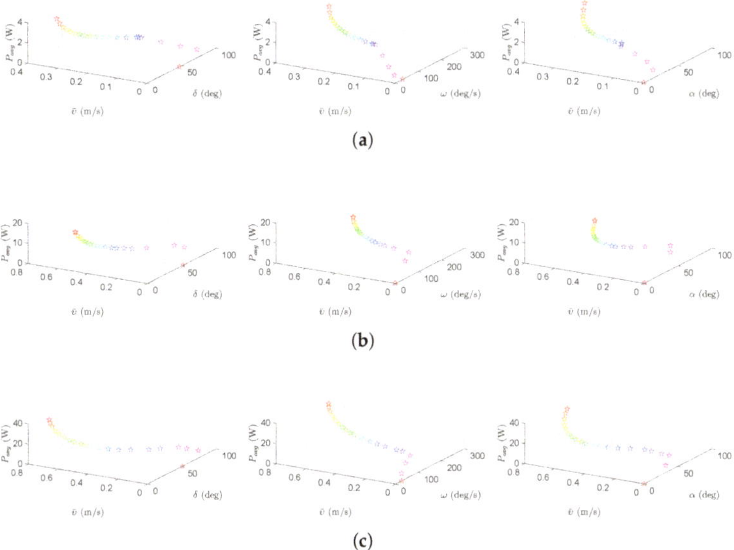

Figure 7. Optimal gait parameters for the eel-like motion pattern for the USR, considering the negative work effect: (**a**) 3D plots of optimal gait parameters for robot with $n = 5$ links; (**b**) 3D plots of optimal gait parameters for robot with $n = 10$ links; and (**c**) 3D plots of optimal gait parameters for robot with $n = 20$ links. Colors illustrate different weights of the objective functions.

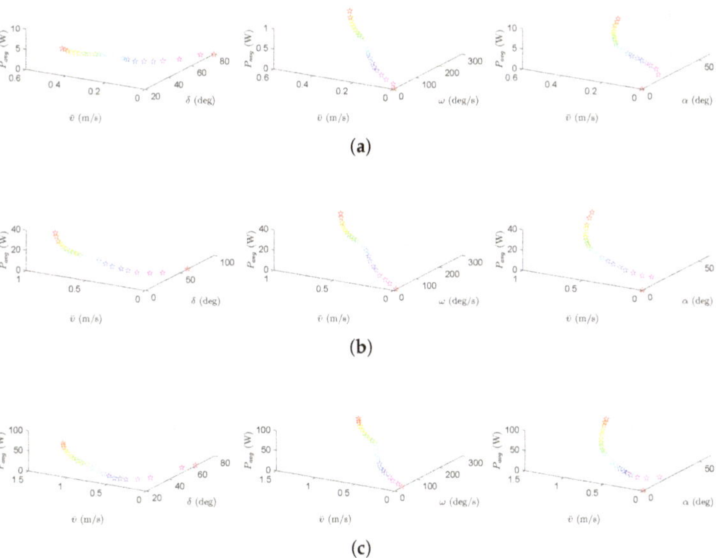

Figure 8. Optimal gait parameters for the lateral undulation motion pattern for the USR, without considering the negative work effect: (**a**) 3D plots of optimal gait parameters for robot with $n = 5$ links; (**b**) 3D plots of optimal gait parameters for robot with $n = 10$ links; and (**c**) 3D plots of optimal gait parameters for robot with $n = 20$ links. Colors illustrate different weights of the objective functions.

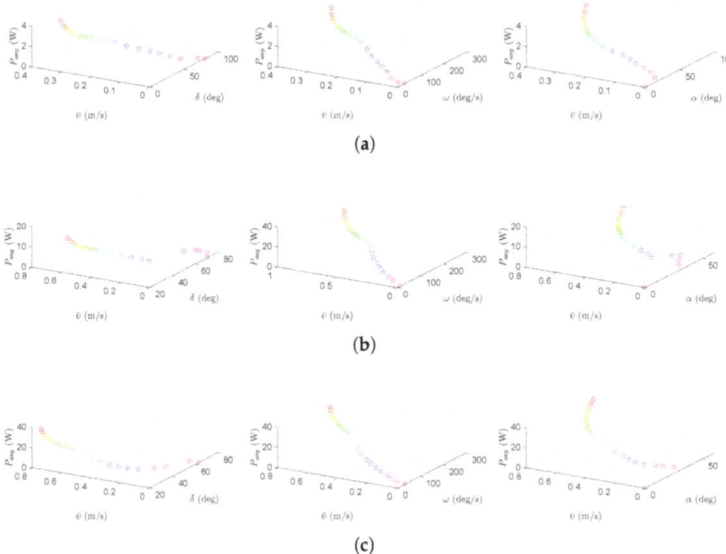

Figure 9. Optimal gait parameters for the eel-like motion pattern for the USR, without considering the negative work effect: (a) 3D plots of optimal gait parameters for robot with $n = 5$ links; (b) 3D plots of optimal gait parameters for robot with $n = 10$ links; and (c) 3D plots of optimal gait parameters for robot with $n = 20$ links. Colors illustrate different weights of the objective functions.

Table 3. Multi-objective optimization results for the lateral undulation motion pattern for the USR with different numbers of links, considering the negative work effect.

n	α (deg)	ω (deg/s)	δ (deg)	$\bar{v}$ (m/s)	P_{avg} (W)
5	57.68	210	26.82	0.46	4.83
5	52.33	210	35.65	0.44	2.68
5	51.68	210	39.33	0.43	2.20
10	44.01	210	15.14	0.84	34.25
10	34.92	210	20.32	0.81	18.92
10	33.42	210	22.97	0.79	15.69
20	41.21	210	20.00	1.06	74.01
20	32.45	210	21.73	1.00	46.74
20	29.16	210	24.01	0.95	36.75

Table 4. Multi-objective optimization results for the eel-like motion pattern for the USR with different numbers of links, considering the negative work effect.

n	α (deg)	ω (deg/s)	δ (deg)	$\bar{v}$ (m/s)	P_{avg} (W)
5	79.21	210	34.23	0.38	3.26
5	72.04	210	43.73	0.37	1.89
5	69.69	210	50.09	0.36	1.39
10	59.24	210	26.20	0.60	13.44
10	52.61	210	33.47	0.58	7.88
10	51.33	210	36.69	0.57	6.37
20	55.13	210	23.97	0.77	36.08
20	43.83	210	29.48	0.73	20.81
20	42.03	210	33.86	0.68	14.71

Table 5. Multi-objective optimization results for the lateral undulation motion pattern for the USR with different numbers of links, without considering the negative work effect.

n	α (deg)	ω (deg/s)	δ (deg)	$\bar{v}$ (m/s)	P_{avg} (W)
5	58.05	210	26.68	0.46	5.33
5	51.16	210	35.17	0.44	3.18
5	48.14	210	39.34	0.42	2.54
10	44.02	210	15.17	0.84	34.92
10	34.76	210	20.98	0.81	19.31
10	32.26	210	24.68	0.77	14.92
20	40.96	210	20.00	1.06	83.45
20	34.03	210	21.07	1.02	57.09
20	29.81	210	22.64	0.97	43.92

Table 6. Multi-objective optimization results for the eel-like motion pattern for the USR with different numbers of links, without considering the negative work effect.

n	α (deg)	ω (deg/s)	δ (deg)	$\bar{v}$ (m/s)	P_{avg} (W)
5	79.21	210	34.23	0.38	3.45
5	69.37	210	45.38	0.37	1.89
5	66.13	210	50.18	0.35	1.52
10	59.24	209.98	26.20	0.60	14.96
10	51.83	210	34.41	0.58	8.53
10	50.17	209.98	37.48	0.56	7.14
20	55.13	210	23.97	0.77	36.78
20	43.49	210	29.28	0.73	22.21
20	40.39	209.95	31.23	0.70	18.62

4.4. Results for Land-Based Snake Robots

In this section, optimization results for the land-based snake robots are discussed. The motion pattern is considered lateral undulation which is the most common motion pattern for land-based snake robots [1]. Figure 10 presents the Pareto fronts (trade off curves) for different numbers of links. Figure 11 illustrates the Pareto optimal average power consumption and forward velocity versus corresponding optimal gait parameters. Table 7 summarizes a selected number of Pareto optimal objective functions and related Pareto optimal gait parameters. For a land-based snake robot with 5, 10, and 20 links, the maximum forward velocity is equal to $\bar{v} = 0.46$ m/s, $\bar{v} = 0.94$ m/s, and $\bar{v} = 1.27$, respectively, and the corresponding average power consumption is equal to $P_{avg} = 0.96$, $P_{avg} = 6.86$, and $P_{avg} = 14.62$, respectively. This means increasing the number of links from 5 to 10 leads to approximately a 104% increase in forward velocity, while the average power consumption is increased approximately seven times. We observe about 35% increment in the forward velocity by changing the number of links from 10 to 20, and as a result the average power consumption is increased approximately two times. From this table, we notice that the optimal frequency related to the maximum forward velocity is equal to $\omega = 210\ ^\circ/\text{s}$ for different numbers of links. The optimal maximum amplitude is equal to $\alpha = 44.32^\circ$, $\alpha = 25.27^\circ$, and $\alpha = 22.22^\circ$, respectively, for 5, 10 and 20 links. The optimal phase shift for 5, 10, and 20 links is defined equal to $\delta = 36.04^\circ$, $\delta = 18.02^\circ$, and $\delta = 20^\circ$, respectively.

We observe from Figure 10 that one can obtain a significant decrease in the average power consumption as a trade-off for a negligible decrease in the achieved forward velocity. This trade-off is also stated in Table 7. For a snake robot with 5 links, the table gives a 40.79% decrease of the average power consumption (from $P_{avg} = 0.96$ W to $P_{avg} = 0.57$ W) results in just 4.35% decrease in the forward velocity ($\bar{v} = 0.46$ m/s to $\bar{v} = 0.44$ m/s). Another alternative is to decrease the average power consumption from $P_{avg} = 0.96$ W to $P_{avg} = 0.44$ W (about 54% increment) by decreasing the forward velocity from $\bar{v} = 0.46$ m/s to $\bar{v} = 0.42$ m/s (8.69% decrease). Similar results are also stated for a robot with 10 and 20 links. For a snake robot with 10 links, we see that the robot is able to consume 45.63% and 56.26% less power by reducing the forward velocity by 4.26% and 7.45%, respectively. Similarly, for a snake robot with 20 links, we can reduce the power consumption by 34.82% and 46.64% by reducing the achieved forward velocity by 4.72% and 10.23%, respectively. Consequently, based on these results, the operators can select the gait parameters in a way that would result in a significant decrease of the consumed power, while giving only a slight decrease of the forward velocity.

As we can see in Figure 11, the forward velocity and the power consumption decrease when the value of the gait parameter δ increases. In addition, we observe that the optimal value of the gait parameter α is greater than 14°, 17° and 13° in all weighting sets for the robot with $n = 5$, $n = 10$ and $n = 20$ links, respectively. In particular, the gait parameter α takes values in the range of $[14°, 45°]$, $[17°, 30°]$ and $[13°, 21°]$ for all optimization trials for the land-based snake robot studied in this paper with $n = 5$, $n = 10$ and $n = 20$ links, respectively. The values of the phase shift parameter δ are in the range of $[36°, 72°]$, $[18°, 78°]$ and $[20°, 60°]$ for the robot with $n = 5$, $n = 10$ and $n = 20$ links, respectively. The maximum forward velocity is achieved for the maximum values of $\omega = 210$ °/s. Note that these insights are essential for control design purposes for bio-inspired snake robots. Furthermore, they can be used to customize the necessary constraints for future optimization investigations.

Remark 3. *Note that the Pareto fronts illustrated in Figures 4, 5 and 10 constitute an informative visualization tool, which helps control system designers to get full information about objective values (power consumption and velocity) and objective tradeoffs. Consequently, depending on the control objectives and the available power of the robot, the operators or designers of bio-inspired snake robots can choose the optimal operational point using the Pareto fronts. Afterward, the corresponding gait parameters can be obtained.*

Remark 4. *We observe in Tables 3–7 that the frequency parameter ω tends to stay at the maximum value. Therefore, there is a possibility to reduce the dimension of the search space, n_c, to 2 by eliminating ω from the decision variables and setting this parameter equal to the maximum possible value. This observation can also be embedded in the design of the bio-inspired snake robots. This means one should consider the maximum possible value for the frequency during actuation mechanism selection. As a result, it motivates designers to investigate possible high frequency actuation solutions in control and design of underwater and land-based snake robot.*

Table 7. Multi-objective optimization results for the land-based snake robot with different numbers of links.

n	α (deg)	ω (deg/s)	δ (deg)	$\bar{v}$ (m/s)	P_{avg} (W)
5	44.32	210	36.04	0.46	0.96
5	37.81	210	43.78	0.44	0.57
5	33.92	210	46.78	0.42	0.44
10	25.27	210	18.02	0.94	6.86
10	20.87	210	23.29	0.90	3.73
10	19.61	210	25.57	0.87	3.00
20	22.22	210	20.00	1.27	14.62
20	15.87	210	20.02	1.21	9.53
20	14.73	210	21.38	1.14	7.80

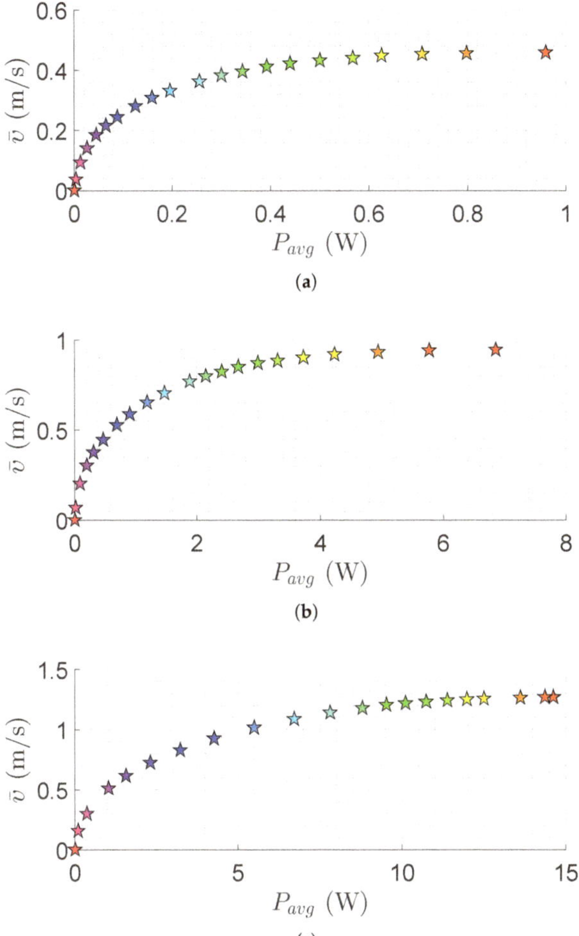

Figure 10. Pareto front for the land-based snake robot using the lateral undulation motion pattern: (**a**) Pareto front for robot with $n = 5$ links; (**b**) Pareto front for robot with $n = 10$ links; and (**c**) Pareto front for robot with $n = 20$ links. Colors illustrate different weights of the objective functions.

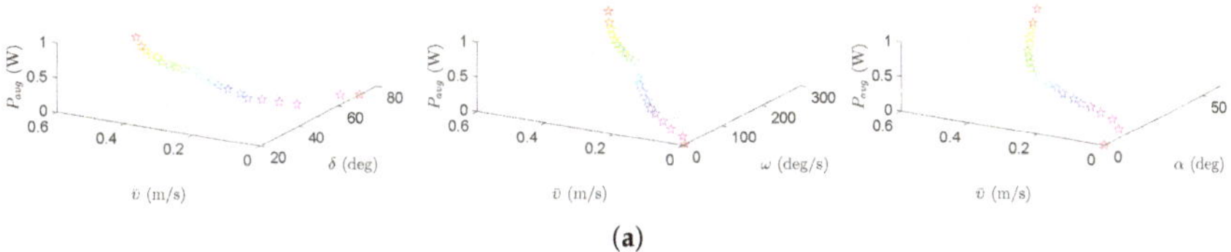

(**a**)

Figure 11. *Cont.*

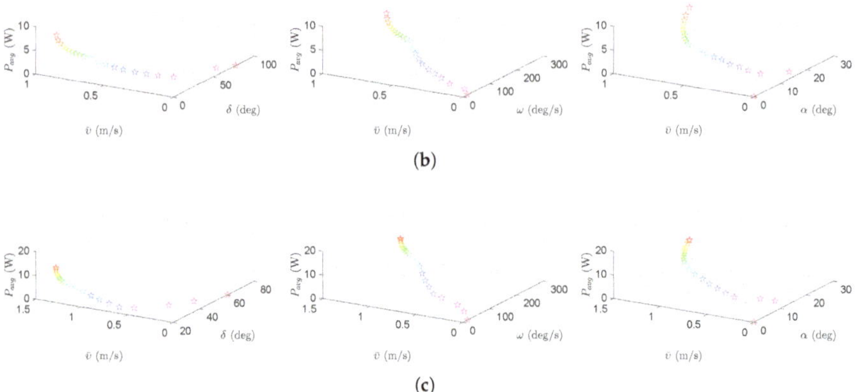

Figure 11. Optimal gait parameters for the lateral undulation motion pattern for the land-based snake robot: (**a**) 3D plots of optimal gait parameters for robot with $n = 5$ links; (**b**) 3D plots of optimal gait parameters for robot with $n = 10$ links; and (**c**) 3D plots of optimal gait parameters for robot with $n = 20$ links. Colors illustrate different weights of the objective functions.

5. Conclusions

A general multi-objective optimization framework was presented in this paper. The proposed framework was applied to study the locomotion efficiency for both swimming and land-based snake robots with different numbers of links. The optimal gait parameters were obtained for different configurations of the snake robots. The proposed optimization scheme is general in the sense that it can be used to study the locomotion efficiency for robotic systems using undulatory motion patterns. Using this framework, locomotion efficiency optimization was analyzed for both land-based snake robots moving according to the gait pattern lateral undulation, and for swimming snake robots, using lateral undulation and eel-like motion. In the latter case, the negative work effect was also taken into account. In particular, the energy efficiency of the underwater snake robots was studied using two different equations to calculate power consumption. In the first case, the absolute value of the theoretical joint power was considered as the power consumption, while, in the second case, the benefit of the negative work effect was also considered, i.e., taking into account that the robot is able to recover energy and thus calculating the power consumption using the net joint power. Optimization results for the land-based snake robot were presented considering only the absolute values of the theoretical joint power because of the energy consumption of the ground friction, which does not give a negative work effect. PSO was chosen as a suitable optimization algorithm to acquire the Pareto optimal solutions taking into account the trade-offs between the consumed power and the achieved forward velocity. From the obtained Pareto fronts, we concluded that improving energy efficiency corresponds to the decrements of the forward velocity. Consequently, depending on the control objectives and the available power of the robot, the operators or designers of bio-inspired snake robots can choose the optimal operational point using the Pareto fronts as an informative tool. Furthermore, based on the multi-objective optimization study, we managed to acquire useful insights regarding the optimal values of the gait parameters for both land-based and swimming snake robots. These observations are important to consider in the control and design of snake robots. Investigating the locomotion efficiency of bio-inspired snake robot using other types of sinusoidal motion patterns will be the subject of future work. It is likely that other design parameters could also improve the motion efficiency of the snake robots, and we plan to study the effect of these in future work. Moreover, the proposed optimization framework can also be used for optimizing gaits of biped robots by introducing newly revised optimization variables.

Acknowledgments: Research partly funded by VISTA a basic research program in collaboration between The Norwegian Academy of Science and Letters, and Statoil, and partly supported by the Research Council of Norway through its Centres of Excellence funding scheme, project no. 223254-NTNU AMOS.

Author Contributions: The authors have contributed equally to this work.

Conflicts of Interest: The authors declare no conflict of interest.

References

1. Liljebäck, P.; Pettersen, K.Y.; Stavdahl, Ø.; Gravdahl, J.T. *Snake Robots: Modelling, Mechatronics, and Control*; Springer: London, UK, 2013.
2. Gray, J. Studies in Animal Locomotion. *J. Exp. Biol.* **1933**, *10*, 88–104.
3. Hirose, S. *Biologically Inspired Robots: Snake-Like Locomotors and Manipulators*; Oxford University Press: Oxford, UK, 1993.
4. Colgate, J.; Lynch, K. Mechanics and control of swimming: A review. *IEEE J. Ocean. Eng.* **2004**, *29*, 660–673.
5. Fossen, T.I. *Handbook of Marine Craft Hydrodynamics and Motion Control*; John Wiley & Sons, Ltd.: Hoboken, NJ, USA, 2011.
6. Kelasidi, E.; Pettersen, K.Y.; Gravdahl, J.T.; Liljebäck, P. Modeling of underwater snake robots. In Proceedings of the IEEE International Conference on Robotics and Automation (ICRA), Hong Kong, China, 31 May–7 June 2014; pp. 4540–4547.
7. Boyer, F.; Porez, M.; Khalil, W. Macro-continuous computed torque algorithm for a three-dimensional eel-like robot. *IEEE Trans. Robot.* **2006**, *22*, 763–775.
8. Wiens, A.; Nahon, M. Optimally efficient swimming in hyper-redundant mechanisms: Control, design, and energy recovery. *Bioinspir. Biomim.* **2012**, *7*, 046016.
9. Khalil, W.; Gallot, G.; Boyer, F. Dynamic modeling and simulation of a 3-D serial eel-like robot. *IEEE Trans. Syst. Man Cybern. Part C Appl. Rev.* **2007**, *37*, 1259–1268.
10. Tesch, M.; Schneider, J.; Choset, H. Expensive multiobjective optimization and validation with a robotics application. In Proceedings of the Neural Information Processing Systems: Workshop on Bayesian Optimization & Decision Making, Lake Tahoe, NV, USA, 7 December 2012.
11. Hasanzadeh, S.; Tootoonchi, A.A. Ground adaptive and optimized locomotion of snake robot moving with a novel gait. *Auton. Robots* **2010**, *28*, 457–470.
12. Crespi, A.; Ijspeert, A.J. Online optimization of swimming and crawling in an amphibious snake robot. *IEEE Trans. Robot.* **2008**, *24*, 75–87.
13. Kelasidi, E.; Pettersen, K.Y.; Gravdahl, J.T. Energy efficiency of underwater snake robot locomotion. In Proceedings of the 23th Mediterranean Conference on Control Automation (MED), Torremolinos, Spain, 16–19 June 2015; pp. 1124–1131.
14. Kelasidi, E.; Pettersen, K.Y.; Gravdahl, J.T. Energy efficiency of underwater robots. In Proceedings of the 10th IFAC Conference on Manoeuvring and Control of Marine Craft (MCMC), Copenhagen, Denmark, 24–26 August 2015; pp. 152–159.
15. Kelasidi, E.; Liljebäck, P.; Pettersen, K.Y.; Gravdahl, J.T. Experimental investigation of efficient locomotion of underwater snake robots for lateral undulation and eel-like motion patterns. *Robot. Biomim.* **2015**, *2*, 1–27.
16. Kelasidi, E.; Jesmani, M.; Pettersen, K.Y.; Gravdahl, J.T. Multi-objective optimization for efficient motion of underwater snake robots. *Artif. Life Robot.* **2016**, *21*, 1–12.
17. McIsaac, K.; Ostrowski, J. Motion planning for anguilliform locomotion. *IEEE Trans. Robot. Autom.* **2003**, *19*, 637–652.
18. Taylor, G. Analysis of the swimming of long and narrow animals. *Proc. R. Soc. Lond. Ser. A Math. Phys. Sci.* **1952**, *214*, 158–183.
19. Lighthill, M.J. Large-amplitude elongated-body theory of fish locomotion. *Proc. R. Soc. Lond. Ser. B. Biol. Sci.* **1971**, *179*, 125–138.
20. Chen, J.; Friesen, W.O.; Iwasaki, T. Mechanisms underlying rhythmic locomotion: Body-fluid interaction in undulatory swimming. *J. Exp. Biol.* **2011**, *214*, 561–574.
21. Kelasidi, E.; Pettersen, K.Y.; Liljebäck, P.; Gravdahl, J.T. Integral line-of-sight for path-following of underwater snake robots. In Proceedings of the IEEE Multi-Conference on Systems and Control, Juan Les Antibes, France, 8–10 October 2014; pp. 1078–1085.

22. Kelasidi, E.; Liljebäck, P.; Pettersen, K.Y.; Gravdahl, J.T. Integral line-of-sight guidance for path following control of underwater snake robots: Theory and experiments. *IEEE Trans. Robot.* **2014**, *33*, 610–628.
23. Koziel, S.; Yang, X.S. (Eds.) *Computational Optimization, Methods and Algorithms*; Springer: Berlin/Heidelberg, Germany, 2011; Volume 356.
24. Kuo, P.D.; Grierson, D. Genetic algorithm optimization of escape and normal swimming gaits for a hydrodynamical model of carangiform locomotion. In Proceedings of the Genetic and Evolutionary Computation Conference (GECCO), Chicago, IL, USA, 12–16 July 2003; pp. 170–177.
25. Kern, S.; Koumoutsakos, P. Simulations of optimized anguilliform swimming. *J. Exp. Biol.* **2006**, *209*, 4841–4857.
26. Kennedy, J.; Eberhart, R. Particle swarm optimization. In Proceedings of the IEEE International Conference on Neural Networks, Perth, Australia, 27 November–1 December 1995; pp. 1942–1948.
27. Kober, J.; Bagnell, J.A.; Peters, J. Reinforcement learning in robotics: A survey. *Int. J. Robot. Res.* **2013**, *32*, 1238–1274.
28. Di Mario, E.; Talebpour, Z.; Martinoli, A. A comparison of PSO and reinforcement learning for multi-robot obstacle avoidance. In Proceedings of the IEEE Congress on Evolutionary Computation (CEC), Cancun, Mexico, 20–23 June 2013; pp. 149–156.
29. Ariizumi, R.; Tesch, M.; Kato, K.; Choset, H.; Matsuno, F. Multiobjective optimization based on expensive robotic experiments under heteroscedastic noise. *IEEE Trans. Robot.* **2017**, *33*, 468–483.
30. Ariizumi, R.; Matsuno, F. Dynamic analysis of three snake robot gaits. *IEEE Trans. Robot.* **2017**, *33*, 1075–1087.
31. Kelasidi, E.; Pettersen, K.Y.; Gravdahl, J.T. Stability analysis of underwater snake robot locomotion based on averaging theory. In Proceedings of the IEEE International Conference on Robotics and Biomimetics (ROBIO), Bali, Indonesia, 5–10 December 2014; pp. 574–581.
32. Pareto, V. *Manual of Political Economy*; A. M. Kelley: New York, NY, USA, 1906.
33. Liljebäck, P.; Stavdahl, Ø.; Pettersen, K.; Gravdahl, J. Mamba—A waterproof snake robot with tactile sensing. In Proceedings of the International Conference on Intelligent Robots and Systems (IROS), Chicago, IL, USA, 14–18 September 2014; pp. 294–301.
34. Jesmani, M.; Bellout, M.C.; Hanea, R.; Foss, B. Well placement optimization subject to realistic field development constraints. *Comput. Geosci.* **2016**, *20*, 1185–1209.
35. Wetter, M. GenOpt—A generic optimization program. In Proceedings of the 7th International Conference of the International Building Performance Simulation Association Conference (IBPSA), Rio de Janeiro, Brazil, 13–15 August 2001; pp. 601–608.
36. Parsopoulos, K.; Vrahatis, M. Recent approaches to global optimization problems through particle swarm optimization. *Nat. Comput.* **2002**, *1*, 235–306.
37. Carlisle, A.; Dozier, G. An off-the-shelf PSO. In Proceedings of the Workshop on Particle Swarm Optimization, Indianapolis, IN, USA, 6–7 April 2001.

Article

Snake-Like Robot with Fusion Gait for High Environmental Adaptability: Design, Modeling, and Experiment

Kundong Wang [1,*], Wencan Gao [1] and Shugen Ma [2]

[1] Department of Instrument Engineering, Shanghai Jiao Tong University, Shanghai 200240, China; vincent_ko@sjtu.edu.cn

[2] Department of Robotics, Ritsumeikan University, Kyoto 525-8577, Japan; shugen@se.ritsumei.ac.jp

* Correspondence: kdwang@sjtu.edu.cn; Tel.: +86-021-34205936

Received: 22 September 2017; Accepted: 31 October 2017; Published: 3 November 2017

Abstract: A snake changes its gait to adapt to different environments. A snake-like robot that is able to perform as many or more gaits than a real-life snake has the potential to successfully adapt to a range of environments, similar to a real-life snake. However, only a few mechanisms in the current snake-like framework can perform common gaits. In this paper, a novel snake-like robot is developed to resolve this problem. A multi-gait is established and used as a reference for the articulation design. A non-snake-like mechanism with linear articulation is combined with the classical swing joint. A prototype is designed and constructed for verification and analysis. Two basic main gaits, namely, serpentine and rectilinear locomotion, are fused, and a novel obstacle-aided locomotion based on rectilinear motion is developed. The experiment demonstrates that the robot can generate all of the expected gaits with high movement efficiency.

Keywords: multi-gait; linear articulation; novel locomotion; movement efficiency

1. Introduction

A snake is a reptile that has been evolving for more than 130 million years. Snakes have good motion adaptability and powerful attack capability on land and in water despite their simple string-like bodies. Researchers have explored and applied smart mechanisms that are based on snakes, including active cord mechanism (ACM) and Crawler robots, to emulate the unique motion of snakes, such as serpentine and rectilinear locomotion. A snake-like robot is often equipped with terrain adaptability by means of wheel or foot locomotion. Additionally, a slender body acts like a string and can enter narrow spaces. The unique motion of snake-like robots can generate the action of a hand and a leg [1]. Thus, snake-like robots can be successfully applied in nonstructural environments, such as disaster rescue, human body cavity examination, and industrial pipe inspection [2,3].

Hirose studied and developed many snake-like robots with ACM, a recent version of which is the ACM-R5 with amphibious features [4]. Ma completed a series of explorations of 3D snake-like locomotion, including works on a 3D joint mechanism and dynamics analysis [5,6]. However, passive wheels can limit adaptability and complicate locomotion, such as during disasters or in mud puddles. Thus, some researchers shifted to wheel-less snake-like robots, whereas others focused on crawler robots with additional functions. Bayraktaroglu reported a wheel-less robot with obstacle-aided locomotion [7]. Transeth presented a system for modeling and controlling a limbless snake-like robot [8]. Kuwada designed a snake-like robot with a rotary connection between swing joints that can be utilized for pipe inspection [9]. Other works focused on wheel-less snake-like robots, such as those by Crespi and Klaassen [10,11]. Crawler robots that can enter highly complicated environments have also been studied [3].

Wheel-less snake-like robots generally have high environmental adaptability but at the cost of motion efficiency. That is, they have high power consumption and low motility. Active crawler robots are bulky and have poor trafficability. However, in the natural world, snakes move with different gaits in different environments. Most of the current research focuses on classical serpentine locomotion, which is the most efficient type of motion. Few works concentrate on side-winding, concertina, and rectilinear types of movement. For example, the ACM-R3 can execute serpentine, rolling, and 3D motion [12]. The SIA (Shenyang Institute of Automation) snake-like robot can move in a helical gait in 3D space [13]. If a snake-like robot is configured with 2-DOF (Degree of Freedom) swing joints, then it can perform serpentine and side-winding motions but not concertina and rectilinear motions related to contractions and extensions. Meanwhile, a worm moves by shortening and elongating its body. In special cases, worm-like motion is superior, such as in terms of adaptability to narrow spaces, as it can avoid regular shape points. Thus, a 2-DOF joint with linear translation was developed for configuring a snake-like robot. This robot can easily realize serpentine and rectilinear gaits. Moreover, a solution related to multi-gait realization and kinematics analysis was discussed in our laboratory. As preliminary work, a prototype was proposed and experiments were performed [14,15].

On the basis of the preliminary findings, a 2-DOF joint with linear translation was developed and a snake-like robot was configured for the present study. A prototype was then designed and constructed for verification and analysis. Two basic main gaits, namely, serpentine and rectilinear locomotion, were fused, and a novel obstacle-aided locomotion based on rectilinear motion was developed.

This paper is organized as follows. Section 2 analyzes the classical gait of snakes and determines the DOF design of joints. Then, a method for realizing these gaits is introduced. On this basis, the details of the joint mechanism for the prototype and its control system are discussed. Section 3 explains the experiment verification and results. Finally, the conclusions and the discussion are outlined in Section 4.

2. Methods

2.1. Gait Analysis

The anatomy of a snake features hundreds of short globe articulations; this structure is difficult to fully reproduce in mechanical systems given the current technical limitations. Most snake-like robots are composed of 1- or 2-DOF swing joint serials. However, a simplistic mechanism leads to limited functions (i.e., few gaits). In nature, snakes commonly have four types of locomotion gaits: serpentine, side-winding, rectilinear, and concertina. However, locomotion gaits that have not been observed in nature have been realized. This paper excludes non-natural gaits and instead focuses on serpentine and rectilinear locomotion, which are the basic gaits that allow a robot to adopt different gaits according to the environment.

Snakes adopt the graceful and aesthetically pleasing serpentine locomotion when they are on grass or irregular plains. The whole body is thrust into several sinusoidal curves, with some body parts following the path of the head. The serpentine motion is shown in Figure 1a.

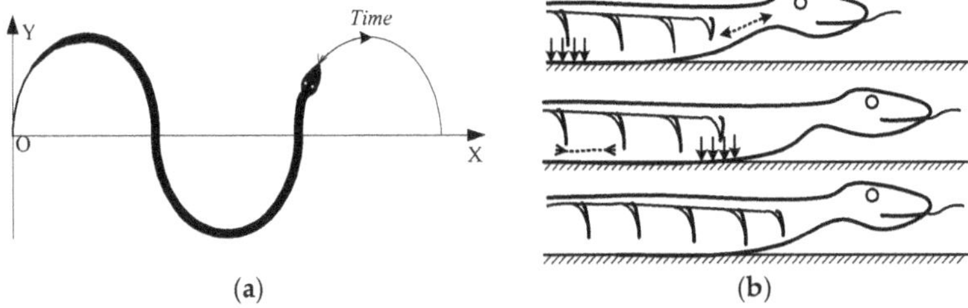

Figure 1. (**a**) Serpentine locomotion. (**b**) Rectilinear locomotion.

Rectilinear motion is common among boas, which have big bodies. As shown in Figure 1b, the front of the body is pressed against the ground as the rear part is extended by muscle release. In addition, the front is fixed on the ground while the remaining parts are pulled forward by muscle contraction. Although rectilinear locomotion is slow and motion-inefficient, it allows a snake to pass through narrow line-typed or circular-shaped pipes, in which other gaits cannot be realized.

2.2. Kinematics of Locomotion Gaits

The articulation modules in the snake-like robot are connected in serial, as shown in Figure 2. The world coordinate system xyz is fixed in extern with origin O. Absolute position and attitude are denoted by the head or tail point coordinate (x, y, z) in xyz and the inner body shape, respectively. To describe shape, a series of moving coordinates $x_1y_1z_1$, $x_1y_1z_1$, ..., $x_ny_nz_n$ is designated at every module from tail to head. Variables O_{n+1} and O_0 represent the head and the tail endpoints, respectively. The i-th moving coordinate $x_iy_iz_i$ in the i-th module is constructed similar to Figure 2. Variable y_i is the pitch joint axis, and z_i is the roll joint axis. The origin is designated as the crossing point of y_i and z_i. The positive direction of z_i is determined on the basis of the contraction of the linear joint. Variable x_i is determined by the left-hand law of the Cartesian coordinate system. If the robot is shaped into a line, then x_i and y_i (i = 1, 2, ..., n) are parallel to each other and all z_i (i = 1, 2, ..., n) are collinear. Variables l_i, γ_i, and θ_i denote the i-th translational displacement, roll angle, and pitch angle, respectively. The transformation of two adjacent module coordinates can be calculated as follows:

$$T_{i-1}^{i} = Ry \cdot Rz \cdot P_z = \begin{bmatrix} C\theta_i & 0 & S\theta_i & 0 \\ 0 & 1 & 0 & 0 \\ -S\theta_i & 0 & C\theta_i & 0 \\ 0 & 0 & 0 & 1 \end{bmatrix} \cdot \begin{bmatrix} C\gamma_i & S\gamma_i & 0 & 0 \\ -S\gamma_i & C\gamma_i & 0 & 0 \\ 0 & 0 & 1 & 0 \\ 0 & 0 & 0 & 1 \end{bmatrix} \cdot \begin{bmatrix} 1 & 0 & 0 & 0 \\ 0 & 1 & 0 & 0 \\ 0 & 0 & 1 & -l_i \\ 0 & 0 & 0 & 1 \end{bmatrix} = \begin{bmatrix} C\theta_i \cdot C\gamma_i & C\theta_i \cdot S\gamma_i & S\theta_i & -l_i \cdot S\theta_i \\ -S\gamma_i & C\gamma_i & 0 & 0 \\ -S\theta_i \cdot C\gamma_i & -S\theta_i \cdot S\gamma_i & C\theta_i & -l_i \cdot C\theta_i \\ 0 & 0 & 0 & 1 \end{bmatrix} \quad (1)$$

Equation (1) can be used to compute the coordinates of every joint point. Variables l_i, γ_i, and θ_i are used as inputs for the robot shape, which is related to gait. Kinematic analysis was performed to identify the best control input law and to establish the snake gait. The succeeding section shows the creation of the mathematical model of each gait on the basis of the aforementioned coordinate system and the mechanism.

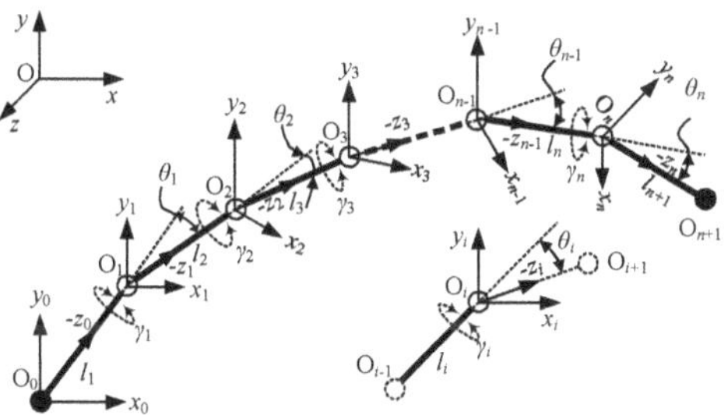

Figure 2. Coordinate system and joint parameters.

2.2.1. Serpentine Gait

Serpentine gait is obtained by tracing the movement of a snake on a plane. The snake body is shaped into a serpentine curve. If the joint swings are modulated by the curvature function of the serpentine curve, then the snake can move forward continuously and gracefully. The serpentine curve is given by

$$\begin{cases} x(s) = \int_0^s cos(\xi_\sigma)d\sigma \\ y(s) = \int_0^s sin(\xi_\sigma)d\sigma \end{cases} \quad (2)$$

where $\xi_\sigma = a \cdot \cos(b \cdot \sigma) + c \cdot \sigma$, a, b, and c are the parameters used to determine the amplitude, frequency, and direction of the curve, respectively, and s is the curve length from the origin to the point (x, y). The curvature is expressed by

$$\kappa(s) = \sqrt{\left(\frac{d^2x}{ds^2}\right)^2 + \left(\frac{d^2y}{ds^2}\right)^2} = |a \cdot b \cdot sin(b \cdot s) - c| \,. \tag{3}$$

Snake-like robots differ from natural snakes because the rigid links in robots are longer than those in real snake joints. In many prototypes, a serpentine is often approximated by discrete fold lines. Consequently, two solutions are suggested in Figure 3. As shown by the solid fold line, all joint points are located along the curve when passive wheels are installed on the joint point; this method is called the section method. When the wheels are defined along the link tangent center with the serpentine, the abovementioned discrete method is denoted by an imaginary fold line; this process is then called the tangent method. During joint point motion, the joint angle is designated as the control input. Assuming that the robot moves on a plane, our design should therefore determine θ_i ($i = 1, 2, \ldots, n$).

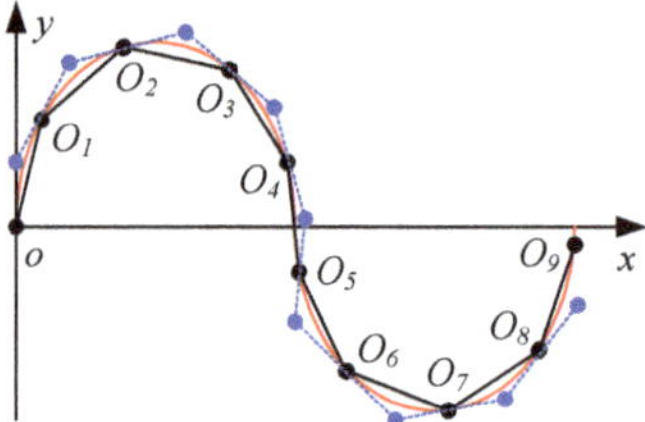

Figure 3. Ideal joint point distribution.

In the section method, the total snake length is normalized to 1 such that the link length l is $1/\text{n}$. A discrete equation of (2) is used to calculate the angle.

$$\begin{cases} x_i = \sum\limits_{k=1}^{i} \frac{1}{n} \cdot cos(a \cdot cos\left(\frac{k \cdot b}{n}\right) + \frac{k \cdot c}{n}) \\ y_i = \sum\limits_{k=1}^{i} \frac{1}{n} \cdot sin(a \cdot cos\left(\frac{k \cdot b}{n}\right) + \frac{k \cdot c}{n}) \end{cases} \tag{4}$$

where (x_i, y_i) is the i-th coordinate of the joint in xoy and l is the link length. If we normalize l to 1, the joint angle can be easily calculated as follows:

$$\theta_i = arctan\frac{y_i - y_{i-1}}{x_i - x_{i-1}} - arctan\frac{y_{i+1} - y_i}{x_{i+1} - x_i} = \alpha \cdot sin\left(i\beta + \frac{\beta}{2}\right) + \gamma \tag{5}$$

where

$$\alpha = 2asin\left(\frac{\beta}{2}\right),\ \beta = \frac{b}{n},\ \gamma = -\frac{c}{n}.$$

2.2.2. Rectilinear Gait

The locomotion of a snake-like robot with swing joints is characterized by motion singularity when its body is shaped into a line or an arc. The robot gait presented in this paper can resolve the aforementioned limitation. Linear joints play an important role in gaits. The locomotion of a robot composed of three articulation modules and a head cabin is shown in Figure 4. In the figure, 1, 2, and 3 denote pitch joints; I, II, and III represent links that change lengths as linear joints; and 0 is the head cabin. Only 0, 1, 2, and 3 touch the ground. In the same figure, the stroke along the line indicates motion from Phase A to E. First, Link I is extended while others are kept static. Then, the head cabin,

0, is pushed forward in Phase A. Link I contracts and Link II extends, thus pushing Joint 1 to Phase B. A similar action happens in Phase C, and Joint 2 advances. Finally, Joint 3 is pulled forward as Link III contracts. The robot advances Δs in this case. Phase E has the same configuration as A. Thus, the robot is continuously thrust forward through repetition.

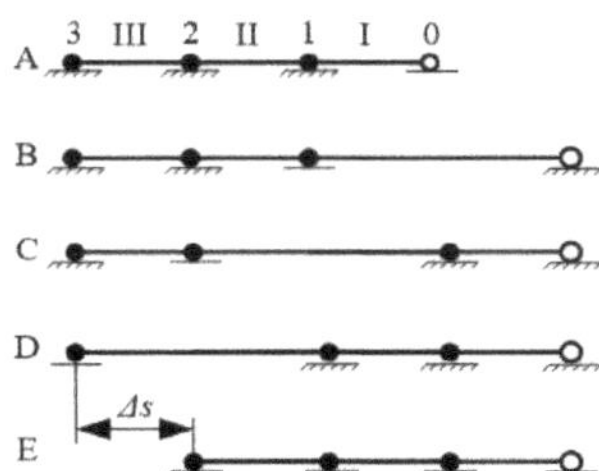

Figure 4. Rectilinear gait on a line.

The rectilinear gait is similar to the inchworm-like motion adopted by many worms. The force balance condition should be analyzed because gait can be realized only when partial articulation modules move while others are kept stationary. We assume that a joint in any articulation module is locked until the motors are powered. In the present study, analysis is performed in the quasi-static condition. The robot is composed of n articulation modules with identical masses. Subsequently, a general case is selected for the analysis. The i-th module moves at the speed of v, while others are kept stationary. Then, the i-th module is isolated from the mechanism. The equation for rectilinear gait in linear motion is as follows:

$$\begin{cases} F_{i-1} = \sum_{k=1}^{i-1} f_k \leq (i-1)\cdot mg\cdot\mu_s \\ F_{i-1} + F_i = f_i = mg\cdot\mu_d \\ F_i' = \sum_{k=i+1}^{n} f_k \leq (n-1)\cdot mg\cdot\mu_s \end{cases} \tag{6}$$

where F_i is the pulling force that $i-1$-th module applies on the i-th module, F_{i+1} is the pulling force that $i+1$-th module applies on the i-th module, F_i' and F_{i+1}' are the counteracting forces of F_i and F_{i+1}, respectively, f_k is the static friction of the rolling shaft in the passive wheels if $k \neq I$, f_i is the kinetic friction force of the rolling shaft, and μ_s and μ_d are the static and kinetic friction coefficients of the rolling shaft, respectively. The equation can be simplified by

$$\mu_d \leq (n-1)\cdot\mu_s. \tag{7}$$

Equation (7) can be satisfied theoretically when $n \geq 2$. In real motion, the i-th module always starts with acceleration; in this case, the robot may slip because the condition cannot be assured. However, if we control acceleration and configure additional modules for the robot, adequate locomotion may still be achieved.

2.2.3. Obstacle-Aided Locomotion

To further improve the motion efficiency of snake-like robots in narrow and limited spaces, a novel gait called obstacle-aided locomotion is proposed for rectilinear locomotion. Obstacle-aided locomotion is an important way for biological snakes moving through an ill-conditioned environment and was explained in [8] as follows. "The fastest biological snakes exploit roughness in the terrain for locomotion. They may push against rocks, branches, or other obstacles to move forward more efficiently." In obstacle-aided locomotion, the snake-like robot utilizes external objects, such as walls, stones, and other obstacles, for propulsion [16]. In narrow paths, both side walls serve as a fulcrum. Following earlier discussions on rectilinear locomotion, the novel gait is then represented visually, as shown in Figure 5.

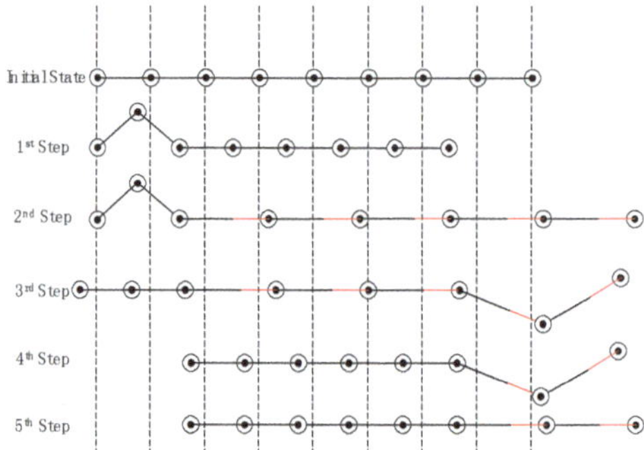

Figure 5. Obstacle-aided locomotion model.

In the initial state, each joint of the snake-like robot is set in a state of contraction. Then, the tail of the robot curves against the sides of the wall while supporting the body to avoid slipping. The other five joints at the front of the robot stretch simultaneously and slide forward in the second step, which is a bigger development than multi-wave rectilinear locomotion. The third step is symmetric with the first step, with the snake-like robot straightening out its tail and curving its front joints. The joints are contracted simultaneously to efficiently complete the rectilinear motion in the fourth step. Finally, the snake-like robot straightens out all its joints to complete a cycle of motion.

2.3. *Articulation Modular Design*

The independent articulation modular integrates the mechanical design and the electrical system. The control system includes communication and power supply components, which are useful for mobility and agility.

2.3.1. Joint Mechanism

The articulation unit of the snake-like robot is similar to a 2-DOF joint with two orthogonal swing axes. In a simplified mechanism, 1-DOF hinges are used to connect the links in serial. This configuration can realize planar locomotion. For linear translational joints, a gear-rack drive is used to realize rectilinear locomotion, whereas passive wheels are mounted at the bottom of each module for motion support.

The articulation modular consists of a swing joint for serpentine locomotion and a linear translational joint for rectilinear locomotion. An exploded-view of the design is presented in Figure 6a.

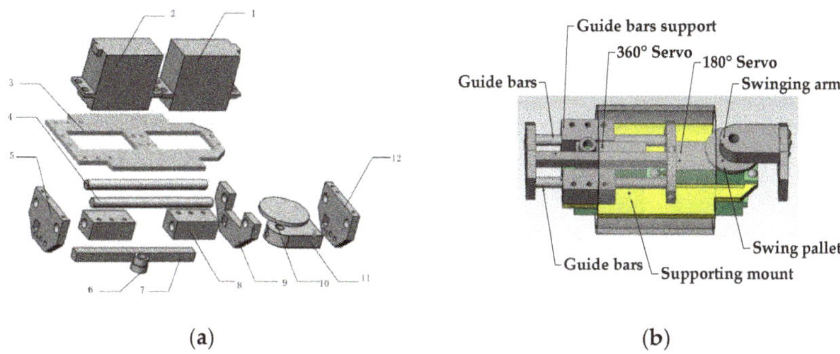

Figure 6. (**a**) Exploded view of snake-like robot. (**b**) Assembly drawing of snake-like robot.

The swing joint consists of a 180° servo and a swing arm. The swing arm is affixed to the swing pallet, which is connected to the servo. Thus, when the module is fully connected, the arm swings with the servo and generates serpentine gait.

The linear translational joint consists of guide bars, guide supports, gears, racks, and end caps at both ends. The transmission bar of the 360° servo is affixed to the gear that engages with the rack to provide linear motion. The rack is connected to the two guide bars and the end cover, thus forming a linear translation joint that moves on the guide rod support seat. The name of each part and its properties are shown in Table 1.

Table 1. Components of the snake-like robot.

Labels	Part Name	Properties
1	Servo 1	±90°
2	Servo 2	360°
3	Supporting mount	Aluminum alloy
4	Guide bars	Carbon steel
5	Left end cover	Aluminum alloy
6	Gear	Alloy steel, 0.5 module, 20 teeth
7	Rack	Alloy steel, 0.5 module, 6 mm height and width
8	Guide bars support	Aluminum alloy
9	Right end cover	Aluminum alloy
10	Swing pallet	Plastic
11	Swinging arm	Aluminum alloy
12	End cover of joint	Aluminum alloy

The assembly drawing of the snake-like robot is shown in Figure 6b. The apparatus is compact, light, and easy to assemble–disassemble. Limited by the mechanism design of the snake-like robot, the rotation of robot module is −60° to 60°. Details of the parameters are shown in Table 2.

Table 2. Parameters of the snake-like robot.

Number of Module	7
Overall length (mm)	1280
Individual module diameter	70
Single module quality (g)	350
Joint rotation angle	−60°–60°
Step distance of linear translational joint (mm)	34.4

2.3.2. Electrical System

The complete setup of the articulation unit of the independent snake should integrate not only the mechanism but also the power supply and its actuating and control components. The mobility and reliability of disconnected cables in a single modular should be improved. In the design of the control system (Figure 7), *Ci* represents the controller of the *i*-th modular. The micro-processing unit (MPU) is used to control two servo motors driven by the pulse–width–modulation (PWM) signal and to communicate with the external controller by using a wireless communication chip (CC1101). A computer is used to control the robot and switch the gait. The instruments connect to every articulation unit by an external wireless communication circuit affixed to the PC by a serial port. All chips are powered by a rechargeable battery in an articulation unit. Thus, cables do not need to be installed in-between adjacent articulation units.

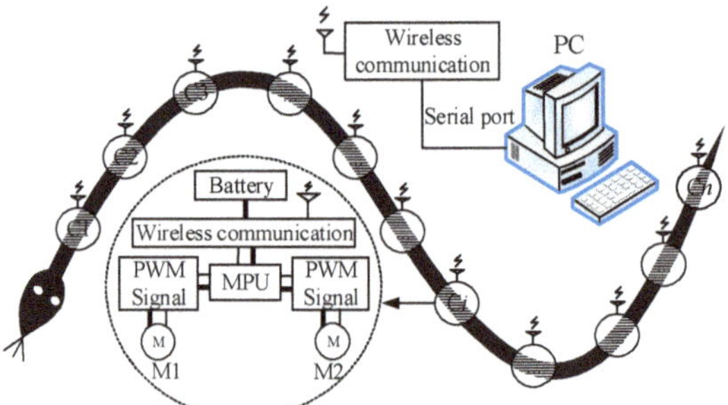

Figure 7. Control system.

2.3.3. Articulation Module and Prototype

Figure 8a shows an assembled articulation unit, which includes two servos, an MPU, a battery, and a wireless communication chip. A passive wheel is mounted at the bottom of the articulation unit for simulating the skin of the snake. In comparison with the two-wheeled module, a single-wheel module can reduce the friction coefficient at the longitudinal direction and improve movement efficiency. Figure 8b shows the prototype of the snake-like robot, which comprises eight connected articulation units. Each articulation unit is independent and follows instructions from the PC without affecting the other units. This mechanical structural design allows the body length to be changed easily according to various applications and environments.

(a) **(b)**

Figure 8. (**a**) Single snake-like robot module. (**b**) Connection of the snake-like robot module.

2.3.4. Control System

Serpentine Locomotion

The coordinate of the i-th joint in xoy (x_i, y_i) is calculated from the discrete equation of the serpenoid curve (4). When the snake-like robots move, the serpenoid curve changes at a radian frequency ω. Equation (4) can then be obtained by Formula (8) in the extended time domain.

$$\begin{cases} x_i(t) = \sum_{k-1}^{i} \frac{1}{n} cos\left(a \cdot cos\left(\frac{k \cdot b}{n}\right) + \omega \cdot t\right) \cdot + \frac{k \cdot c}{n}) \ (i = 1, 2 \sim n) \\ y_i(t) = \sum_{k-1}^{i} \frac{1}{n} sin\left(a \cdot cos\left(\frac{k \cdot b}{n}\right) + \omega \cdot t\right) \cdot + \frac{k \cdot c}{n}) \ (i = 1, 2 \sim n) \end{cases} . \quad (8)$$

Then, θ_i can be calculated by Equation (5).

The specific position of each swing joint can be determined at any time during locomotion, and angle data are stored into the MPU for gait generation. A set of data for Module 1 is shown in Table 3. However, the step distances of the swing joints vary with time. Changes in each angle are divided into 10-step datasets to achieve a regulated PWM duty cycle, thereby improving smoothness and stability. Thus, a timer is used with the MPU to interrupt subroutines corresponding to changes in the PWM. In these subroutines, the angle value of t_{i-1} is labeled "*i*Value," whereas "*f*Value" is used to retain the angle value of t_i. Subsequently, the degree of change in the control of the PWM can be calculated as

$$Command = \frac{iValue + (fValue - iValue)}{10} \cdot n \quad (9)$$

where n is a variable that will self-increase at each time point until it equals 10. The time for "*Command*" is equal to "*fValue*", which indicates the terminal position of t_i.

Table 3. Position of the Module 1 servo at each time [1].

Time (s)	**0**	**1**	**2**	**3**	**4**	**5**	**6**	**7**	**8**
Rotation angle	−42	−13	24	47	42	13	−24	−47	−42

[1] The parameters are $a = \frac{\pi}{3}, b = 2.4 \cdot \pi,\ c = 0$.

Rectilinear Locomotion

The module actuator (360° servo) is driven by the PWM. A duty cycle of 0.15 with 50 Hz PWM implies forward-driving cycles, whereas 0.05 indicates backward-driving cycles. Each module of the snake-like robot is set in an initial state of contraction under free state. Then, Module 1 contracts with a backward-driving input to the actuator while the other modules remain static. Module 2 extends as soon as Module 1 arrives and stops moving. A similar action is accomplished by the next module. Thus, we can easily implement the mode introduced in Section 2.2.2. However, as shown by the movement of earthworms in nature, adopting multiple standing waves can improve movement efficiency. This motor pattern can also be applied to snake-like robots. Subsequently, we add multiple standing waves to the forward cycle. The moving model of multi-wave rectilinear locomotion is shown in Figure 9, in which the deepening unit is a moving module. At high levels, the servo is in positive rotation; at low levels, the servo is in reversed rotation. The timing diagram of the servo is illustrated by Figure 9.

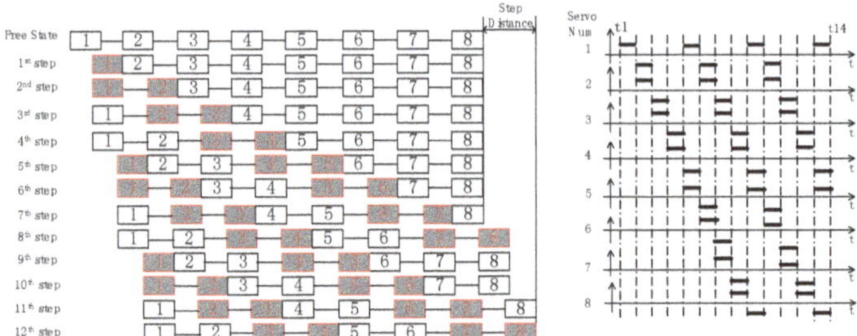

Figure 9. Multi-wave rectilinear movement gait and timing of the servo.

3. Experiment and Results

On the basis of the above analysis, we build a prototype for the 8-module 7-joint snake-like robot. The two experiments for serpentine and rectilinear locomotion are presented in this section. Furthermore, a video demo named "Snake-Like Robot with Fusion Gait" is supplied to demonstrate these performance of locomotion (see video S1).

3.1. Serpentine Movement

We have previously completed the kinematics calculation of the snake-like robot and determined the specific position of every module at each time for the theoretical analysis. Consequently, the serpentine locomotion experiment can be conducted using the calculated data.

The values for a, b, and c in the serpenoid are known parameters for the amplitude, frequency, and direction of a curve, respectively. Subsequently, a favorable parameter can be selected. In the present study, we set $a = \frac{\pi}{3}, b = 2.4 \cdot \pi,\ c = 0$. Using the aforementioned control algorithm in Section 2.3.3, a smooth and well-simulated serpentine gait is implemented, as shown in Figure 10.

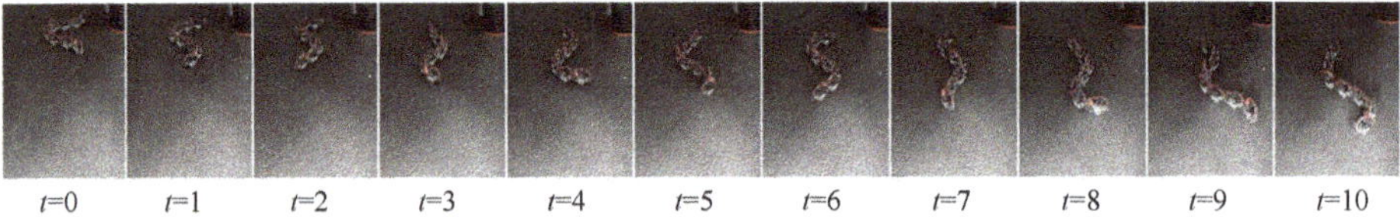

Figure 10. Snapshots of forward serpentine locomotion.

Furthermore, the left and right turns of the serpentine locomotion are implemented by changing parameter c in the locomotion equation. In this paper, we select $a = \frac{\pi}{3}, b = 2.4 \cdot \pi,\ c = \frac{\pi}{4}$ for the left turn and $a = \frac{\pi}{3}, b = 2.4 \cdot \pi, c = -\frac{\pi}{6}$ for the right turn. The simulation is favorable, and the motion track of the snake-like robot is shown in Figure 11.

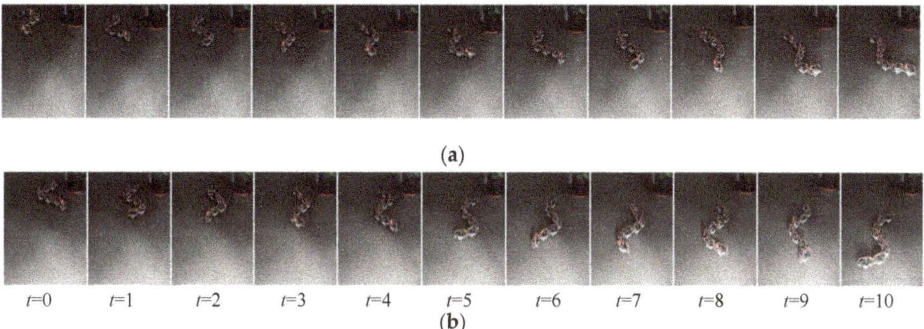

Figure 11. (**a**) Experiment on left-turn serpentine locomotion at $a = \frac{\pi}{3}, b = 2.4 \cdot \pi, c = \frac{\pi}{4}$. (**b**) Experiment on right-turn serpentine locomotion at $a = \frac{\pi}{3}, b = 2.4 \cdot \pi, c = -\frac{\pi}{6}$.

3.2. Rectilinear Locomotion and Its Fusion Gaits

The snake-like robot is placed against a wall to simulate actual environments. Figure 12a shows one cycle of the multi-wave rectilinear locomotion. The multi-wave rectilinear locomotion is more efficient than the single-wave gait. To assess performance and its improvement, we measure the speed of the snake-like robot by counting the time it takes to move 100 mm forward. By referring to the gait in Figure 8 for analysis, the speed can therefore reach approximately 8.6 mm/s, which is approximately two times faster than that of single-wave locomotion.

Obstacle-aided locomotion tests are also performed. To simulate the narrow and limited space environments, the snake-like robot is placed between two parallel vertical walls. Figure 12b shows one cycle of obstacle-aided locomotion. The width of the narrow path is 18 mm, whereas the onward direction is from right to left. The results show that the robot can reach a speed of 20 mm/s, which is significantly faster than the 8.6 mm/s speed in rectilinear locomotion.

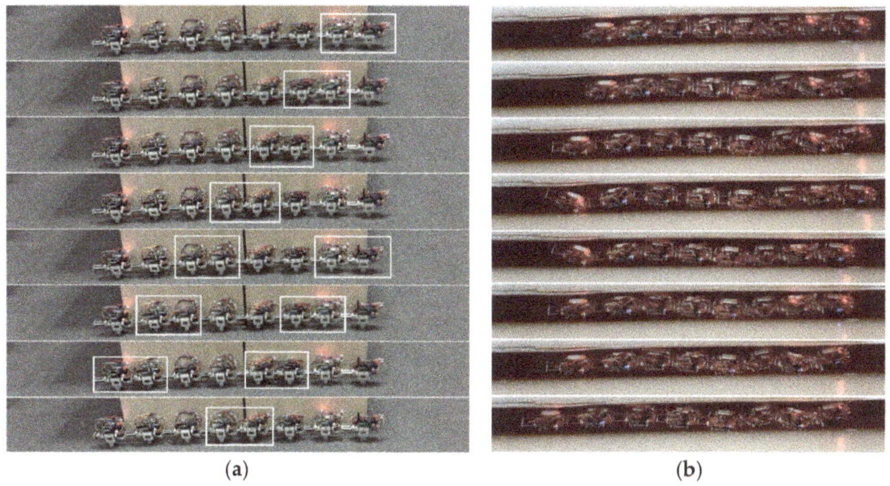

Figure 12. (**a**) Multi-wave rectilinear locomotion. (**b**) Obstacle-aided locomotion.

4. Conclusions and Discussion

In this paper, a novel solution is presented for resolving the problem of multi-gaits for snake-like robots. Serpentine locomotion is the most efficient movement, but snake-like robots cannot move in narrow spaces. Several works have been conducted on serpentine and rectilinear locomotion. However, only a few mechanisms have been studied for adopting all of the common gaits into the current snake-like framework. To allow a snake-like robot to traverse a wide range of different complex environments, an integrated robot that can perform multi-gait locomotion is proposed. Consequently, a linear translational joint is combined with the swing joint and then applied to rectilinear and serpentine locomotion, respectively. The design of the serpentine locomotion is based on the serpenoid curve, and a distributed control system is used to control the multi-modular robotic unit. Experimental results show that our design has an acceptable performance as both serpentine and rectilinear locomotion correspond well with theoretical analyses. Finally, and perhaps most importantly, a novel obstacle-aided locomotion based on rectilinear locomotion that significantly improves motion efficiency is proposed.

However, the present study has certain limitations. First, full gait is needed in robot design to achieve the expected performance in a variety of complex environments. We integrated only two basic gaits, and these are insufficient in fulfilling the objective. Second, the gait transition of snake-like robots is a considerable problem in complex environments. Biological snakes can use the most suitable gaits for different environments; however, those utilized by robots are mostly determined by PCs. Thus, locomotion can be invalid when a robot encounters an obstacle while performing the serpentine gait. A sensory–perceptual system is required to help snake-like robots perceive environments and determine locomotion gait, thereby improving the adaptability of the robots. [16] Another issue that is worth studying is the performance of snake-like robots on different surface types. When such a robot moves on smooth ground, the serpentine locomotion may be invalid because of the insufficient radial friction. In this case, utilizing irregularities in the terrain to avoid skidding is a solution for efficient locomotion. Moreover, using the linear translational joint in our design to change body length and shape may allow snake-like robots to adapt to different ground surfaces. In the future, we will integrate concertina and side-winding locomotion to the snake-like robotic gait. We will also adopt the full gait and perception-driven obstacle-aided locomotion for snake-like robots.

Supplementary Materials: zenodo DOI:10.5281/zenodo.1041033 (https://zenodo.org/record/1041033#.WfvNTYSGMcY). Video S1: Snake-Like Robot with Fusion Gait.

Acknowledgments: The study was funded by the Shanghai Qiming Star Program of the Shanghai Committee of Science and Technology (13QA1402200). The funding covers the costs of open access publishing.

Author Contributions: Kundong Wang and Shugen Ma co-organized the work and analyzed the data. Kundong Wang and Wencan Gao designed and performed the experiments and wrote the paper.

Conflicts of Interest: The authors declare no conflict of interest.

References

1. Xuesu, X.; Cappo, E.; Weikun, Z.; Jin, D.; Ke, S.; Chaohui, G.; Travers, M.; Choset, H. Locomotive reduction for snake robots. In Proceedings of the 2015 IEEE International Conference on Robotics and Automation (ICRA), Seattle, WA, USA, 26–30 May 2015.
2. Neumann, M.; Predki, T.; Heckes, L.; Labenda, P. Snake-like, tracked, mobile robot with active flippers for urban search-and-rescue tasks. *Ind. Robot Int. J.* **2013**, *40*, 246–250. [CrossRef]
3. Liljebäck, P.; Pettersen, K.; Stavdahl, Ø.; Gravdahl, J. A review on modelling, implementation, and control of snake robots. *Robot. Auton. Syst.* **2012**, *60*, 29–40. [CrossRef]
4. Hirose, S.; Yamada, H. Snake-Like Robots Machine Design of Biologically Inspired Robots. *IEEE Robot. Autom. Mag.* **2009**, *16*, 88–98. [CrossRef]
5. Ma, S.; Tadokoro, N.; Inoue, K. Influence of the gradient of a slope on optimal locomotion curves of a snake-like robot. *Adv. Robot.* **2006**, *20*, 413–428. [CrossRef]
6. Ye, C.; Ma, S.; Li, B.; Liu, H.; Wang, H. Development of a 3D Snake-like Robot: Perambulator-II. In Proceedings of the 2007 International Conference on Mechatronics and Automation, Harbin, China, 5–8 August 2007.

7. Bayraktaroglu, Z. Snake-like locomotion: Experimentations with a biologically inspired wheel-less snake robot. *Mech. Mach. Theory* **2009**, *44*, 591–602. [CrossRef]
8. Transeth, A.; Leine, R.; Glocker, C.; Pettersen, K.; Liljeback, P. Snake Robot Obstacle-Aided Locomotion: Modeling, Simulations, and Experiments. *IEEE Trans. Robot.* **2008**, *24*, 88–104. [CrossRef]
9. Kuwada, A.; Wakimoto, S.; Suzumori, K.; Adomi, Y. Automatic pipe negotiation control for snake-like robot. In Proceedings of the 2008 IEEE/ASME International Conference on Advanced Intelligent Mechatronics, Xian, China, 2–5 July 2008.
10. Crespi, A.; Ijspeert, A. Online Optimization of Swimming and Crawling in an Amphibious Snake Robot. *IEEE Trans. Robot.* **2008**, *24*, 75–87. [CrossRef]
11. Klaassen, B.; Paap, K. GMD-SNAKE2: A snake-like robot driven by wheels and a method for motion control. In Proceedings of the 1999 IEEE International Conference on Robotics and Automation (Cat. No.99CH36288C), Detroit, MI, USA, 10–15 May 1999.
12. Mori, M.; Hirose, S. Locomotion of 3D Snake-Like Robots—Shifting and Rolling Control of Active Cord Mechanism ACM-R3. *J. Robot. Mechatron.* **2006**, *18*, 521–528. [CrossRef]
13. Shumei, Y.; Shugen, M.; Bin, L.; Yuechao, W. Analysis of helical gait of a snake-like robot. In Proceedings of the 2008 IEEE/ASME International Conference on Advanced Intelligent Mechatronics, Xian, China, 2–5 July 2008.
14. Wang, K.; Ma, S. Analysis to Serpentine Locomotion Efficiency of Snake-like Robot. *Abstr. Int. Conf. Adv. Mechatron.* **2010**, *5*, 43–48. [CrossRef]
15. Wang, K.; Ma, S. Kinematic analysis of snake-like robot using sliding joints. In Proceedings of the 2010 IEEE International Conference on Robotics and Biomimetics, Tianjin, China, 14–18 December 2010.
16. Sanfilippo, F.; Azpiazu, J.; Marafioti, G.; Transeth, A.; Stavdahl, Y.; Liljebäck, P. Perception-Driven Obstacle-Aided Locomotion for Snake Robots: The State of the Art, Challenges and Possibilities. *Appl. Sci.* **2017**, *7*, 336. [CrossRef]

Article

Numerical Simulation of an Oscillatory-Type Tidal Current Powered Generator Based on Robotic Fish Technology

Ikuo Yamamoto [1,*], Guiming Rong [2], Yoichi Shimomoto [2] and Murray Lawn [3]

[1] Organization for Marine Science and Technology, Nagasaki University, Nagasaki 852-8521, Japan
[2] Graduate School of Engineering, Nagasaki University, Nagasaki 852-8521, Japan; rong@nagasaki-u.ac.jp (G.R.); goma@nagasaki-u.ac.jp (Y.S.)
[3] Surgical Oncology, School of Medicine, Nagasaki University, Nagasaki 852-8521, Japan; lawnmj@nagasaki-u.ac.jp
* Correspondence: iyamamoto@nagasaki-u.ac.jp; Tel.: +81-95-819-2512

Received: 18 August 2017; Accepted: 12 October 2017; Published: 16 October 2017

Abstract: The generation of clean renewable energy is becoming increasingly critical, as pollution and global warming threaten the environment in which we live. While there are many different kinds of natural energy that can be harnessed, marine tidal energy offers reliability and predictability. However, harnessing energy from tidal flows is inherently difficult, due to the harsh environment. Current mechanisms used to harness tidal flows center around propeller-based solutions but are particularly prone to failure due to marine fouling from such as encrustations and seaweed entanglement and the corrosion that naturally occurs in sea water. In order to efficiently harness tidal flow energy in a cost-efficient manner, development of a mechanism that is inherently resistant to these harsh conditions is required. One such mechanism is a simple oscillatory-type mechanism based on robotic fish tail fin technology. This uses the physical phenomenon of vortex-induced oscillation, in which water currents flowing around an object induce transverse motion. We consider two specific types of oscillators, firstly a wing-type oscillator, in which the optimal elastic modulus is being sort. Secondly, the optimal selection of shape from 6 basic shapes for a reciprocating oscillating head-type oscillator. A numerical analysis tool for fluid structure-coupled problems—ANSYS—was used to select the optimum softness of material for the first type of oscillator and the best shape for the second type of oscillator, based on the exhibition of high lift coefficients. For a wing-type oscillator, an optimum elastic modulus for an air-foil was found. For a self-induced vibration-type mechanism, based on analysis of vorticity and velocity distribution, a square-shaped head exhibited a lift coefficient of more than two times that of a cylindrically shaped head. Analysis of the flow field clearly showed that the discontinuous flow caused by a square-headed oscillator results in higher lift coefficients due to intense vortex shedding, and that stable operation can be achieved by selecting the optimum length to width ratio.

Keywords: oscillatory-type tidal current powered generator; elastic wing; coefficient of lift (Cl); self-induced oscillation; two way Fluid-Structure Interaction (FSI) problem

1. Introduction

In recent years, we have been faced with issues such as global warming and depletion of fossil fuels. Particularly in Japan, since the Eastern Japan earthquake disaster and the stopping of many nuclear power plants, there has been a significant effort to move towards renewable clean energy generation. As Japan is surrounded by the sea which provides stable ocean currents throughout the year, we propose an oscillatory-type tidal current-powered generator based on robotic fish technology.

A sea bream robotic fish was developed in 1995, as shown in Figure 1. A flexure like a fish fin was produced from variously changing elastic modules of the flexible oscillating fin [1–3]. The author found that the fin had significant potential as an oscillatory-type tidal current-powered generator.

Figure 1. Sea bream robotic fish.

In tidal power generation, propeller-based mechanisms are predominantly used. However, maintenance costs are high, due to marine fouling from incrustations and entanglement by seaweed; the rotating propeller itself also negatively impacts the marine environment in various ways. Furthermore, the relatively high complexity of a propeller combined, with the intense corrosion that occurs in sea water, results in high manufacturing and maintenance costs. Therefore, there have been many studies on fundamentally different approaches to marine energy generation. Among them is the VIVACE system of the University of Michigan [4]. This system uses the physical phenomenon of vortex-induced oscillation, in which water currents flowing around cylinders induce transverse motion. The energy resulting from the movement of the cylinders is then converted to electricity. There is also a small reciprocating-type generator in Japan, made by Abiru at Fukuoka Industrial University [5], in which the fluttering behavior of a wing is used. Further, a water flow-powered generator using pendulum-oscillation has been developed at Okayama University [6,7], which uses the same principle as the VIVACE system.

The authors' aim is to provide some assistance toward developing a more efficient reciprocating-type power generating system. Specifically, with regard to determining the optimal elastic modulus of a wing foil-type oscillator, and the optimum shape of a reciprocating oscillator head (from a group of six common shapes) to provide high lift coefficients when these oscillating heads are placed in transverse flows. The focus of this paper is on the optimization of this lift force. The magnitude of the "coefficient of lift" (Cl) varies with the shape of the oscillating mechanism. Therefore, as the oscillator is the main part of a reciprocating type generator, it's design is critical. The purpose of this study is to find the shape of oscillator head that produces the largest Cl.

To begin with, the optimum shape of the oscillator is found through numerical analysis; then, a physical machine using the oscillator will be created for further testing. A problem with coupling fluid flow to an object/mechanism must be solved, because the flow pressure causes deformation and displacement of the object (in this case, the oscillator), and the movement of the object in turn affects the flow (its velocity, pressure, etc.). The interdependent fluid-mechanical coupled system is numerically simulated in ANSYS.

The fundamental equations of the two-way coupled problem are as follows:

The unsteady flow field (incompressible) around the oscillator is governed by the following Navier-Stokes equation, and the equation of continuity, where t is the time, ρ is the mass of the flow, $\boldsymbol{V}$ is the velocity, $\boldsymbol{F}$ is the body force and p is the pressure:

$$\rho\frac{d\boldsymbol{V}}{dt} = \boldsymbol{F} - \nabla p + \mu\nabla^2\boldsymbol{V},\ \nabla\cdot V = 0 \tag{1}$$

And the equation governing the solid is derived from Newton's second law:

$$a_s = \frac{1}{m_s}\nabla\cdot\sigma_s + g_s \tag{2}$$

Here m_s is the mass of the solid, a_s is the acceleration, σ_s is the stress tensor on the solid and g_s is the body force.

The coupling equations are (displacement d, fluid subscript f, and solid s):

$$\tau_f n_f = \tau_s n_s \qquad d_f = d_s \tag{3}$$

where τ is the stress and n the normal vector. The lift force L is defined as

$$L(t) = Cl\frac{1}{2}v^2\rho A \tag{4}$$

where ρ is the density of the flow and A is the projected area of the body in the direction perpendicular to the flow. It is clear from the equation that, in a determined flow with a determined velocity, the magnitude of lift force is determined by Cl and A. That is, a shape with a large Cl, as well as an appropriately large A, is required. Furthermore, a lift that consistently varies is desirable, in order to facilitate reciprocation.

Regarding the calculation parameters, a flow velocity of Vx = 0.5 m/s [8] is used, because this is the slowest tidal speed that can be used for generating power. The Reynold's number is approximately 7.5×10^4, because the oscillator size against the flow is 0.15 m. A uniform flow field with Vx = 0.5 m/s (flow from the left side—x direction) is used in all of the following calculations [9].

2. Study on an Airfoil-Type Oscillator

An airfoil shape would naturally be thought of as a logical shape for generating a large Cl. However, it does not oscillate unless it is exposed to an appropriate turbulent flow. Furthermore, based on the authors' efforts to mechanically emulate the propulsion mechanism that can be observed in a regularly shaped fish as it swims at very high speeds, it was considered to be a natural selection for the oscillator [10,11]. On the assumption that Cl would increase if there were some elasticity in the oscillating wing, a numerical experiment to find the optimum elastic modulus on a symmetric foil was carried out.

As shown in Figure 2, an NACA0015 model foil, with a shaft located at 0.1 m from the front leading edge is used (the shaft is not considered in the calculation). A reciprocating displacement of $y = 0.001\sin(2\pi t)$ m is given to the shaft, and the elastic modulus E of the foil is varied stepwise from $E = 10^8$ to 5×10^4 Pa to investigate the change of Cl, details of the simulation conditions are provided in Table 1.

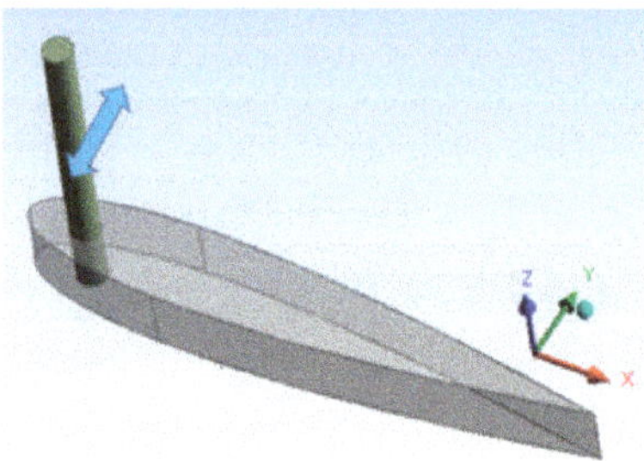

Figure 2. NACA0015 foil simulation model.

This two way FSI problem was solved using ANSYS Mechanical and ANSYS FLUENT (14.5, Ansys inc., Canonsburg, PA, USA) [12] controlled by System Coupling. The results are plotted in Figure 3, from which significant variation of *Cl* due to variation in *E* can be clearly seen. When the foil's $E = 10^8$ Pa, there is almost no deformation; however, when $E = 10^6$ Pa, the maximum *Cl* is obtained, where *Cl* is approximately three times greater. However, the *Cl* will also drop when the value of *E* is decreased. Therefore, it can be concluded that the optimum elastic modulus is $E = 10^6$ Pa for this shape of foil, in the case of different cambers the optimum value of *E* will most likely vary.

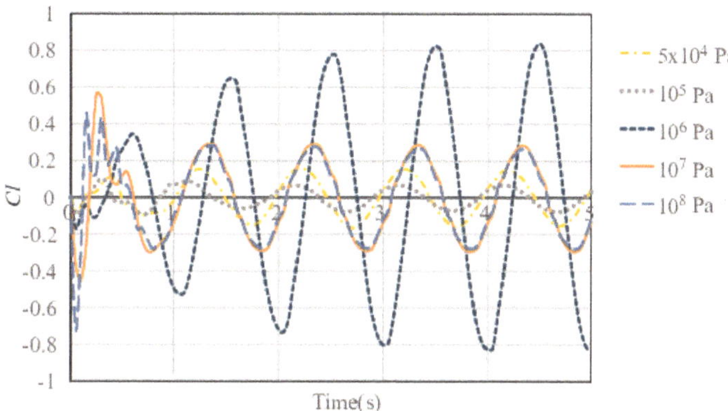

Figure 3. Variation of coefficient of lift (*Cl*) versus rigidity.

Table 1. Simulation conditions.

Time Step Size for Flow Calc. (s)	0.005
Scale of flow domain (m)	$6 \times 3 \times 0.1$
Type of mesh	prism
Min. size of mesh (m)	0.007
Density of structure (kg/m^3)	1000
Length of structure (m)	1
Depth of structure (m)	0.1

The overall deformation of the foil for $E = 5 \times 10^4$, 10^6 Pa, and 10^8 Pa has been examined, providing an optimal *Cl* value for $E = 10^6$ Pa. The scale in the *y* direction is scaled up as the deformation is very small. $E = 10^8$ is not shown here, as the deformation is insignificant. Instantaneous deformations of the foil for $E = 5 \times 10^4$ and $E = 10^6$ are shown in Figures 4 and 5, respectively. It is recognized that, compared to the case of $E = 5 \times 10^4$, in which the body curves with more than one center of curvature, in the case of $E = 10^6$, the foil is singularly concave or convex, providing an asymmetric foil. This forms an attack angle for the foil, and is the reason for the increase of *Cl*.

Clearly, the lift force of a foil will change according to its attack angle. The lift increases as the attack angle increases within the range of $\pm 15°$. Facilitation of this lift angle must be considered when manufacturing an elastic oscillator-based power generator; however, this study only investigated the influence of the rigidity. In conclusion, this study has shown that the optimum value of the elastic modulus is $E = 10^6$ Pa; the lift will decrease with any increase or reduction of the value of *E*.

A reciprocating displacement of $y = 0.002 \sin(2\pi t)$ m occurs when $E = 10^6$ Pa. The *Cl* is three times that of the previous $y = 0.001 \sin(2\pi t)$ m. However, the calculation failed due to the occurrence of negative meshes when a larger displacement was used. The authors will investigate the reason for this in future research.

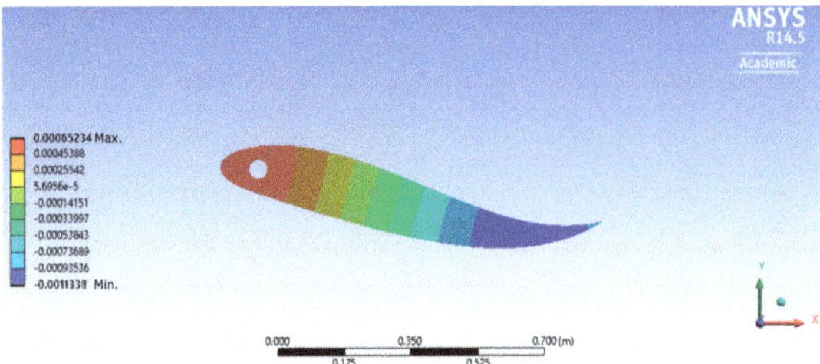

Figure 4. Deformation in case of $E = 5 \times 10^4$ Pa at t = 2.1 s, Cl is reduced at this instant. Scale in y direction is scaled up.

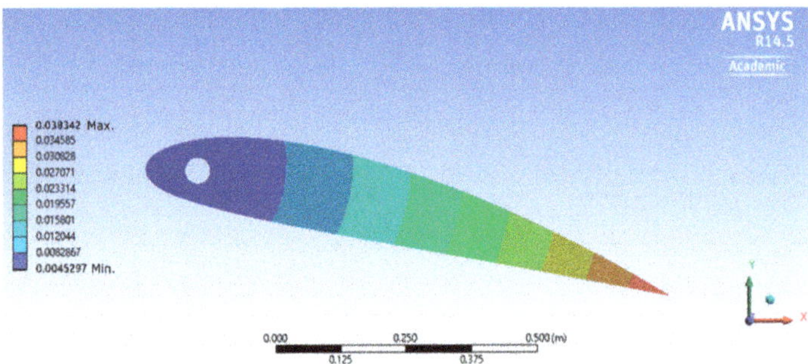

Figure 5. Deformation in case of $E = 10^6$ Pa at t = 2.1 s, Cl will increase after this instant. Scale in y direction is scaled up.

3. The Optimum Shape of a Reciprocating Oscillator for Self-Induced Oscillation

3.1. The Pre-Test in 2D

An essential requirement for a self-induced oscillator like VIVACE's is the stable shedding of vortices. For this reason, the authors attempted to find a shape that could stably shed vortices more effectively than that of a circular shaped interface, as well as providing an increase in Cl [13]. According to experiments by Hiejima in Okayama University, the oscillation of a semicircular-shaped oscillator is better than that of a circle when the plane face of the semicircle is against the flow. Also, according to Equation (4), the dimension parallel to the flow is proportional to the lift force. Based on the above two facts, the authors experimented with an elongated semicircle (rectangle plus semicircle). As the pre-test, models of various shapes, including those in Table 2, were made for two-dimensional calculations to investigate their Cl values. The mesh resolution around the model was set to 240 in order to see the vortex clearly. ICEM CFD (16.1, Ansys inc., Canonsburg, PA, USA) was used for meshing and FLUENT was used for calculating the flow. A 2D flow domain with a size of 4 × 3 (m) ($x \times y$) was used, and the time step size was 0.05 s. The convergence of the calculation was acceptable, as the maximum residuals of iterations were under 0.01.

Table 2 shows part of the results. It is clear that the semicircle has a higher Cl value than the circular shape. The elongated semicircle also has a higher Cl value; however, the oscillation is not stable, making it impractical for power generation. The reason for this phenomenon of instability in the case of the elongated semicircle was investigated by checking the flow field and vorticity for each case. The reason for this instability was the discontinuous flow resulting from a higher flow rate, and the ensuing reattachment of the separated flow.

Table 2. Results of the pre-test.

	Shape	x/y	Max Cl to Circle's	Periodic	Stability	Cd * to Circle
1	semicircle	0.5	1.53	good	not good	1.67
2	rect.+half cir.1	1	1.92	good	not good	1.67
3	rect.+half cir.2	1.4	3.2	detected	Very poor	1.67
4	oval-circle	2.5	1.3	excellent	excellent	0.5
5	triangle	1	1.53	excellent	good	1.67
6	square	1	1.97	excellent	excellent	1.67

* Cd coefficient of drag.

3.1.1. Reason for a Higher Cl Value in the Case of a Semicircle

From Figure 6, it can be seen that, in the case of a circle, the flow velocity changes from zero to its maximum gradually, from the forefront to a separation point, this is also reflected in the vorticity (Figure 7) and there is only a slight variation in angle between the stream in the x direction. In contrast, in the case of the semicircle, as shown in Figure 8 rapid streams occur at the edges and separate at an angle of about 45 degrees to the x direction, as the watercourse suddenly narrows. The maximum velocity of this case is 1.44 times that of the circle, and the negative pressure becomes about three times greater.

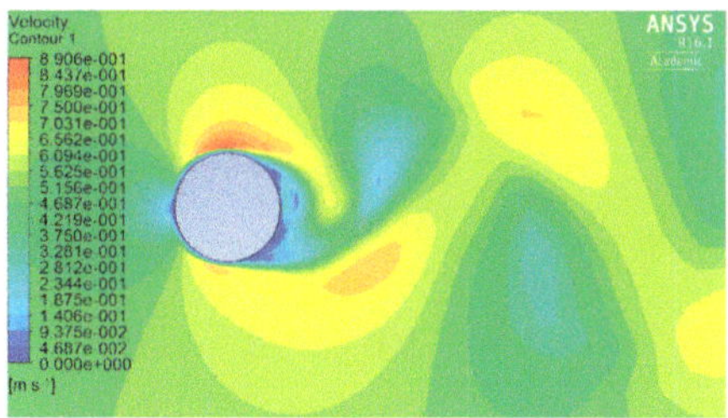

Figure 6. Distribution of velocity for a circle at t = 15.4 s. Descending Cl.

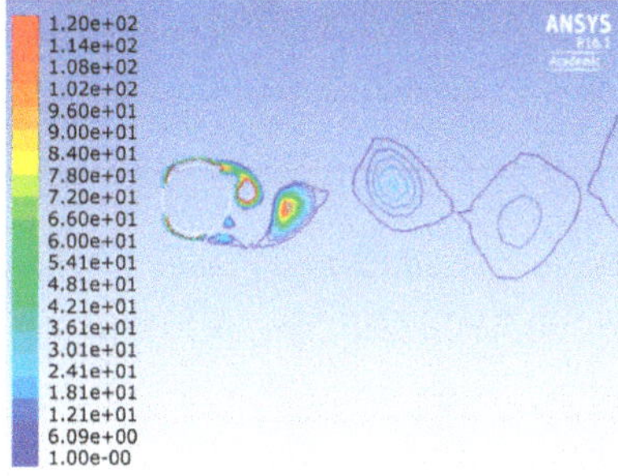

Figure 7. Distribution of vorticity for a circle at t = 15.4 s. Descending Cl.

However, it was noted that a trace of the oscillation of *Cl* showed slight irregularity, indicating some instability, as shown in Figure 9 The reason for this can be seen in the distribution of velocity and vorticity. Figures 8 and 10 show $t = 18$ s when a vortex is shedding and the lift is downward. The back of the rapid stream produces a dead water region which extends in the x direction up to half the width of the object (see Figure 8). The flow will then reattach to the object and reconnect to the rapid flow (this phenomenon can be observed in both Figures 8 and 11). In the case of the semicircle, reattachment occurs to the rear on the circular side (see Figure 8), the gradually increasing region of rapid velocity due to reattachment, together with the rapid stream from the edge, forms a strong vortex, which is then shed (see Figure 10), providing the object with a strong lifting force. As a result, the object will oscillate as the vortex makes contact with it.

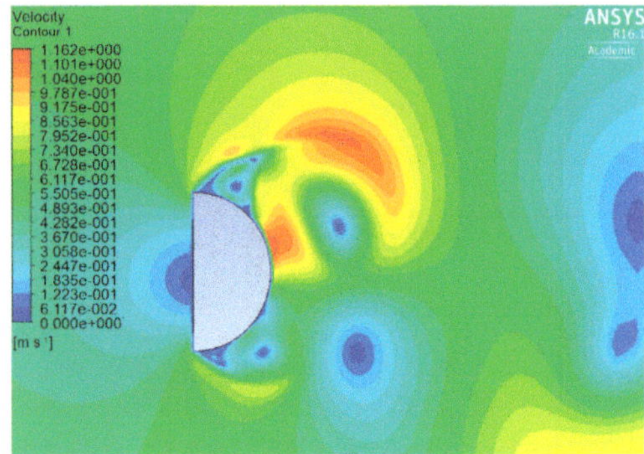

Figure 8. Distribution of velocity for a semicircle at $t = 18$ s. Descending *Cl*.

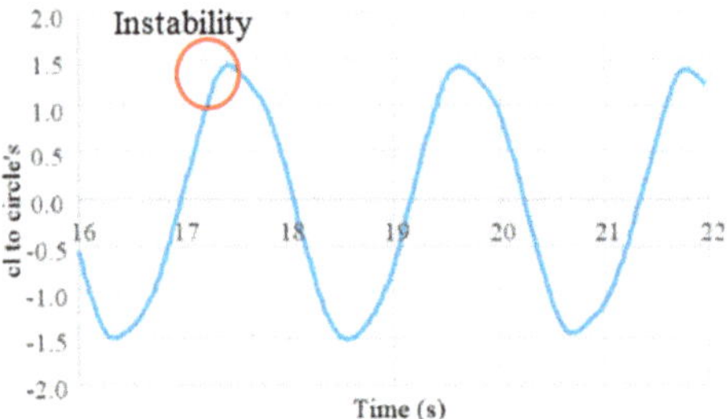

Figure 9. Sinusoidal trace of *Cl* for a semicircle.

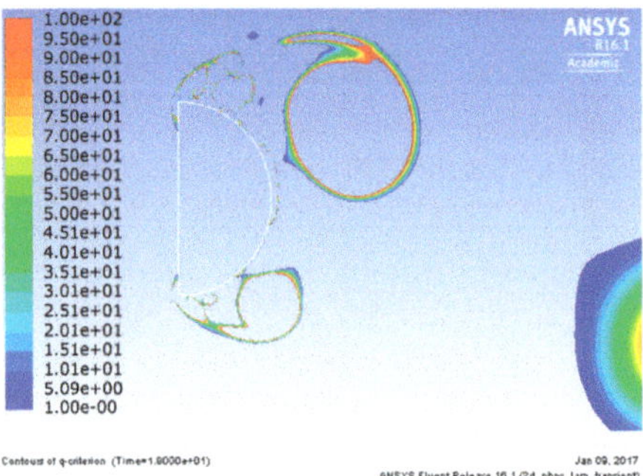

Figure 10. Distribution of vorticity for a semicircle at $t = 18$ s. Descending *Cl*.

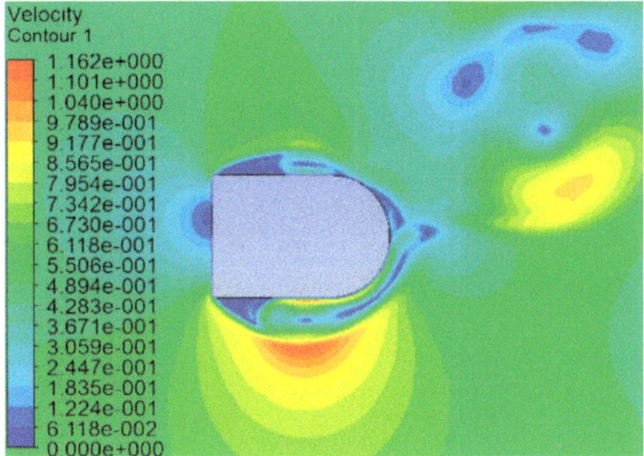

Figure 11. Distribution of velocity for an extended semicircle (model 3) at t = 15.5 s. Oscillation is unstable.

3.1.2. Reason for Increasing Instability with Elongation

The lift force is mainly produced by the vortex shedding on the opposite side of the rapid stream separating from the edge, in the case of an extended semicircle. The main difference is that the region of reattachment occurs in a plane parallel to the object's body, as shown in Figure 11; this rapid flow region will move to the rear with time, producing a moving negative-pressure zone. This kind of reattachment causes unstable oscillation, since its distribution of vorticity is irregular, as shown in Figure 12.

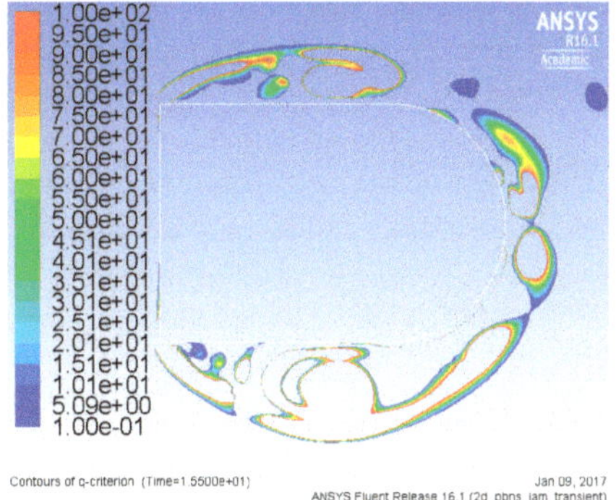

Figure 12. Distribution of vorticity for an extended semicircle (model 3) at t = 15.5 s. Oscillation is unstable.

On the other hand, an elliptically headed shape with its major axis parallel to the x direction produced a regular sinusoidal oscillation as is reflected in the stability data in Table 2, even when extended in the x direction. Its drag force reduces, and the maximum velocity of flow is 0.9 times that of a circle. Therefore, the period for vortex shedding is a little longer, and the intensity of the vortex also reduces compared to the circle. Elongation in the x direction increases Cl; however, the oscillation becomes increasingly unstable. This occurs because the shed vortex reattaches to the object. This analysis confirmed that a regular oscillation of Cl can be obtained if the ratio of length to width is under two.

3.1.3. Finding a Model with Optimal Length in the x-Direction

Comparing models 2 and 3, model 2 is more stable. The distribution of model 2's velocity is shown in Figure 13; this is similar to that of model 1, shown in Figure 6. It is therefore also expected to provide stable oscillation.

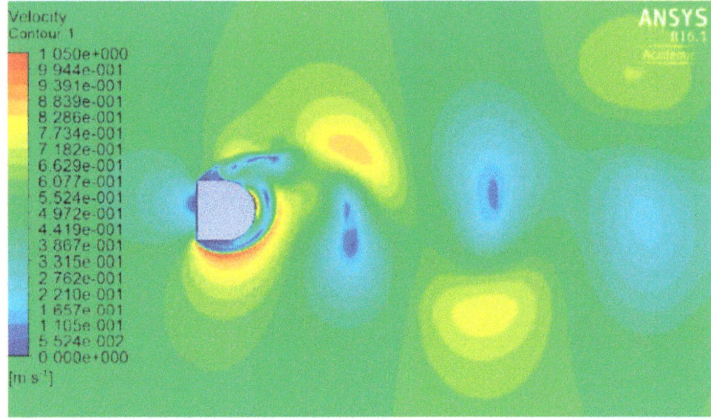

Figure 13. Distribution of velocity for an extended semicircle (model 2) at t = 23.75 s. The reattached region is moving to the rear on the circular side of the object.

From the results in Table 2, it is clear that the square-headed shapes produce greater Cl, except in cases where oscillation becomes unstable. However, the reattachment of a shed vortex must be avoided. Based on the above 2D analysis results, a 3D simulation of the fluid-structure coupling of a square-headed shaped oscillator was carried out.

3.1.4. 3D Fluid-Structure Coupled Analysis

The simulation conditions for the coupled problem are listed in Table 3. Based on the previously discussed results, the authors consider that model 2 (ref. to Table 2) may be optimal, as it produces higher Cl and is relatively stable. Model 1, a semicircle, was analyzed for comparison. Model 3 was an elongated version of model 2. In order to compare their maximum Cl, the weight of the three models with varying lengths in the x direction are the same, as is the support constraint. (Elastic base setting to move 0.01 m under a force of 1.92 N). The mechanics are shown in Figure 14, in which the reciprocating motion in the y direction from the oscillator parallel to z direction will be transmitted to a shaft, then the motion will be converted into rotation by an appropriate mechanism in order to generate electricity. Care was required in order to avoid negative meshes, the mesh type and its size, together with the total number of meshes, including the scale of the fluid domain, were iterated, as dynamic mesh is required for three-dimensional simulation; furthermore, an academic version of ANSYS was used, which involves some limitations. The convergence of the calculation was acceptable. However, the residues were a little higher than for the 2D case, indicating the presence of some error. The computing precision is currently limited by the use of a regular PC for simulation; in the future, this will be addressed by an upgrade of the processing environment.

The simulation results are shown in Table 4. From the results, it can be concluded that Cl increases with elongation in the x direction, as was observed in the 2D model. However, the maximum displacement of model 3 is less than that of model 2. Based on the previously observed data, this reduction of output will be due to instability, resulting in an inability to meet the conditions required to move the object (oscillate). It can be concluded that model 2 is the optimum shape from among those investigated in this study.

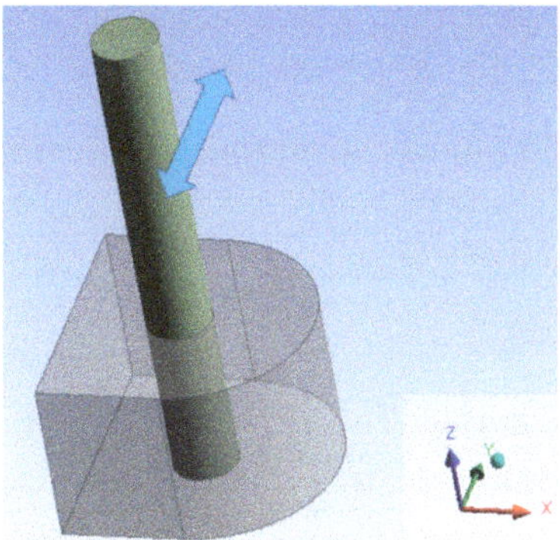

Figure 14. Model for coupling calculation.

Table 3. Conditions for the fluid-structure coupled problem.

Flow time step size (s)	0.005
Flow domain scale (m)	3 × 2 × 0.1
Flow domain mesh type	prism
Min. size of mesh (m)	0.017
Weight of struc. model (kg)	1.62
Size of model 1 (m)	0.075 × 0.15 × 0.1
Size of model 2 (m)	0.15 × 0.15 × 0.1
Size of model 3 (m)	0.18 × 0.15 × 0.1
Dia. of hole in models (m)	0.04
Elastic base (N/m^3)	48,000

Table 4. Results from coupling calculation.

Model	Size in x-direc. (m)	Max Cl	Max Disp. (m)
No.1	0.075	7	0.0120
No.2	0.15	8.8	0.0213
No.3	0.18	10	0.0198

Model 6 in Table 2 may also be optimal based on its stability and high Cl. However, the coupled simulation result cannot be compared with those in Table 4, as the simulation failed under these specific conditions; the simulation, however, was successful with different weight and support parameters. Clearly the relationships between Cl, model weight, supporting conditions, and optimum length require further research.

The square-headed model has a higher Cl compared to the circular model. However, a very important aspect regarding installation of this mechanism is that, if it is not perpendicularly installed to the flow, reciprocation will not occur.

4. Conclusions

(1) An airfoil-shaped oscillator with optimal elasticity effectively increases lift, and we found that the elastic modulus E = 106 Pa is the best for a NACA0015 model foil.

(2) Analysis of the flow field for six common head shapes clearly showed that the discontinuous flow caused by a square-headed oscillator results in higher Cl due to intense vortex shedding. Stable operation can be achieved by selecting the optimum length to width ratio, and this ratio is confirmed to be one (square) by our simulation.

(3) A shape with a higher *Cl* than the semicircle has been identified, and the efficacy of this shape was confirmed in the fluid-structure coupled analysis.

As was mentioned in the introduction, this study is limited to the numerical simulation of two oscillator based generators. Clearly, more extensive simulations are needed on which to base potential relationships between the parameters involved. Such optimization will require iterative design, simulation, prototyping and testing of physical models before a comparison with existing technologies can be objectively made.

Acknowledgments: The authors express their sincere gratitude to Wei Deng, Jincheng Xu, Qiangqiang Ren, and Joshua Lawn for supporting the research.

Author Contributions: I.Y. supervised this research, G.R. planned this paper, G.R. and Y.S. carried out the simulations, I.Y., G.R., Y.S. and M.L. together wrote up the paper. The English was checked by M.L.

Conflicts of Interest: The authors declare no conflict of interest.

References

1. Yamamoto, I. *Practical Robotics and Mechatronics: Marine, Space and Medical Applications*; IET: Croydon, UK, 2016; pp. 1–192.
2. Yamamoto, I. Propulsion system with flexible/rigid oscillating fin. *IEEE J. Ocean. Eng.* **1995**, *20*, 23–30. [CrossRef]
3. Yamamoto, I.; Terada, Y.; Nagamatu, T.; Imaizumi, Y. Research on an oscillating fin propulsion control system. In Proceedings of the OCEANS '93: Engineering in Harmony with Ocean, Victoria, BC, Canada, 18–21 October 1993. [CrossRef]
4. University of Michigan. Available online: http://www.vortexhydroenergy.com/ (accessed on 30 November 2016).
5. Abiru, H.; Yoshitake, A. Experimental Study on a Cascade Flapping Wing Hydroelectric Power Generator. *J. Energy Power Eng.* **2012**, *6*, 1429–1436.
6. Hiejima, S.; Nakano, S. Experimental study on control parameters for aerodynamic vibration-based power generation using feedback amplification. *J. Jpn. Soc. Civ. Eng.* **2012**, *68*, 88–97. [CrossRef]
7. Hiejima, S.; Hiyoshi, Y. Vibrational amplification technique for power generation using wind-induced vibration. *Wind Energy* **2010**, *34*, 135–141.
8. Annual Report on the Environment in Japan 2016. Available online: http://www.env.go.jp/en/wpaper/2016/pdf/2016_all.pdf (accessed on 14 October 2017).
9. Rodríguez, I.; Lehmkuhl, O.; Chiva, J.; Borrell, R.; Oliva, A. On the flow past a circular cylinder from critical to super-critical Reynolds numbers: Wake topology and vortex shedding. *Int. J. Heat Fluid Flow* **2015**, *55*, 91–103. [CrossRef]
10. Hu, W. Hydrodynamic study on a pectoral fin rowing model of a fish. *J. Hydrodyn.* **2009**, *21*, 463–472. [CrossRef]
11. Xu, Y.; Wan, D. Numerical simulations of fish swimming with rigid pectoral fins. *J. Hydrodyn.* **2012**, *24*, 263–272. [CrossRef]
12. ANSYS (Flunet, Mechanical) User's Manual in Japanese. Available online: http://tsubame.gsic.titech.ac.jp/docs/guides/isv-apps/fluent/pdf/Fluent.pdf (accessed on 14 October 2017).
13. Sun, X.; Kato, N.; Li, H. Effects of cross section and flexibility of pectoral fins on the swimming performance of biomimetic underwater vehicles. *J. Jpn. Soc. Nav. Archit. Ocean Eng.* **2012**, *15*, 175–189. [CrossRef]

Article

Three-Dimensional Modeling of a Robotic Fish Based on Real Carp Locomotion

Gonca Ozmen Koca [1,*], Cafer Bal [1], Deniz Korkmaz [2], Mustafa Can Bingol [1], Mustafa Ay [1], Zuhtu Hakan Akpolat [1] and Seda Yetkin [3]

[1] Department of Mechatronics Engineering, University of Firat, Elazig 23119, Turkey; caferbal@gmail.com (C.B.); mustafacanbingol@gmail.com (M.C.B.); mustafaay023@gmail.com (M.A.); z.h.akpolat@gmail.com (Z.H.A.)

[2] Department of Electrical and Electronics Engineering, University of Firat, Elazig 23119, Turkey; denizkorkmaz17@gmail.com

[3] Department of Electronics and Automation, University of Bitlis Eren, Bitlis 13000, Turkey; syetkin@beu.edu.tr

* Correspondence: gonca.ozmen@gmail.com; Tel.: +90-424-237-0000

Received: 9 November 2017; Accepted: 22 January 2018; Published: 26 January 2018

Abstract: This work focuses on developing a complete non-linear dynamic model comprising entirely kinematic and hydrodynamic effects of Carangiform locomotion based on the Lagrange approach by adapting the parameters and behaviors of a real carp. In order to imitate biological features, swimming patterns of a real carp for forward, turning and up-down motions are analyzed by using the Kineova 8.20 software. The proportional optimum link lengths according to actual size, swimming speed, flapping frequency, proportional physical parameters and different swimming motions of the real carp are investigated with the designed robotic fish model. Three-dimensional (3D) locomotion is evaluated by tracking two trajectories in a MATLAB environment. A Reaching Law Control (RLC) approach for inner loop (Euler angles-speed control) and a guidance system for the outer loop (orientation control) are proposed to provide an effective closed-loop control performance. In order to illustrate the 3D performance of the proposed closed loop control system in a virtual reality platform, the designed robotic fish model is also implemented using the Virtual Reality Modeling Language (VRML). Simulation and experimental analysis show that the proposed model gives us significant key solutions to design a fish-like robotic prototype.

Keywords: dynamic model; robotic fish; trajectory tracking; biomimetic modeling; fish-like motion

---

1. Introduction

In recent years, biologically inspired behavior-based systems have been more and more popular topic in underwater vehicles. The goal of the bio-inspired approach is to adapt the biological features of underwater creatures such as fish to Autonomous Underwater Vehicle (AUV) designs and also imitate the aquatic locomotion abilities of them. Biomimetic modeling also provides enhanced performance and increased efficiency, maneuverability and acceleration for novel AUV designs [1–5]. As it is known, many kinds of fish perform high-efficiently locomotion and maneuvering in the water. Propulsion efficiencies of the rotary propeller AUVs are limited below 70%, while the swimming mechanism of a real fish is 20% more efficient than rotary propellers [6]. In addition, the turning radius of propellers is big and speed performances are low. Therefore, this kind of propulsion is more noisy and ineffective than bio-inspired systems [6,7]. These advantages have great benefit for marine applications and fish-like robots have evolved to provide versatile solutions for a wide variety of underwater applications, such as undersea investigation, pollution detection, deep sea monitoring and mapping, military operations and protections, etc. [8–11].

An elaborate observation of the whole biological structure of a fish is necessary to identify the principles of a bio-inspired approach to the fish. In this way, significant limitations and constraints of any biological feature are determined before modeling the robotic fish. A bio-inspired robotic fish is defined as an aquatic vehicle, which is propelled by undulatory or oscillatory body motions [10,12]. In nature, most fish exhibit various swimming behaviors by bending their bodies and/or their caudal fins (BCF). Alternatively, some fish use their median and/or pectoral fins (MPF) [13]. Commonly, more than 85% of fish swim by BCF locomotion and on average 15% of fish swim in the world by MPF. Carangiform locomotion, the most common type of BCF, generates undulatory swimming patterns, high speed performance and low noise [11,13]. Thus, biomimetic Carangiform design is an appropriate solution for AUVs [14–17].

There have been some studies about robotic fish modeling in the literature. The first mathematical models including fish swimming were proposed by Wu and Lighthill in 1960s. Wu suggested simplified two-dimensional waving plate theory and Lighthill defined the body-traveling wave, which takes a sinusoidal form of undulatory motion [1,18]. Sfakiotakis explained the fish-swimming modes for oscillatory and undulatory motions of fins and bodies [14]. McIsaac and Ostrowski established a model for Anguilliform locomotion to give swimming gaits [19], while Morgansen et al. derived 3D models for pectoral fins and tails [20]. Also, Yu et al. applied the Central Pattern Generator (CPG)-based locomotor controller to their dynamic model. CPG is a neural circuit diagram inspired from biological musculoskeletal systems to perform gait generation of bio-inspired robots [21–23]. Liu and Hu built a simulator for the autonomous Carangiform motion of a four-joint robotic fish, but up-down motions were not modeled [24]. Yu et al. also modeled up-down motions [25,26]. Unlike these, Zhou et al. presented a dynamic backward analysis for a biomimetic Carangiform robotic fish [27]. Furthermore, Nakashima and Ohgishi presented a dynamic model of the double-jointed robotic fish to examine self-exciting conditions and propulsive characteristics [28]. Kim et al. presented the four-link motion mechanism of fish [29]. Masoomi and colleagues modeled a four degrees of freedom (4-DoF) mathematical model including hydrodynamic forces inspired by tuna fish [10]. A multi-DoF robotic fish model was suggested by Suebsaiprom and Lin to determine the position and orientation of the rigid body. They used a state feedback controller in the simulation [15,30]. According to the above literature survey, a 3D dynamic model including all kinematic and hydrodynamic effects by analyzing realistic parameters of a real fish would be a significant reference while developing a realistic prototype.

The contribution of this paper is to develop a complete dynamic model including kinematic and hydrodynamic effects of Carangiform locomotion based on the Lagrange approach by adapting the parameters and behaviors obtained from real carp. This model provides appropriate solutions for biomimetic design of fish-like motions with 3D carp gait patterns unlike the dynamic robotic fish models available in the literature with non-realistic parameters. In order to mimic real fish-swimming abilities, the designed robotic fish model has been focused on the proportional optimum link lengths according to actual size, swimming speed, flapping frequency, proportional physical parameters and different swimming motions of the real carp. 3D locomotion is evaluated by applying various trajectories in different MATLAB environments. The closed loop performance is provided by using the Reaching Law Control (RLC) approach in the inner loop and orientation control is also achieved with a guidance system in the outer loop. These simulations and experimental analyses show that the proposed model gives us significant key solutions to design a fish-like robotic prototype.

The rest of this paper is organized as follows: in the next section, Carangiform locomotion is described and a complete dynamic model, including kinematic and hydrodynamic effects, is given. Then the maneuverability of the robotic fish model is performed and simulation results are given. Finally, conclusions are presented.

2. Three Dimensional Dynamic Model and Motion Equations of the Robotic Fish

The fish model analyzed in this study is inspired by a Carangiform-mode fish based on the BCF locomotion. A Carangiform mode exhibits more significant swimming actions such as fast start, rapid turning, C-shape turning, cruise swimming and high accelerations [14]. A Carangiform fish swims on the lateral sinusoidal motion, which increases the amplitude from nose to caudal fin in a spine. This kinematic motion can be described using the body-traveling wave function, suggested by Lighthill [18,31]. However, this function only includes the simple kinematic equality for the fish tail and there is no whole dynamic effect to understand the fish locomotion. Based on the biological features of Carangiform locomotion, the non-linear dynamic model of the robotic fish should be derived, including 6-DoF motion. This study focuses on deriving a complete non-linear mathematical model of a Carangiform robotic fish by analyzing real carp locomotion. The dynamic model is performed by using realistic parameters from 3D swimming patterns of the carp. The proposed non-linear mathematical model of the robotic fish includes the body and tail kinematics, hydrodynamic effects acting on each link and rigid body motions. This model behaves like a real fish with complete non-linear derivations. In this way, biomimetic modeling of the robotic fish prototype can be easily implemented in future work.

2.1. Carp Locomotion

In this study, forward, turning and ascending-descending swimming patterns are investigated with top- and side-view cameras for 3D records in a 120 cm length, 80 cm width and 60 cm depth test tank to analyze the maneuverability of real carp. The resolution of the high-definition video camera is 1920 × 1080 p, the data rate is 19,949 kb/s, the total bit rate is 20,205 kb/s and frame rate is 29 fps, respectively. Stabilized swimming over a distance of 60 cm is selected for testing. Snapshots are also captured and examined by using the Kinovea 8.20 software. The sampling time is set to 0.05 s.

The main idea for kinematic modeling of the robotic fish is to assume the robotic fish to have a multi-link mechanism. The equivalent model of the robotic fish and description of the one link are given in Figure 1.

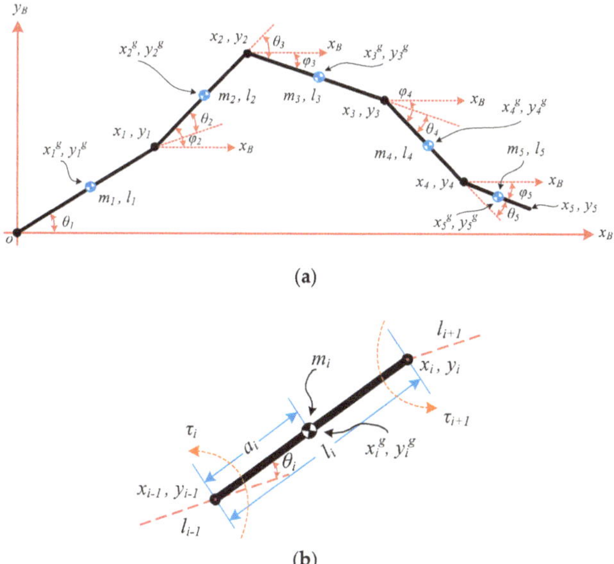

Figure 1. Kinematic scheme of the robotic fish: (a) Equivalent model; (b) Description of the one link.

The designed robotic fish consists of five links. We will define $[l_1, \ldots, l_i] \to (i = 1, 2, \ldots, N)$ as one body link, three active tail links and one passive oscillating caudal fin link. $N = 5$ is the number of links. θ_1 is the angle of head during swimming, $\theta_k (k = 2, 3, 4)$ is the each active link angle of the propulsion mechanism and θ_5 is angle of the caudal fin. m_1 is the body mass and each link mass of the propulsion mechanism equals from m_2 to m_5. (x_i^g, y_i^g) and (x_i, y_i) are link coordinates of the center of mass and end points of the link coordinates, respectively. Also, $\varphi_i = \theta_i + \theta_{i-1}$ is the orientation angle of the joints in the horizontal plane. It can be seen from Figure 2b that each link is homogeneous and the middle points of the links are regarded as the center of mass. a_i is the distance from joint to center of the link mass and τ_i is the turning moment of the each joint.

Motion of a real fish depends on the body-traveling wave. In order to imitate the body-traveling wave, there are two general methods: sine-based joint location or intersection methods [19,32], and CPGs [21,33]. In this paper, the improved intersection method with the Big Bang–Big Crunch optimization algorithm, which is also used in our previous work [32], is applied to the link points of the real carp. In this method, locations of links are determined within a scanning area to achieve the minimum error of the fitting body-travelling wave. Thus, optimization criteria is chosen as error of the envelop area with optimum link lengths [32,34]. In the analysis of the carp, five points are defined to verify optimized link lengths on forward and turning patterns obtained from the top view of the carp. The locations of the points and optimized link lengths on forward and the turning patterns are shown in Figure 2.

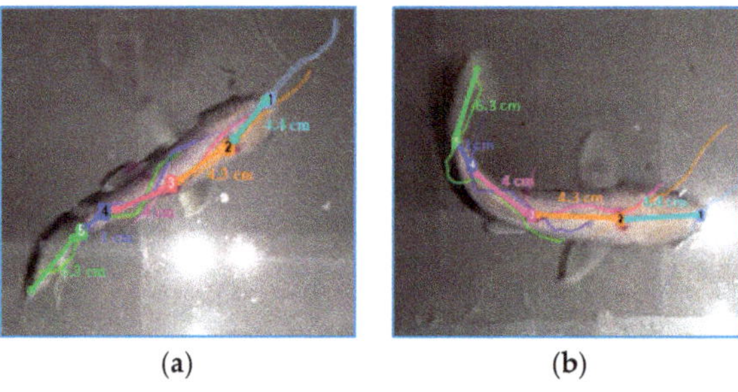

Figure 2. Trajectory study of two swimming patterns of carp. The fish is simplified as a 5-joint linkage represented by different colors. The light color curves indicate a trajectory of the corresponding joint within a period of swimming pattern: (**a**) Forward swimming pattern, period 350 ms; (**b**) Turning swimming pattern, period 550 ms.

In the experiments, five points are marked on the real fish in order to define link locations according to optimized link lengths of the robotic fish. Locations of the points are tested by applying links between every two points to ensure a minimum area.

As seen from Figure 2, light blue line is the anterior body, orange, pink and deep blue lines are active links and the green line is the passive caudal fin, respectively.

2.2. Swimming Motion of the Robotic Fish

Motions of the carp obtained from experimental analysis according to all joints for forward and turning swimming modes are given in Figure 3.

The average speed of a carp ranges from 0 to 2.5 m/s in the ichthyology. In the experimental studies, speeds of forward swimming are recorded for some different ranges from 0.105 m/s, 0.125 m/s, 0.144 m/s, 0.233 m/s to 0.41 m/s and turning speed is obtained as 0.24 m/s, approximately. In order to determine flapping frequency of the carp, motion of the caudal fin is analyzed as shown in Figure 4 during different periods. The average recorded flapping frequencies of the carp are measured as 1.818 Hz, 2 Hz, 2.439 Hz and 2.857 Hz, approximately. In order to calculate one sample flapping frequency for 2.857 Hz, minimum and maximum angles of the caudal fin from the center of gravity are measured and time of the fin motion is determined. In this sample, motion time for one period is measured as 350 ms.

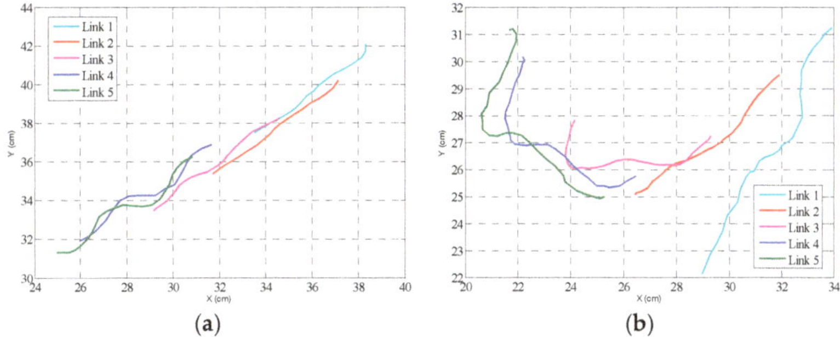

Figure 3. Path tracking of the all links: (**a**) Forward swimming pattern; (**b**) Turning swimming pattern.

Figure 4. One period of the tail: (**a**) Trace of the travelling wave for 350 ms; (**b**) Caudal fin angles for forward swimming pattern: $\theta_5 = [\pi - 0.6\pi, \pi - 1.15\pi]$.

Angles of the active links are also shown in Figure 5 for forward and turning swimming modes during one period.

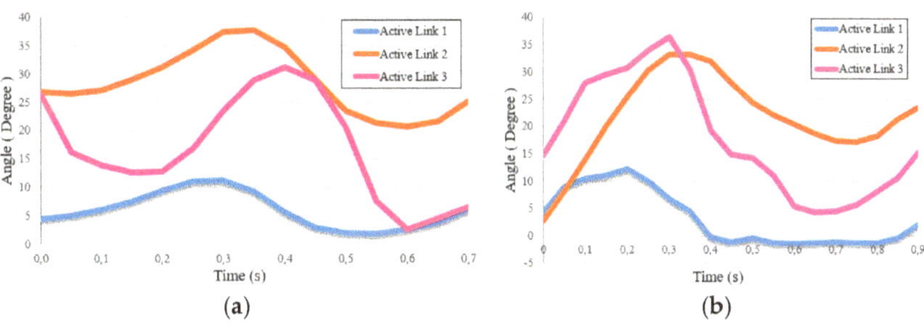

Figure 5. Real fish angles according to the active links: (**a**) Forward swimming mode; (**b**) Turning swimming mode.

Motion of the robotic fish is realized by the powerful spine that generates the lateral sinusoidal undulations, while the anterior body is relatively rigid and determines the swimming direction. Active tail links are actuated by servomotors and the passive caudal fin is connected to the fourth link with a peduncle. It is noted that the thrust force is generated by the caudal fin.

In order to obtain real fish-like motions, the dynamic model of the robotic fish is derived by using following second order form:

$$M(\theta_i)\ddot{\theta} + C(\theta_i, \dot{\theta}_i)\dot{\theta}_i + B(\dot{\theta}_i) + K(\theta_i) = \tau_i \tag{1}$$

$M(\theta_i)$ is the moment of inertia, $C(\theta_i, \dot{\theta}_i)$ is the Coriolis matrix, $B(\dot{\theta}_i)$ is the damping coefficient matrix, $K(\theta_i)$ is the spring coefficient matrix and τ_i is the force vector that includes hydrodynamic forces.

2.3. Hydrodynamic Forces Acting on the Tail

As shown in Figure 6, hydrodynamic forces depend on the tail motion and there are five main hydrodynamic forces acting on the tail. Inertial fluid force is F_V, lift force is F_J, thrust force is F_F, lateral force is F_C and drag force is F_D.

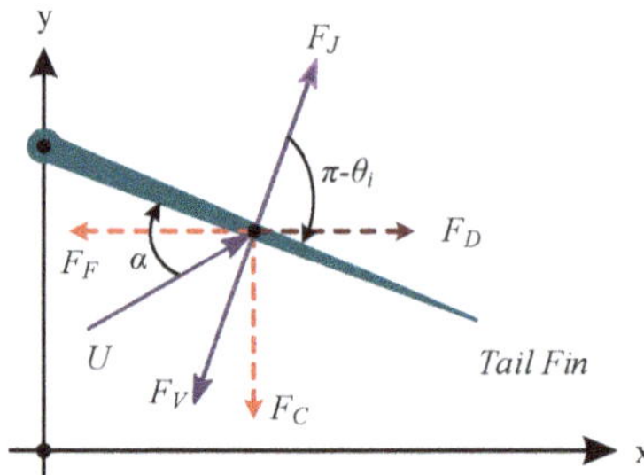

Figure 6. Hydrodynamic force distributions acting on the tail.

It is considered that tail fin of the robotic fish flaps in constant flow representing by U_m so that the inertial and lift forces influence the caudal fin. Afterwards, thrust force and lateral force components are calculated by using inertial fluid and lift forces. The value of constant flow is determined as 0.08 m/s to reduce the turbulent effects [35]. According to Figure 6, F_V is a proportional force to the acceleration and it is calculated by Equation (2) and the lift force is also vertical to the flow and it is defined as Equation (3):

$$F_V = \pi\rho LC^2\dot{U}\sin(\alpha) + \pi\rho LC^2U\dot{\alpha}\cos(\alpha) \tag{2}$$

$$F_J = 2\pi\rho LCU^2\sin(\alpha)\cos(\alpha) \tag{3}$$

here, the chord length is $2C$, the span of the caudal fin is L, density of the water is ρ, relative speed at the center of caudal fin is U and the angle of attack is α. Relative speed and angle of attack variables are time dependent and the first derivative of these should be obtained, smoothly. These forces are separated into the thrust force on the x-axis and lateral force on the y-axis components:

$$F_F = (F_V - F_J)\sin(2\pi - \sum_{i=1}^{N}\theta_i) \tag{4}$$

$$F_C = (F_V - F_J)\cos(2\pi - \sum_{i=1}^{N}\theta_i) \tag{5}$$

If motion of the robotic fish is considered in the x-axis direction, the relative speed on the center of caudal fin in the y-axis direction is expressed by the following equation:

$$\begin{aligned} u_f = \; & l_1\dot{\theta}_1\cos(\theta_1) + l_2(\dot{\theta}_1 + \dot{\theta}_2)\cos(\theta_1 + \theta_2) \\ & + l_3(\dot{\theta}_1 + \dot{\theta}_2 + \dot{\theta}_3)\cos(\theta_1 + \theta_2 + \theta_3) \\ & + l_4(\dot{\theta}_1 + \dot{\theta}_2 + \dot{\theta}_3 + \dot{\theta}_4)\cos(\theta_1 + \theta_2 + \theta_3 + \theta_4) \\ & + a_5(\dot{\theta}_1 + \dot{\theta}_2 + \dot{\theta}_3 + \dot{\theta}_4 + \dot{\theta}_5)\cos(\theta_1 + \theta_2 + \theta_3 + \theta_4 + \theta_5) \end{aligned} \tag{6}$$

Relative speed on the center of caudal fin U is determined by $\sqrt{U_m{}^2 + u_f{}^2}$, because U_m and u_f are perpendicular to each other. The thrust force F_F pushes the robotic fish in the forward direction.

The drag force F_D acts on the fish body in the flow direction by the friction. The definition of the drag force is presented in Equation (7):

$$F_D = \frac{1}{2}\rho C_D V_f^2 S \quad (7)$$

where, C_D is the drag coefficient, V_f is the relative speed of the robotic fish and S is the area of the main body. The head of the robotic fish is assumed as conic and the value of drag coefficient is defined as 0.5. Both hydrodynamic and external forces act on the robotic fish tail. These forces are called with $\tau_i = [\tau_1, \tau_2, \tau_3, \tau_4, \tau_5]^T$ which affect to the ith link, and can be expressed by:

$$\text{if}\begin{cases} active\_link \rightarrow \tau_i = F_C l_i \cos(\sum_{i=1}^{N} \theta_i) - F_F l_i \sin(\sum_{i=1}^{N} \theta_i) + T_{i-1}^u \\ passive\_link \rightarrow \tau_i = F_C l_i \cos(\sum_{i=1}^{N} \theta_i) - F_F l_i \sin(\sum_{i=1}^{N} \theta_i) \end{cases} \Bigg|_{last\_link \rightarrow l_i = a_i} \quad (8)$$

The designed robotic fish is driven by three active joints that generate the thrust force. These joints are actuated by three DC servo motors with input torques $T_i^u = A_{\max}\sin(2\pi f_T t \pm \Delta_i)$. $A_{\max}$ is the amplitude of input torque, f_T is the frequency of input torque and $\pm\Delta_i$ is the phase angle between the two input torques. By using Lagrange's equation, the dynamic model of the robotic fish is described briefly in Equation (9):

$$\begin{bmatrix} M_{11} & M_{12} & M_{13} & M_{14} & M_{15} \\ M_{21} & M_{22} & M_{23} & M_{24} & M_{25} \\ M_{31} & M_{32} & M_{33} & M_{34} & M_{35} \\ M_{41} & M_{42} & M_{43} & M_{44} & M_{45} \\ M_{51} & M_{52} & M_{53} & M_{54} & M_{55} \end{bmatrix} \begin{bmatrix} \ddot{\theta}_1 \\ \ddot{\theta}_2 \\ \ddot{\theta}_3 \\ \ddot{\theta}_4 \\ \ddot{\theta}_5 \end{bmatrix} = \begin{bmatrix} N_1 \\ N_2 \\ N_3 \\ N_4 \\ N_5 \end{bmatrix} \quad (9)$$

M_{ij} is the moment element of the inertia matrix and each value of θ_i can be obtained from Equation (9).

2.4. Modeling of the Up-Down Motion Mechanism

In nature, up-down motions such as ascending and descending are another important criterion to exhibit high swimming performance. In order to generate ascending and descending motions, a fish changes its center of mass (COM) position and uses the air bladder. Figure 7 gives the descending pattern of the carp during 1.05 s. In this analysis, pitch angle of the carp is measured as 35°. This performance is the high motion ability for a real fish and similar values of this angle are validated by using the derived model.

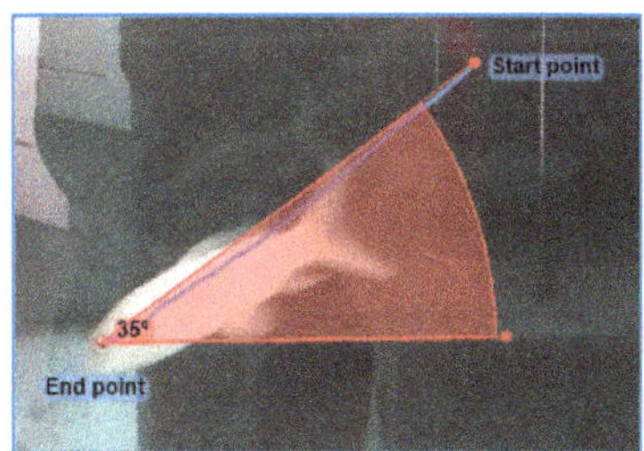

Figure 7. Descending pattern of the carp during 1.05 s.

In Figure 8a,b, the model of the up-down motion mechanism and flow chart, which represents the working principle of the mechanism, are given, respectively. An up-down mechanism is designed and modeled by using the Lagrange approach to generate and control pitch and roll motions.

This mechanism is placed inside the anterior robotic fish body. While the robotic fish is swimming under water, swimming speed is not zero and the depth position of the robotic fish in the vertical plane can be controlled. Control of the orientation angles are performed by changing COM position. There are two masses (m_x, m_y) moving on the shaft. Shafts are connected to the pulley sets with DC motors. By moving the masses along the x and y-axis, the up-down mechanism changes the pitch and roll angles.

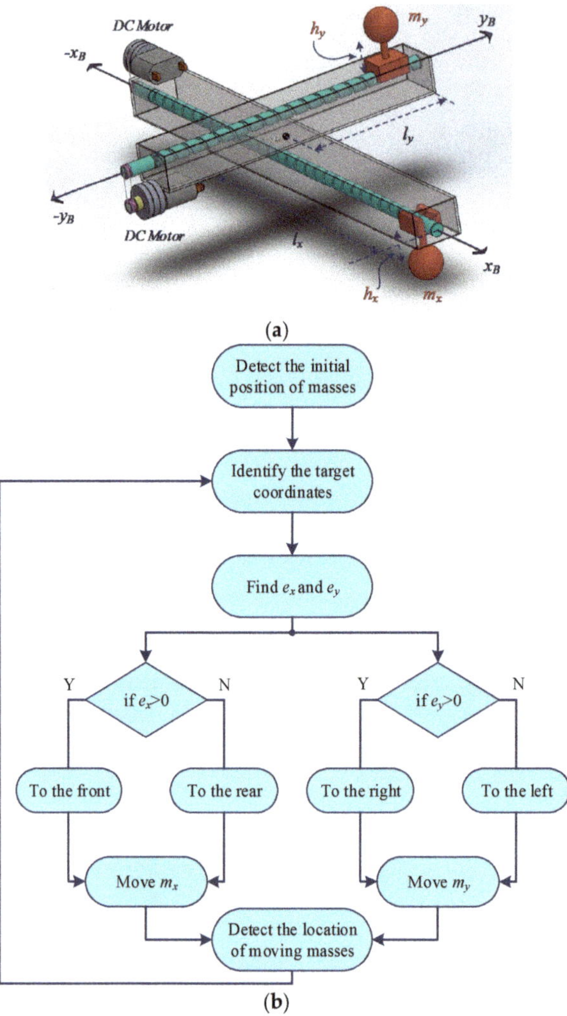

Figure 8. (**a**) Model of the up-down motion mechanism; (**b**) Flow chart of the up-down motion.

When mass positions change, pitch and roll torques are generated. These mentioned torques are considered in Equations (10) and (11), respectively.

$$\tau_x = g((l_x \sin(\phi)\sin(\theta) - h_x \sin(\phi)\cos(\theta))m_x + (l_y \cos(\phi) - h_y \sin(\phi)\cos(\theta))m_y) \tag{10}$$

$$\tau_y = g((l_x \cos(\theta) - h_x \sin(\theta))m_x + (h_y \sin(\theta))m_y) \tag{11}$$

Here, g is the acceleration of the gravity, (l_x, l_y) are the distance from the origin and (h_x, h_y) are the heights from the axis.

2.5. Three Dimensional Model Equations of Motion

Based on the ichthyology of carp, a multi-link and autonomous-swimming biomimetic robotic fish prototype is designed. The prototype consists of a rigid body, three active tail links and a passive caudal fin. In order to analyze the guidance states of the robotic fish, 6-DoF body motion equations are presented in this paper. These equalities define the kinematic and dynamic behaviors of the robotic fish in two coordinate frames and also include the hydrodynamic forces and moments acting on the robotic fish body. The moving coordinate frame is fixed to the body and it is expressed as a body-fixed reference frame. The origin of the body-fixed frame is pointed with the center of gravity. Also, the motion of the body-fixed frame depends on the inertial reference frame that is named the earth-fixed frame. The 6-DoF motion of the robotic fish prototype in the Earth-Fixed frame with translation and rotation relations is shown in Figure 9.

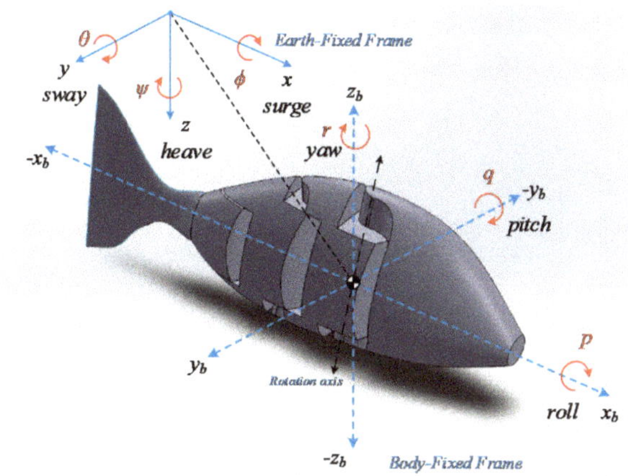

Figure 9. The designed multi-link robotic fish prototype on reference frames.

The generalized speed and positions of the robotic fish are completely given by the state variable vector $s = \begin{bmatrix} u & v & w & p & q & r & x & y & z & \phi & \theta & \psi \end{bmatrix}^T$. The linear speeds (u, v, w) are surge, sway and heave, respectively. The angular speeds (p, q, r) are roll, pitch and yaw speeds, respectively. Surge, sway and heave positions (x, y, z) are determined with the earth-fixed frame. Roll, pitch and yaw angles (ϕ, θ, ψ) are the rigid body orientations with respect to the earth-fixed frame.

The kinetic translation between the body-fixed and the earth-fixed frames for the robotic fish motion is expressed as:

$$\dot{\eta} = \begin{bmatrix} R_\Theta(\eta) & 0_{3\times3} \\ 0_{3\times3} & T_\Theta(\eta) \end{bmatrix} v \tag{12}$$

where, $\eta = \begin{bmatrix} x & y & z & \phi & \theta & \psi \end{bmatrix}^T$ and $v = \begin{bmatrix} u & v & w & p & q & r \end{bmatrix}^T$ are the generalized positions and speeds, respectively. η is relative with the body-fixed frame and speed vector depends on the earth-fixed frame. $R_\Theta(\eta)$ is the rotation matrix and $T_\Theta(\eta)$ is the transformation matrix. In this way, the 3D equation of the rigid body motion is given by:

$$M\dot{v} + C(v)v + D(v)v + g(\eta) = \tau_f \tag{13}$$

where, $M = M_{RB} + M_A$ is the mass matrix, M_{RB} is the rigid body mass matrix and M_A is the added mass matrix. $C(v) = C_{RB}(v) + C_A(v)$ is the Coriolis and centripetal matrix, $C_{RB}(v)$ is the rigid body Coriolis matrix and $C_A(v)$ is the hydrodynamic Coriolis matrix. $D(v)$ is the hydrodynamic damping matrix including linear and non-linear damping effects. τ_f is the vector that includes the hydrodynamic forces. Rigid body mass matrix and added mass matrix are defined as:

$$M_{RB} = \begin{bmatrix} mI_{3\times3} & -mS(r_G) \\ mS(r_G) & I_0 \end{bmatrix} \tag{14}$$

$$M_A = \begin{bmatrix} X_{\dot{u}} & 0 & 0 & 0 & 0 & 0 \\ 0 & Y_{\dot{v}} & 0 & 0 & 0 & Y_{\dot{r}} \\ 0 & 0 & Z_{\dot{w}} & 0 & Z_{\dot{q}} & 0 \\ 0 & 0 & 0 & K_{\dot{p}} & 0 & 0 \\ 0 & 0 & M_{\dot{w}} & 0 & M_{\dot{q}} & 0 \\ 0 & N_{\dot{v}} & 0 & 0 & 0 & N_{\dot{r}} \end{bmatrix} \tag{15}$$

In these equations, m is the total body mass. The position of the center of gravity is equal to $r_G = \begin{bmatrix} x_G & y_G & z_G \end{bmatrix}^T$. The moments of inertia are defined as I_0. Asymmetric matrix S is used for easier computation. $X_{\dot{u}}$ and so forth are the zero-frequency added mass coefficients. $C_{RB}(v)$ matrix indicates the Coriolis matrix $\begin{bmatrix} p & q & r \end{bmatrix}^T \times \begin{bmatrix} u & v & w \end{bmatrix}^T$ and it is expressed by:

$$C_{RB}(v) = \begin{bmatrix} 0_{3\times3} & -S(v_1 mI_{3\times3} - v_2 mS(r_G)) \\ -S(v_1 mI_{3\times3} + v_2 - mS(r_G)) & -S(v_1 mS(r_G) + v_2 I_0) \end{bmatrix} \tag{16}$$

$C_A(v)$ matrix contains Munk moments and it is given by:

$$C_A(v) = \begin{bmatrix} 0 & 0 & 0 & 0 & -Z_{\dot{w}}w & Y_{\dot{v}}v \\ 0 & 0 & 0 & Z_{\dot{w}}w & 0 & -X_{\dot{u}}u \\ 0 & 0 & 0 & -Y_{\dot{v}}v & X_{\dot{u}}u & 0 \\ 0 & -Z_{\dot{w}}w & Y_{\dot{v}}v & 0 & -N_{\dot{r}}r & M_{\dot{q}}q \\ Z_{\dot{w}}w & 0 & -X_{\dot{u}}u & N_{\dot{r}}r & 0 & -K_{\dot{p}}p \\ -Y_{\dot{v}}v & X_{\dot{u}}u & 0 & -M_{\dot{q}}q & K_{\dot{p}}p & 0 \end{bmatrix} \tag{17}$$

Hydrodynamic damping effects of the robotic fish are separated into linear and non-linear terms. These matrixes can be defined as below equations:

$$D_L = \begin{bmatrix} X_u & 0 & 0 & 0 & 0 & 0 \\ 0 & Y_v & 0 & 0 & 0 & -Y_r \\ 0 & 0 & Z_w & 0 & Z_q & 0 \\ 0 & 0 & 0 & K_p & 0 & 0 \\ 0 & 0 & M_w & 0 & M_q & 0 \\ 0 & -N_v & 0 & 0 & 0 & N_r \end{bmatrix} \tag{18}$$

$$D_{NL} = \begin{bmatrix} X_{|u|u}\left|u\right| & 0 & 0 & 0 & 0 & 0 \\ 0 & Y_{|v|v}\left|v\right| & 0 & 0 & 0 & -Y_{|v|r}\left|v\right| \\ 0 & 0 & Z_{|w|w}\left|w\right| & 0 & Y_{|w|q}\left|w\right| & 0 \\ 0 & 0 & 0 & K_{|p|p}\left|p\right| & 0 & 0 \\ 0 & 0 & M_{|q|w}\left|q\right| & 0 & M_{|q|q}\left|q\right| & 0 \\ 0 & -N_{|r|v}\left|r\right| & 0 & 0 & 0 & N_{|r|r}\left|r\right| \end{bmatrix} \tag{19}$$

D_L and D_{NL} are the diagonal matrixes and these matrixes depend on the body shape and speed of the robotic fish, respectively. X_u, $X_{|u|u}$ and so forth terms are linear and non-linear positive damping coefficients. In marine applications, underwater vehicles move at low speed and adding mass terms can be neglected. However, these effects are used to obtain complete motion model.

$W = mg$ is the gravity force and $B = \rho g \Delta$ is the buoyancy force. Here, Δ is the submerged volume. Thus, hydrostatic effect $g(\eta)$ is calculated as:

$$g(\eta) = \begin{bmatrix} (W-B)s(\theta) \\ -(W-B)c(\theta)s(\phi) \\ -(W-B)c(\theta)c(\phi) \\ -(y_gW - y_bB)c(\theta)c(\phi) + (z_gW - z_bB)c(\theta)s(\phi) \\ -(z_gW - z_bB)s(\theta) + (x_gW - x_bB)c(\theta)c(\phi) \\ -(x_gW - x_bB)c(\theta)s(\phi) - (y_gW - y_bB)s(\theta) \end{bmatrix} \tag{20}$$

Note that (*s*(), *c*(), *t*()) are equal to (*sin*(), *cos*(), *tan*()). $r_B = \begin{bmatrix} x_B & y_B & z_B \end{bmatrix}^T$ is the position vector from the body origin to center of buoyancy. The force vector τ_f includes the hydrodynamic forces and moments acting on the fish body and it is expressed as:

$$\tau_f = \begin{bmatrix} F_x & F_y & F_z & \tau_x & \tau_y & \tau_z \end{bmatrix}^T \tag{21}$$

where, (F_x, F_y, F_z) represent the surge, sway and heave forces. (τ_x, τ_y, τ_z) are the yaw, pitch and roll moments, respectively. It is considered that $F_x = F_F \cos(\theta_2 + \theta_3 + \theta_4)$. F_y and F_z are chosen as zero, and $\tau_z = ((l_1 + l_2)\cos(\theta_2) + l_3\cos(\theta_2 + \theta_3) + l_4\cos(\theta_2 + \theta_3 + \theta_4))F_F \sin(\theta_2 + \theta_3 + \theta)$. Finally, 6-DoF motion block diagram is given in Figure 10 that includes the fish body and fish tail equations both body-fixed frame and earth-fixed frame.

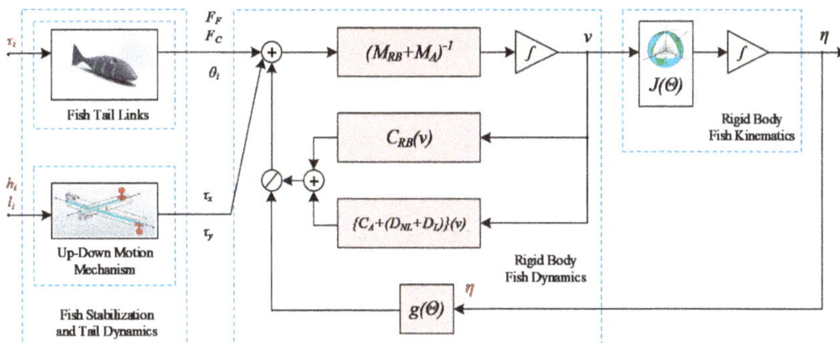

Figure 10. 6-DoF motion block diagram.

By solving these equations, whole 3D non-linear dynamic model of the robotic fish is obtained.

3. Implementations of the Fish-Like Motion

In order to mimic real fish swimming abilities, the robotic fish model focuses on five special points according to the real carp:

1. Proportional optimum link lengths according to actual size,
2. Flapping frequency,
3. Swimming speed performance,
4. Proportional physical parameters according to the carp,
5. Trajectory tracking.

As can be observed from the experiments, proportional link lengths generate sinusoidal angles while the robotic fish propels itself. This situation causes the sinusoidal undulation motions. Secondly, obtained flapping frequency from the observations is an appropriate value and it is applied to the robotic fish model. Forward and turning speeds of the carp are also examined in the simulations. Finally, guidance trajectory tracking is performed to validate the dynamic model.

3.1. Ability of the Fish-Like Motion

In this study, numerical analyses are performed in a MATLAB environment. The flapping frequency of the carp is observed from the experiments and it is determined as 2.857 Hz. This value is also applied to the robotic fish model to validate the ability of the fish-like motion. In Figure 11, free-swimming motion of the robotic fish is illustrated during an 80 s simulation time.

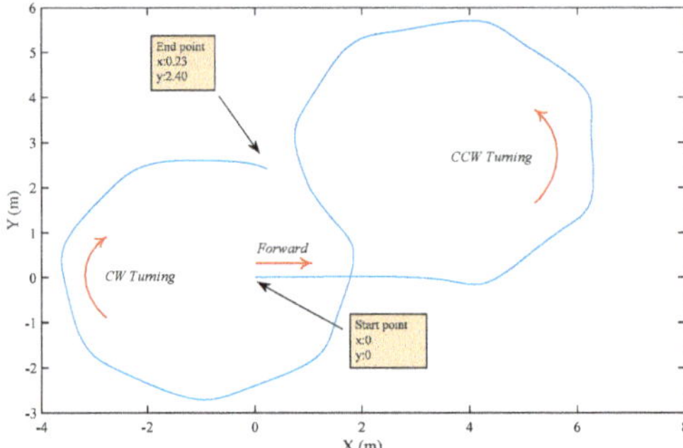

Figure 11. Free-swimming motion of the robotic fish. It is noted that there are three swimming modes: Forward, Counter Clock Wise (CCW) and Clock Wise (CW) turning.

Figure 11 shows open loop free-swimming behavior on a two-dimensional (2D) reference frame. The start position of the robotic fish is X = 0 m, Y = 0 m. After 80 s simulation time, the robotic fish reaches the final position at X = 0.23 m, Y = 2.40 m. The speed of the robotic fish is measured as 0.184 m/s and 0.474 m/s for forward and turning swimming motions, respectively. Figure 12 shows the link angles and Figure 13 presents the thrust force on the x-axis.

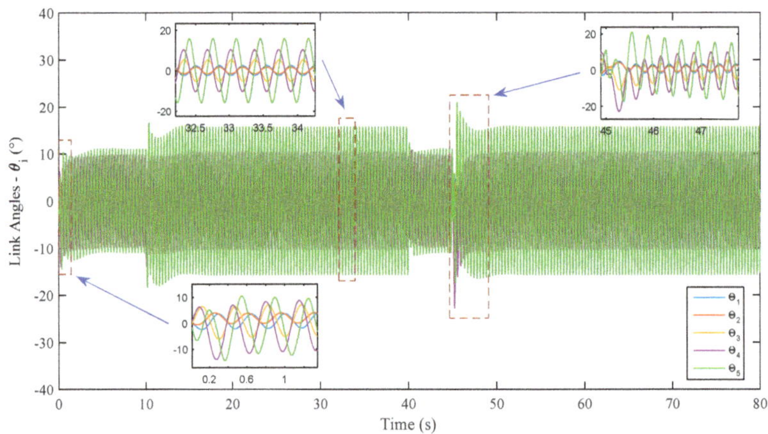

Figure 12. Link angles of the robotic fish for free-swimming motion during 80 s simulation time.

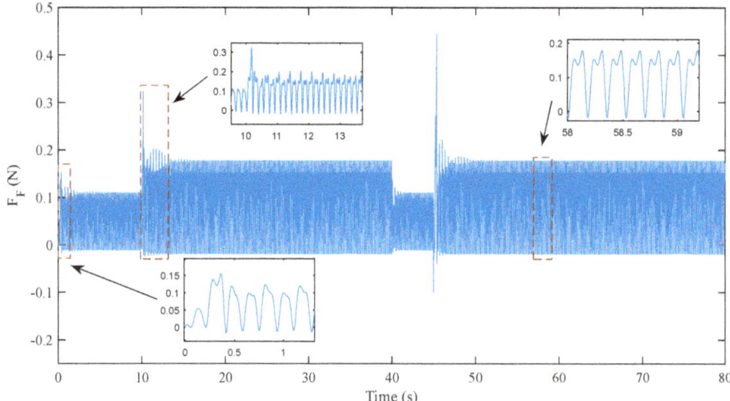

Figure 13. Thrust force of the robotic fish for free-swimming motion during 80s simulation time.

In Figure 12, tail link behaviors are illustrated for different swimming motions. The robotic fish tail swings itself with sinusoidal undulations with pure sine waves.

The thrust force of the tail is determined as nearly 0.110 N and 0.177 N for forward and turning swimming modes.

3.2. Guidance and Trajectory Tracking

The proposed model of the robotic fish including guidance and trajectory tracking consists of a three-layered hierarchical architecture: inner loop, outer loop and trajectory generator. The inner loop is controlled with the simple and practical control structure inspired by Reaching Law Control (RLC) approach. The inner loop occurs in the body frame of the robotic fish. The outer loop deals with the guidance control structure. On the other hand, the trajectory generator builds up the 3D path tracking in earth-fixed frame. The block diagram of the closed loop system is shown in Figure 14.

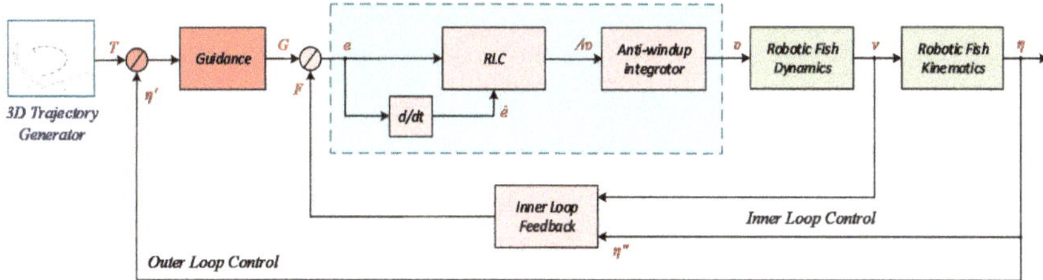

Figure 14. Block diagram of the proposed guidance and trajectory tracking with RLC.

In Figure 14, T, η, G and F indicate reference trajectory, current positions, reference of the guidance, and current orientations and speed, respectively. These parameters are defined as below:

$$\begin{cases} T = \begin{bmatrix} X_{ref} & Y_{ref} & Z_{ref} \end{bmatrix}^T \\ \eta' = \begin{bmatrix} x & y & z \end{bmatrix}^T \\ \eta'' = \begin{bmatrix} \phi & \theta & \psi \end{bmatrix}^T \\ G = \begin{bmatrix} \theta_{ref} & \psi_{ref} & v_{ref} \end{bmatrix}^T \\ F = \begin{bmatrix} \theta & \psi & v \end{bmatrix}^T \end{cases} \tag{22}$$

It is noted that the reference matrix of the guidance is calculated by using a trigonometrical approach and that each element of the matrix can be given by Equation (23):

$$\begin{cases} \theta_{ref} = -\tan^{-1}\left(\left(z_{ref}-z\right)/\left(\sqrt{\left(x_{ref}-x\right)^2+\left(y_{ref}-y\right)^2}\right)\right) \\ \psi_{ref} = \tan^{-1}\left(\left(y_{ref}-y\right)/\left(x_{ref}-x\right)\right) \\ v_{ref} = \zeta\sqrt{\left(x_{ref}-x\right)^2+\left(y_{ref}-y\right)^2+\left(z_{ref}-z\right)^2} \end{cases} \tag{23}$$

here, ζ is a constant speed control gain. In the general RLC approach, the discrete time reaching law [36,37] can be given by:

$$\Delta S(k+1) = -\sigma \mathrm{sgn}(S(k)) - \alpha S(k) \tag{24}$$

where σ and α are constant parameters. $\Delta S(k+1)$ is then expressed by:

$$\Delta S(k+1) = \frac{S(k+1) - S(k)}{T_s} \tag{25}$$

T_s is the sampling time. With the assumption of $\{\Delta S(0) = 0\}$, discrete time switching function is given as:

$$S(k) = \lambda e(k) + \Delta e(k) \tag{26}$$

here, $e(k)$ is the error of the guidance and $\Delta e(k)$ is the derivative of the error in the kth sampling step. λ is the slope of the switching line and indicates a first-order dynamics of the error. The high-frequency oscillation problem known as chattering is occurred if Equation (24) is used directly in practice [37,38]. This chattering problem is presented in Figure 15 and it can be solved by using $sat(S(k))$ function instead of $\mathrm{sgn}(S(k))$. $sat(\cdot)$ function is expressed by:

$$sat(\Omega) = \begin{cases} \Omega & \text{if}|\Omega| \le 1 \\ \mathrm{sgn}(\Omega) & \text{if}|\Omega| > 1 \end{cases} \tag{27}$$

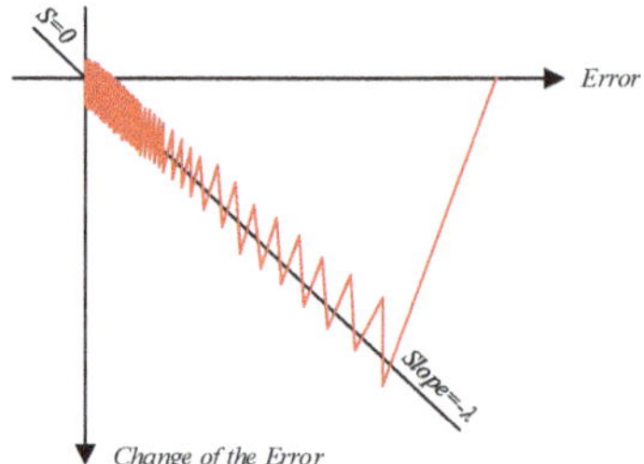

Figure 15. A typical high-frequency oscillation problem.

The new reaching law can be obtained as:

$$\Delta S(k+1) = -\sigma sat(S(k)) - \alpha S(k) \tag{28}$$

The reaching law is also rewritten as below with this $sat(\cdot)$ function in the Boundary Layer (BL):

$$\Delta S(k+1) = -\beta S(k) \tag{29}$$

where $\beta = (\sigma/BL) + \alpha$. Thus, the control output becomes:

$$\Delta v(k) = KS(k) \tag{30}$$

Since S is forced to reach zero with this RLC approach, exponentially reducing of the error to zero with time constant of $(1/\lambda)$ is provided.

Also, an anti-windup integrator is added to stop over-integration for the protection of the designed model as seen in Figure 16.

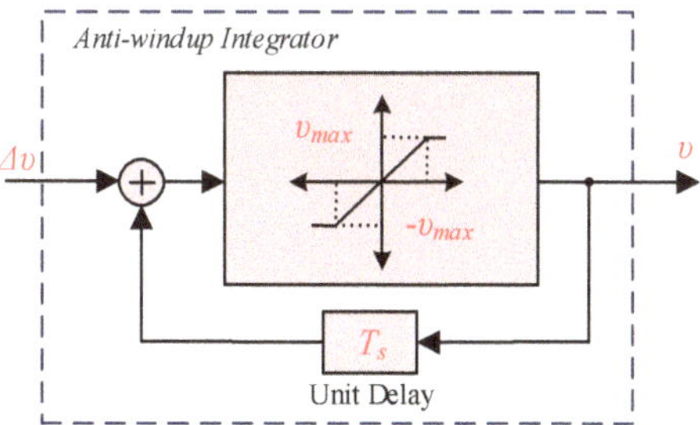

Figure 16. Thrust force of the robotic fish for free-swimming motion during 80 s.

6-DoF motion responses completely include the generalized linear and angular positions and speeds of the robotic fish. Positions (X, Y, Z) are determined with the earth-fixed frame. Roll, pitch and yaw angles (ϕ, θ, ψ) are also rigid body orientations with respect to the earth-fixed frame. In order to validate the accuracy of the designed model, simulations are examined in this section.

Figure 17a shows the closed loop forward speed responses of the robotic fish from 0.1 m/s to 0.4 m/s. These speed values are chosen as observed speed ranges of the carp. Figure 17b shows the orientation control that includes yaw angles of the robotic fish model during 40 s simulation time. The control performance of the closed loop system is summarized in Table 1.

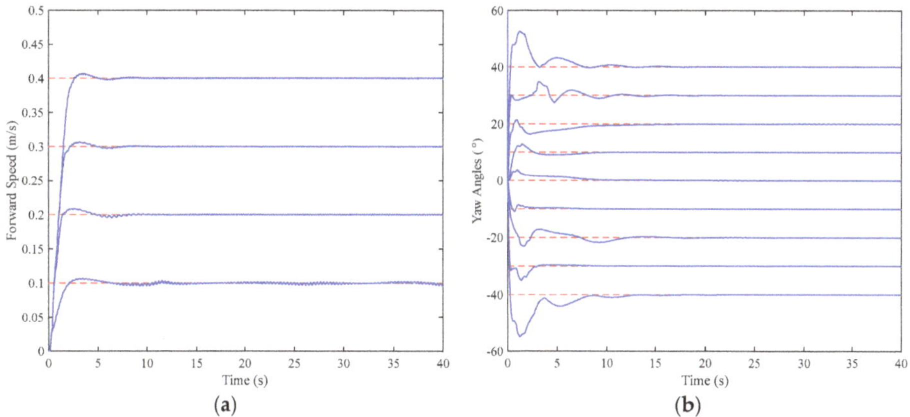

Figure 17. Closed loop responses of the 6-DoF system during 40 s simulation time: (**a**) Forward speed control; (**b**) Yaw angles tracking control.

Another key point is to realize pitch and roll motions together. In Figure 18, these motions are performed for step and stairs references. The closed loop control performance of the roll and pitch motions are illustrated in Table 2.

In order to validate the closed loop system, two different path trajectories are also generated in earth-fixed frame for 3D trajectory tracking. First path is a 3D sinusoidal path (P_1) reference and the second path is a 3D CCW circular path (P_2) reference. Trajectory tracking responses indicate the time-dependent performances of the robotic fish model. Figure 19 shows the closed loop P_1 tracking performance during a 60 s simulation time.

Amplitude and frequency of the P_1 reference are ± 0.18 m and 0.2 Hz, respectively. The depth of P_1 is 3 m. The robotic fish tracks the P_1 reference with a speed of 0.42 m/s. Figure 20 presents the P_1 tracking errors of speed and orientation angles.

Table 1. The closed loop control performance: Maximum overshoot (%) and settling time (s) of forward speed for (0.1 m/s)–(0.4 m/s) and yaw angles for (−40°)–(+40°).

	Interval	Maximum Overshoot (%)	Settling Time (s)
Forward Speed (m/s)	0.1	6.80	16.30
	0.2	4.40	11.80
	0.3	2.23	15.40
	0.4	1.65	12.20
Yaw Angles (°)	−40	37.50	18.30
	−30	16.66	8.80
	−20	15	16.65
	−10	10	8.20
	0	36	14.50
	10	27.50	10.80
	20	6.50	21.50
	30	15	17.60
	40	31.25	18.90

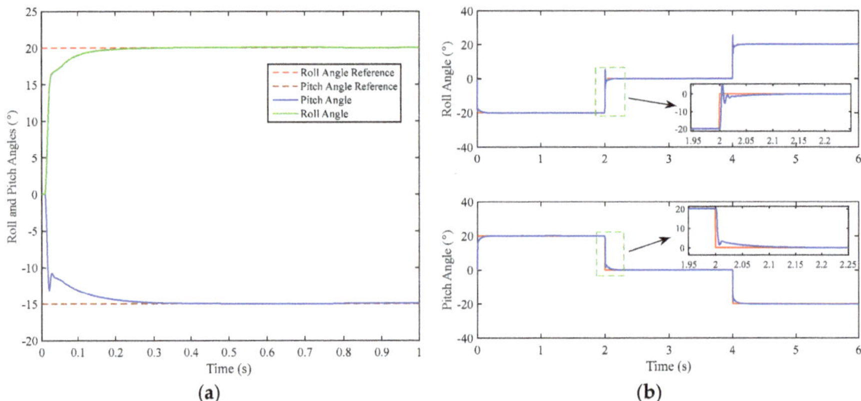

Figure 18. Roll and pitch motions together: (**a**) Step response; (**b**) Stairs response.

Table 2. The simultaneously closed loop control performance of the roll and pitch motions for step and stairs references.

Criteria	Step		Stairs					
Value (°)	20	−15	−20		0		20	
Motion	*Roll*	*Pitch*	*Roll*	*Pitch*	*Roll*	*Pitch*	*Roll*	*Pitch*
Maximum Overshoot (%)	-	-	-	-	27	-	27.50	-
Settling Time (s)	0.38	0.37	0.20	0.24	0.18	0.26	0.19	0.24

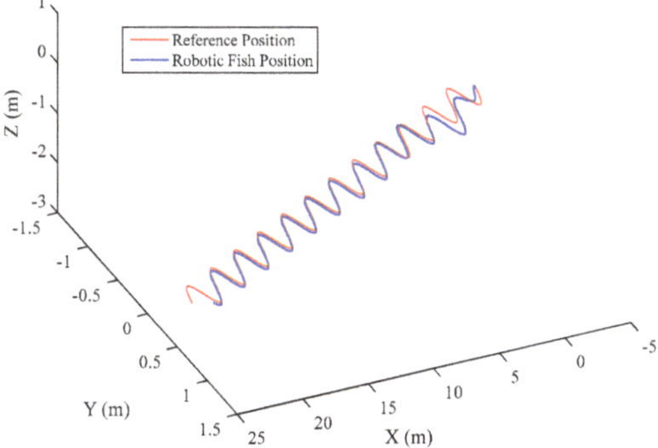

Figure 19. 3D P_1 tracking control performance.

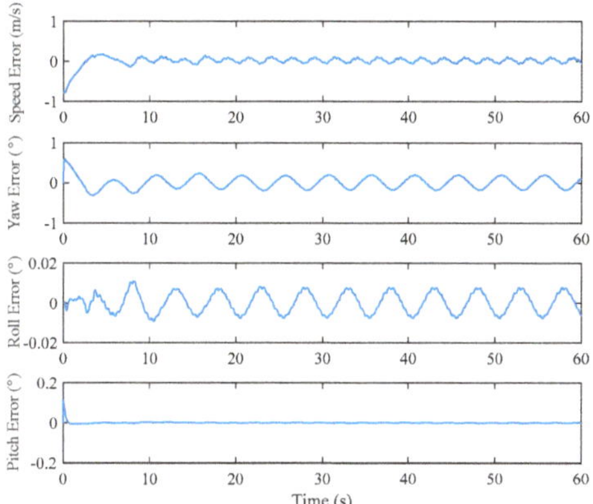

Figure 20. Errors for P_1 tracking.

6-DoF motion responses including positions, yaw, roll, pitch and also link angles are given in Figure 21. The robotic fish swings the flexible tail to keep the P_1 reference during closed loop tracking.

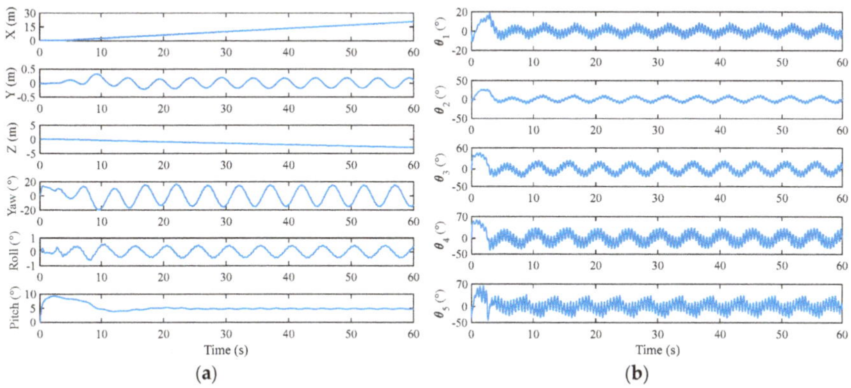

Figure 21. P_1 tracking responses: (**a**) 6-DoF responses; (**b**) Link angles.

Figure 22 shows the closed loop P_2 tracking performance during a 64 s simulation time. Diameter of the P_2 is 8 m and depth is 1.28 m. Initial positions of the robotic fish and P_2 reference begin the origin at X = 0 m, Y = 0 m, Z = 0 m.

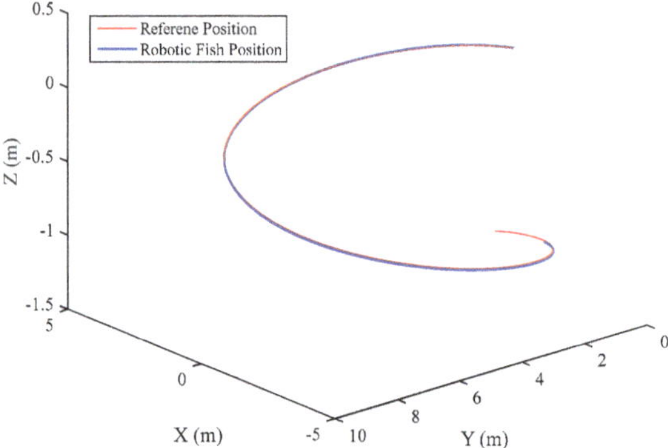

Figure 22. 3D P_2 tracking control performance.

The robotic fish turns around the CCW direction to track the P_2 reference with a speed of 0.4 m/s. Figure 23 also presents the P_2 tracking errors of speed and orientation angles.

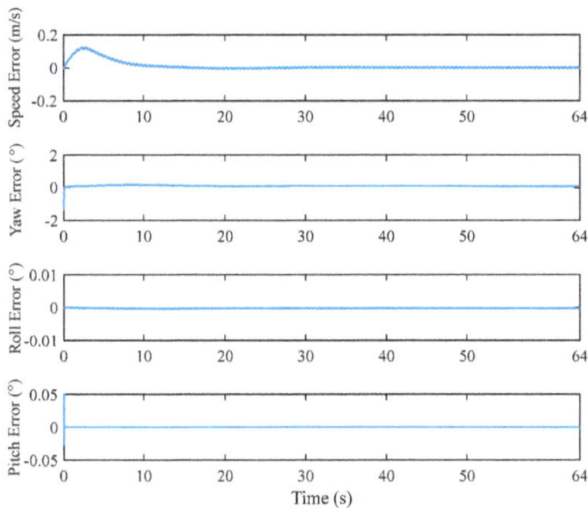

Figure 23. Errors for P_2 tracking.

6-DoF motion responses are given in Figure 24. When the robotic fish reaches the steady state, link angles become stable and generate small amplitudes.

As shown in Table 3, the tracking errors of P_1 and P_2 references on the x, y and z axis are evaluated by Root Mean Square Error (RMSE) performance index and given as:

$$RMSE = \sqrt{\frac{\sum_{k=1}^{n} e(k)^2}{n}} \tag{31}$$

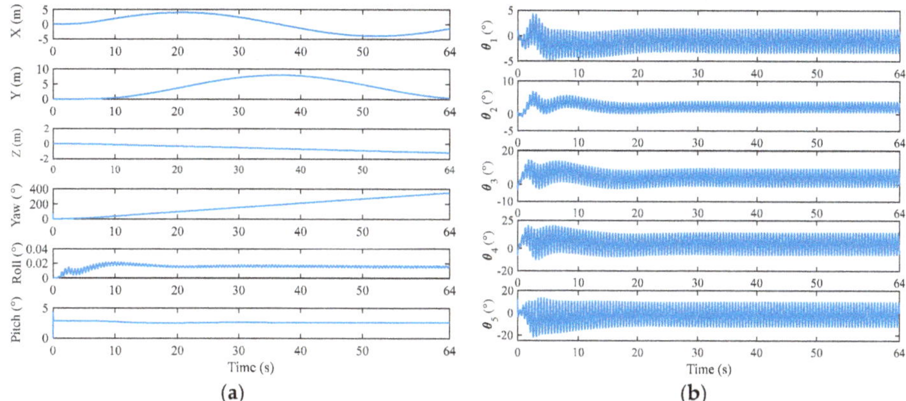

Figure 24. P_2 tracking responses: (**a**) 6-DOF responses; (**b**) Link angles.

Table 3. The tracking errors for proposed closed loop control system: P_1 and P_2 indicate the generated 3D path trajectories.

Trajectory	e_x (m)	e_y (m)	e_z (m)
P_1	0.0968	0.0131	0.1058
P_2	0.0827	0.0107	0.0920

In order to visualize the performance of the proposed closed loop control system in a virtual reality platform, the designed robotic fish model is also adopted to Virtual Reality Modeling Language (VRML). A simulation scenario is constructed consisting of a marine environment. The goal is to perform the P_2 tracking underwater. Figure 25 shows the top view above the sea, and the isometric view of the underwater environment is also shown in Figure 26.

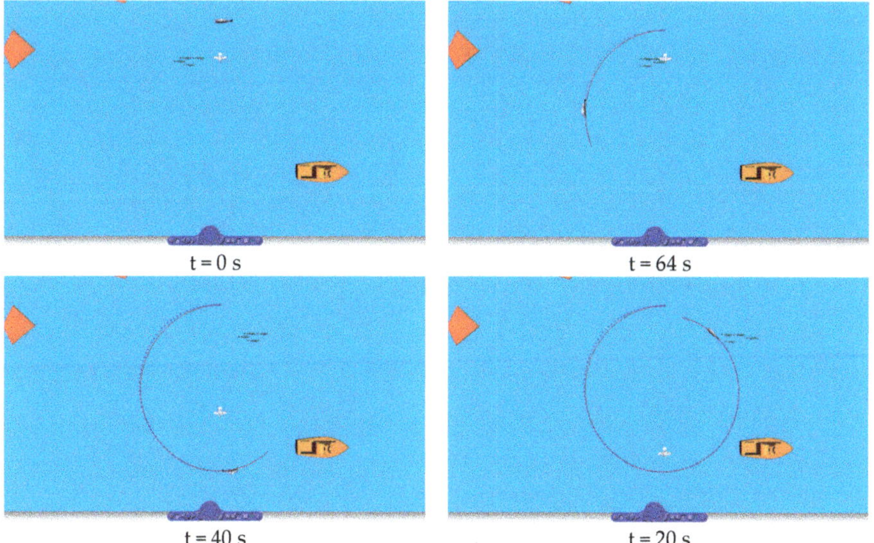

Figure 25. Top view above the sea for closed loop P_2 tracking simulation.

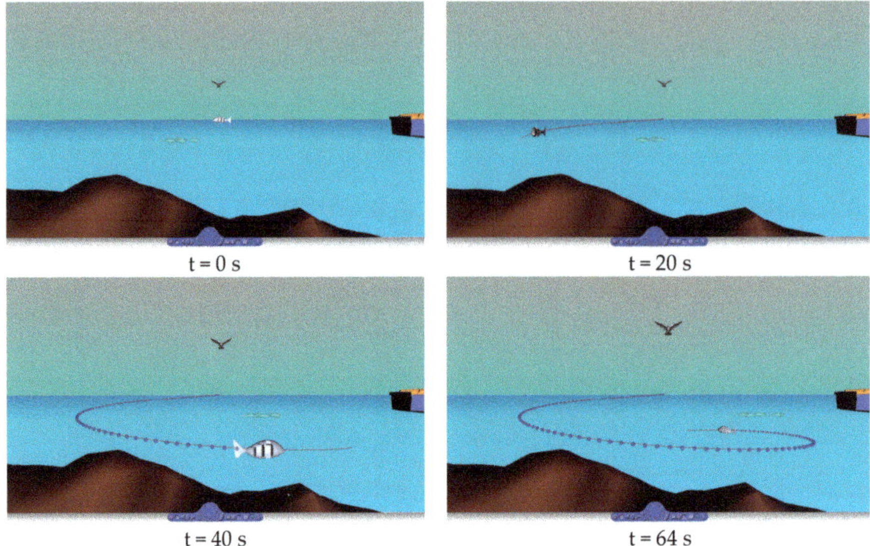

Figure 26. Isometric view of the underwater environment for closed loop P_2 tracking.

All of the simulation results show that the bio-inspired non-linear dynamic model of the robotic fish is suitable to mimic biological features of Carangiform carp. In this way, biomimetic modeling can be easily implemented to the robotic fish prototype in the future work.

4. Conclusions

This study presents a complete 3D non-linear dynamic model of a biomimetic robotic fish based on real carp locomotion. The designed model is derived by using the Lagrange formulation approach and includes both kinematic and hydrodynamic effects acting on each link. In order to achieve six degrees of freedom motions in the earth fixed frame, an up-down mechanism is designed that controls the influence of pitch and roll. This mechanism gives the position and orientation of the robotic fish in 3D space. Moreover, linear and rotational behavior responses are obtained and controlled in the simulations. In order to mimic biological features of fish with the dynamic model, various swimming patterns of a real carp are observed in the experimental pool. Link length ratio, flapping frequency, speed range, displacement and Euler angles are analyzed to determine appropriate desired inputs for the designed model. According to optimized real carp values, link lengths of the anterior body, active links and passive caudal fins are determined as 0.044 m, [0.043 m 0.040 m 0.010 m] and 0.063 m, respectively. Flapping frequency of carp is observed from 1.818 Hz to 2.857 Hz and it is also chosen as 2.857 Hz in the simulations. Average speed of a carp ranges from 0 to 2.5 m/s in the ichthyology. In the experimental studies, the speed range of forward swimming is recorded from 0.105 m/s to 0.41 m/s and turning speed is obtained as 0.24 m/s, approximately. It is also seen that the simulation speed values remained in these ranges. The pitch angle of the descending pattern of the carp is measured as 35° during 1.05 s. According to simulation results, the descending motion of the robotic fish is performed with 0.24 s settling time for 20° pitch angle, which has very similar performance to carp.

Forward and turning speeds, descending-ascending motions and yaw, pitch, roll angles are also controlled to achieve real fish behaviors. Figures 11, 17 and 18 show the maneuverability of the fish-like motions. Similarly, two different reference trajectories (P_1 and P_2) are tracked to evaluate sufficient swimming performances in Figures 19 and 22. The proposed closed loop control system is also adapted to VRML in order to validate the P_2 tracking in virtual marine environment. These simulation and experimental analyses show that the derived model achieves real carp motions to implement robotic fish prototypes for biomimetic design.

In future work, these experiences should be transferred to robotic fish prototypes in experimental platforms and path-tracking performance should be examined with various intelligent control algorithms.

Acknowledgments: This research was supported by the 114E652 TUBITAK 1001 project. We thank because of the financial support and guiding reports.

Author Contributions: Gonca Ozmen Koca, Cafer Bal, Deniz Korkmaz, Mustafa Can Bingol, Mustafa Ay, Zuhtu Hakan Akpolat and Seda Yetkin conceived all ideas and worked together to achieve this work.

Conflicts of Interest: The authors declare no conflict of interest.

References

1. Yu, J.; Wang, M.; Su, Z.; Tan, M.; Zhang, J. Dynamic modeling of a CPG-governed multijoint robotic fish. *Adv. Robot.* **2013**, *27*, 275–285. [CrossRef]
2. Ding, R.; Yu, J.; Yang, Q.; Tan, M.; Zhang, J. Dolphin-like swimming modeling for a biomimetic amphibious robot. *IFAC Proc. Vol.* **2011**, *18*, 9367–9372. [CrossRef]
3. Zhou, C.; Cao, Z.; Wang, S.; Tan, M. A marsupial robotic fish team: Design, motion and cooperation. *Sci. China Technol. Sci.* **2010**, *53*, 2896–2904. [CrossRef]
4. Lauder, G.V.; Madden, P.G.A. Learning from fish: Kinematics and experimental hydrodynamics for roboticists. *Int. J. Autom. Comput.* **2006**, *3*, 325–335. [CrossRef]
5. Zhang, F.; Tan, X. Three-Dimensional Spiral Tracking Control for Gliding Robotic Fish. In Proceedings of the 2014 IEEE 53rd Annual Conference on Decision and Control (CDC), Los Angeles, CA, USA, 15–17 December 2014; pp. 5340–5345.
6. Yang, G.H.; Choi, W.; Lee, S.H.; Kim, K.S.; Lee, H.J.; Choi, H.S.; Ryuh, Y.S. Control and Design of a 3 DOF Fish Robot 'ICHTUS'. In Proceedings of the 2011 IEEE International Conference on Robotics and Biomimetics (ROBIO), Karon Beach, Phuket, Thailand, 7–11 December 2011; pp. 2108–2113.
7. Zhou, C.; Tan, M.; Cao, Z.; Wang, S.; Creighton, D.; Gu, N.; Nahavandi, S. Kinematic Modeling of a Bio-Inspired Robotic Fish. In Proceedings of the IEEE International Conference on Robotics and Automation, Pasadena, CA, USA, 19–23 May 2008; pp. 695–699.
8. Suebsaiprom, P.; Lin, C.L. Fish-Tail Modeling for Fish Robot. In Proceedings of the 2012 International Symposium on Computer, Consumer and Control (IS3C), Taichung, Taiwan, 4–6 June 2012; pp. 548–551.
9. Ryuh, Y.S.; Yang, G.H.; Liu, J.; Hu, H. A school of robotic fish for mariculture monitoring in the sea coast. *J. Bionic Eng.* **2015**, *12*, 37–46. [CrossRef]
10. Masoomi, S.F.; Gutschmidt, S.; Chen, X.; Sellier, M. The Kinematics and Dynamics of Undulatory Motion of a Tuna-mimetic Robot. *Int. J. Adv. Robot. Syst.* **2015**, *12*, 83. [CrossRef]
11. Liu, J.; Hu, H. Biological Inspiration: From Carangiform Fish to Multi-Joint Robotic Fish. *J. Bionic Eng.* **2010**, *7*, 35–48. [CrossRef]
12. Niu, X.; Xu, J.; Ren, Q.; Wang, Q. Locomotion learning for an anguilliform robotic fish using central pattern generator approach. *IEEE Trans. Ind. Electron.* **2014**, *61*, 4780–4787. [CrossRef]
13. Li, G.; Deng, Y.; Osen, O.L.; Bi, S.; Zhang, H. A Bio-Inspired Swimming Robot for Marine Aquaculture Applications: From Concept-Design to Simulation. In Proceedings of the Oceans 2016—Shanghai, Shanghai, China, 10–13 April 2016; pp. 1–7.
14. Sfakiotakis, M. Review of Fish Swimming Modes for Aquatic Locomotion. *IEEE J. Ocean. Eng.* **1998**, *24*, 237–252. [CrossRef]
15. Suebsaiprom, P.; Lin, C.L. Maneuverability modeling and trajectory tracking for fish robot. *Control Eng. Pract.* **2015**, *45*, 22–36. [CrossRef]
16. Chowdhury, A.R.; Prasad, B.; Kumar, V.; Kumar, R.; Panda, S.K. Design, Modeling and Open-Loop Control of a BCF Mode Bio-Mimetic Robotic Fish. In Proceedings of the 2011 International Siberian Conference on Control and Communications (SIBCON), Krasnoyarsk, Russia, 15–16 September 2011; pp. 226–231.
17. Hu, Y.H.; Zhao, W.; Wang, L. Vision-Based Target Tracking and Collision Avoidance for Two Autonomous Robotic Fish. *IEEE Trans. Ind. Electron.* **2009**, *56*, 1401–1410.

18. Yu, J.; Wang, S.; Tan, M. A simplified propulsive model of bio-mimetic robot fish and its realization. *Robotica* **2005**, *23*, 101–107. [CrossRef]
19. Yu, J.Z.; Liu, L.Z.; Tan, M. Three-dimensional dynamic modelling of robotic fish: simulations and experiments. *Trans. Inst. Meas. Control* **2008**, *30*, 239–258.
20. Wang, W.; Xie, G.; Shi, H. Dynamic Modeling of an Ostraciiform Robotic Fish Based on Angle of Attack Theory. In Proceedings of the 2014 International Joint Conference on Neural Networks (IJCNN), Beijing, China, 6–11 July 2014; pp. 3944–3949.
21. Li, G.; Zhang, H.; Zhang, J.; Bye, R.T. Development of adaptive locomotion of a caterpillar-like robot based on a sensory feedback CPG model. *Adv. Robot.* **2014**, *28*, 389–401. [CrossRef]
22. Li, G.; Zhang, H.; Li, W.; Hildre, H.P.; Zhang, J. Design of Neural Circuit for Sidewinding of Snake-Like Robots. In Proceedings of the 2014 IEEE International Conference on Information and Automation (ICIA), Hailar, China, 28–30 July 2014; pp. 333–338.
23. Ijspeert, A.J. Central pattern generators for locomotion control in animals and robots: A review. *Neural Netw.* **2008**, *21*, 642–653. [CrossRef] [PubMed]
24. Liu, J.; Hu, H. A 3D simulator for autonomous robotic fish. *Int. J. Autom. Comput.* **2004**, *1*, 42–50. [CrossRef]
25. Yu, J.Y.J.; Liu, L.L.L.; Wang, L.W.L. Dynamic Modeling of Robotic Fish Using Schiehien's Method. In Proceedings of the 2006 IEEE International Conference on Robotics and Biomimetics, Kunming, China, 17–20 December 2006; pp. 457–462.
26. Liu, L.; Yu, J.; Wang, L. Dynamic Modeling of Three-Dimensional Swimming for Biomimetic Robotic Fish. In Proceedings of the IEEE International Conference on Intelligent Robots and Systems, Beijing, China, 9–15 October 2006; pp. 3916–3921.
27. Zhou, C.; Cao, Z.; Wang, S.; Tan, M. The Dynamic Analysis of the Backward Swimming Mode for Biomimetic Carangiform Robotic Fish. In Proceedings of the IROS 2008 IEEE/RSJ International Conference on Intelligent Robots and Systems, Nice, France, 22–26 September 2008; pp. 3072–3076.
28. Nakashima, M.; Ohgishi, N.; Ono, K. A study on the propulsive mechanism of a double jointed fish robot utilizing self-excitation control. *JSME Int. J. Ser. C Mech. Syst. Mach. Elem. Manuf.* **2003**, *46*, 982–990. [CrossRef]
29. Kim, H.; Lee, B.; Kim, R. A Study on the motion mechanism of articulated fish robot. In Proceedings of the 2007 IEEE International Conference on Mechatronics and Automation (ICMA), Harbin, China, 5–8 August 2007; pp. 485–490.
30. Suebsaiprom, P.; Lin, C.L. 2-DOF Barycenter Mechanism for Stabilization of Fish-Robots. In Proceedings of the 2013 IEEE 8th Conference on Industrial Electronics and Applications (ICIEA), Melbourne, VIC, Australia, 19–21 June 2013; pp. 1119–1122.
31. Yu, J.; Wang, L.; Tan, M. Geometric optimization of relative link lengths for biomimetic robotic fish. *IEEE Trans. Robot.* **2007**, *23*, 382–386. [CrossRef]
32. Bal, C.; Korkmaz, D.; Koca, G.O.; Ay, M.; Akpolat, Z.H. Link Length Optimization of A Biomimetic Robotic Fish Based on Big Bang—Big Crunch Algorithm. In Proceedings of the 21st International Conference on Methods and Models in Automation and Robotics (MMAR), Miedzyzdroje, Poland, 29 August–1 September 2016; pp. 189–193.
33. Yu, J.; Tan, M.; Chen, J.; Zhang, J. A survey on CPG-inspired control models and system implementation. *IEEE Trans. Neural Netw. Learn. Syst.* **2014**, *25*, 441–456. [PubMed]
34. Ozmen Koca, G.; Korkmaz, D.; Bal, C.; Akpolat, Z.H.; Ay, M. Implementations of the route planning scenarios for the autonomous robotic fish with the optimized propulsion mechanism. *Measurement* **2016**, *93*, 232–242. [CrossRef]
35. Vo, T.Q.; Kim, H.S.; Lee, B.R. Smooth gait optimization of a fish robot using the genetic-hill climbing algorithm. *Robotica* **2012**, *30*, 257–278. [CrossRef]
36. Gao, W.B.; Wang, Y.F.; Homaifa, A. Discrete-Time Variable-Structure Control-Systems. *IEEE Trans. Ind. Electron.* **1995**, *42*, 117–122.

37. Akpolat, Z.; Gokbulut, M. Discrete time adaptive reaching law speed control of electrical drives. *Electr. Eng.* **2003**, *85*, 53–58. [CrossRef]
38. Akpolat, Z.H.; Guldemir, H. Trajectory following sliding mode control of induction motors. *Electr. Eng.* **2003**, *85*, 205–209. [CrossRef]

Article

Hybrid Locomotion Evaluation for a Novel Amphibious Spherical Robot

Huiming Xing [1], Shuxiang Guo [1,2,*], Liwei Shi [1,*], Yanlin He [1], Shuxiang Su [1], Zhan Chen [1] and Xihuan Hou [1]

[1] Key Laboratory of Convergence Medical Engineering System and Healthcare Technology, the Ministry of Industry and Information Technology, School of Life Science, Beijing Institute of Technology, Beijing 100081, China; xinghuiming@bit.edu.cn (H.X.); heyanlin@bit.edu.cn (Y.H.); sushuxiang@bit.edu.cn (S.S.); chenzhan@bit.edu.cn (Z.C.); houxihuan@bit.edu.cn (X.H.)

[2] Faculty of Engineering, Kagawa University, 2217-20 Hayashicho, Takamatsu, Kagawa 761-0396, Japan

* Correspondence: guoshuxiang@bit.edu.cn (S.G.); shiliwei@bit.edu.cn (L.S.); Tel.: +86-159-1102-2732 (S.G.); +86-10-6891-5908 (L.S.)

Received: 14 November 2017; Accepted: 10 January 2018; Published: 24 January 2018

Abstract: We describe the novel, multiply gaited, vectored water-jet, hybrid locomotion-capable, amphibious spherical robot III (termed ASR-III) featuring a wheel-legged, water-jet composite driving system incorporating a lifting and supporting wheel mechanism (LSWM) and mechanical legs with a water-jet thruster. The LSWM allows the ASR-III to support the body and slide flexibly on smooth (flat) terrain. The composite driving system facilitates two on-land locomotion modes (sliding and walking) and underwater locomotion mode with vectored thrusters, improving adaptability to the amphibious environment. Sliding locomotion improves the stability and maneuverability of ASR-III on smooth flat terrain, whereas walking locomotion allows ASR-III to conquer rough terrain. We used both forward and reverse kinematic models to evaluate the walking and sliding gait efficiency. The robot can also realize underwater locomotion with four vectored water-jet thrusters, and is capable of forward motion, heading angle control and depth control. We evaluated LSWM efficiency and the sliding velocities associated with varying extensions of the LSWM. To explore gait stability and mobility, we performed on-land experiments on smooth flat terrain to define the optimal stride length and frequency. We also evaluated the efficacy of waypoint tracking when the sliding gait was employed, using a closed-loop proportional-integral-derivative (PID) control mechanism. Moreover, experiments of forward locomotion, heading angle control and depth control were conducted to verify the underwater performance of ASR-III. Comparison of the previous robot and ASR-III demonstrated the ASR-III had better amphibious motion performance.

Keywords: the amphibious spherical robot; the lifting and supporting wheel mechanism; sliding locomotion; wheel-legged robot; waypoints tracking; hybrid locomotion

---

1. Introduction

Amphibious robots are currently under development. Such robots can both walk on rough terrain and swim, and are reliable, stable, able to work rapidly and carry large loads, and display intelligence. These robots have been widely and successfully used for monitoring, recovery, detection of pollution, submarine sampling, data collection, video-mapping, vision perception [1], exploration of unstructured amphibious environments, object recovery under challenging circumstances, and other tasks [2–5].

Recently, many robotic designs [6–8] have been inspired by biology, especially amphibious robots differing in terms of propulsion method and shape. Inspired by the crab and the lobster, Jun et al.

developed a multi-legged underwater-walking robot (CR200) [9,10] to survey underwater structures and shipwrecks off the coast of the Korean Peninsula; the robot is slow because it features legged propulsion. In 2013, Crespi designed a bio-amphibious salamander robot (Salamandra robotica II) [11] that both walked and swam; the robot has four legs and an active spine allowing anguilliform swimming in water and on-ground walking with various gaits, featuring body-limb coordination and bodily undulation. Earlier, in 2006, Crespi et al. created an amphibious snake robot termed AmphiBot II [12] that crawled and swam using a central-pattern generator. In 2009, Tang developed a wheel-propeller-leg-integrated amphibious robot [13] that crawled on land, swam underwater (at certain depths), and crept on the ocean floor. In 2016, Zhang developed the amphibious robot AmphiHex-I [14] which walked on rough terrain, maneuvered underwater, and used specially designed transformable flipper legs to traverse soft muddy or sandy substrates in littoral areas between land and water. All of these amphibious robots have particular characteristics and advantages. In terms of on-land locomotion, legged robots [15–18] deal better with uneven terrain, but are slow, whereas wheeled robots [19,20] move rapidly but cope poorly with rough ground; serpentined robots [21,22] (inspired by snakes) work well on flat terrain but it is difficult to control the direction and speed of motion. In water, robots with screw propellers are more stable and mobile than those with oscillatory and undulatory capacities.

Earlier, inspired by amphibious turtles, the amphibious spherical robot I (ASR-I) [23,24] that could move both on land and underwater was constructed. The ASR-I diameter (including its four legs) was only 25 cm. To improve stability while retaining precision, we built the ASR-II [25–28] using 3-D printing. This robot could walk at up to 8.5 cm/s (0.34 the body length) and climb a slope of 5°. To improve locomotion on land, Li added a propulsion mechanism featuring active wheels [29]. The direct current (DC) motors driving the wheels increased the speed. An active wheel-legged robot [30] with 6 legs was developed to get better actions in unknown environment with reinforcement learning. The robot used values of external force measured on the robot's legs as states and rewards, and it can adapt to different terrains in real time using on-line learning. However, propulsion system redundancy was evident, and the active wheels (with DC motors) increased the size of the robot, causing problems during walking.

The passive wheel-legged robot named Roller-walker [31–34] with two locomotion modes was early proposed. Roller-walker can transform into sole mode by rotating the ankle roll joint with the passive wheel on each leg. Using the transformation mechanism, the robot can switch between the quadruped walking and roller-skating on smooth flat terrain. The transformation of the actuation modes can be realized by one degree of freedom of each leg. The mechanical legs were redesigned to feature four passive wheels [35], allowing the amphibious robot to move rapidly on smooth flat ground. The robot exhibited both roller-skating and walking locomotion. Compared with active-wheeled robots, passive wheel-legged robots are of lower weight, consume less energy, and are smaller. However, directional control of passive wheel-legged robots is difficult; such robots swing while walking and skating.

Therefore, to enhance the space available for sensors, and to improve on-land stability and velocity, we developed a sliding mode-based, amphibious spherical robot employing a wheel-legged, water-jet, composite driving mechanism (including a lifting and supporting wheel mechanism (LSWM)) and mechanical legs with water-jet thruster. To allow the robot to readily move in any direction, four omni-directional passive wheels were added to the LSWM mobile platform. The LSWM lifts and drops the mobile platform. The robot undergoes sliding locomotion when the platform is lowered, and walking locomotion when the platform is raised. Therefore, the robot combines the advantages of locomotion afforded by legs and wheels when walking on rough terrain, and sliding locomotion on flat ground or slopes. We conducted a waypoint tracking experiment using a vision localization system to evaluate stability, velocity, and directional control. Finally, the underwater locomotion experiments with four vectored water-jet thrusters was conducted to evaluate the performance of underwater locomotion.

The remainder of this paper is organized as follows: In Section 2, we describe our advanced design of such a robot, the wheel-legged water-jet composite driving mechanism, the new LSWM assisting the robot to stand or slide, and the electrical system delivering either strong or weak power. Section 3 presents the forward and inverse kinematic models of the mechanical legs, the novel sliding and walking locomotors gait based on the LSWM, the underwater locomotion. In Section 4, we describe experiments verifying the LSWM, the sliding and walking gaits using the LSWM, and the comparison of two on-land locomotion on different terrain. A waypoint tracking experiment evaluates the reliability and feasibility of the robot in real work using a closed-loop proportional-integral-derivative (PID) control mechanism. We conducted forward motion experiment and heading angle control experiment in Section 5. Sections 6 and 7 conclude the paper.

2. The Amphibious Spherical Robot (ASR) III

2.1. The Mechanism

To create more space for sensors, and to improve stability and velocity on land, we designed an on-land, multiple-locomotion-based, amphibious spherical robot. As shown in Figure 1, the robot has a hemispherical waterproof hull, two quarter-spherical hulls, a central plant, four mechanical legs, a mechanism for lifting and supporting the wheels, an electrical circuit, and five batteries. The thickness of the hemispherical hull, which is made of Acrylonitrile Butadiene Styrene (ABS) plastic, is about 6 mm, adequate to allow the robot to operate up to 10 m underwater. The top of the hull has a waterproof plug; when this is removed, the control system of the robot can be connected to a remote computer via an optical fiber cable, allowing software debugging. An O-ring is placed between the hemispherical hull and the central plant to ensure waterproofing; again, the manual indicates that the robot can dive to a maximum of 10 m. The actuator is a wheel-legged composite mechanism including the LSWM and mechanical legs. In the water, the two quarter-spherical hulls close up like a ball and the robot moves using four vectored water-jet thrusters. On land, the two hulls open and the robot can slide or walk using the wheel-legged composite driving mechanism. To deal with the extra weight, five batteries, one of which contains 6000 mAh, are used to extend the working time.

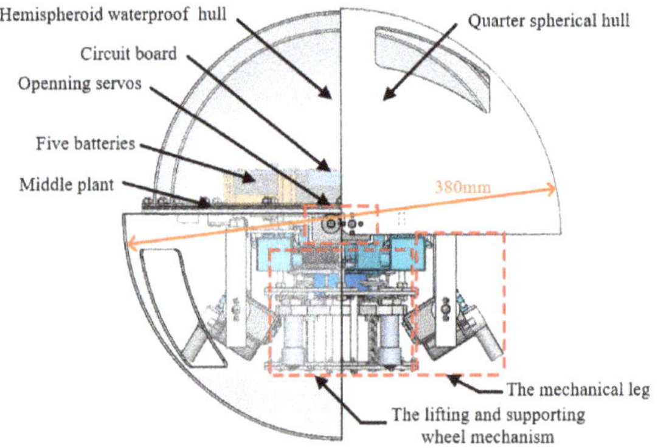

Figure 1. The novel amphibious spherical robot III (ASR-III).

2.2. The Wheel Legged Water-Jet Composite Driving Mechanism

As more multifunctional sensors are required when a robot increases in size and weight, the actuation mechanism becomes challenging. The robot diameter was 35 cm following a redesign to include more sensors. Buoyancy calculations indicated that the weight needed to be 11.22 kg before the robot sank in water. When standing, the four mechanical legs must support the robot and the numerous sensors; there is a danger that a sudden current surge might burn out the electronic circuit. Thus, we developed a wheel-legged and water-jet composite driving mechanism including the LSWM (Figure 2a) and mechanical legs (Figure 2b). The LSWM assists the robot to stand and perform on-land locomotion that is rapid on smooth, flat terrain. The mechanism reduces wear on the actuating mechanism (the steering gear, etc.), extending the working life of the robot. Also, the robot can slide down slopes, which saves a great deal of power.

The LSWM is shown in Figure 2a. At the top, two servos can rotate through 360 degrees. The mechanism includes an upper support platform with four top beams and a lower support platform with four lower beams. A mobile platform with four supporting legs allows automatic transition between the sliding and walking modes. The ends of the legs bear four omni-directional wheels, allowing the robot to slide in any direction on flat, smooth terrain. As shown in Figure 2b, the mechanism also includes a mechanical leg with two joints, two servos, two omni-directional wheels, an upper holder, a lower holder, and a water-jet thruster. In previous work [36], the thruster was just used to spray water in one direction. Now, by the LM298N, a motor driving module, the thruster can spray the water in two directions. Therefore, for the forward motion and rotary motion in water, four water-jet thrusters all can be used simultaneously. The details are described in Section 3.

When the robot moves from rough to smooth terrain, the two servos rotate in the same direction, simultaneously and synchronously, and the mobile platform drops vertically along the guide bar. Then, using the four mechanical legs, the robot switches to sliding locomotion. When the robot walks on rough terrain, the two servos rotate in the opposite direction simultaneously and synchronously, lifting the mobile vertically.

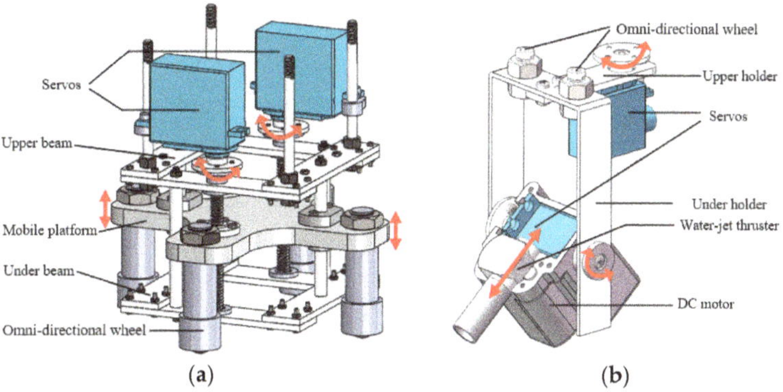

Figure 2. The wheel-legged, water-jet composite driving mechanism. (**a**) The lifting and supporting wheel mechanism; (**b**) The mechanical leg of ASR-III.

The frequency of the control signal is 50 Hz, and the duty factor S_{PWM} of the pulse width modulation (PWM) signal can be set to obtain the desired angular velocity w; the relationship is described by Equation (1).

$$w = kS_{PWM} \tag{1}$$

where, k is the scaling factor between the angular velocity and the duty factor of PWM signal.

The LSWM is modeled using Equation (2).

$$H = \Delta l \times \frac{n \times 2\pi}{kS_{PWM}} \tag{2}$$

where H is the distance through which the support platform travels, Δl is the length travelled as the servo rotates once, n is the number of rotations, S_{PWM} is the PWM signal value, and k is as described above.

2.3. The Electrical Circuit

To improve the stability and logic of the circuit (Figure 3), we divided it into four parts: the power supply unit (PSU), the decision center, and sensor and driver layers. A stable, powerful, electrically isolated PSU is essential. When the robot begins to stand, eight servos require electricity simultaneously, triggering a surge current that challenges the controller and sensors. Hence, we used five 8.4-V batteries, two of which (Battery II) power the decision center and the sensor layer, and three of which (Battery I) power the controls and the driver layer. To isolate the control and dynamic electricity, they are not common-grounded; the control system outputs PWM signals to the motors of the dynamic system via optocouplers that transform photoelectric signals. We used a Zynq-7000 SoC system as the "brain", to offer a platform which has the computational capabilities to realize the intelligence and function of the robot. When exploring and sensing amphibiously, the robot uses many sensors, such as a Global Positioning System (GPS), an inertial navigation system (INS), cameras, pressure sensors, and a communication modem; all signals are sent to the decision center. In this paper, the JY901 IMU module with high-precision is a key sensor giving the angle feedback for motion control, and installed in the center of the PCB board. IIC bus is used to produce data and make it available to the Zynq-7000 SoC system.

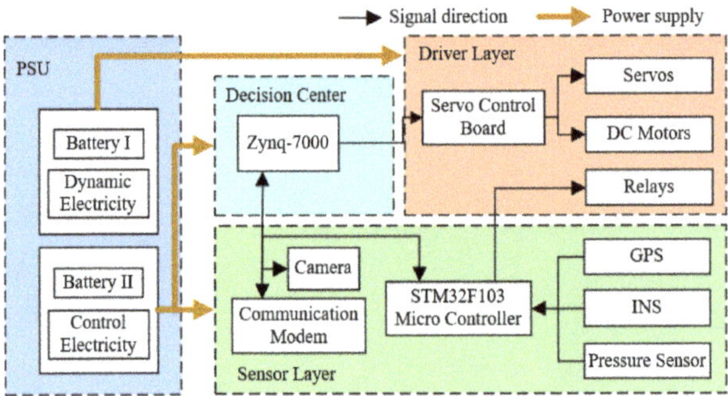

Figure 3. The circuit system of ASR-III. DC: direct current; GPS: Global Positioning System; INS: inertial navigation system.

3. The Hybrid Locomotion

A multiply gaited, vectored water-jet, hybrid locomotion was designed for ASR-III to improve the performance in amphibious environment. Different gaits on land afford distinct advantages in terms of speed, stability and energy consumption; specific substrates require different gaits. For example, walking on mechanical legs with multiple degrees of freedom is optimal on rough terrain. However, on smooth flat terrain and slopes, sliding locomotion for ASR-III is much better in terms of velocity, energy efficiency and stability. Besides on-land locomotion, underwater locomotion is actuated by four vectored water-jet thrusters, and the robot is capable of forward motion, heading angle control, and depth control. Therefore, a multi-locomotion capacity greatly improves robotic performance and practicability in amphibious environment.

3.1. Forward Kinematic Model of ASR-III

To describe the locomotion of the leg-wheeled robot simply, we term the four legs LF, LH, RH, and RF (the left foreleg, left hind leg, right hind leg, and right foreleg, respectively; Figure 4). The Figure 4 shows the coordinate system $\{O_B\}$ of the trunk and assumes that the coordinate lies in the geometrical center of the trunk. The X_B axis represents the forward direction, the Z_B axis the vertical direction (perpendicular to the trunk's horizontal plane), and the Y_B axis the direction to the right of the trunk. Figure 4 shows the coordinates of the legs $\{O_0^i\}$ where i represents a leg number ($i = 1$ for LF, $i = 2$

for LH, $i = 3$ for RH, and $i = 4$ for RF). The coordinates of each leg are similar; we describe only the coordinates of the left foreleg in Figure 5. Each leg has a hip joint, a knee joint, and a toe, corresponding to the coordinates $\{O_1\}$, $\{O_2\}$, and $\{O_3\}$, respectively.

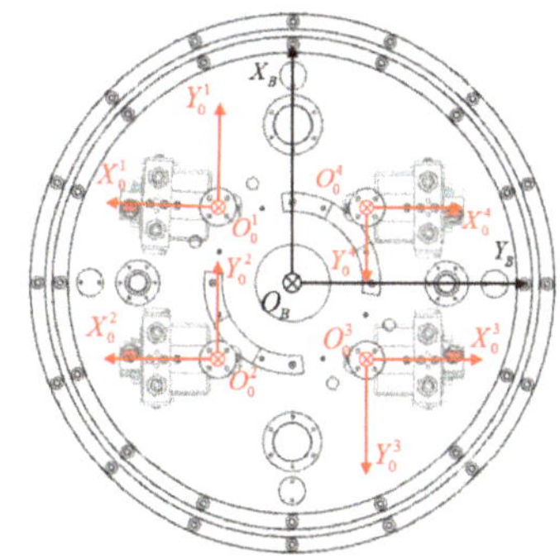

Figure 4. The coordinates of the robot and the legs (top view).

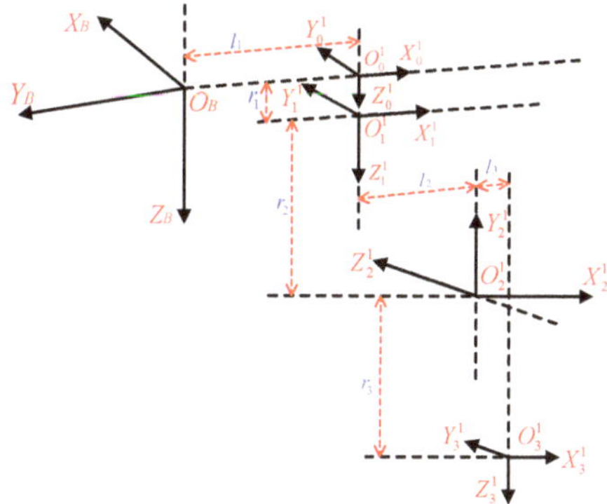

Figure 5. Kinematics model of the left foreleg of ASR-III.

For on-land locomotion of the leg-wheeled robot, it can be known that two parameters (the stride and frequency) influence the velocity of the robot. In order to get a high velocity, the stride length and frequency need to be increased. Considering the compact structure of ASR-III, the design using servo motors is a nice choice to the mechanical leg. Every leg has two degrees of freedom with two servo motors. The bottom and center of the middle plant is a good site to install the communication module. In order to hold the communication module and lower the barycenter of the robot, the height of the LSWM was decided to 12.1 cm. As shown in Figure 2a, the distance from the upper state to the lower state was 2.5 cm. With the increasing of the under holder, the robot gets a higher barycenter, which decreases the stability of ASR-III. Therefore, to obtain a lower barycenter and make the robot stand up at the lower state of LSWM, the length of the under holder shown in Figure 2b was decided to 9.5 cm. In our robot, the longer the upper holder, the longer the stride length. However, for keeping the spherical boundary, the longest length of the upper holder was 4 cm. The robot was designed using Solidworks 2017. In addition, avoiding interference of parts in assembly, all the parameters of the leg were optimized to get a higher velocity of sliding locomotion and walking locomotion, and a lower barycenter for ASR-III.

Table 1 lists all of the Denavit-Hartenberg (D-H) parameters of the left foreleg. θ_j^i and d_j^i are the joint angle and the distance between the joint $j-1$ and j of leg i; and α_{j-1}^i and a_{j-1}^i are the torsional angle and the length of the bar $j-1$ of leg i, respectively. The five principal physical parameters of the leg are shown in Table 1. The parameter l_1, the horizontal distance from the center of the plant to the hip joint, is 70 mm.

Table 1. The Denavit-Hartenberg (D-H) parameters of the left foreleg.

Joint j	ˋ^i_j	ff$^i_{j-1}$	ff$^i_{j-1}$ (mm)	d^i_j (mm)	ˋ^i_j
1	$\theta^1_1(0)$	0	0	$r_1(10)$	$(0, \pi/2)$
2	$\theta^1_2(0)$	0	$l_2(40)$	$r_2(95)$	$(-\pi/6, \pi/4)$
3	0	$\pi/2$	$l_3(20)$	$r_3(60)$	0

Using the D-H homogeneous transformation formula in the driven-axis context, we obtain Equation (3).

$$ {}^{B}\mathbf{T}^1_3 = \begin{bmatrix} s^1_1c^1_2 & c^1_1 & s^1_1s^1_2 & l_3s^1_1c^1_2 + r_3s^1_1s^1_2 + l_2s^1_1 + l_1 \\ -c^1_1c^1_2 & s^1_1 & -c^1_1s^1_2 & -l_3c^1_1c^1_2 - r_3c^1_1s^1_2 - l_2c^1_1 - l_1 \\ -s^1_2 & 0 & c^1_2 & -l_3s^1_2 + r_3c^1_2 + r_3 + r_1 \\ 0 & 0 & 0 & 1 \end{bmatrix} \quad (3) $$

where, $s^i_j = \sin\theta^i_j$ and $c^i_j = \cos\theta^i_j$.

The position of the toe of the left foreleg in terms of the trunk coordinates can be obtained as follows Equation (4).

$$ \mathbf{p}^1 = \begin{bmatrix} p^1_x \\ p^1_y \\ p^1_z \end{bmatrix} = \begin{bmatrix} l_3s^1_1c^1_2 + r_3s^1_1s^1_2 + l_2s^1_1 + l_1 \\ -l_3c^1_1c^1_2 - r_3c^1_1s^1_2 - l_2c^1_1 - l_1 \\ -l_3s^1_2 + r_3c^1_2 + r_3 + r_1 \end{bmatrix} \quad (4) $$

Similarly, the equations of the left hind leg, the right hind leg and the right foreleg also can be obtained by the same procedure of the left front leg. The forward kinematic model can be succinctly written as Equation (5).

$$ {}^{B}\mathbf{p}_{toe} = FK(\boldsymbol{\theta}) \quad (5) $$

where, FK represents the inverse kinematics, allowing mapping from the joint space to the Cartesian space.

3.2. Inverse Kinematic Model

The movement required to control the robot can be calculated using the inverse kinematic model. We derive inverse kinematic equations employing the forward kinematic model.

When the left forward toe is in position $\mathbf{p}^1_{toe} = [\ p^1_x \ \ p^1_y \ \ p^1_z\ ]^T$ of $\{O_B\}$, its position in other coordinate systems can be obtained by inverse transformation. The position in $\{O_1\}$ is expressed by Equation (6).

$$ \begin{aligned} {}^{1}\mathbf{T}^1_3 &= ({}^{0}\mathbf{T}^1_1)^{-1}({}^{B}\mathbf{T}^1_0)^{-1}{}^{B}\mathbf{T}^1_3 \\ &= \begin{bmatrix} c^1_1n^1_x + s^1_1n^1_y & c^1_1o^1_x + s^1_1o^1_y & c^1_1a^1_x + s^1_1a^1_y & s^1_1(p^1_x - l_1) - c^1_1(p^1_y + l_1) \\ -s^1_1n^1_x + c^1_1n^1_y & -s^1_1n^1_x + c^1_1n^1_y & -s^1_1n^1_x + c^1_1n^1_y & c^1_1(p^1_x - l_1) + s^1_1(p^1_y + l_1) \\ n^1_z & o^1_z & a^1_z & p^1_z - r_1 \\ 0 & 0 & 0 & 1 \end{bmatrix} \end{aligned} \quad (6) $$

Then, the position also can be obtained as Equation (7).

$$ {}^{1}\mathbf{T}^1_3 = {}^{1}\mathbf{T}^{1}_2{}^{2}\mathbf{T}^1_3 = \begin{bmatrix} c^1_2 & 0 & -s^1_2 & l_3c^1_2 + r_3s^1_2 + l_2 \\ 0 & -1 & 0 & 0 \\ -s^1_2 & 0 & -c^1_2 & -l_3s^1_2 + r_3c^1_2 + r_2 \\ 0 & 0 & 0 & 1 \end{bmatrix} \quad (7) $$

Employing Equations (6) and (7), we obtain Equation (8) containing two joint variables,

$$\begin{cases} s_1^1(p_y^1 + l_1) + c_1^1(p_x^1 - l_1) = 0 \\ -l_3 s_2^1 + r_3 c_2^1 + r_2 = p_z^1 - r_1 \end{cases} \tag{8}$$

Define $t = p_z^1 - r_2 - r_1$. Then, we obtain Equation (9).

$$\begin{cases} \theta_1^1 = \text{atan2}(l_1 - p_x^1, l_1 + p_y^1) \\ \theta_2^1 = \text{atan2}(r_3, l_3) - \text{atan2}(t, \pm\sqrt{r_3^2 + l_3^2 - t^2}) \end{cases} \tag{9}$$

Similarly, the equations of the left hind leg, the right hind leg and the right foreleg also can be obtained by the same procedure. The inverse kinematic model can be succinctly written as Equation (10).

$$\theta = IK(^B\mathbf{p}_{toe}) \tag{10}$$

where IK refers to the inverse kinematics, allowing mapping from the Cartesian space to the joint space.

3.3. Multiple On-Land Locomotion

Our earlier robot had a diameter of 20 cm and used four legs for support. The new robot is 35 cm in diameter (enlarged to carry more sensors), rendering it much heavier. It is challenging to support the body with four mechanical legs. Therefore, we designed the LSWM to allow the robot to slide on land like a turtle. We carefully observed how a turtle moved, and used this principle to design the gaits. To move flexibly and steadily, the turtle uses both symmetrical and asymmetrical gaits. The symmetrical gait is basically a trot (a two-beat gait) (Figure 6b) and the asymmetrical gait is basically a tripod walk (a four-beat gait) (Figure 6a); the synchronous gaits are shown in Figure 6c–e. When trotting, the four legs move in the sequence "LH, RF → LF, RH → LH, RF", as shown in Figure 7b. We set the duty factors during the transfer phase to 50%. The sequence of the tripod walk shown in Figure 7a is "LF → RH → RF → LH", and the duty factor is about 75%. During synchronous sliding with two legs (Figure 7c), the sequence is "LF, RF → LF, LH", and the duty factor is 50%. The turtle also performs synchronous sliding using four legs, as shown in Figure 6d, to achieve both forward and rotatory locomotion. In the real world, turtles often move from one point to another. When moving short distances using little energy, the turtle rotates with a zero radius if it wants to go back. Therefore, we implemented a rotary gait with zero radius as the directional control; the duty factor was 50%. Thus, like the turtle, the robot can turn left and right.

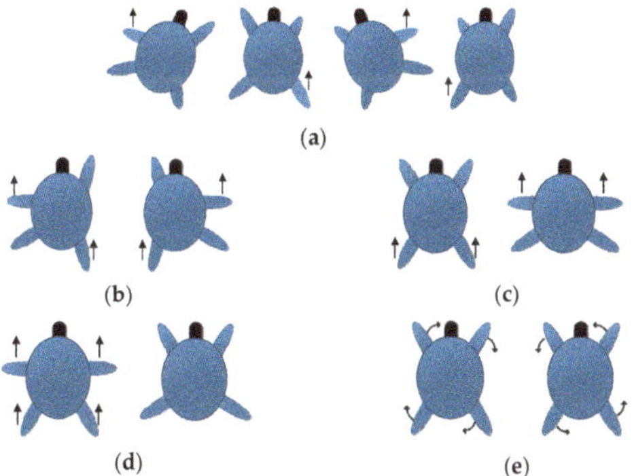

Figure 6. Turtle gaits (The black arrow indicates that the leg is in transfer phase). (**a**) The tripod walking gait (TPWG); (**b**) the trotting walking gait (TTWG); (**c**) the synchronous sliding gait with two legs (SSGTL); (**d**) the synchronous sliding gait with four legs; (**e**) the rotary sliding gait with zero radius.

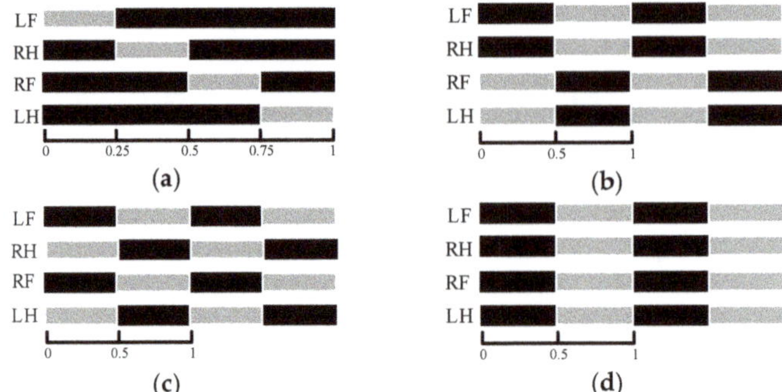

Figure 7. The sequence of the gaits. (**a**) TPWG; (**b**) TTWG; (**c**) SSGTL; (**d**) the synchronous sliding gait with four legs. The gray and the black bars show that the leg is in the transfer and support phase, respectively. Over single cycles of the trotting and synchronous gaits, the robot simultaneously has two legs in the transfer phase and two legs in the support phase. When in the synchronous sliding gait, the four legs of the robot swing synchronously with the same stride, generating different types of locomotion (rotatory or forward, based on different leg swing directions). Over a single cycle of TPWG, only one leg is in the transfer phase and, under the same conditions, the cycle time is twice that of the other three synchronous gaits.

3.4. Underwater Locomotion

Every water-jet thruster of four legs can rotate in the vertical and horizontal planes. With the directions and propulsive forces of four vectored water-jet thrusters changing, the robot can move forward or backward, turn left or right, float or sink in the underwater environment. Figure 8 illustrates the underwater locomotion of the robot.

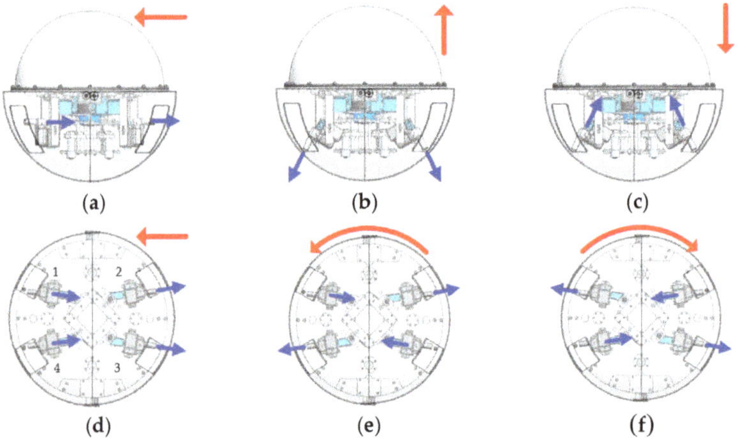

Figure 8. Underwater locomotion. Side view: (**a**) forward motion; (**b**) floating motion; (**c**) sinking motion. Bottom view: (**d**) forward motion; (**e**) left turning motion; (**f**) right turning motion. The blue arrow indicates the water-jet direction of the thruster, and the red arrow represents the motion direction of the robot.

The four legs are denoted by number 1, 2, 3 and 4, as shown in Figure 8d. Leg 1 and 2 are on the left of the robot, and leg 3 and 4 on the right. As depicted in Figure 8a,d, the robot can adjust the water-jet direction of thrusters in the same horizontal plane to realize forward or backward motion.

In addition, by the difference of the left water-jet force and the right water-jet force, the robot can implement the left and right turning motion, as shown in Figure 8e,f. Besides, as shown in Figure 8b,c, with adjusting the direction and force of four water-jet thrusters in the vertical plane, the robot can float upward and sink down. With the flexible motion, our robot can be used for environment monitoring, object tracking and detection of pollution with proper sensors.

4. The On-Land Locomotion Experiments

We performed experiments to verify the reliability and feasibility of the LSWM, an essential feature of sliding locomotion. We measured the distance that the LSWM traveled, and acquired the optimal length and frequency of the stride for the sliding and walking locomotion modes on different terrain.

4.1. The Lifting and Supporting Wheel Mechanism (LSWM)

To reduce the load on the actuator and extend its working life, we developed an LSWM driven by two 360-degree servos. Before the robot moved, the mobile platform descended, powered by the two servos rotating synchronously in the same direction. The PWM control signals were sent to the servos by the motor control board. We used waterproof servos (model HDKJ D3009) with a working voltage of 4.8–7.2 V. To maximize force and torque, we selected 7.2 V as the working voltage. The stall torque was 10 kg·cm and the operating speed was 0.14 s/60°. The frequency of the control signal was 50 Hz, and the duty factor s of the PWM signal ranged from 6.9 to 8.1%. We set the step size of the PWM signal to 0.05%. Thus, we obtained 20 data groups (from the lower to the upper and from the upper to the lower states, as shown in Figure 9); the distance Δh between the upper and lower state was 40 mm. All data groups were collected 10 times and averaged. The rising and falling velocities were as Equation (11).

$$v = \frac{\Delta h}{t} \tag{11}$$

(a)

(b)

Figure 9. (**a**) The upper state of lifting and supporting wheel mechanism (LSWM); (**b**) the lower state of LSWM.

The nonlinear relationship between the rising and falling velocity v and the duty factor s is shown in Figure 10. We considered the falling velocity positive and the rising velocity negative. When the duty factor ranged between 7.41 and 7.51%, the servo torque was very low and the LSWM velocity was almost zero. When the duty factor was more than 7.9% or less than 7.05%, the servo angular velocity, rendering both the rising and falling velocity relatively stable. Finally, based on the variation in velocity as the duty factor ranged from 7 to 8%, the nonlinear relationship can be described by the sum of sines (Equation (12)).

$$v = \begin{cases} -2.1 & s < 7.1 \\ 0 & 7.41 < s < 7.51 \\ 2.1 & s > 7.9 \\ 1.968\sin(3.448s + 18.22) + 0.1923\sin(9.059s + 23.33) & else \end{cases} \quad (12)$$

When the robot moves from flat, smooth terrain to rough terrain and vice versa, autonomous stable transition from sliding to walking (or vice versa) is achieved by varying the duty factor, dramatically improving performance on different substrates.

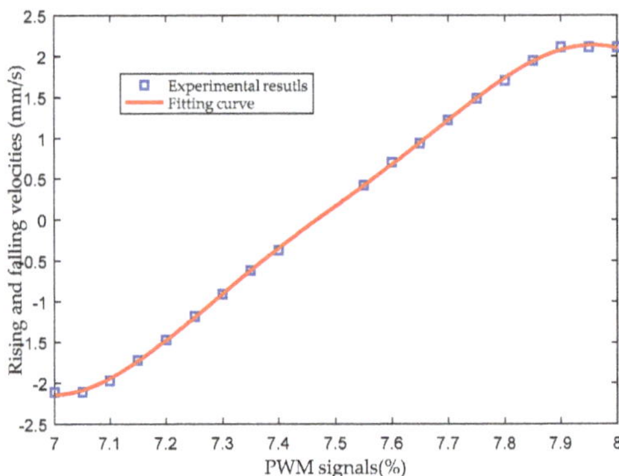

Figure 10. The relationship between the duty factor and the velocity. The falling velocity is positive, and the rising velocity negative.

4.2. Sliding Locomotion Using the LSWM

Sliding is appropriate on smooth flat terrain. Below a certain frequency, the velocity of the sliding gait varies with the distance by which the LSWM falls. As the mobile platform descends, the toe can touch the ground with the knee joint rotated clockwise, reducing the horizontal distance Δl from the knee joint to the toe. As shown in Figure 11, the blue state ($\Delta h = 11$ mm) of the leg reflects the situation where the toe only just touches the ground, and the red state ($\Delta h = 36$ mm) reflects the situation when the stance is vertical. Therefore, the fall distance Δh is 11–36 mm. We used HS-5646 WP servos, rotating at a rate of 0.18 s/60° at 7.4 V. We set the rotary angle θ_1^i and the rotary speed w_s of the horizontal servo to 90° and 0.5 s/60° respectively, in case a slice developed. We obtained the velocity of the sliding gait by Equations (13) and (14).

$$v = \frac{l}{t} = \frac{l_2 + \Delta l}{t} = \frac{l_2 + \sqrt{R^2 - [r_3 - (36 - \Delta h)]^2}}{t} \quad (13)$$

$$t = 2\theta_1^i w_s \quad (14)$$

where R is the distance from the knee joint to the toe ($R = 63.25$ mm).

To evaluate sliding locomotion using the LSWM on smooth flat ground, we modeled a synchronous sliding gait using two legs. We measured the sliding velocities at different descent values Δh of the mobile platform. In Figure 12, the blue curve shows a simulation of sliding velocity based on Equations (13) and (14). The deep-sky-blue circles indicate the experimental sliding velocities. We set 1 mm as the step size, and obtained 25 groups of experimental results from 11 to 36 mm. The maximum velocity was 10.2 cm/s at $\theta_2^i = 37.97°$ and $\Delta h = 11$ mm. As the descent distance of the mobile platform increased, the velocity declined to 66.5 mm/s. The experimental and simulated results differed by about 10 mm, which may reflect installation error. However, the trends of the two curves were similar.

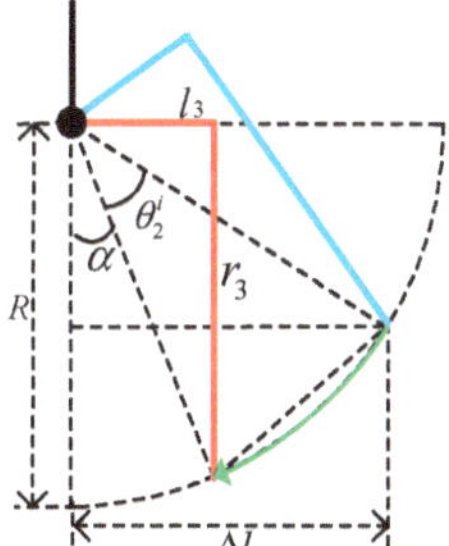

Figure 11. Extent of rotation of the knee joint by LSWM descent distance. The blue state ($\Delta h = 11$ mm) indicates that the toe only just touches the ground, and the red state ($\Delta h = 36$ mm) indicates the vertical stance. The green arrow indicates that the knee joint rotates from the blue to the red state.

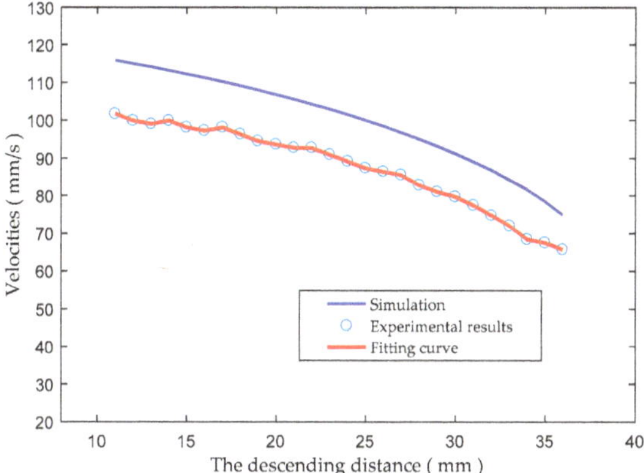

Figure 12. The sliding velocities at different descent distances.

4.3. Two On-Land Locomotion Experiments

To measure sliding mode velocities, we evaluated the relationships between robot velocity, and the swing frequency and angle of the legs. For sliding locomotion and walking locomotion of ASR-III, two key parameters affecting velocity are the stride length and frequency. To evaluate performance on land, we conducted sliding and walking experiments exploring these parameters.

When walking, the velocity is given by Equation (15).

$$v = s_L \times f \tag{15}$$

where, s_L is the length of one stride in the x direction, and f is the stride frequency. Using the ASR-III kinematic model, the stride in the x direction is given by Equation (16).

$$s_L = l \times \sin\theta_1^i \tag{16}$$

where l is the horizontal length of the leg (related to the extent of rotation of the knee joint), and can be obtained by Equation (17).

$$l = l_2 + \Delta l = l_2 + R\sin(\theta_2^i + \alpha) \tag{17}$$

where, R is the distance from the knee joint to the toe, and α is the leg angle ($\alpha = 18.43°$ in Figure 11).

The stride frequency depends on the rotation rate of the servo (w_s); this was 0.18 s/60° at 7.4 V. A short time Δt elapses between the support and the transfer phase of the leg, maintaining stability. Therefore, the stride frequency is as Equation (18).

$$f = \frac{1}{2\theta_1^i w_s + \Delta t} \tag{18}$$

Finally, the velocity is given by Equation (19).

$$v = \frac{[l_2 + R\sin(\theta_2^i + \alpha)]\sin\theta_1^i}{2\theta_1^i w_s + \Delta t} \tag{19}$$

From Equation (19), as the angle θ_2^i increases, the velocity increases when $\theta_2^i + \alpha < 90°$. When θ_2^i is 37.97°, the leg length is maximal (up to 52.68 mm). Then, the stride length depends only on the angle θ_1^i. To maximize velocity, we sought to define the optimal angle θ_1^i of the horizontal servo.

As the rotatory range of the servo is 0–90°, we evaluated nine angles (10, 20, 30, 40, 50, 60, 70, 80, and 90°). In Figure 13, the green curve represents the simulated results; the maximum velocity was about 12.0 cm/s when the rotary angle was less than 50°. The deep-sky-blue asterisks indicate the experimental results; the red curve fits these results. The maximum simulated and experimental velocities differed, perhaps because of slipping between the toe and the ground. However, the trends of the two curves were similar. Servo wear was more severe at higher frequencies. We set the optimal angle θ_1^i to 55°.

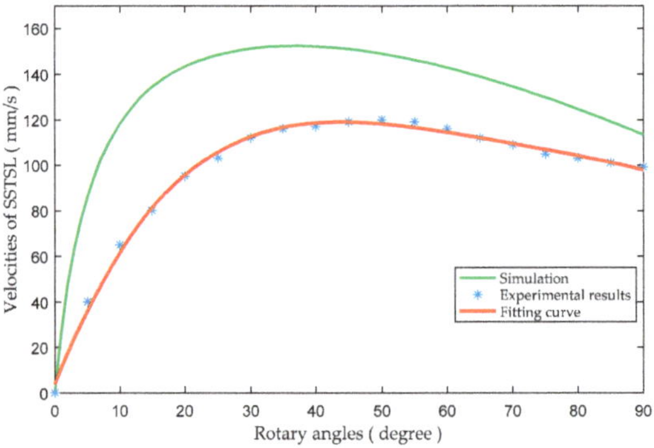

Figure 13. The velocities at different rotary angles of the horizontal servos.

The other parameter affecting velocity is the stride frequency. Ideally, the higher the frequency, the greater the velocity. However, stride frequency cannot be infinite, and slipping will develop at higher frequencies. To evaluate the sliding gaits, we measured velocities when the horizontal servo (set to 55°) operated at various frequencies. On smooth flat terrain, we tested three gaits (i.e., a synchronous gait using two legs, a trot, and a tripod walk). Given the limited performance of the servo, the frequencies of the synchronous sliding gait with two legs (SSGTL) and the trotting walking gait (TTWG) ranged from 0–2.5 Hz, and that of the tripod walking gait (TPWG) from 0–1.7 Hz. Figure 14 shows the velocities at different frequencies on smooth flat terrain. With the SSGTL and TPWG, as the gait frequency increased, the velocity initially increased linearly at low frequencies and then declined at high frequencies. For the TTWG, at relatively low frequencies (under 0.5 Hz), the robot could not maintain balance. The robot had only two legs on the ground, so the pressure associated with TTWG was larger than that associated with other gaits, so the velocity slowly declined. With the TPWG (a four-beat gait), the velocity was lower than those of the other gaits, not exceeding 5.2 cm/s. For SSGTL, the maximum velocity of 16.7 cm/s on smooth flat terrain was attained at 1.5 Hz and increased 13.6% than TTWG.

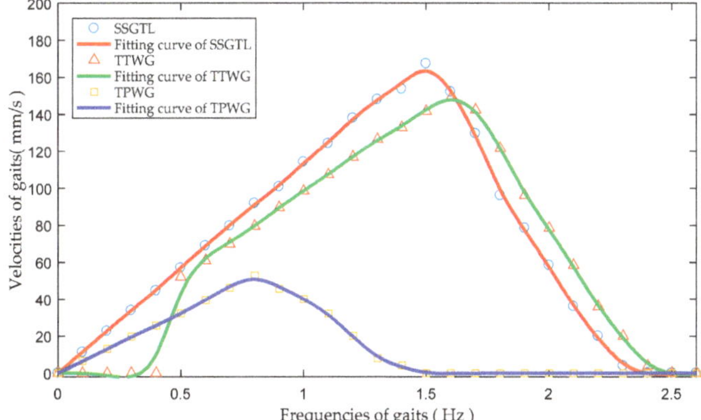

Figure 14. The velocities of gaits with different frequencies on smooth flat terrain. The deep-sky-blue circles, the coral triangles and the orange squares indicate the experimental synchronous sliding gait with two legs (SSGTL), TTWG, and TPWG data. The red, green and blue curves are the fitting curves for these data, respectively.

Figure 15a,b represent the roll, pitch, yaw angle of SSGTL and TTWG, respectively. For both gaits, the roll and pitch angle change slightly while moving. ASR-III with TTWG always swung around the z axes and the swing angle was about 5.1°, and the robot with SSGTL had a slight offset (2.2°). Consequently, compared to the walking gaits, the sliding gait for ASR-III on smooth flat terrain was better in terms of stability.

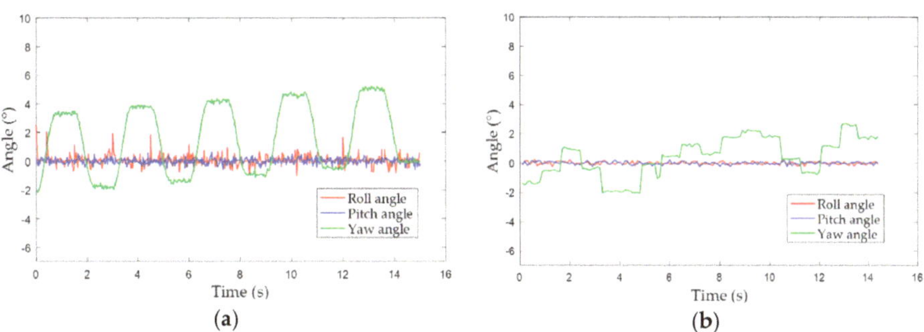

Figure 15. Roll, pitch, yaw angle of WWTG and SSTGL. (**a**) WWTG; (**b**) SSTGL.

Figure 16 shows the velocities of TTWG and TPWG on rough terrain (an earth road) at different frequencies. Such terrain is characterized by variations in the friction coefficient and roughness. With increasing frequency, the velocity increased linearly and then fell to zero. For both TTWG and TTPG, the velocity on rough terrain was greater than on smooth flat terrain, attaining 18.5 cm/s at 1.9 Hz and 7.0 cm/s at 1.1 Hz, and increased 20.5% and 28.6%, respectively. Besides, the mean power with SSTGL and TTWG measured approximately at the maximum velocity were 40.5 W and 56.7 W, respectively. According to the customary, we set the output factor of batteries as 80%. The robot can work 3h in the on-land range of the radius of 2.0 km.

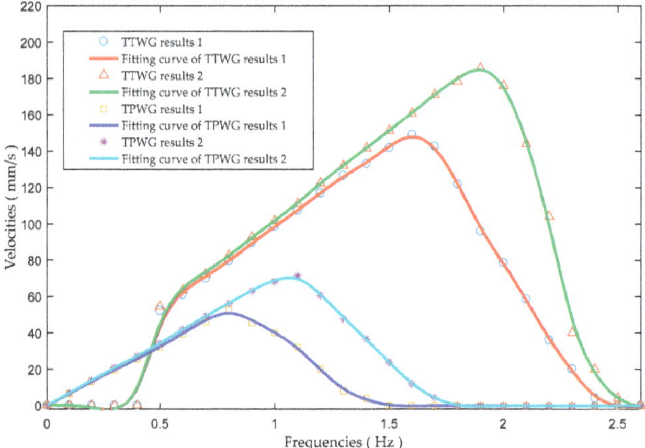

Figure 16. The velocities of gaits of different frequency on smooth flat, and rough, substrates. The deep-sky-blue circles and the coral triangles indicate the experimental TTWG data on smooth, flat and rough terrain, and the orange squares and the purple asterisks indicate the experimental TPWG data on smooth, flat and rough terrains. The red, green, blue and cyan curves are the corresponding fitting curves.

We evaluated rotatory motion using the sliding rotatory gait (SRG) and walking rotatory gait (WRG). In Figure 17, the red and blue curves show the rotatory velocities of the SRG and WRG at different frequencies, respectively. For both rotatory gaits, the rotary angle of a single step was 90° and the rotatory velocity increased linearly with increasing frequency, to attain a maximum. As the frequency increased further, the rotatory velocity declined rapidly because of slipping. Compared to the WRG, the SRG had a higher rotatory velocity.

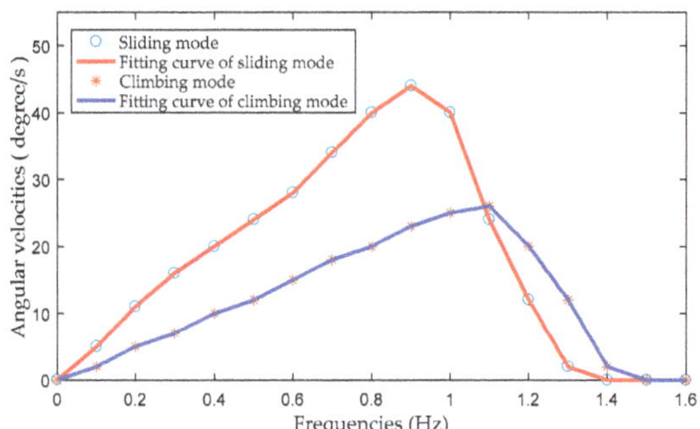

Figure 17. Angular velocity of rotary movement of the robot on different frequency.

4.4. A Waypoint Tracking Experiment Featuring Closed-Loop Proportional-Integral-Derivative (PID) Control

When an intelligent and multi-functional robot performs real work, a selection of gaits affording excellent performance is essential, as is the ability to track waypoints flexibly and autonomously. A real-time vision localization system is required. Thus, we defined three coordinate systems: a pixel system $\{O^{pixel}\}$, an image system $\{O^{i}\}$, and a world system $\{O^{p}\}$. To optimize performance, the optical center axis of the camera lay perpendicular to the object plane. As shown in Figure 18, the robot positions detected by the global camera running the Kalman Consensus Filter (KCF) algorithm in the image and object planes are (x_r^i, y_r^i) and (x_r^p, y_r^p) respectively. Using the principle of pinhole imaging, we obtain Equations (20) and (21).

$$\begin{bmatrix} x_r^i \\ y_r^i \\ 1 \end{bmatrix} = \begin{bmatrix} \frac{1}{d_x} & 0 & x_0^{pixel} \\ 0 & -\frac{1}{d_y} & y_0^{pixel} \\ 0 & 0 & 1 \end{bmatrix} \begin{bmatrix} x^{pixel} \\ y^{pixel} \\ 1 \end{bmatrix} \tag{20}$$

$$d \begin{bmatrix} x^{pixel} \\ y^{pixel} \\ 1 \end{bmatrix} = \begin{bmatrix} f & 0 & 0 & 0 \\ 0 & f & 0 & 0 \\ 0 & 0 & 1 & 0 \end{bmatrix} \begin{bmatrix} x_r^p \\ y_r^p \\ d \\ 1 \end{bmatrix} \tag{21}$$

Thus, the position of the robot in the experimental scenario is yielded by Equation (22).

$$\begin{cases} x_r^p = \frac{d(x_r^i - x_0^{pixel})d_x}{f} = \frac{d(x_r^i - x_0^{pixel})}{f_x} \\ y_r^p = -\frac{d(y_r^i - y_0^{pixel})d_y}{f} = -\frac{d(y_r^i - y_0^{pixel})}{f_y} \end{cases} \tag{22}$$

where $(x_0^{pixel}, y_0^{pixel})$ are the principal point coordinates relative to the image plane, and f_x and f_y are the focal lengths of the camera in the x and y directions, respectively.

We established four waypoints (A, B, C, and D; the corners of a rectangle). Figure 19 shows the experimental scenario that the ASR-III is on the blue ground. The camera software calculates the position of the robot using Equation (22) and sends the information to the robot. As descripted in Appendix A, we designed a waypoint tracking algorithm featuring precise positional and heading angle control, and allowed flexible locomotion (as described above). In waypoints tracking algorithm (Algorithm A1 in Appendix A), the parameter δ_d is the distance threshold; when the distance from the current position to the next waypoint is less than δ_d, we consider that the robot has arrived at the waypoint.

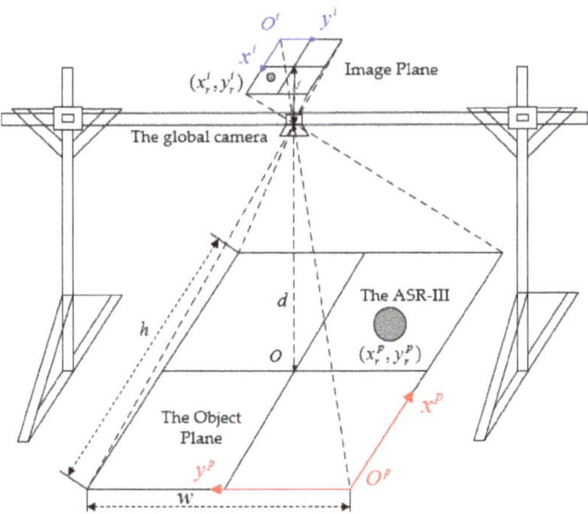

Figure 18. The vision localization system and the global camera. The gray circle indicates the ASR-III.

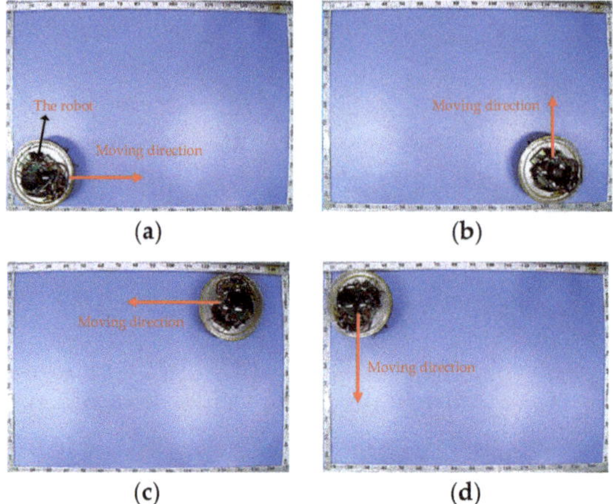

Figure 19. Snapshots of waypoint tracking (view of the global camera). (**a**) at point A; (**b**) at point B; (**c**) at point C; (**d**) at point D.

A common steering method used for legged robots on both sides is known as differential drive, where the swing angle α_{robot} is proportional to the difference between left α_l and right α_r leg angles. In the heading angle control algorithm (Algorithm A2 in Appendix A), we implemented a closed loop PID controller that ran at 50 Hz and an open loop controller. In the closed loop PID controller, IMU (JY901 IMU module) readings of yaw angle provided feedback, and the PID controller monitors proportional, integral, and derivative errors in order to output the PWM values required for achieving the desired swing angles of legs. As described in Algorithm A2 in Appendix A, when orientation errors appear, the robot need to adjust it direction by swing angles of both sides of legs. The parameter δ_a is an angle threshold; when the orientation error is less than δ_a, the robot goes forward with different swing angles of legs. When the orientation error is larger than δ_a, the robot stops to rotate with the rotary gait, and goes forward only until the orientation decline under δ_a. In the open loop controller, the difference of left legs and right legs was calculated by the distance from the current position to the reference trajectory. The longer the distance, the larger the difference.

The waypoint tracking experiment is shown in Figure 19, as captured by the global camera at A, B, C, and D. In Figure 20, the four red stars are the four waypoints, and the blue lines are the reference trajectory. The vision localization system of the global camera records the trajectory (black curves) when the parameters δ_d and δ_a are 3 cm and 8°, respectively. With the experiment's data of ASR-III trajectory, the maximum offsets with closed-loop PID and an open loop PID were about 7.0 cm and 10 cm, and the mean offset were up to 3.42 cm and 4.16 cm. Compared to the open loop trajectory, the maximum and mean offset with closed-loop PID reduced 30% and 17.9%, respectively. In order to evaluate the performance of the waypoints tracking with closed-loop PID, we also recorded the heading angle as shown in Figure 21. The maximum offsets with closed-loop PID and open loop manner were 8.6° and 10.4°. The mean offset with closed-loop PID and open loop manner were about 1.86° and 4.15°. Compared to the heading angle with open loop manner, the maximum and mean offset with closed-loop PID reduced 17.3% and 55.2%. In addition to the imperfect motion caused by the manufacturing tolerance, the robot location obtained by the vision localization system also had a slightly error because of the variations in illumination. However, compared to the size (35 cm) of the robot, the maximum trajectory offset with closed-loop PID only occupied 1/5, which was really slight offset. The heading angle exhibited several fluctuations, but the robot effectively corrected them. As a result, apart from several slight drifts, the robot tracked the task with less great precision, guided by the vision localization system.

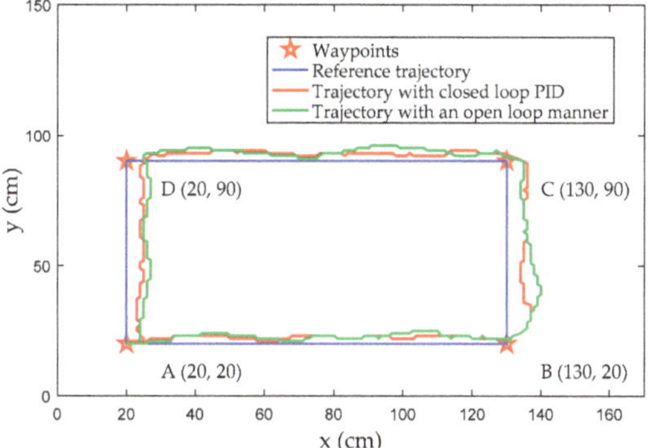

Figure 20. The experimental waypoint trajectory. The red stars represent the waypoints. We show the four waypoints A (20, 20), B (130, 20), C (130, 90), and D (20, 90), and the reference trajectory (blue lines; a precise rectangle). The red and green curves are the trajectory of the robot with closed loop PID and an open loop manner as recorded by the vision localization system, respectively.

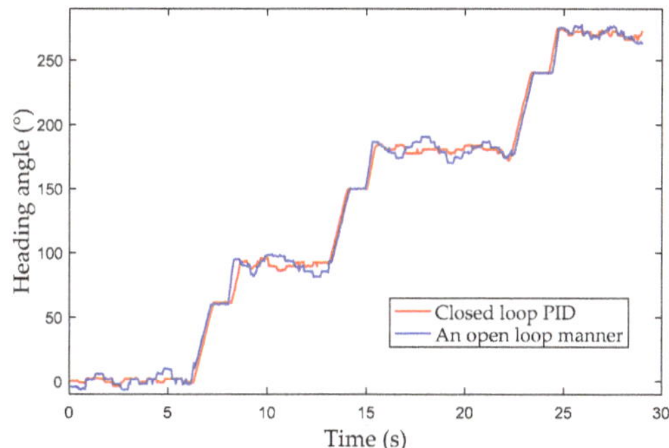

Figure 21. The heading angle in waypoint trajectory experiment. The red and blue curves represent heading angle with closed loop PID and an open loop manner, respectively.

5. The Underwater Locomotion Experiments

Due to the symmetrical spherical design, the robot has the same model for forward motion and backward motion. Hence experiments in forward motion were carried out to evaluate the performance of ASR-III underwater motions. Figure 22 shows the video sequence of the forward motion of ASR-III on the water surface of the pool with the size of 3 m × 2 m × 1 m (height × width × depth). In the pool, there were two vertical rulers (2 m) and one horizontal ruler (2.5 m). We assumed that the water was static. As depicted in Section 3, we adapted the forward motion strategy with four thrusters. The motor of thruster worked under the voltage of 20 V. Initially the robot was put at one side of the horizontal ruler, and then swam along the ruler to other side. We recorded the time and the displacement of forward motion, and then calculate the velocity. As shown in Figure 23, the red and blue curves indicate the displacement and the velocity of the robot varying with time. The robot speeded up to the maximum velocity (21.5 cm/s, 0.61 body/s) at about 6.5 s, and then the robot swam at the constant speed. From the start to the end, the average velocity of the robot was 18.1 cm/s (0.52 body/s). Furthermore, the mean power measured approximately at maximum velocity was 64.5 W, and the robot can cruise 2.5 h in the range of the radius of 1.5 km.

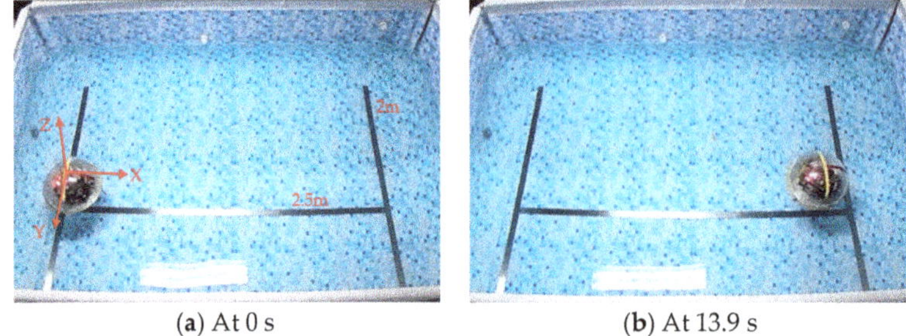

(**a**) At 0 s (**b**) At 13.9 s

Figure 22. Underwater forward motion on the water surface.

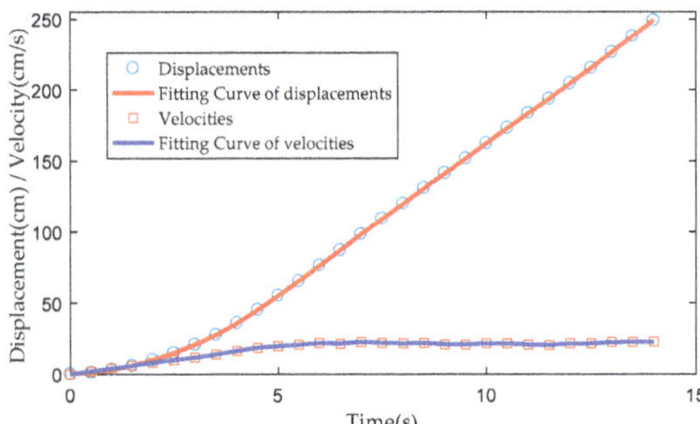

Figure 23. Experimental results for forward motion.

The heading angle control experiment was conducted to evaluate the performance of ASR-III with the strategy of the left turning and right turning. The heading angle measured by the JY901 IMU module was regarded as the feedback for the closed-loop control. The experiment was carried out in the same pool. Figure 24 shows four snapshots of the heading angle control motion. The initial angle of the robot was set to 0°, and the excepted heading angle was set to 90°. As shown in Figure 25a, the robot can keep stable at 16.5 s.

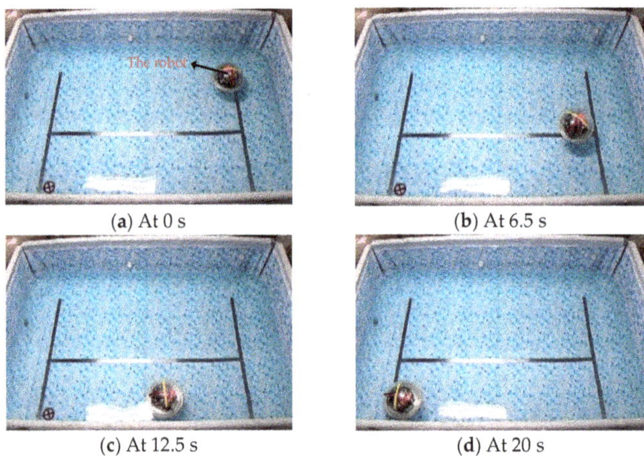

(**a**) At 0 s (**b**) At 6.5 s (**c**) At 12.5 s (**d**) At 20 s

Figure 24. Heading angle control underwater motion on the water surface.

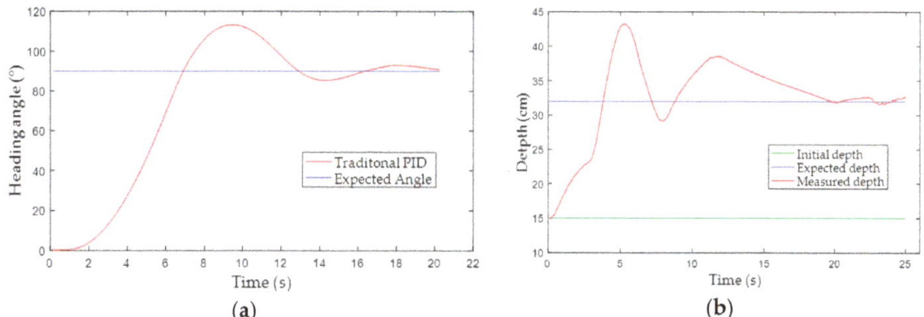

Figure 25. (**a**) Experimental results for heading angle control; (**b**) experimental results for depth control.

Besides the forward motion and the heading angle control experiment, the depth control experiment was carried out to evaluate the performance of ASR-III with the strategy of floating and sinking. The depth values were collected by IIC bus from the pressure sensor (MS5803-01BA module) installed under the middle plant, and regarded as the feedback for the closed-loop PID control. The robot started at a depth of 15 cm, and the expected depth was 32 cm. The results of depth control were illustrated in Figure 25b. The robot reached the expected depth after 20 s. In the steady state, the mean depth error was 0.56 cm.

6. Discussion

Based on the results obtained from the experiments, on land locomotion and underwater locomotion were concluded. First, the LSWM was evaluated to make the robot stand up and slide on smooth flat ground. We obtained the relationship between the rising and falling velocity v and the duty factor s, which assisted the robot realizing the automatic lifting and dropping of the LSWM. Second, the two parameters which affected the velocity of the robot on land were the stride length and frequency. After experiments, we found that the optimal rotary angle was 55°. On smooth flat terrain, the maximum velocity of SSTGL (sliding locomotion) was 16.7 cm/s at 1.5 Hz and increased 13.6% than TTWG (walking locomotion), which proved that SSTGL on smooth flat terrain had higher velocity performance than TTWG for ASR-III. On rough terrain, the maximum velocity of TTWG is 18.5 cm/s at 1.9 Hz. In addition, the mean power with SSTGL and TTWG measured approximately at the maximum velocity were 40.5 W and 56.7 W, respectively, which testified the low power loss of SSTGL. Compared to the maximum velocity (8.5 cm/s) of the previous robot [37], the maximum velocity of TTWG was about twice higher. For the rotatory gait, the maximum rotatory velocity in previous robot [38] was up to 67°/s while the weight of the robot was only 2.1 KG. However, Now ASR-III's weight was up to 8.2 KG (about four times heavier than the previous robot), and the rotatory velocity was 44.5°/s (decline 33.6%). Third, for the waypoints tracking, the maximum trajectory offset with closed-loop PID only occupied 1/5 to the size (35) of the robot, which was really slight offset. Finally, underwater locomotion, such as forward motion, backward motion, turn left or right, floating and sinking were realized. Compared to the previous work [36] in forward locomotion, the average velocity of the robot was just 10 cm/s (0.29 body/s), and increased about 81%. For the heading angle control, the robot can keep the expected angle at 25 s, and the time was shortened about 34% compared to the previous work [36]. For the depth control, the robot has a sensitive response, and the control errors are in the acceptable range. Besides, with the comparison to the related works, Locomotion speed of ASR-III on ground was slower than AmphiBot II [12], and locomotion speed of ASR-III in water was almost equal to AmphiBot II. However, the robot can rotate with zero-radius both on land and in water, and have a stable heading angle control. Compared to the hexapedal robot [16], ASR-III on ground and in water was slower than the hexapedal robot. However, our robot can realize multi-locomotion, such as sinking down and floating up in water, rotating with zero-radius both on land and in water. In addition, the robot can keep a stable depth in water for some complex tasks.

7. Conclusions

We describe the novel, multiply gaited, vectored water-jet hybrid locomotion-capable, amphibious, spherical robot III. To improve on-land velocity on different substrates and decrease the payload on the leg servos, we designed an automated lifting and supporting mechanism with four omni-directional wheels to facilitate sliding locomotion, and also the transition between smooth flat and rough terrain; two platform servos were synchronously controlled. The efficacy of the LSWM was verified experimentally.

To improve stability and mobility on land, we designed sliding gaits for smooth flat terrain and walking gaits for rough terrain. We experimented with different gaits on different substrates to define the optimal gaits. Compared with walking gaits, sliding locomotion using SSGTL or SRG afforded good stability and mobility on smooth flat terrain, and relatively higher velocities. Walking gaits caused the swing of the robot, which reduces stability. However, TTWG and TPWG can be used on rough terrain, and the speed increased 20.5% and 28.6% than on smooth terrain. The maximum speed is almost double that of the previous work. In terms of rotation, SRG exhibited a higher rotatory velocity (up to 44.5°/s) than WRG. Although the weight of the robot increased almost four times, the robot also rotates smoothly with a high angular velocity. In addition, we developed a waypoint tracking algorithm combining a vision localization system with a global camera. The maximum offset of this trajectory is just 1/5 of the diameter of the robot, which showed experimentally that the robot performed well.

The robot can also realize the underwater locomotion with four vectored water-jet thrusters. Compared to the previous work, the maximum velocity of forward motion increased 81%, and the time to be stable in heading angle control was shortened 34%. Besides, the robot also has a sensitive response in depth control, and the control errors are in the acceptable range. Therefore, the performance of underwater locomotion was improved.

With the high performance of the robot in amphibious environment, we will build multiple robot system in future.

Acknowledgments: This work was supported by National Natural Science Foundation of China (61503028, 61773064, 61375094), and Excellent Young Scholars Research Fund of Beijing Institute of Technology (2014YG1611). This research project was also partly supported by National High Tech. Research and Development Program of China (No. 2015AA043202).

Author Contributions: Huiming Xing conceived and designed the experiments, and then wrote the paper; Shuxiang Guo and Liwei Shi guided the system design and revised the manuscript. Yanlin He, Shuxiang Su and Zhan Chen performed the experiments. Xihuan Hou analyzed the data.

Conflicts of Interest: The authors declare no conflict of interest.

Appendix A. Waypoints Tracking Algorithm

As mentioned in Section 5, we designed a waypoint tracking algorithm including position and heading angle control. The following three algorithms are the position control algorithm, the heading angle control algorithm, and the waypoint tracking algorithm, respectively.

Algorithm A1 Waypoints tracking algorithm with Closed-loop PID Control

```
System Initialization
Waypoints set up
Go to the next waypoint (w_x^i, w_y^i)
Calculate the distance d = sqrt((w_x^i − x_r)^2 + (w_y^i − y_r)^2)
IF d < δ_d
IF this waypoint is the last one
Waypoints tracking ends
ELSE
Go to the next waypoint
ELSE
IF Road is smooth
Do ASR-III sliding motion with heading angle control
ELSE
Do ASR-III climbing motion with heading angle control
```

Algorithm A2 Heading Angle control algorithm

```
Initial the parameters
Read the current heading angle
Calculate the differential angle α = α_expected − α_heading
While
IF α ≥ δ_a or α ≤ −δ_a
Get Leg PID angle α_p
Swing angle of legs α_legs ← α_i + α_p
Stop and execute the sliding rotary gait
ELSE
Get leg PID angle α_p
IF 0 < α < δ_a (turn left)
Left legs α_l ← α_i − α_p
Right legs α_r ← α_i + α_p
ELSE
Left legs α_l ← α_i + α_p
Right legs α_r ← α_i − α_p
Go forward with different swing angles
End while
```

References

1. Shi, Q.; Li, C.; Wang, C.B.; Luo, H.B.; Huang, Q.; Fukuda, T. Design and Implementation of an Omnidirectional Vision System for Robot Perception. *Mechatronics* **2017**, *41*, 58–66. [CrossRef]
2. He, Y.L.; Guo, S.X.; Shi, L.W. Preliminary Mechanical Analysis of an Improved Amphibious Spherical Father Robot. *Microsyst. Technol.* **2016**, *22*, 2051–2066. [CrossRef]
3. Guo, S.X.; Pan, S.W.; Li, X.Q.; Shi, L.W. A System on Chip-based Real-time Tracking System for Amphibious Spherical Robots. *Int. J. Adv. Robot. Syst.* **2017**, *14*, 1–19. [CrossRef]
4. Guo, S.X.; Pan, S.W.; Shi, L.W. Visual Detection and Tracking System for an Amphibious Spherical Robot. *Sensors* **2017**, *17*, 870. [CrossRef] [PubMed]
5. Kaznov, V.; Seeman, M. Outdoor Navigation with a Spherical Amphibious Robot. In Proceedings of the IEEE/RSJ International Conference on Intelligent Robots and Systems, Taipei, Taiwan, 18–22 October 2010.
6. Shi, Q.; Ishii, H.; Sugahara, Y.; Sugita, H.; Takanishi, A.; Huang, Q.; Fukuda, T. Design and Control of a Biomimetic Robotic Rat for Interaction with Laboratory Rats. *IEEE/ASME Trans. Mechatron.* **2015**, *20*, 1832–1842. [CrossRef]

7. Shi, Q.; Ishii, H.; Kinoshita, S.; Konno, S.; Takanishi, A.; Okabayashi, S.; Iida, N.; Kimura, H.; Shibata, S. Modulation of Rat Behaviour by Using a Rat-like Robot. *Bioinspir. Biomim.* **2013**, *8*, 046002. [CrossRef] [PubMed]
8. Shi, Q.; Ishii, H.; Tanaka, K.; Sugahara, Y.; Takanishi, A.; Okabayashi, S.; Huang, Q.; Fukuda, T. Behavior Modulation of Rats to a Robotic Rat in Multi-rat Interaction. *Bioinspir. Biomim.* **2015**, *10*, 050611. [CrossRef] [PubMed]
9. Jun, B.H.; Shim, H.; Kim, B.; Park, J.Y.; Baek, H.; Lee, P.M.; Kim, W.J.; Park, Y.S. Preliminary Design of the Multi-Legged Underwater Walking Robot CR200. In Proceedings of the Oceans MTS/IEEE Conference, Yeosu, Korea, 21–24 May 2012.
10. Shim, H.; Yoo, S.Y.; Kang, H.; Jun, B.H. Development of Arm and Leg for Seabed Walking Robot CRABSTER200. *Ocean Eng.* **2016**, *116*, 55–67. [CrossRef]
11. Crespi, A.; Karakasiliotis, K.; Guignard, A.; Ijspeert, A.J. Salamandra Robotica II: An Amphibious Robot to Study Salamander-Like Swimming and Walking Gaits. *IEEE Trans. Robot.* **2013**, *29*, 308–320. [CrossRef]
12. Crespi, A.; Ijspeert, A.J. AmphiBot II: An Amphibious Snake Robot that Crawls and Swims using a Central Pattern Generator. *Color Res. Appl.* **2006**, *27*, 130–135.
13. Tang, Y.G.; Zhang, A.Q.; Yu, J.C. Modeling and Optimization of Wheel-Propeller-Leg Integrated Driving Mechanism for an Amphibious Robot. In Proceedings of the 2nd International Conference on Information and Computing Science, Manchester, UK, 21–22 May 2009; pp. 73–76.
14. Zhang, S.W.; Zhou, Y.C.; Xu, M.; Liang, X.; Liu, L.; Yang, J. AmphiHex-I: Locomotory Performance in Amphibious Environments with Specially Designed Transformable Flipper Legs. *IEEE-ASME Trans. Mechatron.* **2016**, *21*, 1720–1731. [CrossRef]
15. Sun, Y. Planning of Legged Racewalking Gait for an Epaddle-Based Amphibious Robot. *IFAC Proc. Vol.* **2012**, *45*, 218–223. [CrossRef]
16. Kim, H.; Lee, D.; Liu, Y.; Seo, T.; Jeong, K. Hexapedal Robot for Amphibious Locomotion on Ground and Water. In Proceedings of the IEEE/ASME International Conference on Advanced Intelligent Mechatronics (AIM), Busan, Korea, 7–11 July 2015; pp. 121–126.
17. Aoki, T.; Ito, S.; Sei, Y. Development of quadruped walking robot with spherical shell-mechanical design for rotational locomotion. In Proceedings of the IEEE/RSJ International Conference on Intelligent Robots and Systems, Hamburg, Germany, 28 September–2 October 2015; pp. 5706–5711.
18. Yin, X.; Wang, C.; Xie, G. A salamander-like amphibious robot: System and control design. In Proceedings of the International Conference on Mechatronics and Automation, Chengdu, China, 5–8 August 2012; pp. 956–961.
19. Boxerbaum, A.S.; Werk, P.; Quinn, R.D.; Vaidyanathan, R. Design of an Autonomous Amphibious Robot for Surf Zone Operation: Part I Mechanical Design for Multi-Mode Mobility. In Proceedings of the IEEE/ASME International Conference on Advanced Intelligent Mechatronics, Monterey, CA, USA, 24–28 July 2005; pp. 1459–1464.
20. Harkins, R.; Ward, J.; Vaidyanathan, R.; Boxerbaum, A.S.; Quinn, R.D. Design of an Autonomous Amphibious Robot for Surf Zone Operations: Part II—Hardware, Control Implementation and Simulation. In Proceedings of the IEEE/ASME International Conference on Advanced Intelligent Mechatronics, Monterey, CA, USA, 4–28 July 2005; pp. 1465–1470.
21. Li, B.; Ma, S.G.; Ye, C.L.; Yu, S.M.; Zhang, G.W.; Gong, H.L. Development of an Amphibious Snake-like Robot. In Proceedings of the 8th World Congress on Intelligent Control and Automation (WCICA), Jinan, China, 6–9 July 2010; pp. 613–618.
22. Hirose, S.; Yamada, H. Snake-like Robots: Machine Design of Biologically Inspired Robots. *IEEE Robot. Autom. Mag.* **2003**, *16*, 88–98. [CrossRef]
23. Pan, S.W.; Guo, S.X.; Shi, L.W.; He, Y.L.; Wang, Z.; Huang, Q. A Spherical Robot based on all Programmable SoC and 3-D Printing. In Proceedings of the 11th IEEE International Conference on Mechatronics and Automation (ICMA), Tianjin, China, 3–6 August 2014; pp. 150–155.
24. He, Y.L.; Guo, S.X.; Shi, L.W.; Pan, S.W.; Wang, Z. 3D Printing Technology-based an Amphibious Spherical Robot. In Proceedings of the 11th IEEE International Conference on Mechatronics and Automation (ICMA), Tianjin, China, 3–6 August 2014; pp. 1382–1387.

25. He, Y.L.; Shi, L.W.; Guo, S.X.; Guo, P.; Xiao, R. Numerical Simulation and Hydrodynamic Analysis of an Amphibious Spherical Robot. In Proceedings of the IEEE International Conference on Mechatronics & Automation, Beijing, China, 2–5 August 2015; pp. 848–853.
26. Pan, S.W.; Shi, L.W.; Guo, S.X.; Guo, P.; He, Y.L.; Xiao, R. A Low-power SoC-based Moving Target Detection System for Amphibious Spherical Robots. In Proceedings of the IEEE International Conference on Mechatronics & Automation, Beijing, China, 2–5 August 2015; pp. 1116–1121.
27. Shi, L.W.; Xiao, R.; Guo, S.X.; Guo, P.; Pan, S.W.; He, Y.L. An Attitude Estimation System for Amphibious Spherical Robots. In Proceedings of the IEEE International Conference on Mechatronics & Automation, Beijing, China, 2–5 August 2015; pp. 2076–2081.
28. Pan, S.W.; Shi, L.W.; Guo, S.X. A Kinect-Based Real-Time Compressive Tracking Prototype System for Amphibious Spherical Robots. *Sensors* **2015**, *15*, 8232–8252. [CrossRef] [PubMed]
29. Guo, J.; Guo, S.X.; Li, L.G. Design and Characteristic Evaluation of a Novel Amphibious Spherical Robot. In Proceedings of the 3rd International Conference on Engineering and Technology Innovation (ICETI), Kenting, Taiwan, 31 October–4 November 2014; Volume 23, p. 6.
30. Nishimura, Y.; Mikami, S. Learning Adaptive Escape Behavior for Wheel-Legged Robot by Inner Torque Information. In Proceedings of the 2016 Joint 8th International Conference on Soft Computing and Intelligent Systems (SCIS) and 17th International Symposium on Advanced Intelligent Systems, Sapporo, Japan, 25–28 August 2016; pp. 10–15.
31. Endo, G.; Hirose, S. Study on Roller-Walker (system integration and basic experiments). In Proceedings of the 1999 IEEE International Conference on Robotics and Automation, Detroit, MI, USA, 10–15 May 1999.
32. Endo, G.; Hirose, S. Study on Roller-Walker—Adaptation of characteristics of the propulsion by a leg trajectory. In Proceedings of the IEEE/RSJ International Conference on Intelligent Robots and Systems, Nice, France, 22–26 September 2008; pp. 1532–1537.
33. Endo, G.; Hirose, S. Study on Roller-Walker—Energy efficiency of Roller-Walk. In Proceedings of the 2011 IEEE International Conference on Robotics and Automation, Shanghai, China, 9–13 May 2011; pp. 5050–5055.
34. Endo, G.; Hirose, S. Study on Roller-Walker (multi-mode steering control and self-contained locomotion). In Proceedings of the IEEE International Conference on Robotics and Automation, San Francisco, CA, USA, 24–28 April 2000; Volume 3, pp. 2808–2814.
35. Li, M.X.; Guo, S.X. A Roller-Skating/Walking Mode-based Amphibious Robot. *Robot. Comput. Integr. Manuf.* **2017**, *44*, 17–29. [CrossRef]
36. Shi, L.W.; Su, S.X.; Guo, S.X. A Fuzzy PID Control Method for the Underwater Spherical Robot. In Proceedings of the IEEE International Conference on Mechatronics & Automation, Takamatsu, Japan, 6–9 August 2017; pp. 626–631.
37. Shi, L.W.; Pan, S.W.; Guo, S.X. Design and Evaluation of Quadruped Gaits for Amphibious Spherical Robots. In Proceedings of the IEEE International Conference on Robotics and Biomimetics, Qingdao, China, 3–7 December 2017; pp. 13–18.
38. Li, M.X.; Guo, S.X. Design and performance evaluation of an amphibious spherical robot. *Robot. Auton. Syst.* **2015**, *64*, 21–34. [CrossRef]

Article

Performance Evaluation of a Novel Propulsion System for the Spherical Underwater Robot (SURIII)

Shuoxin Gu [1] and Shuxiang Guo [2,3,*]

1 Graduate School of Engineering, Kagawa University, Takamatsu, Kagawa 760-0396, Japan; s16d642@stu.kagawa-u.ac.jp

2 Key Laboratory of Convergence Medical Engineering System and Healthcare Technology, The Ministry of Industry and Information Technology, School of Life Science, Beijing Institute of Technology, Beijing 100081, China

3 Department of Intelligent Mechanical Systems Engineering, Kagawa University, Takamatsu, Kagawa 760-0396, Japan

* Correspondence: guo@eng.kagawa-u.ac.jp

Received: 28 September 2017; Accepted: 17 November 2017; Published: 20 November 2017

Abstract: This paper considers a novel propulsion system for the third-generation Spherical Underwater Robot (SURIII), the improved propulsion system is designed and analyzed to verify its increased stability compared to the second-generation Spherical Underwater Robot (SURII). With the new propulsion system, the robot is not only symmetric on the X axis but also on the Y axis, which increases the flexibility of its movement. The new arrangement also reduces the space constraints of servomotors and vectored water-jet thrusters. This paper also aims to the hydrodynamic characteristic of the whole robot. According to the different situations of the surge and heave motion, two kinds of methods are used to calculate the drag coefficient for the SURIII. For surge motion, the drag coefficient can be determined by the Reynolds number. For heave motion, considering about the influences of edges and gaps of the SURIII, the drag coefficient needs to be calculated by the dynamic equation. In addition, the Computational Fluid Dynamics (CFD) simulation is carried out to estimate some parameters which cannot be measured. The pressure contours, velocity vectors and velocity streamlines for different motions are extracted from the post-processor in the CFD simulation. The drag coefficients of surge and heave motion are both calculated by the simulation results and compared with the chosen one by Reynolds number. Finally, an experiment is also conducted for measure the propulsive force of the multi-vectored water-jet thrusters by using a 6-DoF load cell. The experimental results demonstrate the propulsive force is better than a previous version. Thus, the propulsive performance is better than before.

Keywords: spherical underwater robot; hydrodynamic analysis; Computational Fluid Dynamics simulation; propulsion system; vectored water-jet thrusters

1. Introduction

With the unceasing underwater exploration of the ocean, autonomous underwater vehicles (AUVs) have also been improved and developed for underwater detection. If AUVs in the unknown underwater environment can still maintain adequate flexibility and sensitivity, they can be suitable for a variety of complex underwater tasks. The IFREMER L'Eqaulard AUV was probably the first AUV which deployed for marine geoscience [1,2]. Anirban Mazumdar et al. developed a ball shaped underwater robot which was completely smooth and used jets to propel and maneuver through water-filled environments such as the inside of nuclear power plants [3]. MIT researchers also unveiled an oval-shaped submersible robot smaller than a football, flattened on one side and able to slide along an underwater surface to perform ultrasound scans [4]. But these two kinds of robots are both designed for special environments such as pipelines or confined spaces, with relatively simple forward and backward motion. Heejoong Kim and Jihong Lee proposed a swimming pattern generator mimicking locomotion of diving beetles with the viewpoint of biomimetics for legged underwater robots [5]. In our previous study, firstly a spherical underwater robot with multi vectored water-jet thrusters was developed [6,7]. Then a Father-son Underwater Intervention Robotic System (FUIRS) was proposed [8]. In this system, the SURII plays the role of the father robot and the micro robot plays the role of the son robot [9]. The father robot has a spherical shape and a vectored water-jet-based propulsion system, so it can work with high stability and low noise. The FUIRS system is shown in Figure 1. With the development of the previous research, the SURIII which has four vectored water-jet thrusters for the propulsion system is proposed [10–12].

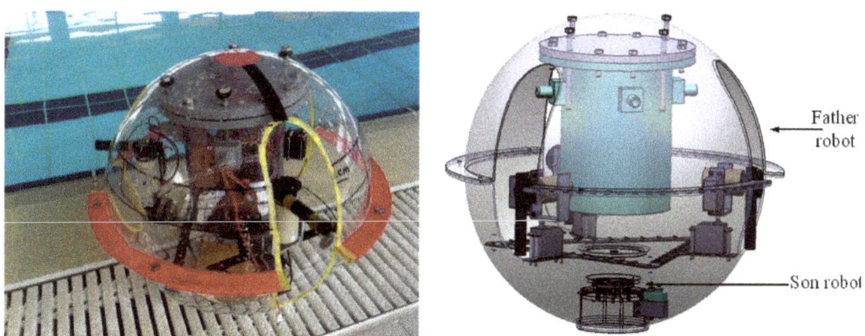

Figure 1. Father-son Underwater Intervention Robot System (FUIRS).

AUVs face different working conditions, requiring different configurations and size, thus many new underwater technologies are being explored and applied to AUVs. Because of the excellent diving capability of underwater creatures, some underwater robots are designed to mimic their characteristics. Hyung-Jung Kim et al. developed a turtle-like swimming robot with biomimetic flippers. This robot uses a smart soft composite structure to realize smooth and soft flapping motion [13]. Guo lab in Kagawa University developed an octopus-inspired micro robot using ionic polymer metal composite (IPMC). It can cooperate with other robot to execute underwater intervention tasks [14]. These above robots imitate the appearance or movement of aquatic creatures to better adapt to the underwater environment. Watson S. A. et al. developed a micro-spherical underwater robot which used six propellers to generate propulsive force and this robot was designed for monitoring the nuclear storage pond and preventing leakage in wastewater treatment facilities [15,16]. However, its propeller thruster cannot adjust direction. MIT researchers designed a robot called "Omni-Egg", which is smooth, spheroidal and completely appendage-free [17]. It used a novel pump-jet system built into the streamlined shell. Researchers at Harbin Engineering University also developed a spherical underwater robot with three water-jet thrusters [18,19].

Spherical Underwater Robot (SUR) is a kind of special structural underwater robot which has better compression performance. Due to its overall symmetry, there are no coupling items in fluid dynamics calculation and the hydrodynamic parameters are also equal. Therefore, it has more extensive application. In this paper, a novel spherical underwater robot (SURIII) which has four vectored water-jet thrusters was developed. A novel annular symmetrical structure was proposed to ensure the stability of the propulsion system. The static analysis proved that this new structure is more stable than before. The improvement of structure increases the utilization of space. Therefore, the rotation angle of vectored water-jet thrusters is more varied and flexible. Hydrodynamic characteristics are the key point and difficulty for underwater robot research. In CFD simulation, the pressure contours, velocity vectors and velocity streamlines are obtained from the post-processor.

This paper is organized as follows. Section 2 presents the mechanical design and analysis of the SURIII. Analysis of the hydromechanics of the vehicle, including hydrodynamic analysis of the impact of propulsion system deformation on vehicle stability as well as calculation of drag characteristics, are presented in Section 3. The results of the CFD simulation carried out by ANSYS-CFX are presented in Section 4. Experimental evaluation of the propulsive force of the propulsion system is described in Section 5. Finally, some conclusions are provided in Section 6.

2. Mechanical Design and Analysis

2.1. Inspiration for Design

The inspiration for the design of the SUR propulsion system came from the propulsive mechanism of the jellyfish. A study indicated that jellyfish are the most energy efficient underwater swimmers [20]. Figure 2 demonstrates how jellyfish, through the alternate contraction and relaxation of their bodies, produce a jet of water whose propulsive force propels them through the water. The design of the vehicle, as described subsequently, is aimed at mimicking the efficient motion of the jellyfish and attaining symmetry, stability and performance.

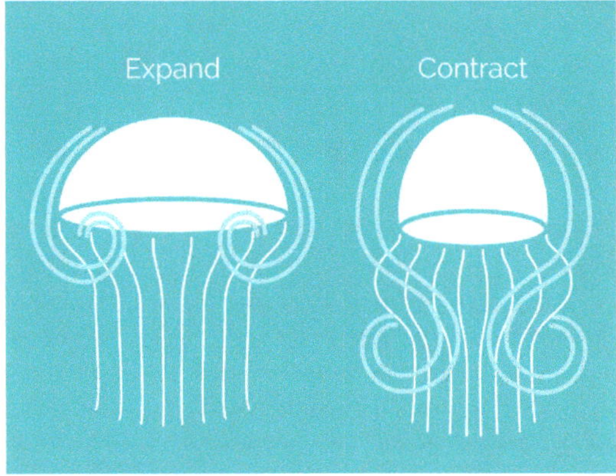

Figure 2. Propulsion mechanism of the Jellyfish.

2.2. Mechanical Design and Analysis of Propulsion System

Taking the requirement of volume into consideration, SUR as a kind of small underwater robot in our lab, its movement is restricted by the structure and space. Thus, its propulsion system needs to be improved in the limited space to enhance the hydrodynamic property. In the propulsion system of the SURII, three water-jet thrusters were homogeneously located on an equilateral triangle support as shown in Figure 3. This kind of layout limited the rotation angle of servo motors, thereby effecting the movement performance, especially the forward movement. Such a structure cannot achieve the maximum propulsive force and the payload capability will be affected. In terms of the current issue, a novel annular symmetrical mechanism is designed for the propulsion system of SURIII as illustrated in Figure 4. This system adopts an annular structure which can provide support for four thrusters. The maximum distance between the two opposite thrusters is 48 cm, the inside diameter of the annular support is 19 cm and its thickness is 0.5 cm. Considering the size of spherical hull, the weight of the robot and advantages of symmetric structure, only one more set thruster is added. Moreover, the stability of robot will be impacted less while improving the dynamic power. In order to enhance the stiffness of the propulsion system, the annular support structure is fixed with the waterproof bin by jackscrews.

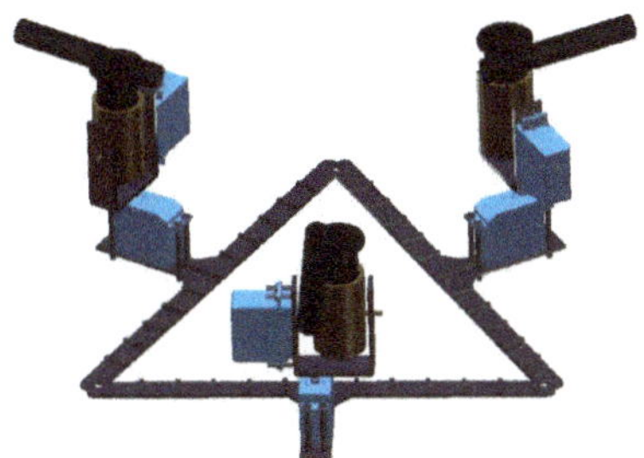

Figure 3. Propulsion system of SURII.

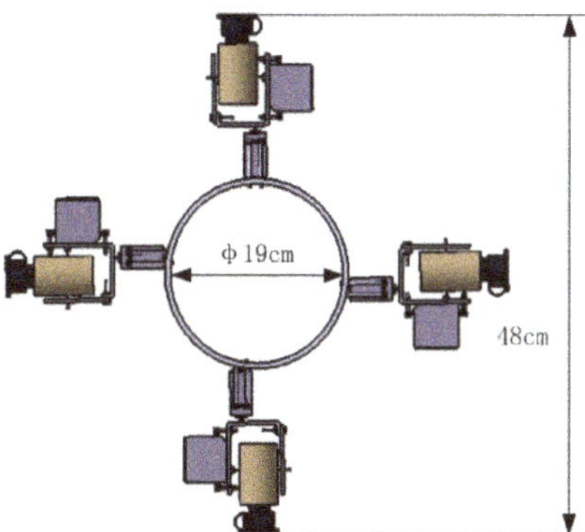

Figure 4. Propulsion system of SURIII.

For the novel propulsion system, the static analysis is necessary to evaluate its performance. Above all, unforeseen deformation will adversely affect the robot, particularly the direction of the nozzle. Although the second-generation robot has already enhanced the rotation support between servo motors and the water-jet thruster and reduced the deformation of the system [21], its triangle support is only suitable for three water-jet thrusters and cannot be used for fours. Therefore, the deformation analysis for the annular support is very important. Moreover, the weight of the propulsion system is increased after adding one more set of vectored water-jet thruster, hence the static analysis is quite essential.

The basic movements of the robot include three motions: up, down and forward motion. Based on this, the static analysis is carried out by the ANSYS WORKBENCH. Figure 5 illustrates the simulation models of the different motions for SURIII. The up and down motion models are shown in Figure 5a,c, the 2 N force is applied to each nozzle, respectively. The forward motion model is shown in Figure 5b, the 2 N force is exerted on three working nozzles. The position and direction of these forces are the same as the static analysis of the SURII. Both the triangle support of the SURII and the annular support of the SURIII are fixed with the ground.

Figure 6 displays the deformation of the propulsion system. Figure 6a,b describe the deformation of the previous and the novel propulsion system, respectively. According to the rainbow map in the left of the interface, red color indicates the maximum deformation in the result. Observing these two propulsion systems, the maximum deformation occurs at the same location, therefore the maximum deformation of these two propulsion systems is compared to evaluate the stiffness. Figure 7 is the comparison of the maximum deformation between these two propulsion systems. From the comparison, it is evident that no matter what kind of motion state, the deformation of propulsion system in SURIII is smaller than that in SURII. However, the maximum deformation (0.045 mm) of all motions for SURIII which generates in the up motion is still far less than the minimum deformation (0.422 mm) of all motions for SURII which generates in the down motion.

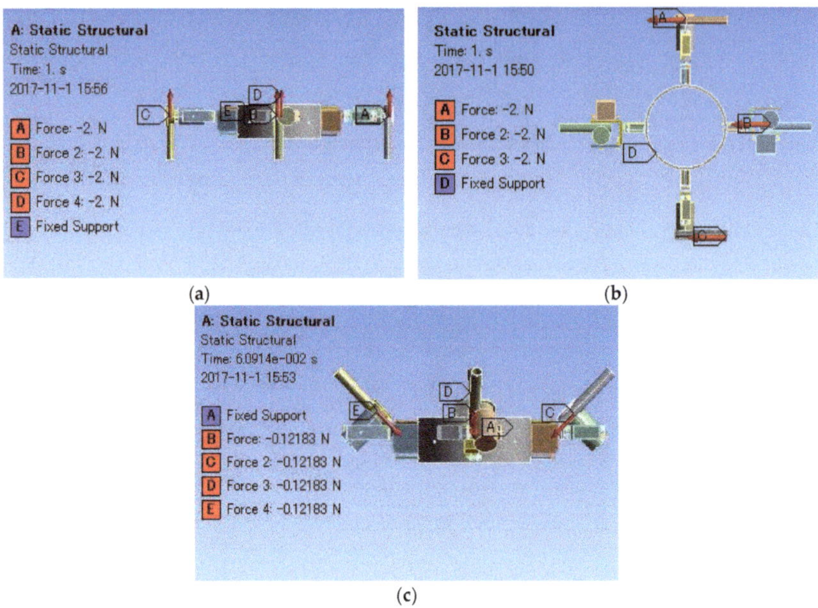

Figure 5. The pre-process of the static analysis. (**a**) Up motion; (**b**) Forward motion; (**c**) Down motion.

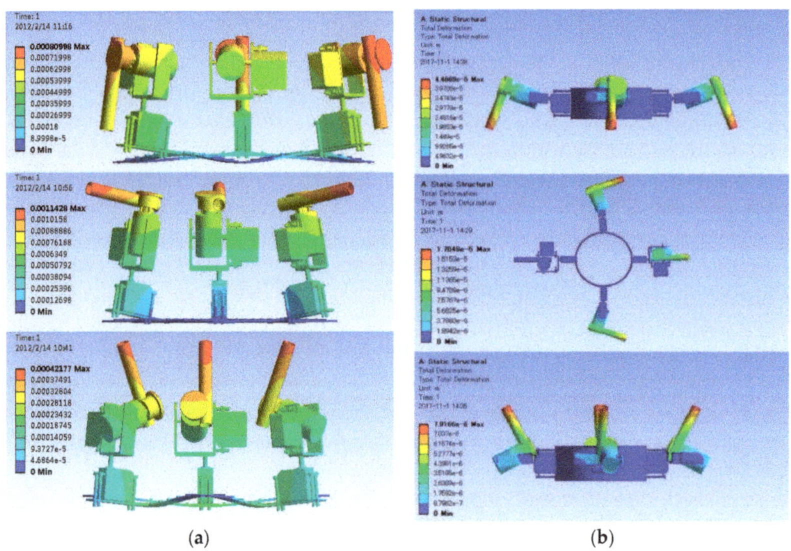

Figure 6. Deformation of the propulsion system. (**a**) Deformation of the propulsion system for SURII; (**b**) Deformation of the propulsion system for SURIII.

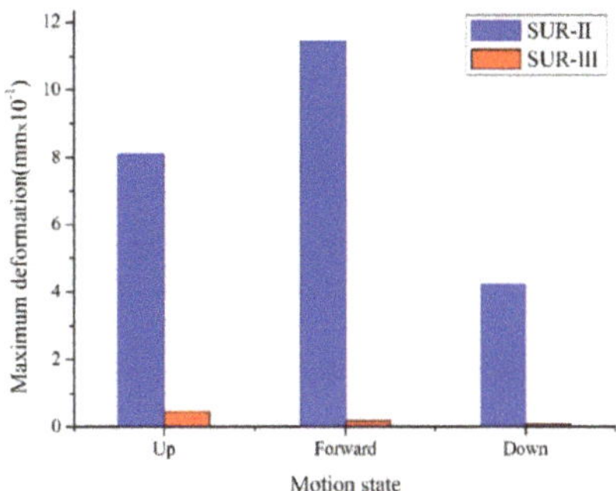

Figure 7. Deformation comparison of SURII and SURIII.

Due to the good characteristics of the vectored water-jet thruster, this part uses the previous design. Each set of vectored water-jet thruster is composed of four parts, which includes a water-jet thruster, a waterproof box for protecting DC motor, two servo motors and a support for them. The propulsive force of the SURIII comes from vectored water-jet thrusters and the orientation of propulsive force can be controlled by the rotation angle of two servo motors.

Two servo motors are rotated in two planes which are vertically intersecting, so they provide two rotational DoFs for the thruster. Figure 8 shows the rotation range of these two servo motors. From the side view of vertical motion in Figure 8a, the nozzle turns 45° counter clockwise, the robot will move down underwater. When the nozzle turns 90° clockwise, the robot will move up to the surface of water. And from the top view of forward motion in Figure 8b, the robot will move forward when the nozzle turns 90° clockwise. If servo motor 1 rotates 180° in the forward motion and then the nozzle turns 90° counter clockwise as shown in Figure 8c, the robot will change from forward motion to backward motion quickly without any change in heading. In other words, through the simultaneous action of these two servo motors, the water-jet thruster can rotate from −90° to 90° in horizontal plane. This angle will affect the orientation of horizontal motions and the orientation can be changed directly by rotating two servo motors. Besides the down motion, all the other motions have the same orientations with the propulsive force and this can maximize the use of it.

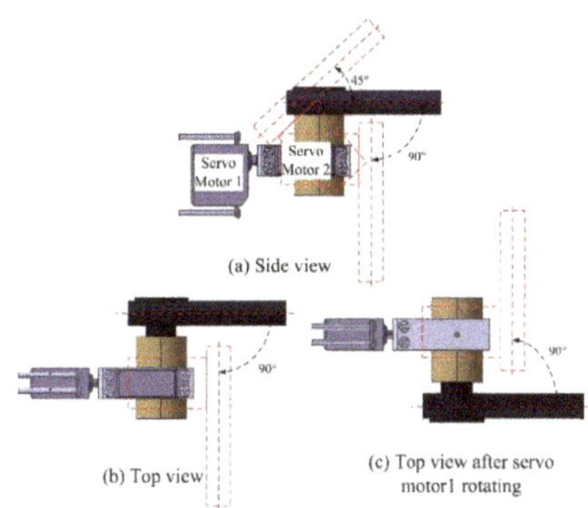

Figure 8. The nozzle angle of different motion states.

2.3. Structure Design of SURIII

In SURIII the design of the propulsion system is improved. The fixed form of the waterproof bin and design of the spherical hull should be changed to fit the new structure. The 3D model of the

SURIII is illustrated in Figure 9. All control circuits are collected in the waterproof bin. Four screws are used to hold up the waterproof bin in the hull. For the spherical hull, it needs to be made four symmetrical openings for four vectored water-jet thrusters. In order to prevent interference between the spherical hull and thrusters and ensure the stiffness of the hull, the opening is designed as shown in Figure 9. The middle part opens wide and both ends open narrowly. Furthermore, on the one hand the hull can block the external damage to protect the internal structure, on the other hand the spherical hull can realize a zero turning radius smoothly.

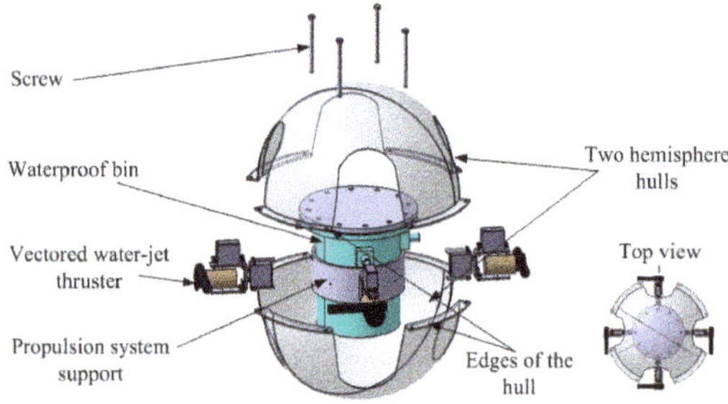

Figure 9. 3D model of SURIII.

2.4. Motions of SURIII

Degrees of freedom (DoF) for the underwater robot include surge, sway, heave, roll, pitch and yaw. Table 1 summarizes the utilization ratios of different degrees of freedom for an underwater robot. DoFs as sway, roll and pitch are rarely used [22].

In SURIII, there are four DoF which involve surge, sway, heave and yaw. These DoFs are shown in Figure 10. There are four water-jet thrusters in the robot and the inclined angle between two thrusters is 90 degrees, so it can be switched between the surge and sway directly by changing the working thrusters.

Table 1. Utilization ratio of different DoFs.

DoF	Surge	Sway	Heave	Roll	Pitch	Yaw
Utilization ratio	100%	31%	96%	33%	7%	100%

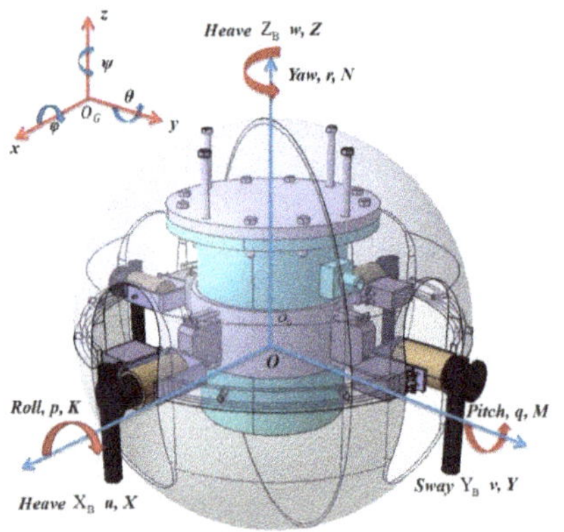

Figure 10. DoFs of SURIII.

Three DoFs which are surge, heave and yaw will be researched as the main target. In surge motion, three thrusters can work simultaneously to achieve a high-speed mode and the remaining thruster can be used to brake or veer, as shown in Figure 11a,b. In addition, two opposite thrusters work together to achieve a low-speed mode, the other two thrusters can be used to brake and veer respectively, as shown in Figure 11c. In yaw motion, a gyroscope will be employed to measure the rotation angle. If all thrusters work at the same time, the robot can be realized a high-speed rotating mode which is shown in Figure 11d, this style is suitable for fast turning. Otherwise working of two opposite thrusters will realize a low-speed rotating mode. Due to the low speed, it is easier to accurately control the rotation. In heave motion, there are two kinds of motions: up and down, corresponding to Figure 11e,f.

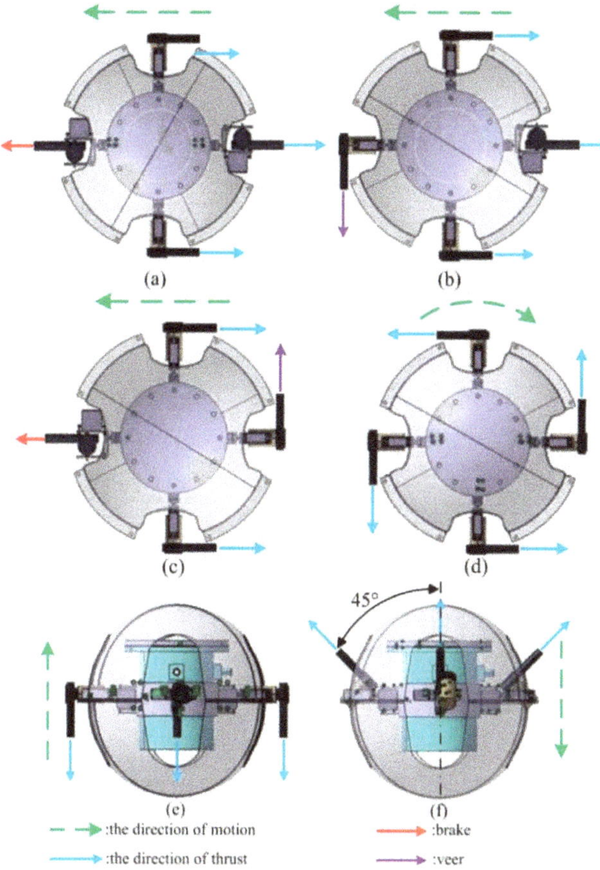

Figure 11. Motion state of SURIII.

3. Hydrodynamic Analysis

The underwater environment is complicated to the AUV. The hydrodynamic analysis of the SURIII described in this section is for getting theoretical basis which will provide initial conditions of the simulation. Before the specific analysis, some assumptions will be given firstly as follows.

1 SURIII is a spherical robot;
2 The flow field is water;
3 The temperature of flow field is 20 °C.

3.1. Dynamic Model of SURIII

Generally, the 6-DoF dynamic equation of underwater robot can be given as [21,23,24],

$$M\dot{v} + C(v)v + D(v)v + G = \tau + \tau_{wind} + \tau_{wave} \quad (1)$$

where $M \in R^{6\times6}$ is the sum of system mass matrix and added mass matrix, $\dot{v}$ is the vector of generalized accelerations, $C(v) \in R^6$ is the Coriolis and centripetal matrix. $D(v) \in R^6$ is the damping matrix, $G \in R^6$ is vector of restoring forces and moments, $\tau \in R^6$ is the vector of generalized forces which input by control system and include the propulsive force and moment, τ_{wind} and τ_{wave} are vectors of forces generated by wind and wave. But effects of wind and wave usually occur in complicated marine environment, for our robot, these two factors can be neglectful. Hence, the dynamic equation can be simplified to Equation (2),

$$M\dot{v} + C(v)v + D(v)v + G = \tau \quad (2)$$

SURIII is a spherical robot and its structure is symmetrical, thus, M can be further simplified as follows,

$$M = \begin{bmatrix} mI_{3\times3} & 0_{3\times3} \\ 0_{3\times3} & I_{zz} \end{bmatrix} \quad (3)$$

where m is the sum of rigid body mass and added mass, I_{zz} is the resultant of inertia tensor. Also, M can be described as follows,

$$M = M_{RB} + M_A \quad (4)$$

where M_{RB} is the mass of the robot, M_A is the added mass. Moreover, $C(v)$ is the sum of rigid body and hydrodynamic Coriolis matrix, so it can be expressed as,

$$C(v) = C_{RB}(v) + C_A(v) \quad (5)$$

The Coriolis force is generated by the rotation of the robot. The rotation speed of the SURIII is minute, so $C(v)$ can be ignored in the hydrodynamic analysis. $D(v)$ is an important parameter for the hydrodynamic analysis of the underwater robot. It contains the linear damping term D_1 and nonlinear damping term $D_q(v)$ and $D(v)$ is give as,

$$D(v) = D_1 + D_q(v) \quad (6)$$

G is the restoring force which contains gravity and buoyancy. It has been mentioned before that the SURIII has four DoFs: surge, sway, heave and yaw. In this research, the SURIII is considered as the suspended status underwater, the gravity is equals to the buoyancy. So, the influence of gravity and buoyancy can be ignored. Finally, the Equation (1) can be simplified as follows,

$$(M_{RB} + M_A)\dot{v} + (D_1 + D_q(v))v = \tau \quad (7)$$

3.2. Related Parameters

M_{RB} is received by weighing the robot in air. The SURIII is 6.9 kg, so the 3D model is:

$$M_{RB} = \begin{bmatrix} m & 0 & 0 \\ 0 & m & 0 \\ 0 & 0 & I_{zz} \end{bmatrix} = \begin{bmatrix} 6.9 & 0 & 0 \\ 0 & 6.9 & 0 \\ 0 & 0 & 0.1615 \end{bmatrix} \quad (8)$$

M_A is the mass of a fluid sphere, so its 3D model is:

$$M_A = \begin{bmatrix} 16.75 & 0 & 0 \\ 0 & 16.75 & 0 \\ 0 & 0 & 0.3864 \end{bmatrix} \quad (9)$$

The coefficients of linear damping force can be expressed as:

$$D_1 v = C_1 \times diag\{u, v, w, r\} \quad (10)$$

The initial conditions stipulate that the temperature of water is 20 °C. At this temperature, the linear viscous coefficient is $C_1 = 1 \times 10^{-4}$. The *diag* $\{u,v,w,r\}$ is the velocity matrix of the SURIII, where u is the velocity of the surge motion, v is the velocity of the sway motion, w is the velocity of heave motion, r is the rotation velocity of the yaw motion. These two parameters are both very small, so $D_1 v$ can be ignored. The Equation (7) may be expressed as:

$$(M_{RB} + M_A)\dot{v} + D_q(v)v = \tau \quad (11)$$

where $D_q(v)v$ equals to the resistance of the robot underwater,

$$D_q(v)v = F_d = \frac{1}{2} C_d \rho V^2 A \quad (12)$$

where F_d is the drag of water, C_d is the drag coefficient, ρ is the density of the fluid, V is the relative speed between SURIII and fluid in different motions, A is the effective cross-sectional area of SURIII.

For surge motion, the SURIII could be approximately as a sphere. In this case, C_d only associated with the Reynolds number [25] and its formula is:

$$R_e = \frac{uD}{v} \quad (13)$$

where R_e is the Reynolds number, u is the velocity of the surge motion, D is the diameter of the robot, ν is the kinematic viscosity of the water and $\nu = 1 \times 10^{-6}$ Pa·s at the temperature of 20 °C. The experimental maximum velocity of surge motion is 0.15 m/s [11], the diameter of the robot is 0.4 m, so the R_e equals to 6×10^4 and this value is more than 10^4 and less than 3×10^5. The drag coefficient in surge motion can be selected from Table 2 [26,27]. If R_e equals to 6×10^4, C_d is 0.4.

Table 2. Relationship between Reynolds number and drag coefficient for the sphere.

R_e	$R_e < 10^4$	$10^4 < R_e < 3 \times 10^5$	$3 \times 10^5 < R_e < 10^6$
C_d	$24/R_e + 6.48 \times R_e^{-0.573} + 0.36$	0.4	0.4
C_d	$30/R_e + 0.46$	0.46	0.46
C_d	$24/R_e + (1 + 0.0654\, R_e^{2/3})^{2/3}$	0.4	0.40
C_d	$(0.325 + (0.124 + 24/R_e^{1/2}))$	-	-
C_d	$(0.63 + 4.8 \times R_e^{-0.5})^2$	0.4	-

In heave motion, the robot cannot be regarded as a sphere. On the one hand, for the case shown in Figure 11a, there are four gaps for the water-jet thrusters and the fluid will flow across these gaps, so that the robot cannot be classified as a closed sphere in this motion. On the other hand, for the case shown in Figure 9, the edge of hull cannot be ignored for heave motion, so that the two hemispheres cannot form a smooth sphere and the diameter of the hull is transformed to 46 cm. According to the Equation (12), the C_d can be expressed as:

$$C_d = \frac{2F_d}{\rho V^2 A} \tag{14}$$

In Equation (14), the maximum experimental velocity of heave motion is 0.083 m/s [11]. The maximum cross-sectional area can be obtained by the projection of 3D model in the vertical direction, so A is 0.1661 m^2. The drag of water cannot be calculated in this case, thus the CFD simulation is conducted to get the estimation of it.

4. CFD Simulation

In this research, all the simulations were calculated by CFX which is associated with ANSYS WORKBENCH. The basic steps of CFD analysis is listed as followings,

A. Analysis of physical problems and pre-processor of the hydrodynamic model;
B. Solver execution;
C. Results of the post-processing.

4.1. Pre-Processor of the CFD Simulation

The pre-processor of the hydrodynamic model includes many contents. Firstly, in the design process, some details need to be added in the 3D model of the robot to make it closer to the prototype. However, it is necessary to simplify the 3D model for CFD simulation, so that it can get effective results and reduce the computational time. The simplified parts are as follows:

(1) The thruster has some complicated surfaces and their area are very small, so these surfaces are pre-processed as regular surfaces;
(2) Some irregular solids have been changed to cylinder or cuboid shape;
(3) Some parts such as screws and nuts have been omitted. And the simulation models are shown in Figure 12.

Secondly, for CFD simulation, another important model is the flow field. The flow field will affect the hydrodynamic characteristic of the robot, so it needs to be established base on the robot. The boundary of the flow field should be large enough, so it will avoid influence the domain around the robot. A cylinder with the diameter of 2 m and the height of 4 m is chosen as the flow field and the robot is put in the center of it. However, in hydrodynamic analysis, it generally needs a computational domain. The computational domain can be obtained by using Boolean operation between 3D model and flow field. The complete hydrodynamic model is shown in Figure 13.

Thirdly, the boundary conditions of the hydrodynamic model are set in the pre-processor. In this research, boundaries make up of inlet, outlet and wall which are shown in Figure 14. For different motions, it only needs to adjust the relationship between the robot and the direction of the inlet. The inlet velocity of the forward motion is 0.15 m/s and the inlet velocity of up and down motion is 0.083 m/s. The flow field is the turbulence handled by the $k - \varepsilon$ model. The mesh is also the crucial step of hydrodynamic analysis. It is generated by using the curvature method. This method can refine the element size of the edge and curve surface. In this hydrodynamic model, the robot is considered as a suspended status. At the present step, the gravity and buoyancy are both ignored. The environment condition is set as the isothermal temperature of 20 °C.

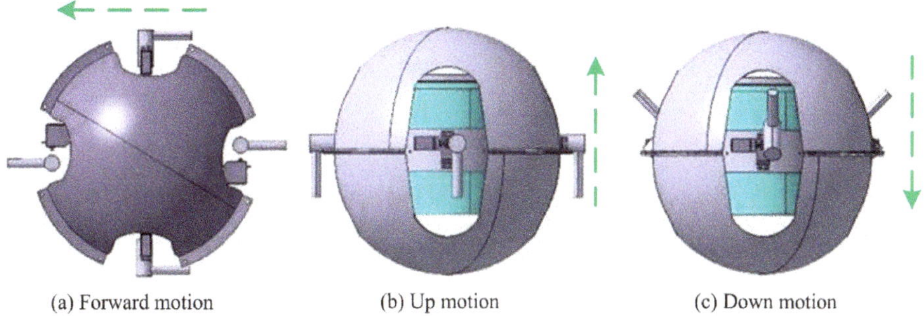

Figure 12. Simulation model of SURIII.

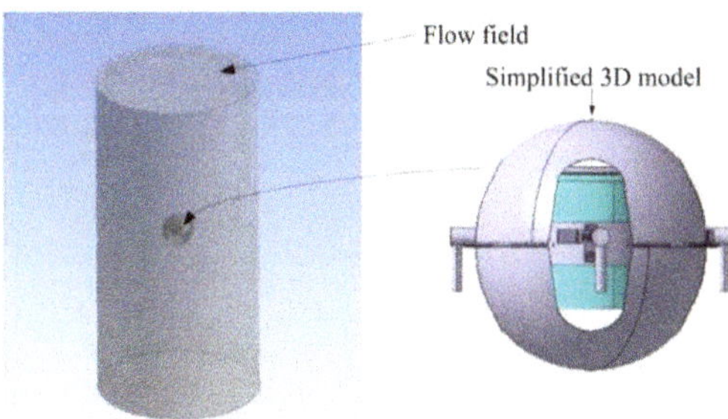

Figure 13. Flow field of SURIII.

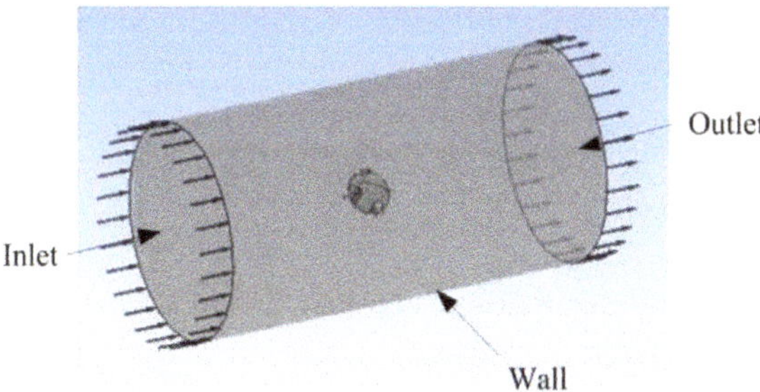

Figure 14. Fluid boundary setting.

Finally, the convergence criterion is set as 1×10^{-4} in the solver control. Three hydrodynamic models (up, down and forward) are calculated for the CFD simulation.

4.2. Results in the Post-Processing

After solver execution, some results can be extracted from the post-processor of ANSYS CFX. In heave motion, there are two kinds of motions: up and down. Figure 15 shows the simulation results of the down motion. Figure 15a is the pressure contour for the down motion, the maximum pressure occurs in the bottom of the robot and near the edge of the robot. This confirms that the influence of the edge in heave motion cannot be ignored. Figure 15b,c display the velocity vector and velocity streamline, respectively. As these two figures shown, the velocity of the down motion also affected by the edge of the robot. Therefore, the influence of the edge needs to be considered when calculating the drag coefficient. As well, the simulation results of up motion are shown in Figure 16. Figure 16a shows the maximum pressure occurs in the top and the edge of the robot. Figure 16b,c are the velocity vector and streamline in the up motion, respectively. As with the down motion, the up motion is also affected by the edge of the robot. For both up and down motions, the water will flow through these gaps and this may have influence on the robot in the simulation. Comparing the simulation results of these two motions, the robot has larger resistance force when it sinks. The resistance of the down motion in the simulation is 0.334 N. The maximum relative velocity in the post-processor is 0.1021 m/s and the maximum cross-sectional area is 0.1661 m^2. According to the Equation (14) and the drag coefficient can be calculated as 0.386. This drag coefficient has a 3.5% error compared to drag coefficient determined by Reynolds number for the sphere. It indicates that edges and gaps in the heave motion effect little.

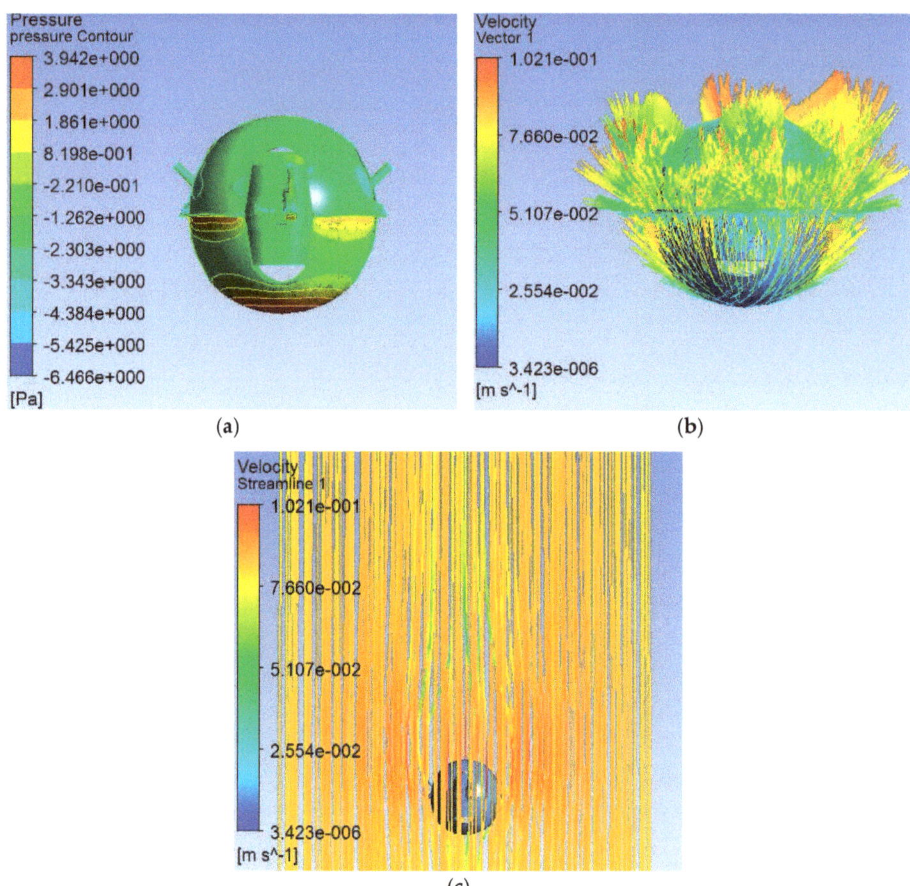

Figure 15. Down motion in heave. (**a**) Pressure contour; (**b**) Velocity vector; (**c**) Velocity streamline.

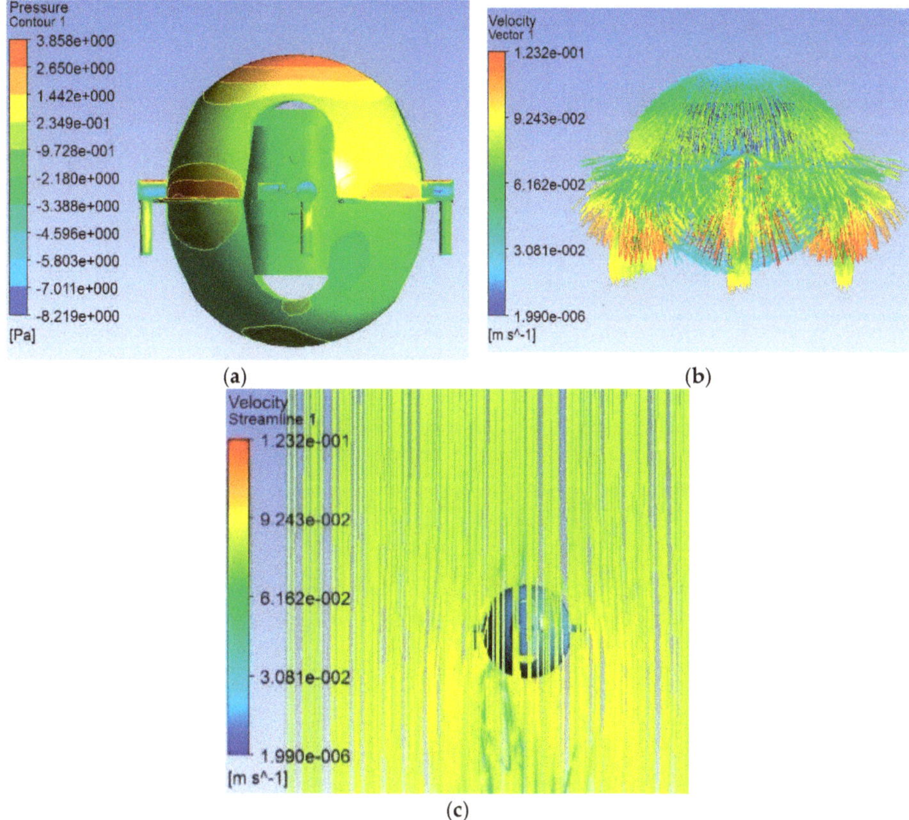

Figure 16. Up motion in heave. (**a**) Pressure contour; (**b**) Velocity vector; (**c**) Velocity streamline.

Figure 17 shows the surge motion for the SURIII, the high-speed mode is analyzed in this research, both the gaps and the edge of the robot cannot be affected to this motion. Figure 17a shows the pressure contour and the maximum pressure occurs near the edge of the robot. Figure 17b,c are the velocity vector and streamline of the high-speed surge motion and the edge can be ignored because of its small impact for the velocity vector. The resistance of the forward motion is also extracted from the post-processor to verify the drag coefficient, the force acting on the robot is 0.988 N in the flow direction, the maximum relative velocity in the post-processor is 0.1998 m/s and the maximum cross-sectional area is 0.1256 m^2. The drag coefficient can be calculated by Equation (14) and is equals to 0.394, it has a 1.5% error compared to the drag coefficient determined by Reynolds number which indicates the robot can be similarly considered as a sphere in surge motion.

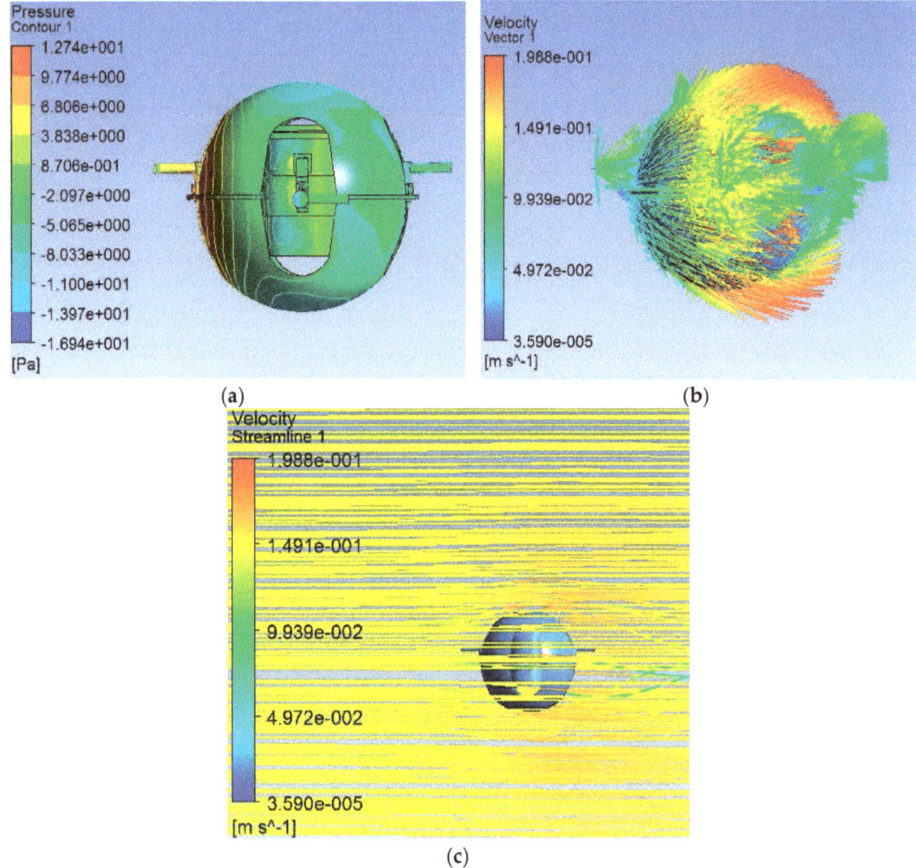

Figure 17. Forward motion in surge. (**a**) Pressure contour; (**b**) Velocity vector; (**c**) Velocity streamline.

5. Experiments and Results

For evaluating the performance of the novel propulsion system, some experiments are conducted in this section. An inflatable swimming pool is prepared with the water at the depth of 25 cm. A load cell with 6-DoF is employed to measure the propulsive force of the multi-vectored water-jet thrusters. Two power supplies are employed as the electric source and each one will give a 7.2 voltage power to two vectored water-jet thrusters. The experimental setup is shown in Figure 18. All vectored water-jet thrusters are connected with the load cell by using the aluminum alloy profile and it can be considered as rigid connection. And all the support parts of the experimental setup are also rigid bodies.

Although the SURIII has 4 DoFs, our experiments only include three different cases as the low-speed mode for the forward motion, high-speed mode for the forward motion and upward motion due to the limitation of experimental conditions. Each case will be carried out for 10 times.

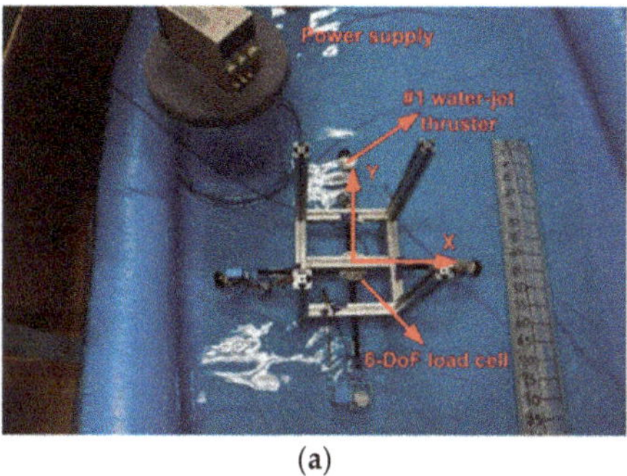

(**a**)

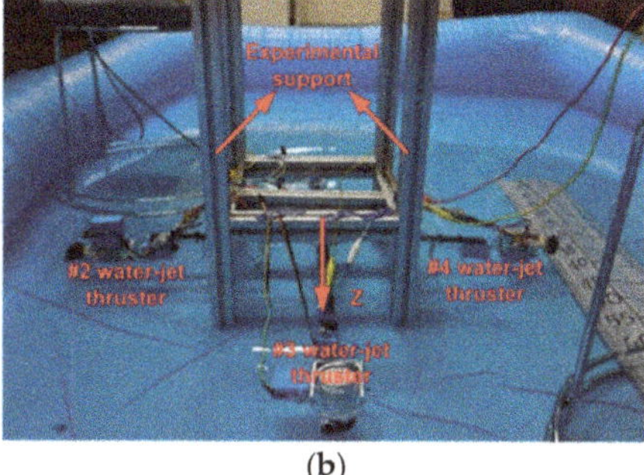

(**b**)

Figure 18. The experimental setup. (**a**) Top view; (**b**) Side view.

Case I: Low-speed mode for forward motion

Step 1: Choose #2 and #4 vectored water-jet thrusters to work (Figure 11c);
Step 2: Adjust the direction of propulsive forces as the Y direction;
Step 3: Set the load cell to obtain 200 values and provide the power supply at 7.2 V;
Step 4: Stop for 30 s and repeat the step 3 and 4 for 10 times.

Case II: High-speed mode for forward motion

Step 1: Choose #2, #3 and #4 vectored water-jet thrusters to work (Figure 11a);
Step 2: Adjust the direction of propulsive forces as Y direction;
Step 3: Set the load cell to obtain 200 values and provide the power supply at 7.2 V;
Step 4: Stop for 30 s and repeat the step 3 and 4 for 10 times.

Case III: Up motion

Step 1: Choose #1, #2, #3 and #4 vectored water-jet thrusters to work (Figure 11e);
Step 2: Adjust the direction of propulsive forces as Z direction;
Step 3: Set the load cell to obtain 200 values and provide the power supply at 7.2 V;
Step 4: Stop for 30 s and repeat the step 3 and 4 for 10 times.

Figure 19 displays the experimental results. The average propulsive force for each time is shown in each case. The average propulsive forces of 10 experiments for each case are 3.57 N, 5.36 N and 7.27 N, respectively. In our previous research, the SURII also has two modes for forward motion. Although an experiment which measured propulsive force of single thruster was conducted, the propulsive force of the high-speed motion and up motion can be calculated by vector synthesis. The maximum propulsive force of forward motion is $2\sqrt{3}$ N. The maximum propulsive force of up motion is 6 N [21]. It indicates that the propulsive force of the novel propulsion system in forward and up motion is better than before. In this experiment, the vectored water-jet thrusters performed in a stable manner. In either case, the changed amplitude of the propulsive force compared with the average value is within 6%.

The Equation (11) shows the relationship between mass, resistance and propulsive force. Although the SURIII is 0.6 kg heavier than SURII in air, the resistance variety is not greater than the previous research [21]. The variety of mass underwater is less than that in air and the propulsive force is improved significantly, so the speed of the SURIII will be enhanced. When the robot moves at a constant speed, the acceleration is equal to 0, then from Equation (11) the resistance can be obtained and is equals to the propulsive force. The movement process of the SURIII can be inferred, the initial stage is an acceleration process, the propulsive force will be greater than the resistance, then the speed of the robot reaches a constant, the propulsive force will be the same as the resistance, finally the robot reaches the target and the speed is slow down, the propulsive force will be less than the resistance.

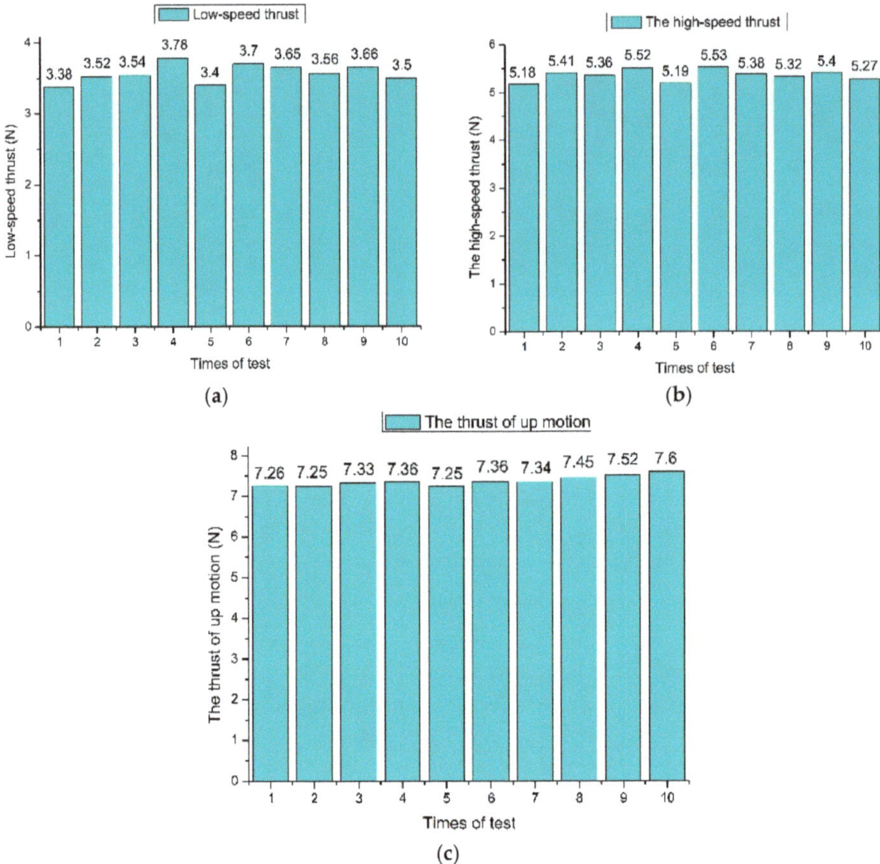

Figure 19. The average propulsive force of different experimental cases. (**a**) Case I; (**b**) Case II; (**c**) Case III.

6. Conclusions

A novel spherical underwater robot which has more stable propulsion system and better hydrodynamic performance than a previous version was developed. Firstly, a novel propulsion system was proposed to improve the dynamic performance and stability of the SURIII. Although one more water-jet thruster increases the weight of the robot from 6.3 kg to 6.9 kg, the improvement of performance for the robot more than offsets the additional weight. In the up and forward motions, the direction of the propulsive force is aligned with the direction of motion, rather than merely having a component along it. Therefore, this kind of propulsion system can provide greater propulsive force for the underwater robot. Secondly, the hydrodynamic analysis of the SURIII was presented as two different motions: surge and heave. According to the influence of edges and gaps in the SURIII, two kinds of methods were given to analyze the drag coefficient. In surge motion, the drag coefficient is 0.4 which can be selected according to the Reynolds number. Thirdly, the hydrodynamic model for CFD simulation was established. Some relevant parameters and conditions were presupposed in the pre-processing. And the pressure contours, velocity vectors and velocity streamlines were obtained from the post-processing. The maximum pressure occurred in the top, bottom and near the edge in heave motion. The maximum resistance was obtained in down motion. The drag coefficient was calculated by hydrodynamic equation and it equals to 0.386. This indicates that edges and gaps effect little. The calculated drag coefficient in surge motion is 0.394, it has a 1.5% error compared with the chosen one, which indicates the robot can be similarly considered as a sphere. Finally, the experiment is conducted for measuring the propulsive force of multi-vectored water-jet thrusters. The experimental results indicate that the propulsive force is greater than that of the previous version. The changed amplitude of the propulsive force is within 6%.

Acknowledgments: This research is partly supported by the National Natural Science Foundation of China (61375094), the National High Tech. Research and Development Program of China (No. 2015AA043202) and SPS KAKENHI Grant Number 15K2120.

Author Contributions: Shuxiang Guo and Shuoxin Gu co-organized the work; Shuxiang Guo and Shuoxin Gu conceived and designed the simulation and experiment work; Shuoxin Gu performed the simulation and experiment work; Shuxiang Guo contributed analysis and experiment tools; Shuoxin Gu wrote the paper; Shuxiang Guo have made substantial contributions to guide and revise the manuscript.

Conflicts of Interest: The authors declare no conflict of interest.

References

1. Galerne, E. Epaulard ROV used in NOAA polymetallic sulfide research. *Sea Technol.* **1983**, *24*, 40–42.
2. Russell, B.W.; Veerle, A.I.H.; Timothy, P.L.B.; Bramley, J.M.; Douglas, P.C.; Brian, J.B.; Henry, A.R.; Kirsty, J.M.; Jeffrey, P.; Daniel, R.P.; et al. Autonomous Underwater Vehicles (AUVs): Their past, present and future contributions to the advancement of marine geoscience. *Mar. Geol.* **2014**, *352*, 451–468.
3. Mazumdar, A.; Fittery, A.; Ubellacker, W.; Asada, H.H. A ball-shaped underwater robot for direct inspection of nuclear reactors and other water-filled infrastructure. In Proceedings of the Robotics and Automation (ICRA) 2013 IEEE International Conference, Karlsruhe, Germany, 6–10 May 2013; pp. 3415–3422.
4. Bhattacharyya, S.; Asada, H.H. Single jet impinging verticial motion analysis of an underwater robot in the vicinity of a submerged surface. In Proceedings of the 2016 IEEE Oceans Conference, Shanghai, China, 10–13 April 2016; pp. 1–8.
5. Kim, H.J.; Lee, J. Swimming Pattern Generator Based on Diving Beetles for Legged Underwater Robots. *Int. J. Mater. Mech. Manuf.* **2014**, *2*, 101–106.
6. Lin, X.; Guo, S.; Tanaka, K.; Hata, S. Underwater experiments of a water-jet-based spherical underwater robot. In Proceedings of the 2011 IEEE International Conference on Mechatronics and Automation, Beijing, China, 7–10 August 2011; pp. 738–742.
7. Lin, X.; Guo, S. Development of a spherical underwater robot equipped with multiple vectored water-jet-based thrusters. *J. Intell. Robot. Syst.* **2012**, *67*, 307–321. [CrossRef]
8. Yue, C.; Guo, S.; Li, M.; Li, Y.; Hirata, H.; Ishihara, H. Mechatronic system and experiments of a spherical underwater robot: SUR-II. *J. Intell. Robot. Syst.* **2015**, *80*, 325. [CrossRef]
9. Yue, C.; Guo, S.; Shi, L. Design and performance evaluation of a biomimetic microrobot for the father-son underwater intervention robotic system. *Microsyst. Technol.* **2016**, *22*, 831–840. [CrossRef]
10. Li, Y.; Guo, S.; Yue, C. Preliminary Concept of a Novel Spherical Underwater Robot. *Int. J. Mechatron. Autom.* **2015**, *5*, 11–21. [CrossRef]
11. Li, Y.; Guo, S.; Wang, Y. Design and Characteristics Evaluation of a Novel Spherical Underwater Robot. *Robot. Auton. Syst.* **2017**, *94*, 61–74. [CrossRef]
12. Li, Y.; Guo, S. Communication between Spherical Underwater Robots Based on the Acoustic Communication Methods. In Proceedings of the 2016 IEEE International Conference on Mechatronics and Automation, Harbin, China, 7–10 August 2016; pp. 403–408.
13. Kim, H.J.; Song, S.H. A turtle-like swimming robot using a smart soft composite (SSC) structure. *Smart Mater. Struct.* **2012**, *22*, 14007. [CrossRef]
14. Yue, C.; Guo, S.; Li, Y.; Li, M. Bio- Inspired robot launching system for a mother-son underwater manipulation task. In Proceedings of the 2014 IEEE International Conference on Mechatronics and Automation, Tianjin, China, 3–6 August 2014; pp. 174–179.
15. Watson, S.A.; Crutchley, D.J.P.; Green, P.N. The design and technical challenges of a micro-autonomous underwater vehicle AUV. In Proceedings of the 2011 IEEE/ICMA International Conference on Mechatronics and Automation, Beijing, China, 7–10 August 2011; pp. 567–572.

16. Waston, S.A.; Green, P.N. Propulsion System for Micro-Autonomous Underwater Vehicles (μAUVs). In Proceedings of the 2010 IEEE Conference on Robotics, Automation and Mechatronics, Singapore, 28–30 June 2010; pp. 435–440.
17. Fittery, A.; Mazumdar, A.; Lozano, M.; Asada, H.H. Omni-Egg: A smooth, spheroidal, appendage free underwater robot capable of 5 DOF motions. In Proceedings of the IEEE Oceans, Hampton Roads, VA, USA, 14–19 October 2012; pp. 1–5.
18. Guo, S.; Du, J.; Ye, X.; Gao, H.; Gu, Y. Realtime adjusting control algorithm for the spherical underwater robot. *Inf. Int. Interdiscip. J.* **2010**, *13*, 2021–2029.
19. Guo, S.; Du, J.; Ye, X.; Yan, R.; Gao, H. The computational design of a water-jet propulsion spherical underwater vehicle. In Proceedings of the 2011 IEEE International Conference on Mechatronics and Automation, Beijing, China, 7–10 August 2011; pp. 2375–2379.
20. Gemmell, B.J.; Costello, J.H.; Colin, S.P.; Stewart, C.J.; Dabiri, J.O.; Tfti, D.; Priya, S. Passive energy recapture in jellyfish contributes to propulsive advantage over other metazoans. *Proc. Natl. Acad. Sci. USA* **2013**, *110*, 17904–17909. [CrossRef] [PubMed]
21. Yue, C.; Guo, S.; Shi, L. Hydrodynamic Analysis of the Spherical Underwater Robot SUR-II. *Int. J. Adv. Robot. Syst.* **2013**, *10*, 1–12. [CrossRef]
22. Jiang, X.; Feng, S.; Wang, L. *Unmanned Underwater Robot*; Liaoning Science and Technology Publishing House: Shengyang, China, 2000.
23. Fossen, T.I. *Guidance and Control of Ocean Vehicles*; John Wiley & Sons Inc.: Hoboken, NJ, USA, 1994.
24. Lin, X.; Guo, S.; Tanaka, K.; Hata, S. Development and Evaluation of a Vectored Water-jet-based Spherical Underwater Vehicle. *Information* **2010**, *13*, 1985–1998.
25. Houghton, E.L.; Carpenter, P.W. *Aerodynamics for Engineering Students*, 5th ed.; Butterworth-Heinemann: Oxford, UK, 2013; pp. 8–15.
26. Ceylan, K.; Altunbas, A.; Kelbaliyez, G. A new model for estimation of drag force in the flow of Newtonian fluids around rigid or deformable particles. *Powder Technol.* **2011**, *119*, 250–256. [CrossRef]
27. Kundu, P.K.; Ira, M.C.; David, R.D. *Fluid Mechanics*, 5th ed.; Academic Press: Cambridge, MA, USA, 2012; pp. 4251–4472.

Article

Swarming Behavior Emerging from the Uptake–Kinetics Feedback Control in a Plant-Root-Inspired Robot

Emanuela Del Dottore *, Alessio Mondini, Ali Sadeghi and Barbara Mazzolai *

Center for Micro-Biorobotics, Istituto Italiano di Tecnologia, 56025 Pontedera, Italy; alessio.mondini@iit.it (A.M.); ali.sadeghi@iit.it (A.S.)

* Correspondence: emanuela.deldottore@iit.it (E.D.D.); barbara.mazzolai@iit.it (B.M.); Tel.: +39-050-883092 (E.D.D.)

Received: 15 November 2017; Accepted: 26 December 2017; Published: 1 January 2018

Abstract: This paper presents a plant root behavior-based approach to defining the control architecture of a plant-root-inspired robot, which is composed of three root-agents for nutrient uptake and one shoot-agent for nutrient redistribution. By taking inspiration and extracting key principles from the uptake of nutrient, movements and communication strategies adopted by plant roots, we developed an uptake–kinetics feedback control for the robotic roots. Exploiting the proposed control, each root is able to regulate the growth direction, towards the nutrients that are most needed, and to adjust nutrient uptake, by decreasing the absorption rate of the most plentiful one. Results from computer simulations and implementation of the proposed control on the robotic platform, Plantoid, demonstrate an emergent swarming behavior aimed at optimizing the internal equilibrium among nutrients through the self-organization of the roots. Plant wellness is improved by dynamically adjusting nutrients priorities only according to local information without the need of a centralized unit delegated for wellness monitoring and task allocation among the agents. Thus, the root-agents can ideally and autonomously grow at the best speed, exploiting nutrient distribution and improving performance, in terms of exploration capabilities and exploitation of resources, with respect to the tropism-inspired control previously proposed by the same authors.

Keywords: bioinspired control; swarm intelligence; plant-inspired robot; emergent behavior

1. Introduction

A decentralized control system is a system in which the components act on the basis of local information when accomplishing global tasks. In such systems, the collaborative behavior emerges from independent local decisions without the need for centralized processing [1]. This definition is intrinsically linked to the idea of self-organization. Many natural systems have been studied in terms of their ability to perform complex tasks without a centralized control but due to simple rules followed by many distributed agents with communication capabilities [2]. Well-known examples include ant colonies that are able to accomplish foraging tasks by following pheromone trails [3], or honeybees that indicate the direction of the nectar source by dancing [4]. However, systems that also have no nervous system are considered, such as bacteria that organize themselves in order to maximize nutrient availability [5].

The analysis of these behaviors has inspired routing algorithms [6], load balance problem solutions [7], ant colony optimization [8] or particle swarm optimization [9] algorithms. This approach typically provides a scalable and robust solution to large-scale complex problems. It has also opened the door to a relatively new discipline, called swarm robotics, which applies swarm intelligence principles to robotics [10]. Collaborative exploration specifically in unstructured environments such as disasters or dangerous areas is a particularly important application of this discipline.

Like animals, plants also need to explore the environment for foraging purposes. They actively interact with the environment perceiving, for instance, the presence of obstacles and adjusting their growth when mechanically stimulated [11,12]. Moreover, they need to optimize their energy due to the uncertainty of nutrient availability. Specifically, production, mobilization and allocation among the tissues of photosynthesis products (e.g., carbon and sugars), that regulate plant growth and development, are highly affected by sugar and hormone signals in response to environmental cues [13–15]; in addition, nutrient uptake and usage are regulated according to the availability of the nutrient [16]. However, unlike animals, plant locomotion is irreversible, since it takes place through organ growth, which suggests that plants should focus more on decision-making activities compared to animals.

Plants have been already taken as source of inspiration in engineering [17–19], and they have been also explored for optimization algorithms [20]. Macro-rules for the design of metaheuristics have been extracted for instance from pollination processes [21], the colonization of invasive weeds [22] or strawberry plant propagation strategies [23]. Plant roots have also been considered, in particular, their distribution in searching for optimal soil, water, and fertilizer conditions. Specifically, Qi et al. [24] proposed the Root Mass Optimization Algorithm (RMO) where the search for optimality is driven by operators inspired by the concepts of growth and branching, and the search evolves through generations of roots, as with a classical genetic algorithm. Roots, which represent different initialization points in the search domain, can grow in conditions of optimal soil impedance, water, and fertilization. This optimal condition is monitored by a fitness function. In addition, roots can decide to generate a branch with a random probability in a random position; each root is then evaluated and the best are selected for the next generation. Similarly, Zhang et al. [25] proposed the Root Growth Algorithm (RGA) based on root branching and root hair growth operators. In this case, the length and distance of hairs and roots are also important in obtaining wider spatial distribution and increasing the diversity of fitness values.

However, to our knowledge, in the robotic community, plants have not yet been explored as swarm intelligent systems, while at the same time extensively analyzing plant root behavior.

In the field of robotics, in a previous paper, we analyzed plant root behavior for the implementation of a plant-inspired control [26]. The previous control was inspired by the tropic responses of plant roots, e.g., attraction to water, attraction to gravity, attraction to an optimal defined temperature, repulsion to obstacles. From the observation of tropisms, we developed a stimulus-oriented control. The direction of growth or bending was defined by combining the preferential direction obtained for each stimulus. Each stimulus had a fixed priority to amplify the attraction or repulsion towards or away from that stimulus, and, by vectorization, we defined the preferred direction of growth or bending. However, at this stage, the chemical signals were neglected with a consequent disregard of the nutrient uptake mechanism and the regulation of internal needs. The robotic roots operated independently on the basis of local perception, without an internal memory and with no inter-agent communication.

This paper presents a plant root behavior-based approach for defining the control architecture of a plant-root-inspired robot. Specifically, we looked at the movements, communication channels and uptake mechanism used by plant roots to explore and exploit the environment for the entire plant survival. We demonstrate that taking inspiration from plants can lead to the extraction of new technologies and control principles that are relevant in robotics as well as in other fields (e.g., optimization problems, traffic management, marketing strategies, etc.), in the same way as ethology did. At the same time, this approach can improve the knowledge on plant behavior by extending the analysis of internal processes.

In this work, we aim to verify the hypothesis that plant roots can be considered as simple agents working as a swarm in order to ensure plant survival, although still acting independently on the basis of local information and perception. With respect to our previous plant-inspired control [26], in this paper, we introduce local memory, inter-agent communication and the dynamic evolution of nutrient priorities.

Section 2 provides an extensive explanation of the biological aspects characterizing plant root behavior from which the main properties for control are extracted (Section 2.1) and from which the plant wellness problem is formalized (Section 2.2). These features are then implemented as a control strategy for a robotic platform, the Plantoid [26], which mimics the key movements of plant roots, i.e., directional bending of the apical part of the root. The robotic system and the simulated environment used for validation are briefly introduced in Section 2.3, followed by a detailed presentation of the implemented control (Section 2.4) and the experiments performed to validate the hypothesis (Section 2.5). Results of simulations and of the experiment performed on the robot are presented in Section 3, and discussed in Section 4.

2. Materials and Methods

2.1. Clues from Plants

2.1.1. Uptake–Kinetics

As plants are sessile organisms, they have to develop a series of strategies for survival. The ability to adapt morphological and physiological properties to environmental stimuli is called plasticity and has enabled plants to explore and exploit the environment [27]. The role of roots is to supply nutrients to the whole plant. In order to maximize the probability of success, they have developed several strategies for nutrient exploitation, e.g., increase in root hairs, growth of lateral roots in patches of soil that is nutrient rich (morphological plasticity) or increasing the nutrient uptake rate (physiological plasticity) in conditions of nutrient deficiency [28].

In soil, nutrients are not always present and distributed in such a way to satisfy the requirements of plants for optimal growth [29]; indeed, plant growth is limited by their availability, quantity, and ratio [30]. Ion charges in plants also need to maintain a balance for the correct evolution of processes such as protein synthesis or ion transportation through the membranes. Experiments evaluating the interaction among ions show in fact that the uptake of Na^+ and K^+ changes according to the presence of calcium in the medium or processes such as the synthesis of organic acids are altered by excessive cation or anion uptake [31]. These observations suggest that the internal concentrations of nutrients in plants need to maintain an equilibrium to prevent process alterations. Various characteristic indices and requirements among macronutrients have also been established [32].

There is thus the first fundamental property of plant behavior:

Property 1: Plant growth is driven more by maintaining a balance of the internal nutrient concentrations rather than by collecting the closest and most available nutrient in the soil.

The consequence of *Property 1* is the selectivity of ion uptake. Uptake rates have been analyzed in both lower and higher plants by comparing the accumulation of nutrients in roots with the concentration remaining in the external solution. Results have shown that ratios among these two quantities differ for each observed nutrient [31], and thus confirm the selective characteristic of nutrient absorption by plants.

Nutrient uptake is known to work as a function of the nutrient concentration in soil (at least up to a limiting threshold) with the saturation kinetics mechanism, which is described by the Michaelis–Menten equation [33], and defines the uptake rate, or absorption velocity, as:

$$I = \frac{I_{max} \times C}{K_m + C}. \tag{1}$$

In Equation (1), I_{max} identifies the capacity factor, maximal rate of absorption, which is approached asymptotically when the ion concentration in the medium increases. K_m represents the concentration with half of the maximal rate of absorption and C is the concentration perceived. It has been shown that the parameters of the uptake–kinetics (I_{max} and K_m) are strongly influenced by the internal concentration status of the plant. For instance, in both *Zea mays* (corn) and soybean, it has been shown that with an increasing concentration of phosphorus in plants, both parameters decrease linearly (I_{max} more rapidly than K_m) [34]. The same behavior has been observed in barley roots when the internal nitrate availability increases [35]. This adjustment of I_{max} and K_m suggests a second property:

Property 2: A feedback control modifies the uptake rate of a nutrient according to its internal state in plants.

2.1.2. Tropisms

Another kind of plasticity shown by plants is the directional response to environmental stimuli, which can be attractive or repulsive (tropism); for instance, gravity is an attractive stimulus in roots, inducing them to bend towards the gravity vector (gravitropism). With thermotropism, temperature has been shown to be attractive (repulsive) when a low (high) threshold temperature is reached [36]. And directed responses towards increasing concentrations of moisture with hydrotropism [37], or towards attractive chemicals in chemotropism [38] (or away from negative chemicals [39]) have also been observed. These and many other tropisms (thigmotropism, phototropism, magnetotropism, etc.) interact with each other, thus leading to a unified directional response [40]. For instance, the interaction between gravitropism and mechanical stimulation has been analyzed observing a modulation of the response to gravity under mechanical stress [41]; analogously, the interaction among gravitropism and hydrotropism showed a reduced response to gravity in the presence of moister gradients [42].

The above observations can be summarized in three main properties of plant root behavior:

Property 3: Roots show directional responses towards or away from a stimulus;
Property 4: The directional response is probably induced by the perception of a gradient;
Property 5: Tropic responses are combined to obtain a single directional response.

2.1.3. Plant Intra-Communication and Local Storage

The sharing of minerals and other substances between roots and shoot is possible in plants thanks to the internal vascular system, which is composed of two channels, called the xylem and phloem (Figure 1). The xylem is the central vessel that runs along the whole structure and represents a direct omnidirectional connection from roots to shoot. Here, water and nutrients are transported by bulk flow, exploiting water pressure, to the aerial parts of the plants. On the other hand, the phloem is a slower connection channel, with an osmotic mechanism, enabling substances to be diffused from source to sink. In this channel, elements move from an area with a high concentration (source) to an area with a lower concentration (sink). In fact, the direction of transport is defined by the nutritional requirements of the different plant organs or tissues, ensuring nutrient cycling and a fair redistribution between shoot and roots. It can thus be considered as an important communication channel of the internal nutritional status [31], e.g., if a required nutrient is not delivered to the sink, it is highly probable that that nutrient is not available.

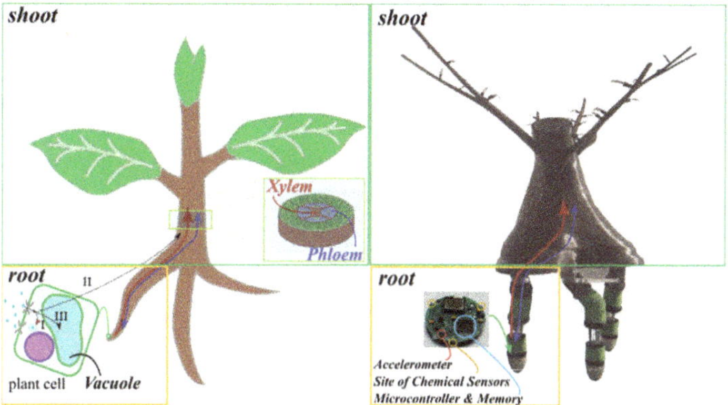

Figure 1. The plant system. On the left, the biological model showing communication channels: the red arrow represents the monodirectional communication from root to shoot (xylem), while the blue arrow is the bidirectional communication along the whole system; on the root module, a schematic of ions uptake is presented with case (I) of internalization, case (II) of immediate sending to the shoot and case (III) storage in vacuole. On the right, there is the analogy with the plant inspired robot (Plantoid).

Once mineral nutrients in the soil reach the root surface and are absorbed by cells, they can (I) be immediately used for local processes (e.g., protein synthesis), (II) sent to the shoot through the xylem or (III) stored in vacuoles, which are cell components that work as pools for substances that need to be readily provided to cell processes [43]. When the request from the shoot is high, nutrients are rapidly pumped up (alternative II), and then redistributed towards the requesting sinks.

Plants thus also have the following three properties:

Property 6: There is a fast and direct highway where nutrients are transported immediately from roots to shoot;

Property 7: Nutrients are distributed among requesting organs according to the strength of their requests;

Property 8: Root tissues and cell vacuoles are local memories storing information on nutrient status.

In fact, regulation of the uptake rate has been correlated to the mineral nutrients stored in vacuoles [44].

2.2. Plant Wellness Problem

The eight properties outlined in Section 2.1 indicate that the interest of the entire plant is to collect nutrients, thus preserving optimal ratios in order not to compromise the correct functioning of internal processes. To confirm this theory, the demand of nutrients in plants was found to reflect specific ratios between nutrients (i.e., N:P and K:P both equal to 10) [45]. We can thus formalize the plant's interest in minimizing the imbalance (ε_P) among nutrients during its life (for every instant of time $\bar{t}$):

$$\min\{\varepsilon_P(\bar{t})\},\ \forall \bar{t}, \tag{2}$$

with the imbalance defined as:

$$\varepsilon_P(\bar{t}) = \sum_{j=1}^{J-1} \sum_{l=j+1}^{J} \left| \frac{c_P^j(\bar{t})}{K_{Ch}^j} - \frac{c_P^l(\bar{t})}{K_{Ch}^l} \right|, \tag{3}$$

where J represents the set of nutrients and K^j_{Ch} represents the ratio between nutrient j and Ch, where Ch is a nutrient chosen as reference among all. $c^j_P(\bar{t})$ is the concentration of nutrient j in the entire plant at a certain instant of time ($\bar{t}$). The concentration in the plant of a single nutrient in a certain instant of time can then be obtained by:

$$c^j_P(\bar{t}) = \int_0^{\bar{t}} (\sum_{r=1}^{R} I^j_r(t) - \sum_{r=1}^{R} O^j_r(t)) \times dt, \quad (4)$$

where R is the number of roots in the apparatus, $I^j_r(t)$ is the uptake of nutrient j from root r at time t and $O^j_r(t)$ is the consumption of nutrient j from root r at time t. In (4), for this work, only uptake and consumption actuated by the root apparatus are considered, neglecting photosynthesis and other transformation processes actuated in the shoot.

2.3. Robotic Architecture, Simulated Environment and Sensing

Root-inspired behavior was implemented first in a simulation and then in a robotic platform called Plantoid, which is a plant-inspired robot with a root apparatus where the robotic roots mimic the bending movements of plant roots thanks to the actuation of three soft springs that can elongate differentially [26]. Each root is endowed with perception capabilities by embedding an accelerometer, to detect gravity, three commercial temperature sensors, placed at 120° from each other, and customized humidity and tactile sensors (for details on the design and sensors, see [26]).

The overall system (Figure 1) can be considered as a multi-agent system with two types of agents: a shoot agent, grouping all aerial elements (trunk, branches and leaves), dedicated in this implementation only to the collection and redistribution of nutrients, plus three root agents that search for and collect nutrients. Root and shoot agents, in the following also only called roots and shoot unless ambiguous, each have their local memory for nutrient storage with a maximal capacity (with variable name storeCapacity) for each nutrient (*Property 8*). Roots have a direct communication with the shoot in both directions to simulate the xylem channel (from roots to shoot) and phloem channel (bidirectionality).

With the three soft spring robotic roots, it is possible to visualize the directional response by bending the root (Supplementary Video S1). While growth is simulated on a virtual environment created in MATLAB (R2016b, The Mathworks, Natick, MA, USA) and containing three roots where only the skeleton is visible. The simulated environment is used to plot tip positions (the apical part is at the beginning oriented downwards), the historical path (previous positions of the point at the back of the tip) and to provide a chemical stimulation to the robotic tips. Roots are stimulated with gravity and chemical stimuli. On the robot, temperature sensors are used instead of chemical sensors, thus the gradient of a nutrient is simulated with a gradient of temperature, while the other two nutrients are provided with the virtual environment to the robot. Consequently, chemical receptors on the simulated roots are localized at the same positions as temperature sensors present in the robotic root. In fact, there is a receptive site for nitrate (N), for potassium (K) and for phosphorus (P) every 120° along the circumference of the tip.

The environment simulates static gradients of the three selected nutrients (Figure 2), drawn with a Gaussian function:

$$G(x) = d \times e^{\frac{-(x-c)^2}{2\sigma^2}}, \quad (5)$$

with σ Gaussian root mean square width, d maximal concentration and c central location of nutrient in the soil.

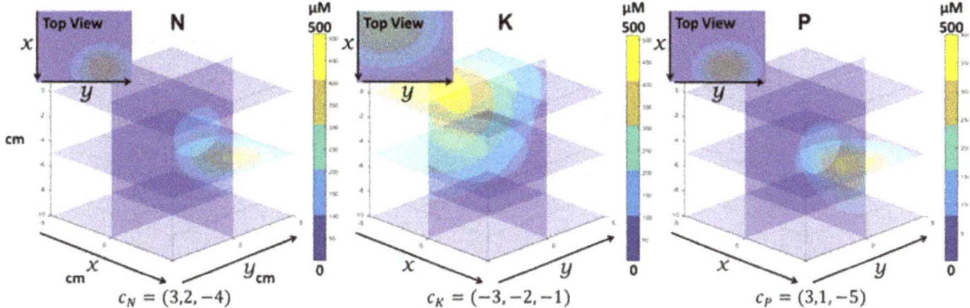

Figure 2. An example of the virtual environment. Nutrients gradients are visualized over three separated spaces, one for each nutrient (N: nitrate, K: potassium, P: phosphorus) where the centers are, respectively, $C_N = (3, 2, -4)$, $C_K = (-3, -2, -1)$, $C_P = (3, 1, -5)$. Roots move in the environment where these three gradients coexist.

The proof of concept works best with a simplified system so several assumptions were made. For instance, it is assumed that all nutrients are highly requested by the shoot to force the immediate sending of the total amount from roots to shoot (using *Property 6*—alternative II in Section 2.1.3). It is assumed that there is always enough water potential to pump ions in the xylem channel, and to always have a constant optimal temperature in order to consider the influence of this factor to be negligible on the uptake–kinetics parameters. In addition, only the interaction between gravity and attractive chemicals are considered here.

The aerial part in this case only works as a gateway, collecting and redistributing the absorbed nutrients. This means that processes such as photosynthesis and nutrient transformation are not modeled. Consequently, energy production, redistribution, and consumption are not considered here.

Taking this simplification into account, to prevent a rapid filling of local storages, a consumption factor was introduced to decrease the root local memory, which should be in the future related to energy consumption. All three nutrients are decreased with an amount equal to the minimum nutrient stored minus a constant threshold.

2.4. Uptake–Kinetics Feedback Control

At each time step, each root takes a decision independently from the other and with only the knowledge of its internal state and environmental perception. Therefore, each robotic root is an autonomous agent that repeatedly performs steps in the following order:

1. Update of internal state and uptake–kinetics parameters;
2. Perception of the environment;
3. Uptake of nutrients;
4. Nutrients sent to shoot;
5. Evaluation of growing direction;
6. Growth.

As the shoot is a collector and distributor of nutrients (satisfying *Property 7*), when it receives all the nutrients from each agent, it sends an amount of nutrients back in proportion to the request received from the root. In fact, together with the uptake, each root also makes a request to the shoot for each nutrient that corresponds to the free internal memory. The request is expressed by the root as a percentage of free memory over the storeCapacity.

The feedback control inspired by the uptake–kinetics of plant roots, called the uptake–kinetics feedback control, is summarized in Figure 3.

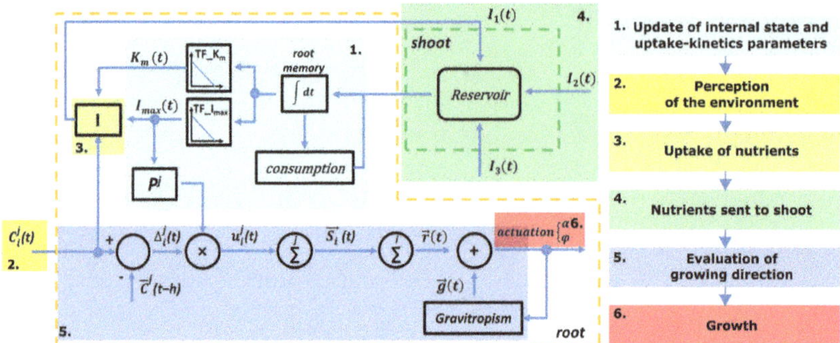

Figure 3. Schematic of the uptake–kinetic feedback control. On the left, a detailed control diagram, and on the right, the workflow of the control showing the with sequential blocks of internal macro operations. Each root sends the quantity of each nutrient acquired from the environment to the shoot agent that redistributes values among all root agents. Internally, each root, after updating its memory, updates Michaelis–Menten parameters and nutrients' priorities. On the second step, each root senses the environment and with the concentration perceived for each nutrient and its corresponding Michaelis–Menten parameters actuates the uptake and sends it to the shoot; with the same perception, each root performs also the evaluation of the next direction of growth that is obtained as combination between gravitropism and chemotropism.

At the beginning of each loop, the root updates the internal status with the nutrients received by the shoot and then proceeds to update the uptake–kinetics parameters accordingly (Step 1). New I_{max} and K_m are obtained as a function of the internal quantity of each nutrient (*Property* 2) with a linear transformation. Coefficients of transformation functions and fitting parameters are reported in Table 1.

Table 1. Transformation functions. Coefficients of the linear functions used to obtain the parameters I_{max} and K_m of the uptake rate, given the internal concentration of each nutrient. The last column reports the points used for polynomial fitting. Considering that Michaelis–Menten parameters have been found to preserve the same ratios of nutrient requirement (N:P and K:P equal to 10 from [45]), points for fitting I_{max} and K_m, for nutrient N and K, were estimated from the values known for nutrient P at different concentration as found in the literature [34].

		p_1	p_2	Points Used for Fitting
Transformation	N	-6.992×10^{-3}	3.672×10^{-6}	$(0, 3.672 \times 10^{-6})$, $(5 \times 10^{-4}, 1.76 \times 10^{-7})$
function for I_{max}	K	-6.992×10^{-3}	3.672×10^{-6}	$(0, 3.672 \times 10^{-6})$, $(5 \times 10^{-4}, 1.76 \times 10^{-7})$
(TF_I_{max}) $y = p_1 \times x + p_2$	P	-6.992×10^{-3}	2.125×10^{-7}	$(0, 3.672 \times 10^{-7})$, $(5 \times 10^{-4}, 1.76 \times 10^{-8})$
Transformation	N	-8.4×10^{-4}	61	$(0, 61)$, $(2 \times 10^{-4}, 19)$
function for K_m	K	-8.4×10^{-4}	61	$(0, 61)$, $(2 \times 10^{-4}, 19)$
(TF_K_{max}) $y = p_1 \times x + p_2$	P	-8.4×10^{-4}	6.1	$(0, 6.1)$, $(2 \times 10^{-5}, 1.9)$

Since the variation of I_{max} reflects the variation in the internal state of nutrients, it can be considered as an estimator of nutrient priorities. In fact, when the internal state of a nutrient increases, its I_{max} decreases, indicating that this nutrient needs to reduce its uptake because of the increase in its internal availability. We mapped I_{max} directly into a priority:

$$P = \frac{1.1 \times {I_{max}}^2}{{I^{+}_{max}}^2} - 0.1, \quad (6)$$

with I^{+}_{max} the maximum value of I_{max} (Table 2) obtained from the literature [34,45]. In Equation (6), we used a quadratic function to speed-up the priority adjustment.

Table 2. Parameters initialization. Initialization of the constants I^{+}_{max} and storeCapacity used in the root agent control. Values of I^{+}_{max} were fixed considering ratios N:P and K:P equal to 10 as found in the literature [34,45]. P was selected as the referential nutrient (*Ch*) and used for conversions and normalizations since it is the nutrient that has the lowest absolute value of the three nutrients required by plants.

	N	K	P
I^{+}_{max}	36.72×10^{-7} μM cm^{-1} s^{-1}	36.72×10^{-7} μM cm^{-1} s^{-1}	36.72×10^{-8} μM cm^{-1} s^{-1}
storeCapacity	0.002 μM	0.002 μM	0.0002 μM

If a nutrient is completely lacking, its priority rises to 1. On the other hand, when the internal memory is full, or, as in our case, when it reaches a maximal filling threshold (we imposed a threshold equal to storeCapacity/4), this nutrient is no longer needed and its priority decreases to 0. *P* becomes negative when a nutrient is accumulated over the filling threshold, transforming that nutrient into a repulsive stimulus.

For each stimulus perceived by receptor site i, the corresponding tropic response (*Property 3*) is defined by the vector whose magnitude expresses the strength of attraction in that direction, similarly to [26], but here, unlike in [26], each nutrient j is weighted with its dynamic priority (P^j) obtained by Equation (6). In addition, while in [26], the instantaneous concentration value was considered, now due to *Property 4*, the variation in concentration of nutrient j perceived in direction i ($\Delta^j_i(t)$) is taken, obtained as the difference between the current concentration at time t with the averaged concentration among all directions at some previous time step ($t - h$). For each stimulus in each direction, the strength of attraction is defined as:

$$u^i_j(t) = P^j \times \Delta^j_i(t). \tag{7}$$

Vector of attraction for chemical stimulation towards each direction i is then defined by:

$$\vec{s_i}(t) = \hat{i} \times \sum_{j=1}^{J} u^i_j(t), \tag{8}$$

where $\hat{i}$ represents the unit vector towards direction i. Thus, for a generic nutrient j that has a positive priority P^j, when the concentration of nutrient j decreases ($\Delta^j_i(t)$ is negative), the chemotropic response becomes repulsive towards direction i for that nutrient, while it is attractive if the concentration increases (positive $\Delta^j_i(t)$).

The directional resulting vector, representing the final chemotropic response, is obtained by:

$$\vec{r}(t) = \sum_{i=1}^{S} \vec{s_i}(t). \tag{9}$$

From $\vec{r}$, the preferential direction of growth (10) and strength of attraction (11) can be extracted for the chemical stimulation:

$$\alpha_{ch} = \tan^{-1}\left(\frac{r_y}{r_x}\right), \tag{10}$$

$$\varnothing_{ch} = \begin{cases} \overline{\varnothing_{ch}}\sqrt{r_x^2 + r_y^2} & \overline{\varnothing_{ch}}\sqrt{r_x^2 + r_y^2} \leq \overline{\varnothing_{ch}} \\ \overline{\varnothing_{ch}} & otherwise \end{cases}. \tag{11}$$

In Equation (11), $\overline{\varnothing_{ch}}$ represents a maximum threshold for the bending angle that can be induced and reached in one single time step by chemical stimulation, fixed in our implementation at 0.5°.

The chemotropic response now needs to be combined with gravitropism to reflect *Property 5*. To find the gravitropic response, the gravity vector is retrieved (in the case of the robot, it is directly obtained by the accelerometer) and its projection on the *x*–*y* plane of the tip is obtained to find the direction of response (α_g), as in [26]. The strength of this signal is known to respond in plant roots with a sin law [46]:

$$\varnothing_g = a \times \sin(\beta - \theta) + b, \tag{12}$$

in which a and b are two constants, θ is the tolerance angle (which can vary from species to species) and β is the inclination of the tip from the gravity. a, b and θ are fixed parameters (we considered $a = 15.9°$, $b = 7.9°$, $\theta = 20°$ as in [46], where values were experimentally obtained from *Arabidopsis thaliana*—arabidopsis).

The combined directional response (*Property 5*) can be obtained by vectorization:

$$\vec{v} = \begin{pmatrix} \varnothing_{ch} \times \cos\alpha_{ch} + \varnothing_g \times \cos\alpha_g \\ \varnothing_{ch} \times \sin\alpha_{ch} + \varnothing_g \times \sin\alpha_g \end{pmatrix}, \tag{12}$$

$$\alpha = \tan^{-1}\left(\frac{v_y}{v_x}\right), \tag{13}$$

$$\varnothing = \sqrt{v_x^2 + v_y^2}. \tag{14}$$

2.5. Experiments

In order to verify the effect of a complexification of the control by introducing a priority adjustment and the uptake–kinetics mechanism, we simulated the evolution of three roots that share nutrients through the shoot, and we implemented the control described above (Section 2.4), hereafter alternative A. We then compared the simulation result with three other alternatives:

B. the same control but with a linear priority adjustment ($P = I_{max} / I^{+}_{max}$);
C. a control without steps 1 and 3, in fact there is no uptake mechanism nor any priority adjustment, priorities of all chemicals are fixed at 1 (similarly to our previous stimulus-oriented control [26]);
D. the uptake mechanism is implemented but is not used for priority adjustment, also in this case priorities are fixed at 1.

For each alternative, we monitored for each root the evolution in time of nutrients perception, their internal memory, the uptake rate and nutrient priority. To verify if the adopted control is able to solve the problem formalized in Equation (2), we also monitored the evolution of the internal nutrients' ratios and the resulting imbalance for the whole plant.

Alternative A is then also tested on the robotic platform. Nutrients N and K are in this case provided through the virtual environment, while P is provided with a halogen lamp and the temperature sensors of the robotic roots are used for stimulus perception (Supplementary Video S1).

3. Results

As shown in Figure 4, while growing roots have a different perception of the environment from each other, and can uptake a different number of nutrients, they end up with an identical internal state thanks to the communication channel that facilitates a complete sharing of the resources collected. On the basis of their local memory, they adjust the Michaelis–Menten parameters (I_{max} in Figure 5) and nutrient priorities (Figure 5). The result of this priority adjustment is that nutrient ratios tend to optimality (Figure 6).

By comparing the control alternative A (Figure 7A) with the others (B, C and D), a different arrangement of the roots in the environment can be observed, suggesting that the use of dynamic priorities for each stimulus can greatly affect plant root architecture. The different arrangement of the roots induces a subsequent different perception and different uptake of nutrients. The curves of the nutrient imbalance highlight the trend of alternative C (Figure 7C) to completely diverge from zero. This thus suggests that the uptake–kinetics mechanism is fundamental for solving the plant wellness problem and the chemotropic response without the adjustment of nutrient priorities (similarly to the stimulus-oriented control previously developed) combined with other tropic responses are insufficient to ensure plant survival. In addition, the conversion function between I_{max} and priorities is fundamental for establishing root architecture and a consequent faster or slower adjustment of nutrient balancing. In fact, the results of alternative B (Figure 7B), where a linear function was used instead of the quadratic function as in alternative A, show a different organization of the roots followed by a slower adjustment of the imbalance (on average, it reached ~4.9×10^{-5} at the end of the simulation, ~1.8 times higher than alternative A) (Table 3). In alternative D (Figure 7D), although the priorities are fixed (as in alternative B), there is an initial decrease in the imbalance. This is due to the variation in Michaelis–Menten parameters, which leads to a dynamic adjustment of the instantaneous uptake of nutrients. However, since this variation is not reflected in stimuli priorities, the root architecture is affected, inducing a deviation in the nutrient balance from the optimal condition.

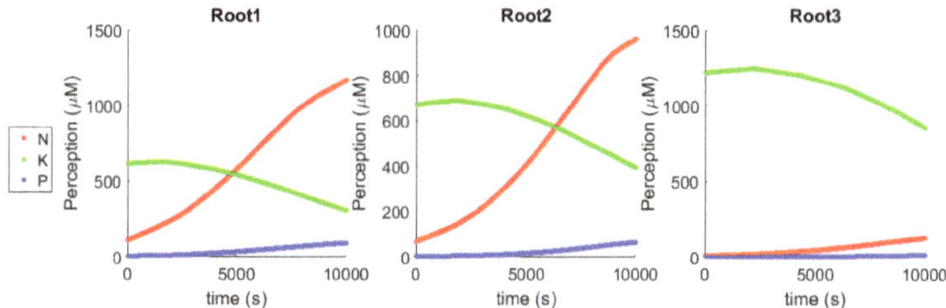

Figure 4. Evolution of nutrients perception. An example of nutrients perception evolution for each root (three roots where placed in the environment) while growing. Case of nutrients with centers $C_N = (3, 2, -4)$, $C_K = (-3, -2, -1)$, $C_P = (3, 1, -5)$.

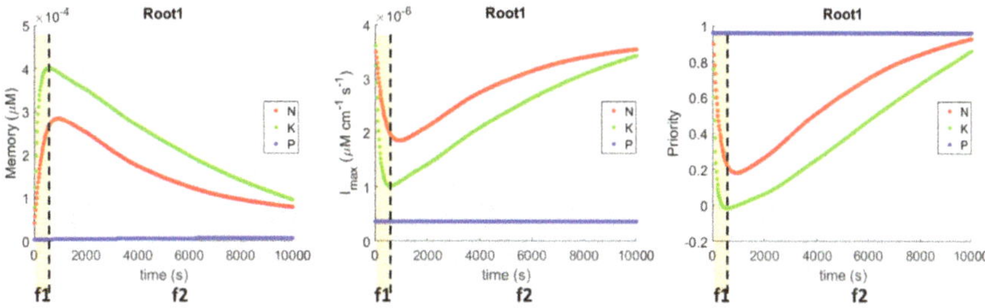

Figure 5. Evolution of memory, nutrient priority and uptake. An example of evolution of internal memory, I_{max} parameter and priority for Root 1. Case of nutrients with centers $C_N = (3, 2, -4)$, $C_K = (-3, -2, -1)$, $C_P = (3, 1, -5)$. The three roots with alternative A result with the same internal memory, and, consequently, they also have the same I_{max} and priorities evolution. In the graphs, we can observe a face f1 where the internal availability of both N and K increases inducing a consequent decreasing of both I_{max} and priority. When the internal availability starts to decrease, there is a consequent increasing of I_{max} and priority (face f2).

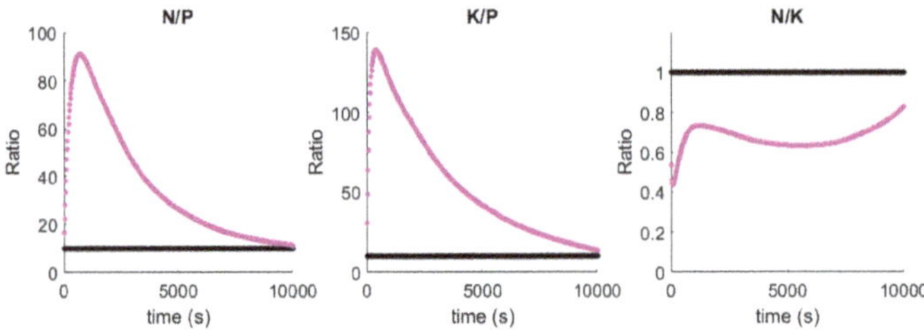

Figure 6. Evolution of nutrients' ratios. The evolution of nutrients ratios (in magenta color) in the entire plant while growing with alternative A in the case of nutrients with centers $C_N = (3, 2, -4)$, $C_K = (-3, -2, -1)$, $C_P = (3, 1, -5)$. The solid black lines represent the optimal nutrient ratios (N:P and K:P equal to 10 from [37]).

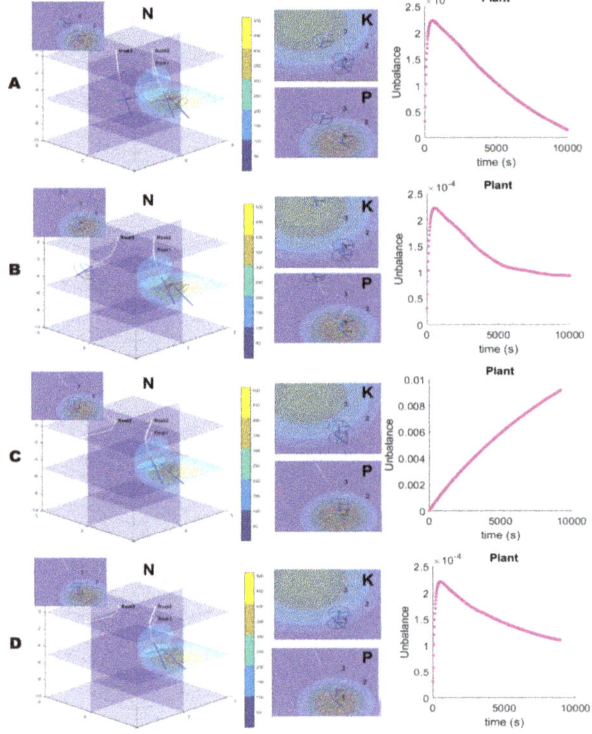

Figure 7. Results of simulations. An example of simulation results after 10,000 s of growth. Roots started from positions: Root1 (0, 2, 0), Root2 (2, 0, −2) and Root3 (0, 0, 0). In the simulated environments, only the skeleton of the roots is visible, with the white lines depicting the region of maturity, blue lines the tip of the root and the circles depict the circumference of the root. The three roots are presented for alternative A: with the uptake–kinetic feedback control as presented in Section 2.5; alternative B: with the uptake kinetic feedback control where priorities function is a linear function; alternative C: where the control does not implement the uptake kinetic mechanism and priority adjustment; and alternative D: where the control implements the uptake kinetic mechanism but not priority adjustment. Each row in the figure presents from left to right: the root architecture at the end of simulation in the environment with nitrate, in the middle, the top view of the roots in the environments with potassium and phosphorus, and, on the right, the curves of nutrients imbalance.

Table 3. Nutrient imbalance. Nutrient imbalance obtained at the end of simulation (~3 h of simulated growth), for each of the control alternatives. Resulting imbalance is obtained as the averaged imbalance over three different runs, placing the centers of nutrients at different random locations: RUN 1 $C_N = (3, 2, -4)$, $C_K = (-3, -2, -1)$, $C_P = (3, 1, -5)$; RUN 2 $C_N = (-3, 3, -4)$, $C_K = (-3, -2, -3)$, $C_P = (1, 1, -5)$; RUN 3 $C_N = (1, 2, -2)$, $C_K = (4, 1, -1)$, $C_P = (2, 1, -2)$.

A	B	C	D
2.66×10^{-5}	4.93×10^{-5}	546×10^{-5}	5.69×10^{-5}

Alternative A provides the best performance in terms of nutrient imbalance and was selected as the control for the robotic platform. The supplementary video (Supplementary Video S1) shows how each agent independently moves according to their internal state and local perception, and the immediate response of the uptake–kinetics mechanism that, as soon as the missing nutrient (P) is inserted in the environment, leads to a decreasing of the imbalance of nutrients in the whole plant.

4. Discussion and Conclusions

In this paper, plant roots are proposed as a source of inspiration for the design and architecture of robotic control. The behavior analysis from the literature led to the extraction of eight fundamental principles, which helped in the creation of a feedback control for a plant-inspired robotic platform (Plantoid).

The proposed control, inspired by the uptake–kinetics of plant roots and by the internal communication system of plants, was implemented in simulation and on the Plantoid, showing the effective directionality of root growth towards attractive stimuli and a natural adjustment of the internal nutrient balance. The control runs independently on each individual root agent and is able to dynamically adjust nutrient priorities for that agent alone on the basis of local memory and to direct its growth on the basis only of local and instantaneous perception.

Looking at the global architecture and internal state, roots are shown to organize themselves, leading to a collaborative behavior among the agents aimed at improving the equilibrium of nutrients and indeed plant wellness, thus autonomously satisfying *Property 1*, without the need of a centralized unit for the control of nutrient status and of tasks' distribution on the agents. The swarming behavior is an emergent result of the auto-regulation adopted by the uptake–kinetics control and the redistribution of nutrients among the agents. We have shown, in fact, that the internal imbalance never seems to converge to optimality with a stimulus-oriented control (alternative C, lacking an uptake mechanism and priority adjustment) compared to the proposed uptake–kinetics feedback control (imbalance with alternative A $2.66 \times 10^{-5} \ll 546 \times 10^{-5}$ imbalance with alternative C after ~3 h of simulated growth).

Even though, there is not confirmation from biology on how uptake and stimuli priorities are related, here, we have proposed the use of a quadratic function to map uptake–kinetics parameters into nutrients' priorities, and by comparing this choice with the alternative that uses a linear function, we demonstrate that a priority function has an important role in defining root architecture and plant wellness. We show that the priority adjustment, independently actuated by each root, is a powerful instrument for the development of collaboration, without the need for a centralized unit to delegate task allocation.

The developed control can be used in artificial root-like robots, which can move for instance in rescue scenarios for survival detection, in unstructured environments for mapping, or in space exploration. Wellness can be defined according to the application, for instance by adjusting ratios among several and different stimuli, not necessarily chemical.

The work proposed here can be considered as a first milestone in plant-root inspired control, which can also contribute to understanding plant-root behavior. In fact, the implementation of hypothesis made on a biological model using a biomimetic platform, such as Plantoid, can help in hypothesis validation. For instance, the specific control can be incrementally adapted to include additional features of plant-root behavior in order to understand their role, e.g., the influence of temperature in Michaelis–Menten parameters and the subsequent influence on root distribution in soil. It would also be interesting to evaluate the energy costs of the biological model and use it as a nutrient consumption component on the control, observing how the behavior is affected. A subsequent evaluation would be how growth velocity is affected as well as an evaluation of the behavior when not all the absorbed nutrients are immediately sent to the shoot.

In terms of exploring the environment, it would be interesting to mimic lateral root growth not only from an engineering point of view for the development of new technological mechanisms but also analyzing how and where plants allocate new resources, i.e., roots, can lead to new ideas for collaborative exploration control strategies as well as for solving optimization problems.

Supplementary Materials: The following are available online at www.mdpi.com/2076-3417/8/1/47/s1, Video S1: Swarming Behavior Emerging from the Uptake–Kinetics Feedback Control in a Plant-Root-Inspired Robot.

Acknowledgments: The study was funded by Istituto Italiano di Tecnologia. The funding covers the costs of open access publishing.

Author Contributions: Emanuela Del Dottore performed the literature analysis, conceived the control, implemented and performed simulations and experiments, and wrote the paper; Ali Sadeghi designed and realized the robotic platform; Alessio Mondini designed and realized the electronics; Emanuela Del Dottore and Alessio Mondini discussed the control, experiments and results; Barbara Mazzolai oversaw, defined and advised on the research; and all the authors revised the paper.

Conflicts of Interest: The authors declare no conflict of interest.

References

1. Bakule, L. Decentralized control: An overview. *Annu. Rev. Control* **2008**, *32*, 87–98. [CrossRef]
2. Bonabeau, E.; Theraulaz, G.; Deneubourg, J.L.; Aron, S.; Camazine, S. Self-organization in social insects. *Trends Ecol. Evol.* **1997**, *12*, 188–193. [CrossRef]
3. Beckers, R.; Goss, S.; Deneubourg, J.L.; Pasteels, J.M. Colony size, communication, and ant foraging strategy. *Psyche* **1989**, *96*, 239–256. [CrossRef]
4. Seeley, T.D.; Camazine, S.; Sneyd, J. Collective decision-making in honey bees: How colonies choose among nectar sources. *Behav. Ecol. Sociobiol.* **1991**, *28*, 277–290. [CrossRef]
5. Passino, K.M. Biomimicry of bacterial foraging for distributed optimization and control. *IEEE Control Syst.* **2002**, *22*, 52–67. [CrossRef]
6. Wedde, H.F.; Farooq, M.; Zhang, Y. BeeHive: An efficient fault-tolerant routing algorithm inspired by honey bee behavior. In Proceedings of the International Workshop on Ant Colony Optimization and Swarm Intelligence, Brussels, Belgium, 5–8 September 2004; Springer: Berlin/Heidelberg, Germany, 2004; pp. 83–94. [CrossRef]
7. Babu, L.D.D.; Krishna, P.V. Honey bee behavior inspired load balancing of tasks in cloud computing environments. *Appl. Soft Comput.* **2013**, *13*, 2292–2303. [CrossRef]
8. Dorigo, M.; Birattari, M.; Stutzle, T. Ant colony optimization. *IEEE Comput. Intell. Mag.* **2006**, *1*, 28–39. [CrossRef]
9. Kennedy, J. Particle swarm optimization. In *Encyclopedia of Machine Learning*; Springer: New York, NY, USA, 2011; pp. 760–766. [CrossRef]
10. Brambilla, M.; Ferrante, E.; Birattari, M.; Dorigo, M. Swarm robotics: A review from the swarm engineering perspective. *Swarm Intell.* **2013**, *7*, 1–41. [CrossRef]
11. Monshausen, G.B.; Gilroy, S. Feeling green: Mechanosensing in plants. *Trends Cell Biol.* **2009**, *19*, 228–235. [CrossRef] [PubMed]
12. Gilroy, S.; Masson, P.H. *Plant Tropisms*; Blackwell Publishing: Ames, IA, USA, 2008; ISBN 9780813823232.
13. Baier, M.; Hemmann, G.; Holman, R.; Corke, F.; Card, R.; Smith, C.; Rook, F.; Bevan, M.W. Characterization of mutants in Arabidopsis showing increased sugar-specific gene expression, growth, and developmental responses. *Plant Physiol.* **2004**, *134*, 81–91. [CrossRef] [PubMed]
14. Rolland, F.; Baena Gonzalez, E.; Sheen, J. Sugar sensing and signaling in plants: Conserved and novel mechanisms. *Annu. Rev. Plant Biol.* **2006**, *57*, 675–709. [CrossRef] [PubMed]

15. Tuteja, N.; Sopory, S.K. Chemical signaling under abiotic stress environment in plants. *Plant Signal. Behav.* **2008**, *3*, 525–536. [CrossRef] [PubMed]
16. Alam, S.M. Nutrient uptake by plants under stress conditions. In *Handbook of Plant and Crop Stress*, 2nd ed.; Pessarakli, M., Ed.; Marcel Dekker, Inc. Publisher: New York, NY, USA, 1999; pp. 285–313, ISBN 9780824719487.
17. Mazzolai, B.; Mondini, A.; Corradi, P.; Laschi, C.; Mattoli, V.; Sinibaldi, E.; Dario, P. A miniaturized mechatronic system inspired by plant roots for soil exploration. *IEEE/ASME Trans. Mechatron.* **2011**, *16*, 201–212. [CrossRef]
18. Kim, S.W.; Koh, J.S.; Lee, J.G.; Ryu, J.; Cho, M.; Cho, K.J. Flytrap-inspired robot using structurally integrated actuation based on bistability and a developable surface. *Bioinspir. Biomim.* **2014**, *9*. [CrossRef] [PubMed]
19. Li, S.; Wang, K.W. Fluidic origami with embedded pressure dependent multi-stability: A plant inspired innovation. *J. R. Soc. Interface* **2015**, *12*. [CrossRef] [PubMed]
20. Akyol, S.; Alatas, B. Plant intelligence based metaheuristic optimization algorithms. *Artif. Intell. Rev.* **2016**, *47*, 417–462. [CrossRef]
21. Yang, X.S. Flower pollination algorithm for global optimization. In Proceedings of the International Conference on Unconventional Computing and Natural Computation (UCNC 2012), Orléans, France, 3–7 September 2012; pp. 240–249. [CrossRef]
22. Mehrabian, A.R.; Caro, L. A novel numerical optimization algorithm inspired from weed colonization. *Ecol. Inform.* **2006**, *1*, 355–366. [CrossRef]
23. Salhi, A.; Eric, S.F. Nature-inspired optimisation approaches and the new plant propagation algorithm. In Proceedings of the International Conference on Numerical Analysis and Optimization (ICeMATH 2011), Yogyakarta, Indonesia, 6–8 June 2011. [CrossRef]
24. Qi, X.; Yunlong, Z.; Hanning, C.; Dingyi, Z.; Ben, N. An idea based on plant root growth for numerical optimization. In Proceedings of the International Conference on Intelligent Computing, Nanning, China, 28–31 July 2013; Springer: Berlin/Heidelberg, Germany, 2013; pp. 571–578. [CrossRef]
25. Zhang, H.; Yunlong, Z.; Hanning, C. Root growth model: A novel approach to numerical function optimization and simulation of plant root system. *Soft Comput.* **2014**, *18*, 521–537. [CrossRef]
26. Sadeghi, A.; Mondini, A.; Del Dottore, E.; Mattoli, V.; Beccai, L.; Taccola, S.; Lucarotti, C.; Totaro, M.; Mazzolai, B. A plant-inspired robot with soft differential bending capabilities. *Bioinspir. Biomim.* **2016**, *12*. [CrossRef] [PubMed]
27. Gruber, B.D.; Giehl, R.F.; Friedel, S.; von Wirén, N. Plasticity of the Arabidopsis root system under nutrient deficiencies. *New Phytol.* **2013**, *163*, 161–179. [CrossRef] [PubMed]
28. Hodge, A. The plastic plant: Root responses to heterogeneous supplies of nutrients. *New Phytol.* **2004**, *162*, 9–24. [CrossRef]
29. Macy, P. The quantitative mineral nutrient requirements of plants. *Plant Physiol.* **1936**, *11*, 749–764. [CrossRef] [PubMed]
30. Van der Ploeg, R.R.; Kirkham, M.B. On the origin of the theory of mineral nutrition of plants and the law of the minimum. *Soil Sci. Soc. Am. J.* **1999**, *63*, 1055–1062. [CrossRef]
31. Marschner, H. *Mineral Nutrition of Higher Plants*, 2nd ed.; Academic Press: London, UK, 1995; ISBN 9780124735439.
32. Sumner, M.E. Application of Beaufils' diagnostic indices to maize data published in the literature irrespective of age and conditions. *Plant Soil* **1977**, *46*, 359–369. [CrossRef]
33. Epstein, E. *Mineral Nutrition of Plants: Principles and Perspectives*; Wiley: London, UK, 1972; ISBN 0-471-24340-X.
34. Jungk, A.; Asher, C.J.; Edwards, D.G.; Meyer, D. Influence of phosphate status on phosphate uptake kinetics of maize (*Zea mays*) and soybean (*Glycine max*). *Plant Soil* **1990**, *124*, 135–142. [CrossRef]
35. Siddiqi, M.Y.; Glass, A.D.; Ruth, T.J.; Rufty, T.W. Studies of the uptake of nitrate in barley I. Kinetics of $^{13}NO_3^-$ influx. *Plant Physiol.* **1990**, *93*, 1426–1432. [CrossRef] [PubMed]
36. Fortin, M.C.; Poff, K.L. Characterization of thermotropism in primary roots of maize: Dependence on temperature and temperature gradient, and interaction with gravitropism. *Planta* **1991**, *184*, 410–414. [CrossRef] [PubMed]
37. Eapen, D.; Barroso, M.L.; Ponce, G.; Campos, M.E.; Cassab, G.I. Hydrotropism: Root growth responses to water. *Trends Plant Sci.* **2005**, *10*, 44–50. [CrossRef] [PubMed]

38. Rhodes, A.L. Chemotropism of roots. *Bot. Gaz.* **1910**, *50*, 71. [CrossRef]
39. Sun, F.; Zhang, W.; Hu, H.; Li, B.; Wang, Y.; Zhao, Y.; Li, K.; Liu, M.; Li, X. Salt modulates gravity signaling pathway to regulate growth direction of primary roots in Arabidopsis. *Plant Physiol.* **2008**, *146*, 178–188. [CrossRef] [PubMed]
40. Hart, J.W. *Plant Tropisms: And Other Growth Movements*; Chapman & Hall: London, UK, 1990; ISBN 978-0-412-53080-7.
41. Massa, G.D.; Gilroy, S. Touch modulates gravity sensing to regulate the growth of primary roots of Arabidopsis thaliana. *Plant J.* **2003**, *33*, 435–445. [CrossRef] [PubMed]
42. Takahashi, N.; Yamazaki, Y.; Kobayashi, A.; Higashitani, A.; Takahashi, H. Hydrotropism interacts with gravitropism by degrading amyloplasts in seedling roots of Arabidopsis and radish. *Plant Physiol.* **2003**, *132*, 805–810. [CrossRef] [PubMed]
43. Marty, F. Plant vacuoles. *Plant J.* **1999**, *11*, 587–599. [CrossRef]
44. Lee, R.B.; Ratcliffe, R.G. Subcellular distribution of inorganic phosphate, and levels of nucleoside triphosphate, in mature maize roots at low external phosphate concentrations: Measurements with 31P-NMR. *J. Exp. Bot.* **1993**, *44*, 587–598. [CrossRef]
45. Schenk, M.K.; Barber, S.A. Potassium and phosphorus uptake by corn genotypes grown in the field as influenced by root characteristics. *Plant Soil* **1980**, *54*, 65–76. [CrossRef]
46. Mullen, J.L.; Wolverton, C.; Ishikawa, H.; Evans, M.L. Kinetics of constant gravitropic stimulus responses in Arabidopsis roots using a feedback system. *Plant Physiol.* **2000**, *123*, 665–670. [CrossRef] [PubMed]

Article

Data-Foraging-Oriented Reconnaissance Based on Bio-Inspired Indirect Communication for Aerial Vehicles

Josué Castañeda Cisneros [1], Saul E. Pomares Hernandez [1,2,3,*], Jose Roberto Perez Cruz [4,5], Lil María Rodríguez-Henríquez [1,4] and Jesus A. Gonzalez Bernal [1]

1 Department of Computer Science, Instituto Nacional de Astrofísica, Óptica y Electrónica (INAOE), Tonantzintla, Puebla 72840, México; josue@ccc.inaoep.mx (J.C.C.); lmrodriguez@inaoep.mx(L.M.R.-H.); jagonzalez@inaoep.mx (J.A.G.B.)
2 CNRS, LAAS, 7 Avenue du Colonel Roche, F-31400 Toulouse, France
3 Université de Toulouse, LAAS, F-31400 Toulouse, France
4 Consejo Nacional de Ciencia y Tecnología (CONACYT), Av. Insurgentes Sur 1582, Col. Crédito Constructor Del. Benito Juárez C.P.: 03940, Ciudad de México; jrperezcr@conacyt.mx
5 Universidad Michoacana de San Nicolás de Hidalgo (UMSNH), Morelia 58040, México
* Correspondence: spomares@inaoep.mx; Tel.: +52-(222)-266-3100

Academic Editor: Fei Chen
Received: 31 May 2017; Accepted: 12 July 2017; Published: 16 July 2017

Abstract: In recent years, aerial vehicles have allowed exploring scenarios with harsh conditions. These can conduct reconnaissance tasks in areas that change periodically and have a high spatial and temporal resolution. The objective of a reconnaissance task is to survey an area and retrieve strategic information. The aerial vehicles, however, have inherent constraints in terms of energy and transmission range due to their mobility. Despite these constraints, the Data Foraging problem requires the aerial vehicles to exchange information about profitable data sources. In Data Foraging, establishing a single path is not viable because of dynamic conditions of the environment. Thus, reconnaissance must be focused on periodically searching profitable environmental data sources, as some animals perform foraging. In this work, a data-foraging-oriented reconnaissance algorithm based on bio-inspired indirect communication for aerial vehicles is presented. The approach establishes several paths that overlap to identify valuable data sources. Inspired by the stigmergy principle, the aerial vehicles indirectly communicate through artificial pheromones. The aerial vehicles traverse the environment using a heuristic algorithm that uses the artificial pheromones as feedback. The solution is formally defined and mathematically evaluated. In addition, we show the viability of the algorithm by simulations which have been tested through various statistical hypothesis.

Keywords: data foraging; reconnaissance; bio-inspired indirect communication; graph exploration; interruptibility

1. Introduction

In recent years, the use of Unmanned Aerial Vehicles (UAVs) has become important in numerous tasks, such as security surveillance, transportation [1], rescue and environmental monitoring. An outstanding capacity of these vehicles is that they can allow not only monitor environments with harsh conditions where humans cannot have access, but they can also monitor scenarios that change periodically, with a high spatial and temporal resolution [2].

For example, in flood monitoring it is necessary to identify the increase in water levels of certain regions in the environment. The water level can move in an uncontrolled way and change with a high frequency. Therefore, to identify these changes it is necessary to perform reconnaissance

(Reconnaissance refers to the task of traveling with the purpose of discovering new territories, unknown spaces, roads and routes.) tasks over an area. This implies frequently collecting and selecting the most relevant data about the current status of the environment. Once the reconnaissance task has been done, oversampling some regions is necessary to determine which regions are relevant despite changes in the environmental conditions.

A feasible solution is to employ UAVs to perform the reconnaissance task. In this sense, some UAVs can sample the area through various flights, exchanging partial views to determine the regions with relevant environmental data. However, due to their mobility, UAVs have inherent constraints in terms of energy and transmission range. Thereby, it is necessary not only to perform the communication among these vehicles even with a lack of direct coupling among senders and receivers, but also to tackle the problem of monitoring a changing environment.

In nature, some animals face similar problems when foraging for food, and the main objective is to retrieve the most profitable food resources by considering various restrictions, including energy. In order to communicate the findings obtained through reconnaissance to other animals, several species use indirect communication such as segregation of pheromones. Data Foraging is related to the selection of profitable data sources in a dynamic environment with mobile sensors.

Many approaches related to reconnaissance with mobile sensors have been proposed, especially in robotics, where the main objective is to find an optimal path to maximize the knowledge over a particular area [3–8].

Finding a single optimal reconnaissance path is not suitable for data foraging, particularly when an operational environment with highly changing attributes is considered, and where the objectives may change dynamically.

In this sense, Data Foraging-Oriented Reconnaissance (DFORE) requires establishing multiple dynamic paths to ensure that a profitable data source can be identified. Figure 1a depicts how a group of ants performs the exploration and the exploitation of their environment. In some species, food collection is achieved by thousands of workers travelling along well-defined foraging trails. These trails emerge from a succession of pheromone deposits that can result in a complex network of interconnected routes [9]. To perform DFORE, a UAV searches for points of interest in an unknown environment, as depicted in Figure 1b. Through various trips, the UAV can identify a region which has something of interest to the application. Both systems are dynamic; therefore, several paths must be explored in order to exploit useful resources.

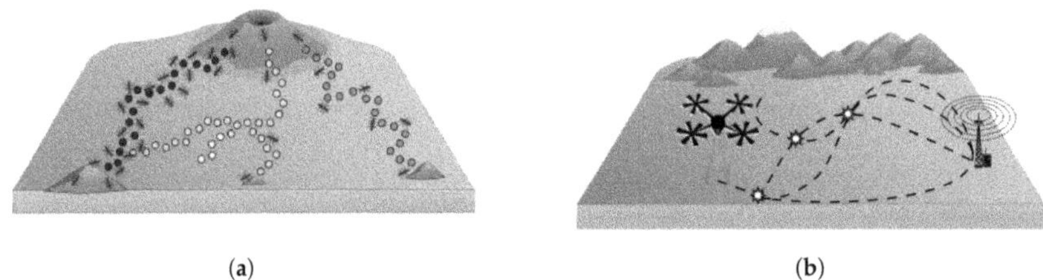

(a) (b)

Figure 1. Both the ants and the Unmanned Aerial Vehicles (UAV) forage resources. They must be able to identify dynamic resources with limited energy and temporal constraints. (**a**) Ants foraging; (**b**) UAV performing Data Foraging-Oriented Reconnaissance (DFORE).

In this work, we propose a Data Foraging-Oriented Reconnaissance algorithm for a single aerial vehicle. Inspired by the stigmergy principle, aerial vehicles communicate indirectly through an artificial pheromone to create several paths and to explore the operational environment with limited movement capabilities; thus, the focus of the research is how these devices through indirect communication can do a reconnaissance task. A hexagonal grid to represent the operational environment is used. Hexagonal models allow for a better movement representation in 2D due to the uniform distance in any direction.

We assume the aerial vehicle has limited movement capabilities, and for this reason, the aerial vehicle needs to be recharged as many times as needed at a base station. The required movement capacity of the aerial vehicle per trip to explore an operational environment is identified based on the size of the hexagonal grid. The algorithm accomplishes the temporal constraints that are defined for DFORE. It is proved in a formal way that the algorithm satisfies such constraints. A computational cost analysis and a simulation are presented to show the viability of our solution. We compare our proposed algorithm to MULES [10], adapted to the foraging reconnaissance task, which is a random walk with uniform distribution algorithm to collect data from the environment. MULES has similarities with our proposed mechanism, they both use indirect communication through an intermediary in our case artificial pheromones and in MULES the use of a mobile data relay. We measured the number of trips by each algorithm in different conditions. The comparison with MULES is justified since this proposal is the baseline algorithm for indirect communication among mobile sensors. Also, this algorithm is extensively used in recent works to solve problems like patrolling, source location privacy, data collection, etc. [11–13].

The organization of this document is as follows: A survey of recent literature is presented in Section 2. In Section 3, the preliminaries are explained along with the system model. Our proposed solution is presented in Section 4. The analysis of the proposed algorithm is shown in Section 5. In Section 6, the proof to validate our proposed algorithm is discussed. A series of experiments are shown in Section 7. The discussion of our algorithm is presented in Section 8. Conclusions are presented in Section 9. As a quick guide to follow this work, the notation is presented in Table 1.

Table 1. Notation table.

U	$\triangleq$	Set of Mobile Data Foragers, where each $u \in U = \{u_1, u_2, \cdots, u_q\}$
R	$\triangleq$	The operational environment represented by a set of regions r
H	$\triangleq$	The Hextille that represents the operational environment
G_h	$\triangleq$	The Data Foraging graph to model Hextille H with radius h_{ex}
n_h	$\triangleq$	Number of cells of G_h
p_h	$\triangleq$	Depth of G_h
ϵ_h	$\triangleq$	Required steps to traverse G_h
e_u	$\triangleq$	Endurance of mobile sensor u
F_c	$\triangleq$	Set of food pheromones
F_t	$\triangleq$	Set of travel pheromones
F	$\triangleq$	Set of pheromones. The union of $F_c \cup F_t$
F_r	$\triangleq$	The set of pheromones present at region r
η	$\triangleq$	The farthest region from the base within the main line
ζ	$\triangleq$	The region r where the base station is located
T_{expMax}	$\triangleq$	Maximum reconnaissance time
$Tstart$	$\triangleq$	Reconnaissance's start time
$Tstamp_t(r)$	$\triangleq$	The time when the region r is stamped

2. Related Work

There are several remarkable works related to our proposal from different perspectives. In this section, the related work for constrained exploration is presented, where a mobile agent must interrupt, return to the base station and refuel before continuing exploration. Next, the works that address the reconnaissance problem are discussed, which is a special type of exploration where the objective is to gain strategic information from uncertain environments with an optimal path. Finally, the differences between reconnaissance and Data-Foraging Oriented Reconnaissance are presented.

2.1. Constrained Exploration

Exploration of an unknown environment has been studied in numerous occasions [14]. In most proposals, the unknown environment is modeled as a graph. For such approaches, the task is to explore a given graph while optimizing the exploration routes. In general, exploration algorithms can be classified into two main types: offline and on-line. Offline exploration occurs when the graph information is known in advance. In contrast, during an on-line exploration, the information about the graph can only be learned in the execution of the exploration algorithm. Based on the concept of on-line exploration, the most used algorithms are the Depth First Search (DFS) and the Breadth First Search (BFS) [15]. A variation of on-line exploration was introduced by Betke et al. [16]. In this variation, called piecemeal search model (PSM), two constraints were added to the problem of graph exploration:

- Continuity: An agent must traverse the graph by passing through incident nodes. There is no teleportation of the agent to any node.
- Interruptibility: The agent must return to the start node s after traversing ϑ steps to recharge energy, where ϑ is a constant . In this sense, for PSM, the agent's energy (required to travel ϑ steps), is set to $2(1+\alpha)r$; where r is the distance to the farthest node from the starting node s and $\alpha > 0$ is a constant. The agent's energy is proportional to α.

With these constraints, BFS and DFS are not able to solve the piecemeal search problem [17]. Thus, several algorithms have been proposed to tackle such a problem. Betke et al. [16] present two algorithms: Wavefront and Ray. The Wavefront algorithm is based on BFS. It expands knowledge in waves from a starting node, just like a pebble expands a wave when thrown in a pond. The graph is decomposed into four regions, and each region is explored through ripples. The authors also present an algorithm based on DFS, which is called Ray and is similar to Wavefront, but it considers the shortest path from the starting node and any point in the ray. The main objective of these algorithms is to reduce the uncertainty of new routes to traverse an area. To the best of our knowledge, there are only three more works that deal with the restrictions proposed by Betke et al. [16]: Argamon et al. [18], Duncan et al. [17], P. B. Sujit and Debasish Ghose [19]. Moreover, in opportunistic routing, Shah et al. [10] presented another work that can be extended to satisfy the PSM constraints.

Argamon et al. [18] present an *on-line exploration while performing* algorithm for a repeated task(s). A repeated task must be done continuously, more than once. An agent needs to go from two points at least r times. The goal of the agent is to minimize the overall cost of performing the task(s). The agent also searches for new paths in the graph that are not explored. Movement is done through the expected utility of each path taken. The path between the two known points, improves over time. This movement is not restricted by energy constraints, and it is assumed that the agent has enough energy to get to the two points.

In Duncan et al. [17], the authors present an optimal constrained graph exploration algorithm called Bounded Deep First Exploration (bDFX), which uses a rope of size $(1+\alpha)r$ for some constant $\alpha > 0$ and a known radius r. To be able to access every node in the graph, bDFX prunes the nodes beyond the rope and maintains a list of disjoint subtrees of the original graph whose union contains all the nodes not visited. After applying a deep first search algorithm to each subtree, an agent can visit all the nodes of that particular subtree.

P. B. Sujit and Debasish Ghose [19] introduce game theory, where two UAVs explore an area in order to minimize the uncertainty of the sampling area. They proposed computing a non-cooperative Nash equilibrium to coordinate the two UAVS. However, it is very expensive to compute it. Furthermore, they have a q-ahead look-up policy, which makes calculating the Nash equilibrium even more costly.

In opportunistic routing, mobile sensors have uncontrolled mobility and move in a random fashion, similar to a random walk. Despite not using the constraints of the PSM, Shah et al. [10] proposed a three-tier architecture with a mobile sensor named Data Mobile Ubiquitous Local Area Network Extensions (MULEs) to collect data from sensors and transfer them to the sink. Thus, the MULEs are a mechanical carrier of information and achieve indirect communication between sensors. In order to include the constraints of the PSM, it is necessary to limit the movement of the MULEs and make them return to a base station after some steps. Using the approach of indirect communication, the network life is extended as indirect communication removes the burden of control information from the sensor, although latency is increased because the sensors have to wait for a MULE to approach before they can transfer data. As a result, high latency is the main disadvantage of such approaches.

2.2. Reconnaissance Problem

There have been many works that tackle the reconnaissance problem with aerial vehicles. Most of them are focused on the path taken by these aerial vehicles; thus, the interest is to find the optimal path under a series of constraints. Strategic information is represented as targets. The targets can remain fixed or change with respect of time, i.e., the operational environment is dynamic and uncertain. Therefore, the task of reconnaissance is divided into two approaches: Static and Dynamic.

Static path optimization relies on knowledge about the operational environment. Traditional approaches such as Particle Swarm Optimization [20], Genetic Algorithms [21] and Ant Colony [22] are used to obtain an optimal path for the aerial vehicles. Several other constraints to find an optimal path have been studied. Time is one of them; thus, the problem of task assignment has been researched in [23,24]. The minimum number of turns required to cover an area is explored in [25]. Formation for several aereal vehicles is also analysed in [26]. There are also others interesting optimization approaches based on techniques like Taguchi-methods, differential evolution, hybrid Taguchi-cuckoo search algorithm [27–29] that have given nice result optimizing objectives in multiples scenarios such as two degrees of freedom compliant mechanism, micro-displacement sensors, and positioning platforms.

The main disadvantage of these approaches regarding an unknown environment is the assumption of information known *a priori*. Another issue is that the optimal path obtained by these approaches must remain constant; however, there are dynamic environments where new conditions must be taken into account in order to get a useful path, e.g. the aerial vehicles must avoid moving obstacles. Dynamic reconnaissance for unknown environments has been studied in the following works. The aerial vehicles must respond to the dynamic changes in the environment. The use of probability distributions with *a priori* information has been addressed in [30,31], where the main idea is to adapt the path taken by the aerial vehicles based on the probability of new threats or emerging targets. To avoid obstacles, a hybrid approach was proposed in [32].

2.3. Differences between Reconnaissance and Data-Foraging Oriented Reconnaissance

There are differences between common reconnaissance and Data Foraging-Oriented Reconnaissance (DFORE). In common reconnaissance, every node must be visited with equal priority, and a single trip is sufficient to gather data from the nodes. However, for DFORE, the priority of every node can change based on the retrieved information of the node; thus, the interest is to overlap several trips. Therefore, in the common reconnaissance problem, the movement capability to traverse the whole graph is greater than the number of nodes: $\epsilon \geq \alpha n$ where ϵ is the movement capability a mobile agent has, $\alpha > 1$ is a constant and n is the number of nodes. Then, the objective is to minimize ϵ with an optimal route because every node has the same priority. On the other hand, DFORE considers multiple trips to explore the whole graph due to the movement constraints of the mobile elements. Thus, the objective is to obtain valuable data sources based on the overlapping paths generated by several trips in unknown environments. Most of the cited works do not meet the constraints imposed by DFORE, with the exception of MULES [10] modified to have movement constraints. The objective of our work is to explore a delimited area with endurance constraints for a dynamic environment by using a single aerial vehicle to obtain valuable data sources, which meet the constraints of the DFORE problem.

3. Preliminaries

In this section, the system model, as well as the formal definition of DFORE with its restrictions, are discussed.

3.1. Problem Definition

The problem of DFORE is related to the reconnaissance task in uncertain environments, where valuable regions can change with respect to time due to their dynamic nature. More precisely, each profitable region has a lifetime associated to it. Each of these regions has different values according to the application. Considering the conditions of the operational environment, it is necessary to oversample it through various trips and to selectively choose the more profitable regions, taking into account the temporal constraints of the regions. Therefore, the reconnaissance step must ensure that the entire sampling area is examined before a maximum time t_1.

3.2. Modeling the Operational Environment

Exploring the operational environment is done through a single aerial vehicle. The aerial vehicle lifts off the base station ζ, explores the operational environment, and returns to the base station to refuel. These three activities, together, determine a trip. Due to the flight endurance, which refers to the amount of time a mobile element spends in flight without landing, it might be impossible to visit all the regions of the operational environment in a single trip. In this work we assume that the mobile element must return to a base station to recharge fuel; that is, the flight endurance of the mobile element is not enough to explore the whole environment. For these reasons, the aerial vehicle needs to perform several trips in order to explore the operational environment.

3.3. System Model

Next, the system model is defined in order to describe and represent our system.

- Mobile Data Foragers. The explorer entities in the system are modeled as MDF. Each MDF belongs to the set $U = \{u_1, u_2, \ldots, u_q\}$. An MDF $u \in U$ represents aerial vehicles flying over the operational environment. Each $u \in U$ has a finite amount of steps it can make, limited storage and computational resources. Every time the MDF moves to an adjacent region, the number of steps of the MDF is reduced by one.
- Pheromones. Since there is no single reconnaissance route, each MDF $u \in U$ is guided by a trail of pheromones. In this work, a pheromone is defined as an abstract data type as follows. A pheromone $f \in F$ is represented as a tuple $f = \{r, counter\}$, where r is the identifier of the region where the pheromone was placed, and *counter* is the number of pheromones placed in such region. There are two types of pheromones: food and travel pheromones. Food pheromones indicate that in a specific region there is something of interest to the application; food pheromones are denoted by the set F_c. On the other hand, travel pheromones indicate that the region has been visited; they are denoted by the set F_t. The set F of pheromones is the union of food and travel pheromones, that is $F = F_c \cup F_t$. Each pheromone belongs to the set $F = \{f_1, f_2, \ldots, f_k\}$. Each pheromone f has a lifetime associated to the maximum time a pheromone can be in a region.
- Operational environment. We represent the operational environment as a Hextille H of radius h_{ex} in the form of a set $R = \{r_1, r_2, \ldots, r_i\}$, where each $r \in R$ is a sampling region. The radius h_{ex} of the Hextille H is defined as the linear distance from the center of the Hextille to the farthest hexagon of the Hextille in any of the six directions. Figure 2 shows an example of the environment as a tilled hexagonal grid. It should be noted that the hexagonal grid is not restricted just to the specific radius shown in Figure 2 and can vary in radius. A two-dimensional space is considered along with the knowledge of the environment in the form of a map, but without the characteristic and conditions of the environment; that is, there is no information about where valuable data sources are located. Exploring the environment is done through one MDF. The MDF starts at the base station, explores the environment, and returns to the base station to refuel; this is called a trip. The sampling area is a subset $S \subseteq R$, where each region $r \in S$ is a hexagon with a diameter equal to the sensing range of an MDF $u \in U$. Only one $u \in U$ can be in the environment at any given time. It should be noted that each region $r \in S$ has dynamic changing conditions, which

means that regions that are valuable do not necessarily remain valuable indefinitely. Each $r \in R$ has a number of pheromones $f \subseteq F_r$, where F_r is the set of pheromones present at region r.

- Base station. The base station ζ is a processing unit, associated to the physical place where each MDF lifts to explore the environment and drops the retrieved data after each expedition. It is assumed that the base station has enough resources to process, and send control messages to the MDFs. There is a unidirectional channel between the base station and the MDF present in the environment.
- Maximum reconnaissance time: This refers to the maximum time to cover the sampling area. It is denoted by T_{expMax}. According to Duncan et al. [17] the upper bound of exploration under energy constraints is $\mathcal{O}(n^2)$, where n is the number of regions.

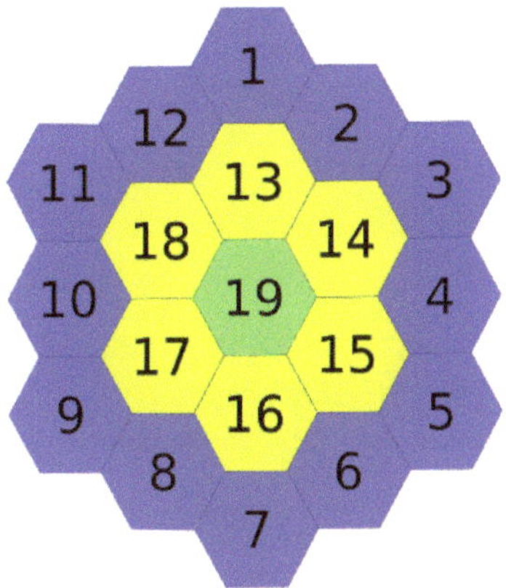

Figure 2. Environment as a Hextille.

4. Data Foraging-Oriented Reconnaissance

In order to explore the whole area, the Hextille first is modeled as a special graph called a Data Foraging Graph (DFG). The graph must be labeled and its properties are analyzed. Our algorithm is designed and implemented based on the properties of the DFG.

4.1. Creating a Data Foraging Graph

We are interested in a graphical representation with morphological properties, such as uniform distance and symmetry. The focus is twofold; first, to reduce the overhead complexity of the algorithm, and second, to understand how Hextilles grow in order to obtain the properties used to propose our solution. Thus, any Hextille H with radius h_{ex} is modeled as a connected undirected graph G called a Data Foraging Graph (DFG) of size h (G_h). The approach to create a DFG is as follows. First, the position of the base station is chosen. In this work, the base station can be located on the border of Hextilles; this means that the base is placed outside the sampling area for practical reasons. Due to the symmetrical properties, any hexagon in the border of a Hextille can be chosen and the DFG will remain the same; for example, in Figure 2, hexagons 1, 3, 5, 7, 9 and 11 can be used interchangeably to place the base station. Figure 3 shows an example with a DFG G_3. For every hexagonal cell, a node is created. Nodes are related with edges if they share a vertex. Therefore, any node has a maximum of six neighbors.

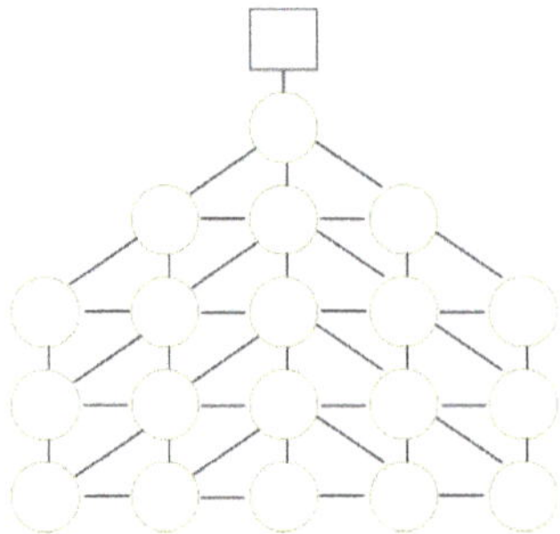

Figure 3. A Data Foraging Graph is created from the Hextille of Figure 2. The square represents the base station while the circles represent the regions of the Hextille. Every region has adjacent regions which are connected with edges between the circles.

4.2. Enumerating a Data Foraging Graph

To identify the nodes in a unique way, it is necessary to label the graph with numbers. There are many ways to enumerate the nodes of a DFG. Our approach is based on the previously defined representation:

- The root of the graph (base station) is numbered as 0. The next node to be numbered is chosen from a clockwise spiral, as shown in Figure 4.
- The process stops when all nodes of the DFG are numbered.

This approach is used because it simplifies comprehension and readability of the DFG.

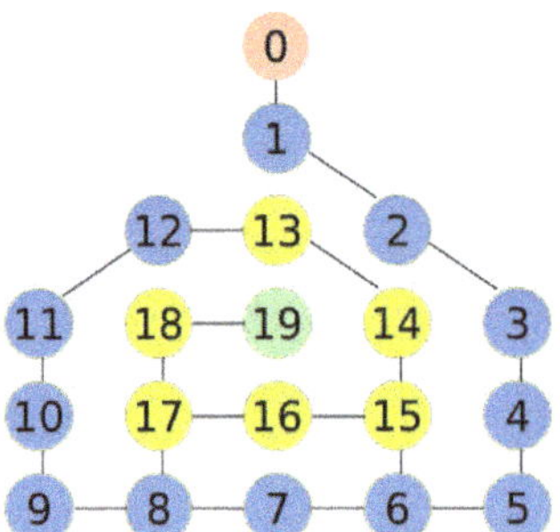

Figure 4. Enumerating the graph.

4.3. Data Foraging Graph Properties

The properties of any given DFG G_h of size h are introduced. The properties help us formally define the problem of Data Foraging-Oriented Reconnaissance (DFORE) for any Hextille of radius h_{ex} and analyze our algorithm to prove it satisfies the restrictions of DFORE.

The linear distance between the base station and the farthest region is called the depth of the DFG.

Property 1. *For any given DFG G_h where $h \geq 1$, its depth p_h is equal to:*

$$p_h = 2h - 1. \tag{1}$$

To show that Property 1 a straight line is drawn from the base station of G_h to the farthest node of the graph and count the number of regions the lines cross are counted.

The hexagonal tilling consists of a number of hexagons bordered by other hexagons. It is necessary to know the number of hexagons for any Hextille with radius h_{ex}. This allows us to know how the sampling area grows and to measure the performance of our algorithm compared to others, in terms of the number of regions they visit. The DFG will also have the same number of nodes as the Hextille.

Property 2. *For any given DFG G_h with depth p_h (see Equation (1)), the number of regions n_h is given by the following recursion:*

$$n_h = (3p_{h-1} + 3) + n_{h-1}$$

The close solution for the recurrence is (see Appendix A):

$$n_h = 3h^2 - 3h + 1 \tag{2}$$

The focus is on the minimum number of steps to travel from the base station to another region of the DFG without having to recharge the MDF.

Property 3. *Let ϵ_h be the required number of steps that are necessary to reach any region from the base station. For any DFG G_h, $h \geq 1$, the required number of steps ϵ_h is:*

$$\epsilon_h = 2h - 1. \tag{3}$$

Since it is necessary to return to the base station, the total number of steps an MDF can make is $2\epsilon_h$.

Property 4. *Let η be the farthest region from the base within the main line. Let ζ be the base station, which is at the root of the DFG.*

It is impossible to get from one region to all the regions of the DFG with the movement capabilities of the MDF. Only the base station can be reached from any region of the DFG with the required number of steps (see Property 3). Therefore, we are interested in defining the set of reachable regions given the remaining energy of the MDF.

Property 5. *Let be two regions, $r, r' \in R, r \neq r'$, the remaining number of steps e_u of an MDF and the physical distance between a pair of regions r and r' noted as $d(r, r')$. A region r' is reachable if and only if there are enough steps Π to visit the region and return to the base station:*

$$\Pi = e_u - d(r, r') - d(r', \zeta) \geq 0 \tag{4}$$

4.4. Problem Definition According to Our Environment Representation

With our environment representation, the problem of Data Foraging-Oriented Reconnaissance (DFORE) can be formally defined using the previously defined properties.

DFORE: The objective of DFORE is to stamp every region in the sampling area, while visiting nodes according to their priority. This will ensure that the algorithm obtains points of attraction based on trip overlaps. Each MDF will stamp regions by retrieving data from the region. Formally, the problem of DFORE must meet the following restrictions:

Restriction 1. *Every region in the sampling area must be visited and stamped with a pheromone before a maximum time. Let $Tstart$ be the reconnaissance's start time, $Tstamp_t(r)$ be the time of the visit and stamping of a region $r \in R$ with a pheromone f in the set F at step t, related to the number of regions visited since $Tstart$. Finally, let T_{expMax} be the maximum reconnaissance time, the reconnaissance step must meet:*

$$\forall r \in R, |Tstamp_t(r) - Tstart| \leq T_{expMax} \tag{5}$$

Restriction 2. *Every MDF must return to the base station ζ within its specified maximum endurance. Let e_u be the endurance of an MDF u*

$$Tstamp_t(\zeta) \leq e_u. \tag{6}$$

Restriction 3. *Continuity: Every move of an MDF must be done only on adjacent regions. That is, an MDF cannot jump from one region to another one which is not adjacent to it.*

Restriction 4. *Interruptibility: The MDF must return to the base station ζ in at most $2\epsilon_h$ steps, where ϵ_h is the required number of steps to arrive from the base station ζ to the farthest region η.*

4.5. Proposed Algorithm

At the beginning of the mission, there is no information about the sampling area. After Hextille H with radius h_{ex} is selected, we construct a DFG G_h of the sampling area and explore it. Once the MDF has visited a region in G_h, the MDF stamps the region. The objective of reconnaissance is to expand the knowledge of the sampling area while visiting nodes based on their priority. In order to explore new nodes getting to farthest and less stamped nodes is preferred. It is necessary to satisfy Restrictions 2, 3 and 4 (see Section 4.4). The following rules are:

Rule 1. *Given the remaining energy e_u of an MDF, the set of potential nodes L_r, the minimum number of steps ϵ_h to traverse the DFG G_h, a region r, its neighbors V_r, the next potential node r' to be visited by an MDF is a $r' \in L_r \subseteq V_r$. The set L_r is determined by:*

- *(a) if $e_u > \epsilon_h$, $L_r \leftarrow \{r' \in V_r : d(r',\zeta) \geq d(r'',\zeta) \forall r'' \in V_r\}$,*
- *(b) otherwise, $L_r \leftarrow \{r' \in V_r : d(r',\zeta) \leq d(r'',\zeta) \forall r'' \in V_r\}$.*

Rule 2. *Given a region r and its neighbors L_r, an MDF can only move if $\exists r' \in L_r$ such that the estimated remaining energy between r and r', $\Pi \geq 0$ (see Property 5).*

The rules are implemented in the algorithm. The main function of the algorithm is shown in Algorithm 1. The detailed description of the DFORE algorithm is presented in Appendix D.

Algorithm 1 Exploration algorithm.

```
function [LIST<CELL>, INT] EXPLORATION(int e_u, List<Cell> area)
    int time ← 0
    Cell r ← 0
    while e_u ≥ 0 do
        if (e_u − 2) − e_base- ≥ 0 then
            Cell r' ← choose(neighbors(r)) /* See Appendix D */
        else
            Cell r' ← traceback(neighbors(r), e_u) /* See Appendix D */
        end if
        e'_u ← e_u − 1
        r'.stamp ← r'.stamp + 1
        time' ← time + 1
        r ← r'
    end while
    List < Cell > foodCells ← getFoodCells(area) /* See Appendix D */
    return [foodCells, time]
end function
```

Next, the DFORE algorithm is described. If the environment has not been visited completely, the reconnaissance continues its execution while the MDF has enough number of steps to continue exploring the DFG. In order to choose the next node to be visited, it is necessary to verify whether the MDF has enough remaining steps to proceed (function EXPLORATION line 4, Rule 2) or needs to return (function EXPLORATION line 6). If the MDF can proceed, the following heuristics are applied. First, the MDF selects the adjacent nodes with the largest distance to the base station ζ (see Rule 1a). Second, the MDF chooses the nodes with fewer stamps (function choose see Appendix D line 6). Third, if all conditions hold, i.e., every node has the same distance to the base station and the nodes have the same number of stamps, then the MDF chooses a node at random with a uniform distribution (function choose see Appendix D line 7). After moving to the last node, the MDF must return to recharge energy. Thus, when the MDF cannot proceed, then it will begin to choose nodes which are nearer to the base station (see Rule 1b); therefore, the MDF will return to the base station satisfying Restriction 4 (see Section 4.4). See Figure 5 for the following example. All red nodes are stamped, the number of stamps is represented by the intensity of the color. The MDF is currently on node 0. At step 0, the MDF chooses node 1 since it has only one choice. At step 1, the MDF chooses node 13, which is not stamped. Node 19 with less stamps is chosen at step 2. The MDF chooses randomly node 15 at step 3. In step 4, the MDF chooses node at random.

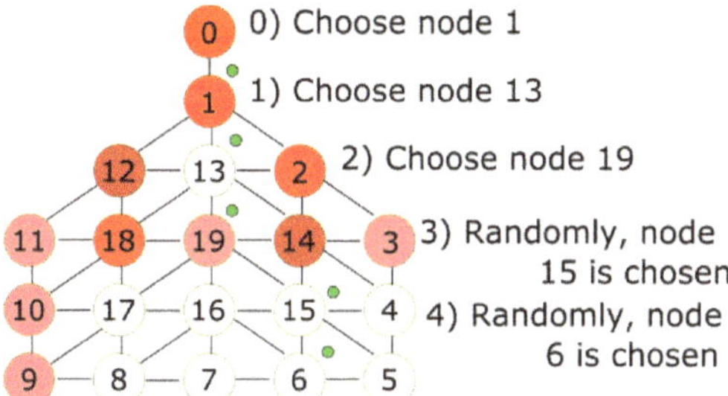

Figure 5. Reconnaissance example using the heuristics. The number of stamps in a node is represented by the intensity of the color.

To show an example of the execution of our algorithm in various trips, Figure 6 depicts an example of the reconnaissance algorithm for a DFG G_3. Each color represents the trip taken. Uncolored nodes are not visited yet. The first trip is colored in green. When the MDF cannot proceed, it returns to recharge energy at the base station. In the second trip, the MDF explores unvisited nodes, puts a pink stamp and returns. Finally, in the last trip, the MDF visits another group of nodes and paints them blue. In order to explore the entire graph, overlaps between trips must occur. It can be noted that in the last trip, the MDF exploits better the movement capabilities since it explores more nodes in one trip.

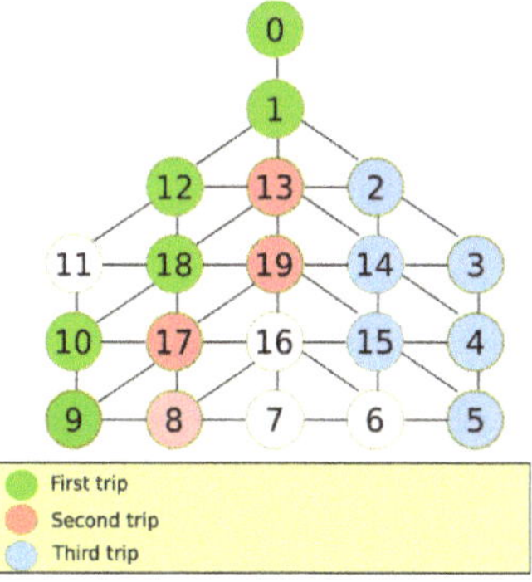

Figure 6. Reconnaissance of the Data Foraging Graph (DFG) G_3. Each trip can be different. The base station is placed at node 0. The Mobile Data Forager (MDF) is restricted to visit 10 nodes.

5. Algorithm Analysis

In this section, the mathematical analysis of our algorithm is discussed to show that it satisfies the restrictions of the problem. The number of trips required to explore any DFG with n regions is analyzed. The first part is devoted to the use of the required number of steps ϵ_h to traverse the whole DFG, both in the best and in the general case of reconnaissance. Based on this analysis, the minimum and expected number of trips for any DFG with n regions is obtained. Finally, the number of trips required to visit every node in the DFG is analyzed with a greater number of steps than ϵ_h.

5.1. Number of Trips Using the Required Number of Steps: Best Case

The best case occurs when there is almost no overlap of paths to visit every node in the DFG. If the graph G_h is divided by a straight line between the base station and the farthest region η, symmetrical sub areas are obtained. Figure 7 shows the environment divided in half. If the process continues, eventually only a straight line is obtained. Thus, it is possible to apply a divide and conquer strategy. Formally, this behavior is defined as follows. Let *f(n)* be the problem of exploring an environment represented by graph G with n regions.

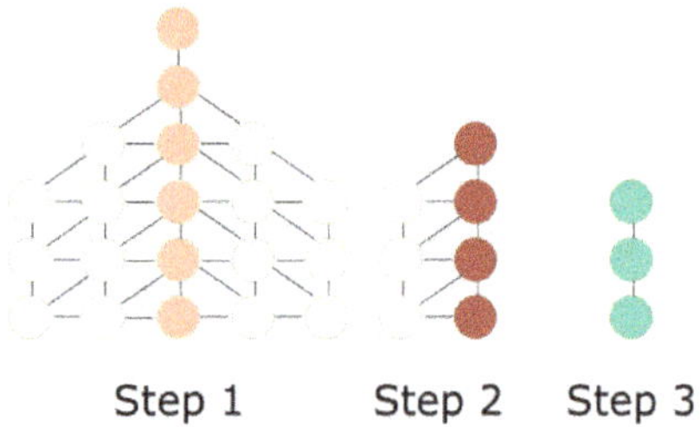

Figure 7. A central line divides the environment in half. Step 1 divides the Hextille in two symmetric parts. Step 2 continues to do this until there is only one straight line at Step 3.

Definition 1. *Each time the DFG G_h is divided, f(n) is split into two equal subsets with the same cardinality. In order to combine the solution, at least n steps need to move towards the base station. Therefore, for any given environment $f(n)$ can be expressed by:*

$$f(n) = 2f(n/2) + c \tag{7}$$

where c is a constant.

We have a linear time to explore n nodes, that is: $f(n) = \mathcal{O}(n)$ (see Appendix B). However, there is a precise way of calculating the minimum number of trips required to explore any DFG G_h, given the required number of steps ϵ_h available. There are two ways that a main line can be traversed: Either by choosing a main line and returning to the base using the same regions, or by backtracking using the next row of regions. However, the farthest region η of the next row will not be marked. Figure 8 shows an example of this situation. It can be seen that the minimum number of trips for any given DFG G_h is equal to its depth p_h.

Definition 2. *The minimum number of trips T_h to explore any given DFG G_h is equal to its depth p_h (see Equation (1)) .*

$$T_h = p_h \tag{8}$$

For any DFG G_h, in every trip, the number of nodes traversed is $2(2h-1)$. Since there are T_h trips, the total number of nodes traversed required to explore the DFG G_h is $2T_h(2h-1)$. Based on Equation (8), the previous equation is $2(2h-1)^2$. Expanding the equation yields:

$$f(n) = 8h^2 - 8h + 2 \tag{9}$$

However, to obtain profitable data sources, various trips must overlap in order to discriminate valuable sources from common ones. Therefore, a general case of reconnaissance where various trips overlap must be addressed.

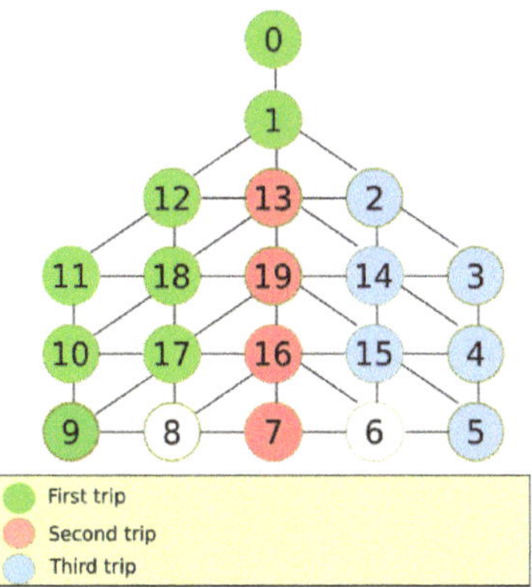

Figure 8. The best way to explore a graph with the required number of steps is by having each trip visits both a line and its adjacent line. Since there are p_h lines, that is the minimum number of trips to explore the whole graph. In this particular example, p_h is equal to five.

5.2. Number of Trips Using the Required Number of Steps: General Case

In the general case, the interest is to visit several nodes repeatedly in order to obtain valuable nodes, contrary to the goal in the best case. Therefore, the focus is to obtain the average number of trips to explore the DFG considering the heuristics of our proposal. In order to calculate the average number of trips, the environment is divided into rows and columns. The rows correspond to the levels in the graph while the columns are represented by the width of the graph. This division is shown in Figure 9. The columns correspond to the blue nodes, and the levels start at the base station. Only the blue nodes are considered because if a blue node is visited, there is a chance to visit all the nodes in the column due to the heuristics of the proposed solution. For example, if the current node is 3, the MDF will choose 4 over 14 and 2 because the distance from 4 to the base station is greater than all the adjacent nodes of the current node.

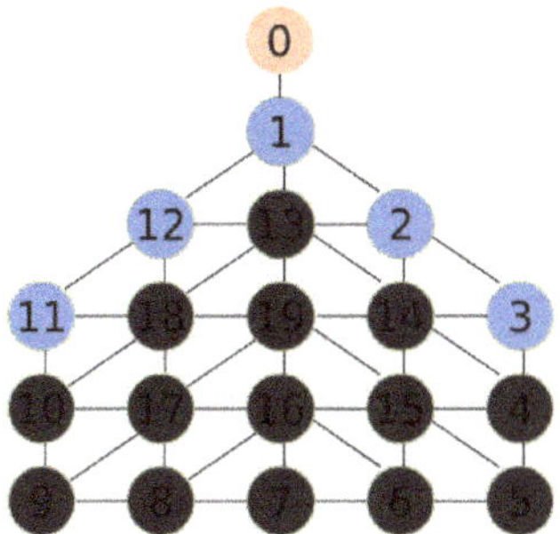

Figure 9. Lines divide the environment in rows and columns.

Since there are many possibilities to travel in the blue nodes, it is necessary to calculate the number of trips on average to stamp every node in the column. Table 2 shows the average ways we need to pass by each blue node in order to stamp every node in that column. The first column of Table 2 contains each blue node, and the second column presents the average number of ways a particular blue node can have all its siblings visited based on the DFORE algorithm. The number of average trips is the number of edges the MDF can take from a blue node using the heuristics of the algorithm. For example, for the blue node 2, there are four possible edges. The first edge is in the line that consists of nodes {14, 15, 6}; the second edge is from node 2 to 3; the third edge is from 14 to 4; and the last edge is from 15 to 5. There is not an edge between 6 and 5 since it is impossible to get from 6 to 5 with the required number of steps. Therefore, the number of edges of all blue nodes is the average number of trips required to explore the graph. We show the equation of the expected number of trips for any given DFG.

Table 2. Average trips to cover all the siblings on the line of each blue node.

Node	Average Ways
1	9
2	4
3	1
12	4
11	1

Definition 3. *Given a DFG G_h with depth p_h (see Equation (1)). The expected average number of trips T_h is:*

$$T_h = 2p_h + 2\sum_{j=1}^{h-2}(p_h - j) + 1. \tag{10}$$

However, there is a simpler way to calculate the expected average number of trips for any given DFG. To obtain the expression, a table for each environment is built. Table 3 shows each DFG G_h with the correspondent variables. In the first column, the size of each DFG is shown. The second column contains the number of nodes in each DFG. The third column shows the depth of the graphs. Finally, the difference between the number of nodes of DFG of size h and $h-1$ is presented in the last column. We note that each successive environment grows by a fixed amount of six as shown in the last row of the table. For example, with the DFG G_3, the number of nodes is 19, while for the DFG it is 7, and their difference is $19-7$, which is equal to 12.

Table 3. Variables for each environment.

DFG size h	Number of nodes n_h	Depth of G_h	$n_h - n_{h-1}$
1	1	1	
2	7	3	6
3	19	5	12
4	37	7	18
5	61	9	24
6	91	11	30
7	127	13	36
8	169	15	42
9	217	17	48
10	271	19	54
11	331	21	60

Taking into consideration the growth of each successive Hextille, the number of regions for any Hextille is given by the following expression: $n_h = 3h^2 - 3h + 1$ (see Appendix B). Finally, the expected average number of trips is $T_h = 3h^2 - 3h + 1$ (see Appendix A). Therefore, $T_h = n_h$.

Furthermore, to get the computational cost of the general case, consider that the number of nodes visited by each trip is $2\epsilon_h$. Therefore, if there are T_h trips, the total number of nodes is $2T_h\epsilon_h$. Since

$T_h = n_h$, the expression is $2n_h\epsilon_h$. Also, notice that $n_h > 2\epsilon_h$. Since $2\epsilon_h$ is a constant, we ignore it. Thus, the general case of exploration is linear $\mathcal{O}(n)$ with respect to the number of nodes in any DFG.

We have calculated the average number of trips for any DFG with the required number of steps ϵ_h. When the number of steps of the MDF e_u is bigger than the required number of steps ϵ_h of a given DFG G_h, that is $e_u > \epsilon_h$, the number of regions visited is increased by a constant factor. Therefore, the expected number of trips remains the same; thus, the reconnaissance time is linear with respect to the number of nodes visited: $f(n) \subseteq \mathcal{O}(n)$.

6. Correctness Proof

Section 5 shows that our algorithm has a linear time $\mathcal{O}(n)$ while the upper bound of exploration under interruption is $\mathcal{O}(n^2)$. Now we prove that our proposal satisfies the DFORE's restrictions. Reconnaissance must satisfy the following restriction: $\forall r \in R, |Tstamp_t(r) - Tstart| \leq T_{expMax}$ (see Section 4.4). The maximum reconnaissance time for any given DFG G_h where $i \in \mathbb{N}$ under interruptibility is n_h^2. This is the time needed to explore the graph using DFS [17]. If the reconnaissance time of our proposed algorithm is greater than n_h^2, then it is not better than DFS, and therefore, our proposed algorithm does not satisfy the restriction of the DFORE problem. For this particular proof, we define the reconnaissance time of our algorithm as the sum of the differences between each stamp of a region with respect to the start time $Tstart$, in other words, the time taken by our algorithm to stamp every region in G_h. By definition, the stamping time at time t is equal to: $Tstamp_t(r) = Tstamp_{t-1}(r') + t$ where $r \neq r'$ and $Tstamp_0(\zeta) = Tstart$.

Definition 4. *For a DFG G_h the number of nodes is $n_h = 3h^2 - 3h + 1$; therefore, the reconnaissance time T_{expMax} is equal to:*

$$T_{expMax} = (3h^2 - 3h + 1)^2. \tag{11}$$

The following restriction should be satisfied:

Restriction 5. *The reconnaissance time of our algorithm $T_{expAlgorithm}$ is less than the maximum reconnaissance time: $T_{expAlgorithm} < T_{expMax}$.*

In order to satisfy Restriction 5, the whole area should be explored within the given time T_{expMax}. Therefore, both the best and general case must be analyzed. The following theorems state that our algorithm satisfies Restriction 5.

Theorem 1. *The divide and conquer reconnaissance algorithm for the best case satisfies Restriction 5 for any given DFG G_h.*

Theorem 2. *The reconnaissance algorithm for the general case satisfies Restriction 5 for any given DFG G_h.*

To prove Theorem 1, the time taken by the reconnaissance step and the time to sample the entire area are calculated. According to Definition 4, the reconnaissance time for any given DFG G_h is $T_{expMax} = (3h^2 - 3h + 1)^2$. The time it takes to explore G_h is calculated using the divide and conquer method $T_{expDivide}$ which is equal to $T_{expAlgorithm}$. We know that $T_{expDivide} = 2\epsilon_h p_h$ (see Section 5.1). The depth of any given DFG G_h where $h \in \mathbb{N}$ is $p_h = 2h - 1$. Therefore, $T_{expDivide} = 2(\epsilon_h)^2$. To prove that $T_{expMax} > T_{expDivide}$, analyzing the inequality:

$$(3h^2 - 3h + 1)^2 > 2(\epsilon_h)^2 \tag{12}$$

$$(9h^4 - 18h^3 + 15h^2 - 6h + 1) > (8h^2 - 8h + 2) \tag{13}$$

The inequation holds if the DFG is greater than one; therefore, $T_{expMax} > T_{expDivide}$ if $h > 1$.

Now, the general case for Theorem 2 is proven. Based on the analysis, the average number of trips is $T_h = 3h^2 - 3h + 1$. Since every trip takes $2\epsilon_h$ of steps, the average reconnaissance time $T_{expAlgorithm}$ is: $2T_h\epsilon_h$. It is necessary to verify that $T_{expAlgorithm} < T_{expMax}$:

$$\begin{aligned} T_{expMax} &> T_{expAlgorithm} \\ 2(T_h)^2 &> 2T_h\epsilon_h \\ T_h &> \epsilon_h \\ 3h^2 - 3h + 1 &>^2 (2h - 1) \\ 3h^2 - 3h + 1 &>^4 h - 2 \\ T_{expMax} &> T_{expAlgorithm}|h > 2 \end{aligned} \tag{14}$$

We have proved that both the best and the general case of our algorithm, for any given DFG, satisfies DFORE's restrictions. In the following section, our theoretical results are compared with the experimental values.

7. Experiments

To determine the performance of our algorithm under various conditions, two experiments were defined.

In the first experiment, the movement range of the MDF was set between ϵ_h and $2\epsilon_h$ to measure the average number of trips required to place a pheromone in every node of the DFG. The second experiment compares the performance of the DFORE algorithm versus the one obtained by MULES [10], adapted to the foraging reconnaissance task. The comparison with MULES is justified since this proposal is the baseline algorithm for indirect communication among mobile sensors.

7.1. Simulation Versus Theoretical Value

This experiment is conducted in two phases. The first phase is to determine the difference between the average number of simulated trips versus the theoretical bound. In the second phase, the experiments are validated through statistical inference.

7.1.1. Experimental Setup

This experiment includes 5,000,000 flights since the number of simulations provided sufficient data to measure the average number of trips, such that the difference among several simulations was not significant. To determine if the proposed algorithm accomplishes the constraint on the average number of trips given by $T_h = 2p_h + 2\sum_{i=1}^{h-2}(p_h - i) + 1$, the number of trips required to travel every DFG G_h were measured. The number of steps was set in the range of $[\epsilon_h, 2\epsilon_h]$, with increments of two units, since a unitary increment does not change the behavior of the algorithm due to Restriction 2 (see Section 4.4). Table 4 shows the results of this experiment.

Table 4. Average number of trips per DFGs with our algorithm.

DFG Depth h	AVG Trips	Std. Deviation	Variance	Max	Min
2	3.8890	0.8315	0.6915	5	3
3	10.6992	2.9336	8.6060	32	5
4	19.7592	5.3702	28.8388	59	7
5	33.2963	9.2975	86.4443	114	11

Figures 10–12 show the distribution of trips for every DFG considering the different number of steps that the MDF can make. Each colored line represents a histogram with the corresponding trips

per steps. The required steps ϵ_h to traverse G_h is colored blue. When the number of steps increases by a factor of two, the data distribution is skewed towards the left around a peak value.

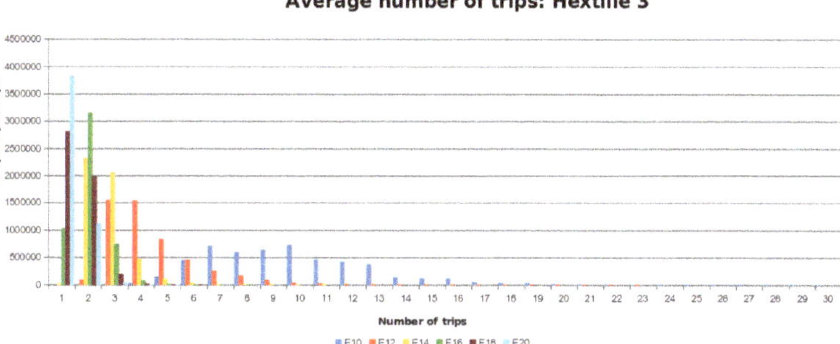

Figure 10. DFG G_3 with the variation of the required number of steps ϵ_h for the 5,000,000 flights. The average number of trips for G_3 with ϵ_h is 10 trips with a frequency between 500,000 and 1,000,000 flights to explore the whole DFG. If the number of steps is incremented, the average number of flights reduces.

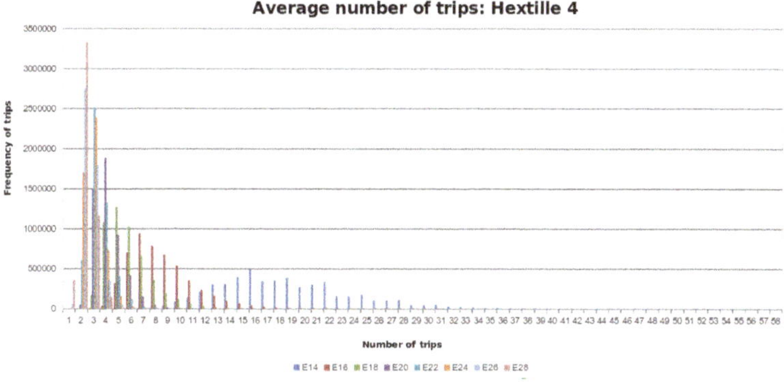

Figure 11. DFG G_4 with the variation of the required number of steps ϵ_h for the 5,000,000 flights. The average number of trips for G_4 with ϵ_h is 19 trips with a frequency close to 500,000 to explore the whole DFG. If the number of steps is incremented, the average number of flights reduces.

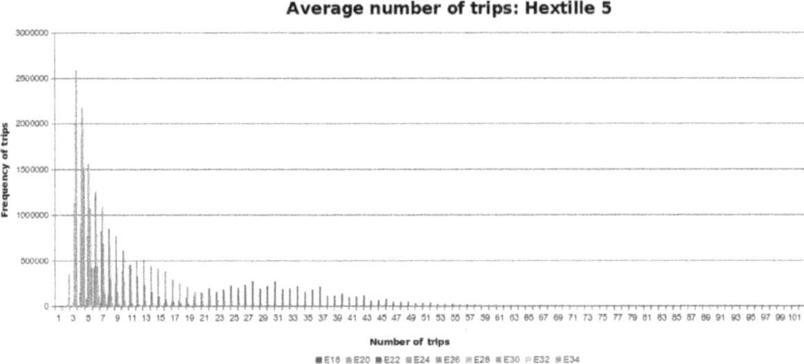

Figure 12. DFG G_5 with the variation of the required number of steps ϵ_h for the 5,000,000 flights. The average number of trips for G_5 with ϵ_h is 33 trips with a frequency between 200,000 and 300,000 flights to explore the whole DFG. If the number of steps is incremented, the average number of flights reduces.

7.1.2. Statistical Inference

A statistical inference test was done to prove that the average number of trips performed by the proposed algorithm is different from the theoretical value T_h. For this, 50 random samples of flights were taken, as statistical sample, for every DFG. We define the null hypothesis H_0 as: *the average number of trips* θ *is equal to* T_h and the alternative hypothesis H_1 as: *the average number of trips* θ *is less than* T_h. Considering the p-value obtained from each set of experiments, the null hypothesis is rejected with a 95% level of confidence, as it can be seen in Table 5.

A *t*-test is applied since we obtain a normal distribution due to the randomness of the movements performed in the experiments. In addition, due to the size of the sample, the variance between trips is homogeneous and there are no significant outliers. Table 6 shows the statistic test. Since the significance level α is 0.05, for every DFG G_h the test passed.

Table 5. T-statistics for each DFGs G_3, G_4, G_5 with the samples. For each row, we have tested H_0 against the results.

DFG Depth h	AVG Trips	T-Statistic	p-Value	H_0
3	10.8	6.82	<0.00001	Reject
4	20.9	4.01	0.002289	Reject
5	33.37	2.84	0.005388	Reject

Table 6. Average trips per DFGs G_3, G_4, G_5 taken from a sample of 50 random flights.

DFG Depth h	AVG Trips	Std. Deviation	Variance	T_h
3	10.8	3.0017	9.0101	19
4	20.9	6.3365	40.1515	37
5	33.37	8.7532	76.6193	61

7.2. DFORE Compared with MULES Reconnaissance

In this section the DFORE algorithm is compared with MULES. The MDF starts at the base station both for the DFORE algorithm and MULES. The MDF explores the whole environment using the two algorithms in different experiments, measuring the average number of trips performed by each one over DFGs G_3, G_4. In this way, considering a sample of 10,000 flights, the difference between the average number of visited regions by each algorithm was measured.

We define the null hypothesis H_0 as: *there is no significant difference between MULES [10] and our proposed algorithm*, i.e., the average number of trips θ of MULES is not different from the average number of trips ϑ of our proposed algorithm. The alternative hypothesis H_1 is: *there is a significant difference between θ and ϑ*. The null hypothesis is rejected with a 95% level of confidence.

Table 7 shows the results of this experiment. There is a clear difference between the average number of trips of the two algorithms. The proposed algorithm has a better performance than MULES, since it performs less trips than MULES.

Table 7. Results of our proposed algorithm compared with MULE for DFGs G_h of size h = (2, 3, 4).

Algorithm	AVG Trips	Std. Deviation	Variance	Max	Min
Proposed alg. G_2	3.8922	0.8326	0.6933	5	3
MULES G_2	13.7152	10.5584	111.4790	122	3
Proposed alg. G_3	10.6784	2.9283	8.5751	26	5
MULES G_3	206.6570	181.1481	32,814.6230	2290	6
Proposed alg. G_4	19.7463	5.3766	28.9077	54	7
MULES G_4	3189.4170	2981.9314	8,891,914.6828	33730	66

8. Discussion

Based on the results obtained from the experiments two facts are concluded. First, the data obtained shows that the average number of trips falls within the mathematical bound obtained theoretically. From Figures 10–12, it can be seen that the data follows a distribution towards the mean, despite the randomness part of our proposed algorithm. In addition, if the movement capabilities are increased by two units, the number of trips decreases, as shown in Figure 12. Furthermore, based on the results of the second experiment (see Section 7.2), we have evidence that our algorithm has better performance than MULES [10]. This is explained by the random movement of MULES against the oriented movement of our proposed algorithm. The orientation towards unexplored regions is done through indirect communication using the artificial pheromones segregated by each mobile sensor in the regions. Therefore, the average number of trips required to deposit at least one pheromone in all the graph using our proposed algorithm is less than that of MULES. The trade-off between computational time and run time of the algorithm is shown in a comparison between a random walk algorithm, such as Data MULES, and our proposed algorithm.

9. Conclusions

We have presented a data-foraging-oriented reconnaissance algorithm based on bio-inspired indirect communication for aerial vehicles. One original contribution is the definition of an artificial pheromone, as an abstract data type, oriented to perform stigmergy-based communications. Through the virtual segregation of such pheromones, the algorithm allows aerial vehicles which sense a given area, to communicate indirectly their findings. In this way, aerial vehicles can create several paths oriented to explore the environment and recognize profitable data sources. By considering the energy constraints of aerial vehicles and their impact on their movement capabilities, the operational environment was discretized in the form of a set of regions organized into a Hextille. Then, based on the Hextille, the environment is formally modeled as a connected undirected graph called Data Foraging Graph (DFG). The artificial pheromones segregated are related to an area that is the region visited, which corresponds to a node in the DFG. The Data Foraging-Oriented Reconnaissance problem has been defined. We identify and define the required and sufficient movement capacity capabilities of the aerial vehicle per trip to explore an environment according to the depth of the DFG. The solution proposed was formally specified and mathematically evaluated. The results prove the viability and efficiency of the solution. Additionally, we have presented a study increasing the aerial vehicle's movement capability. The results of this study show that the average number of trips and the run time to explore the environment highly decrease as the movement capability increases.

Acknowledgments: This work is supported by the National Council for Science and Technology of Mexico (CONACYT) through the Master scholarship number 390398 and the project ID PDCPN2013-01-215421.

Author Contributions: The five authors contributed proportionally in these categories: Conception or design of the work, data analysis and interpretation, drafting the article, critical revision of the article and final approval of the version to be published.

Conflicts of Interest: The authors declare no conflict of interest.

Abbreviations

The following abbreviations are used in this manuscript:

UAV	Unmanned Aerial Vehicle
DFORE	Data Foraging-Oriented Reconnaissance
DFS	Depth First Search
BFS	Breadth First Search
PSM	Piecemeal Search Model
bDFX	Bounded Deep First Exploration
MULE	Mobile Ubiquitous Local Area Network Extension
MDF	Mobile Data Forager
DFG	Data Foraging Graph

Appendix A. General Case of Algorithm

To find the solution of the following recurrence, it should be unfolded:

$$T_h = 2p_h + 2\sum_{j=1}^{h-2}(p_h - j) + 1$$

This expression is in terms of depth p; however, the expression in terms of the number of nodes n is required. It is possible to see that there is a factor of six if we subtract every second row of Table 3. Therefore, we calculate a factor of expansion equal to

$$\frac{n}{6} - 1$$

Also, by checking the table, an expression to calculate the number of regions in a DFG G_h is obtained. Find a close equation for this recurrence is needed; therefore, we try to unfold it to see a pattern.

$$n_h = (3p_{h-1} + 3) + n_{h-1}$$
$$n_h = (3p_{h-1} + 3) + (3p_{h-2} + 3) + n_{h-2}$$
$$n_h = (3p_{h-1} + 3) + (3p_{h-2} + 3) + (3p_{h-3} + 3) + n_{h-3}$$
$$n_h = (3p_{h-1} + 3) + (3p_{h-2} + 3) + (3p_{h-3} + 3) + (3p_{h-4} + 3) + n_{h-4}$$
$$n_h = 4 * 3 + 3(p_{h-1} + p_{h-2} + p_{h-3} + p_{h-4}) + n_{h-4}$$

There is a pattern in the recurrence; every time the recurrence is unfolded a 3 in depth is obtained. If we continue to unfold the recurrence, we get:

$$n_h = 4 * 3 + 3 * (p_{h-1} + p_{h-2} + p_{h-3} + p_{h-4}) + (3 * p_{h-5} + 3) + n_{h-5}$$
$$n_h = 4 * 3 + 3 * (p_{h-1} + p_{h-2} + p_{h-3} + p_{h-4}) + (3 * p_{h-5} + 3) + n_{h-5} + \ldots$$
$$n_h = 3k + 3 * \sum_{j=1}^{k}(p_{h-j}) + n_{h-k} \ \forall k$$

If we set $k = h$:

$$n_h = 3h + 3 * \sum_{j=1}^{h} (p_{h-j}) + n_{h-h} \tag{A1}$$

It is necessary to find $\sum_{j=1}^{h} (p_{h-j})$. We unfold this sum:

$$\sum_{j=1}^{h} (p_{h-j}) = p_{h-1} + p_{h-2} + \ldots + p_{h-h}$$
$$\sum_{j=1}^{h} (p_{h-j}) = p_{h-1} + p_{h-2} + \ldots + p_0$$
$$\sum_{j=1}^{h} (p_{h-j}) = p_{h-1} + \ldots + 7 + 5 + 3 + 1$$
$$\sum_{j=1}^{h} (p_{h-j}) = (h-1)^2$$

Therefore:

$$n_h = h * 3 + 3 * (h-1)^2 + n_{h-h}$$
$$n_h = h * 3 + 3 * (h-1)^2 + n_0$$
$$n_h = h * 3 + 3 * (h-1)^2 + 1$$
$$n_h = 3h^2 - 3h + 1$$

The depth of Hextille with radius h_{ex} is:

$$p_h = 2i - 1$$

Therefore, the average number of trips for any Hextille with radius h_{ex} is:

$$T_h = 2p_h + 2 \sum_{j=1}^{h-2} (p_h - j) + 1$$
$$T_h = 2(2h-1) + 2 \sum_{j=1}^{h-2} ((2h-1) - j) + 1$$
$$T_h = 4h - 2 + 2[(h-2)(2h-1) - \sum_{j=1}^{h-2} (j)] + 1$$
$$T_h = 4h - 2 + 2[(h-2)(2h-1) - \sum_{j=1}^{h-2} (j)] + 1$$
$$T_h = 4h - 1 + 2(2h^2 - 5h + 2) - 2(h-1)(h-2)/2$$
$$T_h = 4h - 1 + 2(2h^2 - 5h + 2) - (h^2 - 3h + 2)$$
$$T_h = 3h^2 - 3h + 1$$

Appendix B. Best Case of Algorithm

To prove the linear time of the best case, a close solution to the following recurrence must be found:

$$f(n) = 2f(n/2) + c$$

where c is a constant.

$$\begin{aligned} f(n) &= 2f(n/2) + c \\ &= 2(2f(n/4) + c) + c = 4f(n/4) + 3c \\ f(n) &= 4(2f(n/8) + c) + 3c = 8f(n/2) + 7c \\ &\dots \\ & 2^k f(n/2^k) + (2^k - 1)c \end{aligned}$$

We need to get rid of $f(n/2^k)$ and reach f(1). If $\log_2 n = k$, a closest solution is possible.

$$\begin{aligned} f(n) &= 2^k f(n/2^k) + (2^k - 1)c \\ &= 2^{\log_2 n} f(n/2^{\log_2 n}) + (2^{\log_2 n} - 1)c \\ &= nf(1) + (n-1)c \\ &= n + (n-1)c \\ &= \mathcal{O}(n) \end{aligned}$$

Therefore, we have a linear time to explore n nodes, that is:

$$f(n) = \mathcal{O}(n)$$

Appendix C. Multiple MDFs at a Time

In this section, the behavior of our proposal is evaluated in presence of multiple mobile elements. Each group of MDFs $u \in U$ takes off sequentially from the base station ζ. When all the MDFs of the group return to the ζ, the next batch of MDFs will lift from the base station, until all regions in the operational environment have pheromones. The MDFs do not have previous knowledge of the deposited pheromones by others MDF; thus, the base station must communicate this information to them. There are two cases to be considered: a) overlapping regions and b) disjoint regions.

1. **Disjoint regions.** At any time that a MDF u_q is sampling a region r_i, denoted as $< u_q, r_i >$ then $\forall u \in U| < u, r_i >$, then $u_q \neq u$ given that there are multiple choices from the adjacent regions.
2. **Overlapping regions:** The mobile elements may share a region at any time.

Disjoint regions. When no overlap exists at any given time, every MDF will explore different regions. Figure A1 shows an example of this. MDF u_q is represented by blue while MDF u_{q-1} is represented by red. In the base station there is no knowledge about the operational environment. At step 0, since there is only one region, both MDFs must share a region. The MDFs follow Rules 1 and 2 from Section 4.5. At step 1, the MDFs have chosen and move to new regions and select new regions following the rules. After several steps, the MDFs must return to the base station.

Each one communicates to the base station the deposited pheromones as shown in Figure A2. Next, the base station combines the information of the two MDFs into one snapshot of the DFG. In the next cycle, two new MDFs will continue the reconnaissance task, however each one of them will have the snapshot from the previous task. Each time new regions are visited, the overall time required to do DFORE will reduce by a certain amount; however, the algorithm for DFORE stays the same. This amount is bounded by the depth of the DFG.

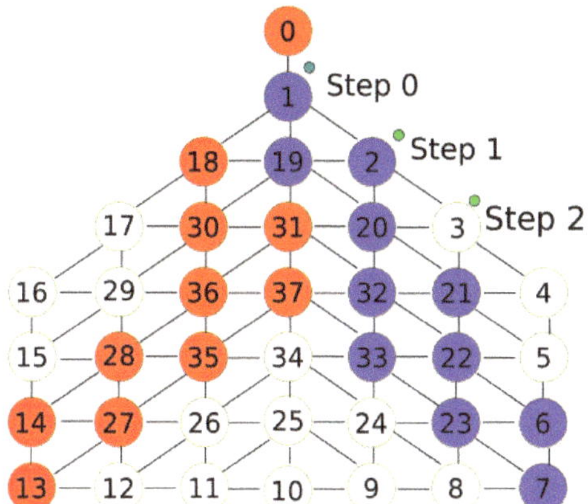

Figure A1. Two MDFs do the recoinnasance task. Each pheromone deposited by the MDFs is colored as red or blue.

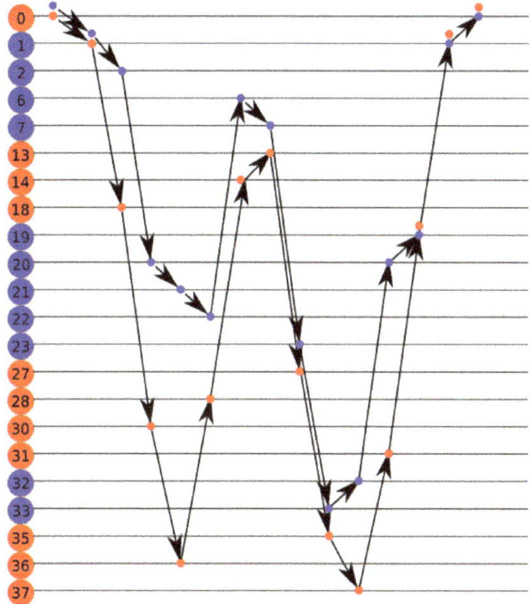

Figure A2. The operational environment as a causal graph. Some regions are visited by multiple MDFs when there is only one possibility.

Overlapping regions. Similarly, when overlap occurs, the information can be shared among the MDFs trough the base station. In the worst case scenario, every MDF will visit the same region at the same time. Thus, every region will be visited multiple times. This is equivalent to the single MDF scenario where the only difference is the amount of pheromones deposited in each region. The number of trips in this scenario is bounded between the single case and the disjoint scenarios. Since the number of trips in the single case is greater than any of the multiple cases, Restriction 1 is satisfied.

Appendix D. Reconnaissance Algorithm

We present the reconnaissance algorithm.

Algorithm A1 Main function.

```
function LIST<CELL> MAIN
    int accumulatedTime ← 0
    List<Cell> inspectionRegions ← ∅
    while check(r ∈ N) do
        [region, time] = reconnaissance(e_u, area)
        accumulatedTime ← accumulatedTime + time
        inspectionRegions = inspectionRegions ∪ region
    end while
    return inspectionRegions
end function
```

Algorithm A2 Reconnaissance algorithm.

```
function [LIST<CELL>, INT] RECONNAISSANCE(int e_u, List<Cell> area)
    int time ← 0
    Cell r ← 0
    while e_u ≥ 0 do
        if (e_u − 2) − e_base- ≥ 0 then
            Cell r' ← choose(neighbors(r))
        else
            Cell r' ← traceback(neighbors(r), e_u)
        end if
        e'_u ← e_u − 1
        r'.stamp ← r'.stamp + 1
        time' ← time + 1
        r ← r'
    end while
    List < Cell > foodCells ← getFoodCells(area)
    return [foodCells, time]
end function
```

Algorithm A3 Obtain regions with food from the Hextille.

```
function LIST<CELL> GETFOODCELLS(List<Cell> area)
    List<Cell> foodCells ← ∅
    for all (r ∈ area) do
        if r.hasFood then
            foodCells ← foodCells ∪ r
        end if
    end for
    return foodCells
end function
```

Algorithm A4 Choose the next cell to visit.

```
function CELL CHOOSE(List<Cell> neighbors)
    List<Cell> markedNeighbors = getMarked(neighbors)
    if markedNeighbors!= ∅ then
        Cell new = random(markedNeighbors)
    else
        Cell lessStamped = selectMinimumStamps(neighbors)
        Cell new = random(farthest(lessStamped))
    end if
    return new
end function
```

Algorithm A5 Get the marked regions.

```
function LIST<CELL> GETMARKED(List<Cell> area)
    List<Cell> markedCells ← ∅
    for all (r ∈ area) do
        if r.stamp > 0 then
            markedCells ← markedCells ∪ r
        end if
    end for
    return markedCells
end function
```

Algorithm A6 Get the minimum number of stamps.

```
function LIST<CELL> SELECTMINUMUMSTAMPS(List<Cell> neighbors)
    List<Cell> leastStampedCells ← ∅
    int minimumStamps ← minStamp(neighbors)
    for all (r ∈ neighbors) do
        if r.stamp == minimumStamps then
            leastStampedCells ← leastStampedCells ∪ r
        end if
    end for
    return leastStampedCells
end function
```

Algorithm A7 Get the farthest regions from my current position.

```
function LIST<CELL> FARTHEST(List<Cell> neighbors)
    List<Cell> farthestCells ← ∅
    int maxDistance ← maxDistance(neighbors)
    for all (r ∈ neighbors) do
        if r.distance >= maxDistance then
            farthestCells ← farthestCells ∪ r
        end if
    end for
    return farthestCells
end function
```

Algorithm A8 Get the maximum distance between the adjacent regions.

```
function INT MAXDISTANCE(List<Cell> neighbors)
    int max ← 0
    for all (r ∈ neighbors) do
        if r.distance > max then
            max ← r.distance
        end if
    end for
    return max
end function
```

Algorithm A9 Return to the base station.

```
function CELL TRACEBACK(List<Cell> neighbors(r), int e_u)
    List < Cell > legalNeighbors = checkSteps(neighbors, e_u)
    Cell new = random(legalNeighbors)
    return new
end function
```

Algorithm A10 Check if a region can be visited.

```
function LIST<CELL> CHECKSTEP(List<Cell> neighbors, int e_u)
    List<Cell> legalNeighbors ← ∅
    for all (r ∈ neighbors) do
        if (e_u − 1) − r.distance >= 0 then
            legalNeighbors ← legalNeighbors ∪ r
        end if
    end for
    return legalNeighbors
end function
```

References

1. Valavanis, K.P.; Vachtsevanos, G.J. *Handbook of Unmanned Aerial Vehicles*; Springer Publishing Company, Incorporated: Dordrecht, The Netherlands, 2014.
2. McGill, P.; Reisenbichler, K.; Etchemendy, S.; Dawe, T.; Hobson, B. Aerial surveys and tagging of free-drifting icebergs using an unmanned aerial vehicle (UAV). *Deep Sea Res. Part II Top. Stud. Oceanogr.* **2011**, *58*, 1318–1326.
3. Burgard, W.; Moors, M.; Stachniss, C.; Schneider, F.E. Coordinated multi-robot exploration. *IEEE Trans. Robot.* **2005**, *21*, 376–386.
4. Choset, H. Coverage for robotics—A survey of recent results. *Ann. Math. Artif. Intell.* **2001**, *31*, 113–126.
5. Bonin-Font, F.; Ortiz, A.; Oliver, G. Visual navigation for mobile robots: A survey. *J. Intell. Robot. Syst.* **2008**, *53*, 263–296.
6. Rooker, M.N.; Birk, A. Multi-robot exploration under the constraints of wireless networking. *Control Eng. Pract.* **2007**, *15*, 435–445.
7. Dessmark, A.; Pelc, A. Optimal graph exploration without good maps. *Theor. Comput. Sci.* **2004**, *326*, 343–362.
8. Fraigniaud, P.; Ilcinkas, D.; Peer, G.; Pelc, A.; Peleg, D. Graph exploration by a finite automaton. *Theor. Comput. Sci.* **2005**, *345*, 331–344.
9. Kramer, D.L. Foraging behavior. In *Evolutionary Ecology: Concepts and Case Studies*; Oxford University Press: New York, NY, USA, 2001; pp. 232–246.
10. Shah, R.C.; Roy, S.; Jain, S.; Brunette, W. Data mules: Modeling and analysis of a three-tier architecture for sparse sensor networks. *Ad Hoc Netw.* **2003**, *1*, 215–233.
11. Chang, C.Y.; Yu, G.J.; Wang, T.L.; Lin, C.Y. Path construction and visit scheduling for targets by using data mules. *IEEE Trans. Syst. Man Cybern. Syst.* **2014**, *44*, 1289–1300.

12. Singh, J.P.; Roy, P.K.; Singh, S.K.; Kumar, P. Source location privacy using data mules in Wireless Sensor Networks. In Proceedings of the 2016 IEEE Region 10 Conference (TENCON), Singapore, 22–25 November 2016; pp. 2743–2747.
13. Das, A.; Mazumder, A.; Sen, A.; Mitton, N. On mobile sensor data collection using data mules. In Proceedings of the 2016 International Conference on Computing, Networking and Communications (ICNC), Kauai, HI, USA, 15–18 February 2016; pp. 1–7.
14. Megow, N.; Mehlhorn, K.; Schweitzer, P. Online graph exploration: New results on old and new algorithms. In Proceedings of the International Colloquium on Automata, Languages, and Programming, Zurich, Switzerland, 4–8 July 2011; Springer: Berlin/Heidelberg, Germany, 2011; pp. 478–489.
15. Albers, S. Online algorithms: A study of graph-theoretic concepts. In Proceedings of the International Workshop on Graph-Theoretic Concepts in Computer Science, Ascona, Switzerland, 17–19 June 1999; Springer: Berlin/Heidelberg, Germany, 1999; pp. 10–26.
16. Betke, M.; Rivest, R.L.; Singh, M. Piecemeal learning of an unknown environment. *Mach. Learn.* **1995**, *18*, 231–254.
17. Duncan, C.A.; Kobourov, S.G.; Kumar, V. Optimal constrained graph exploration. *ACM Trans. Algorithms (TALG)* **2006**, *2*, 380–402.
18. Argamon-Engelson, S.; Kraus, S.; Sina, S. Utility-based on-line exploration for repeated navigation in an embedded graph. *Artif. Intell.* **1998**, *101*, 267–284.
19. Sujit, P.; Ghose, D. Two-agent cooperative search using game models with endurance-time constraints. *Eng. Optim.* **2010**, *42*, 617–639.
20. Bao, Y.; Fu, X.; Gao, X. Path planning for reconnaissance UAV based on particle swarm optimization. In Proceedings of the 2010 Second International Conference on Computational Intelligence and Natural Computing Proceedings (CINC), Wuhan, China, 13–14 September 2010; Volume 2, pp. 28–32.
21. Obermeyer, K.J. Path planning for a UAV performing reconnaissance of static ground targets in terrain. In Proceedings of the AIAA Conference Guidance, Navigation and Control, Chicago, IL, USA, 10–13 August 2009.
22. Huang, L.; Qu, H.; Ji, P.; Liu, X.; Fan, Z. A novel coordinated path planning method using k-degree smoothing for multi-UAVs. *Appl. Soft Comput.* **2016**, *48*, 182–192.
23. Bertuccelli, L.; Alighanbari, M.; How, J. Robust planning for coupled cooperative UAV missions. In Proceedings of the 43rd IEEE Conference on Decision and Control (CDC), Nassau, Bahamas, 14–17 December 2004; Volume 3, pp. 2917–2922.
24. Zhao, J.W.; Zhao, J.J. Study on Multi-UAV Task clustering and Task Planning in Cooperative Reconnaissance. In Proceedings of the Sixth International Conference on Intelligent Human-Machine Systems and Cybernetics (IHMSC), Hangzhou, China, 26–27 August 2014; Volume 2, pp. 392–395.
25. Li, Y.; Chen, H.; Er, M.J.; Wang, X. Coverage path planning for UAVs based on enhanced exact cellular decomposition method. *Mechatronics* **2011**, *21*, 876–885.
26. Shames, I.; Fidan, B.; Anderson, B.D. Close target reconnaissance using autonomous UAV formations. In Proceedings of the 47th IEEE Conference on Decision and Control (CDC), Cancun, Mexico, 9–11 December 2008; pp. 1729–1734.
27. Dao, T.P.; Huang, S.C. Optimization of a two degrees of freedom compliant mechanism using Taguchi method-based grey relational analysis. *Microsyst. Technol.* **2017**, 1–16, doi:10.1007/s00542-017-3292-1.
28. Dao, T.P.; Ho, N.L.; Nguyen, T.T.; Le, H.G.; Thang, P.T.; Pham, H.T.; Do, H.T.; Tran, M.D.; Nguyen, T.T. Analysis and optimization of a micro-displacement sensor for compliant microgripper. *Microsyst. Technol.* **2017**, 1–21, doi:10.1007/s00542-017-3378-9.
29. Dao, T.P.; Huang, S.C.; Thang, P.T. Hybrid Taguchi-cuckoo search algorithm for optimization of a compliant focus positioning platform. *Appl. Soft Comput.* **2017**, *57*, 526–538.
30. Jun, M.; D'Andrea, R. Path planning for unmanned aerial vehicles in uncertain and adversarial environments. In *Cooperative Control: Models, Applications and Algorithms*; Springer: Boston, MA, USA, 2003; pp. 95–110.

31. Fan, Q.; Wang, F.; Shen, X.; Luo, D. Path planning for a reconnaissance UAV in uncertain environment. In Proceedings of the 2016 12th IEEE International Conference on Control and Automation (ICCA), Kathmandu, Nepal, 1–3 June 2016; pp. 248–252.
32. Yao, P.; Wang, H.; Su, Z. Cooperative path planning with applications to target tracking and obstacle avoidance for multi-UAVs. *Aerosp. Sci. Technol.* **2016**, *54*, 10–22.

Article

Characterization of Control-Dependent Variable Stiffness Behavior in Discrete Muscle-Like Actuators

Caleb Fuller †,* **and Joshua Schultz** †

Department of Mechanical Engineering, University of Tulsa, Tulsa, OK 74104, USA; joshua-schultz@utulsa.edu
* Correspondence: caleb-fuller@utulsa.edu; Tel.: +1-970-623-3859
† Current address: 800 S. Tucker Dr. Tulsa, OK 74104, USA.

Received: 5 February 2018; Accepted: 22 February 2018; Published: 28 February 2018

Abstract: This paper presents the modeling, characterization and validation for a discrete muscle-like actuator system composed of individual on–off motor units with complex dynamics inherent to the architecture. The dynamics include innate hardening behavior in the actuator with increased length. A series elastic actuator model is used as the plant model for an observer used in feedback control of the actuator. Simulations are performed showing the nonlinear nature of the changing stiffness as well as how this affects the dynamics, clearly observed in the phase portrait. Variable-stiffness hardening behavior is evaluated in experiment and shows good agreement with the model.

Keywords: actuators; variable stiffness; muscle-like actuators; discrete; control; Series Elastic Actuator (SEA) model

---

1. Introduction

Bio-inspired robotic actuator designs seek to emulate certain desirable traits seen in human skeletal muscle systems, and can sometimes exceed them in performance. The design presented in this article utilizes skeletal muscle characteristics, including: compliance, redundancy to failure, and a modular building block or cellular architecture, see Figure 1. One other concept taken from human skeletal muscle systems is that of motor unit recruitment [1]. Motor unit recruitment is a discretized actuation strategy that provides human skeletal muscle with a plan to generate the particular motion dictated by the nervous system. Each motor unit is responsible for recruiting certain muscle fibers, spread throughout the muscle, to achieve the contraction. In this way, if a contraction of higher strength is needed, more motor units are recruited. An actuation scheme based on this idea can simply change which motor units are activated if one motor unit suddenly is rendered broken or useless. Such an actuation device offers a high resistance to failure, which is a very attractive feature in high-reliability applications. The key advantage is that a component failure results not in a total loss of motion for that particular robotic limb, but only a decrease in performance, meaning a decrease in the maximum force that can be produced. A discrete actuator design that uses "on"–"off" motor units must utilize such a recruitment strategy for its actuation and control. There are challenges associated with quantizing an analog control signal into activating discrete units, which will be addressed by showing two methods of quantizing such a signal. The first method is a constant scaling of an approximation for the force, length and number of total motor units active. The second method is through the use of a pre-computed look-up table referenced during run time.

Compliance from unit-to-unit must also be a design component in order not to violate physical laws, as shown by Mathijssen et al. [2]. By introducing this compliance two things happen, the first is that it makes modeling the actuator very complex as each discrete unit now has an elastic element associated with it, but it also allows for the opportunity to vary the compliance in such a way as to introduce damping-like behavior where there was not before, which will be explained in more detail later on. The units making up discrete actuators offer a redundancy that resists failure, for if one of the units is suddenly broken another unit can be recruited to "pick up the slack". These three characteristics make discrete actuators very desirable for use in robotic applications of various types, but modeling the dynamics of these actuators becomes quite complex because each discrete unit must be treated as a mass-spring system, which results in very high order systems of differential equations. High order systems makes constructing a closed form control law difficult. This paper provides a method for overcoming these hurdles using a Series Elastic Actuator model and an observer system.

When modeling the dynamics of the actuator presented in this work, prior research in Series Elastic Actuators (SEAs) provides guidance in developing a controller to work with the actuator. An SEA was chosen to aid in the model development because its compliant element that operates in series with the motor is similar to the unit-to-unit compliance in a discrete muscle-like actuator. In the design of discrete actuators that are aligned in parallel formations there must be compliance coupling them together in order to adhere to physical laws. See Figure 2 for reference to this concept and read [2,3] for more detail.

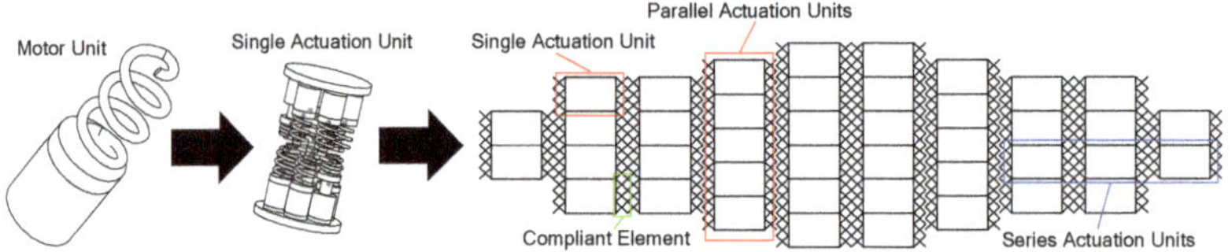

Figure 1. Modular, discrete muscle-like variable-stiffness actuator.

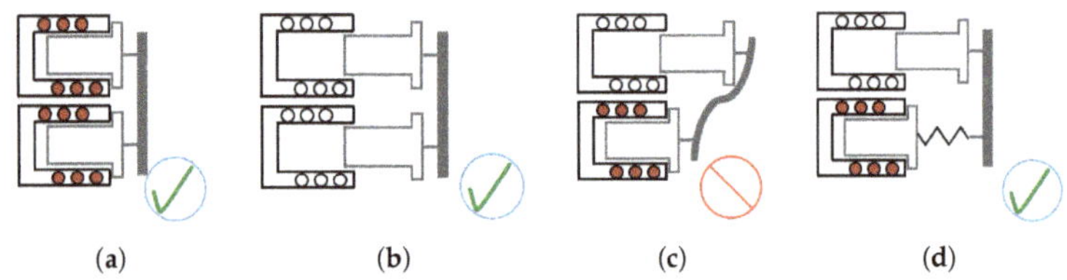

Figure 2. Compliance is necessary for allowing different contraction patterns for parallel motor units. See how in the third figure violation has occurred and is resolved by the addition of a compliant spring element in the fourth figure: (**a**) rigid connection for active units; (**b**) rigid connection for non-active units; (**c**) breaks laws of mechanics of materials; and (**d**) addition of compliant element.

The SEA was one of the earlier developments made in compliant actuators and provided a platform used as a basis for many of the later designs. Pratt and Williamson [4] developed an actuator that included a series elastic element that gives the actuator a compliant nature. More recent work in SEA's has led to the development of Variable Stiffness Actuators (VSA), or Variable Impedance Actuators (VIA's). Vanderborght et al. provide a definition for and discussion of the range of VSA's and many of their characteristics in his overview of Variable Stiffness Actuators [5]. These actuators have numerous advantages (some of which have already been mentioned) including: compliance, the ability to reject disturbances due to the nature of their system dynamics, and the ability to be used in unpredictable environments. In his paper, Vanderborght also categorizes the different ways that variable stiffness is achieved in the many designs that have been developed. The three main categories are: active impedance control, inherent compliance and inherent damping, and inertial actuators. The actuator outlined in this work falls under the first and second categories. The actuator itself is defined by inherent variable stiffness and it can be actively controlled to vary this stiffness even more.

Discrete actuators have been implemented in the past with platforms seeking to control vibration and antenna shaping, see [6–10]. The main types of discrete actuators used in those applications were piezoelectric, but the theoretical work done in this paper is able to be applied to many different types of discrete actuators. Discrete actuators are able to emulate human skeletal muscle with several distinctive characteristics. The first is *compliance*, which is a necessary characteristic in designing the actuator for applications where human–robot interaction is needed [11]. The second characteristic is *redundancy*, which is important because it achieves something that is novel and desirable for actuators, keeping the robot operational despite a component failure, which is an important characteristic when considering actuators that will be used in remote areas making repair difficult or impossible. The third characteristic is a *modular construction*; this is significant because the cost for expanding the actuator to new configurations or larger applications is low and the complex nature of redesigning an actuator for larger applications can be simplified if a framework for one of these actuators is already in place. This article seeks to provide such a framework for discrete actuators. A discussion on the actuator design will lead into showing the variable stiffness behavior seen most vividly in simulation, after which the capability of the SEA model will be shown, leading into a discussion on the control law developed through simulation, and finally an experiment with data showing the variable stiffness behavior.

2. Muscle-Like Actuators Composed of Discrete Building Blocks

Human skeletal muscle is made up of small discrete building blocks, which create the larger actuator. Coming from the perspective of control, our application begins with the motor units as the most fundamental. An actuation unit is the next step up from a motor unit, and is the reconfigurable unit for the actuator as a whole. A motor unit is defined as the "smallest individually activated force producing device", while the actuation unit is defined as the "smallest building block of the actuator that can be inserted or removed" [2]. Putting these actuation units together into series and parallel bundles results in different arrays and arrangements providing the necessary configuration for the performance requirements of the task (Figure 1).

2.1. Actuation Unit Design and Definition

A discrete actuator is "a motion or force producing device composed of more than one actuation unit and containing more than one motor unit" [2]. The control paradigms presented in this article are based on the physical parameters of the platform built by Mathijssen et al., some details of which will be provided in Section 3. Other platforms will have many of the same features displayed in this design including: compliant spring element, modular actuation units, and mechanical stop/limiting component.

The motor units provide a small amount of displacement, Δx, extending the spring element thus producing a restoring force. To increase the force ,one simply activates more motor units. See Figure 3 for an annotated graph depicting the various components of a discrete actuation unit.

Creating a usable lumped-parameter model of a compliant muscle-like discrete actuator that is able to be used in control development is not a simple problem. Schultz et al. [7] define a "class" structure to enumerate the types of different stiffness behaviors exhibited by the actuator at different lengths. However, this concept was only shown in simulation and the development of the idea did not explain what the actuator behavior is beyond the Critical Length. Thus, the work by Schultz et al. [7] did not provide a framework for the hardening behavior beyond this "Critical Length".

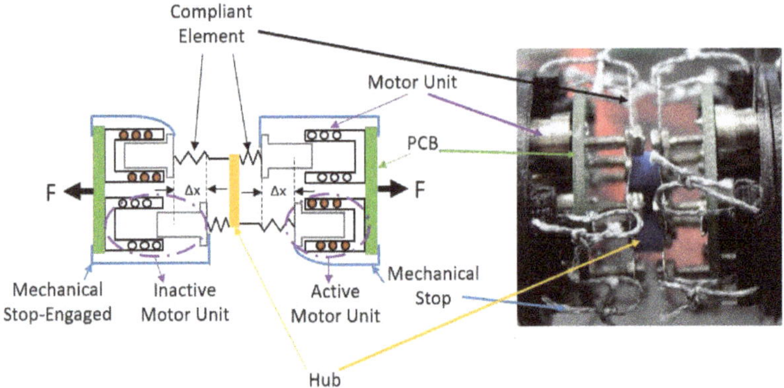

Figure 3. Actuation unit with annotated components.

2.2. The Mechanical Stop and Limiting Condition

To properly analyze and understand the hardening behavior of a discrete actuator one must understand how *mechanical stops* contribute to the system. The stops engage with the springs when an external force pulls on the units and extends them. In this actuator design the stops are attached to each lobe of the leaf spring. These stops give the actuator its variable impedance nature. When a load is placed across an actuation unit the spring begins pulling on the contraction mechanism. The springs on this actuator engage with the mechanical stops, shown in Figure 4, which prevent the inactive motor units from being dislodged and rendered useless. The mechanical stops keep the motor unit contraction mechanism (in this case, solenoid plungers) from pulling out of the motor unit housing (in this case, within the solenoid flux gap). However, when the springs encounter the mechanical stop the system dynamics of the actuator deviate from those of a typical spring-mass model. However, if the length of the actuator is stretched beyond this point then the spring deforms from the mechanical stops meaning the actuation units stiffness is a maximum stiffness of all the spring lobes in parallel. In this way before the particular spring lobe attached to the inactive motor units encounters the mechanical stop that spring element does not contribute to the overall stiffness of the actuator. That is the reason this actuator can vary the impedance through control of the different motor units.

(**a**)

(**b**)

Figure 4. Demonstration of straps engaged with leaf-spring: (**a**) mechanical stop not engaged; and (**b**) mechanical stop engaged.

2.3. Class Structure

Four different classes of an actuation unit will now be defined. These class definitions help to understand how the dynamics of the actuator can be altered with changes of the various elements: motor units active, extension of the compliant element and the mechanical stop. After the definition and explanation of each Class, simulations showing the nonlinear variable stiffness qualities inherent in the design of the actuator will be shown, which exemplify how class affects the system dynamics. To begin a discussion of the class architecture describing the length-hardening behavior of the actuator, the resting length of the spring at two states must be defined, one when the actuation unit is active and the other when the actuation unit is inactive. At any given instant, an active actuation unit is a unit with *one or more* active motor units, while an inactive actuation unit is one with *no* active motor units. It is important to note that the actuation units move from one class to another during operation. It is assumed that there is at least some small pre-load ϵ on the actuator at all time. The actuation unit can assume one of two distinct resting lengths. When an actuation unit has an active motor unit, whether it is one or n units active, the resting length is defined to be ℓ^a. When an actuation unit has no active motor units, then the pre-load extends the actuation unit until the mechanical stops are encountered, which is defined as, ℓ^o. The difference between the active resting length and the inactive resting length is the stroke length of the motor units, $\ell_t = \ell^o - \ell^a$. It is critical to note that the resting length for an actuation unit is the same regardless of whether one, two, three, or more motor units are active.

2.3.1. Class 0

Class Zero is defined as being the state of the actuation unit when no motor units are active and the length of the actuation unit is less than the resting length, ℓ^o, of the inactive actuation unit. This means the motor units are free to slide and move in and out of the metal casing of the motor units. For quasi-static motion (under the assumption of a small pre-load), Class Zero does not appear on any unit. Figure 5a shows the actuation cell in Class Zero.

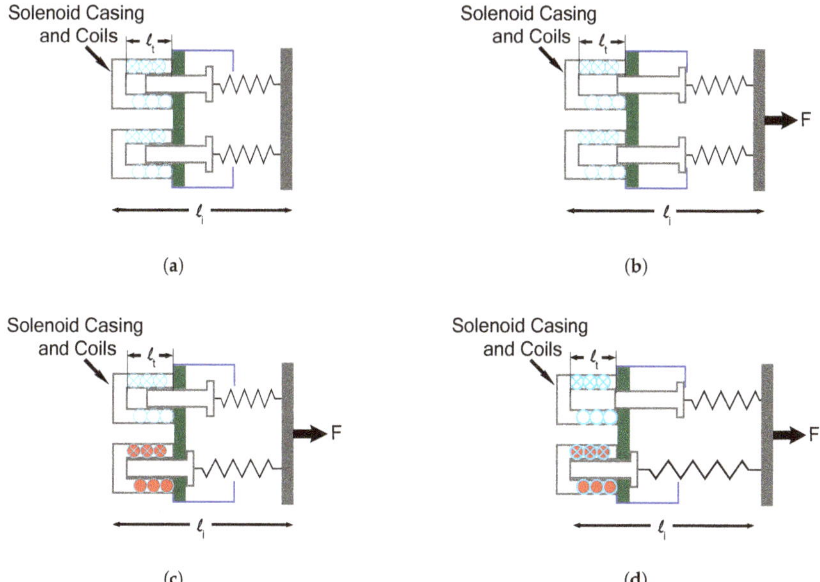

Figure 5. Variety in arrangements of actuators: (**a**) Class Zero; (**b**) Class I; (**c**) Class II; and (**d**) Class III.

2.3.2. Class I

Class I is defined as being the state of the actuation unit when no motor units are active and the length of the actuator has been extended beyond ℓ^o. With an initial infinitesimal pre-load on the system this actuation unit will assume a resting length ℓ^o, which is the length of the unit when resting against the mechanical stops. This can be seen in Figure 5b.

2.3.3. Class II

Class II is defined as being the state of the actuation unit when at least one motor unit is active and the length of the unit, ℓ_i, is greater than the resting length of an active unit, ℓ^a, but less than the length at which the mechanical stops are engaged, $\ell^a < \ell_i < \ell^o$. In this state the active motor units carry the load, while the inactive motor units are free to slide and move without carrying any load. In Class II, as the load approaches zero, the actuation unit will assume a resting length of ℓ^a. Figure 5c shows a representation of an actuation unit in Class II.

2.3.4. Class III

Class III is defined for when the actuation unit has M active motor units, where $0 < M < N$ ($M = N \Longrightarrow$ fully activated), and the unit has been stretched to a length that is beyond the mechanical stops, ℓ^o. (For the platform presented in [2] the total number of motor units is $N = 6$). Thus, for Class III both the active and inactive motor units are carrying the load for the actuator. However, this would manifest itself as a different stiffness than simply having 6 of the units in parallel because the compliant spring attached to the active motor units will be stretched an extra solenoid stroke length, ℓ_t, further than the inactive motor units. Figure 5d shows a representation of an actuation unit in Class III.

2.3.5. Critical Length

Imagine that the actuator load quasi-statically stretches the actuator p units resting length to some length L. If the number of active motor units for each actuation unit i is given as p_i, then p_{min} is the minimum active over all i units. In [7], the term "Critical Length" is coined and is defined as the overall actuator length, L, when p_{min}, the weakest actuation unit (actuation unit with the fewest active motor units, and thereby the lowest stiffness), experiences extension and moves from Class II to Class III, where all the spring lobes of the inactive motor units engage with the mechanical stops and increase the overall spring constant of that actuation unit. It is important to realize that this will happen to each actuation unit individually at a different value of L, depending on how many motor units are active. One of the actuation units could be in Class III while another is in Class II and yet another could be in Class I. In extending the actuator's length, the first actuation unit that will reach Class III is p_{min}. It should also be noted and is outlined in the three spring model from [7], that if there are more than one actuation units at p_{min} then these units will extend in length at the same rate and both move from Class II to Class III at the same value of L. Furthermore, given a total number of motor units active for the entire actuator, distributing them differently over the actuation units will result in a different Critical Length.

2.4. Activation Patterns and Activation Levels

There are many ways that the active and inactive motor units can be distributed over the actuation units that will total the same number of motor units active for the entire actuator, see Figure 6. The ***activation level*** is defined as the sum of active motor units on each actuation unit. Thus, if four actuation units are in series, two motor units could be activated on the first actuation unit, five motor units on the second, one motor unit on the third and zero on the fourth. This can be represented by the vector [2 5 1 0]. An example of a different ***activation pattern*** while keeping the same activation level would be [2 2 2 2]. However, another activation pattern that could be realized would be [6 2 0 0], each of these patterns totalling eight units active. However, specifying an activation pattern does not determine *which* of the motor units on a given actuation unit are active, simply that two of the $N = 6$ motor units are active. It should be evident that a large finite number of different stiffness values that depends on the number of actuation units and the number of motor units used to make up the actuator can be achieved. The following equations represent the activation level with M being the number of active motor units per actuation unit. In the example illustrated in this section, four actuation units $P = 4$, connected in series are used. Each actuation unit consists of one PCB with six motor units, $N = 6$, mounted on each PCB. The number of different stiffness levels that can be selected via the control input can be calculated using binomial coefficients and the equation for combinations seen in Equation (1).

$$\binom{n}{k} = \frac{n!}{k!(n-k)!} \tag{1}$$

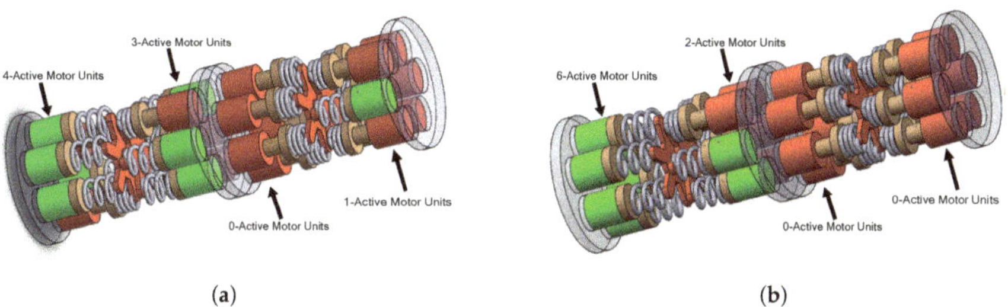

(a) (b)

Figure 6. Different activation patters while maintaining same activation level: (**a**) eight motor units active (highlighted in green) over two actuation units—shorthand written as a vector [4 3 0 1]; and (**b**) eight motor units active (highlighted in green) over two actuation units—Shorthand written as a vector [6 2 0 0].

When calculating the number of stiffness levels ,substitutions for n and k can be made, as shown in Equation (2). In the substitution, N stands for the number of motor units used and P denotes the number of actuation units for the actuator following similar notation to [7]. This equation can be simplified to Equation (3).

$$\binom{N-1+P}{N-1} = \frac{(N-1+P)!}{(N-1)!((N-1+P)-(N-1))!} \tag{2}$$

$$\frac{(N-1+P)!}{(N-1)!(P)!} \tag{3}$$

For the length based hardening simulation that will be discussed later, if 0 is included as a state showing that there are no active motor units for that actuation cell, then four actuation units in series results in a total of 210 different stiffness levels that can be achieved through different activation patterns. For longer chains, it can easily be shown that by substituting the number of units in the equation for P and evaluating gives a determination for the new number of total activation levels.

As an individual actuation cell lengthens and moves from Class II into Class III, the overall stiffness of the actuator should increase. This is because the actuation cell moves from having less than the maximum number of motor units $(M < N)$ providing stiffness, to a point where the springs of the inactive motor units $(N - M)$ engage with the mechanical stops and begin providing stiffness to the overall spring constant. The exception is if all the motor units are active $(M = N)$ or all the motor units are inactive $(M = 0)$, in these cases there will not be a Class III distinction for this actuator or a critical length.

The stiffness of an actuator, made up of four actuation cells, increases in a nonlinear fashion as each of the cells moves from Class II into Class III. This happens in distinct steps because of the difference in activation level between the four cells. The least active cell will reach Class III first, followed by the next least active cell and so on until the cell with the most active motor units reaches Class III. Once all the actuation cells have reached Class III and none are left in Class II the stiffness will remain constant with increasing length.

3. Complex System Dynamics of a Variable Impedance Discrete Actuator with a Simple SEA Observer Model

This section contains the simulations, results and discussion of driving an inertial load mass with an actuator made up of a chain of actuation units connected in series. Building upon the knowledge of the system variable stiffness behavior gained from the simulations, the equations of motion describing the actuator dynamics can be derived. These equations of motion include the hardening behavior presented in Section 2. This complex plant model is then simulated numerically under feedback control of the inertial load, with discretization of the control input, allowing us to observe the effects of the innate variable impedance behavior. However, before properly implementing the observer/controller, one needs to understand key ideas regarding a control system with discrete actuators.

1. A typical continuously variable signal $u(t)$ will not be sufficient for driving a discrete actuator.
2. The control signal must be transformed into an integer number of motor units "on".
3. This number of motor units "on" must be selected based upon the closest value to the control signal that is possible.
4. One must decide which units will lead to the most desired affect on the system dynamics.
5. These steps are symplified by treating the entire actuator as an SEA.

In this section, there are three governing assumptions made about the system.

1. The contraction of the motor units is frictionless.
2. The mass of the actuation units is negligible as compared to the driven inertial mass.
3. The mechanical stops at each spring lobe engage at same length.

3.1. Mathematica Formulation

Changing between the classes of the actuation units, which depends on the overall length of the actuator, is a nonlinear switching condition. The different classes of the actuator, discussed earlier in Section 2.3, give definition to the nonlinear phase portraits, a plot of the position versus velocity for the system, that will be shown and discussed in detail in this section. The setup in Figure 7 shows the simple actuator–mass–return spring configuration that was used in the simulations to be discussed shortly.

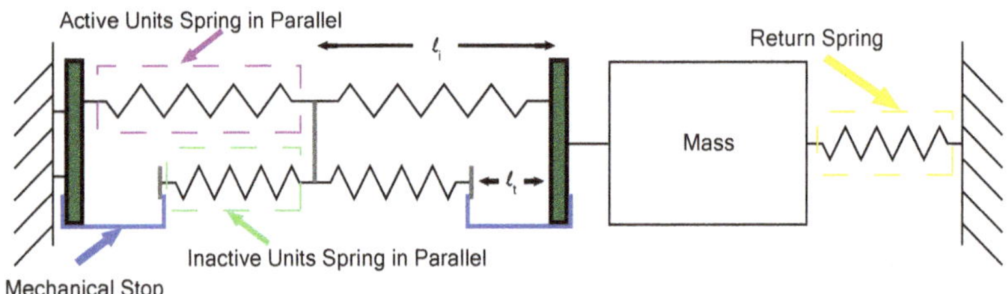

Figure 7. Horizontal oscillatory motion.

The equations describing the variable stiffness are not easily described in closed form due to the piecewise-linear behavior described in Section 2. Equation (4) gives the sum of the forces about the common node for two actuation units in a series chain connection, i and $i+1$, as shown in Figure 7.

Table 1 displays the terms in the equation. The equation provides the force developed by a chain of actuation units based upon displacement and the stiffness, however this equation does not divulge which Class the actuation units are in, which is critical to knowing what the stiffness is for the actuator. The displacement of the actuation units, ℓ_t, will be changing rapidly and the stiffness of the actuator will change according to which Class it is in, thus conditional statements must be used to check the Class of the actuator before computing force across the actuator.

$$\sum_i F_i : k_i^a(\ell_i - \ell_i^a) + k_i^{in}(\ell_i - \ell_i^o) = k_{i+1}^a(\ell_{i+1} - \ell_{i+1}^a) + k_{i+1}^{in}(\ell_{i+1} - \ell_{i+1}^o) \tag{4}$$

Table 1. Nomenclature for Equation (4).

Term	Description
k_i^a	spring constant active motor units in parallel
k_{i+1}^a	spring constant active motor units in parallel for next node in chain
k_i^{in}	spring constant inactive motor units in parallel
k_{i+1}^{in}	spring constant inactive motor units in parallel for next node in chain
ℓ_i	length of actuation unit
ℓ_{i+1}	length of next actuation unit in chain
ℓ_i^a	resting length of active actuation unit
ℓ_{i+1}^a	resting length of next active actuation unit in chain
ℓ_i^{in}	resting length of inactive actuation unit
ℓ_{i+1}^{in}	resting length of next inactive actuation unit in chain

In Equation (4), it should be noted that all ℓ_i are unknown variables, and all ℓ_i^a and ℓ_i^o are known system constants. To properly identify the Class that each actuator resides in, the individual length of each actuator must be known. Therefore, the goal in arranging Equation (4) is to solve for those lengths, ℓ_i and ℓ_{i+1}. The total length of the chain of actuation units can easily be measured empirically by an encoder or similar sensor. In practice the ℓ_i are not easily measured. A record of which motor units are active can also be obtained from the operating firmware. Therefore, the first step is to expand and collect the terms in Equation (4) about ℓ_i and ℓ_{i+1}. This gives the equation seen in Equation (5).

$$\ell_i(k_i^a + k_i^{in}) - \ell_{i+1}(k_{i+1}^a + k_{i+1}^{in}) = k_i^a\ell_i^a + k_i^{in}\ell_i^o - k_{i+1}^a\ell_{i+1}^a - k_{i+1}^{in}\ell_{i+1}^o \tag{5}$$

This will be simplified further by applying assumption (3), which says that all actuation units have the same active and inactive resting length, $\ell_i^a = \ell_{i+1}^a = \ell^a$ and $\ell_i^o = \ell_{i+1}^o = \ell^o$, in practice there are slight manufacturing variations, but implementing this simplification results in Equation (6).

$$\ell_i(k_i^a + k_i^{in}) - \ell_{i+1}(k_{i+1}^a + k_{i+1}^{in}) = (k_i^a - k_{i+1}^a)\ell^a + (k_i^{in} - k_{i+1}^{in})\ell^o \tag{6}$$

Dividing by $k_i^a + k_i^{in}$ gives Equation (7). Then, a change of variable can be made setting $a_i = \frac{(k_{i+1}^a + k_{i+1}^{in})}{(k_i^a + k_i^{in})}$, which gives Equation (8).

$$\ell_i - \ell_{i+1}\frac{(k_{i+1}^a + k_{i+1}^{in})}{(k_i^a + k_i^{in})} = \frac{(k_i^a - k_{i+1}^a)}{(k_i^a + k_i^{in})}\ell^a + \frac{(k_i^{in} - k_{i+1}^{in})}{(k_i^a + k_i^{in})}\ell^o \tag{7}$$

$$\ell_i - \ell_{i+1}a_i = \frac{(k_i^a - k_{i+1}^a)}{(k_i^a + k_i^{in})}\ell^a + \frac{(k_i^{in} - k_{i+1}^{in})}{(k_i^a + k_i^{in})}\ell^o \Longrightarrow \begin{cases} \text{Class 0,} & undefined \\ \text{Class I,} & k_i^a = 0 \\ \text{Class II,} & k_i^{in} = 0 \\ \text{Class III,} & k_i^a \neq 0 \ \& \ k_i^{in} \neq 0 \end{cases} \tag{8}$$

Notice the conditional statement introduced that allows this general equation to be used for Class Zero–Class III (definitions of which can be found in Section 2.3). Equation (8) represents the force–length relationship for actuation units in Class III. However, in order to adjust this to be used with a unit in Class II the term k_i^{in} must be set equal to 0. This is the term quantifying the spring constant once the units have reached critical length and extend beyond the mechanical stops. Setting this equal to zero and substituting into Equation (8) will give a correct representation of the sum of forces for this Class. a_i will change to: $a_i = \frac{(k_{i+1}^a + k_{i+1}^{in})}{k_i^a}$. If, however, the unit is in Class I, then the motor units on board the actuation unit are all contributing to the stiffness and are all inactive and the length has been extended beyond the inactive resting length, $\ell_i > \ell^o$. If this is the case, then substituting $k_i^a = 0$ into the equation for a_i and Equation (8) will correctly categorize the sum of the forces for this state of the actuator. The last Class is Class Zero for which the force across the actuator is zero and neither the inactive nor the active motor units contribute to carrying the load. In this case, the load is only across the return spring seen in Figure 7, which corresponds to Equation (7) being undefined for Class Zero.

These equations and conditions can then be written for any number of actuators in series and put in matrix form as seen in Equations (9)–(12). Whereas the a_i matrix is defined in Equation (10), c_i is defined in Equation (12), and the individual actuation unit lengths are given by Equation (11).

$$A \cdot \ell = C \tag{9}$$

$$A = \begin{pmatrix} 1 & -\frac{k_{i+1}^a + k_{i+1}^{in}}{k_i^a + k_i^{in}} & 0 & \cdots & 0 \\ 0 & 1 & -\frac{k_{i+2}^a + k_{i+2}^{in}}{k_{i+1}^a + k_{i+1}^{in}} & \ddots & \vdots \\ \vdots & \ddots & \ddots & \ddots & 0 \\ 0 & \cdots & 0 & 1 & -\frac{k_n^a + k_n^{in}}{k_{n-1}^a + k_{n-1}^{in}} \end{pmatrix} \tag{10}$$

$$\ell = \begin{pmatrix} \ell_i \\ \ell_{i+1} \\ \vdots \\ \ell_n \end{pmatrix} \tag{11}$$

$$C = \begin{pmatrix} \frac{k_i^a - k_{i+1}^a}{k_i^a + k_i^{in}} \cdot \ell^a + \frac{k_i^{in} - k_{i+1}^{in}}{k_i^a + k_i^{in}} \cdot \ell^o \\ \frac{k_{i+1}^a - k_{i+2}^a}{k_{i+1}^a + k_{i+1}^{in}} \cdot \ell^a + \frac{k_{i+1}^{in} - k_{i+2}^{in}}{k_{i+1}^a + k_{i+1}^{in}} \cdot \ell^o \\ \vdots \\ \frac{k_{n-1}^a - k_n^a}{k_{n-1}^a + k_{n-1}^{in}} \cdot \ell^a + \frac{k_{n-1}^{in} - k_n^{in}}{k_{n-1}^a + k_{n-1}^{in}} \cdot \ell^o \end{pmatrix} \tag{12}$$

Looking at these equations, one can see that deliberately picking which motor units are activated alters the distribution of active motor units across the actuation units, allowing a tunable hardening behavior for the system dynamics, or an introduction of damping-like behavior without any damping in the system (see Corr and Clark [8]). This is demonstrated in the following section as the phase portraits are examined, which is a common tool for analyzing nonlinear dynamic systems.

3.2. Simulations and Illustrative Examples of Discrete Muscle-Like Actuator Dynamics

The best way to present the information described in the previous section is to contrast a single-ended actuator with a return spring in an antagonist setup, as shown in Figure 7. Pure simple harmonic motion for such an antagonist setup would give a uniform ellipsoid.

Using the matrix definitions in Equations (10)–(12), the equations of motion can be simulated to analyze the oscillatory motion of a mass attached to the end of a muscle-like actuator as seen in the setup of Figure 7. In these simulations, a setup of two actuation units was chosen each with six motor units. Choosing two actuation units allows the actuator to display the discretization effects seen in the phase portraits and control input, however, the theoretical formulation developed in the previous section is valid for n units in a chain and for any number of motor units per actuation unit, and it can also extend easily to bundles of chains. Adding more actuation units decreases the discretization effects making it appear more and more linear until at some point the quantization is so linear it can be more or less considered continuously variable. However, most practical applications will have a small enough number that quantization is necessary. There are four cases shown in Figure 8: (1) a simple harmonic motion between a linear spring and our actuator; (2) a switching method based on time; (3) a switching method based upon actuator position, right or left of the origin; and (4) a switching method based on the sign of the velocity, so above or below the origin. Thus, it can be said that the third and fourth methods are a type of quadrant based switching, but with a limited number of activation patterns and activation levels dictated by the number of actuation units and their arrangement within the actuator. In all of these figures, activation level is held constant but the activation pattern is varied.

1. Figure 8a shows the phase portrait of the natural response of the muscle-like actuator due to an initial condition (blue), contrasting a linear spring setup showing the familiar symmetrical ellipse pattern. The "D" shape curve clearly shows the hardening behavior. The three vertical lines show the points at which the actuation units move from Class II to Class III, the first vertical line closest to the origin being the Critical Length of the actuator.
2. Figure 8b looks at the phase portrait in which the actuator switches between two different activation patterns halfway through the simulation time. It is evident the phase portrait changes shape, and is easy to tell that the second pattern has a higher stiffness than the first.
3. Figure 8c displays a switching pattern that changes the activation pattern at equilibrium position. It is evident that this switching pattern does not offer the stark difference between activation patterns, and thus it should be noted that as long as the actuator is in Class Zero it has no hardening behavior, this also means that as long as the oscillation has a negative position on the phase portrait it is only affected by the return spring and carries none of the aspects of the variable stiffness muscle-like actuator.
4. Figure 8d shows a switching condition based upon the velocity of the load mass, which can easily be measured in practice. Velocity switching occurs by moving from an actuation pattern that is less "stiff" to an actuation pattern that is more "stiff" at the point when the velocity sign changes; it can be noted that this can be done with the same total number of motor units active across the actuation units (but changing the distribution) or by increasing the number of total active motor units. The results from this simulation, see Figure 8d, are quite interesting. For the mass fixed to the end of this actuator and a return spring it exhibits damping-like behavior without there being any damping modeled in the system.

Switching control is a method of control used in practice and has been shown by Gasparri et al. to be a feasible control scheme for his variable stiffness actuator [9].

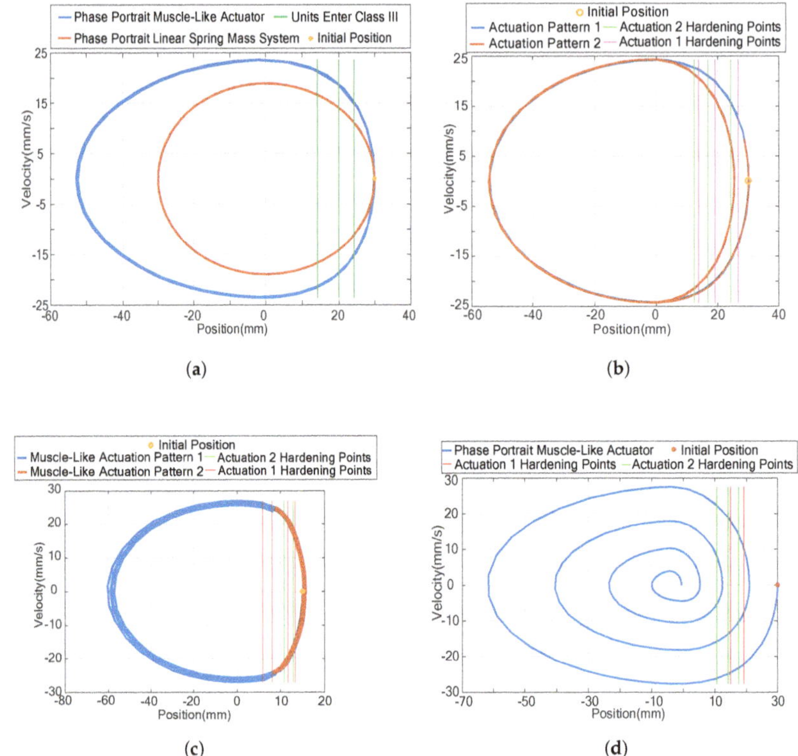

Figure 8. Actuator phase portraits of different switching patterns: (**a**) simple-harmonic motion of variable stiffness actuator; (**b**) switching based upon time; (**c**) switching based upon position; and (**d**) switching activation based upon velocity.

3.3. SEA Model Implementation

The motivation behind the development of the SEA model is to use standard state space control methods in designing an linear observer to reconstruct the state for full state feedback control. The observer captures the nominal linear behavior of the average stiffness; it does not account for the nonlinearities in the plant due to the hardening behavior. The crucial aspect of using this model, however, is how control of a discretized plant with all of its complexity is accomplished using a simple linear observer architecture based only on measurement of the load position.

The Series Elastic Actuator was originally developed by Pratt and Williamson (see [4] for simple model development and control designed for use with their actuator). The model used in the work for this project borrows concepts from Pratt and other developments in the SEA [10,11]. For this project, a model was developed using a linear motor, a spring and a load mass. This greatly simplifies a model in which each motor unit is treated as its own mass-spring-damper system coupled together in series-parallel configurations. Such a model would reach extremely high order state spaces with even one or two actuation units coupled together. The SEA model sufficiently captures all the dominant dynamic effects of the actuator system.

Following the procedure for designing an observer as presented in Brogan [12] and Sinha [13], the necessary steps were taken to create an observer using the SEA model discussed previously, see Figure 9a,b.

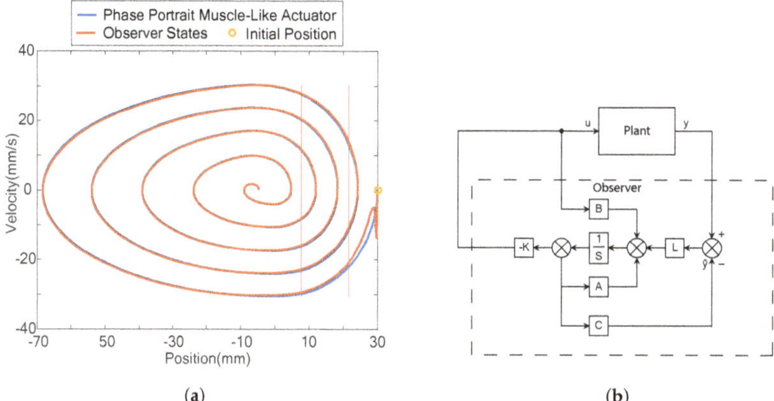

(a)

(b)

Figure 9. Observer feedback loop and phase portrait: (**a**) phase portrait with observer; and (**b**) observer with full state feedback control.

For any linear system, this would give the controller for the system, however, the discrete muscle-like actuator has one last important characteristic which has not yet been discussed thoroughly. The muscle-like discrete actuator must receive a quantization of the control input u in order to actuate the particular motor units needed to generate the force dictated by the controller. The output, u, of the controller is a continuously real-valued variable which must be translated to a discrete number of motor units "on", an integer value [14]. This section examines this problem and shows the novel solution implemented and the initial value simulations that resulted from this solution.

3.3.1. Quantization of the Observer Control Signal to Discrete Control Signal

The produced force of the actuator at any given length can be obtained using Equation (4) after using the matrix Equations (9)–(12) to solve for the individual actuation unit lengths. Two similar methods of quantizing the control input from the controller are explained. Method I is by constant scaling of the force, length and approximating the number of total motor units active; this method computes the force available for the current actuator length that is closest to the control input, u. Method II is a lookup table that is generated off-line and referenced during run-time; this method searches for the force in the table that is closest to the control input and selects the activation pattern associated with the force. We must perform these steps separately from the control because the individual actuation unit lengths, ℓ_i, are unknown. Figure 10a shows the muscle-like actuator given an initial position and zero initial velocity and displays how the controller is able to converge on the equilibrium point. It must be noted that given the nature of quantized control the system will not converge exactly to the equilibrium point but will instead converge within a certain radius R of a sphere inscribed within an ellipsoid [15]. Using the second method of pre-computing, a lookup table that is referenced in run-time gives the results seen in Figure 10b.

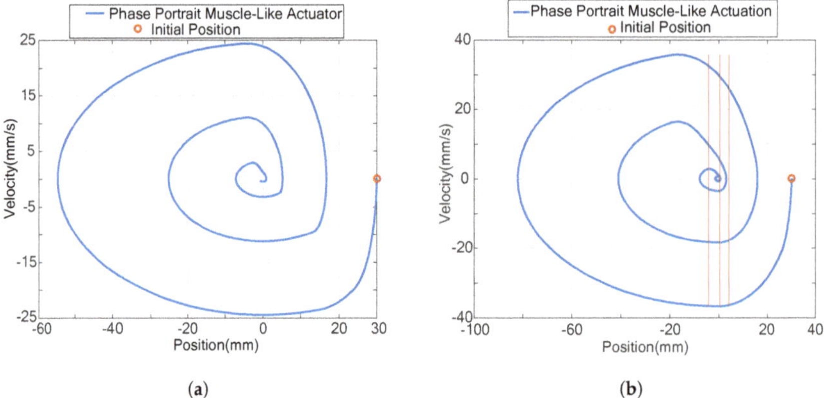

Figure 10. Observer control phase portraits with two different discrete quantization methods: (**a**) quantization Method I; and (**b**) quantization Method II.

3.3.2. Introducing a Disturbance

To observe how the actuator behaves when disturbances perturb the system, a constant disturbance force is introduced. This was analyzed for the open loop velocity switching condition and for both of the two methods discussed above for the quantization of the control signal. The results can be seen in Figures 11 and 12b. Notice in the two figures with the Observer control methods that both of them only converge to a certain point but then enter a cycle of constant oscillations. This region of the phase portrait shows that the actuator does not receive valid control input for this region. The reason for this is that when the actuator is in Class Zero, it does not matter what pattern or activation level is switched to; in this mode of the actuator, it can impose no force and therefore is not controllable.

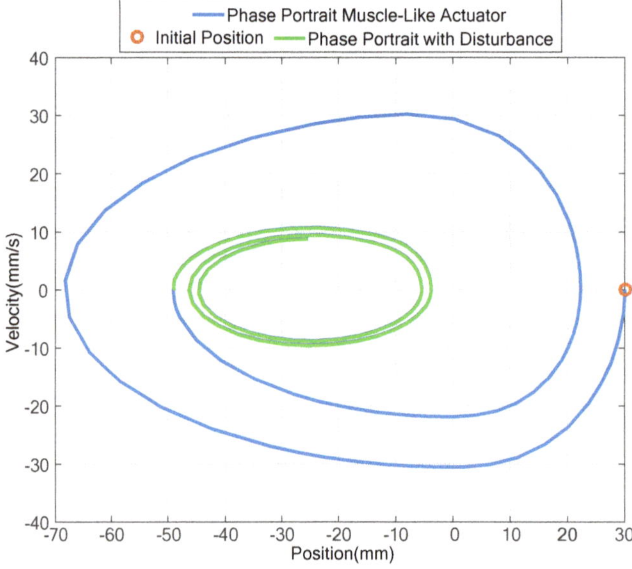

Figure 11. Velocity switching with disturbance force.

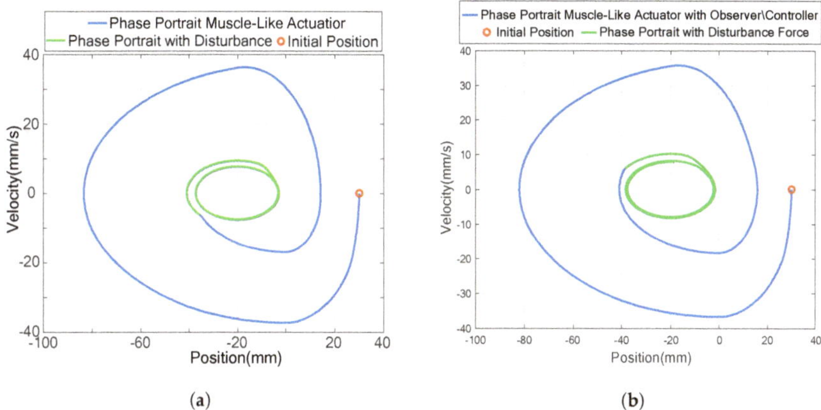

Figure 12. Introduction of a disturbance force into the system: (**a**) control Method I with disturbance (green); and (**b**) control Method II with disturbance (green).

3.3.3. Muscle-Like Actuator in Antagonistic Setup

In order not to lose controllability in certain regions of the statespace, an antagonist actuator setup using two variable stiffness discrete muscle-like actuators can be utilized (see Figure 13). In this scenario, the SEA model used as the observer maintains a strong performance and gives a good estimation of the states of the antagonistic actuator setup. Here it must be explained why the observer model was not reconfigured for an antagonistic SEA model and how the antagonistic muscle-like actuator is superior to the antagonistic SEA. The answer is really quite simple, as stated in [16] and also proven through analysis of the controllability matrix. An antagonistic SEA model with linear springs of the same spring constant is not controllable because the stiffness of the SEA model becomes independent of the controllable parameters. Figure 14 shows such a setup. Summing the forces acting on the mass in the center gives Equation (13). The stiffness of this setup is given by Equation (14) [16] and shows that the result for the stiffness is dependent only on the constant k which is not variable and therefore renders this setup uncontrollable. Thus, considering that argument forward progress can be made with the same observer used in the previous section and shown in Figure 15a, the result of using an antagonist actuator in place of the return spring. This result is easily seen to be much improved from the results of the actuator with a return spring, which should make sense because in an antagonistic setup there is no longer a Class Zero region where the actuator is uncontrollable.

$$\begin{aligned} F &= -k(x - x_{0A}) + k(x_{0B} - x) \\ &= -2k + k(x_{0A} - x_{0B}) \end{aligned} \tag{13}$$

$$K = \frac{dF}{dx} = -2k \tag{14}$$

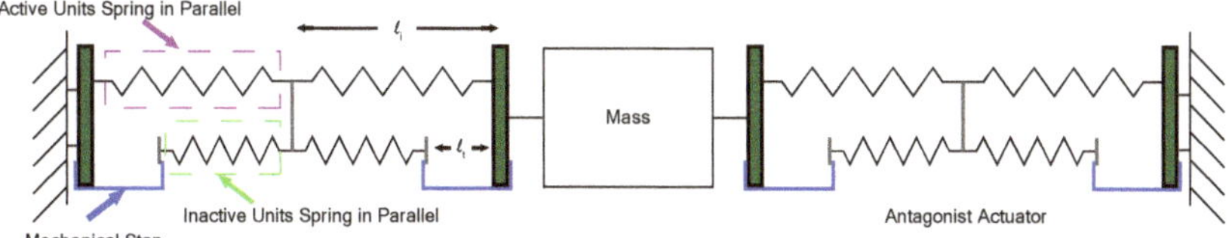

Figure 13. Antagonistic actuator setup.

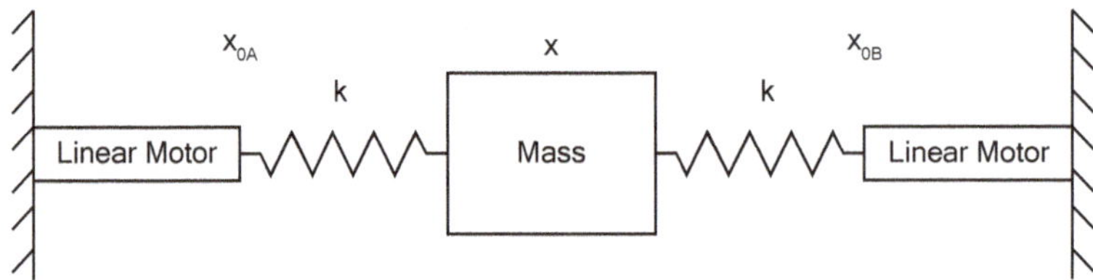

Figure 14. Demonstration of the uncontrollable linear springs.

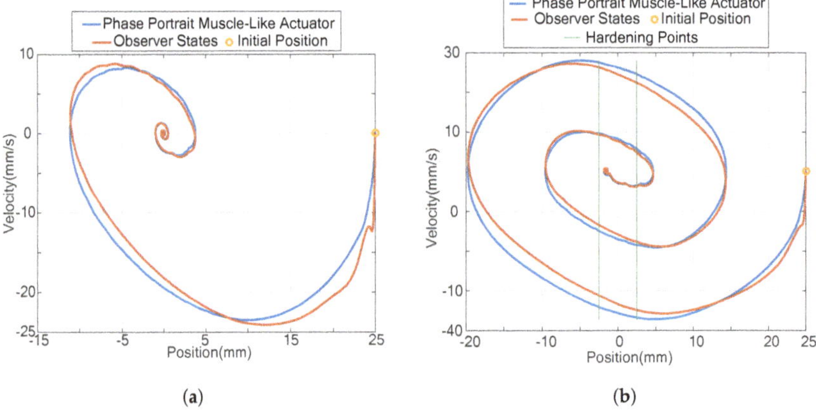

Figure 15. Antagonist control phase portraits: (**a**) antagonistic control with observer; and (**b**) antagonistic control with observer and disturbance Force.

4. Experimental Verification of Length-Based Hardening

To verify the variable stiffness characteristics discussed in Sections 2.1–2.3, an experiment was designed that would enable the actuator to be given an actuation pattern and then steadily lengthened while keeping the activation pattern constant throughout. The setup of the experiment can be seen in Figure 16 and is described as follows. A chain of four actuation units of the type described in Mathijssen et al. [2] (see Figure 16) makes up an actuator mounted between two fixtures where one of the fixtures is secured to an optical table while the other fixture is mounted to a micrometer stage. In series with the actuator is a Futek 1lb force cell whose voltage output was measured using a LeCroy (WaveRunner 604Zi, Teledyne LeCroy, Rockland County, NY, United States) oscilloscope. The fixture mounted on the micrometer stage gave the ability to lengthen the actuator in steps as fine as 0.025 mm, however, for this experiment steps of approximately 1.2 mm were sufficient. The drive circuitry supplying current to each motor unit is a custom design, described in Mathijssen et al. [2]. The digital signal supplied to the circuitry is from an National Instruments DAQ and manually activated through a National Instruments LabView Virtual Instrument. Taking data sets was performed by setting a constant activation pattern and then increasing the length of the actuator in steps of approximately 1.2 mm, the actual excursion recorded directly from the micrometer stage, and measuring the force at each step. This was repeated with a different activation pattern after resetting the stage. At the extremes of extension, the restoring force of the spring overcame the magnetic force holding the solenoid plunger in the solenoid housing, and data points could not be collected beyond this ultimate length, which varied with the activation pattern (in practice, this ultimate length constitutes failure and should be avoided). Table 2 shows the activation patterns that were used in the experiment and their corresponding activation levels and the figure where their data is graphed and compared to the theoretical model.

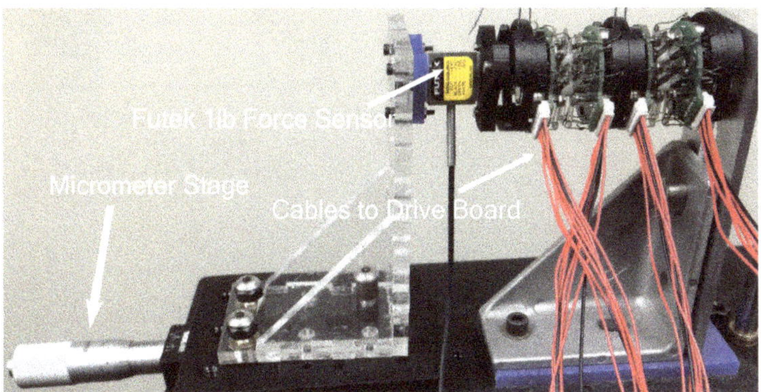

Figure 16. Experimental hardware setup.

Table 2. Experimental actuation patterns.

Pattern	Activation Level	Figure
[4 0 1 3]	8	Figure 17a
[6 2 0 0]	8	Figure 17b

The experiment was performed for different actuation patterns and levels. The experiment was also run for the same actuation level repeatedly, while varying which motor unit was used; to evaluate the effects of manufacturing variations, the motor units were activated in different combinations. For example, if the activation level is 4 for a certain actuation cell, then given a total number of motor units of six, the different combinations for activation four motor units is given $\binom{n}{k} = \frac{n!}{k!(n-k)!}$ for a total of 15. Figure 17a shows the experimental data plotted with standard deviation error bars with an activation pattern of [4 0 1 3]. This plot includes four data sets for the same activation pattern just with different choices for which solenoid to activate from those available on each cell. Although there is some deviation in the slope of each segment due to manufacturing variation, the hardening trend described in Section 2 is evident. The data stop before the actuator reaches the last hardening point of the actuator, this happens because the restoring force of the springs became too great and overcame the magnetic field of one or more of the motor units. The plot shown in Figure 17b shows a different activation pattern but the same activation level, eight motor units. Note that this activation pattern has different hardening behavior than the previous one and the data follows the trend predicted by the model. Again, the data do not go to excursions as far as the model predicts because the spring force overcame the magnetic field force. This is a limitation of Mathijssen and coworkers' platform (or the line electric solenoid/drive board), not the theory in and of itself.

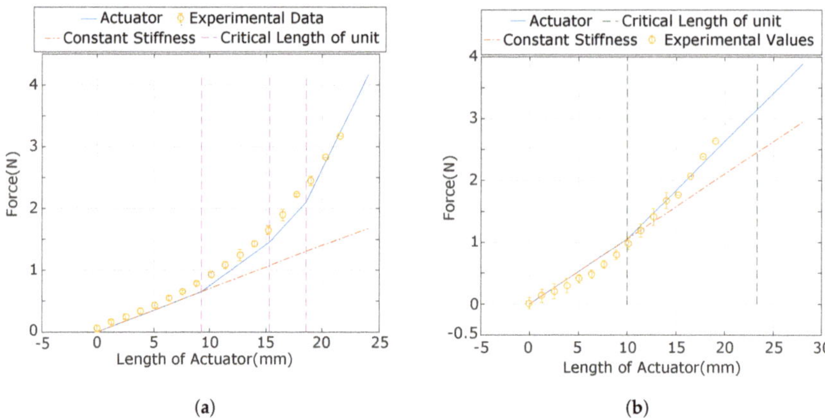

Figure 17. Length dependent hardening: (**a**) experimental hardening vs. theoretical hardening—Pattern: 4013; and (**b**) experimental hardening vs. theoretical hardening—Pattern: 6200.

5. Conclusions and Future Work

The work presented in this article characterizes the length dependent hardening behaviors encountered in the use of discrete muscle-like actuators. This characterization is done through the introduction of a Class system that defines the regions for which an actuation unit will experience different stiffness values. It is noted that different stiffness values for the actuator can be achieved while keeping the activation level constant, through a simple change in the activation pattern. This behavior of length dependent hardening is validated through the experiment shown in Section 4.

A control method that simplifies the complex nature of discrete actuator control, by exploiting the similar qualities of an SEA and implementing a linear observer to reconstruct the state and deploy a full-state feedback control law is also presented in this article. The challenges associated with applying a continuously varied control signal to a discrete number of motor units "on" necessitates the development of a quantization method that approximates the best activation level and pattern for the control signal.

Future work for this concept includes conducting dynamic experiments with a moving mass while using the observer architecture as the hardware improves with better design and manufacturing processes for discrete actuators incorporating the three main properties: compliance, modularity, and redundancy. We also desire to research a more general framework and control strategy for more complex hierarchical arrangements of discrete actuators.

Acknowledgments: This work was partially supported by NSF CNIC: US-Belgium Engineering Planning Visit: Simplified Models for Bio-inspired Actuators Grant 1427787. The authors would like to express their thanks to Glenn Mathijssen and Dr. Bram Vanderborght in Brussels, Belgium for the helpful conversations that spurred on the work shown in this paper.

Author Contributions: Caleb Fuller and Joshua Schultz conceived and designed the experiment; Caleb Fuller performed the experiment and analyzed the data; Caleb Fuller wrote the paper; Joshua Schultz provided edits and ideas for writing the paper.

Conflicts of Interest: The authors declare no conflicts of interest.

References

1. Enoka, C. Motor Unit. *Compr. Physiol.* **2012**, *2*, 2629–2682.
2. Mathijssen, G.; Schultz, J.; Vanderborght, B.; Bicchi, A. A muscle-like recruitment actuator with modular redundant actuation units for soft robotics. *Robot. Auton. Syst.* **2015**, *74*, 40–50.
3. Gere, J.; Goodno, B. *Mechanics of Materials*; Cengage Learning: Boston, MA, USA, 2008.

4. Pratt, G.A.; Williamson, M.M. Series elastic actuators. In Proceedings of the 1995 IEEE/RSJ International Conference on Intelligent Robots and Systems 95. "Human Robot Interaction and Cooperative Robots", Pittsburgh, PA, USA, 5–9 August 1995; Volume 1, pp. 399–406.
5. Vanderborght, B.; Albu-Schaeffer, A.; Bicchi, A.; Burdet, E.; Caldwell, D.G.; Carloni, R.; Catalano, M.; Eiberger, O.; Friedl, W.; Ganesh, G.; et al. Variable impedance actuators: A review. *Robot. Auton. Syst.* **2013**, *61*, 1601–1614.
6. Tonietti, G.; Schiavi, R.; Bicchi, A. Design and Control of a Variable Stiffness Actuator for Safe and Fast Physical Human/Robot Interaction. In Proceedings of the 2005 IEEE International Conference on Robotics and Automation, Barcelona, Spain, 18–22 April 2005.
7. Schultz, J.; Mathijssen, G.; Vanderborght, B.; Bicchi, A. A selective recruitment strategy for exploiting muscle-like actuator impedance properties. In Proceedings of the 2015 IEEE/RSJ Intelligent Robots and Systems Conference, Hamburg, Germany, 28 September–2 October 2015; pp. 40–50.
8. Corr, L.R.; Clark, W.W. Comparison of low-frequency piezoelectric switching shunt techniques for structural damping. *Smart Mater. Struct.* **2002**, *11*, 307, doi:10.1088/0964-1726/11/3/307.
9. Gasparri, G.M.; Garabini, M.; Pallottino, L.; Malagia, L.; Catalano, M.; Grioli, G.; Bicchi, A. Variable Stiffness Control for Oscillation Damping. In Proceedings of the 2015 IEEE/RSJ Intelligent Robots and Systems Conference, Hamburg, Germany, 28 September–2 October 2015; pp. 6543–6550.
10. Rahman, S.M.M. A novel variable impedance compact compliant series elastic actuator for human-friendly soft robotics applications. In Proceedings of the IEEE International Workshop on Robot and Human Interactive Communication, Paris, France, 9–13 September 2012; pp. 19–24.
11. Braun, D.J.; Petit, F.; Huber, F.; Haddadin, S.; Van Der Smagt, P.; Albu-Schaffer, A.; Vijayakumar, S. Robots driven by compliant actuators: Optimal control under actuation constraints. *IEEE Trans. Robot.* **2013**, *29*, 1085–1101.
12. Brogan, W. *Modern Control Theory*, 3rd ed.; Prentice Hall: Upper Saddle River, NJ, USA, 1991; pp. 532–533.
13. Sinha, A. *Linear Systems: Optimal and Robust Control*; CRC Press: Boca Raton, FL, USA, 2007.
14. Schultz, J.A.; Ueda, J. Intersample discretization of control inputs for flexible systems with quantized cellular actuation. In Proceedings of the ASME 2010 Dynamic Systems and Control Conference, Cambridge, MA, USA, 12–15 September 2010; American Society of Mechanical Engineers: New York, NY, USA, 2010; pp. 421–428.
15. Corradini, M.L.; Orlando, G. Robust quantized feedback stabilization of linear systems. *Automatica* **2008**, *44*, 2458–2462.
16. Van Ham, R.; Thomas, S.; Vanderborght, B.; Hollander, K.; Lefeber, D. Review of Actuators with Passive Adjustable Compliance/Controllable Stiffness for Robotic Applications. *IEEE Robot. Autom. Mag.* **2009**, doi:10.1109/MRA.2009.933629.

Article

Bio-Inspired Adhesive Footpad for Legged Robot Climbing under Reduced Gravity: Multiple Toes Facilitate Stable Attachment

Zhongyuan Wang [1,2], Zhouyi Wang [2], Zhendong Dai [2,*] and Stanislav N. Gorb [3]

1 The 28th Research Institute of China Electronics Technology Group Corporation, Nanjing 210007, China; wangzy051@163.com

2 Institute of Bio-inspired Structure and Surface Engineering, Nanjing University of Aeronautics and Astronautics, Nanjing 210016, China; wzyxml@nuaa.edu.cn

3 Department Functional Morphology and Biomechanics, Zoological Institute of the University of Kiel, Kiel 24118, Germany; sgorb@zoologie.uni-kiel.de

* Correspondence: zddai@nuaa.edu.cn

Received: 13 November 2017; Accepted: 11 January 2018; Published: 15 January 2018

Abstract: This paper presents the design of a legged robot with gecko-mimicking mechanism and mushroom-shaped adhesive microstructure (MSAMS) that can climb surfaces under reduced gravity. The design principle, adhesion performance and roles of different toes of footpad are explored and discussed in this paper. The effect of the preload velocity, peeling velocity and thickness of backing layering on the reliability of the robot are investigated. Results show that pull-force is independent of preload velocity, while the peeling force is relying on peeling velocity, and the peel strength increased with the increasing thickness of the backing layer. The climbing experiments show that the robot can climb under mimic zero gravity by using multiple toes facilitating adhesion. The robot with new type of footpads also provides a good platform for testing different adhesive materials for the future space applications.

Keywords: legged robot; mushroom-shaped adhesive microstructure; adhesion performance; reduced gravity; climbing

---

1. Introduction

The extraordinary locomotory abilities of geckos, insects and spiders, which are attributed to striking adhesive setae segmented into scansors or pads on the undersides of the toes or legs, have been well-known for many centuries [1]. The outstanding climbing performances of animals inspired the engineers and researchers for the design of artificial systems, such as adhesive materials, which in turn enabled the construction of climbing robots that are capable of climbing smooth walls and ceiling [2–4].

Inspired by hairy footpads of various animals in nature, fiber arrays with different end shapes such as mushrooms, asymmetric spatulae and concave structures have been developed [5–11]. By using the fibrillar surfaces, several robotic prototypes with tank-like treads and rotary legs have been reported. Mini Whegs robot [12] was equipped with mushroom-shaped adhesive microstructure (MSAMS) made of polyvinyl siloxane (PVS), using four mulit-spoke wheel-leg appendages. This 100–120 g heavy robot easily climbs smooth vertical and even inverted surfaces. Waalbot [13], equipped with rotating compliant legs, is able to climb on smooth vertical walls using angled polyurethane fiber arrays.

Due to the limitation of the mechanism these robots can only work on relatively flat areas of the surface using relatively uniform motion pattern. However, legged climbing robots may increase their potential to adapt their gaits to uneven surfaces and allow for more instantaneous control of stability, if we can push or pull them effectively during the intermittent contact with the substrate at each step. Several groups have developed legged robots that use dry adhesives and capable of such a contact formation/breakage control [14–16]. One of the most impressive robots is the Stickybot [17], which utilizes directional polymer stalks (DPS) to effectively climb vertical surfaces, such as glass, plastic, and polished wood panels. Although the climbing abilities of the robot on ground and wall have been reported, its ability to perform locomotion under zero gravity, when contact formation by DPS is not supported by the gravity force, is still unknown.

Legged climbing robot involves the controlled application of forces during the leg-substrate contact and release, in order to propel the body forwards. The achievement of strong attachment and easy removal of the adhesive is the critical issues, when we apply the adhesive to the robot's footpad. Previous studies have considered the effect of the peel angle, the length and the shape of the peel region of the adhesive itself on the adhesion performance. However, the influences of the shape and stiffness of the footpad itself are rarely studied.

IBSS_Gecko is a legged mechanism for a gecko-inspired robot, which is first introduced by Dai and Sun (Figure 1a) [18]. Using a single toe which is composed of a spring-loaded runner sheet coated with pressure-sensitive adhesive at each foot, IBSS_Gecko_6 was able to climb on smooth vertical surface [19]. The improvements presented in this work are built upon the basic design principles of the robot. Inspired by the geckos, we introduce here a footpad with multiple toes (Figure 1b) and biomimetic MSAMS (Figure 1c–e) [10], to increase adaptability and stability of the climbing robot. The legged robot with only one toe is difficult to cope with the different environments. And the novel fibrillar adhesives may enable the robot to climb on surfaces regardless of the gravity.

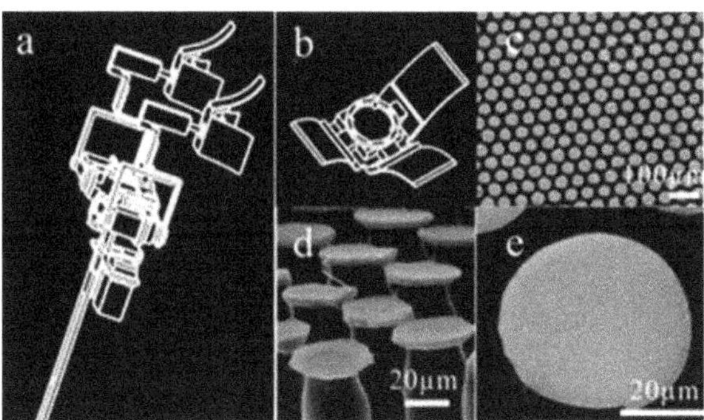

Figure 1. Structural hierarchy of the biologically inspired robot adhesive system. (**a**) Gecko-inspired robot; (**b**) Footpad of the robot; (**c–e**) Biomimetic mushroom-shaped adhesive microstructure (MSAMS) made of PVS (reproduced with permission from [10]. Copyright the Royal Society, 2007).

This paper is organized in the following manner. Section 2 briefly reviews the biomechanics of a gecko climbing on a surface. Based on the biological principle in micro and macro scales, we develop a gecko-inspired footpad. The manner of footpad with strong attachment and easy-removal properties is designed and tested in Section 3. Furthermore, experiments on the IBSS_Gecko on the challenged condition-reduced gravity is designed, analyzed and discussed in Section 4. Finally, conclusions and future work are reported in Section 5.

2. Footpad Design

Biomechanics of a gecko climbing on a surface reveals the utilization of a particular Y-configuration in both the first and fifth toes and opposite feet (Figure 2a) [20–26]. The opposite feet are pulling inward toward the center of mass, generating large in-plane forces to ensure the angle of pulling force vector towards the substrate remains small. The geckos favor setal attachment at two different hierarchical characteristic sizes of toes and feet. Additionally, geckos with multiple toes can avoid detachment by aligning some of their toes opposite to the force vector [24,27].

IBSS_Gecko_6 robot consists of a rigid body and four legs (Figure 2b). The mass of the robot is about 500 g, while the length and width is 240 mm and 100 mm, respectively. Each leg has two linkages and three degrees of freedom. Three motors drive the leg to move it in the three dimensional space. The first motor (Hitec HS-85MG, Chungcheongbuk-do, South Korea) controls the rotational motion. The second and the third motors (Hitec HS-65MG, Chungcheongbuk-do, South Korea) control the motion and position of the foot. An integrated flexure in the foot (ball hinge and four springs) allows for up to 15 degrees of pitch, roll and yaw misalignment to obtain alignment and full contact with substrate. This design largely reduces required precision of footpad's trajectory and greatly increases the adaptability of the robot to different surfaces.

To mimic the function of the gecko's footpad, a biomimetic footpad was developed (Figure 2b). It has similar geometric appearance as the gecko's footpad, including three rectangular toes pointing to three directions. Every toe is composed of supporting layer, backing layer and adhesive layer. The adhesive layer is attached to one side of the supporting layer, while the backing layer is attached to the other side of the supporting layer. The supporting layer was superimposed with the backing layer, so that it can prevent the adhesive from deforming irreversibly. Obviously, the thickness of the backing layer is largely influence adhesion strength of the footpad. However, the role of the backing layer thickness on adhesion performance for legged climbing robot has not been investigated in detail so far.

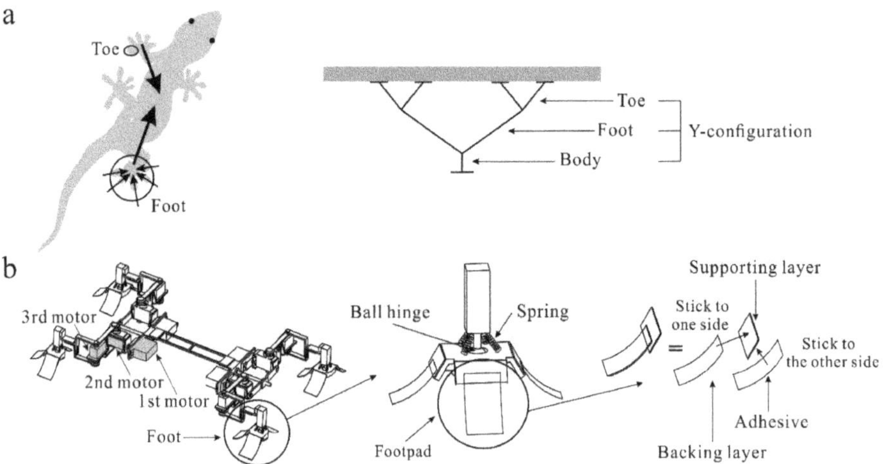

Figure 2. (**a**) Arrangement of toes in the gecko walking on a ceiling surface reveals the utilization of a particular Y-configuration in both toes and feet; (**b**) Mechanical structure of the legged robot and its footpad.

3. Adhesion Performance

The footpad movements consist of two phases: attachment and detachment. Attachment and detachment time determine the speed and influence stability of the robot. For example, for pressure-sensitive adhesives (PSA), if the footpad attaches just for a short time, it may only generate a small adhesion force. Furthermore, if the PSA-footpad tries to detach at once, it must overcome a strong adhesion force. Reducing the peeling force effectively at detachment is essential for keeping the stability of the robot. In this part of the locomotory cycle, both the pull-off test and peeling test will examine how attachment and detachment times influence adhesion performance. Furthermore, we study here the backing layer thickness effect on the adhesion performance of the pad.

3.1. Pull-Off Test

A pull-off test was conducted using Nano and Micro Tribometer UMT-2 (Campbell, CA, USA). MSAMS specimen in the form of discs of 3 mm diameter was used as a dry adhesive. The specimen was attached to an aluminum block, fixed on the force transducer. The specimen and block were perpendicular to the load direction. The specimen was brought into contact and detached by linear motion of the translation stage, which was oriented perpendicularly to the glass surface (average roughness $Ra = 3.3 \pm 2.1$ nm) (Figure 3a). The pull-off force was measured, while the translation stage moved at different attach velocities (V_0) with the same preload and the same detach velocity ($V_t = 0.2$ mm/s). Contact time between the MSAMS specimen and the glass substrate was 5 s.

The pull-off force measured between the PVS specimen and the glass substrate is presented in Figure 3b. The result reveals that the pull-off force on the glass substrate is not only independent of the load force [10], but also relies on the load velocity. The result indicates that for the robot, the footpad of this kind does not need much load control: the pad just needs to contact the surface, in order to produce substantial adhesion force and meet the requirement of the attaching phase. This characteristic of MSAMS is extremely beneficial for applying it at zero gravity, when the robot cannot rely on gravity to load the adhesive pad.

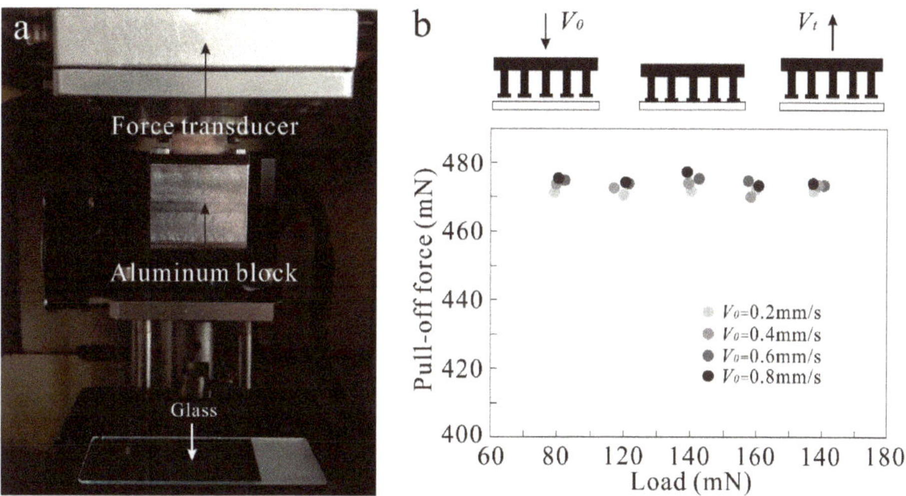

Figure 3. (**a**) Image of the setup used to measure the pull-off force; (**b**) The pull-off force measured between mushroom-shaped adhesive microstructure (MSAMS) and glass as a function of load at different load velocities.

3.2. Peeling Test

A peeling strength analysis is performed according to the Kendall's peeling model [28]:

$$(\frac{F}{b})^2 \frac{1}{2Ed} + (\frac{F}{b})(1 - \cos\theta) - R = 0 \tag{1}$$

where F is the peeling force, d is the thickness of the adhesive, b is the width of the adhesive, E is the Young's modulus of the adhesive, θ is the peeling angle, and R is the energy required to fracture a unit area of interface.

Kendall's peeling model demonstrates that peel strength increases with increasing Young's modulus of the backing layer. When the PVS is peeled off from a substrate, fibrills are stretched and deformed until the tension stress surpasses the pull-off stress. Under this condition, detachment is started. For the fibrillar surface, the stretching velocity of the fibril influences the resulting peel force [29]. Previous studies have shown that the relationship between the peel force and the peel velocity at the same peel angle can be described by [30–32]

$$\frac{F}{b} = kV^n \quad (2)$$

where k is a function of the peel angle and is related to the thickness of the adhesive, V is peel velocity, and n is a constant related to the intrinsic properties of the adhesive: this parameter is often called adhesive energy.

Our testing platform was set up to estimate the effect of peel velocity and the thickness of the adhesive on the peel strength (Figure 4). Pieces of the MSAMS dry adhesive (20 mm × 100 mm) with different thickness of backing layer (Polyvinyl chloride plastic) were attached to a smooth glass surface (average roughness Ra = 3.2 ± 1.9 nm) and preloaded with different weights. The peel angle was adjusted by fixing the glass surface at various tilted angles. Different peel velocities are obtained by changing the loading weights.

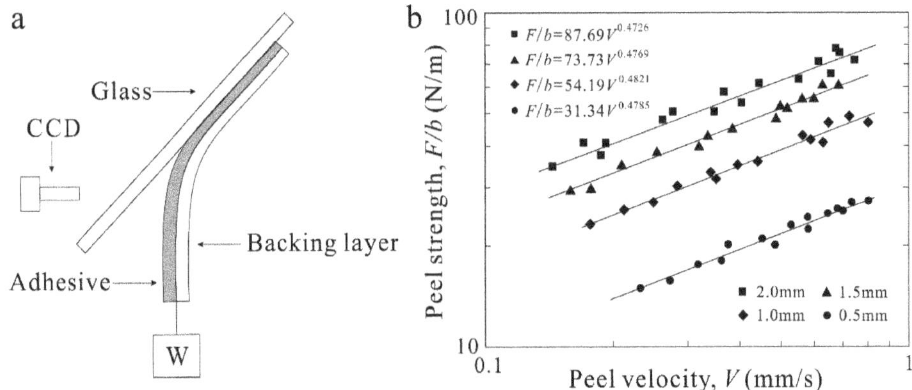

Figure 4. Results of the peeling test. (**a**) Schematic diagram of the peeling experiment; (**b**) The relationship between peel strength and peel velocity at a peeling angle of 30° with different thicknesses of backing layer in log-log coordinates.

A linear relationship is found between the peel strength and the peel velocity in the log-log plot. Peel strength keeps increasing while thickness of the backing layer increases. The slopes of the four curves are almost the same, indicating that the physical properties of the adhesive are identical, while the differences between the intercepts indicate the different stiffness of the backing layer. The results agree with the experimental results obtained by Blum et al. [31] and Zhou et al. [33]. When the thickness of the backing layer is varied, the stiffness of the backing layer is changed, resulting in the increasing of the peel strength as stiffness increasing up until a saturation point (Figure 5). For the robot's footpad, the backing layer is indispensable, because of the stronger peeling strength. Thin backing layer effects not only in low peeling strength, which might be advantageous for contact breakage, but also in strong toe deformation during contact formation, which might prevent proper contact formation.

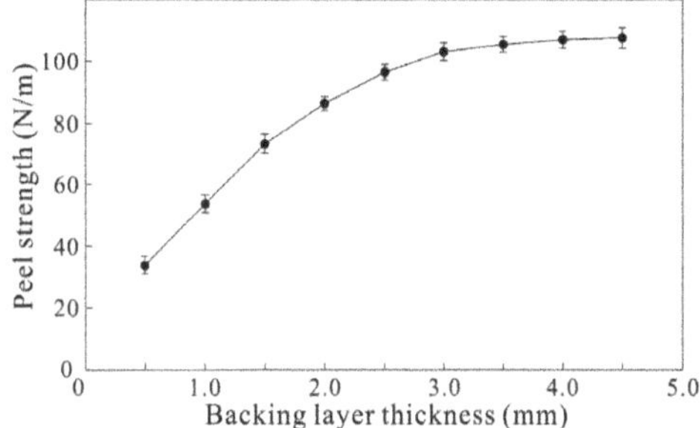

Figure 5. The influences of backing layer thickness on the peel strength at a peeling angle of 30° and at a peeling velocity of 2 mm/s.

4. Experiments on Robots

4.1. Climbing Ability with Different Toes

Helium balloons which are fixed on the center of mass of the robot were used to reduce gravity. The climbing abilities of the robot under different conditions under reduced gravity were tested to verify the performance of the toes (Figure 6a). Here we define a ratio (μ) of reduced gravity (RRG):

$$\mu = \frac{G - G_r}{G} \tag{3}$$

Here G is the weight of the robot, G_r is the weight of the robot under reduced gravity conditions. All experiments were conducted on a glass surface, the size of each toe is 2 cm × 3 cm.

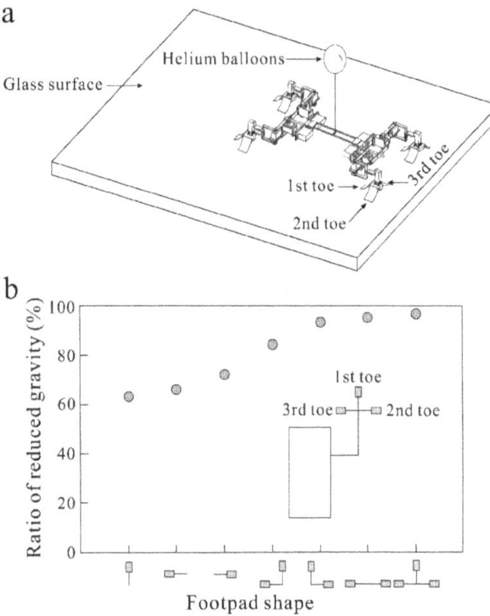

Figure 6. (**a**) Schematic of the robot climbing under reduced gravity; (**b**) Climbing ability with different footpad shape at different RRG.

The results show the synergistic effect of different toes in one leg (Figure 6b). The 2nd and 3rd toes are mainly responsible for the improvement of the lateral balance of the robot by pulling inward toward the center of the leg. As the RRG increases, the role of lateral toes becomes even more important, because the robot needs to make transition from a crouched posture to a sprawled one, to ensure smaller peeling angle and prevent failure. This splaying behaviour is also observed in geckos [24] and tree frogs [34] when they face a stepper surface. The legs should generate large inward forces, in order to ensure that the angle of the pulling force remains small. Thus, the outward oriented toe, which can generate reliable shear force, is essential under a challenging condition, such as wall, ceiling and zero-g environment.

Figure 7 shows the robot climbing at a RRG of 94% by using the 1st and 2nd toes at each foot (Supplementary video 1). Even without the 3rd toe, the robot can climb stably. The robot can climb under a lower ratio of the reduced gravity, when the 3rd toe is added, because an extra toe leads to the peeling force increase, in turn leading to an increased peeling time that ensure stability of locomotion. The ratio of reduced weight increases only slightly with an addition of the 3rd toe (Figure 6b).

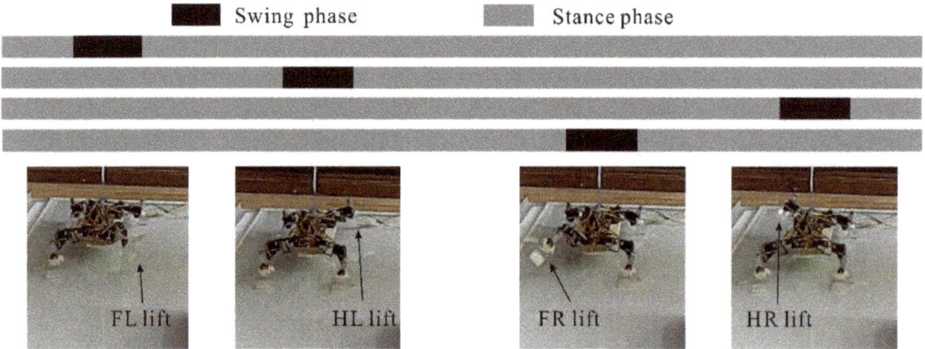

Figure 7. Robot climbing on a glass surface at a RRG of 94%. Photos show different legs in swing phase. FL: Front left, HL: Hind left, FR: Front right, HR: Hind right.

4.2. Climbing Performance with Lateral Impact

We push the balloons in side direction slowly and quickly, respectively, to test the climbing performance at the lateral impact. When we pushed the balloons slowly, there was no slip observed during climbing. The climbing performance was not affected by such a lateral impact, and the robot could continue to move forward (Figure 8a) (Supplementary video 2). When we pushed balloons quickly, the robot slipped off, but did not fall down within a short distance in lateral direction (Figure 8b). Then, the robot recovered quickly and continued forward motion. These results demonstrate an enhanced self-balancing ability of the robot with multiple toes. It demonstrated feasibility of the multi-toe footpad and its higher reliability in the climbing performance under reduced gravity.

Kendall's peeling model demonstrates that peel strength increases with increasing Young's modulus of the backing layer. When the PVS is peeled off from a substrate, fibrills are stretched and deformed until the tension stress surpasses the pull-off stress. Under this condition, detachment is started. For the fibrillar surface, the stretching velocity of the fibril influences the resulting peel force [29]. Previous studies have shown that the relationship between the peel force and the peel velocity at the same peel angle can be described by [30–32].

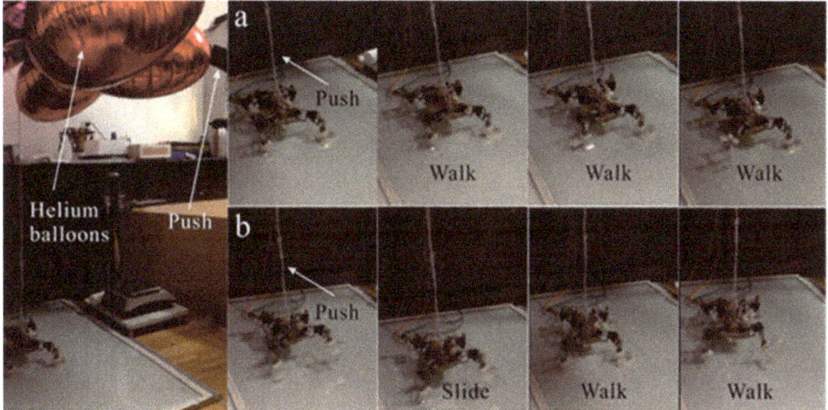

Figure 8. Robot climbing on glass surface with lateral impact at a RRG of 94%.

4.3. Climbing Ability of the Robot

In order to test climbing ability of the robot under reduced gravity, the balloons were used to fully balance the robot's weight, as shown in Figure 9. The size of the toe is 2 cm × 4 cm. Our experiment shows that the robot can climb under mimic zero gravity at the velocity of up to 1 cm/s with diagonal gait (Duty factor is 0.6). When the velocity of the increased up to 2 cm/s, the robot was slipped and become unstable. This is because as the velocity increases, peel strength also increases, which leads to larger peeling force.

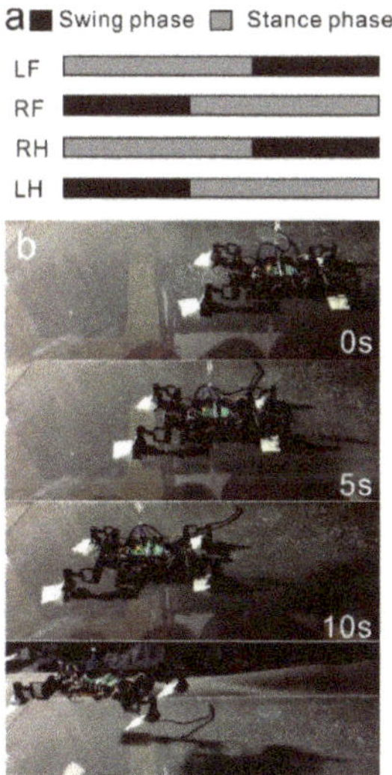

Figure 9. Experiments of robot under micro-gravity environment with diagonal gait.

5. Conclusions and Future Work

This paper presents the biologically inspired design of the footpad for a legged climbing robot. The biomimetic MSAMS-based tape was used here to design a bio-inspired footpad with multiple toes. The experimental tests demonstrated that the robot can climb undermimic zero gravity at the velocity of up to 1 cm/s, and even resist lateral impact. The major contribution of this work is the application of the biological principle of multi-toe footpad to the legged robot, and the demonstration of the feasibility and reliability of this design. To the best of our knowledge, the legged robot, climbing under nearly zero gravity, was demonstrated here for the first time. We also explored the effect of the (1) preload, (2) preload velocity, (3) peeling velocity, (4) thickness of the backing layer, and (5) collective effect of multiple toes in one leg on the reliability of the robot locomotion.

Further optimization of the robot is planned to improve its performance in climbing on vertical and inverted surfaces. Our preliminary studies have shown that it also can climb on incline surface of 75° without tail, but failed to climb the inverted surface due to its heavy weight. The decrease of the robot mass will aid in an increase of the climbing ability to more challenged surfaces. In addition, the climbing speed is planned to be improved by using dynamical trotting gait.

Supplementary Materials: The following are available online at http://www.mdpi.com/2076-3417/8/1/114/s1.

Acknowledgments: This work is supported by the National Natural Science Foundation of China (No. 51435008 and 31601870), the Doctoral Fund of Ministry of Education of China (No. 20123218110031), the Natural Science Foundation of Jiangsu Province, China (No. SBK2016040649) and the National Defense Pre-Research Foundation of China (No. B2520110013). This work is also partially supported by Jiangsu Province Friendship Award to S N Gorb.

Author Contributions: Zhongyuan Wang and Zhendong Dai conceived and designed the experiments; Zhongyuan Wang performed adhesion performances test, Zhongyuan Wang and Stanislav N. Gorb performed robot experiment, Zhongyuan Wang analyzed the data; Zhongyuan Wang, Zhouyi Wang, Zhendong Dai and Stanislav N. Gorb wrote the paper.

Conflicts of Interest: The authors declare no conflict of interest. The founding sponsors had no role in the design of the study; in the collection, analyses, or interpretation of data; in the writing of the manuscript, and in the decision to publish the results.

References

1. Arzt, E.; Gorb, S.; Spolenak, R. From micro to nano contacts in biological attachment devices. *Proc. Natl. Acad. Sci. USA* **2003**, *100*, 10603–10606. [CrossRef] [PubMed]
2. Cutkosky, M.R.; Kim, S. Design and fabrication of multi-material structures for bioinspired robots. *Philos. Trans. R. Soc. A* **2009**, *367*, 1799–1813. [CrossRef] [PubMed]
3. Gorb, S.N.; Sinha, M.; Peressadko, A.; Daltorio, K.A.; Quinn, R.D. Insects did it first: A micropatterned adhesive tape for robotic applications. *Bioinspir. Biomim.* **2007**, *2*, S117. [CrossRef] [PubMed]
4. Sitti, M.; Fearing, R.S. Synthetic gecko foot-hair micro/nano-structures for future wall-climbing robots. In Proceedings of the IEEE International Conference on Robotics and Automation (ICRA '03), Taipei, Taiwan, 14–19 September 2003.
5. Campo, D.A.; Greiner, C.; Alvarez, I.; Arzt, E. Patterned surfaces with pillars with controlled 3D tip geometry mimicking bio-attachment devices. *Adv. Mater.* **2007**, *19*, 1973–1977. [CrossRef]
6. Lee, D.Y.; Lee, D.H.; Lee, S.G.; Cho, K. Hierarchical geckoinspired nanohairs with a high aspect ratio induced by nanoyielding. *Soft Matter* **2007**, *8*, 4905–4910. [CrossRef]
7. Davies, J.; Haq, S.; Hawke, T.; Sargent, J.P. A practical approach to the development of a synthetic gecko tape. *Int. J. Adhes. Adhes.* **2009**, *29*, 380–390. [CrossRef]
8. Haefliger, D.; Boisen, A. Three-dimensional microfabrication in negative resist using printed masks. *J. Micromech. Microeng.* **2006**, *16*, 951–957. [CrossRef]
9. Sameoto, D.; Menon, C. Direct molding of dry adhesives with anisotropic peel strength using an offset lift-off photoresist mold. *J. Micromech. Microeng.* **2009**, *19*, 115026. [CrossRef]
10. Gorb, S.; Varenberg, M.; Peressadko, A.; Tuma, J. Biomimetic mushroom-shaped fibrillar adhesive microstructure. *J. R. Soc. Interface* **2007**, *4*, 271–275. [CrossRef] [PubMed]
11. Kim, S.; Sitti, M. Biologically inspired polymer microfibers with spatulate tips as repeatable fibrillar adhesives. *Appl. Phys. Lett.* **2006**, *89*, 261911. [CrossRef]
12. Daltorio, K.A.; Wei, T.E.; Horchler, A.D.; Southard, L.; Wile, G.D.; Quinn, R.D.; Gorb, S.N.; Ritzmann, R.E. Mini-whegs TM climbs steep surfaces using insect-inspired attachment mechanisms. *Int. J. Robot. Res.* **2009**, *28*, 285–302. [CrossRef]

13. Murphy, M.P.; Kute, C.; Mengüç, Y.; Sitti, M. Waalbot II: Adhesion recovery and improved performance of a climbing robot using fibrillar adhesives. *Int. J. Robot. Res.* **2011**, *30*, 118–133. [CrossRef]
14. Unver, O.; Uneri, A.; Aydemir, A.; Sitti, M. Geckobot: A gecko inspired climbing robot using elastomer adhesives. In Proceedings of the 2006 IEEE International Conference on Robotics and Automation (ICRA '06), Orlando, FL, USA, 15–19 May 2006.
15. Li, Y.; Ahmed, A.; Sameoto, D.; Menon, C. Abigaille II: Toward the development of a spider-inspired climbing robot. *Robotica* **2012**, *30*, 79–89. [CrossRef]
16. Henrey, M.; Ahmed, A.; Boscariol, P.; Shannon, L.; Menon, C. Abigaille-III: A versatile, bioinspired hexapod for scaling smooth vertical surfaces. *J. Bionic Eng.* **2014**, *11*, 1–17. [CrossRef]
17. Kim, S.; Spenko, M.; Trujillo, S.; Heyneman, B.; Santos, D.; Cutkosky, M.R. Smooth vertical surface climbing with directional adhesion. *IEEE Trans. Robot.* **2008**, *24*, 65–74.
18. Dai, Z.; Sun, J. A biomimetic study of discontinuous-constraint metamorphic mechanism for gecko-like robot. *J. Bionic Eng.* **2007**, *4*, 91–95. [CrossRef]
19. Wang, Z.; Dai, Z.; Yu, Z.; Shen, D. Optimal attaching and detaching trajectory for bio-inspired climbing robot using dry adhesive. In Proceedings of the 2014 IEEE/ASME International Conference on Advanced Intelligent Mechatronics (AIM 2014), Besançon, France, 8–11 July 2014.
20. Autumn, K.; Niewiarowski, P.H.; Puthoff, J.B. Gecko adhesion as a model system for integrative biology, interdisciplinary science, and bioinspired engineering. *Annu. Rev. Evol. Syst.* **2014**, *45*, 445–470. [CrossRef]
21. Wang, Z.; Dai, Z.; Ji, A.; Ren, L.; Xing, Q.; Dai, L. Biomechanics of gecko locomotion: The patterns of reaction forces on inverted, vertical and horizontal substrates. *Bioinspir. Biomim.* **2015**, *10*, 016019. [CrossRef] [PubMed]
22. Tian, Y.; Pesika, N.; Zeng, H.; Rosenberg, K.; Zhao, B.; McGuiggan, P.; Israelachvili, J. Adhesion and friction in gecko toe attachment and detachment. *Proc. Natl. Acad. Sci. USA* **2006**, *103*, 19320–19325. [CrossRef] [PubMed]
23. Autumn, K.; Dittmore, A.; Santos, D.; Spenko, M.; Cutkosky, M. Frictional adhesion: A new angle on gecko attachment. *J. Exp. Biol.* **2006**, *209*, 3569–3579. [CrossRef] [PubMed]
24. Autumn, K.; Hsieh, S.T.; Dudek, D.M.; Chen, J.; Chitaphan, C.; Full, R.J. Dynamics of geckos running vertically. *J. Exp. Biol.* **2006**, *209*, 260–272. [CrossRef] [PubMed]
25. Chen, J.J.; Peattie, A.M.; Autumn, K.; Full, R.J. Differential leg function in a sprawled-posture quadrupedal trotter. *J. Exp. Biol.* **2006**, *209*, 249–259. [CrossRef] [PubMed]
26. Lepore, E.; Pugno, F.; Pugno, N.M. Optimal angles for maximal adhesion in living tokay geckos. *J. Adhes.* **2012**, *88*, 820–830. [CrossRef]
27. Goldman, D.I.; Chen, T.S.; Dudek, D.M.; Full, R.J. Dynamics of rapid vertical climbing in cockroaches reveals a template. *J. Exp. Biol.* **2006**, *209*, 2990–3000. [CrossRef] [PubMed]
28. Kendall, K. Thin-film peeling: The elastic term. *J. Phy. D Appl. Phys.* **1975**, *8*, 1449–1452. [CrossRef]
29. Zhou, M.; Pesika, N.; Zeng, H.; Wan, J.; Zhang, X.; Meng, Y.; Wen, S.; Tian, Y. Design of gecko-inspired fibrillar surfaces with strong attachment and easy-removal properties: A numerical analysis of peel-zone. *J. R. Soc. Interface* **2012**, *9*, 2424–2436. [CrossRef] [PubMed]
30. Amouroux, N.; Petit, J.; Léger, L. Role of interfacial resistance to shear stress on adhesive peel strength. *Langmuir* **2001**, *17*, 6510–6517. [CrossRef]
31. Blum, F.D.; Metin, B.; Vohra, R.; Sitton, O.C. Surface segmental mobility and adhesion-Effects of filler and molecular mass. *J. Adhes.* **2006**, *82*, 903–917. [CrossRef]
32. Marin, G.; Derail, C. Rheology and adherence of pressure-sensitive adhesives. *J. Adhes.* **2006**, *82*, 469–485. [CrossRef]
33. Zhou, M.; Tian, Y.; Pesika, N.; Zeng, H.; Wan, J.; Meng, Y.; Wen, S. The extended peel zone model: Effect of peeling velocity. *J. Adhes.* **2011**, *87*, 1045–1058. [CrossRef]
34. Endlein, T.; Ji, A.; Samuel, D.; Yao, N.; Wang, Z.; Barnes, W.J.P.; Federle, W.; Kappl, M.; Dai, Z. Sticking like sticky tape: Tree frogs use friction forces to enhance attachment on overhanging surfaces. *J. R. Soc. Interface* **2013**, *10*, 20120838. [CrossRef] [PubMed]

Article

Grasping Claws of Bionic Climbing Robot for Rough Wall Surface: Modeling and Analysis

Quansheng Jiang [1] and Fengyu Xu [2,*]

[1] School of Mechanical Engineering, Suzhou University of Science and Technology, Suzhou 215009, China; qschiang@163.com

[2] College of Automation, Nanjing University of Posts and Telecommunications, Nanjing 210003, China

* Correspondence: xufengyu598@163.com; Tel.: +86-25-8586-6512

Received: 16 November 2017; Accepted: 21 December 2017; Published: 22 December 2017

Abstract: Aiming at the inspection of rough stone and concrete wall surfaces, a grasping module of cross-arranged claw is designed. It can attach onto rough wall surfaces by hooking or grasping walls. First, based on the interaction mechanism of hooks and rough wall surfaces, the hook structures in claw tips are developed. Then, the size of the hook tip is calculated and the failure mode is analyzed. The effectiveness and reliability of the mechanism are verified through simulation and finite element analysis. Afterwards, the prototype of the grasping module of claw is established to carry out grasping experiment on vibrating walls. Finally, the experimental results demonstrate that the proposed cross-arranged claw is able to stably grasp static wall surfaces and perform well in grasping vibrating walls, with certain anti-rollover capability. This research lays a foundation for future researches on wall climbing robots with vibrating rough wall surfaces.

Keywords: grasping claws; bionic climbing robot; cross-arranged claw; rough wall surface; finite element analysis

1. Introduction

Climbing robot is a kind of special robot, which can climb and walk on a rough vertical wall or spherical surface. Installed with related equipment, it can clean the wall, detect flaws, lay pipes, and paint, etc.

Wall climbing robots, which are a special operation mechanism in extreme environment generally adhere to walls by virtue of negative suckers, permanent magnets, or bio-inspired adhesion [1–6]. However, magnetic adhesion fails for wall surfaces under long-term vibrations, such as diagonal cable bridge towers and viaduct bridge piers. Vacuum adhesion becomes instable due to the influence of vibrating wall surfaces. Bio-inspired viscous materials tend to detach from dusty and rough walls.

The climbing robot with grippers is a kind of climbing mechanism utilizing gripper for grasping and also an important part of climbing robot in the field of mobile robot. These climbing mechanisms are used widely in the domain of robotics. Bartsch et al. [7] designed a hexapod climbing robot based on the bionics principle. The robot could walk freely in an extreme complex environment with some slopes. The climbing robot can also imitate the climbing actions of human being and animals in some specific surfaces. Lam et al. [8] developed a kind of tree-climbing robot with a new flexible movement structure and analyzed the wormlike peristaltic movement to design grippers. They presented an algorithm to increase the motion stability. The robot could stably grasp some irregular objects. Asbeck et al. [9] designed a kind of climbing robot for hard vertical surfaces, which is inspired by the mechanisms that were observed in some climbing insects and spiders. The basic idea is to involve arrays of microspines that could catch on surface asperities.

Sintov et al. [10] proposed a wall-climbing robot with grippers. The robot was similar in structure to the bio-mimetic climbing robot in Reference [9]. There were extremely small barb-like spines on its feet to hook the uplifts on the rough surface. Parness et al. [11] developed a rock-climbing robot with grasping grippers using hook array. Each gripper contained hundreds of flexible hooks. There were distinct differences between bio-mimetic climbing robot and conventional climbing robots in the adsorption principle, movement mode, and appearance, where the former had a strong adaptive capacity on the wall. Liu et al. [12] proposed a climbing robot by imitating tarsi on the leg of Maladera orientalis. Furthermore, the robots were able to climb along the rough wall using the ratchet wheels with hooks arranged into circumference array as driving wheels. Guan et al. [13] desinged a biped modular robot with climbing capacity to replace workers in high risk environments. Wang et al. [14] presented a novel solution for the problem of multi-gaits for snake-like robots. Jiang et al. [15] designed a double claw modular biomimetic robot with climbing and operating functions for aerial work in agriculture, forestry, architecture, and other fields. Kalind et al. [16] proposed a microspine wheel consisting of multiple independent compliant hooks, which can climb up stairs, steps, and rough vertical walls. Arena et al. [17] presented a control system scheme for a biologically inspired walking hexapod robot. The control system is structured as an analog control system realized by Cellular Non-linear Networks (CNNs) generating the locomotion pattern as a function of the sensorial stimuli from the environment.

The power supply is one of the fundamental problems in climbing robots. Buscarino et al. [18] presented a realization and control scheme of a walking microrobot actuated by piezoelectric actuators. The control signals are generated by performing a frequency modulation driven by the chaotic evolution of Chua's circuit state variables. The role of passive and active vibrations for the control of nonlinear large-scale electromechanical systems is analyzed in Reference [19], which makes it possible to regularize imperfect uncertain large-scale systems.

Applying the mode of gripping walls with tiny hooks, Xu et al. [20] designed a multi-legged wall climbing robot with claws and the mechanical model for the climbing of the robot. The climbing experiment showed that the robot could climb upward on walls. However, when there are large external disturbances (wind loads and wall vibrations), the hooks are likely to detach and thereby result in instable attachment onto the wall. This is because multiple hooks are not flexibly connected and the stress is distributed unevenly. In our work, the bionic claws model of "grabbing the wall" is designed to replace the existing form of hooking claws. The presented grasping claws cause the hook to form a force closure between the hooks and uplifts on the wall, and the active force can be exerted on each hook, so as to overcome the disadvantage of unequal force distribution in the hooking claws. Therefore, the grasping claws can improve the adsorption stability of the claws mechanism.

In this paper, by analyzing the interaction between hooks and rough wall surfaces, a grasping module of cross-arranged claws is designed, the grasping ability and disturbance resistance of wall climbing robots to rough walls are improved. The rest of the paper is organized as follows. The grasping module of cross-arranged claws is designed in Section 2. The minute claw is designed and analyzed in Section 3. The grasping claws module based on ANSYS Workbench is analyzed in Section 4. The experiments for climbing model of bionic robot are carried out in Section 5. Conclusions and future work are discussed in Section 6.

2. Design of the Grasping Module of Cross-Arranged Claws

Based on the designed spine structure in Reference [20], two groups of spines were employed to collaboratively realize the grasping, and the grasping module of cross-arranged claws was proposed (Figure 1). Two pairs of secondary claws were installed on the main body, each of which was composed of two pairs of tiny hooks, screw nuts, elastic steel sheets, spring washers, and connecting plates. The two pairs of tiny hooks were symmetrically distributed on the tips of the secondary claws, and were fixed with the screw nuts and the elastic steel sheets. The elastic steel sheets were connected with the secondary claws and a connecting rod.

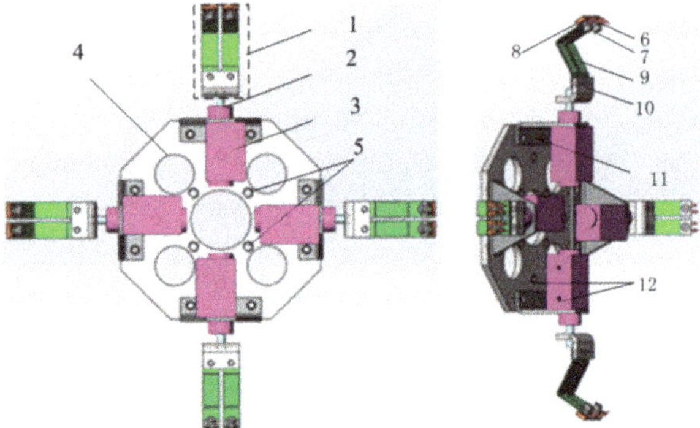

Figure 1. The grasping module of cross-arranged claws. 1—Secondary claws, 2—connecting rod, 3—cylinder, 4—main body, 5—fixture screw thread, 6—single tiny hook, 7—crew nut, 8—spring washer, 9—elastic steel sheet, 10—connecting plate, 11—support frame, 12—hose insertion of the cylinder.

The grasping module works in accordance with the following principle: the two pairs of secondary claws are symmetrically fixed around the claw module and are connected with the cylinders with connecting rods. Two air injecting holes are drilled on each cylinder and connected with the external air source via hoses. When the air source inflates the cylinder through the inside gas hole, the pressure in the cylinder increases, pushes the connecting rod, and thereby drives the motion of the claws. In this way, it controls the outward stretching of the four claws. When the air source inflates via the outside gas hole, it increases the pressure outside the cylinder, thus pushing the connecting rod to draw back the claws. By doing so, it controls the claws to draw back. Through the stretching or draw back of the connecting rod, the hooks at the tips of claws can slip on rough wall surfaces to search stable points that can be grasped. When large enough friction force is produced between the hook and the rough wall, the elastic steel sheet undergoes elastic deformation on the contact wall surface, which enables the robot to adapt to uneven rough wall surfaces.

The designed cross-arranged claws allow for the robot to more stably attach to wall surfaces, prevent sideslip overturn, and have certain vibration resistance. It dramatically improves the adaptability of robots to walls.

3. Design and Analysis of the Miniature Claws

The interaction mechanism of the micro-hook with the microbulges on wall surfaces suggests that the grabbing effect of the bionic hook on the bulges is directly correlated with the wall surface roughness and the top size of the bionic hook.

3.1. Calculation of Hook Top Size

The maximum top size of the micro-hook is associated with the linear profile roughness of wall surface R_a, the frictional coefficient between hook and wall μ, and the average width of wall profile unit S_m. The maximum top size of the micro-hook is derived from Reference [21]. The top size of the micro-hook should be lower than the maximum size. Asbeck et al. [9] established the models of hook and surface profile, and experimentally measured the relationships of the number of unit centimeter micro-bulges with the hook top size, hook load angle, and surface roughness. SpinybotII, the wall-climbing robot, as proposed by Asbeck, had a hook top diameter of 25 μm and a hook neck diameter of 200 μm. Dai et al. [22] established the relationship between the hook top sizes with the surface roughness of objects and pointed out that, the stable stressing point could be achieved as long as the hook top radius was lower than the average radius of the micro-bulges on the wall surface. The smaller that the hook top size is, the stronger is the adhering ability on the rough wall surface.

In present study, the concrete surface was taken as an example. Related parameters were measured as follows: surface profile linear roughness $R_a = 93.0$ μm, frictional factor between hook and concrete surface $\mu = 0.5$, average width of wall profile unit $S_m = 100$ μm, distance from the centroid of the robot to the wall $L = 30$ mm, and distance of the upper and lower claws of the robot $D = 120$ mm. We derived $\tan\alpha = 2L/D$ and the contact angle $\alpha = \arctan(\frac{2\times30}{120}) = 26.5651° < 45°$. Using MATLAB simulation, we calculated that the maximum radius of the micro-hook top (Figure 2) was $r_{\max} = 43.1397$ μm. We set the radius of the micro-hook top as $r_s = 20$ μm, which was approximately equivalent to the size calculated by Kim. The accuracy of the interaction model of the micro-hook and the micro-bulges on the wall was also validated.

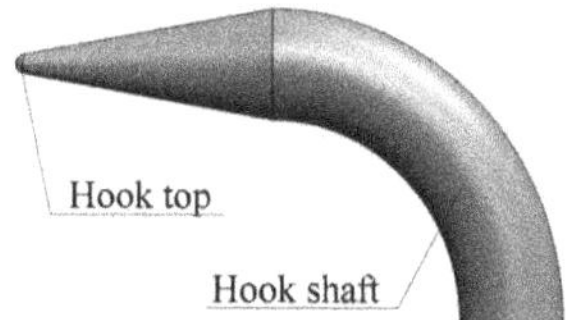

Figure 2. Microhook.

3.2. Failure Form of Hook

The hook should maintain sufficient strength during the hook–wall surface interaction, which is the basic condition to ensure stable grabbing. The stress model of the hook is equivalent to a bended suspension beam (Figure 3), where R is the curvature radius of the hook, d is the diameter of the hook section, β is the angle of the hook top with the y axis, F is the stress on the hook top, and α is the rotation angle of the hook top with the z axis under the effect of force F.

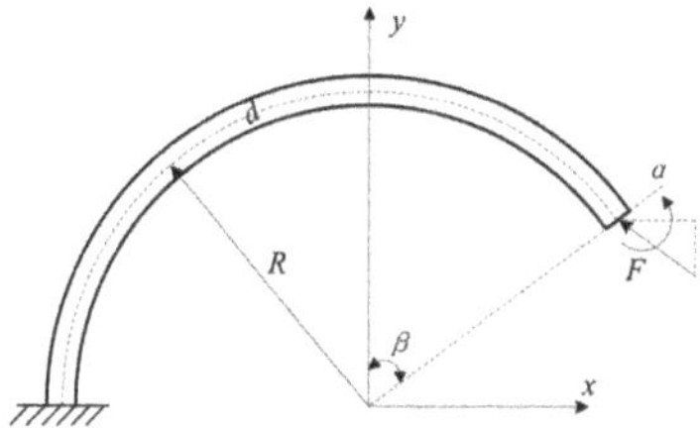

Figure 3. Bending beam model of micro-hook.

The following failure forms were concluded in accordance with the interaction mechanism of the hook with bulges on the wall surface:

(1) failure of the bending stress on the hook base;
(2) bending deformation of the hook and separation from micro-bulges; and,
(3) separation induced by the strength deficiency of the micro-bulge.

In the first failure form, the bending stress on the arbitrary section of the hook can be obtained according to the mechanics of the materials. Failure is avoidable as long as the maximum bending stress is less than the allowable stress of the hook and a certain safety factor can be ensured.

In the second failure form, a virtual torque M can be added to the beam end when calculating the deflection deformation and rotation deformation on the end of the bending beam, according to Castigliano's theorem of mechanics of materials. The rotational angle α is calculated is as follows:

$$\alpha = \partial U/\partial M = \frac{R^2}{2EI}[-2F_y + (2F_x + F_y(\pi + 2\beta))\cos\beta + (-2F_y + F_x\pi + 2F_x\beta)\sin\beta] \propto \frac{1}{d^2} \quad (1)$$

where U is the strain energy of beam, E is the elastic coefficient of material, and R/d is a constant.

In the third failure form, failure is generally analyzed through the Hertz stress distribution on the contact point for the surface made of cementitious materials such as concrete and irregular water brush stone. The maximum pressure at the center of the contact point is achieved by the following equations:

$$\begin{array}{c} p_{\max} = 3F_n/2\pi a^2 = \left(6fE^2/\pi^3R^2\right)^{1/3} \\ a = \left(3F_nR/4E\right)^{1/3}, E = (1-v_s^2)/E_s + (1-v_a^2)/E_a, 1/R = 1/r_s + 1/r_a, \end{array} \tag{2}$$

where v is the material Poisson's ratio, a is the radius of contact spot, F_n is the positive pressure acting on the contact point of the hook and the microbulge, and subscripts s and a represent the hook top and the microbulge on wall surfaces, respectively.

The maximum tensile stress occurs at the edge of the contact spots, i.e.,

$$\sigma_T = ((1-2v_a)p_{\max})/3. \tag{3}$$

Failure depends on the local strain condition, fracture number, and fracture toughness. The positive pressure on the contact spot is the function of the maximum positive stress.

$$F_{n\max} = \left[(\pi\sigma_{\max}/(1-2v_a))^3(9/2E^2)\right]R^2 \tag{4}$$

where the part in the square brackets is constant depending on the properties of the material. Therefore, the maximum tolerable load is determined by the hook top size and the square of the curvature radius of the micro-bulges on the surface.

3.3. Calculation of Supporting Shaft Size

We should calculate the supporting shaft of the hook to determine the cross-sectional diameter of the hook shaft and satisfy the requirements of bending stress. The internal forces on the cross-sectional generally contain moment M, shear force F_S, and axial force F_N during the bending deformation of the hook. We used the section method to analyze the stresses on the curved hook. The curved bars were divided into two parts by the cross-sectional (radial section) $m-m$, with an angle θ with a hook top. Stresses on curved bar of hook using section method are illustrated in Figure 4.

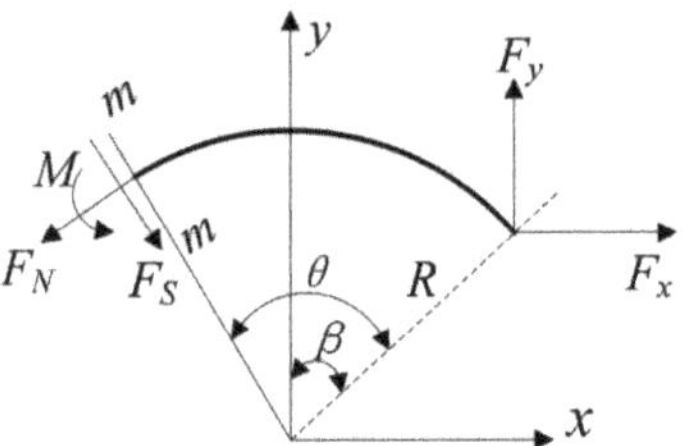

Figure 4. Stresses on curved bar of the hook using section method.

Stressing analysis of $m-m$ section shows that

$$\begin{cases} F_N = F_x\cos(\theta-\beta) + F_y\sin(\theta-\beta) \\ F_s = -F_x\sin(\theta-\beta) + F_y\cos(\theta-\beta) \\ M = F_xR(\cos\beta - \cos(\theta-\beta)) - F_yR(\sin\beta + \sin(\theta-\beta)) \end{cases} \tag{5}$$

where $F_x = F_v/n$ and $F_y = F_s/n$. n are the hook number on a single claw.

For the whole grasping claw, the claw with minimum hooks (such as two hooks arranged on the upper and lower limbs) is taken as an example to examine the interactions between the grasping claws and the wall surface. Figure 5 shows the forces exerting on the hooked claw by the climbing models constructed in the previous section. L, L_1, L_2 and D are the dimensions of the claw. The direction of friction F_{f2} on the lower claw tip is directly related to the torque magnitude of the torsional spring. When $\tau_2 > \tau_0$, F_{f2} points downward along the tangent plane of the micro-protuberance. When $\tau_2 < \tau_0$, F_{f2} points upward along the tangent plane of the micro-protuberance. When $\tau_2 = \tau_0$, $F_{f2} = 0$. In this study, τ_0 is the critical value of the torque of torsional spring.

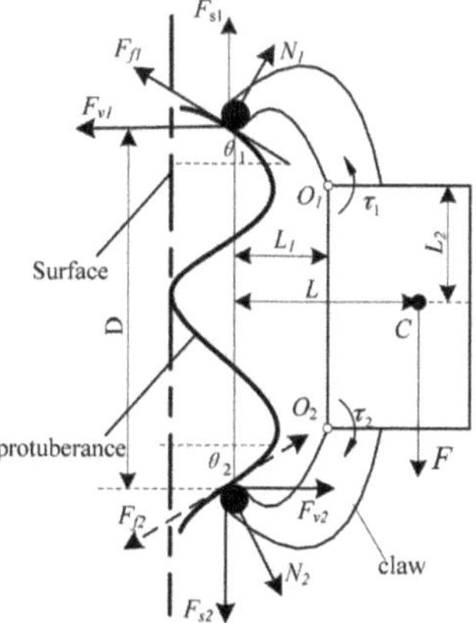

Figure 5. Analysis of stresses on hooked claw.

The angles of the upper and lower micro-protuberances are set to be equal to simplify the problem; that is, $\theta_1 = \theta_2 = \theta$, and $F_{f1}\cos\theta > N_1 \sin\theta$ for the upper claw. Thus, $F_{f1} > N_1 \tan\theta$. Therefore, the angles of micro-protuberances must be smaller than the self-locking angle between the claw tip and the micro-protuberance, i.e., $\theta < \arctan(\mu)$. Otherwise, the upper claw slides away and does not grasp the micro-protuberances steadily.

In Figure 5, N_1 and F_{f1} are the support force and friction exerted on the upper claw tip by the micro-protuberance, respectively; F_{s1} and F_{v1} are the tangential and normal forces along the wall surface of the support force and friction exerted on the upper claw tip, respectively; N_2 and F_{f2} are the support force and friction exerted by the micro-protuberance on the lower claw tip, respectively; and, F_{s2} and F_{v2} are the tangential and normal forces along the wall surface of the support force and friction exerted on the lower claw tip, respectively. C is the mass center of the whole hooked claw; τ_1 and τ_2 are the torques of torsional spring for the upper and lower hooked claws, respectively; θ_1 and θ_2 are the angles of the upper and lower micro-protuberances, respectively; and, G is the gravity of the claw.

Therefore, the normal and tangential forces of the hook along the wall surface are directly correlated with the torque of the coiling spring. The following equations are obtained, according to Equations (1)–(5):

$$\begin{cases} F_{s1} = (M_1 + \frac{GL}{2} - \frac{GLL_2}{D})/L_1 \\ F_{s2} = (M_1 + \frac{GL}{2} - \frac{GLL_2}{D})/L_1 - G \\ F_{v1} = F_{v2} = GL/D \end{cases},$$
$$\sigma_{\max} = \frac{M_{\max}}{W} = \frac{F_x R(\cos\beta - 1) - F_y R\sin\beta}{\frac{\pi}{32}d^3} = \frac{F_{v1} R(\cos\beta - 1) - F_{s1} R\sin\beta}{\frac{n\pi}{32}d^3} \leq [\sigma], \quad (6)$$

In Equation (6), $R = 1.5$ mm, $\beta = 80°$, and $n = 3$. Figure 6 shows the relationship between the maximum bending stress of the hook and the hook shaft diameter. The hook was made of 45 steel and was subjected to modulation processing. The top was unitized to increase wear resistance.

The diameter of the hook neck shaft was set as $d = 0.7$ mm by comprehensively considering the influences of strength, stiffness, and safety.

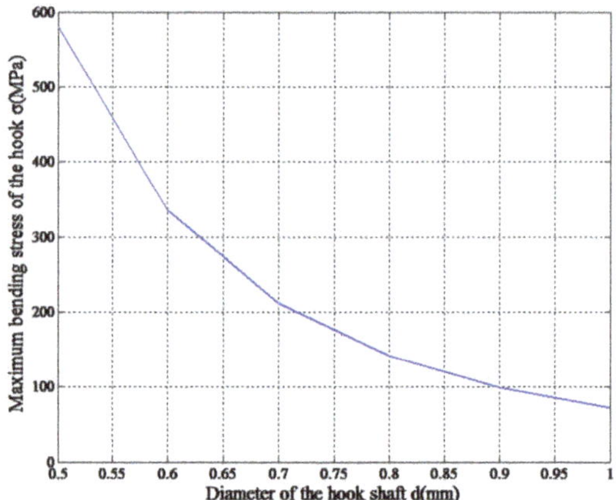

Figure 6. Relationship of maximum bending stress of hook with hook shaft diameter.

4. Analysis of the Grasping Claws Module Based on ANSYS Workbench

4.1. Secondary Claws

An analysis of the claw structure of reptiles revealed that the claws need to meet the following requirements: (1) loading forces applied on hook tips are reasonably distributed to prevent failure of elastic joints caused by stress concentration; and, (2) multiple hooks are independent. That is, while slipping on a rough wall surface, if a hook grasps a protuberance on the wall and forms mechanical interlocking, other hooks can still search regions that can be grasped along the slipping direction.

In order to analyze whether the design of the elastic claws meets the above requirements and verify the operating safety of the hooks, Solidworks was adopted for modeling, and the model was simplified according to the actual demand (Figure 7). After being imported in the ANSYS Workbench for mesh generation (Figure 8), 11,306 nodes and 5229 elements were attained. The properties of the materials that were used for producing the claws are listed in Table 1.

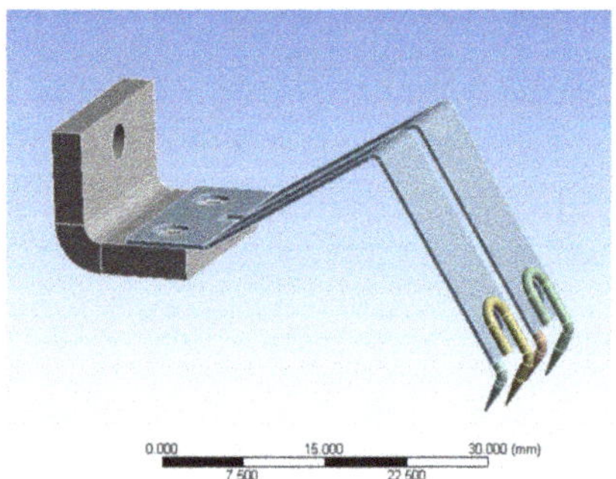

Figure 7. Simplified geometric model of the spiny foot.

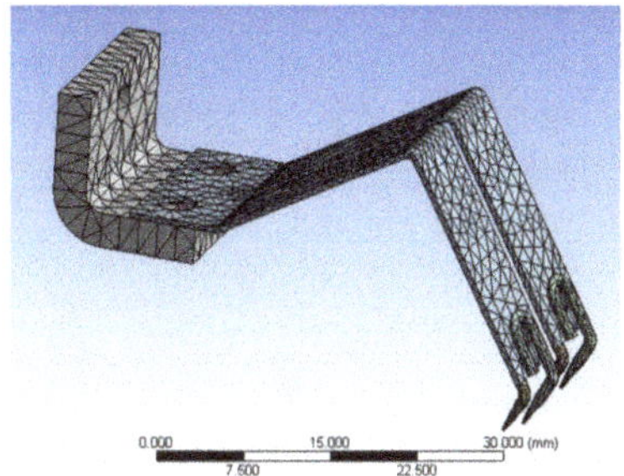

Figure 8. Mesh model of the spiny foot.

Table 1. Properties of materials used for the grasping claws module.

Item	Density [kg/m^3]	Young's Modulus [MPa]	Poisson's Ratio	Yield Strength [MPa]	Tensile Strength [MPa]
45#	7890	2×10^5	0.3	355	610
65 Mn	7810	1.98×10^5	0.26	784	980
Al	2770	7.1×10^4	0.33	250	250

The connecting plate was fabricated with 65 Mn steel with large yield strength, while the hooks were manufactured with #45 steel. When the two pairs of tiny hooks were mechanically interlocked with the wall surface, the hooks and the elastic steel sheets were deformed and the resulting grasping force pointed to the interior of the wall surface. Suppose that the mass of the grasping module of the claws was 0.5 kg, then 5 N pointing to the interior of the wall and vertical to the tips hooks was applied to the claws in each simulation.

The simulation results are shown in Figure 9.

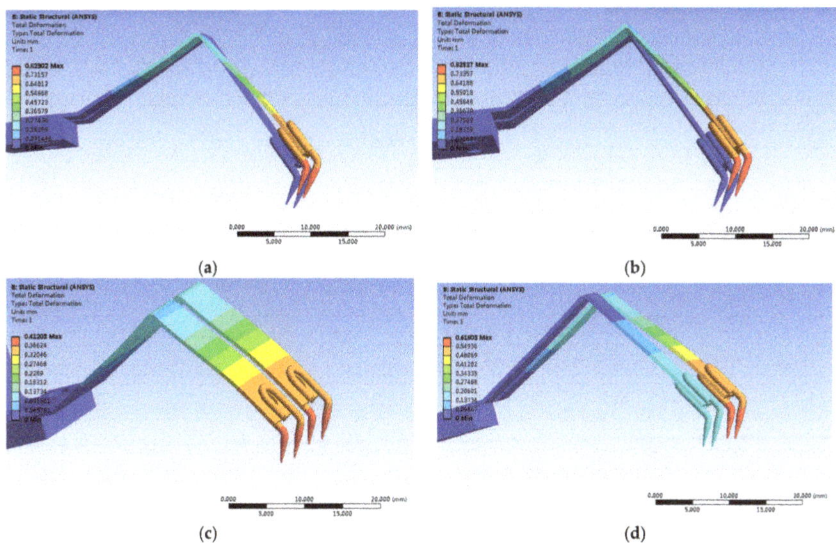

Figure 9. Displacement of the spiny feet. (**a**) Cloud picture for the displacement of a single hook under 5 N; (**b**) Cloud picture for the displacement of a pair of hooks each bearing 2.5 N; (**c**) Cloud picture for the displacement of a pair of hooks each bearing 1.25 N; (**d**) Cloud picture for the displacement of a pair of hooks bearing 3.75 and 1.25 N respectively.

Figure 9a illustrates the displacement when a hook detaches from the wall surface, while the other hooks the wall. The largest displacement (0.82302 mm) was found at the tip hook. Figure 9b demonstrates the displacement when each of the pair of hooks bears 2.5 N. Under this condition, the largest displacement was found to be 0.82527 mm at the tip of the hooks. It can be seen from Figure 9a,b that the displacement of the other pair of tiny hooks was 0. It indicates that even though a pair of hooks grasp the wall, the other pair of hooks can still search positions that can be grasped along the slipping direction.

Figure 9c shows the displacement when the four hooks of two pairs of claws grip the wall. The displacement of the hooks was 0.41203 mm. The displacement under the condition that one pair of the hooks bear 3.75 N, while the other pair is subjected to 1.25 N, as illustrated in Figure 9d. The largest displacement (0.61803 mm) was observed at the tip hook that bore 3.75 N. The above two conditions reveal that, when the two pairs of tiny hooks simultaneously hook the wall, they are independent and not affected by each other.

4.2. Modal Analysis of the Grasping Claws Module

In order to analyze the vibration characteristics of the grasping claws module and provide the theoretical basis for fault diagnosis and optimal design, this section investigated the modal of the grasping claws module.

At first, the claws are modeled and simplified by (1) eliminating small holes having little influences on the whole structure, and (2) using the lumped mass to replace the cylinder. According to the material properties in Table 1, constraints are applied to the fixture screw thread and the tips of the hooks. Then, mesh generation was carried out (shown in Figure 10).

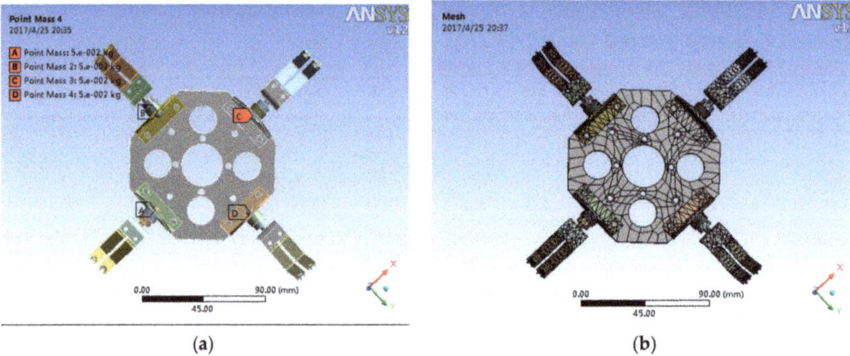

Figure 10. Finite element models. (**a**) Finite element model of the grasping claws module; (**b**) Mesh plot of the grasping claws module.

Based on the design of the pre-processing module, loads were applied for solution. The first six orders of modal were set and solved in the solution item max models to find, and the total deformation was inserted into each order of modal. Through solution, the vibration modes and natural frequencies of the first six orders of the grasping module were obtained. The cloud pictures of the displacement are illustrated in Figure 11.

The analysis from Figure 11 reveals that:

The natural frequencies of the first six orders of the grasping claws module mainly range from 394 Hz to 728 Hz. Among them, the natural frequencies of first four orders are close to each other, and those of the last two orders are approximate.

The vibration modes of the first four orders are all the upward and downward flexural vibrations of the corresponding secondary claws and the corresponding hooks have larger amplitudes, while the central part is stable. The vibration modes of the fifth and sixth orders are the folded vibrations of the two ends of the main body, with obvious amplitudes.

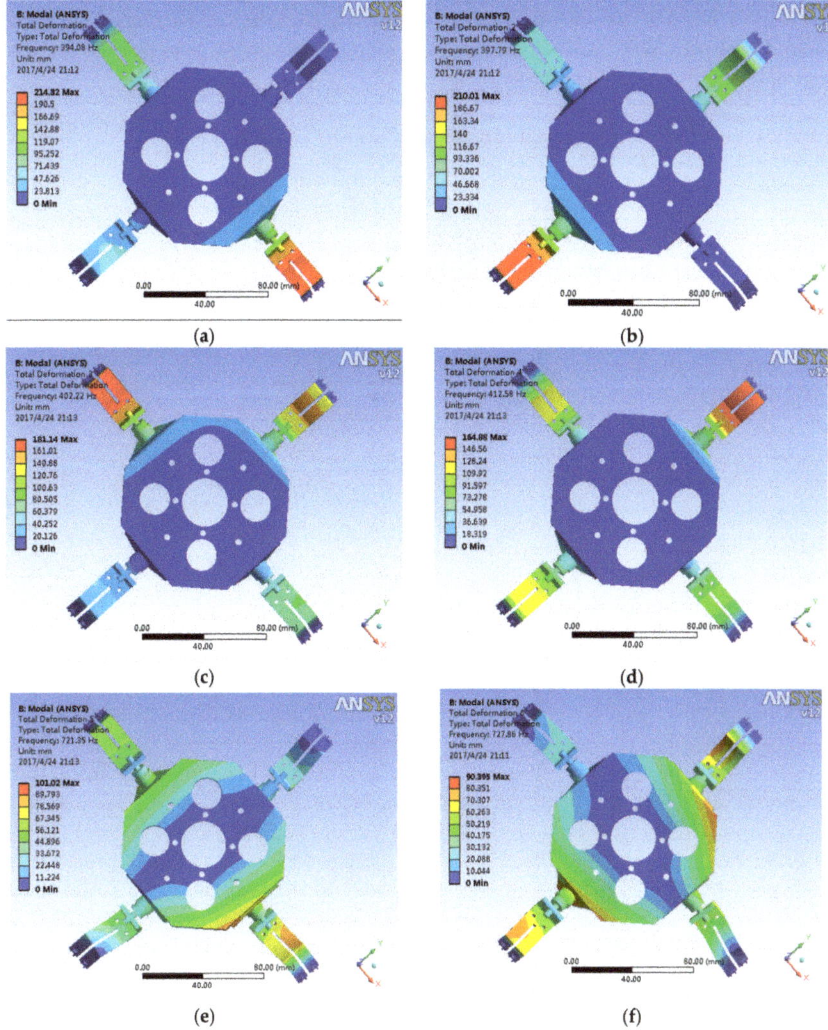

Figure 11. Displacements under vibration modes of the first six orders. (**a**) Vibration mode of the first order; (**b**) Vibration mode of the second order; (**c**) Vibration mode of the third order; (**d**) Vibration mode of the fourth order; (**e**) Vibration mode of the fifth order; (**f**) Vibration mode of the sixth order.

When the external exciting frequency approaches the natural frequency in Figure 8, it is likely to incur the resonance of the grasping claws module, leading to large amplitudes, and finally damaging the grasping claws module.

5. Experiments of Bionic Robot

We propose the climbing model of the bionic robot and design a simple prototype. On this basis, the climbing condition of the micro-hook is investigated experimentally.

5.1. Grasping Claws Module Structure

According to the results of finite element analysis in Section 4.2, #45 steel was used to produce the hooks, and the tips were subjected to nitriding treatment to enhance the wear resistance. The elastic steel sheets made of 65 Mn steel underwent surface finish and the holes were smoothed to avoid stress concentration. The main body was manufactured with aluminum alloy. The mechanism is illustrated in Figure 12.

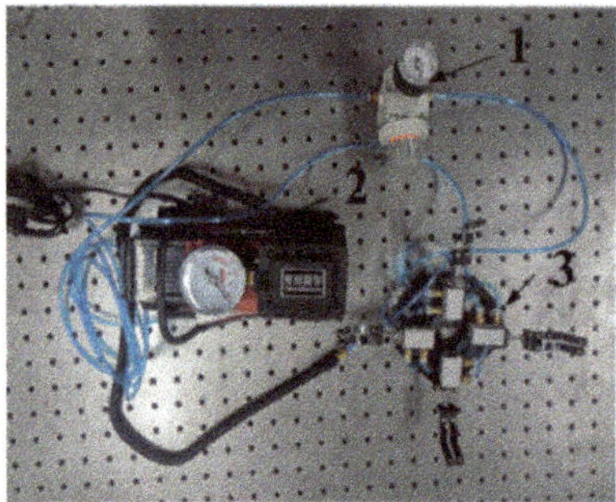

Figure 12. The grasping claws system prototype. 1—Pressure control valve, 2—vehicle mounted pump, 3—grasping module of the claws.

According to the results of modal analysis of claws, the natural frequencies of the grasping claws module are 394 Hz and 728 Hz, so these frequency bands need to be avoided during the simulation experiment. The experiment utilized a vehicle mounted pump as the power source, which was connected with each actuating cylinder through a pneumatic circuit of hoses. A pressure control valve was installed between the actuating cylinder of the secondary claws and the vehicle mounted pump. The pressure inside the cylinder and the driving force can be controlled by adjusting the pressure control valve. So, in the experiment, the pressure control valve mainly played its role according to turning on or off the driving force. During the operation of the vehicle mounted pump, the atmospheric pressure inside the cylinder of the grasping module can be quickly reduced by unplugging the quick connector in one side of the pressure control valve. The total mass of the grasping module of the claws was 431 g. The maximum stroke of the cylinder was 1.75 cm, which proves that the claw can search the grasping rough wall surface in this scope. During the normal work of the vehicle mounted pump in the experiment, the driving force applied by the cylinder to the spiny feet was about 12.7 N.

Figure 13 shows the grabbing states of the grasping claws and the working conditions of each spiny foot during the experiment. In the experiment, when the claws contact the wall, the spiny feet had slight reciprocating vibrations in the direction vertical to the wall under the effect of the driving force applied by the connecting rod. Then, the elastic steel sheets were elastically deformed, resulting in the mutual transformation between and the final stabilization of the kinetic energy and elastic potential energy, allowing for the spiny feet to stably hook the rough wall surface. As the spiny feet in the left and the right can also firmly hook and attach onto the rough wall surface, the grasping claws module is able to stably grab the wall when the wall was slightly vibrated.

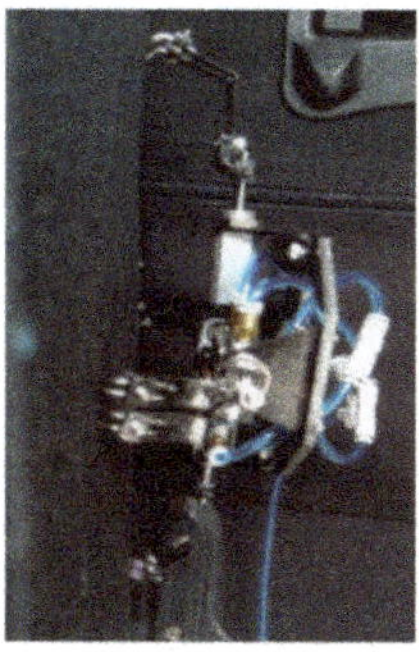

Figure 13. The grabbing posture.

5.2. Indoor Grabbing Experiments of the Grasping Claws

We carry out the bending experiment for the grasping claws in the laboratory, and verify the strength of single claw, as well as bending property. Moreover, we carry out the vibration experiment to verify the vibration resistance performance of the grabbing method.

(1) Experiment of bending performance for tip claw

In this paper, pneumatic system is used in the innovative mechanism. We carry out some experiments to test whether the pneumatic drive is feasible when the precision in demand is not high. Meanwhile, the new designed pneumatic system can be used with the in-car air pump, so as to carry out some field experiments conveniently.

The grasping process is controlled with a pneumatically driven system to provide adequate grasping force for the robot so it can stably attach to the wall. This method has the advantages of low-energy consumption and low noise. In the laboratory, a test system was established using two types of processed rough walls to perform a grasping test for the designed grippers of the climbing robot. One is the extra rough surface and the other is a random surface. The wall's dimension (especial the thickness) is too large, which is too heavy to adjust the slanting angles. We also tested the grasping claw on several vertical walls besides the inclined wall. In an effective stroking range of a 1.75 cm air cylinder, if the pressure of 5–10 MPa (it can produce up to 80 N forces on the tiny hook) is controlled and an appropriate angle of the claws are adjusted, then the appropriate uplifts could be found.

The experimental results show that the tip claw is not flexed when sliding on the wall, and there is no obvious deformation and fracture when various forces are applied to the cylinder.

(2) Vibration experiment of the grasping claws

As shown in Figure 14, the grasping claws were attached to the grabbing position. With grabbing force, the claw can withstand the maximum frequency of 12.3 Hz applied through the test platform and the maximum amplitude vibration intensity were 76.2% (1.5875 mm). When without grabbing force, the values were 12.3 Hz, 20.1% (0.41875 mm). The result indicated that the grabbing mode could significantly enhance the vibration resistance performance.

In order to better verify this problem, we divide the claws grabbing experiment with driving force into two stages: the stage of stability grasping when the driving force is applied, and the phase of the removal of the driving force. The transition between two phases is achieved by closing the pressure control valve.

In the experiment, we select output frequency as 12 Hz and intensity as 60% in the vibration platform. For a more intuitive and accurate description of the claw desorption process, we used high-speed camera to capture the image around closing the cylinder, and introduce the image analysis software Image Pro. The experimental results show that the grasp mode has a better vibration resistance performance than the hanging method after the closure of the cylinder.

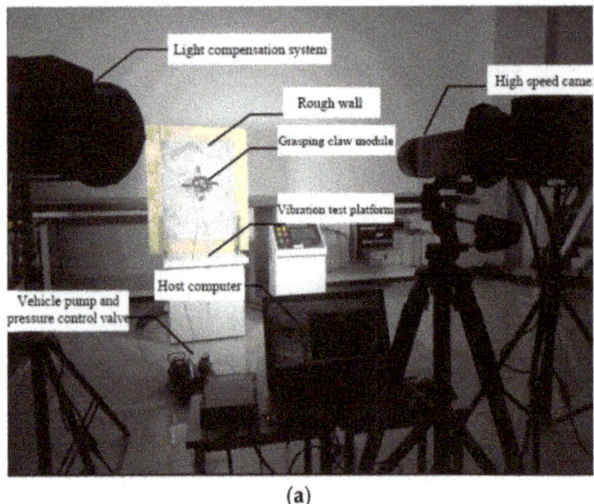

(a)

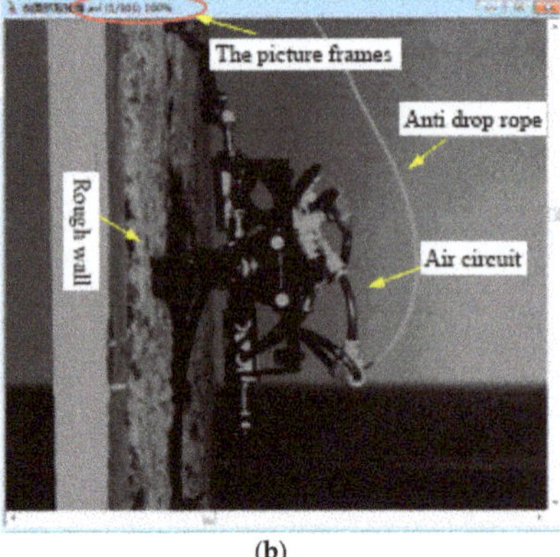

(b)

Figure 14. Indoor grabbing experiments of the grasping claws. (**a**) Vibration experiment of the grasping claws; (**b**) The picture of grabbing process.

5.3. Outdoor Grabbing Experiments of the Grasping Claws

To verify the grabbing ability of the designed mechanism for all kinds of rough wall surfaces, large amounts of outdoor grasping experiments were carried out. Figure 15 shows the grasping claws can achieve stability in the granitic plaster wall. Figure 16 illustrates that the claws can stably grab the rough rock wall. Therefore, the grasping module of cross-arranged claws shows superior robustness against the wall surfaces in the above two cases, indicating that it can work on various rough wall surfaces.

Figure 15. Stably grabbing the granite wall.

Figure 16. Stably grabbing the rough rock wall.

6. Conclusions

Based on the analysis of the grasping mechanism of cockroaches, this paper proposed a bio-inspired compliant spiny feet mechanism, and designed the grasping claws module. By studying the interaction between tiny hooks and rough wall surfaces, the size of miniature hooks was analyzed according to the results of the discriminant algorithm for graspable positions on three-dimensional (3D) rough wall surfaces. Furthermore, the proposed grasping module of the cross-arranged claws was verified based on the grasping model of the claws.

By analyzing the designed mechanism with ANSYS Workbench, the reasonability and reliability of the design of the spiny feet were verified. Meanwhile, the modal of the grasping module of claw was analyzed, which yielded the natural frequencies and vibration modes of the first six orders. On this basis, the authors found that the resonance can be avoided by changing the mass and mass distribution of the grasping claws module. This provides an optimization direction for the overall design of the grasping claws module.

Acknowledgments: This work was supported by the National Natural Science Foundation of China (No. 51775284, 51005025), and Natural Science Foundation of Jiangsu Province (BK20151505, BK20151199). The authors would like to appreciate all the reviewers for their constructive suggestions and comments.

Author Contributions: Quansheng Jiang and Fengyu Xu co-organized the work, designed the experiments, and wrote the paper.

Conflicts of Interest: The authors declare no conflict of interest.

References

1. Koo, I.M.; Trong, T.D.; Lee, Y.H.; Moon, H.; Koo, J.; Park, S.K.; Choi, H.R. Development of Wall Climbing Robot System by Using Impeller Type Adhesion Mechanism. *J. Intell. Robot. Syst.* **2013**, *72*, 57–72. [CrossRef]
2. He, B.; Wang, Z.; Li, M. Wet Adhesion Inspired Bionic Climbing Robot. *IEEE/ASME Trans. Mechatron.* **2014**, *19*, 312–320. [CrossRef]
3. Liu, R.; Chen, R.; Shen, H.; Zhang, R. Wall climbing robot using electrostatic adhesion force generated by flexible interdigital electrodes. *Int. J. Adv. Robot. Syst.* **2013**, *10*, 1–9. [CrossRef]
4. Wang, W.; Wang, K.; Zong, G.; Li, D. Principle and experiment of vibrating suction method for wall-climbing robot. *Vacuum* **2010**, *85*, 107–112. [CrossRef]
5. Gao, X.; Xu, D.; Wang, Y. Multifunctional robot to maintain boiler water-cooling tubes. *Robotica* **2009**, *27*, 941–948. [CrossRef]
6. Guo, C.; Sun, J.; Ge, Y.; Wang, W. Biomechanism of Adhesion in Gecko Setae. *Chin. Sci. Life Sci.* **2012**, *42*, 135–142.
7. Bartsch, S.; Birnschein, T.; Rommermann, M.; Hilljegerdes, J.; Kuhn, D. Development of the Six-Legged Walking and Climbing Robot SpaceClimber. *J. Field Robot.* **2012**, *29*, 506–532. [CrossRef]
8. Lam, T.L.; Xu, Y.S. Motion planning for tree climbing with inchworm-like robots. *J. Field Robot.* **2013**, *30*, 87–101. [CrossRef]
9. Asbeck, A.T.; Kim, S.; Cutkosky, M.R.; Provancher, W.R.; Lanzetta, M. Scaling hard vertical surfaces with compliant microspine array. *Int. J. Robot. Res.* **2006**, *25*, 1165–1179. [CrossRef]
10. Sintov, A.; Avramovich, T.; Shapiro, A. Design and motion planning of an autonomous climbing robot with claws. *Robot. Auton. Syst.* **2011**, *59*, 1008–1019. [CrossRef]
11. Parness, A.; Frost, M.; Thatte, N. Gravity-independent Rock-climbing Robot and a Sample Acquisition Tool with Microspine Grippers. *J. Field Robot.* **2013**, *30*, 897–915. [CrossRef]
12. Liu, Y.; Sun, S.; Wu, X.; Mei, T. A Wheeled Wall-Climbing Robot with Bio-Inspired Spine Mechanisms. *J. Bion. Eng.* **2015**, *12*, 17–28. [CrossRef]
13. Guan, Y.; Zhu, H.; Wu, W. A Modular Biped Wall-Climbing Robot with High Mobility and Manipulating Function. *IEEE/ASME Trans. Mechatron.* **2013**, *18*, 1787–1798. [CrossRef]
14. Wang, K.; Gao, W.; Ma, S. Snake-Like Robot with Fusion Gait for High Environmental Adaptability: Design, Modeling, and Experiment. *Appl. Sci.* **2017**, *7*, 1133. [CrossRef]

15. Jiang, L.; Guan, Y.; Cai, C.; Zhu, H. Gait Analysis of a Novel Biomimetic Climbing Robot. *J. Mech. Eng.* **2010**, *46*, 17–22. [CrossRef]
16. Kalind, C.; Nick, W.; Aaron, P. Rotary Microspine Rough Surface Mobility. *IEEE/ASME Trans. Mechatron.* **2016**, *21*, 2378–2390.
17. Arena, P.; Fortuna, L.; Frasca, M. Attitude control in walking hexapod robots: An analogic spatio-temporal approach. *Int. J. Circuit Theory Appl.* **2002**, *30*, 349–362. [CrossRef]
18. Buscarino, A.; Fortuna, L.; Frasca, M.; Muscato, G. Chaos does help motion control. *Int. J. Bifurc. Chaos* **2007**, *17*, 3577–3581. [CrossRef]
19. Buscarino, A.; Fortuna, C.F.L.; Frasca, M. Passive and active vibrations allow self-organization in large-scale electromechanical systems. *Int. J. Bifurc. Chaos* **2016**, *26*, 1650123. [CrossRef]
20. Xu, F.; Wang, X. Design Method and Analysis for Wall-climbing Robot based on Hooked-claws. *Int. J. Adv. Robot. Syst.* **2012**, *9*, 1–12. [CrossRef]
21. Xu, F.; Shen, J.; Hu, J.; Jiang, G. A Rough Concrete Wall Climbing Robot Based on Grasping Claws: Mechanical Design, Analysis and Laboratory Experiments. *Int. J. Adv. Robot. Syst.* **2016**, *13*. [CrossRef]
22. Dai, Z.; Stanislav, N.; Uli, S. Roughness-dependent friction force of the tarsal claw system in the beetle Pachnoda marginata. *J. Exp. Biol.* **2002**, *205*, 2479–2488. [PubMed]

Article

A Precise Positioning Method for a Puncture Robot Based on a PSO-Optimized BP Neural Network Algorithm

Guanwu Jiang [1,2,3,4,*], Minzhou Luo [1,2,4,*], Keqiang Bai [1,3,*] and Saixuan Chen [1,2,4]

[1] The Department of Automation, University of Science and Technology of China, Hefei 230026, China; chensx@iimt.org.cn

[2] Key Laboratory of Special Robot Technology of Jiangsu Province, Hohai University, Changzhou 213000, China

[3] School of Information Engineering, Southwest University of Science and Technology, Mianyang 621010, China

[4] Institute of Intelligent Manufacturing Technology, Jiangsu Industrial Technology Research Institute, Nanjing 211800, China

* Correspondence: jgwu0816@mail.ustc.edu.cn (G.J.); luomz@iimt.org.cn (M.L.); baisir@mail.ustc.edu.cn (K.B.); Tel.: +86-181-8177-0661 (G.J.); +86-186-6118-1682 (M.L.); +86-159-8365-8733 (K.B.)

Received: 18 July 2017; Accepted: 18 September 2017; Published: 21 September 2017

Abstract: The problem of inverse kinematics is fundamental in robot control. Many traditional inverse kinematics solutions, such as geometry, iteration, and algebraic methods, are inadequate in high-speed solutions and accurate positioning. In recent years, the problem of robot inverse kinematics based on neural networks has received extensive attention, but its precision control is convenient and needs to be improved. This paper studies a particle swarm optimization (PSO) back propagation (BP) neural network algorithm to solve the inverse kinematics problem of a UR3 robot based on six degrees of freedom, overcoming some disadvantages of BP neural networks. The BP neural network improves the convergence precision, convergence speed, and generalization ability. The results show that the position error is solved by the research method with respect to the UR3 robot inverse kinematics with the joint angle less than 0.1 degrees and the output end tool less than 0.1 mm, achieving the required positioning for medical puncture surgery, which demands precise positioning of the robot to less than 1 mm. Aiming at the precise application of the puncturing robot, the preliminary experiment has been conducted and the preliminary results have been obtained, which lays the foundation for the popularization of the robot in the medical field.

Keywords: inverse kinematics; PSO algorithm; BP neural network; precise localization; puncturing robot

---

1. Introduction

Robots are currently used in industrial and medical applications where high accuracy, repeatability, and stability of the operations are required [1]. With the development of modern control technology, robot technology has been widely used in new fields, such as in medical robots. A surgical robot operating system is a collection of a number of modern, complex, high technologies, and the doctor, through the robot system, can perform surgical operations without touching patients. A minimally-invasive surgical robot is a combination of medical image processing technology and the operation of the mechanical arm to perform puncture surgery on the patient, to achieve minimal invasiveness, accuracy, efficiency, and stability.

The most important problem of the serial robot, which is the solution of the kinematics of the manipulator, can be successfully implemented. Robot kinematics handles the mapping between joint space (h) and Cartesian space (x, y, z), where h represents the positions of the joints of a robotic manipulator and (x, y, z) represent the position of the end effector of the manipulator [2]. Kinematics analysis of the robot includes two aspects: the forward kinematics and the inverse kinematics. The forward kinematics are the mappings from the joint angle to the Cartesian coordinate system. The inverse kinematics are known to solve the joint variables under the position and posture of the end effector.

Traditionally, there are three methods to solve the inverse kinematics problem of the robot: the geometric method, the algebraic method, and the iterative method. Any method has its own shortcomings in solving the inverse kinematics. For instance, closed-form solutions are not guaranteed for the algebraic methods, and closed-form solutions for the first three joints of the robot must exist geometrically when the geometric method is used. Similarly, the iterative inverse kinematics solution method converges to only one solution that depends on the starting point [1]. These methods often require high-performance computer hardware, and the calculation accuracy cannot be guaranteed. For these reasons, researchers have begun to focus on the application of artificial neural networks to the kinematics of the robot.

The inverse kinematics analysis of the six degrees of freedom (DOF) industrial robot is carried out by using the back propagation neural network algorithm [3], but this method cannot solve the problem when the joint angle error is too large. In [4], a method is presented for solving the inverse kinematics of redundant robots and the prevention of singular points. In [5], for the singular series robot arm configuration and uncertainty, a method was proposed based on an artificial neural network, and the training process is very difficult, and needs sensors added to each joint.

There are many researchers focusing on the genetic algorithm to obtain the inverse kinematics of the robot [6–8]. Kamal and Djamel [6] researched particle swarm optimization (PSO) and genetic algorithms (GA) for finite impulse response (FIR) filter design. Kalra and colleagues [7] used an evolutionary approach based on a real-coded genetic algorithm to obtain the multimodal inverse kinematics problem of industrial robots. In their method, the fitness function is defined in a manner that requires separate evaluation of the positional error of the robot and the total joint displacement. These two approaches can be used together to solve some specific problems. Mustafa and Kerim [8] used four different optimization algorithms (the genetic algorithm (GA), the particle swarm optimization (PSO) algorithm, the quantum particle swarm optimization (QPSO) algorithm, and the gravitational search algorithm (GSA)) for solving the inverse kinematics problem of a four DOF serial robot manipulator.

The main purpose of this paper is to improve the precision of the inverse kinematics solution from a particle swarm optimization (PSO) back propagation (BP) neural network algorithm, especially for processing data in a short period of time, and the time in which the robot is in motion. The particle swarm optimization algorithm is employed to find the global advantage with the BP neural network to find the optimal solution, overcome some inherent defects (easy to fall into local minimum, slow convergence and poor generalization ability etc.) of the BP neural network, and thus further improve the convergence precision of BP neural network, the convergence speed, and generalization ability. The main contribution is that the algorithm is applied to the inverse kinematics of UR six degree of freedom manipulator, which guarantees the accuracy of the end position accuracy of the robot with six degrees of freedom within 0.1 mm, and the robot joint angle at 0.01 degrees. The main innovation of this study is the application of this technique in the precise localization of medical needle surgery. In the experimental part, we achieved very good results and achieved high-precision positioning of the puncture operation. We had to ensure that the experimental puncture accuracy was less than 1 mm, which can meet the needs of medical needle surgery.

2. Research and Methods

2.1. The Principle of Precision Positioning of a Puncture Robot

Minimally-invasive surgery (MIS) is a cost-effective alternative to open surgery whereby essentially the same operations are performed using specialized instruments designed to fit into the body through several tiny punctures instead of one large incision [9]. The principle of the operation of the puncture robot is shown in Figure 1.

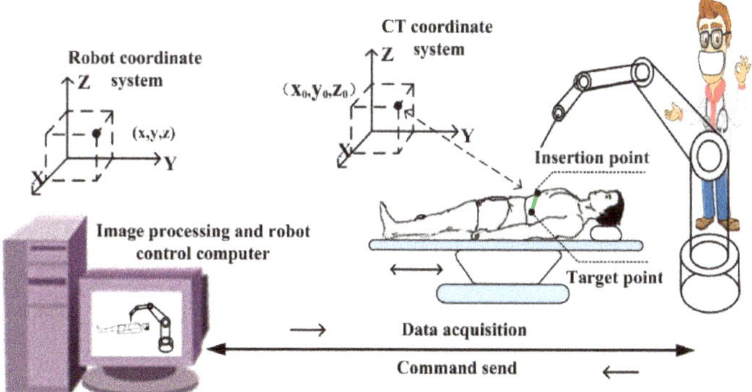

Figure 1. Position principle of the puncturing robot.

Firstly, patients require a CT scan, and the medical image will be generated by the computer to complete the three-dimensional reconstruction. Then a doctor can observe and diagnose the disease according to the three-dimensional model. Finally, the doctor determines the position of the puncture target point coordinates, and determines the insertion point through the analysis of the patients' skin. Between the target point and insertion point of connection is the puncture route (the green line in Figure 1). The route must avoid the patient's bones, blood vessels, and other organs. The accurate puncture route directly determines the quality of the puncturing operation. Research on precise positioning technology of the manipulator used in this study was conducted to establish the series robot puncture route. The puncture route target point and insertion point coordinates are sent to the robot through the data processing computer, and at the end-effector of the robot, the puncture guide tube accurately positions the needle into the patient's skin, the puncture route keeping with the robot end position and posture.

2.2. Analysis of the UR3 Manipulator

The UR3 manipulator (Figure 2) is a new and small six DOF collaborative robotic by the Universal Robots Company (Odense, Denmark). The key features of the UR3 manipulator are that it is a flexible, lightweight, collaborative, and safe table-top robot. The UR3's six joints contribute to the transformational and rotational movements of its end effector. The kinematics analysis of the UR3 is more complex than other manipulators. The Schematic and frame assignment of UR3 is shown in Figure 3.

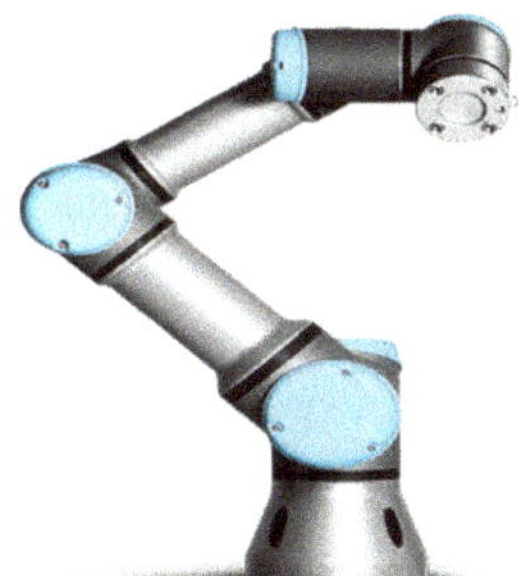

Figure 2. The real UR3 manipulator.

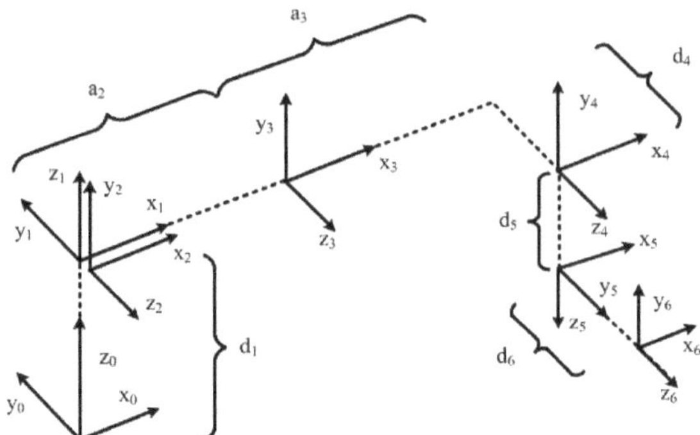

Figure 3. Schematic and frame assignment of UR3 ($\theta_i = 0$ for $1 \leq i \leq 6$).

Currently, there are no exact Matlab models for this robot. In this paper, we researched the precise positioning technique using the UR3 for medical puncture surgery. Kinematic modeling of the robot based on the Denavit-Hartenberg (D-H) parameters provided by the UR3 manual are shown in Table 1.

Table 1. Denavit-Hartenberg (D-H) parameters for the UR3 robot.

Link	θ (rad)	a (mm)	d (mm)	α (rad)
Joint 1	θ_1	0	151.9	$\pi/2$
Joint 2	θ_2	-243.65	0	0
Joint 3	θ_3	-213.25	0	0
Joint 4	θ_4	0	112.35	$\pi/2$
Joint 5	θ_5	0	85.35	$-\pi/2$
Joint 6	θ_6	0	81.9	0

The robot homogeneous transformation matrix ${}^{i-1}_{i}T$ for a single joint is expressed in Equations (1) and (2), which uses four link parameters [10]. This transformation is known as the D-H notation:

$$T = T_{tran}(z_{i-1}, d_i)T_{tran}(x_{i-1}, a_i)T_{rot}(x_{i-1}, a_i) \tag{1}$$

$${}^{i-1}_{i}T = \begin{bmatrix} c\theta_i & -c\alpha_i s\theta_i & s\alpha_i s\theta_i & a_i c\theta_i \\ s\theta_i & c\alpha_i c\theta_i & -s\alpha_i c\theta_i & a_i s\theta_i \\ 0 & s\theta_i & c\alpha_i & d_i \\ 0 & 0 & 0 & 1 \end{bmatrix} \tag{2}$$

based on the transformation matrix between adjacent links ${}^{i-1}_{i}T(i = 1, 2, \ldots, 6)$. Among them, a_i, d_i, α_i depend on the constant of robot structure parameters; $\theta_i(i = 1, 2, \ldots 6)$ represent the joint variables, $c\theta_i = \cos\theta_i$, $s\theta_i = \sin\theta_i$, $s\alpha_i = \sin\alpha_i$, and $c\alpha_i = \cos\alpha_i$. Thus, we can obtain the transformation matrix from the base to the end effector, whose position matrix by the homogeneous coordinate transformation is given by Equation (3):

$${}^{0}_{6}T = \prod_{i=1}^{6} {}^{i-1}_{i}T = \begin{bmatrix} {}^{0}_{6}R_{3\times3} & {}^{0}_{6}P_{3\times1} \\ 0 & 1 \end{bmatrix} = \begin{bmatrix} r_{11} & r_{12} & r_{13} & p_x \\ r_{21} & r_{22} & r_{23} & p_y \\ r_{31} & r_{32} & r_{33} & p_z \\ 0 & 0 & 0 & 1 \end{bmatrix} \tag{3}$$

Among them, ${}^{0}_{6}R_{3\times3}$ is the rotation matrix of the robot end effector, and ${}^{0}_{6}P_{3\times1}$ is the position matrix of the robot end effector, where r_{ij} represents the rotational elements of the transformation matrix (i and j = 1, 2, 3) and p_x, p_y, p_z are the elements of the position vector.

A UR3 robot with six DOF was used in this study. The manipulator has a six DOF Cartesian position of the end effector (x, y, z), which is obtained directly from the ${}^{0}_{6}T$ matrix [11]. The orientation of the end effector is described according to the RPY (roll-pitch-yaw) rotation. These rotations are the angles around the Z-Y-Z axis, as shown in Equation (4):

$$R_{Z'Y'Z'}(\alpha,\beta,\gamma) = \begin{bmatrix} c\alpha c\beta c\gamma - s\alpha s\gamma & -c\alpha c\beta s\gamma - s\alpha c\gamma & c\alpha s\beta \\ s\alpha c\beta c\gamma + c\alpha s\gamma & -s\alpha c\beta s\gamma + c\alpha c\gamma & s\alpha s\beta \\ -s\beta c\gamma & s\beta s\gamma & c\beta \end{bmatrix} \tag{4}$$

Solving the ${}^{0}_{6}T$ matrix, we can obtain the angles, which are calculated by Equations (5)–(7):

$$\alpha = \text{Atan2}(r_{23},\ r_{13}) \tag{5}$$

$$\beta = \text{Atan2}(r_{13}\cos\alpha + r_{23}\sin\alpha,\ r_{33}) \tag{6}$$

$$\gamma = \text{Atan2}(-r_{11}\sin\alpha + r_{21}\cos\alpha, r_{22}\cos\alpha - r_{12}\sin\alpha) \tag{7}$$

These equations can provide the robot position, which is relative to the universe coordinate system [12]. The coordinates of each joint are used to describe the position and orientation of the robot. The forward kinematics equation of the robot is described by Equation (8):

$$\text{F}_{\text{forward kinematics}}(\theta_1,\theta_2,\theta_3,\theta_4,\theta_5,\theta_6) = (p_x,p_y,p_z,\alpha,\beta,\gamma) \tag{8}$$

As shown in Equation (8), when the six joint angles of the robot are known, the Cartesian coordinate system of the robot can be calculated according to the forward kinematics [13]. However, the six joint angles of the robot must be computed in an industrial application, so the inverse kinematics equation is solved by Equation (9):

$$\text{F}_{\text{inverse kinematics}}(p_x,p_y,p_z,\alpha,\beta,\gamma) = (\theta_1,\theta_2,\theta_3,\theta_4,\theta_5,\theta_6) \tag{9}$$

In the next part, $\alpha,\beta,\gamma,p_x,p_y,p_z$ will be used as the input variables of the BP neural network model, and the joint angles $\theta_i(i = 1,2,\ldots,6)$ will be used as the output variables of the BP neural network.

2.3. BP Neural Network

A BP (back propagation) network, which was proposed by Rumelhart and McCelland in 1986, is a type of error back propagation training algorithm for the multilayer feedforward network. It consists of two processes: the forward spread of information and the reverse propagation of error; the neural network model is one of the most widely used. BP neural networks can be compared to the input and output of the highly nonlinear mapping, which is characterized by the spread of the error to correct the weights and thresholds of the network. By approximating the nonlinear function several times, the BP neural network can approximate the complex function. A BP neural network model of a single neuron is shown in Figure 4.

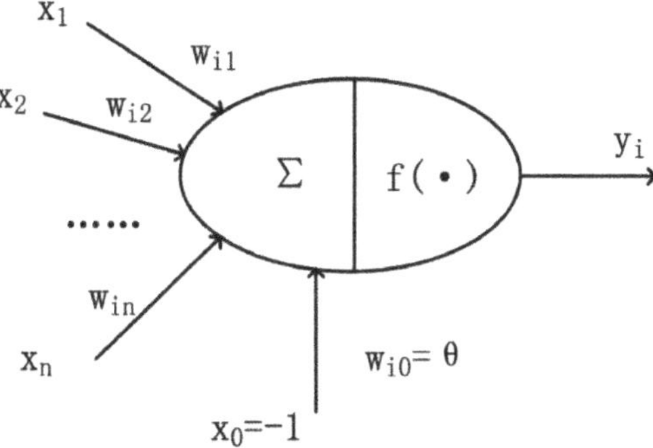

Figure 4. Model of an artificial neuron.

For a given artificial neuron, $x_j(j = 1, 2, \cdots, n)$ is the component of the input vector of the neuron, $\omega_{ij}(j = 1, 2, \cdots, n)$ is the weight component of the input vector of the neuron, and $\omega_{i0} = \theta$ represents a threshold or bias. Usually, the x_0 input is assigned the value -1, which makes it a bias input with $w_{k0} = \theta$. This leaves only actual inputs to the neuron: from x_1 to x_n.

The output of the y_i neuron is expressed in Equation (10):

$$y_i = f(net_i) = f(xw) = f(\sum_{j=1}^{n} w_{ij}x_j - \theta) \tag{10}$$

where f is the transfer function. We approximate the function by the sigmoid function. If the weight is positive, it means that the corresponding input point is in a state of excitement and has a strengthening effect; if the weight is negative, it has an inhibiting effect. The function is expressed in Equation (11):

$$f(x) = \frac{1}{1 + e^{-ax}}, \qquad (0 < f(x) < 1) \tag{11}$$

In this paper, according to the network structure, the input layer has six neurons, respectively, corresponding to $\alpha, \beta, \gamma, p_x, p_y, p_z$. The output layer has six neurons: the six joint angles corresponding to the UR3 robot's six angles θ_1, θ_2, θ_3, θ_4, θ_5, and θ_6. The number of hidden layer units is a very complicated problem: its determination needs to agree with the experience of the designer and several tests. The hidden layer unit number with the number of input/output units has a direct relationship; if the number is too great, not only will the training time increase, causing the learning time to become too long, but the error may be large. If the number is too low, then the neural network may have too little information to solve the problem. Therefore, it is very important to select a suitable number of hidden layers [14,15]. In this paper, after a large number of comparisons and experiments, two hidden layers are selected: 24 and 18 units, respectively. In this paper, the design of the BP neural network topology is shown in Figure 5.

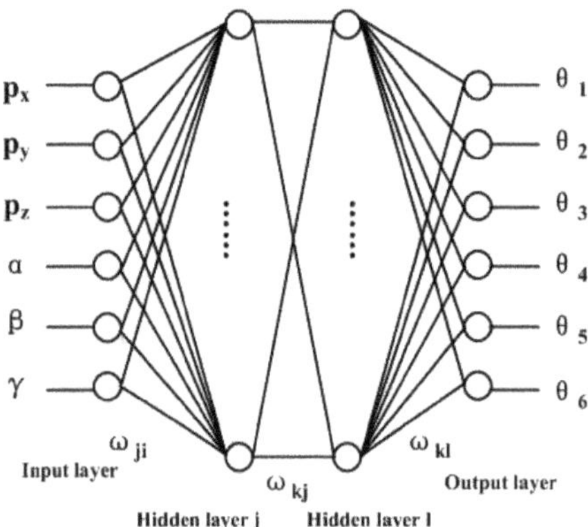

Figure 5. BP (back propagation) neural network topology structure.

Figure 5 illustrates the topology of the designed network which consists of three layers, one of which is a hidden layer, presented to store the internal representation. The fundamental idea underlying the design of the network is that the information entering the input layer is mapped as an internal representation in the units of the hidden layer (s) and the outputs are generated by this internal representation, rather than by the input vector. Given that there are enough hidden neurons, input vectors can always be encoded in such a form that the appropriate output vector can be generated from any input vector [16].

Although BP neural networks have many remarkable characteristics, they also have some inherent defects: easily falling into a local minimum, slow convergence speed, and weak generalization ability. Therefore, they require some improvement by combining with an optimization algorithm.

2.4. PSO Optimization Combined with the BP Neural Network

Particle swarm optimization (PSO) was proposed by Kennedy and Eberhart [17], and is derived from the study of the behavior of bird predation combined with the regularity of the collective activity of birds to establish a simplified model by the population. The PSO algorithm also belongs to the area of evolutionary algorithms (EAs).

PSO, which is similar to a genetic algorithm, is an iterative optimization algorithm [18,19]. The system is initialized with a set of random solutions. PSO does not have a genetic algorithm "crossover" and "mutation" operation, rather, it follows the current search to find the optimal value of the global optimum. The PSO algorithm was brought to the attention of academic circles as it has the advantages of easy implementation, high precision, fast convergence, easy adjustment parameters, and so on. PSO has been extensively used in function optimization, neural network training, fuzzy control systems, and genetic algorithms.

Particle swarm optimization (PSO) is a parallel algorithm, and finds the optimal solution by iteration. In each iteration, the particle updates itself by tracking two extremes [20]. The first extreme is the optimal solution of the particle itself, called the personal best value (pbest); the other is the group best value (gbest). The target value is set as the best value of the fitness function, and i particles can be expressed as a D-dimensional vector $X_i = [X_{i1}, X_{i2}, \ldots, X_{id}]$, and the velocity of particles can be expressed as $V_i = [X_{i1}, X_{i2}, \ldots, X_{id}]$. The current best position of the particle is represented as $X_{pbesti} = X_{pbesti1}, X_{pbesti2}, \ldots, X_{pbestid}$, and the current optimal position of the individual is represented by X_{pbesti}. The best position of the particle group is represented as $X_{gbesti} = X_{gbest1}, X_{gbest2}, \ldots, X_{gbestd}$, X_{gbest}, representing the historical optimal location of a group. In each iteration, the particle swarm updates its speed and position by the individual extreme value, and the group extreme value. The updated formula is shown in Equations (12) and (13):

$$V_{id}^{k+1} = \omega V_{id}^{k} + c_1 r_1 \left(P_{ib}^{k} - X_{id}^{k}\right) + c_2 r_2 \left(P_{gd}^{k} - X_{id}^{k}\right) \quad (12)$$

$$X_{id}^{k+1} = X_{id}^{k} + V_{id}^{k+1} \quad (13)$$

where X_{id}^{k} and V_{id}^{k} are, respectively, the position and velocity of the i-th particle at the k-th iteration. V is the velocity of the particle, ω is the weighting function, r_1 and r_2 are random numbers between 0 and 1, and c_1 and c_2 are learning factors; usually, $c_1 = c_2 = 2$. $P_{ib}^{k} - X_{id}^{k}$ is the deviation of the individual extreme value, and $P_{gd}^{k} - X_{id}^{k}$ is the deviation of the group extreme value. In the process of optimization, the velocity of each particle is limited to the maximum speed of $V_{\max}$; if the update speed is greater than the set $V_{\max}$, then the one-dimensional velocity is limited to $V_{\max}$.

This part of the research setup follows the flowchart based on the PSO-BP neural network as shown in Figure 6, and the steps are as follows:

1. Firstly, one needs to determine the topology of the neural network, initialize the BP neural network, and determine the initial value and threshold.
2. In the PSO algorithm, one needs to initialize the particle velocity, calculate the corresponding fitness function, and create the individual extreme and extreme groups.
3. Then update the particle's velocity, position, and fitness function; update the individual extreme and extreme groups, and determine whether it meets the conditions. If the conditions are not satisfied, the parameter is updated again. If the conditions are satisfied, the optimal solution is obtained.
4. Finally, the optimal solution of the PSO algorithm is given by the trained BP neural network, and then the weights and thresholds are updated to determine whether the training results meet the termination condition. If the condition is not satisfied, then the neural network is trained again; if it is satisfied, then finish the network testing and output the result.

The fitness function of the PSO algorithm is defined as the square sum of the difference between the joint angle and the desired joint angle when the BP neural network is trained, expressed in Equation (14):

$$J_{fitness}(k) = \sum_{i=1}^{N} (\theta_k - \theta_j)^2 \tag{14}$$

The network output θ_k is a function of the weight of ω: $\theta_k = f(\omega)$.

This part studies the PSO algorithm and BP neural network, using the PSO algorithm to find the global advantages combined with the BP neural network to find the optimal local solution, so as to further improve the BP neural network convergence precision, convergence speed, and generalization ability.

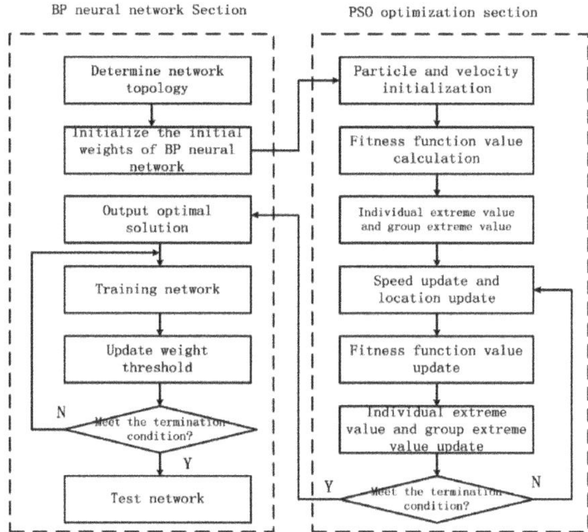

Figure 6. PSO (particle swarm optimization) optimization and BP neural network flowchart.

3. Experimental Results and Validation

Combined with the BP neural network optimized by the PSO algorithm, the algorithm is trained by using the sample data [21]. In the simulation experiment, 950 groups of data were selected as the training samples, and 50 groups of data were used as the test samples by using the UR3 manipulator in the working space of 1000 groups of data. The samples of the UR3 robot end position coordinates and Euler angles were used as the input nodes of the BP neural network, and the UR3 robot's six joint angles for the 50 sets of data output the forecast sample; then, using the PSO algorithm to optimize the convergence, the BP neural network weights and thresholds repeatedly trained the six robot joint angles, providing 50 sets of data for the output prediction samples.

In the course of training, the position vector and the rotation angle of the UR3 robot are used as the input points of the BP neural network, and the values of the six joint angles are the output points. The PSO algorithm is used as the optimization of the fitness function. The weights and thresholds of the BP neural network are obtained by particle swarm optimization, and the output value of each joint variable is obtained by a simulated test.

For the network after training, we get the error range of the test set and the range of the mean square error (MSE) which is shown in Table 2. The error range of 50 sets of joint angles, the error range of the joint angle by the BP neural network was calculated as [−0.3059, 0.4130], and the mean square error (MSE) was in the range of [6.72×10^{-3}, 8.12×10^{-2}]; the error range of the joint angles by the BP neural network-optimized PSO was calculated as [−0.1859, 0.1079], the mean square error (MSE) was in the range of [2.39×10^{-4}, 6.42×10^{-3}]. Figures 7–12 show the contrast of the BP neural network, and the PSO-BP neural network for the output robot joint angle error $\triangle\theta$. Figure 13 is the comparison of the mean square error (MSE) of the two algorithms.

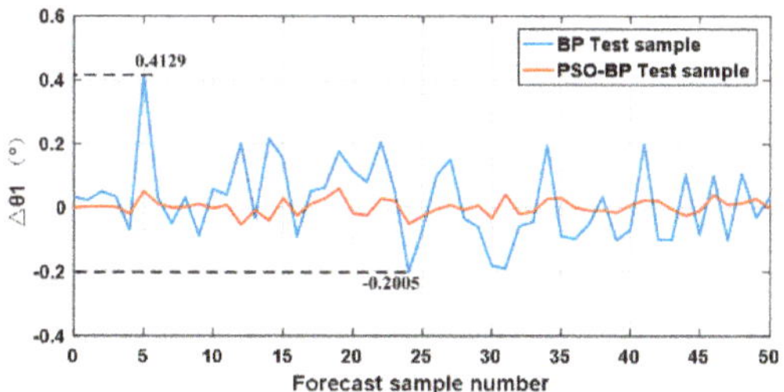

Figure 7. Error contrast of test set's θ_1.

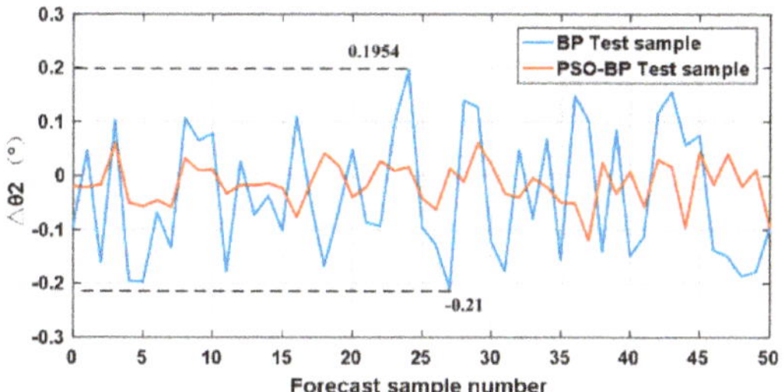

Figure 8. Error contrast of test set's θ_2.

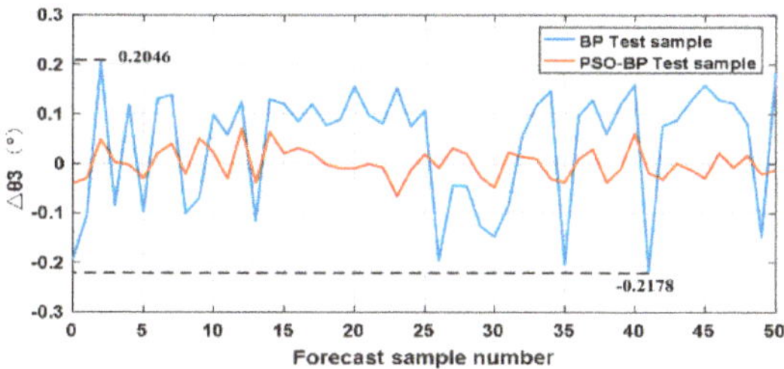

Figure 9. Error contrast of test set's θ_3.

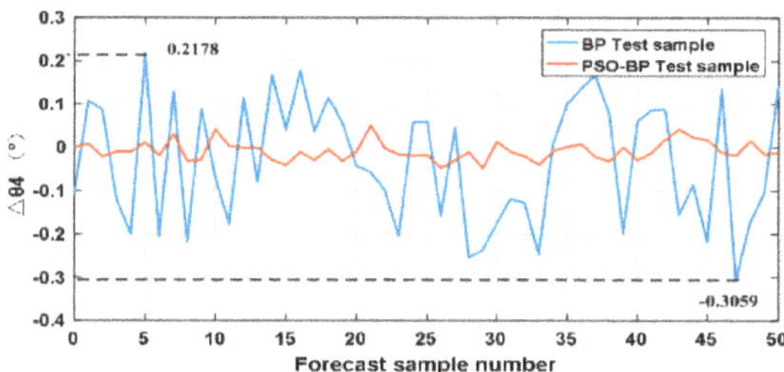

Figure 10. Error contrast of test set's θ_4.

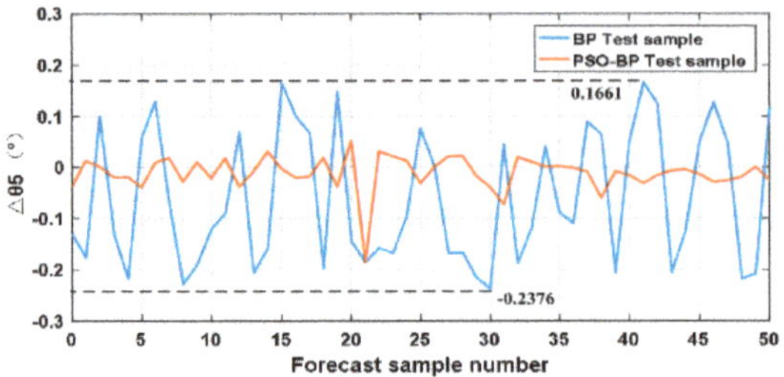

Figure 11. Error contrast of test set's θ_5.

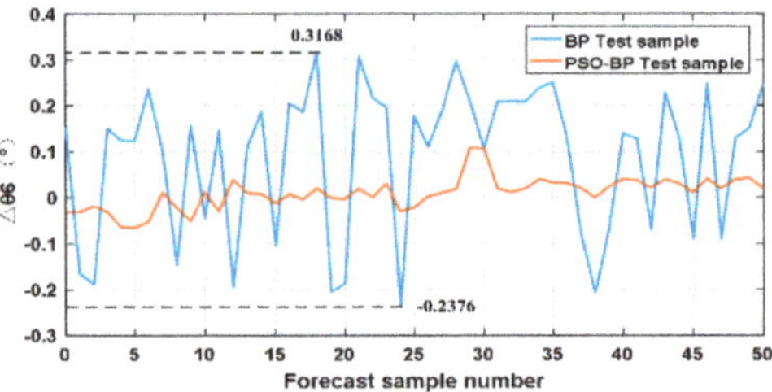

Figure 12. Error contrast of test set's θ_6.

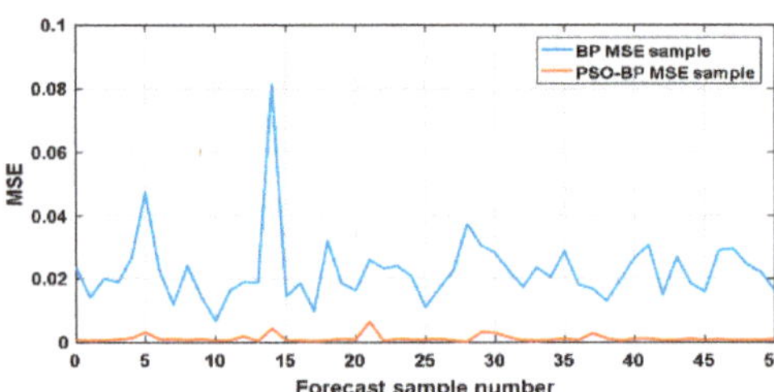

Figure 13. Mean square error (MSE) contrast of test set's θ.

Table 2. Error range test set and mean square error.

Joint Angles and MSE (Mean Square Error)	BP (Back Propagation) Neural Network	PSO (Particle Swarm Optimization)-BP Neural Network
$\Delta\theta_1$	−0.2005–0.4130	−0.0534–0.065
$\Delta\theta_2$	−0.2100–0.1954	−0.1199–0.0605
$\Delta\theta_3$	−0.2178–0.2046	−0.0672–0.0704
$\Delta\theta_4$	−0.3059–0.2178	−0.0486–0.0508
$\Delta\theta_5$	−0.2376–0.1661	−0.1859–0.0526
$\Delta\theta_6$	−0.2376–0.3168	−0.0658–0.1079
MSE	6.72×10^{-3}–8.12×10^{-2}	2.39×10^{-4}–6.42×10^{-3}

The experimental results show that, after the machine arm BP optimization PSO neural network obtained by the inverse kinematics of the robot is compared with the BP neural network joint angle error to a high level, the value of the mean square error (MSE) is also enhanced by an order of magnitude. The BP neural network based on PSO optimization is used to solve the inverse kinematics of the manipulator and to verify its effectiveness.

4. Application of the Puncture Robot

The technology of the puncture robot is new in many research fields, such as medicine, mechanics, imaging, robotics, and so on. This part is based on CT image processing and three-dimensional reconstruction for the guidance and the six degrees of freedom for robot precision positioning technology research; the final step towards completion of medical minimally-invasive surgery with precise positioning control (the principle is shown in Figure 1). The brief process of the experiment is as follows:

(1) Firstly, a piece of porcine spine meat is affixed to some fixed marked points (Figure 14a), and then the DICOM (Digital Imaging and Communications in Medicine) image is obtained by CT scanning.

(2) The CT image is copied to the image processing computer, and the 3D reconstruction is performed on the computer (Figure 14b). Then the coordinates of the marked points in the 3D model are selected.

(3) We can then use, as a coordinate measuring arm, the mechanical measurement of spinal meat marked points on the robot with six degrees of freedom, combined with the coordinates of the marked points in the CT 3D model. Through the space registration conversion, we can obtain the position relationship between the spine meat in the CT 3D model coordinate system, and the robot coordinate system.

(4) The coordinates of the target points and the coordinates of the needle insertion points are selected in the CT 3D model, which determines the route of the needle insertion. The position and rotation in the robot coordinate system are obtained according to the needle insertion route, such as the green line in Figure 14b.

(5) Through the POS-BP neural network algorithm to train the data, we obtain the corresponding joint angles of the robot's end puncture needle guide pipe when required to reach the specified position and rotation. After the command is sent, the robot moves to the position and ensures its rotation (Figure 14c).

(6) The doctor confirms the positioning of the robot and inserts the needle into the porcine spine (Figure 14d). Once again, through the CT scan of the porcine spine meat, and the reconstruction of the three-dimensional image, a comparison of the needle in the three-dimensional model and the surgical planning of the needle route is conducted, so as to determine the success of the puncture operation.

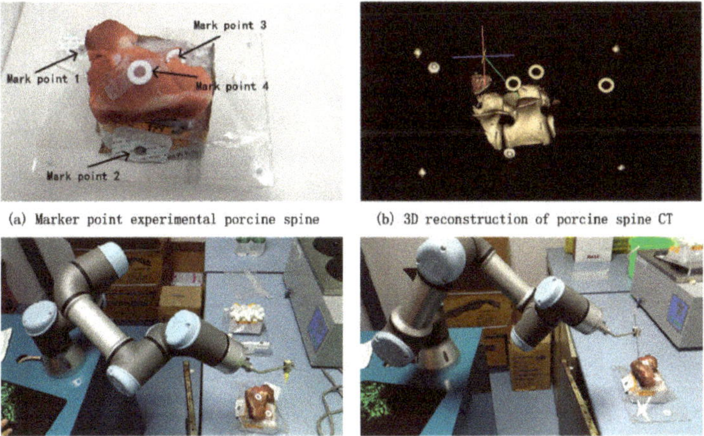

Figure 14. UR3 robot puncture process chart.

In this part, we researched the puncture surgery experiment built on porcine spine meat, and verified the PSO-optimized BP neural network algorithm for a robot based on the inverse kinematics problem. The experiment has achieved good results, the accurate positioning of the UR3 robot was ensured within 0.1 mm, and the precision of the robot joint angle was ensured at 0.01 degrees. The results fully met the requirements of the puncture positioning accuracy requirements within 1 mm.

5. Conclusions

In this paper, a hybrid approach has been presented which combines the particle swarm optimization algorithm and BP neural network algorithms to solve the inverse kinematics problem for a six degree manipulator based on end-effector error minimization. The proposed approach combines the characteristics of particle swarm optimization to overcome some inherent defects of BP neural network, and improve the convergence precision of the BP neural network, the convergence speed, and generalization ability. It is difficult to obtain high-precision inverse kinematics solutions because the six DOF robot arm is complex, and there are many problems such as large computational complexity, no guarantee of accuracy, long computation time and so on. In this case, the proposed method will be particularly useful to improve the precision of the result obtained from the PSO-optimized BP neural network algorithm. The joint error and mean square error performance of the PSO-optimized BP neural network are improved by an order of magnitude over the pure BP neural network.

The innovative application of this technology is used in the precise positioning technology of medical puncture surgery; to perform the traditional puncture operation in low precision; to solve poor reliability; to reduce the pain of the patient; to tackle the large labor intensity of the doctor; to eliminate over reliance on the experience of doctors and to address other issues. After obtaining a large number of experimental data, the accuracy can meet the requirements of precise positioning in medical needle surgery, and it has important theoretical and practical value in the research of the key technologies of precise puncturing robots.

Acknowledgments: The authors would like to thank all the reviews for their constructive comments. This work was supported by University of Science and Technology of China, Southwest University of Science and Technology, Key Laboratory of special robot technology of Jiangsu Province for Hohai University, the Institute of Intelligent Manufacturing Technology for Jiangsu Industrial Technology Research Institute.

Author Contributions: Guanwu Jiang and Minzhou Luo conceived and designed the experiments; Keqiang Bai performed the experments; Guanwu Jiang and Saixuan Chen analyzed the data. Guanwu Jiang wrote the paper.

Conflicts of Interest: The authors declare no conflicts of interest.

References

1. Yang, G.L.; Kurbanhusen, S.; Yeo, S.H.; Lin, W.; Bin, W. Kinematic design of an anthropomimetic 7-DOF cable-driven robotic arm. *Front. Mech. Eng.* **2011**, *6*, 45–60. [CrossRef]
2. Köker, R. A genetic algorithm approach to a neural-network-based inverse kinematics solution of robotic manipulators based on error minimization. *Inf. Sci.* **2013**, *222*, 528–543. [CrossRef]
3. Bingul, Z.; Ertunc, H.M.; Oysu, C. *Applying Neural Network to Inverse Kinematics Problem for 6R Robot Manipulator with Offset Wrist. Coimbra, Portugal, 2005*; Springer: Vienna, Austria, 2005; pp. 112–115.
4. Tabandeh, S.; Clark, C.; Melek, W. A genetic algorithm approach to solve for multiple solutions of inverse kinematics using adaptive niching and clustering. *Comput. Sci. Softw. Eng.* **2006**, *63*, 1815–1822.
5. Hasan, A.T.; Ismail, N.; Hamouda, A.M.S.; Aris, I.; Marhaban, M.H.; Al-Assadi, H.M.A.A. Artificial neural network-based kinematics Jacobian solution for serial manipulator passing through singular configurations. *Adv. Eng. Softw.* **2010**, *41*, 359–367. [CrossRef]
6. Boudjelaba, K.; Ros, F.; Chikouche, D. Potential of Particle Swarm Optimization and Genetic Algorithms for FIR Filter Design. *Circuits Syst. Signal Process.* **2014**, *33*, 3195–3222. [CrossRef]
7. Kalra, P.; Mahapatra, P.B.; Aggarwal, D.K. An evolutionary approach for solving the multimodal inverse kinematics problem of industrial robots. *Mech. Mach. Theory* **2006**, *41*, 1213–1229. [CrossRef]

8. Ayyıldız, M.; Cetinkaya, K. Comparison of four different heuristic optimization algorithms for the inverse kinematics solution of a real 4-DOF serial robot manipulator. *Neural Comput. Appl.* **2016**, *27*, 825–836. [CrossRef]
9. Bernard, C.; Kang, H.; Singh, S.K.; Wen, J.T. Robotic system for collaborative control in minimally invasive surgery. *Ind. Robot* **1999**, *26*, 476–484. [CrossRef]
10. Kebria, P.M.; Al-wais, S.; Abdi, H.; Nahavandi, S. Kinematic and Dynamic Modelling of UR5 Manipulator. In Proceedings of the 2016 IEEE International Conference on Systems, Man, and Cybernetics, Budapest, Hungary, 9–12 October 2016; IEEE: Washington, DC, USA, 2017; pp. 4229–4234.
11. Hasan, A.T.; Hamouda, A.M.S.; Ismail, N.; Al-Assadi, H.M.A.A. An adaptive-learning algorithm to solve the inverse kinematics problem of a 6 DOF serial robot manipulator. *Adv. Eng. Softw.* **2006**, *37*, 432–438. [CrossRef]
12. Rubio, J.J.; Aquino, V.; Figueroa, M. Inverse kinematics of a mobile robot. *Neural Comput. Appl.* **2013**, *23*, 187–194. [CrossRef]
13. Duguleana, M.; Barbuceanu, F.G.; Teirelbar, A.; Mogan, G. Obstacle avoidance of redundant manipulators using neural networks based reinforcement learning. *Robot. Comput. Intergr. Manuf.* **2012**, *28*, 132–146. [CrossRef]
14. Liu, Y.; Wang, D.Q.; Sun, J.; Chang, L.; Ma, C.X.; Ge, Y.J.; Gao, L.F. Geometric Approach for Inverse Kinematics Analysis of 6-Dof Serial Robot. In Proceedings of the 2015 IEEE International Conference on Information and Automation, Lijiang, China, 8–10 August 2015; IEEE: Washington, DC, USA, 2015; pp. 852–855.
15. Ma, C.; Zhang, Y.; Cheng, J.; Wang, B.; Zhao, Q.J. Inverse Kinematics Solution for 6R Serial Manipulator Based on RBF Neural Network. In Proceedings of the 2016 International Conference on Advanced Mechatronic Systems, Melbourne, VIC, Australia, 30 November–3 December 2016; IEEE: Washington, DC, USA, 2016; pp. 350–355.
16. Mayorga, R.V.; Sanongboon, P. Inverse kinematics and geometrically bounded singularities prevention of redundant manipulators An Artificial Neural Network approach. *Robot. Auton. Syst.* **2005**, *53*, 164–176. [CrossRef]
17. Kennedy, J.; Eberhart, R. Particle Swarm Optimization. In Proceedings of the IEEE International Conference on Neural Networks IV, Perth, Australia, 27 November–1 December; IEEE: Piscataway, NJ, USA, 1995; pp. 1942–1948.
18. Chen, C.C.; Li, J.S.; Luo, J.; Xie, S.R.; Li, H.Y. Seeker optimization algorithm for optimal control of manipulator. *Ind. Robot* **2016**, *43*, 677–686. [CrossRef]
19. Si, L.; Wang, Z.W.; Liu, Z.; Liu, X.H.; Tan, C.; Xu, R.X. Health Condition Evaluation for a Shearer through the Integration of a Fuzzy Neural Network and Improved Particle Swarm Optimization Algorithm. *Appl. Sci.-Basel* **2016**, *6*, 171. [CrossRef]
20. Falconi, R.; Grandi, R.; Melchiorri, C. Inverse Kinematics of Serial Manipulators in Cluttered Environments using a new Paradigm of Particle Swarm Optimization. *IFAC Proc. Vol.* **2014**, *47*, 8475–8480. [CrossRef]
21. Kuo, P.H.; Liu, G.H.; Ho, Y.F.; Li, T.H.S. PSO and Neural Network based Intelligent Posture Calibration Method for Robot Arm. In Proceedings of the 2016 IEEE International Conference Systems, Man, and Cybernetics (SMC), Budapest, Hungary, 9–12 October 2016; IEEE: Washington, DC, USA, 2016; pp. 003095–003100.

Article

Bio-Inspired Real-Time Prediction of Human Locomotion for Exoskeletal Robot Control

Pu Duan [1], Shilei Li [2], Zhuoping Duan [1,*] and Yawen Chen [3]

[1] The State Key Laboratory of Explosion Science and Technology, Beijing Institute of Technology, Beijing 100081, China; duanpu@bit.edu.cn
[2] School of Mechatronics Engineering and Automation, Harbin Institute of Technology Shenzhen Graduate School, Shenzhen 518055, China; lishilei@xeno.com
[3] Xeno Dynamics Co. Ltd., Shenzhen 518055, China; chenyawen@xeno.com
* Correspondence: duanzp@bit.edu.cn; Tel.: +86-138-011-661-79

Received: 7 September 2017; Accepted: 31 October 2017; Published: 2 November 2017

Abstract: Human motion detection is of fundamental importance for control of human–robot coupled systems such as exoskeletal robots. Inertial measurement units have been widely used for this purpose, although delay is a major challenge for inertial measurement unit-based motion capture systems. In this paper, we use previous and current inertial measurement unit readings to predict human locomotion based on their kinematic properties. Human locomotion is a synergetic process of the musculoskeletal system characterized by smoothness, high nonlinearity, and quasi-periodicity. Takens' reconstruction method can well characterize quasi-periodicity and nonlinear systems. With Takens' reconstruction framework, we developed improving methods, including Gaussian coefficient weighting and offset correction (which is based on the smoothness of human locomotion), Kalman fusion with complementary joint data prediction and united source of historical embedding generation (which is synergy-inspired), and Kalman fusion with the Newton-based method with a velocity and acceleration high-gain observer (also based on smoothness). After thorough analysis of the parameters and the effect of these improving techniques, we propose a novel prediction method that possesses the combined advantages of parameter robustness, high accuracy, trajectory smoothness, zero dead time, and adaptability to irregularities. The proposed methods were tested and validated by experiments, and the real-time applicability in a human locomotion capture system was also validated.

Keywords: real-time prediction; Takens' reconstruction method; human locomotion; synergy; data fusion

---

1. Introduction

Human motion detection is of fundamental importance for control of human–robot coupled systems such as exoskeletal robots, and for many other applications. In recent years, inertial measurement unit (IMU) technologies have quickly evolved, and complex mechanisms for tracking the operator's movement [1,2] have gradually been replaced by motion capture systems comprising IMUs [3–7]. IMU-based motion capture systems have unique advantages, such as multidimensional information richness, compactness, and light weight. However, compared with biosensors for which signals appears earlier than in human body motion [8–10], human motion capture systems bear the apparent disadvantage of asynchronization with execution delay, which is a critical challenge for applications in exoskeletal robot control.

Human locomotion is a nonlinear quasi-periodic process with chaos [11] resulting from latent human musculoskeletal dynamics. During human walking, the two lower limbs operate synergistically [12–15]. Both the quasi-periodicity and the synergy lead to the development of several efficient prediction methods. Firstly, by drawing evidence from the lower limb synergy, echo control has long been used in powered prosthesis for generating joint trajectories of one limb by a phase shift of the opposite limb [16]. To apply echo control in prediction, the time shift needs to be correctly identified. Secondly, the complementary limb motion estimation (CLME) method [8] improves the echo control method by assuming a coupling relationship between the two lower limbs. However, its performance is not always consistent due to a significant influence from the underlying regression method [17]. Thirdly, the aforementioned quasi-periodic nonlinear behaviour of walking has led to Takens' reconstruction method, which is closely related to Takens' reconstruction theorem. It is essentially a nonlinear time series dynamic factor analysis method for predicting joint trajectories [18,19]. However, it requires historical system data to achieve a better performance which brings in inevitable dead time. There are also some other prediction methods, such as the autoregressive integrated model method, autoregressive integrated moving average method, autoregressive moving average model method [20], and modified recursive least square methods [21], which are also time series analysis methods but with undesired computational expense, dead time, or complex parameter tuning processes. Besides, the consistency and smoothness of human joint motion should not be ignored as human joints possess apparent biological damping [22–24] which damps out the high frequency vibration. Such observations indicate that the traditional Newton's method, in the short term for smoothing or correction techniques, might be beneficial. For Newton's method, estimation of the velocity and acceleration with noise rejecting ability is needed. High-gain observers could be exploited, which are usually used in robot or automation applications [25–27].

In this work, we consider, without loss of generality, a series of prediction methods and its improving techniques utilizing the biological features of human locomotion. Based on progressive analysis, we introduce a novel method which fuses the advantages of different approaches to achieve an even better performance verified by a daily motion dataset of a normal subject, including irregular motion. The method is also applicable to real-time embedded systems.

This paper is organized as follows. Section 2 introduces the IMU-based motion capture system. Section 3 describes different prediction methods, assessment measures, and the used dataset. Section 4 evaluates the influence of several common parameters and the prediction performance with different prediction methods by simulation, elaborates their characteristics, and then proposes our best method. Also, the real-time applicability of the proposed algorithm is demonstrated. Section 5 concludes the paper.

2. The IMU-Based Motion Capture System

A motion capture system using IMUs has been developed by Xeno Dynamics Co., Ltd. (Shenzhen, Guangdong, China), as shown in Figure 1. In each IMU, the gyroscope, accelerometer and magnetometer data are fused with a gradient descent algorithm with magnetic distortion compensation [28]. The compact rigid container with elastic straps could be easily attached to human lower limbs. The data collected by serial communication could either be stored and processed in a standalone central control device as shown in Figure 1b or sent into the developing exoskeletal robots with a sampling frequency of 50 Hz. The test subject can move freely when wearing the motion capture system. In the experiment designed in this research, only the left hip, right hip, left knee, and right knee joint data in the human sagittal plane were collected, as illustrated in Figure 1b.

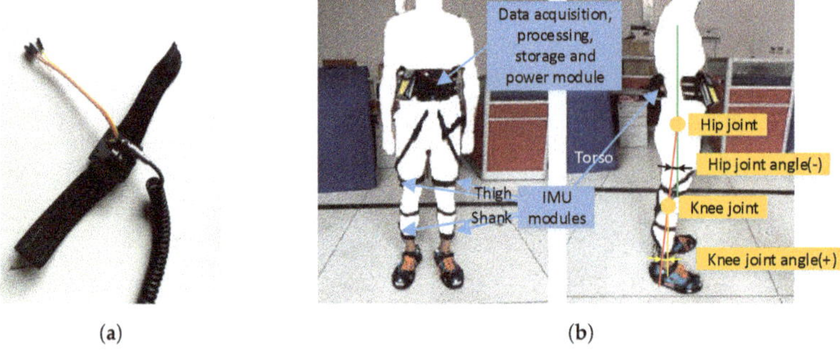

Figure 1. The inertial measurement unit (IMU)-based lower limb motion capture system. (**a**) One IMU module. (**b**) The system worn on a testing subject and the illustration of joint angles in the human sagittal plane collected in this research.

3. Methods

The prediction methods starts with Takens' reconstruction prediction and its enhancement with Gaussian coefficient weighting, offset correction, and a feedback scheme. Synergy of a complementary joint is exploited with two methods producing the synergy-enhanced Takens' reconstruction predictions. To further improve the prediction performance during the short term, and adaptability to irregularities, the classic Newton's method with a high-gain observer is brought in and some new methods appear.

3.1. Takens' Reconstruction Prediction and Modifications Based on Motion Smoothness

Takens' reconstruction method is correct for predicting the joint trajectories of human lower limbs, which present quasi-periodicity and nonlinearity. According to Takens' reconstruction theorem, it is possible to reconstruct the dynamics of a nonlinear system based on its past measurements. For this purpose, a current embedding vector is defined:

$$D(t) = \begin{bmatrix} y(t) & y(t-\Delta t) & \cdots & y(t-(p-1)\Delta t) \end{bmatrix}^T, \quad (1)$$

where $y(t)$ stands for the current state which might be position, velocity, acceleration, etc., t is the current sampling instant, p is the embedding vector length, and the sampling constant is Δt. Takens' original reconstruction algorithm proposed by Herrmann [19] is illustrated in Figure 2:

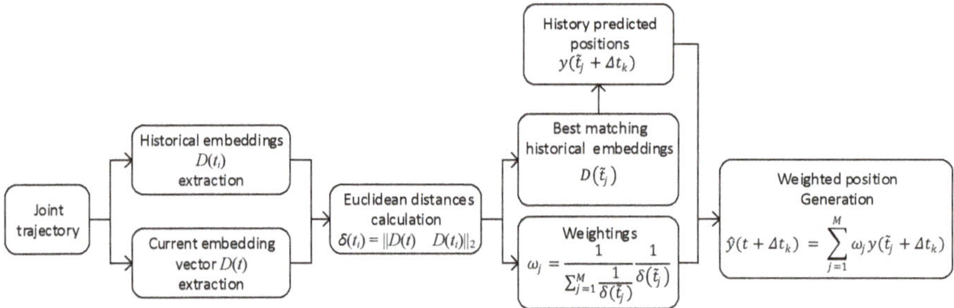

Figure 2. Takens' basic reconstruction method workflow.

At each sampling instant t with $t \geq l\Delta t$, an embedding vector $D(t)$ is acquired. The current embedding vector $D(t)$ is then compared to all collected historical embeddings $D(t_i)$ with $l\Delta t \leq t_i < t$ by determining the Euclidean distances

$$\delta(i) = \sqrt{\sum_{x=1}^{p}(y(t_i - \Delta t_{x-1}) - y(t - \Delta t_{x-1}))^2}, \tag{2}$$

In Equation (2) i is the index of the ith historical embedding $D(t_i)$ and $1 \leq i \leq l$. l is also called the history data length. As $D(t_i)$ is a vector composed of a series of states, the first component of $D(t_i)$ is labeled as $y(t_i)$ and the components are defined as $y(t_i - \Delta t_{x-1})$ where $\Delta t_{x-1} = (x-1)\Delta t$ and $1 \leq x \leq p$.

The M best matching embeddings $D(\tilde{t}_1), D(\tilde{t}_2), \ldots, D(\tilde{t}_M)$ from the sampling instants $\tilde{t}_j$ with $1 \leq j \leq M$ are used to calculate a k-step prediction $\hat{y}(t + \Delta t_k|t)$:

$$\hat{y}(t + \Delta t_k|t) = \sum_{j=1}^{M} \omega_j y(\tilde{t}_j + \Delta t_k), \tag{3}$$

with the weightings

$$\omega_j = \frac{1}{N}\frac{1}{\delta(\tilde{t}_j)}, \tag{4}$$

and

$$N = \sum_{j=1}^{M} \frac{1}{\delta(\tilde{t}_j)}, \tag{5}$$

In Equation (3) $\Delta t_k = k\Delta t$. When $k = 1$, the subscript is dropped as Δt. k is also called prediction horizon.

Although Takens' reconstruction method bears many practical advantages such as easy parameter calibration, result smoothness, accuracy and so on, its improvement is still quite possible. The natural smoothness of motion brought by the human body indicates whether high frequency jumps and sharp spikes in trajectories were initially predicted, as correction and smoothing techniques would be used to produce a more realistic result.

First, we assumed that the $y(t_i - \Delta t_{x-1})$ in $D(t_i)$ closer to $y(t_i)$ had more of an influence on Euclidean distance determination. Thus, the Euclidean distances was calculated with the addition of Gaussian coefficients $G(x)$:

$$\delta(i) = \sqrt{\sum_{x=1}^{p}((y(t_i - \Delta t_{x-1}) - y(t - \Delta t_{x-1}))G(x))^2}, \tag{6}$$

and

$$G(x) = e^{-\frac{1}{2}(\frac{x-1}{\frac{Q}{p}})^2}, \tag{7}$$

where Q is constant and p is the embedding vector length. As the prediction result was not very sensitive to the value of Q, it was chosen to be 4 in this research, which was justified by several items of gait data.

Second, the offset between $y(\tilde{t}_j)$ in a best matching embedding of $D(\tilde{t}_j)$ and $y(t)$ in the current embedding vector was decided during each vector match by

$$\Delta y_j = y(\tilde{t}_j) - y(t), \tag{8}$$

where $y(\tilde{t}_j)$ means the past observation of the current state $y(t)$ in the jth best matching vector $D(\tilde{t}_j)$. The obtained Δy_j is used for correction during the weighting procedure with the equation

$$\hat{y}(t+\Delta t_k|t) = \sum_{j=1}^{M} \omega_j(y(\tilde{t}_j+\Delta t_k) - \Delta y_j), \quad (9)$$

Third, the feedback scheme is integrated and tested as illustrated in Figure 3.

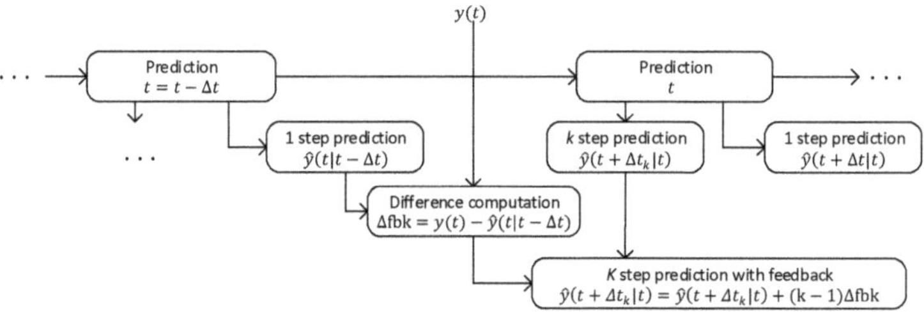

Figure 3. The feedback scheme integrated into Takens' reconstruction-based methods.

At time t, the 1 step prediction $\hat{y}(t|t-\Delta t)$ of the current time is achieved at $t-\Delta t$. The difference between the current position $y(t)$ and $\hat{y}(t|t-\Delta t)$, Δfbk, is added to the current k step prediction $\hat{y}(t+\Delta t_k|t)$ with a factor of $k-1$. At the same time a new 1 step prediction is completed as $\hat{y}(t+\Delta t|t)$, which will be used in the feedback determination in the next time step.

3.2. Takens' Reconstruction Method's Synergy Improvement

The synergy feature of the lower limb joints could be used in a simple way for Takens' reconstruction-based methods, making use of the complementary joint motion data of the other limb. Basically, considering the complementary joint data as another source of history or taking it as an indirect measurement of the original joint motion should be able to provide help. This results in two distinct methods:

1. While searching for the best matching vectors, the embedding vectors are generated with the united source of the two joints' history data. The detailed process is shown in Figure 4.

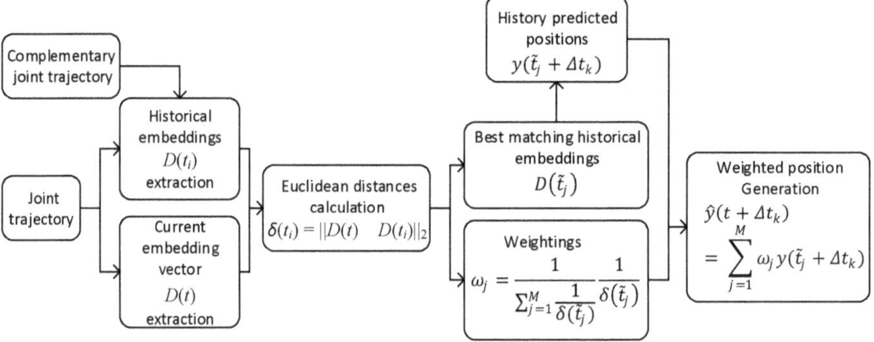

Figure 4. Synergy improvement method (1) of Takens' reconstruction-based methods.

2. Predict k step position ahead as usual and get $\hat{y}(t+\Delta t_k|t)$. Then take the complementary joint as the source of embedding vector generation and predict $\hat{y}_c(t+\Delta t_k|t)$. Finally, fuse $\hat{y}(t+\Delta t_k|t)$ and

$\hat{y}_c(t+\Delta t_k|t)$ with the Kalman filter and obtain $\hat{y}_{fuse}(t+\Delta t_k|t)$. The detailed process is shown in Figure 5.

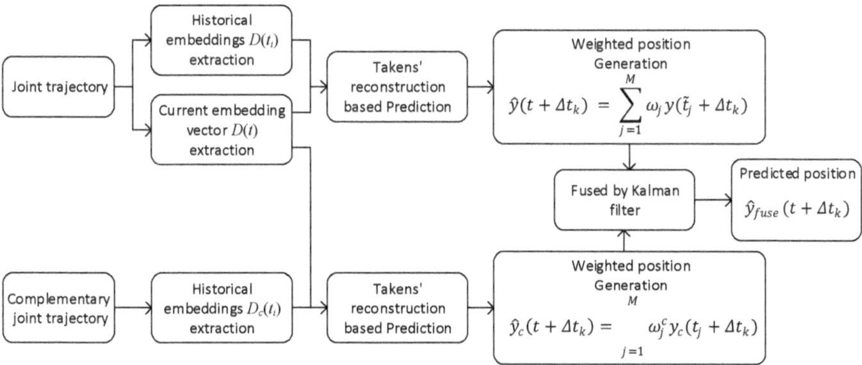

Figure 5. Synergy improvement method (2) of Takens' reconstruction-based methods.

The two means represented different directions. They have limited theoretical foundation as the synergy mechanism is still under research and could not be commented by comparative analysis without being applied to assessment on real data.

3.3. Takens' Reconstruction Adaptability Enhancement with Newton's Method

Motion smoothness also means the assumption of local linearity is appropriate, thus the traditional Newton's method in the short term might be beneficial. A high-gain observer [25–27] is formulated to obtain the required velocity and acceleration for a second-order Newton's method.

$$y_{NHgo}(t+\Delta t_k|t) = y(t) + \hat{\dot{y}}(t)\Delta t_{k+1} + \frac{1}{2}\hat{\ddot{y}}(t)\Delta t_{k+1}^2, \tag{10}$$

where $\hat{\dot{y}}$ and $\hat{\ddot{y}}$ are computed with

$$\hat{\dot{y}} = \frac{\epsilon s}{\frac{\epsilon^3 s^2}{\alpha_2} + \frac{\alpha_1 \epsilon^2 s}{\alpha_2} + \epsilon} y, \tag{11}$$

and

$$\hat{\ddot{y}} = \frac{\epsilon s}{\frac{\epsilon^3 s^2}{\alpha_2} + \frac{\alpha_1 \epsilon^2 s}{\alpha_2} + \epsilon} \hat{\dot{y}}, \tag{12}$$

where $\alpha_1 = 100$, $\alpha_2 = 80$ and ϵ = sampling constant which was 0.02 s in this context.

As a high-gain observer could be taken as a non-linear filter, α_1 and α_2 were chosen so that the amplitude of the data after the filter did not change and the phase delay in time domain was constant. It usually brought in a delay of about one sampling cycle. Equation (10) roughly compensates the delay by taking a one step longer prediction at $k+1$ as the k step result. Finally, the Newton-based and Takens' reconstruction-based methods could be fused to achieve some methods with unique adaptability enhancement.

3.4. Methods Summary

Summarizing the above methods, besides the fundamental Takens' reconstruction and Newton's method, there are 12 new prediction methods that are introduced and analysed which are listed in Table 1.

Table 1. Prediction methods in this article and their abbreviations.

Abbreviation	Method Description	Proposed by Author
Takens'	Takens' reconstruction method	No
NHgo	Newton's method with velocity and acceleration observed by high gain observer	Yes
Gw	Takens' reconstruction with Gaussian weightings	Yes
GwOc	Takens' reconstruction with Gaussian weightings and offset correction scheme	Yes
GwOcFb	Takens' reconstruction with Gaussian weightings, offset correction and feedback scheme	Yes
GwOcFb-KC	GwOcFb predicted result Kalman fusion with the complementary joint GwOcFb predicted	Yes
GwOcFb-US	GwOcFb with coupled joints' united source of historical embedding generation	Yes
GwOc-KC	GwOc predicted result Kalman fusion with the complementary joint GwOc-KC predicted	Yes
GwOc-US	GwOc with coupled joints' united source of historical embedding generation	Yes
GwOcFb-US-NHgo	GwOcFb-US predicted result Kalman fusion with the NHgo predicted	Yes
GwOc-US-NHgo	GwOcFb predicted result Kalman fusion with the NHgo predicted	Yes
GwOcFb-KC-NHgo	GwOcFb-KC predicted result Kalman fusion with the NHgo predicted	Yes
GwOc-KC-NHgo	GwOc-KC predicted result Kalman fusion with the NHgo predicted	Yes

The relationship between different prediction methods with different improving approaches is illustrated in Figure 6.

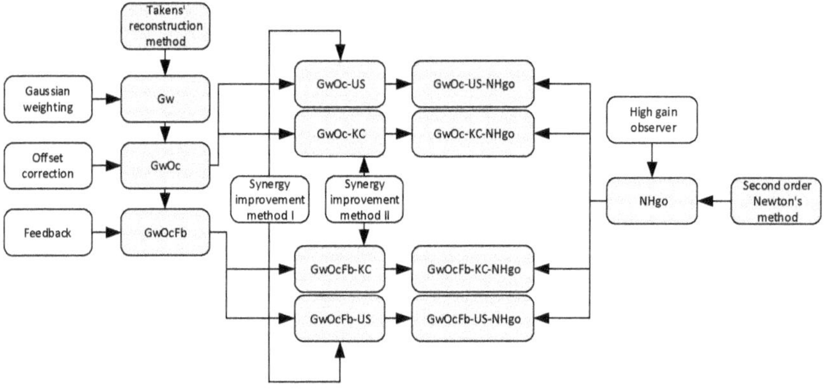

Figure 6. The relationship map of different methods and improving approaches.

3.5. Assessment Data

The dataset used to assess the performance of different prediction methods is 208.5 s long with a sampling rate of 50 Hz, including the joint angles of the knees and hips of a test subject in a sagittal plane. It is comprised of the following continuous motion process segmented and highlighted in Figure 7: 5 stand–sits, 5 steps walking, 2 left turns, 5 steps walking, 2 left turns, 5 steps walking, stepping onto a treadmill, 1 km/h, 2 km/h and 3 km/h walking on a treadmill, stopping and leaving the treadmill, 5 steps walking and sitting down. In order to assess the adaptability of different methods the following segments were defined as irregularities: stand–walk transition, turns, stepping on a treadmill, stopping and leaving the treadmill, and stand–sit.

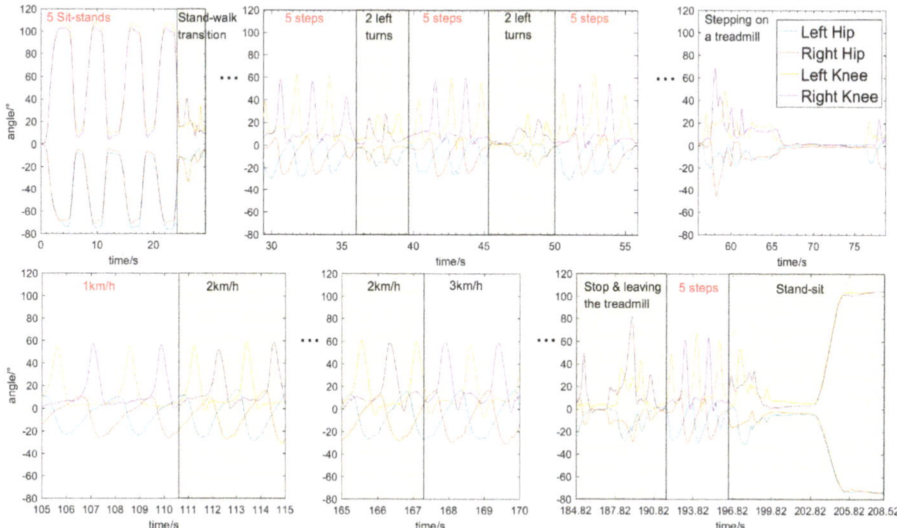

Figure 7. Dataset for prediction method assessment (the joint angles of the knees and hips of a test subject in a sagittal plane collected by the inertial measurement unit-based motion capture system).

3.6. Assessment Measures

To quantitatively evaluate the algorithms' prediction performance, the prediction ratio (*PR*) representing prediction errors, and the smoothness factor (*SF*) measuring predicted trajectory smoothness are calculated by the following equations [19]:

$$PR(y(t), e(t)) = 1 - \frac{RMS(e_{PR}(t))}{RMS(y(t))} \times 100\%, \tag{13}$$

$$e_{PR}(t) = \hat{y}(t|t - \Delta t_k) - y(t), \tag{14}$$

where $\hat{y}(t|t - \Delta t_k)$ represents the predicted current position $\hat{y}(t)$ at $t - \Delta t_k$, which is obtained by shifting the predicted results with Δt_k and

$$SF(\hat{y}(t)) = \frac{1}{t_{End}(max(\hat{y}(t)) - min(\hat{y}(t))} \int_0^{t_{End}} |f(t) - \hat{y}(t)|, \tag{15}$$

where t_{End} is the time when prediction ends and $f(t)$ is the filtered result of $y(t)$ by a zero-phase filter with a cut-off frequency of 5 Hz.

A higher *PR* value means higher accuracy, while a lower *SF* value means smoother motion. In order to provide an overall relative comparison among different methods, *PR* and *SF* are combined into an overall performance index (*OPI*), which is computed with

$$OPI = \alpha PR(i)/max(PR(i)) + (1 - \alpha)min(SF(i))/SF(i), \tag{16}$$

where i represents the index of a specific method involved in the comparison. It should be emphasized that such an index is not an absolute criterion but a relative one. The preference over prediction accuracy and smoothness could be modified by the coefficient α.

In addition, the adaptability of prediction methods to irregularities happening from time to time is very essential. Such capability could be assessed by isolating the irregularities, including the beginning and termination of walking and the transition between different motion speed and modes.

4. Implementation and Assessment

In practical applications, prediction horizon may change in real-time, so a good algorithm does not only perform well at a fixed prediction horizon, but also over a certain range. Before the performance evaluation over different prediction horizons, the common parameters used in Takens' reconstruction-based algorithms must be reasonably selected.

4.1. Parameter Selection for Takens' Reconstruction-Based Methods

For Takens' reconstruction-based methods common parameters were evaluated, including the history data length, embedding vector length, and the number of weighted best matching vectors. A group of basic parameters are listed in Table 2. During the evaluation the evaluated parameters were taken as variables while others were fixed.

Table 2. Basic parameters for parameter evaluation.

Parameter Name	Symbol	Value	Variation Range	Variable
History data length	l	360	200–400	Yes
Embedding vector length	p	20	6–20	Yes
Number of weighted best matching vectors	M	5	1–8	Yes
Prediction horizon	k	5	/	No

Only the left hip data were shown in the following paragraphs because the four joints results exhibited the same varying patterns which meant the ruling laws were the same for different methods.

Theoretically, a longer history is beneficial and Figure 8 gave the proof with continuous escalation of PRs. A steep rise ended at the value of 360, which was the chosen value in Table 2. The history data length had limited influence on SFs after 360, because a longer history could not generate better historical embedding vectors.

The length of embedding vectors represents the orders of the reconstructed system. Excessive higher order might cause overfitting, which reduces the prediction accuracy and increases unnecessary computational expense. Gaussian weightings mitigate the problem by reducing the influence of the history data far away and the offset correction and feedback scheme also provide powerful fixation. Figure 9 proved the analysis. For the SF values, the expending of embedding vectors could improve them more or less for all methods. Thus, considering that a length of 20 is not an burden to the embedded system and it has limited influence on most of the improved methods, the embedded vector length was chosen to be 20.

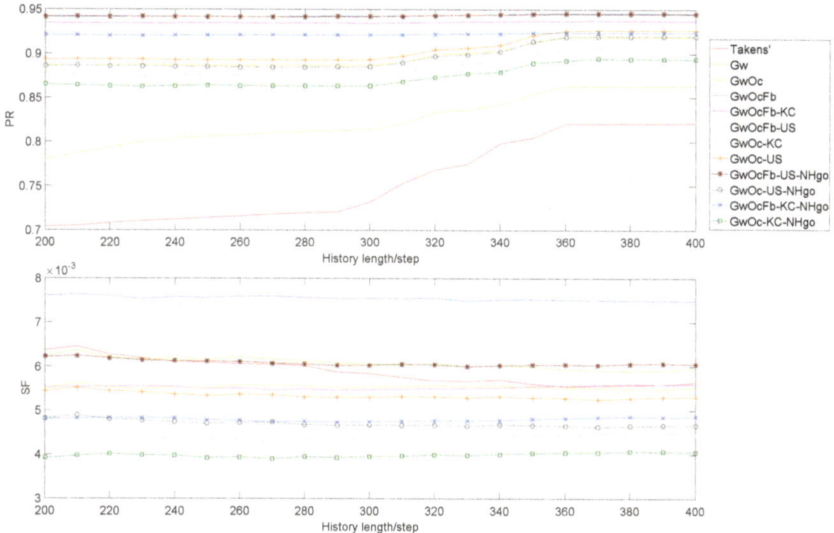

Figure 8. Influence of history data length used in embedding vector generation (left hip joint data in the sagittal plane). *SF*: smoothness factor; *PR*: prediction ratio.

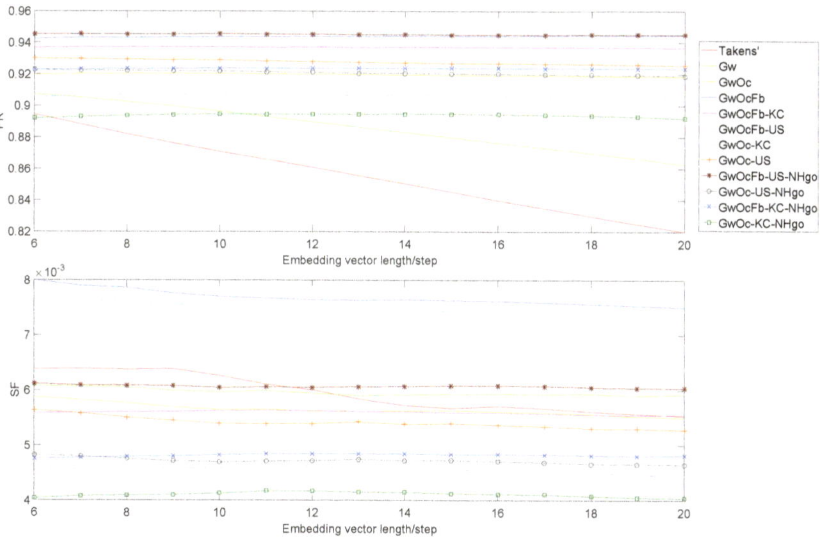

Figure 9. Influence of embedding vector length (left hip joint data in the sagittal plane).

As shown in Figure 10, when the weighted best matching vector number increased, the *PR*s of different methods rose to the highest at 5 and the trend then descended afterwards. Meanwhile, for the *FS*s after 5, they also entered an slow falling state. This was reasonable because the matched vector quality degraded when the number increased to a certain level, and their weighting values fell quickly. At the same time, the increasing number of weighted history embedding vectors worked as a smoother by averaging down the oscillations.

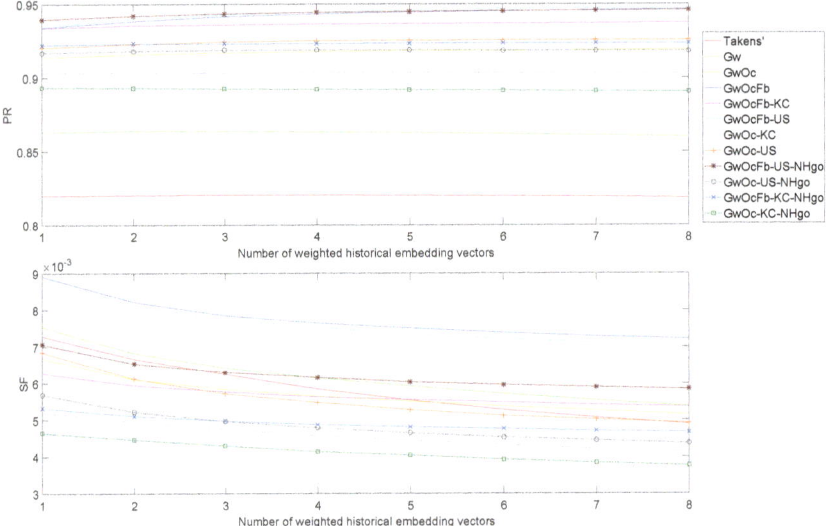

Figure 10. Influence of weighted historical embedding vector number (left hip joint data in the sagittal plane).

Surprisingly, most newly-introduced Takens' reconstruction-based methods exhibited robust performance with parameter variation. Thus, the parameter robustness of the newly introduced methods was discovered.

Summarizing the previous analysis, the basic parameters are selected with enough confidence, and could be applied to algorithm performance assessment.

4.2. Results and Assessment

The prediction methods listed in Table 1 and evaluated (Figure 7) demonstrated the dataset with respect to an extending prediction horizon from 1 to 20, which was the requirement of exoskeletal robot sensing for control. For OPI calculation, α in Equation (16) was chosen as 0.8, which emphasized accuracy over smoothness. The evaluations were performed with the simulation of real-time operation by pushing the data one by one in a time sequence into the last position of a data pool, which was fair for all the methods.

Studying the results by comparison the pros and cons of different methods and improving techniques were clear:

1. It can seen more clearly in Figure 11 that when compared with the original Takens' reconstruction method, the PR and SF values of Gw and GwOC had significant improvements, especially during the lower prediction horizon range, which meant the Gaussian weighting and offset correction were effective and the assumption of smoothness and consistency of human motion was reasonable.
2. Also, it can be inferred by comparing GwOc, GwOc-KC, and GwOc-US that for prediction accuracy the synergy improvement of united source of history embedding vector generation had a positive effect on smoothness improvement and a negative effect on accuracy. In contrast, the improvement of Kalman fusion with complementary joint prediction played an opposite role. Hence, the two synergy improvements would be chosen based on application preference.
3. Figures 11–13 show that feedback scheme would push up the PR value and hold back the FS value. The FS curves of the methods with feedback scheme deviated a lot from the other methods with the extension of the prediction horizon. This was always true by comparative inspection.

Hence, if prediction accuracy is expected with highest priority, the feedback scheme should be considered.

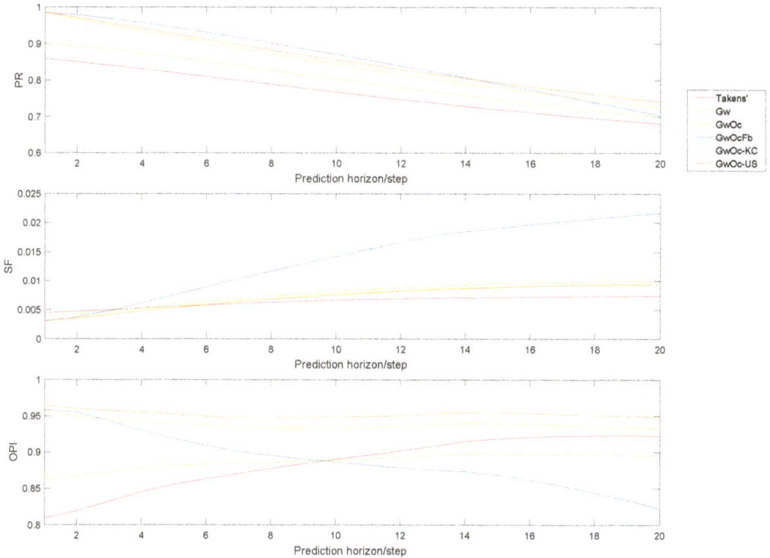

Figure 11. Comparison of prediction performance between the Takens', Gw, GwOc, GwOcFb, GwOc-KC, and GwOc-US methods (left hip joint data in the sagittal plane). *OPI*: overall performance index.

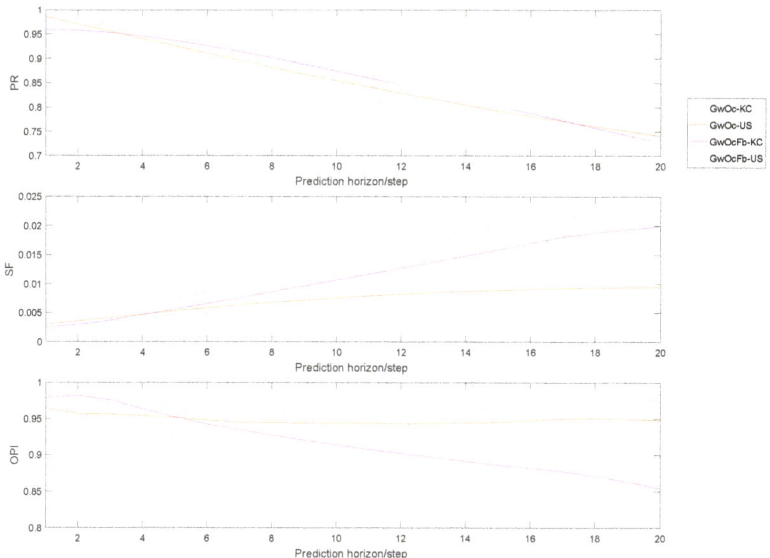

Figure 12. Comparison of the prediction performance of the GwOc-KC, GwOc-US, GwOcFb-KC, and GwOcFb-US methods (left hip joint data in sagittal plane).

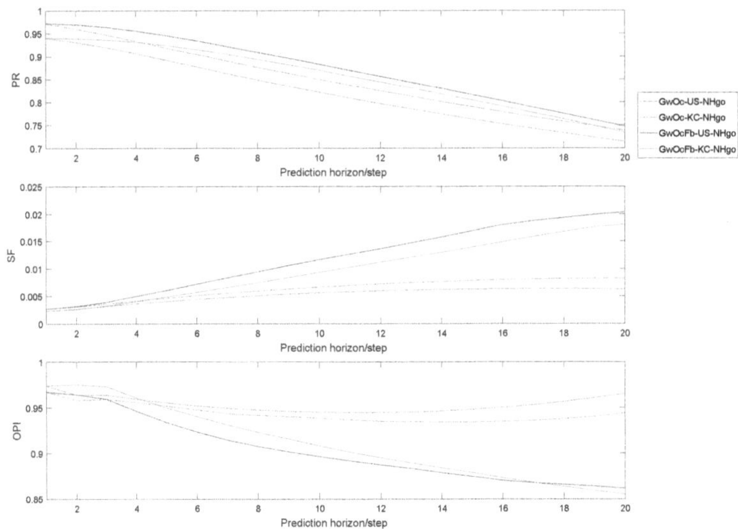

Figure 13. Comparison of the prediction performance of the GwOc-KC-NHgo, GwOc-US-NHgo, GwOcFb-KC-NHgo, GwOcFb-US-NHgo methods (left hip joint data in the sagittal plane).

4. In Figure 14, initially in the prediction horizon range of 1–3, the Newton's method with a high-gain observer had a medium accuracy and also adequate smoothness, but after 3, the Newton's method-based predictions degraded the fastest in both accuracy and smoothness. Meanwhile, for Takens' reconstruction-based methods, the curves changed moderately. The NHgo-fused methods, GwOC-US-NHgo and GwOC-US-NHgo, exhibited a slight drop in *PR* values compared to the methods without fusion, but the smoothness was more elevated. As a result, their overall performance was increased.

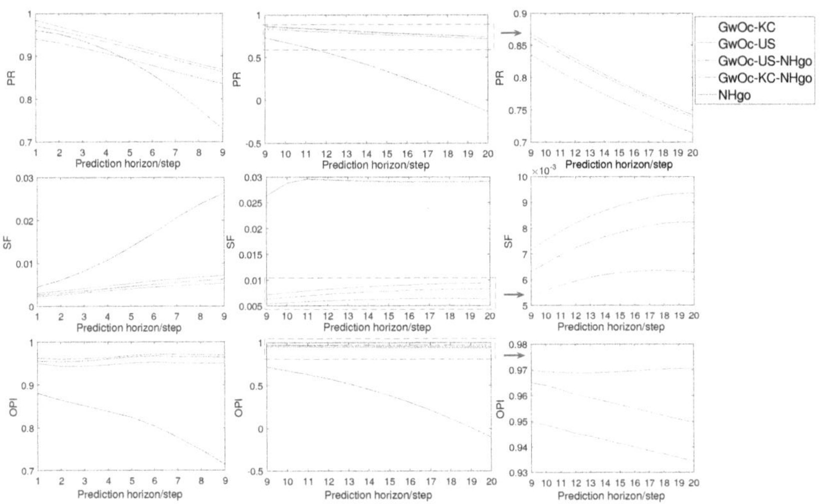

Figure 14. Comparison of the prediction performance of the GwOc-KC, GwOc-US, GwOc-KC-NHgo, GwOc-US-NHgo, and NHgo methods (left hip joint data in the sagittal plane).

5. By observing Figure 15 for short-term prediction such as in steps 1–3, the basic Takens' reconstruction method's performance was unsatisfactory, but its OPI value rose gradually, which meant its relative performance over other methods increased. The OPI curve with the KC-NHgo method changed gradually from an descending trend to ascending, which meant it was superior to other methods with respect to an extended prediction horizon. Considering the OPI values with designed weighting, the fully improved method, GwOc-KC-NHgo, was apparently better across the prediction horizon range. The OPI values are relative criterion and are significantly affected by the weighting value. If smoothness is preferred, the GwOc-KC-NHgo method is superior.

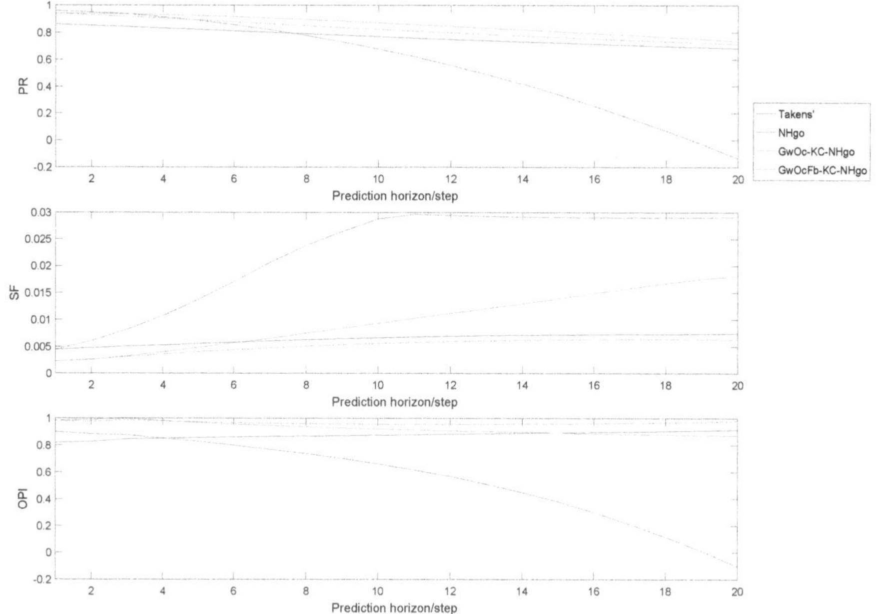

Figure 15. The comparison between Takens', NHgo, GwOc-KC-NHgo and GwOcFb-KC-NHgo methods (left hip joint data in the sagittal plane).

The adaptability to irregularities could be assessed visually and quantitatively by local irregularity tracking performance observations. The segments considered as irregularities in the assessment dataset were isolated and the predictions with different methods based on the parameters in Table 2 were executed. Some typical local trajectories are shown in Figure 16.

It can be seen in Figure 16a that a flat curve with zero values from 0 to 0.4 s exists at the very beginning of the predicted curves of Takens' reconstruction methods. Noting that there are no human data (zero), this meant dead time existed. Zero dead time was gained by other methods via offset correction, feedback, and fusion with the Newton-based methods.

Figure 16b presents the performance of different methods during transitions. Except for the basic Takens' reconstruction method, other methods were able to provide more or less prediction. The NHgo method-predicted curve fluctuated severely. Although the methods fusing the NHgo method, GwOc-KC-NHgo and GwOcFb-KC-NHgo, were influenced by these fluctuations locally, the fusion provided a better inhibiting effect. As the NHgo method exhibited outstanding prediction capability during periodicity loss, the methods fused with it acquired better prediction ability than without it.

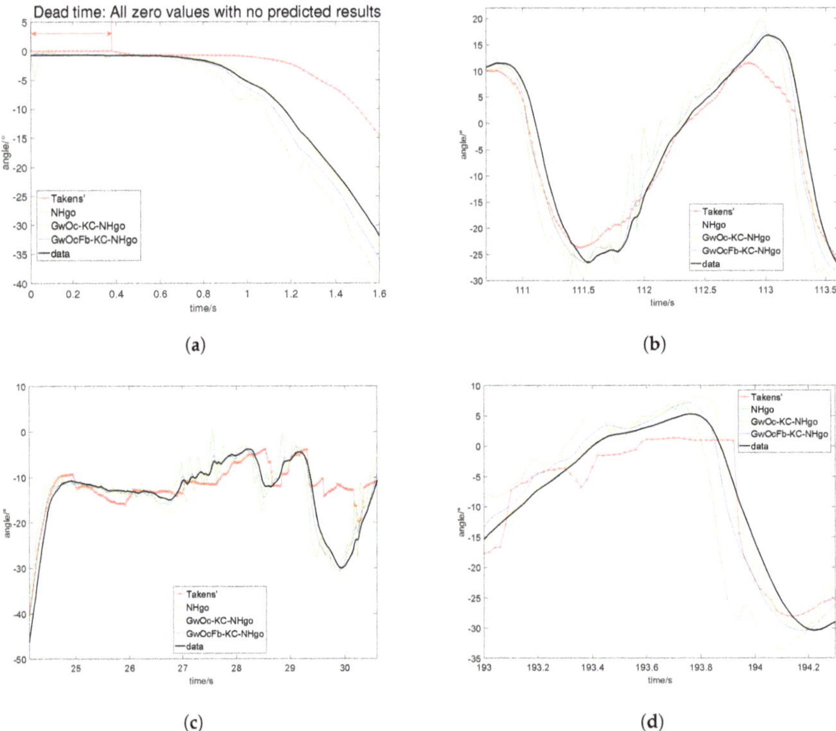

(a) (b) (c) (d)

Figure 16. Typical adaptability of different methods to irregularities (left hip joint angle in sagittal plane). (**a**) Dead time during initiation; (**b**) Speed transition from 1 km/h to 2 km/h; (**c**) Stand–walk transition; (**d**) Non-periodic motion during "stopping and leaving the treadmill".

Figure 16c,d were segments where periodicity was invisible, thus it was quite obvious that Takens' reconstruction method could hardly provide prediction, while other methods still worked more or less.

Quantitatively, the numerical criteria could gave a more objective judgement. Figure 17 compared the PR, OPI, and SF values between the whole dataset and the dataset during irregularities only. There are three phenomena worth noting in the figure:

1. The closing up of the assessment measure value gaps between the whole dataset and dataset during irregularities
2. The improvement in the values' changing trends.
3. The signs of differences in assessment measure values between the whole dataset and dataset during irregularities.

Compared with the basic Takens' method, the GwOc method closed up the gaps of PR, SF, and OPI values between the whole dataset and the dataset during irregularities. Also, all the values were improved with a positive trend. Compared with the GwOc method, the GwOcFb method also made the gap even smaller, but the SF values rose, which brought down the OPI values. With a similar analysis, the GwOc-KC and GwOc-US methods elevated the smoothness with prediction accuracy degradation, and also narrowed the gaps, which means the synergy improvement was effective in improving the prediction performance.

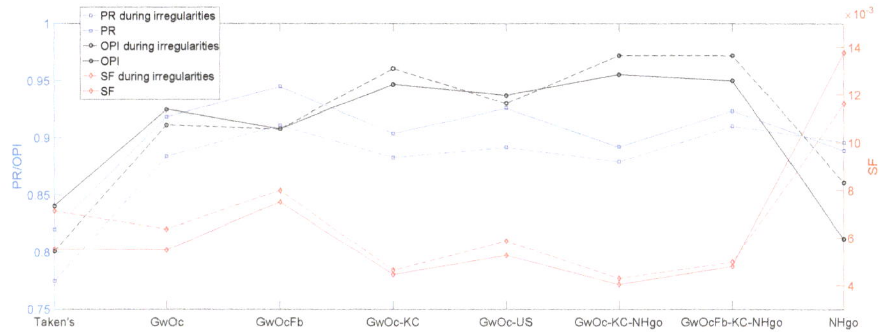

Figure 17. Performance of different methods during irregularities (left hip data).

The *PR* value difference using the NHgo method with respect to the whole dataset and the dataset of irregularities is negative, which means the prediction accuracy adapted better to irregularities, while for the *SF* values the difference is positive, which means the smoothness of predicted trajectories also adapted to irregularities better than periodicity. In contrast, the Takens', GwOc, GwOc-KC, and GwOc-US methods adapted to the whole dataset better.

Thus, by fusing the two types of prediction methods, the GwOcFb-KC-NHgo and GwOc-KC-NHgo methods achieved acceptable accuracy and smoothness, and their adaptability to irregularities was improved. The OPI values for irregularities also demonstrated better adaptability of GwOcFb-KC-NHgo and GwOc-KC-NHgo over other methods, performing extremely well with the whole dataset and the irregular segments.

Based on the previous analysis, the GwOc-KC-NHgo method was the first choice for prediction because of its high prediction accuracy, smoothness, and adaptability to irregularities with no dead time.

4.3. Real-Time Demonstration

The GwOc-KC-NHgo methods were implemented on the human lower limb motion capture system in Figure 1 with the basic parameters designed as in Table 2. The algorithm's computational time was between 3.2 and 3.3 ms (Figure 18) with tiny fluctuations which could satisfy the real-time execution requirements. The computational expenses of the other methods were similar, but could not be shown here.

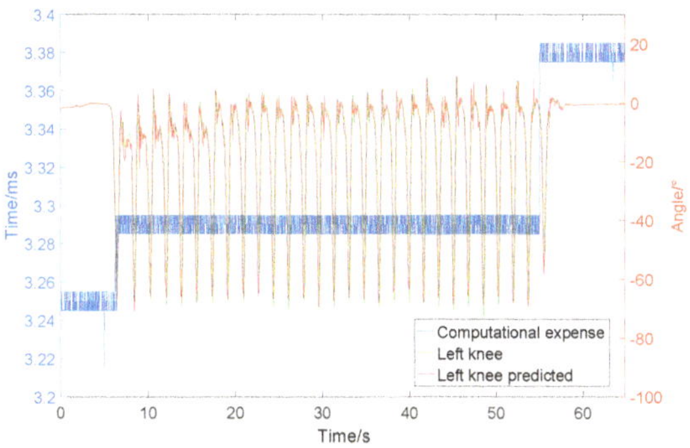

Figure 18. Real-time computational expense of the GwOc-KC-NHgo method (left hip data at 2 km/h walking).

5. Conclusions

In exoskeletal robot control, it would be much easier to handle the human robot coordination problem if the human motion was known in advance. Prediction is a helpful approach. Based on the biological features of human walking, a series of bio-inspired prediction methods were proposed and implemented. The parameters of Takens' reconstruction-based methods were analysed and selected with the parameter robustness of newly introduced methods demonstrated as well. The comparative assessment demonstrated the offset correction and that Gaussian weighting techniques could effectively improve the prediction performance of Takens' reconstruction method in both prediction accuracy and trajectory smoothness. The feedback scheme was useful in prediction accuracy improvement but degraded the smoothness. The synergy improvement of the coupled joints' united source of historical embedding generation exhibited better performance in prediction accuracy than Kalman fusion with the complementary joint prediction and poorer performance in trajectory smoothness. The second-order Newton's method with a high-gain observer gave acceptable performance with short prediction horizons. When considering improving the adaptability to irregularities, the Takens' reconstruction-based methods were fused with Newton-based methods. On comparison with their contending methods, the new hybrid methods showed mixed features with nearly zero dead time, a slight deterioration in prediction accuracy and smoothness in short prediction horizons, and improvement in longer prediction horizons. The methods should be tailored for the detailed requirements of exoskeletal robot applications. When accuracy is emphasized, the GwOc-KC-NHgo method is recommended. The real-time computational expense would not be a burden for the methods introduced.

Acknowledgments: We would like to thank Zhengya Zhang and Xin Huang from Xeno Dynamics Co., Ltd. for the hardware design, fabrication, and algorithm-embedded system realization.

Author Contributions: Pu Duan and Zhuoping Duan conceived of the presented idea. Pu Duan and Shilei Li developed the theoretical formalism. Shilei Li and Yawen Chen collected and analysed the data. Shilei Li performed the numerical simulations. Pu Duan discussed the result and wrote the paper. Zhuoping Duan supervised the entire work.

Conflicts of Interest: The authors declare no conflict of interest.

References

1. Sheridan, T.B. Telerobotics. *Automatica* **1989**, *25*, 487–507.
2. Huang, L. Robotics Locomotion Control. Ph.D. Thesis, University of California, Berkeley, CA, USA, 2005.
3. Novak, D.; Reberšek, P.; De Rossi, S.M.M.; Donati, M.; Podobnik, J.; Beravs, T.; Lenzi, T.; Vitiello, N.; Carrozza, M.C.; Munih, M. Automated detection of gait initiation and termination using wearable sensors. *Med. Eng. Phys.* **2013**, *35*, 1713–1720.
4. Kanjanapas, K.; Wang, Y.; Zhang, W.; Whittingham, L.; Tomizuka, M. A Human Motion Capture System Based on Inertial Sensing and a Complementary Filter. In Proceedings of the ASME 2013 Dynamic Systems and Control Conference on American Society of Mechanical Engineers, Palo Alto, CA, USA, 21–23 October 2013; p. V003T40A004
5. Wang, M.; Wu, X.; Liu, D.; Wang, C.; Zhang, T.; Wang, P. A human motion prediction algorithm for non-binding lower extremity exoskeleton. In Proceedings of the 2015 IEEE International Conference on Information and Automation, IEEE, Lijiang, China, 8–10 August 2015; pp. 369–374.
6. Zhang, W.; Tomizuka, M.; Byl, N. A wireless human motion monitoring system for smart rehabilitation. *J. Dyn. Syst. Meas. Control* **2016**, *138*, 111004.
7. Lin, Y.; Min, H.; Wei, H. Inertial measurement unit-based iterative pose compensation algorithm for low-cost modular manipulator. *Adv. Mech. Eng.* **2016**, *8*, doi:10.1177/1687814015626850.
8. Novak, D.; Riener, R. A survey of sensor fusion methods in wearable robotics. *Robot. Auton. Syst.* **2015**, *73*, 155–170.
9. Fleischer, C.; Hommel, G. A Human–Exoskeleton Interface Utilizing Electromyography. *IEEE Trans. Robot.* **2008**, *24*, 872–882.

10. Li, Z.; He, W.; Yang, C.; Qiu, S.; Zhang, L.; Su, C.Y. Teleoperation control of an exoskeleton robot using brain machine interface and visual compressive sensing. In Proceedings of the 2016 12th World Congress on Intelligent Control and Automation (WCICA), Guilin, China, 12–15 June 2016; pp. 1550–1555.
11. Iqbal, S.; Zang, X.; Zhu, Y.; Saad, H.M.A.A.; Zhao, J. Nonlinear time-series analysis of different human walking gaits. In Proceedings of the 2015 IEEE International Conference on Electro/Information Technology (EIT), Dekalb, IL, USA, 21–23 May 2015; pp. 25–30.
12. Pitkin, M.R. Effects of Design Variants in Lower-Limb Prostheses on Gait Synergy. *JPO J. Prosthet. Orthot.* **1997**, *9*, 113–122.
13. Senda, S.; Takata, N.; Tsujita, K. A study on lower limb's joint synergy in human locomotion with physical constraints on the knee. In Proceedings of the 2012 IEEE/SICE International Symposium on System Integration (SII), Fukuoka, Japan, 16–18 December 2012; pp. 349–354.
14. Sanzmerodio, D.; Cestari, M.; Arevalo, J.C.; Garcia, E. Exploiting joint synergy for actuation in a lower-limb active orthosis. *Ind. Robot. Int. J.* **2013**, *40*, 224–228.
15. Rosenblatt, N.J.; Latash, M.L.; Hurt, C.P.; Grabiner, M.D. Challenging gait leads to stronger lower-limb kinematic synergies: The effects of walking within a more narrow pathway. *Neurosci. Lett.* **2015**, *600*, 110–114.
16. Grimes, D.L.; Flowers, W.C.; Donath, M. Feasibility of an Active Control Scheme for Above Knee Prostheses. *J. Biomech. Eng.* **1977**, *99*, 215–221, doi:10.1115/1.3426293.
17. Vallery, H.; van Asseldonk, E.H.; Buss, M.; van der Kooij, H. Reference trajectory generation for rehabilitation robots: complementary limb motion estimation. *IEEE Trans. Neural Syst. Rehabil. Eng.* **2009**, *17*, 23–30.
18. Broer, H.; Takens, F. Reconstruction and time series analysis. In *Dynamical Systems and Chaos*; Springer: New York, NY, USA, 2011; pp. 205–242.
19. Herrmann, C. Robotic Motion Compensation for Applications in Radiation Oncology. Ph.D. Thesis, University of Würzburg, Würzburg, Germany, 2013.
20. Zhang, W.; Tomizuka, M.; Bae, J. Time series prediction of knee joint movement and its application to a network-based rehabilitation system. In Proceedings of the IEEE 2014 American Control Conference, Portland, OR, USA, 4–6 June 2014; pp. 4810–4815.
21. Zhang, W.; Chen, X.; Bae, J.; Tomizuka, M. Real-time kinematic modeling and prediction of human joint motion in a networked rehabilitation system. In Proceedings of the 2015 American Control Conference (ACC), Chicago, IL, USA, 1–3 July 2015; pp. 5800–5805.
22. Crowninshield, R.; Pope, M.H.; Johnson, R.; Miller, R. The impedance of the human knee. *J. Biomech.* **1976**, *9*, 529–535.
23. Yangming, X.; Hollerbach, J.M. Identification of human joint mechanical properties from single trial data. *IEEE Trans. Biomed. Eng.* **1998**, *45*, 1051–1060.
24. Zhang, L.Q.; Nuber, G.; Butler, J.; Bowen, M.; Rymer, W.Z. In vivo human knee joint dynamic properties as functions of muscle contraction and joint position. *J. Biomech.* **1997**, *31*, 71–76.
25. Andrea, T.; Marcello, M. A low-noise estimator of angular speed and acceleration from shaft encoder measurements. *J. Autom.* **2001**, *42*, 169–176.
26. Ahrens, J.H.; Khalil, H.K. High-gain observers in the presence of measurement noise: A switched-gain approach. *Automatica* **2009**, *45*, 936–943.
27. Nicosia, S.; Tornambe, A. High-gain observers in the state and parameter estimation of robots having elastic joints. *Syst. Control Lett.* **1989**, *13*, 331–337.
28. Madgwick, S.O.H.; Harrison, A.J.L.; Vaidyanathan, R. Estimation of IMU and MARG orientation using a gradient descent algorithm. In Proceedings of the 2011 IEEE International Conference on Rehabilitation Robotics, Zurich, Switzerland, 29 June–1 July 2011; pp. 1–7.

Article

Effective Biopotential Signal Acquisition: Comparison of Different Shielded Drive Technologies

Yanbing Jiang [1,2], Oluwarotimi Williams Samuel [1,2], Xueyu Liu [3], Xin Wang [1,2], Paul Oluwagbenga Idowu [1,2], Peng Li [4], Fei Chen [5], Mingxing Zhu [1], Yanjuan Geng [1], Fengxia Wu [6], Shixiong Chen [1,*] and Guanglin Li [1,*]

1 CAS Key Laboratory of Human-Machine Intelligence-Synergy Systems, Shenzhen Institutes of Advanced Technology, Chinese Academy of Sciences, Shenzhen 518055, China; yb.jiang@siat.ac.cn (Y.J.); samuel@siat.ac.cn (O.W.S.); wangxin@siat.ac.cn (X.W.); Paul@siat.ac.cn (P.O.I.); mx.zhu@siat.ac.cn (M.Z.); yj.geng@siat.ac.cn (Y.G.)

2 Shenzhen College of Advanced Technology, University of Chinese Academy of Sciences, Shenzhen 518055, China

3 Institute of Pharmacy and Bioengineering, Chongqing University of Technology, Chongqing 400054, China; xy.liu2@siat.ac.cn

4 The Third Affiliated Hospital of Sun Yat-Sen University, Guangzhou 510630, China; lp76@163.net

5 Southern University of Science and Technology, Shenzhen 518055, China; fchen@sustc.edu.cn

6 Department of Anatomy, School of Medicine, Shandong University, Jinan 250100, China; wufengxia@sdu.edu.cn

* Correspondence: sx.chen@siat.ac.cn (S.C.); gl.li@siat.ac.cn (G.L.)

Received: 15 November 2017; Accepted: 6 February 2018; Published: 12 February 2018

Abstract: Biopotential signals are mainly characterized by low amplitude and thus often distorted by extraneous interferences, such as power line interference in the recording environment and movement artifacts during the acquisition process. With the presence of such large-amplitude interferences, subsequent processing and analysis of the acquired signals becomes quite a challenging task that has been reported by many previous studies. A number of software-based filtering techniques have been proposed, with most of them being able to minimize the interferences but at the expense of distorting the useful components of the target signal. Therefore, this study proposes a hardware-based method that utilizes a shielded drive circuit to eliminate extraneous interferences on biopotential signal recordings, while also preserving all useful components of the target signal. The performance of the proposed method was evaluated by comparing the results with conventional hardware and software filtering methods in three different biopotential signal recording experiments (electrocardiogram (ECG), electro-oculogram (EOG), and electromyography (EMG)) on an ADS1299EEG-FE platform. The results showed that the proposed method could effectively suppress power line interference as well as its harmonic components, and it could also significantly eliminate the influence of unwanted electrode lead jitter interference. Findings from this study suggest that the proposed method may provide potential insight into high quality acquisition of different biopotential signals to greatly ease subsequent processing in various biomedical applications.

Keywords: shielded drive; power line interference; electrode lead jitter; ECG; EOG; EMG

---

1. Introduction

As important indicators of various physiological parameters of the human body, biopotential signals can directly reflect the physical condition as well as the health status of an individual [1]. With the increase in demand for technologically driven medical devices and the advancements in neuroscience, cognitive psychology, and artificial intelligence research, biopotential signals are being increasingly applied to real-time health monitoring, disease diagnoses, control rehabilitation devices, and brain–computer interfaces among others [2,3]. Generally, biopotential signals such as electrocardiogram (ECG), electromyography (EMG), electroencephalography (EEG), and electro-oculogram (EOG) are characterized by high impedance, low frequency, low amplitude, and strong background noise [4,5]. For instance, the amplitude and frequency of an ECG signal typically lie within the range of 0.05–4 mV and 0.05–150 Hz, respectively [6]; the EMG signal has an amplitude that is relatively weaker, in the range of 0.1–2 mV, and a frequency in the range of 0–500 Hz [7,8]; and the EOG signal's amplitude is in the range of 0.4–10 mV with a spectrum in the range of 0.1–38 Hz [9].

When a biopotential signal is recorded, the target/output of the recordings is almost completely submerged by the background noise/interference inherent in the recordings because the signal is often characterized by low amplitude [10]. Thus, such noise/interference is majorly classified into three categories as follows: Firstly, power line interference, which is composed of a 50 Hz or 60 Hz frequency and its harmonics, commonly defined as the electromagnetic noise produced from the power supply circuitry and electronic equipment [11]. Secondly, movement artifacts caused by poor contact between the electrodes and the surface of an individual's skin [12], or as a result of movement of the wires/cables between the electrodes and the amplifier [13]. Thirdly, physiological artifacts in the form of other signal(s) recorded alongside the originally desired biopotential signals [14,15].

To date, several methods have been proposed to eliminate the above-described interferences/noises from a biopotential signal of interest. To that end, most of the methods focused on improving hardware circuitry and using software filtering techniques [16–18]. There are many commonly applied hardware-based methods for minimizing the interferences inherent in biopotential signal recordings, such as twisting input leads together to reduce the area of the loop formed by the wires [19], or making the acquisition devices dependent on batteries (direct current) for their power supply [20]. Adopting a high common-mode rejection ratio amplifier, a driven-right-leg circuit, active electrodes, and isolation can further suppress interferences [21–23]. The conventional hardware-based methods are limited since they can only reduce the extraneous interferences to a certain extent. Furthermore, Chimeno et al. proposed an interference model that considers interference directly coupled to the measuring electrodes and adds an internal interference arising from the amplifier's power supply [19]. Their study suggested that internal interference from the power supply can affect any measuring instrument with either differential or single-ended input. While the model mainly focused on the case of using analog amplifiers powered by power line transformers, battery-supplied biopotential recording systems did not suffer from such problems with power-supply interference.

In addition, shielded technology is also one of the commonly used hardware methods to suppress power line interference. The most popular shielded technology is the use of a small coax cable, where the outer layer is connected to a certain type of driven signal to prevent the inner wire signal from being contaminated by extraneous interferences. Alnasser showed that unshielded electrodes account for most power line interference, and electrode shielding, together with driven-right-leg, is quite effective in reducing that interference [24]. Sudirman et al. proposed a method that utilized a faraday shield in which the subjects were required to be in a faraday metal cage to eliminate 50 Hz power line interference in an ECG signal, whereby the whole size of the cage must be significantly smaller than the wavelength of the noises [25]. Lee et al. developed a flexible active electrode that contained a shielding metal plate with guarding feedback to protect the ECG signals from extraneous noises [26]. Sullivan et al. used a shielding layer on printed circuit board and implemented a non-contact sensor to record clear EEG and ECG signals [27]. With so many shielded technologies coexisting, it is necessary to compare the performance of different shielding methods so that a general guideline can be established when selecting shielded technologies; it would be best to propose an improved method that could be used reliably to achieve effective biopotential acquisition in clinical or academic practices.

On the other hand, various software-based filtering techniques have also been utilized to minimize the power line noise/interference inherent in biopotential signal recordings [28]. The most common approach is to use a digital notch filter at the power line frequency (50 Hz/60 Hz), which could be easily implemented with low computational cost [29,30]. However, there are often cases when the power line interference contains not only the 50 Hz/60 Hz component but also multiple harmonic components, as well as other interferences caused by the instability of the power supply system [31–33]. Hence, filter banks with notch frequencies located at the fundamental and harmonic components of power line interference were also employed [34]. However, such filter banks would cause information loss of target signals since the frequency range largely overlaps for the biopotential signals and power line interference [35]. Hence, a number of improved filtering techniques have been proposed over the years to minimize interferences in biopotential recordings [36–40]. For instance, Tomasini et al. used an adaptive power line interference filter to estimate the fundamental frequency and harmonics of power line interference, and the estimated interference was subtracted from the noise-affected biosignal [22]. Keshtkaran et al. presented a scalable very-large-scale integration architecture of a robust algorithm for power line interference cancelation in multichannel biopotential recordings [23], and they also proposed an adaptive notch filter to estimate the contents of power line interference with a modified recursive least squares algorithm [41]. Although these filtering technologies showed great performance and robustness, the real time computation load is still high for low-power embedded systems, especially when the sampling frequency is high. Therefore, instead of removing the interferences from the contaminated signal using digital filtering, a robust solution to prevent power line interference from contaminating the biopotential signal during the recording stage would be preferred for long-term healthcare monitoring applications.

During biopotential signal acquisition, device movements, wire jitter, or subject involuntary twitching are inevitable, thus affecting the quality of the recorded signals. The standard and commonly applied method for eliminating interferences resulting from electrode lead jitter is to apply high-pass filtering to the recorded signals [42]. This is the case because the frequencies of the signals are usually very low, and the high-pass filter will cut off the frequencies of interest containing useful information in the desired signal. With regards to thechallenges posed by extraneous interferences in biopotential signal recording, the existing hardware-based methods have limited effect while the traditional software filtering techniques inevitably render the target signal distorted after reducing the interferences. If such interferences can be prevented from mixing up with the desired signal before the raw signal gets into the acquisition device, a quality raw signal could be obtained, thus making the subsequent signal processing and analysis tasks accurate and easier.

The current study proposes an improved hardware-based shielded technology aimed at effectively attenuating the interferences associated with biopotential signal recordings. The proposed technology attempts to resolve the adverse effects of the distributed resistance and capacitance of shielded cables by introducing a shielded signal to suppress resulting interferences as much as possible, and also to prevent the target signal from being distorted. Subsequently, three different methods for the proposed shielded technology were built and applied individually to ECG, EOG, and EMG recordings obtained in our laboratory. The experimental results showed that the proposed method can effectively suppress power line interference, and also shield the target signal. Furthermore, by recording the EMG signal in a shielded room, it was verified that the proposed hardware-based shielded drive circuit can also eliminate electrode lead jitter interference.

2. Methods

2.1. Subjects

Twenty healthy subjects including 12 males and 8 females were recruited, and all participated in the biopotential signal acquisition experiments designed for the current study. The mean age of the subjects was 26.3 years with an age range of 21–31. Prior to the data acquisition experiments, each participant was carefully examined to be sure that they had no impairment with respect to their heart function (for quality ECG signal), eyelid function (for proper EOG recordings), and limb muscle function (for a normal EMG signal). Additionally, the subjects were adequately informed about the objective as well as the purpose of the study, after which they all gave written informed consent and provided permission for the publication of their data. The experimental protocols were approved by the Institutional Review Board of the Shenzhen Institutes of Advanced Technology, Chinese Academy of Sciences (SIAT-IRB-130124-H0015).

2.2. Equipment Setup and Signal Recording

In this study, the ADS1299EEG-FE kit of Texas Instruments (Dallas, Texas, United States) was used as the biopotential signal collection device and was connected to an Arduino microcontroller. The ADS1299EEG-FE demonstration kit has obvious advantages in the collection of biopotential signals such as eight channels of low-noise synchronous EEG recording (which can collect multiple signals simultaneously); a 24-bit high resolution analog to digital converter (ADC), −110 dB common-mode rejection ratio, and very low internal noise (which ensures the accuracy of biopotential signal detection); a 1–24 times adjustable programmable gain amplifier (PGA), which meets the needs of different occasions; a single channel 16 kHz sampling rate (which ensures that the biopotential signals bandwidth is sufficient); and an integrated driven-right-leg circuit (which further reduces environmental interference) [43]. In addition, its flexible connection with the Arduino microcontroller (Arduino is an electronic product development platform based on the microcontroller systems) ensures the development and integration of hardware and software components [44].

As shown in Figure 1, the three previously described biopotential signals (ECG, EOG, and EMG) were collected by the aid of electrodes attached to the subject's (①) skin through the shielded cables into the ADS1299EEG-FE hardware (③), then the recorded signals went through the Arduino microcontroller (④) via a USB interface to the laptop (⑤), and finally the waveforms and spectra components of the collected signals were analyzed using the MATLAB programming tool. Additionally, shielded cables were used as electrode leads, so as to protect the signal in the core from extraneous interferences.

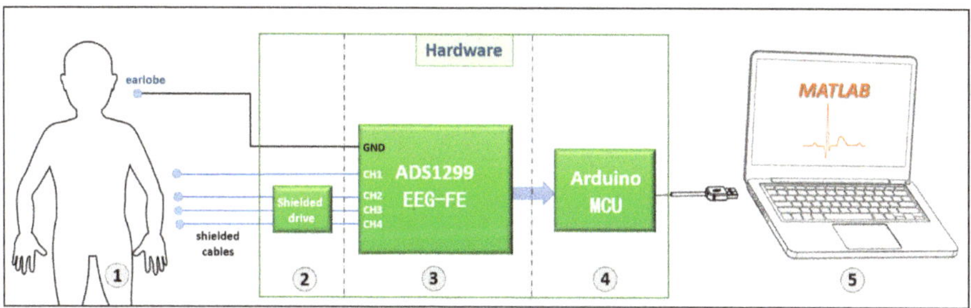

Figure 1. Schematic diagram of the data acquisition system, ① the subject; ② different shielded drive methods proposed in this study; ③ ADS1299EEG-FE kit, where EEG_FE refers to analog front-end designed for electroencephalography, GND is the ground of the power supply, and CH means the input channel index of the kit; ④ Arduino Microcontroller Unit (MCU); ⑤ the laptop.

The recorded signals were sampled at the rate of 2000 Hz and subsequently amplified 24 times. To compare the effects of the hardware-based shielded drive technology with the software-based method, this study also designed a 3-order Butterworth software notch filter with a bandwidth of 6 Hz and an attenuation of 120 dB at the 50 Hz center frequency. The software filter was applied to the raw data of Method 1 (No-Shield), and the performance was evaluated by comparing the time waveforms and spectra of the biopotential signals with different hardware-shielded methods.

2.3. Experimental Principle

In order to prevent the adverse effects of extraneous electromagnetic fields while transmitting the recorded signal (from ① to ③ in Figure 1), shielded cables were utilized as the electrode leads. When the signal is recorded, the shield is usually grounded, thus avoiding the effects of extraneous interferences on the signal in the inner core wire. Although the shielded cables protect the signals in the core from extraneous electromagnetic fields, a distributed resistance and capacitance often exists between the shield and the core wire, thus affecting the quality of recorded biopotential signals that are generally characterized by low amplitude and high impedance [45]. Therefore, this study implemented a shielded drive technology, in which the signal in the core drives the signal in the shield, as shown in Figure 2D. When the output of the amplifier is connected to the shield, which means that the shield is connected to a low internal resistance voltage source, the effect of the shielded cable against interference remains unchanged. However, since the voltage of the core is equal to the voltage of the shield, according to Ohm's law [46], the current in the core automatically becomes zero which then nullifies the resistance and capacitance that once existed between the shield and the core. Hence, the biopotential signal passing through the shielded cable is no longer affected by the distributed resistance and capacitance.

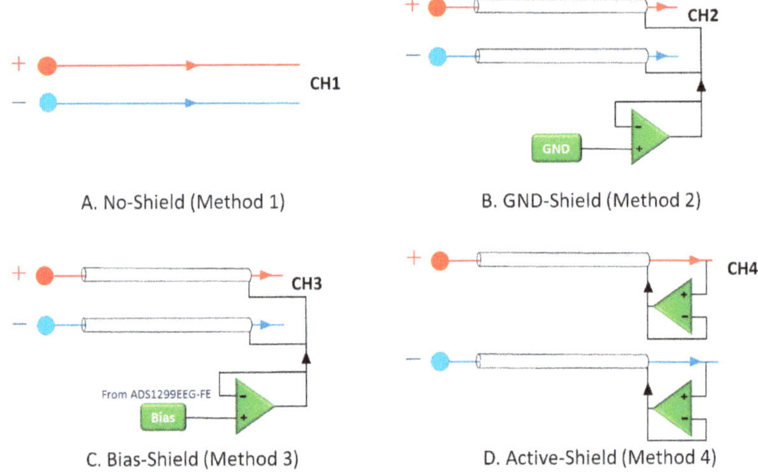

Figure 2. Schematic diagram of the different shielded drive methods investigated in this study.

During the experiments, a total of four electrode channels were simultaneously used to record the ECG, EOG, and EMG signals across all 20 of the recruited subjects. Meanwhile, the first electrode channel denoted as Channel 1 was set as the control channel, thus it was not implemented on the proposed shielded drive circuit. In addition, we implemented three different shielded drive techniques on the other channels (Channel 2, Channel 3 and Channel 4). As shown in Figure 2A, the first channel adopted the conventional method that has no shield (herein referred to as Method 1). In Figure 2B, the second channel connected the outer shields of both electrode cables to the buffered ground signal of the ADS1299EEG-FE, as shown in Figure 3 (referred to as Method 2). Figure 2C shows that the outer shields of the electrode cables of the third channel were connected to the buffered BIAS_SHD signal (Figure 3) generated by the integrated circuit of ADS1299EEG-FE (referred to as Method 3). In Figure 2D, the signal of the inner wire was fed back to the shield using a low impedance output of an amplifier (herein referred to as Method 4). And thus, the voltage of the shielded cable with respect to the shield became zero. Furthermore, the four channels shared the same driven-right-leg (DRL) electrode placed on the earlobe, as the ground and the DRL signal were generated from the integrated circuit of the ADS1299EEG-FE kit. The DRL circuit sensed the common-mode of a selected set of electrodes and created a negative feedback loop by driving the body with an inverted common-mode signal [47].

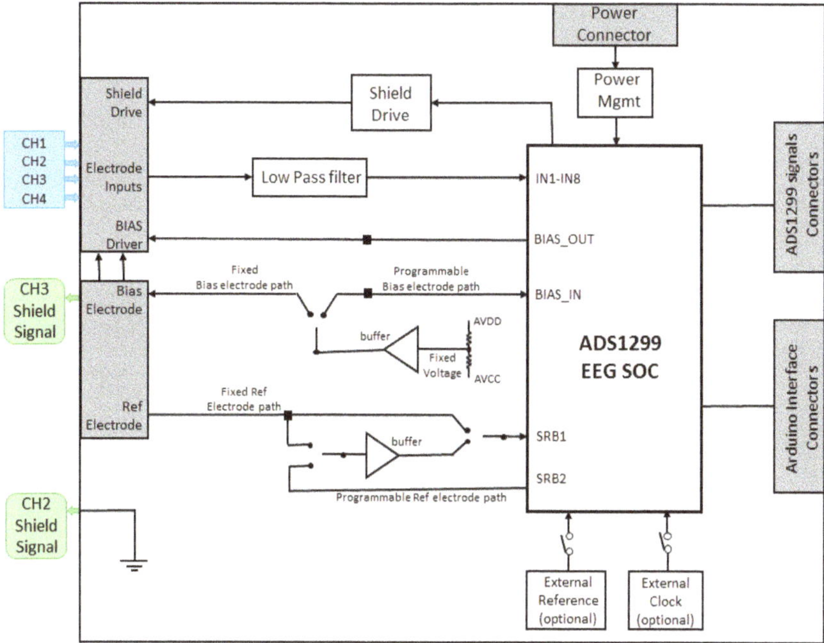

Figure 3. ADS1299EEG-FE functional block diagram.

2.4. Experimental Procedures

In the current study, a total of four different experiments were designed. To investigate the effectiveness of the developed shielded drive circuit with respect to reducing power line interference, three of the experiments were performed under normal laboratory conditions (NLC) (that is, the laboratory was unshielded from sound and electro-magnetic interference) for the recording of ECG, EOG, and EMG signals from the subjects. To equally examine if the proposed method could eliminate the interference of electrode lead jitter, the fourth experiment was designed to collect an EMG signal in a sound-proofed and electro-magnetically shielded laboratory designated as shielded laboratory conditions (SLC). For all of the experiments, the same hardware platform of ADS1299 was used, and only the locations of the electrodes were different for different types of biopotential signals. Prior to the commencement of the experiments, the surface of the subjects' skin was properly cleaned with the aid of an alcohol. Subsequently, conductive gel was applied to the electrodes before attaching them to the skin to minimize the impedance between the skin surface and the electrode. During the experiments, the laptop was completely powered by battery to reduce all forms of power line interference that may result from using an alternating current (AC) supply. The experiments are described in the following subsections.

2.4.1. ECG Data Acquisition Experiment

In the normal laboratory condition, the subjects were instructed to sit comfortably on a chair in a quiet and relaxed manner. Then, the positive and negative nodes of the four electrode channels were placed on the left and right forearm of each subject, respectively, as shown in Figure 4a. It is noteworthy that the electrodes on both sides of the forearms were placed as symmetrically as possible to minimize the impedance mismatch of the positive and negative inputs of each channel. Afterwards, each subject completed three experimental trials with each trial lasting for 5 min, and they all maintained a quiet mode during the signal recording. Lastly, the time waveform and spectra of the recorded signals were monitored in real time via a graphical user interface (GUI) module in the acquisition. Note that the data collection scenario described here is designated as Experiment 1.

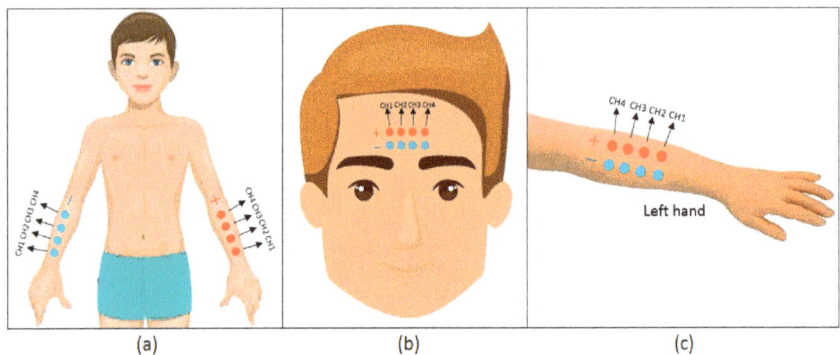

Figure 4. Electrodes location diagram: (**a**) shows the location of the electrodes for electrocardiogram (ECG) acquisition; (**b**) shows the location of the electrodes for electro-oculogram (EOG) acquisition; (**c**) shows the location of the electrodes for electromyography (EMG) acquisition.

2.4.2. EOG Data Acquisition Experiment

By placing the positive and negative nodes of the four electrodes on the forehead of each subject, the EOG signals were recorded under normal laboratory conditions. The placement of the electrode nodes on the forehead of each subject was done in a two by four (consisting of two rows and four columns) grid, as shown in Figure 4b. The first row of the grid represents the positive nodes of the channels while the second row consists of the negative nodes of the channels. The electrode channels were placed close to each other as much as possible to ensure that similar ocular signals were obtained across the channels. The EOG recording session also lasted for 5 min and each subject performed 3 trials. During each experimental trial, the subjects actually blinked their eyes every 2 s with a consistent intensity all through the recordings. Furthermore, the characteristics (in terms of waveform and spectrum) of the recorded EOG signals were observed through a GUI module to assess the recorded signals. Note that the data collection scenario described here is designated as Experiment 2.

2.4.3. EMG Data Acquisition Experiment

The EMG recordings were acquired under normal laboratory conditions by placing the positive and negative nodes of each of the four electrode channels on the left forearm of the subjects, as shown in Figure 4c. The electrodes were placed in a relatively close position to one another to ensure that the recorded EMG signals were identical during muscle contraction. Then, the subjects completed three trials with each trial lasting for 5 min, and they all maintained a quiet mode during the signal recording session. During the experiment, the subject's arm was initially relaxed and his/her hand remained in an open state. After about a minute, the subject clenched his/her fist and kept it closed for 3 s. Then the subject opened his/her hand and returned to the relaxed state. The EMG signal was recorded during the process and the temporal waveforms were observed via a GUI module to assess the signal quality. Note that the data collection scenario described here is designated as Experiment 3.

To avoid the effect of power line interference, the EMG signal recording experiment was performed in the shielded room. In the shielded room, up to 30 dB acoustic noise could be avoided; this is in line with the national standards which conform to GB/T16403-1996 and GB/T16296-1996. Under these laboratory conditions, each subject sat in a comfortable manner while the positive and negative ends of four electrodes were attached to the muscles on their left forearm. To prevent the electrodes from being displaced during the experiment, a bandage was used to firmly hold the electrodes to their skin, as shown in Figure 5. During the experiment, the subjects' forearms with the electrodes were maintained in a stationary position, along with the electrode lead cables, for the first minute. Then, the electrode lead cables were individually shaken at about the same amplitude for 10 s and then returned to their initial positions. Afterwards, each subject repeated the experiment three trials while each trial lasted for 3 min and the recorded signals were examined through a GUI. Note that the data collection scenario described here is designated as Experiment 4.

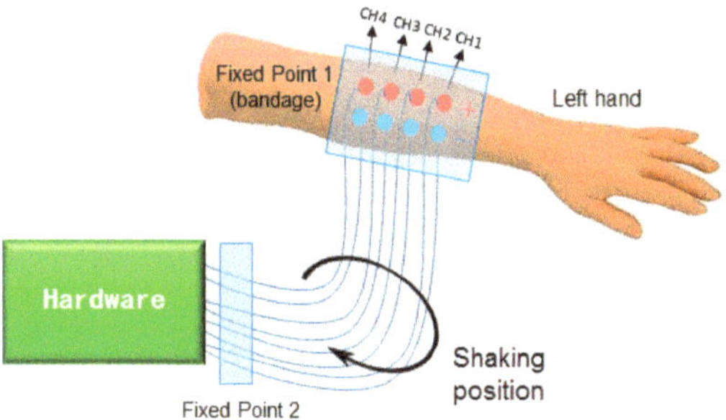

Figure 5. EMG acquisition setup with the introduction of electrode lead jitter.

3. Results and Analysis

By utilizing the biopotential signals obtained from all the recruited subjects, results were obtained for each of the above-described experimental scenarios. For each experimental scenario, the obtained results were compared amongst all subjects, and they were observed to be consistent. Hence, the results obtained for a representative subject are presented and analyzed as follows.

3.1. Reduction of Power Line Interference

3.1.1. Analysis of ECG Recordings

The time-domain waveform and spectrum of the ECG signal of a representative subject collected under NLC are shown in Figure 6. It can be seen from the ECG waveform that Method 1 reflects a series of sine wave that superimposes the ECG signals and thus prevents the ECG recordings from being obvious. However, Methods 2, 3 and 4 produced obvious ECG signals, whose time-domain waveforms are better than that of Method 1. Furthermore, the difference amongst Methods 2, 3 and 4 is not obvious, probably because of their mixture with other biopotential signals such as EMG. This finding suggests that the waveforms of ECG signals acquired with the proposed shielded drive technology would be better than that of the conventional unshielded drive approach.

By analyzing the ECG spectrum, it can be seen that the 50 Hz frequency in the ECG signal collected by Method 1 is as high as 13.6 dB; the 50 Hz power line frequency amplitude of Method 2 is −5.1 dB, which is 18.7 dB lower than that of Method 1; the 50 Hz power frequency amplitude of Method 3 is −9.6 dB, which is 23.2 dB lower than that of Method 1; while Method 4 records a 50 Hz frequency amplitude of −17.0 dB which achieves a reduction of 30.6 dB in comparison to Method 1, and it is much better. In addition, the harmonics of 50 Hz in the spectrum of Method 4 are also eliminated. So it could be concluded that the use of shielded drive can effectively suppress power line interference during ECG signal recording, especially when utilizing Method 4.

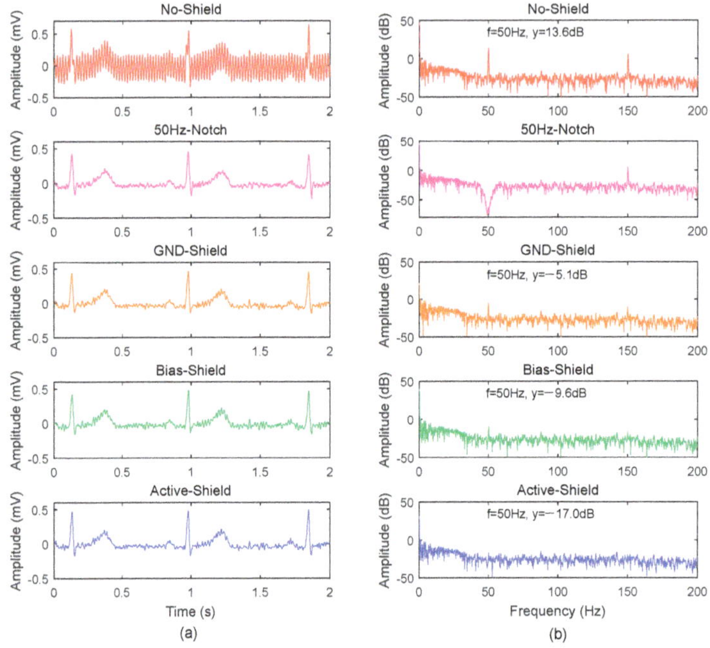

Figure 6. The comparison of the time waveforms (**a**) and spectra (**b**) of the ECG signals (subject #11) under different conditions. The 50Hz notch condition (2nd row) was obtained by applying 50 Hz notch filter (software) to raw data of the No-Shield condition (1st row).

The plot on the second row in Figure 6 shows the time domain waveform and spectrum of the raw signal (the signal gotten from Method 1) after undergoing filtering via a 50 Hz notch filter. Its waveform is similar to those of the shielded drive methods, but the spectrum has a large depression near 50 Hz. Because ECG signal frequency lies in the range of 0.05–100 Hz, the useful components of the recordings would be lost. In contrast, the hardware-shielded results are seen to be better.

3.1.2. Analysis of EOG Recordings

The analysis of spectrum and waveforms of the EOG signal for a representative subject acquired from Experiment 2 is shown in Figure 7. The waveform of the signal obtained by applying Method 1 is a series of approximate sine waves with a peak-to-peak value of 0.5 mV, and blink signals can be seen by the signal envelope; the signal obtained by Method 2 is approximately sinusoidal with a reduced peak-to-peak value of 0.1 mV; the peak-to-peak value of the curve of Method 3 is further reduced to 0.06 mV, and the EOG signal becomes more obvious; In Method 4, the obtained EOG signal has almost no sine wave sequence with a fairly clean blink signal. Thus, the EOG signals obtained via the proposed shielded drive methods are better in comparison to those obtained via the unshielded drive approach (Method 1), and the EOG signal of Method 4 is considered to be the best.

Furthermore, it could be seen from the spectrum analysis results that the 50 Hz frequency of EOG signal acquired by Method 1 is as high as 23.1 dB; the 50 Hz frequency amplitude of Method 2 is 10.3 dB, which is 12.8 dB lower than that of Method 1; Method 3 has a 50 Hz frequency amplitude of 5.7 dB, which is 17.4 dB lower than that of Method 1; Method 4's 50 Hz frequency amplitude of

−23.4 dB is 46.5 dB lower than that of Method 1, which is considered very good. Moreover, the 50 Hz harmonics in the EOG spectrum of Method 4, such as 100 Hz and 150 Hz, also disappeared. We can also conclude that using the shielded drive approach could significantly reduce power line interference when recording EOG signal especially with Method 4.

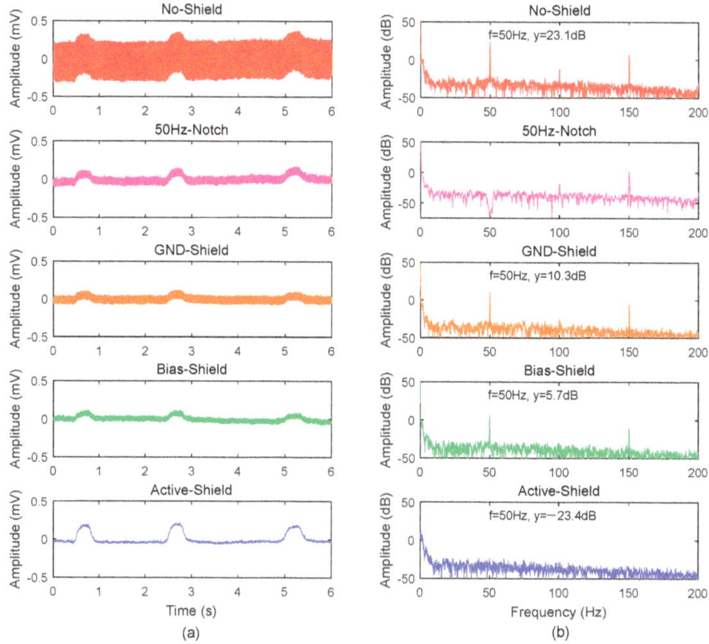

Figure 7. The comparison of the time waveforms (**a**) and spectra (**b**) of the EOG signals (subject #7) under different conditions. The 50Hz notch condition (2nd row) was obtained by applying 50 Hz notch filter (software) to raw data of the No-Shield condition (1st row).

It can also be seen from Figure 7 that the graph obtained after applying the 50 Hz-Notch filter reflects the waveform as well as spectrum of the raw signal (the output of Method 1) which is characterized by a relatively much lower amplitude. The waveform is similar to that of Method 2, but the spectrum has a large depression around 50 Hz, which eventually led to the loss of useful components of the target signal. Hence, the proposed hardware-based method would be better.

3.1.3. Analysis of EMG Recordings

Figure 8 shows the results obtained after analyzing the EMG signals of a representative subject collected in Experiment 3. In the picture, the forearm muscles of the representative subject were initially in their rest state (without activation/contraction), and a second later the muscles were activated. The waveform of the non-activated muscles as obtained by Method 1 represents a series of sine waves that jitters up and down after the activation of the arm muscles; the waveform of the non-activated muscles as obtained by Method 2 denotes a noisy line, and when the muscle is contracted, the EMG signal patterns become obvious. The waveform obtained after applying Method 3 is similar to that of Method 2, but with relatively smaller amplitude; the waveform of non-activated muscles obtained by applying Method 4 appears to be cleaner compared to Method 3, and there is a significant EMG signal burst when the forearm muscles are activated. Therefore, it can be deduced that the conclusion of Experiment 3 is consistent with those of Experiment 1 and Experiment 2, which further proves the potential of the proposed shielded drive technology.

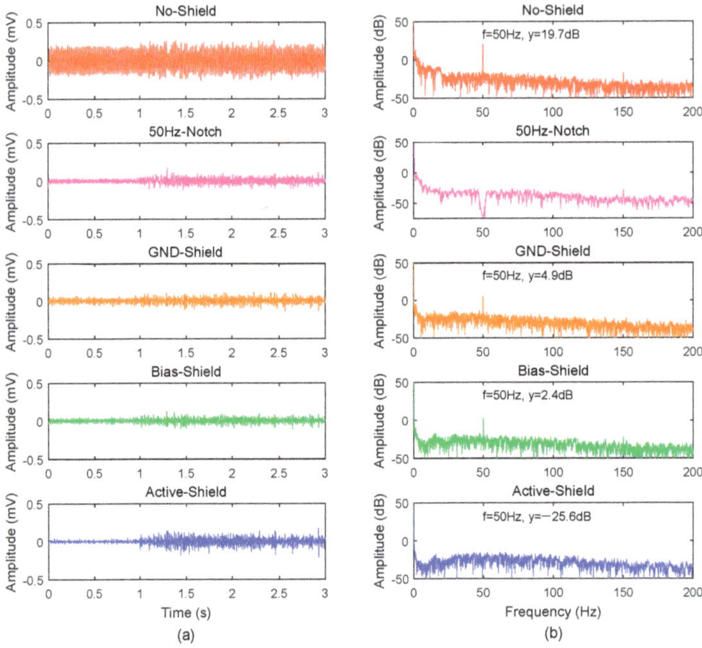

Figure 8. The comparison of the time waveforms (**a**) and spectra (**b**) of the EMG signals (subject #4) under different conditions. The 50 Hz notch condition (2nd row) was obtained by applying 50 Hz notch filter (software) to raw data of the No-Shield condition (1st row).

By analyzing the EMG signal spectrum at 50 Hz, Method 1 achieved a high amplitude of 19.7 dB; Method 2 recorded an amplitude of 4.9 dB, which is 14.8 dB lower than that of Method 1; Method 3 achieved an amplitude of 2.4 dB, which is 17.3 dB lower than that of Method 1. Meanwhile, Method 4 recorded an amplitude of −25.6 dB, which is 45.3 dB lower than that of Method 1, which appeared to be very efficient in comparison to the other methods.

From Figure 8, the graph on the second row denotes the time domain waveform and spectrum after the raw signal (the output of Method 1) was subjected to a 50 Hz notch filtering. Its time domain waveform is similar to that of Method 3, but the spectrum is depressed around 50 Hz. That is, the frequency range of the EMG signal is 0–500 Hz, and the useful components of the signal are lost. In contrast, the hardware-shielded drive processing has a better effect.

Figure 9 shows the mean values of the spectral values at 50 Hz averaged across all the 20 subjects for the ECG, EOG, and EMG experiments under NLC. It could be observed that the spectral amplitude of the 50 Hz interference gradually decreased from Method 1 to Method 4 for all three different experiments, with Method 4 achieving the lowest power line interference that almost fell below the noise floor level. The averaged results in Figure 9 were in accordance with the observations of each individual subject from Figures 6–8.

3.2. Elimination of Electrode Lead Jitter Interference

The results which were obtained based on the EMG recordings (muscles relaxed) acquired in the shielded room session are presented in Figure 10. By closely observing the plots in Figure 10, one can see that the waveform across all four of the tested methods exhibited almost the same characteristics in the first second, which is because the electrode leads were maintained in a fixed state while recording the signals. After the first second, the electrode lead cables were individually shaken at about the same amplitude and the waveform of Method 1 changed greatly while that of Method 2 is changed slightly in comparison to that of the first second. Additionally, Method 3 could be seen to exhibit a slightly lower waveform compared to Method 1 while the waveform of Method 4 has almost no change with respect to that of the first second. This implies that Method 4 has substantially eliminated the interference of the electrode lead jitter on the target signal. Therefore, it could be said that the proposed shielded drive approach would significantly eliminate the interference of the electrode lead jitter, especially when implemented with Method 4.

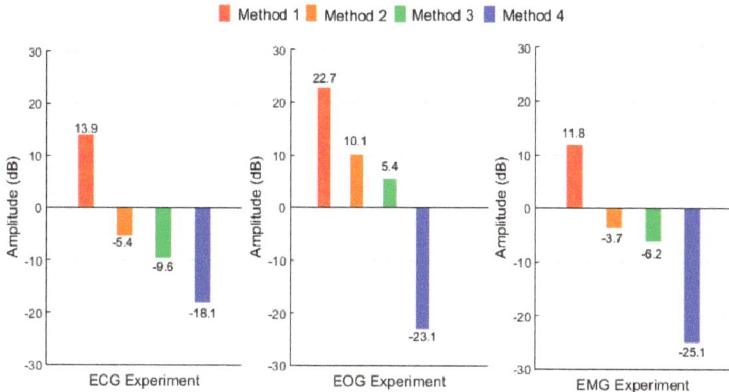

Figure 9. Mean spectral amplitude of 50 Hz interference (averaged across all subjects) of the four different shielded methods under three different experimental sessions: ECG, EOG and EMG experiments.

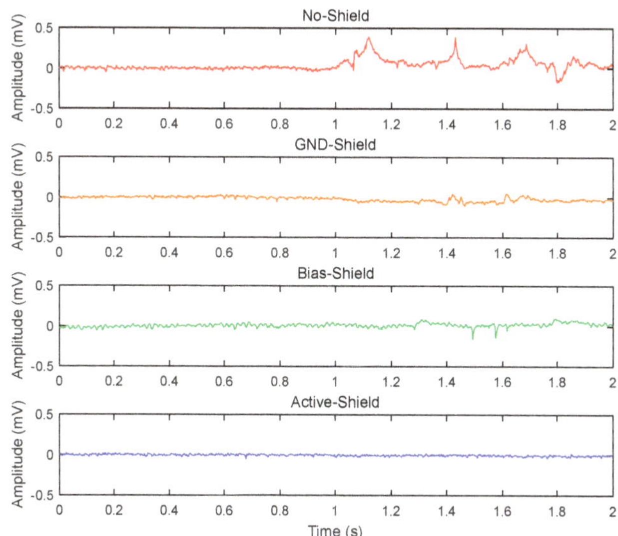

Figure 10. The time waveforms of the recorded signals (subject #15) with the electrode lead cables initially kept still (0–1 s) and then shaken simultaneously (1–2 s) for the four different shielded conditions.

4. Discussion

Towards developing an efficient method for optimal attenuation of multi-source noise/interference in biopotential signal acquisition, this study proposed hardware-based shielded drive technology. Importantly, the proposed method could efficiently shield the recorded biopotential signals from extraneous interferences as much as possible and also preserve the signal quality during the transmission process to obtain a high quality target signal prior to entering the acquisition device. In this regard, three different types of shielded drive methods were developed and their performances were evaluated. The evaluation was done with respect to the capability of methods to attenuate the effect of two kinds of extraneous interferences using a four-electrode channel recording system across three kinds of biopotential signals (ECG, EOG and EMG).

4.1. Reduction of Power Line Interference

With respect to the attenuation of power line interference, three key biopotential signals (ECG, EOG, and EMG) were considered and recorded under the same laboratory settings which have been designated as Experiment 1, Experiment 2 and Experiment 3 in the current study. The analyses of the biopotential signals obtained have been shown in Figures 6–8. The waveforms as well as spectrum characteristics shown in the Figures indicate that the proposed shielded drive methods could effectively suppress power line interference. Also, the effect of Method 4 (Active-Shield) is considered to be the best since it suppressed almost all of the power line interference of 50 Hz as well as its harmonics in the ECG, EOG, and EMG to 32.0 dB, 45.8 dB, and 36.9 dB respectively, as shown in Table 1. Table 1 shows the average dB values across subjects for the three different shielded methods (GND-Shield, Bias-Shield, and Active-Shield) against No-Shield method in suppressing the 50 Hz power line frequency in Experiment 1, Experiment 2, and Experiment 3. It is noteworthy that the greater the number, the better the performance of the methods in suppressing the 50 Hz power line interference. Although the designed notch filter was able to eliminate the 50 Hz frequency inherent in the raw signal's (i.e., No-Shield signal) spectrum, useful components of the target signal near 50 Hz were destroyed, and also the associated harmonic components were still present in the target signal. Adli et al. designed an automatic interference control device (AICD) for the elimination of power line interference, and the AICD was able to suppress the 60 Hz frequency in biopotential signal recordings to 30 dB, yet their method could not eliminate the associated harmonic components [4]. However, by utilizing the proposed shielded drive method, the 50 Hz frequency and its harmonic components were significantly reduced, thus leading to improved quality of the target signals.

Table 1. Performance of three shielded methods (Method 2, Method 3, and Method 4) relative to the unshielded method (Method 1) in suppressing the 50 Hz power line interference for different experiments, the ECG, EOG and EMG Experiments represent recording sessions of electrocardiogram, electro-oculogram, and electromyography signals, respectively.

	M1–M2 (dB)	M1–M3 (dB)	M1–M4 (dB)
ECG Experiment	19.3	23.5	32.0
EOG Experiment	12.6	17.3	45.8
EMG Experiment	15.5	18.0	36.9

Note: M1, M2, M3 and M4 represent the spectrum values at the 50 Hz power line interference for the signal recordings of Method 1, Method 2, Method 3 and Method 4, respectively.

Figure 11A showed an intensive model of the No-Shield method for measuring ECG signals, similar to the interference model in Chimene's study [19]. Because of the capacitive coupling to the mains power (C_{pow}) and the capacitive coupling to the safety earth ground (C_{body}), the power line interference is present at the patient's body as a common-mode signal and it might not be the same as the isolated ground of the power supply, causing a leakage current (I_{diff}) between the patient and the power-supply ground. When the leakage current flowed through the contact impedance of Z_{pgn}, it would introduce a 50 Hz signal visible on both inputs of the amplifier. However, this is not the only source of the 50 Hz disturbance. In case the cables all have a different capacitive coupling to the mains and to ground, the voltage difference between the two cables of the positive and negative inputs will not be zero, causing a current (I_{dcable}) from one cable to the other through the electrodes. For example, even if the current is as low as 0.1 μA, and the two electrodes have an impedance of only 5 kΩ, the differential 50 Hz signal on the bipolar input will be as large as 1 mV, which is a considerable amount of disturbance with the amplitude even larger than that of the ECG signal. In contrast, Figure 11B shows the capacitive coupling model of the Active-Shield method. In this case, the signal of the inner wire is fed back to the shield using a low impedance output of an amplifier. Note that there is still capacitive coupling between the mains power and the outer shields of both electrodes.

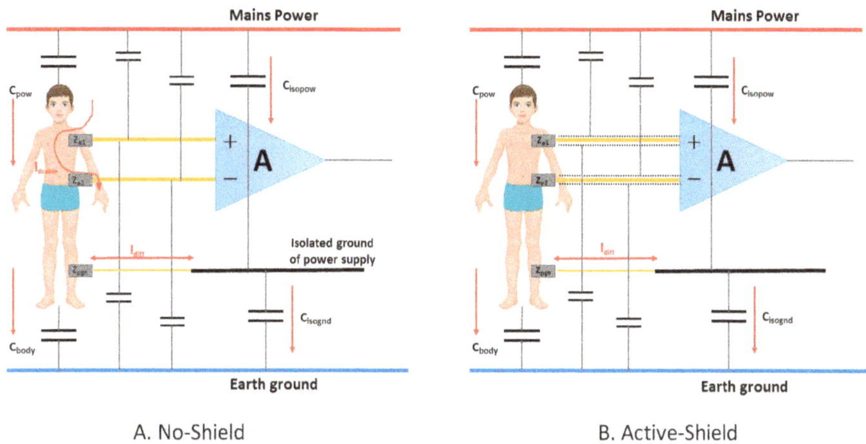

Figure 11. The capacitive coupling model for the power line interference for GND-Shield and Active-Shield methods.

Based on the Figure 11B, the equivalent circuit of the amplifier, as well as the capacitive coupling, is shown in Figure 12a below (also in the revised manuscript) to examine whether the capacitive coupling between the mains power and the shields would introduce significant flux to the loop for the Active-Shield method. The amplifier that drives the outer shield could be equalized to a high input impedance Z_{in}, a low output impedance Z_0 (usually less than 100 Ω), and a voltage source with the same voltage with the input V_i (in reference to the isolated ground of the power supply). The output voltage V_o was composed of two independent components V_1 and V_2, where V_1 is attributed to the voltage source V_i (voltage follower) and V_2 is caused by the capacitive coupling between the mains power and the outer shield. When considering the second part V_2, the voltage source could be regarded as a short circuit according to Thevenin's Theorem and therefore the equivalent circuit is shown in Figure 12b, which is actually a voltage divider circuit between the mains power voltage V_{pow} and the isolated ground, with a series circuit by Z_1 and Z_0. According to the voltage dividing rule, the V_2 voltage could be written as:

$$V_2 = \frac{Z_0}{Z_0 + Z_1}(V_{pow} - V_{isognd})$$

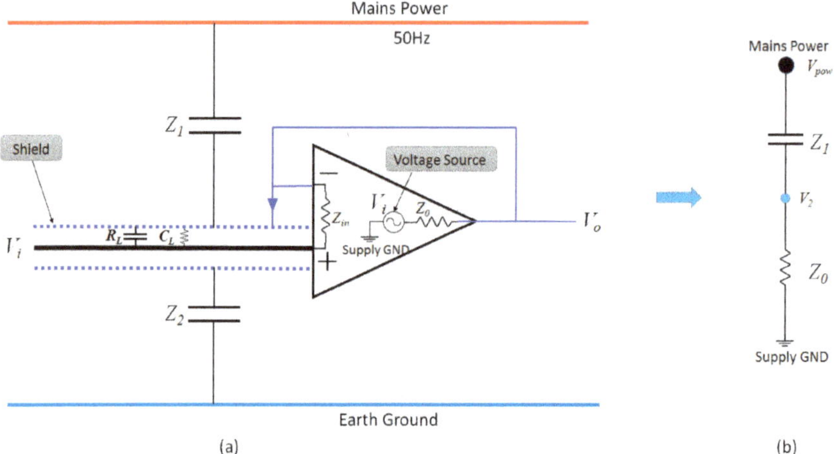

Figure 12. Equivalent circuit of one electrode for the Active-Shield method: (**a**) equivalent circuit of the amplifier and the capacitive coupling introduced by the mains power, (**b**) the equivalent voltage divider circuit of the amplifier.

Given that Z_1 is the coupling impedance between the power line and the shield and could be a few hundred MΩ or even larger, whereas the amplifier output impedance Z_0 is usually less than 100 Ω, the corresponding V_2 is usually a few dozen μV or even less in reference to the isolated ground. That is to say, the power line only introduces an interference V_2 of a few μV for the Active-Shield method, because of the low output impedance of the amplifier. Similar analysis could be applied to the other electrode of the Active-Shield method, and it could also be concluded that the presence of the power line only introduces a few dozen μV (or less) of interference V_2' to the other outer shield. Although the potential of the two shields are different for the Active-Shield method, we can see that the actual potential difference ($V_2 - V_2'$) caused by the power line interference is quite limited, since both V_2 and V_2' are the order of μV and their difference is much smaller.

On the other hand, since the impedance Z_L between the outer shield and the inner core wire is usually quite high (in the order of TΩ) and the other ends of both shields are open (not connected to human body or electrodes), little leakage current would flow from the outer shield to the inner wire or elsewhere, leading to negligible effects of the power line induced V_2 on the voltage of the inner wire (Figure 12). Therefore, the differential voltage between the two inner core wires introduced by the power line interference is also negligible.

As for the electric flux induced by the magnetic field of the power line, we tried to minimize the area of the loop enclosed by the electrode wires during the experiment. For example, we kept the wires as short as possible and made sure that the two electrode cables were as close as possible by twisting them together. In this way, the effective area exposed to the power line magnetic field was largely reduced and the effects of the loop flux induced by the magnetic field were also rather limited. The limited power-line induced flux through the loop could be further confirmed by our experimental observations from Figures 6–9 for the case of Active-Shield method.

Besides, it should be noted that although the electrodes of the four channels were attached as closely as possible in Experiment 2 and Experiment 3, there were still slight differences in the electrode positions, as shown in Figure 4b,c. Therefore, the amplitude of the EOG and EMG signals of four channels might have shown slight differences, as observed in Figures 7a and 8a. Moreover, although the EOG signal frequency is in the range of 0.1–38 Hz, which has no overlap with the spectrum of power line interference, common software bandpass filter methods cannot obtain satisfactory filtered results in routine practices. The reason for this is that the amplitude of the 50 Hz interference is much stronger than EOG signal, and the limited stop-band attenuation at 50 Hz is not enough to bring power line interference below the noise floor level. Hence, our experimental results obviously show that the proposed shielded drive method would potentially be effective for the acquisition of multiple biopotential signals.

4.2. Elimination of Electrode Lead Jitter Interference

By conducting an additional experiment that involved the recording of EMG signals in an electromagnetic shielded room designated as Experiment 4, the effect of electrode lead jitter was studied as shown in Figure 10. By examining the results in Figure 10, one can see that all the shielded drive methods could significantly eliminate the interference in the EMG recordings caused by electrode lead jitter. Of the three shielded drive methods, the Active-Shield (Method 4) had the best performance, which almost completely eliminated the electrode lead jitter interference. It is conceivable that the Active-Shield method could be applied not only to EMG signals but also to the other biopotential signals.

For shielded technologies, the electrode cable consists of a small coax cable, where the outer layer is shielding the inner signal wire. Figure 13 shows the transmission line models of the electrode cable of the two different shielded methods, in which the outer shield is connected to different shielding signals (only one of the two input electrodes of each channel is plotted for simplicity).

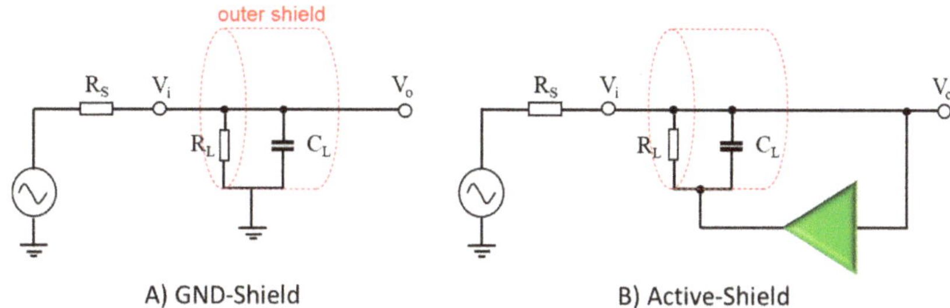

Figure 13. Transmission line model of the electrode cable for GND-Shield and Active-Shield methods.

As seen from Figure 13, the outer shield is connected to the isolated ground of the power supply for the GND-Shield method. In this way, the inner signal wire is protected against the capacitive coupling from the mains power, since the potential between the cables becomes zero and there is no current from one cable to the other through the electrodes, leading to no differential interference input. However, there is a considerable capacitance (C_L) and resistance (R_L) between the inner signal wire and the outer shield. Consequently, when there were some movements of the electrode cables, a large amount of movement artifacts will be seen in the output, caused by variation in the impedance (C_L and R_L) of the cable. The situation was similar for the case of Bias-Shield method. In contrast, for the Active-Shield method, the signal of the inner wire is put on the outer shield using a low impedance output of an amplifier, so the shield has almost the same potential as the inner signal wire in this case. Therefore, the change in C_L and R_L caused by cable movements will no longer have any influence on the potential of the inner wire and the system output. The model in Figure 13 provides the explanation for why the proposed Active-Shield circuit is effective in eliminating the electrode lead jitter interference (Figure 10).

The developed shielded drive circuit in the current study exhibits a simple structure and low cost, and is effective in suppressing power line interference as well as eliminating electrode lead jitter interference which simplifies the subsequent signal processing and analysis tasks. It could prevent the signal from extraneous noises without any requirement of hardware or software filtering. These characteristics of the proposed method are significantly important in developing biopotential signal acquisition and processing methods needed in wearable devices and other types of portable devices. It is noteworthy that the proposed shielded drive circuit built in the current study is independent of the acquisition system and thus we could integrate the shielded drive circuit into different acquisition devices to improve the signal quality in our future work.

5. Conclusions

In this study, the performances of different shielded technologies in reducing power line interference were comprehensively compared for effective biopotential acquisition. By investigating the performance of the proposed Active-Shield circuit with respect to ECG, EOG, and EMG signals acquired under normal laboratory settings, experimental results revealed that the proposed shielded drive method could effectively suppress power line interference including its harmonic components, and greatly improve the signal quality compared to conventional hardware-shielded and software filtering methods. The results also showed that the shielded drive circuit could significantly eliminate the influence of electrode lead jitter interference on the acquired signals, so it could be a great candidate to take care of the noises introduced by body movements of the subjects. In conclusion, the findings of the present study could be quite useful for providing a simple and effective solution to reduce power line interference and movement artifacts to obtain high-quality biopotential signals.

Acknowledgments: This work was supported in part by the Shenzhen High-level Overseas Talent Program (Peacock Plan) (#KQJSCX20160301141522), the National Natural Science Foundation of China under grants (#61771462, #81572233, #61403367, #61650305, #81760416, #81501620, #U1613222,# 61603375), and the Shenzhen Basic Research Grant (#JCYJ20150401145529005, #JCYJ20160331185848286, #GJHS20160329112841007).

Author Contributions: Yanbing Jiang made substantial contributions and wrote the paper; Shixiong Chen conceived and designed the experiments; Xueyu Liu and Xin Wang performed the experiments to acquire the data; Mingxing Zhu and Yanjuan Geng analyzed the data; Peng Li, Fei Chen and Fengxia Wu revised it critically for important intellectual content; Oluwarotimi Williams Samuel and Paul Oluwagbenga Idowu revised the manuscript; and Guanglin Li gave the general supervision of the research group.

Conflicts of Interest: The authors declare no conflict of interest.

References

1. Pu, X.; Wan, L.; Sheng, Y.; Chiang, P.; Qin, Y.; Hong, Z. A Wireless 8-Channel ECG Biopotential Acquisition System for Dry Electrodes. In Proceedings of the 2012 IEEE International Symposium on Radio-Frequency Integration Technology (RFIT), Singapore, 21–23 November 2012; pp. 140–142.
2. Burns, A.; Doheny, E.P.; Greene, B.R.; Foran, T.; Leahy, D.; O'Donovan, K.; Mcgrath, M.J. Shimmer: An Extensible Platform for Physiological Signal Capture. In Proceedings of the 2010 Annual International Conference of the IEEE Engineering in Medicine and Biology, Buenos Aires, Argentina, 31 August–4 September 2010; p. 3759.
3. Pradhan, A.; Nayak, S.K.; Pande, K.; Ray, S.S.; Pal, K.; Champaty, B.; Anis, A.; Tibarewala, D.N. Acquisition and Classification of Emg Using a Dual-Channel Emg Biopotential Amplifier for Controlling Assistive Devices. In Proceedings of the 2016 IEEE Annual India Conference (INDICON), Bangalore, India, 16–18 December 2016; pp. 1–5.
4. Adli, Y.Y.; Nakamura, T.; Kitaoka, K. Automatic Interference Controller Device for Eliminating the Power-Line Interference in Biopotential Signals. In Proceedings of the 17th IEEE Instrumentation and Measurement Technology Conference, Baltimore, MD, USA, 1–4 May 2000; Volume 3, pp. 1358–1362.
5. Samuel, O.W.; Geng, Y.; Li, X.; Li, G. Towards Efficient Decoding of Multiple Classes of Motor Imagery Limb Movements Based on Eeg Spectral and Time Domain Descriptors. *J. Med. Syst.* **2017**, *41*, 194. [CrossRef] [PubMed]
6. Duskalov, I.K.; Dotsinsky, I.A.; Christov, I.I. Developments in ECG Acquisition, Preprocessing, Parameter Measurement, and Recording. *IEEE Eng. Med. Biol. Mag.* **1998**, *17*, 50–58. [CrossRef]

7. Samuel, O.W.; Li, X.; Geng, Y.; Asogbon, M.G.; Fang, P.; Huang, Z.; Li, G. Resolving the Adverse Impact of Mobility on Myoelectric Pattern Recognition in Upper-Limb Multifunctional Prostheses. *Comput. Biol. Med.* **2017**, *90*, 76–87. [CrossRef] [PubMed]
8. Samuel, O.W.; Zhou, H.; Li, X.; Wang, H.; Zhang, H.; Sangaiah, A.K.; Li, G. Pattern Recognition of Electromyography Signals Based on Novel Time Domain Features for Amputees' Limb Motion Classification. *Comput. Electr. Eng.* **2017**, *2017*, 1–10. [CrossRef]
9. Schlögl, A.; Keinrath, C.; Zimmermann, D.; Scherer, R.; Leeb, R.; Pfurtscheller, G. A Fully Automated Correction Method of Eog Artifacts in EEG Recordings. *Clin. Neurophysiol.* **2007**, *118*, 98–104. [CrossRef] [PubMed]
10. Ferdi, Y. Improved Lowpass Differentiator for Physiological Signal Processing. In Proceedings of the International Symposium on Communication Systems Networks and Digital Signal Processing, Newcastle upon Tyne, UK, 21–23 July 2010; pp. 747–750.
11. Chavdar, L.; Mihov, G.; Ivanov, R.; Daskalov, I.; Christov, I.; Dotsinsky, I. Removal of Power-Line Interference from the ECG: A Review of the Subtraction Procedure. *BioMed. Eng. OnLine* **2005**, *4*, 50. [CrossRef]
12. Zhong, Y.; Zhong, P.; Wang, J. The Research of Removing Baseline Wander for ECG. *Comput. Appl. Chem.* **2007**, *24*, 465–468.
13. De Luca, C.J.; Gilmore, L.D.; Kuznetsov, M.; Roy, S.H. Filtering the Surface EMG Signal: Movement Artifact and Baseline Noise Contamination. *J. Biomech.* **2010**, *43*, 1573–1579. [CrossRef] [PubMed]
14. Fatourechi, M.; Bashashati, A.; Ward, R.K.; Birch, G.E. EMG and EOG Artifacts in Brain Computer Interface Systems: A Survey. *Clin. Neurophysiol.* **2007**, *118*, 480–494. [CrossRef] [PubMed]
15. Moretti, D.V.; Babiloni, F.; Carducci, F.; Cincotti, F.; Remondini, E.; Rossini, P.M.; Salinari, S.; Babiloni, C. Computerized Processing of EEG-EOG-EMG Artifacts for Multi-Centric Studies in Eeg Oscillations and Event-Related Potentials. *Int. J. Psychophysiol.* **2003**, *47*, 199–216. [CrossRef]
16. Magri, J.; Grech, I.; Casha, O.; Gatt, E.; Micallef, J. Design of Cmos Front-End Circuitry for the Acquisition of Biopotential Signals. In Proceedings of the 2016 IEEE International Conference on Electronics, Circuits and Systems, Monte Carlo, Monaco, 11–14 December 2016; pp. 161–164.
17. Costa, H.M.; Tavares, M.C. Removing Harmonic Power Line Interference from Biopotential Signals in Low Cost Acquisition Systems. *Comput. Biol. Med.* **2009**, *39*, 519–526. [CrossRef] [PubMed]
18. Mneimneh, M.A.; Yaz, E.E.; Johnson, M.T.; Povinelli, R.J. An Adaptive Kalman Filter for Removing Baseline Wandering in ECG Signals. In Proceedings of the Computers in Cardiology, Valencia, Spain, 17–20 September 2006; pp. 253–256.
19. Chimene, M.F.; Pallas-Areny, R. A Comprehensive Model for Power-Line Interference in Biopotential Measurements. In Proceedings of the 16th IEEE Instrumentation and Measurement Technology Conference, Venice, Italy, 24–26 May 1999; Volume 1, pp. 573–578.
20. Zhang, J.; Wang, L.; Yu, L.; Yang, Y.; Zhang, Y.; Li, B. A Low-Offset Analogue Front-End IC for Multi-Channel Physiological Signal Acquisition. In Proceedings of the 2009 Annual International Conference of the IEEE Engineering in Medicine and Biology Society, Minneapolis, MN, USA, 3–6 September 2009; pp. 4473–4476.
21. Spinelli, E.M.; Martinez, N.H.; Mayosky, M.A. A Transconductance Driven-Right-Leg Circuit. *IEEE Trans. Biomed. Eng.* **1999**, *46*, 1466–1470. [CrossRef] [PubMed]
22. Tomasini, M.; Benatti, S.; Milosevic, B.; Farella, E.; Benini, L. Power Line Interference Removal for High-Quality Continuous Biosignal Monitoring with Low-Power Wearable Devices. *IEEE Sens. J.* **2016**, *10*, 3887–3895. [CrossRef]
23. Keshtkaran, M.R.; Yang, Z. A Robust Adaptive Power Line Interference Canceler VLSI Architecture and Asic for Multichannel Biopotential Recording Applications. *IEEE Trans. Circuits Syst. II Exp. Briefs* **2014**, *61*, 788–792. [CrossRef]
24. Alnasser, E. The Stability Analysis of a Biopotential Measurement System Equipped with Driven-Right-Leg and Shield-Driver Circuits. *IEEE Trans. Instrum. Meas.* **2014**, *63*, 1731–1738. [CrossRef]
25. Sudirman, R.; Zakaria, N.A.; Jamaluddin, M.N.; Mohamed, M.R.; Khalid, K.N. Study of Electromagnetic Interference to ECG Using Faraday Shield. In Proceedings of the Third Asia International Conference on Modelling & Simulation, Bali, Indonesia, 25–29 May 2009; pp. 745–750.
26. Lee, S.M.; Sim, K.S.; Kim, K.K.; Lim, Y.G.; Park, K.S. Thin and Flexible Active Electrodes with Shield for Capacitive Electrocardiogram Measurement. *Med. Biol. Eng. Comput.* **2010**, *48*, 447–457. [CrossRef] [PubMed]

27. Sullivan, T.J.; Deiss, S.R.; Cauwenberghs, G. A Low-Noise, Non-Contact Eeg/Ecg Sensor. In Proceedings of the Biomedical Circuits and Systems Conference, Montreal, QC, Canada, 27–30 November 2007; pp. 154–157.
28. Yacoub, S.; Raoof, K. Power Line Interference Rejection from Surface Electromyography Signal Using an Adaptive Algorithm. *IRBM* **2008**, *29*, 231–238. [CrossRef]
29. Jayant, H.K.; Rana, K.P.S.; Kumar, V.; Nair, S.S.; Mishra, P. Efficient Iir Notch Filter Design Using Minimax Optimization for 50 Hz Noise Suppression in ECG. In Proceedings of the 2015 International Conference on Signal Processing, Computing and Control (ISPCC), Waknaghat, India, 24–26 September 2015; pp. 290–295.
30. Liang, Q.; Ming, Y.E. Design of Digital Trap Filter for Reducing Power Line Interference in Semg. *Comput. Eng. Appl.* **2009**, *45*, 61–63.
31. Spinelli, E.M.; Mayosky, M.A. Two-Electrode Biopotential Measurements: Power Line Interference Analysis. *IEEE Trans. Biomed. Eng.* **2005**, *52*, 1436–1442. [CrossRef] [PubMed]
32. Mitov, I.P. A Method for Reduction of Power Line Interference in the ECG. *Med. Eng. Phys.* **2004**, *26*, 879. [CrossRef] [PubMed]
33. Dotsinsky, I.; Stoyanov, T. Power-Line Interference Removal from ECG in Case of Power-Line Frequency Variations. *Int. J. Bioautom.* **2008**, *3*, 334–340.
34. Hamilton, P.S. A Comparison of Adaptive and Nonadaptive Filters for Reduction of Power Line Interference in the ECG. *IEEE Trans. Biomed. Eng.* **1996**, *43*, 105–109. [CrossRef] [PubMed]
35. Kaur, M.; Singh, B. Powerline Interference Reduction in ECG Using Combination of Ma Method and Iir Notch. *Int. J. Recent Trends Eng.* **2009**, *2*, 125–129.
36. Huang, C.-C.; Liang, S.-F.; Young, M.-S.; Shaw, F.-Z. A Novel Application of the S-Transform in Removing Powerline Interference from Biomedical Signals. *Physiol. Meas.* **2008**, *30*, 13–27. [CrossRef] [PubMed]
37. Avendano-Valencia, L.D.; Avendano, L.E.; Ferrero, J.M.; Castellanos-Dominguez, G. Improvement of an Extended Kalman Filter Power Line Interference Suppressor for ECG Signals. In *2007 Computers in Cardiology*; IEEE: Piscataway, NJ, USA, 2007; pp. 553–556.
38. Mewett, D.T.; Reynolds, K.J.; Nazeran, H. Reducing Power Line Interference in Digitised Electromyogram Recordings by Spectrum Interpolation. *Med. Biol. Eng. Comput.* **2004**, *42*, 524–531. [CrossRef] [PubMed]
39. Ziarani, A.K.; Konrad, A. A Nonlinear Adaptive Method of Elimination of Power Line Interference in ECG Signals. *IEEE Trans. Biomed. Eng.* **2002**, *49*, 540. [CrossRef] [PubMed]
40. Weiting, Y.; Runjing, Z. An Improved Self-Adaptive Filter Based on Lms Algorithm for Filtering 50 Hz Interference in ECG Signals. In Proceedings of the 2007 8th International Conference on Electronic Measurement and Instruments, Xi'an, China, 16–18 August 2007; IEEE: Piscataway, NJ, USA, 2007; pp. 3-874–3-878.
41. Keshtkaran, M.R.; Yang, Z. A Fast, Robust Algorithm for Power Line Interference Cancellation in Neural Recording. *J. Neural Eng.* **2014**, *11*, 026017. [CrossRef] [PubMed]
42. Isaksen, J.; Leber, R.; Schmid, R.; Schmid, H.J.; Generali, G.; Abächerli, R. The First-Order High-Pass Filter Influences the Automatic Measurements of the Electrocardiogram. In Proceedings of the 2016 IEEE International Conference on Acoustics, Speech and Signal Processing (ICASSP), Shanghai, China, 20–25 March 2016; pp. 784–788.
43. Acharya, D.; Rani, A.; Agarwal, S. EEG Data Acquisition Circuit System Based on ADS1299EEG-FE. In Proceedings of the 2015 4th International Conference on Reliability, Infocom Technologies and Optimization (ICRITO), Noida, India, 2–4 September 2015; IEEE: Piscataway, NJ, USA, 2015; pp. 1–5.
44. D'Ausilio, A. Arduino: A Low-Cost Multipurpose Lab Equipment. *Behav. Res. Methods* **2012**, *44*, 305–313. [CrossRef] [PubMed]
45. Reverter, F.; Li, X.; Meijer, G.C.M. Stability and Accuracy of Active Shielding for Grounded Capacitive Sensors. *Meas. Sci. Technol.* **2006**, *17*, 2884. [CrossRef]
46. Schagrin, M.L. Resistance to Ohm's Law. *Am. J. Phys.* **1963**, *31*, 536–547. [CrossRef]
47. Haberman, M.A.; Spinelli, E.M.; García, P.A.; Guerrero, F.N. Capacitive Driven-Right-Leg Circuit Design. *Int. J. Biomed. Eng. Technol.* **2015**, *17*, 115. [CrossRef]

MDPI
St. Alban-Anlage 66
4052 Basel
Switzerland
Tel. +41 61 683 77 34
Fax +41 61 302 89 18
www.mdpi.com

Applied Sciences Editorial Office
E-mail: applsci@mdpi.com
www.mdpi.com/journal/applsci

www.ingramcontent.com/pod-product-compliance
Lightning Source LLC
Chambersburg PA
CBHW051700210326
41597CB00032B/5323

* 9 7 8 3 0 3 8 9 7 0 4 5 3 *